Transducers and Instrumentation

Second Edition

D.V.S. MURTY

Formerly Professor of Electrical Engineering
Indian Institute of Technology Kharagpur

PHI Learning Private Limited

Delhi-110092

2021

₹ 1195.00

TRANSDUCERS AND INSTRUMENTATION, Second Edition
D.V.S. Murty

© 2021 by PHI Learning Private Limited, Delhi. Previous edition © 2002.

ISBN-978-81-203-3569-1 (Print Book)
ISBN-978-93-90669-10-3 (e-Book)

The export rights of this book are vested solely with the publisher.

Published by Asoke K. Ghosh, PHI Learning Private Limited, Rimjhim House, 111, Patparganj Industrial Estate, Delhi-110092 and Printed by Syndicate Binders, A-20, Hosiery Complex, Noida, Phase-II Extension, Noida-201305 (N.C.R. Delhi).

To

The Students of Instrumentation Engineering

Contents

7   Active Electrical Transducers   369–458

8 Feedback Transducer Systems

9 Signal Processing Circuits

13   Sophistication in Instrumentation   **635–684**

14 Process Control Instrumentation

Preface

Since the publication of the first edition of the book in 1995, many remarkable developments have taken place in the science and technology of sensors and instrumentation. The practice of making devices smaller and smaller, yet more powerful, has never been so pervasive as it is today. The materials, production processes and the integrated technology have enabled production of a new breed of sensors, smart in many ways. These developments and advances have prompted me to bring out this new edition. To make the text up to date, I have added three new chapters—Chapters 12, 13, and 14.

Ever since the publication of the first edition, the response to the book from students and the teaching community has been overwhelming, and their continued patronage has been so fulfilling and gratifying to me. What strikes me from the increasing readership of the book is that this book competes well with the existing books in the market, both Indian and foreign ones. However, with the advancement in technology, it is felt necessary to keep the readers abreast of the many changes that have taken place. Hence the addition of new chapters and the publication of the new edition. While preparing the new edition, I have not deleted any chapter or section from the first edition since the readers had given 'thumbs up' to it.

Science is the mother of inventions and, hence, the basic phenomena presented in the book (Chapters 5–8) remain applicable for the development of the microsensors. The development of sensors and the associated instrumentation is mainly geared to meet the new needs of applications of technology while, at the same time, providing safety, security and immunity to electromagnetic interference. One cannot miss the percolation of these modern sensing systems into one's own workplace, home, hospital and, of course, the industrial environment.

The subtle distinction between transducers and sensors has been, of late, ignored by one and all and, therefore, the students undergoing the course on transducers may do well by going through the additional Chapters 12 and 13. Others may utilize these chapters or select topics according to their interest and syllabi requirements. The nature of the subject is such that it embraces all branches of science and engineering and so one has to wade through the extensive literature—books as well as other publications—for getting a deep insight into one particular topic or application.

I am indebted to the authorities of the Indian Institute of Technology Kharagpur for inspiring me to delve into this specialization of Instrumentation Engineering. In many ways, during my service in the Institute from 1953 to 1989, I was fortunate enough to have the ambience necessary for pursuing my areas of interest. It was the spirit of association with all colleagues of the Department of Electrical Engineering and the highly conducive academic atmosphere—I can ill afford to forget it in my life—that has spurred me on to write this book. It is gratifying to learn that the book is prescribed as a textbook to the students of instrumentation engineering in many universities and also as a reference book to students of other disciplines.

I sincerely thank PHI Learning, for bringing out the Second Edition. I am extremely grateful to all the readers of the first edition for their magnanimous response and I fervently hope that they will continue to respond with equal magnanimity to the new edition.

D.V.S. Murty

Preface to the First Edition

In recent times, the importance of technical education in general and training in specific areas in particular has been recognized by people all over the world; India is no exception to this. This is evident from the rapidly increasing number of institutions of scientific and technical education and the diversity of programmes offered at these institutions at the undergraduate, postgraduate and research levels. Instrumentation Engineering has been focussed as an area that needs special emphasis by the Ministry of Human Resources, Government of India. The dearth of manpower demands that the educational institutions urgently meet this need as the requirements of the users industries for sophisticated instrumentation systems cannot be met by the instrumentation industry in India. It is heartening to note that the IITs, Regional Engineering Colleges as well as some colleges of engineering are offering programmes in this discipline.

Today there are very few books available that deal with transducer systems though K.S. Lion and H.K.P. Neubert have made significant contribution in this respect. Students of Instrumentation Engineering should first get acquainted with all the processes and phenomena leading to the development of transducer systems; then they will be able to interface them intelligently with appropriate electronics circuitry. Just like a physician, an instrumentation engineer has to fully understand the system under investigation before suggesting suitable transducers and then utilize the electronic support that not only upholds the basic accuracy of measurement but also enables him to achieve his objectives of on-line instrumentation and control.

The primary aim of this book is to acquaint the students with the basic principles of sensors and transducer systems and show how each of them can be exploited for the measurement of a large number of variables. Chapters 1–4 are devoted to characteristics and principles of electrical measurements and dynamics of measuring systems. Chapters 5–7 discuss the various types of transducers—mechanical, passive electrical, and active electrical—and show how these can be employed for the measurement of a large number of variables, small or large in value. Transducer systems utilizing negative feedback are treated in Chapter 8. The scope of electronic circuitry in dealing with the output signals of transducers and furthering the objectives is described in Chapters 9–11. Chapter 11, in particular, explains the techniques of telemetry and data transmission. Indepth analysis of physical systems and other details pertaining to electronic circuits have been kept to a minimum.

A large number of diagrams have been provided to illustrate the concepts introduced. Also, the numerical examples and the exercises given should be of considerable assistance to the students.

The text is intended for a two/three-semester course on "Transducer Systems" and "Measuring Systems and Instrumentation" in a four-year undergraduate programme in Instrumentation Engineering. By a judicious selection of topics, the book can also be used for a one-semester course for students of other disciplines at both undergraduate and postgraduate levels.

I am indebted to the Director and other authorities of the Indian Institute of Technology Kharagpur for the encouragement they gave by sending me for training to UK for specializing in instrumentation engineering and providing me full opportunity to organize several continuing education programmes on the subject since 1967, thereby inspiring me to write this book. Also, the financial support provided by the Quality-Improvement Programme at IIT Kharagpur is gratefully acknowledged. I also wish to thank my colleagues and other faculty members for their encouragement, guidance and cooperation; indeed, they have been a source of great inspiration to me. Besides, I would like to express my thanks to the typist and the draughtsman for the excellent work they have done. Finally, I sincerely thank my wife for her patience, understanding and encouragement.

Any suggestions for improving the contents will be warmly appreciated.

D.V.S. Murty

CHAPTER 1

Measurement, Instrumentation and Calibration

1.0 INTRODUCTION

Human endeavour is a direct outcome of man's curiosity to learn and understand the environment in which he lives, and his achievements in any field bear witness to the successful applications of his imagination and skills. By his imaginative skills he has been able to develop mathematical strategies and transform them into operative physical systems which enable him to probe into the secrets of nature as well as meet the demands of his necessities for better living. All the successful achievements in science and technology are entirely due to his ability to measure the state, condition, or characteristics of the physical systems, in quantitative terms with sufficient accuracy.

Lord Kelvin stressed the importance of measurement in this context, by saying: "When you can measure what you are speaking about, and express it in numbers, you know something about it".

When information is presented quantitatively, intuitive and intelligent minds experience greater sense of relief and appreciation, resulting in greater enthusiasm for a concerted effort and action in a more befitting and exact manner. We are able to witness the advances in the fields of science and technology in many ways, whether it is at home, in the laboratory, or industry.

Exploration of space goes on simultaneously with the exploration of land and sea for mineral deposits. Day-by-day higher limits of productivity and product quality are set for industrial production. There is a search for the development of new materials with different physical properties. Manufacturing techniques and technologies are undergoing transformation with the development of robotics and nanotechnology. Continuous condition-monitoring and study of a variety of complex systems, e.g. monitoring a patient in the hospital, an aircraft or a spacecraft on its course of motion and the environment around the earth, is increasingly becoming common for obtaining all the information necessary for predicting their behaviour or for deciding the next course of action.

Since World War II, automation has made inroads into almost all scientific and technological fields, leading to ultimate control of physical systems by computers. Aided by the developments in electronic circuits, computers are now capable of achieving goals unheard of and unthinkable earlier, in all the scientific and technological ventures. To exploit the potential of the computers to the maximum, it is imperative that the computer be provided with all the information related to the system under study or control. Most of the information pertaining to a physical system is obtained in a quantitative manner; to be more exact, by means of measurements carried out on the system. To cope with the demands on the variety of measured values that make up the information and the 'quality' and 'integrity' with which they have to be obtained, a large number of physical phenomena are being studied and utilized for development of the basic measuring systems. Measuring systems and techniques have to be designed for each situation so that output signals represent truthfully the state, condition or characteristic of the system under study. The primary or the basic measuring element or the combination of such elements, which 'sense' or 'measure' the quantity sets the limit for the 'integrity' of the measured value and hence any subsequent finding or achievement governed by it cannot possess greater integrity. The accuracy of indicated or displayed value may be less if the output signal of the basic measuring element is handled by some intermediate systems. Similarly, the accuracy with which the condition of a system is controlled by an automatic control system can be no better than the accuracy with which it is measured. Moreover, computer control of industrial processes and systems demands that the process of measurement be fast enough to result in real-time instrumentation and control. The output signals from the basic measuring systems are subjected to signal processing so as to bring them into a form or description compatible with the computer. Thus, whatever be the nature of the output signals of the primary measuring systems, they are converted into proportionate analog electrical signals which in turn, are, converted into digital signals. The developments in integrated-circuit technology and the digital-signal processing techniques have largely been responsible for the development of sophisticated instrumentation systems with multifunction capability. Once the basic measuring element provides an output signal, it should be possible to process it suitably so that its measured value is displayed and recorded in both analog and digital fashion. Digital signals can be stored and reproduced when required for indication and record. They can be transmitted to a distant and remote station, with greater accuracy and integrity than in the case of analog signals. Electronic data handling systems tackle the signals obtained after suitably measuring them, processing them for display, and recording them. Data acquisition systems (DAS) collect the data from all the basic measuring systems, and present the same in the required form for the subsequent operation. The data may be transmitted over a long distance where the data is reduced and reconstructed to the original form. Thus the data handling operations constitute acquisition, transmission and reduction of the data. Data logging operation is done by computers which have the provision for storage of the data. With the increasing role played by digital computers and data loggers, greater attention is being paid to the development of measuring systems that directly deliver digital signals related to the quantity under measurement.

Figure 1.1 shows the various stages of operations that can be carried out on the *measured value,* once the measurement process is completed. Control function generation is the next logical operation after the measured value is compared with the desired or reference value and standardization of the output signals of the control function generators has been accorded greater significance in the field of automatic control. When the system is complex, all the measured data is acquired and pooled at a central location, known as centralized control room, where the data is subjected to further processing in a more convenient 'atmosphere' for display, record or decision-making. The scope of instrumentation activity, thus, encompasses the entire data-handling systems, starting with the basic function of measurement to that of releasing the control function signals, with the computer replacing the human operator, when the system under study and control is complex. Automatic data processing and automatic computation are thus seen to be the logical components of the modern sophisticated instrumentation systems. This sometimes gives an impression that an electronic engineer is an instrumentation engineer. It is important to realize that an instrumentation engineer is one who is conversant with successful application of all the basic measuring systems as well as the processing of the measured data till the end, with no loss of accuracy and integrity of the measured value at any stage. A successful instrumentation engineer has to understand fully the system he investigates, the measuring systems he chooses, and all the means he adopts for the processing of the measured data for the desired end functions. Increasing demands on what the instrumentation system is desired to accomplish make them costly and they are increasingly expected to provide rich dividends in industrial productivity.

In the following sections, we describe the characteristics of the basic measuring systems and how the measured values are assessed for their accuracy.

1.1   MEASUREMENT

The measurement is usually undertaken to ascertain and present the state, condition or characteristic of a system in quantitative terms. It enables the experimenter to understand the state of the system under which it exists or to distinguish the transition of the system from one state to another. To reveal the performance of a physical or chemical system, the first operation carried out on it is measurement. To measure is to determine the magnitude, extent, degree etc. of the condition of a system in terms of some standard. Measurement is the basic and primary operation, the result of which is used only to describe the system and hence treated as an independent operation with no ulterior motive other than to understand it.

Man has always extended his imaginative skills to identify physical phenomena and later developed and utilized the means for confirmation of his understanding of the phenomena. Thus the measuring techniques and systems developed are based on the basic laws of nature and the evolution of these techniques, in course of time, led to the development of indicating systems for some of the physical quantities. These can be seen to consist of mechanisms incorporated inside a housing so as to be coupled to the system under study. Identification of physical phenomena results in the development of measuring systems that interact with the quantity under measurement and develop output signals in the form of linear or angular displacements that can be transmitted to a pointer on a scale. The measuring system may be considered to consist of three basic stages as shown in Figure 1.2:

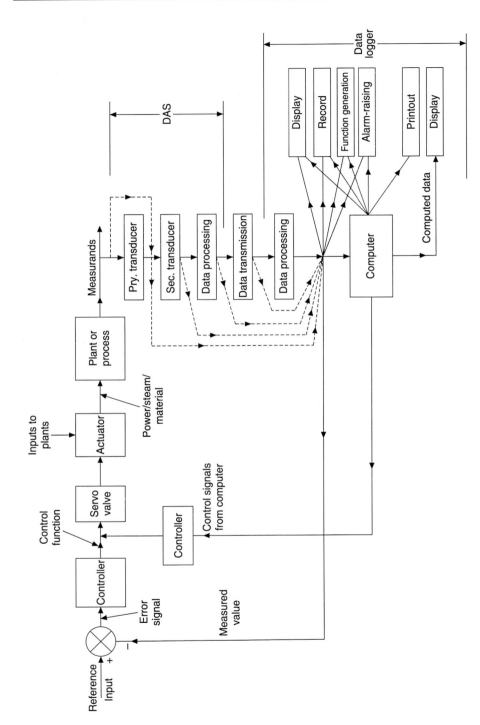

Figure 1.1   Block diagram of an instrumentation system.

1. The primary element that senses the quantity under measurement
2. The intermediate means that modifies suitably the output of the primary element
3. The end device that renders the indication possible on a calibrated scale.

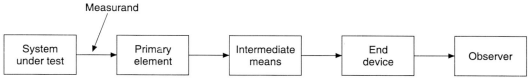

Figure 1.2 A simple measuring system.

The bimetallic thermometer, the Bourdon-tube pressure gauge, and the permanent magnet moving-coil ammeter are all examples of such basic indicating systems. In the bimetallic thermometer, the bimetallic element is the primary sensing element, and the mechanical levers serve as intermediate means for amplifying the displacement of the free tip of the bimetallic element. The pointer and the calibrated scale provide the means for indication. All these stages may be housed in an outfit that becomes the thermometer. It is not always possible to distinguish all the three stages separately in some of the indicating systems which are said to be analog indicators as the pointer can assume any position on the scale in a contiguous manner without displaying any jerky motion from one value to another, however close these may be. The measuring systems are devised to enable the measurement of a quantity that remains at the same value during the process of measurement. In other words, measurement of steady values is carried out by suitable measuring techniques, some of which help the development of ready-reckoning systems that directly indicate the measured value. Such indicating systems are known as *meters* and are calibrated to indicate the measured value. Indication is invariably by means of a pointer on a scale and hence the output signal is a mechanical quantity. Thus the meter can be treated as a combination of certain basic elements, which measures a physical quantity and provides an output signal in the form of a mechanical displacement. It can be considered to be a system that converts a physical quantity under measurement into a displacement signal.

'Metrology' is the science of precise and accurate measurement of the various physical quantities, whereas the 'meter' is understood to be an instrument for indicating or recording the measured value. The variable under measurement is usually referred to as *measurand*. The pointer of the basic meter carries a pen that moves on a chart of paper and plots the variation of the quantity with time, when the paper is moved at constant speed. In such a case, the rate at which the measured quantity varies should be low enough to be followed by the meter. Thus the limitation is set by the inertia and the natural frequency of the mechanism employed for the meter. These indicating and recording meters draw the necessary power for indication, from the system under examination, and the important criterion in the design of such meters is that the meter should not load or burden the system under study and disturb it from its original state. The variable under measurement should have the same value before and after coupling the meter with the system under examination. Other important considerations are the *accuracy* and *precision* of the meter. Meters and indicating instruments of analog type, based on several physical phenomena, have been in existence, since long, with slightly differing accuracy and precision, and are classified as *precision* or *industrial grade instruments*. A variety of such indicating and recording instruments are used to fill the conventional centralized control rooms where decisions are made by the operators for effecting control of a process or plant.

With the developments in electronic circuits, it has become possible to render conversion of analog electrical voltages into digital signals so that a digital readout can be obtained in the desired number of digits with sufficient accuracy. The solid-state electronic circuits, being available in chip form for discrete functions and possessing a high degree of reliability apart from other advantages of low power consumption and size, have been extensively used for the development of digital indicators, not only for all the electrical quantities but also for several non-electrical quantities. Although the advantage of converting a non-electrical signal such as displacement or temperature into a proportionate electrical signal has been recognized earlier, the facility of digital readout systems has been extended to measure every non-electrical quantity. The speed of digital indication may result in continuous change in the display of the least significant digit if the measured quantity is not dead-steady, but it is no major disqualification, in the sense that the display of any quantity of interest could be commanded by the operator in the modern set-up of central control rooms where the data loggers and computer control systems are installed for the monitoring and control of the process. The sizes of alphanumeric and neon display devices can be varied, depending on the distance of the observer from the indicator. Sophisticated electronic instruments have been developed to facilitate the display of not only amplitude of the measured quantity but also other associated characteristics of the input signal.

The absorption of electronics in the field of instrument technology and indicating meters has replaced some of the conventional mechanical and electromechanical systems used extensively so far in the indicating meters, resulting in ease and speed of repair. However, the need for frequent calibration of the electronic indicators persists. The internal calibration facility of many digital instruments has been possible by the use of Zener-diode voltage reference inside the instrument, which serves as an equivalent of the standard Weston cadmium cell.

Certain information pertaining to the characteristics of the indicating instruments and meters is presented in Section 1.5.

1.2   INSTRUMENT

The science and the art of measurement of physical quantities has a long history. Initially, measurements were mainly concerned with the basic and common quantities such as length, time, mass, and force, and meters were developed for their measurement and direct indication under steady state conditions. Indicating systems so developed were regarded as instruments. With the passage of time, instruments came to be understood as tools in the hands of man, which were useful in accomplishing the objectives of sensing, detecting, measuring, recording, controlling, computing or communicating. Extensive modifications in the design of instruments were incorporated to achieve these objectives, while at the same time giving higher accuracy and precision where measurement was involved. Development of a variety of materials for the electrical and mechanical elements and associated solid-state electronic circuits resulted in the design of a fascinating array of instruments with multifunction capability. Considerable improvements in the characteristics of instruments, with regard to reliability, speed of response, visual display of measured quantity etc. are noticeable in the present-day instruments.

Sophisticated instruments have considerably enhanced and refined the sensory perception of the measured data and its usefulness. Although a meter is also treated as a simple instrument, a sophisticated instrument is really a complex system, in which the primary signal obtained after sensing or measurement, undergoes extensive processing before it is presented for display or recording. The primary quantity under measurement may be a non-electrical quantity, in which case means should be provided for conversion of the quantity into a properly related or proportionate electrical signal. Additional signal processing such as amplification, ac to dc conversion, filtering of unwanted signals or analog-to-digital conversion etc. should not entail any errors other than those due to the basic operation of measurement. In other words, the final output of the instrument should possess the desired 'integrity' and 'reliability', however complex the instrument may be in its design. Improving the accuracy of the instrument from ± 1% to ± 0.1% requires utilization of more sophisticated techniques, circuits or systems, thereby making the instrument costlier. Similarly, instruments such as magnetometers, radiation detectors and X-ray fiuoroscopes enable sensing and detection of physical quantities, which cannot be sensed or detected by human beings. For every non-electrical variable, a variety of instruments based on different principles of operation exist, each specifically designed to suit the conditions under which the measurement is made. Instruments possessing additional capabilities such as storage of measured data, correlation, computation and control function generation have been developed for application in situations in which human skills and proficiency were limited. Thus instruments play a significant role in all the activities of man.

1.3 INSTRUMENTATION

Measurement of non-electrical quantities by electrical methods was given greater attention since the middle of the 19th century and this marked the beginning of one of the most fruitful and vigorous areas of activity concerning the scope of their application. Simultaneously, the advances made in the field of electronics substantially contributed to the birth of a new discipline commonly known as *instrumentation*. If sensing a variable draws on the fundamental phenomena provided by the natural sciences of physics, chemistry and biology, the success of design and fabrication of sensors has been possible through the contribution of all engineering disciplines by way of materials, manufacturing processes and integration of several functions on a microscopic composite, the ic chip. The sensors and transducers providing electrical outputs have been favoured since long for reasons of the verstility they offer for conducting investigations of systems, analysis of products and control of an industrial process or plant. Extension of electrical techniques and systems for measuring various nonelectrical quantities necessitated the development of devices known as *sensors* and *transducers* that transform or convert the measurand into another quantity, electrical or mechanical, for which a standard of comparison readily exists. Direct and alternating current voltages and currents stand unique in respect of the ease with which they can be processed or modified in a number of ways. When the output quantity of the sensor or transducer has an electrical property, it is referred to as *signal,* representative of the measurand, and is usually processed by a large number of electrical and electronic signal conditioning circuitry for effecting improvement in the sensitivity of measurement, speed of response, linearity etc. Some of the common signal conditioning operations are amplification, attenuation, comparison, differentiation, integration, dimension and domain conversion, coding, correcting, analyzing, correlating, programming, computing etc.

Instrumentation Engineering and instrument technology may be treated as a multidisciplinary branch of engineering where additional aspects covering the design and manufacture of instrument systems, apart from their utilization, are discussed in depth. Signals obtained after measurement need to be transmitted to a distance, stored, or used for automatic computation of the performance of the systems under study. A variety of theories and techniques making up the disciplines of telecommunication and computer engineering are utilized, leading to significant achievements in on-line instrumentation and control.

The signals obtained from the transducers and associated devices are utilized for condition-monitoring and/or control of industrial processes and plants. Successful application of digital computers for increasing productivity and enhancing product quality is only possible if all information concerning the system is made available to the computer by means of a large number of sensors and transducers. Most of the signals furnished by transducers are analog in nature. But digital techniques and circuits are widely recognized to provide easy and more precise readout facility, greater flexibility of data reduction and, most important of all, better reliability. Where instrumentation systems are required to deal with a large number of signals from transducers and, where extensive data processing needs to be carried out, digital signal approach is considered most effective and best. Transmission and telemetry of measured data to a remote point is possible with greater reliability and accuracy if the measured data is brought out in digital form. Since, most sensors and transducers are analog devices, their output signals require certain analog signal processing before they are converted into digital signals. Of course, certain transducer systems are designed to provide signals in time domain (frequency or pulse train) so that counting circuits may be used to provide digital outputs. Very few transducers develop digital signals in coded form. Coded digital signals are better processed by digital computers. A microprocessor is a simple but versatile computer in a single integrated-circuit package. Developments in microprocessor technology resulted in the evolution of dedicated instrumentation systems in which a microcomputer is used to coordinate the measured data and effect the desired operations on them.

The essential characteristics of an instrumentation system remain the same as those of a measuring system. In fact, it is possible to attain greater sensitivity, precision and response speed, with a proper choice of signal conditioning circuitry. Enough safeguards are provided for preserving the integrity of data from the effects of external disturbances and noise. In fact, it should be understood that the performance of a digital instrument is limited by the characteristics of the primary transducer which is invariably analog in nature. A variety of transducers and measuring systems are being developed for providing digital output signals at the outset with a view to cut down and eliminate analog signal conditioning needs, as far as possible. While the common building blocks for signal conditioning and processing are available as standard integrated circuit chips, the primary measuring transducers and measuring systems constitute a wide variety based on several different phenomena, necessitating intelligent selection of the same, suitable for each measurand and for each system under study.

Since all branches of science and technology utilize instruments for measurement of quantities pertaining to their discipline, instrumentation systems are classified as belonging to chemical, aeronautical, medical, meteorological or optical. The basic differences of one from the other are due to the nature and range of the measurand and the transducer system used to develop output signals which are electrical by nature. Subsequent signal processing circuitry assumes standard pattern for most of the instrumentation systems. The wide range of transducers commonly utilized for development of such instrumentation systems are described in Chapters 5–8 and those modern sensors enabling sophistication in the present day instrumentation are dealt in Chapter 12. Signal processing circuitry and common display and telemetry systems are described in Chapter 9–11. Chapters 13 and 14 are intended to highlight the scope of application of sensors.

1.4   CLASSIFICATION OF TRANSDUCERS

The term 'transducer' has come to be understood as a device or a combination of elements, which responds to the physical condition or chemical state of a substance and develops an output functionally related to the magnitude of the stimulus. Although the word 'transducer' was initially used in the context of instrumentation, subsequently it assumed a wide connotation. In the field of electrical and mechanical measurements, one comes across a range of devices that respond to the physical condition or state of a system, thereby enabling measurement of the same. These devices are treated as mechanical transducers if their output signals are mechanical by nature, and electrical transducers if their output signals are electrical by nature. Where a transducer is composed of more than one element, it can be treated as an open-loop system with the elements connected in cascade as shown in Figure 1.3. The first element which is directly coupled to the system under study is known as the *primary transducer, primary sensing element,* or simply a *sensor. A* sensor may be employed for detecting the existence or absence of a condition, or sensing the degree or level of a physical condition, apart from responding to the condition under measurement. A transducer may be considered to have accomplished the function of measurement by drawing an insignificant amount of power and energy from the system under study whereas a sensor does it by standing aloof without getting into physical contact with the medium or system under examination. Intensity of radiation and luminance of a source may be measured by sensors, whereas the temperature of a fluid is measured by transducers. Both the sensor and the transducer convert one quantity into another quantity of different dimensions. A broader interpretation of these has been found in literature about the function of a transducer, i.e. conversion of one physical quantity into another related quantity. The measurand may be converted directly into a quantity that serves the function of indication or record, in which case comparison with a standard is possible. But in most cases, the measurement is indirect as transduction utilizes certain basic functional relationship between the input and output quantities.

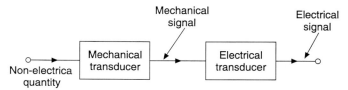

Figure 1.3   Transducer elements in cascade.

Classification of transducers is possible in several ways depending on the role they play, their structure, or the basic phenomena on which they are based.

Treating the transducers as signal converters, it is easy to classify them as input and output transducers. *Input transducers* are meant for measurement of non-electrical quantities, and hence they are usually converters of non-electrical quantities into quantities, which are electrical by nature. They may be also known as instrument transducers. *Output transducers* operate in reverse fashion and deliver output signals of either force, torque, pressure or displacement when the input signals are electrical by nature. They may also be known as power transducers if the output power is considerable and as inverse transducers if the signal power at either end is small and negligible.

Depending on the basic principles of operation, transducers may be known as mechanical, electrical, thermal, magnetic or radiation type transducers.

A transducer is known to consist of more than one physical element, in which case it is treated as a system. The pattern of connection of the several elements may result in an open-loop system as in Figure 1.3 or a closed-loop system. The closed-loop configuration offers several advantages in the measurement and hence feedback-type transducers are used in many situations.

In the field of measurement, the terms *sensors* and *transducers* are still used interchangeably. In Chapters 5-8, primary transducers are treated as 'sensors' if they enable 'ideal' measurement. The primary transducer may be connected in cascade to one or more transducers so as to finally produce an output signal suitable for indication, record or even for transmission to a distant location. *Transmitters* are transducer systems that convert the measurand into a proportionate signal suitable for transmission over a pair of wires (4–20 mA dc) or by means of pneumatic lines (2.07–13.8 N/cm^2).

Transponders are transducers which automatically release an output signal for transmission to a distant point, when interrogated by another signal.

Most of the transducers can be classified as either mechanical or electrical type. A large number of mechanical transducers are in use for measuring the process quantities, either because of their low cost or their inherent safety. But many of them are now used in conjunction with the electrical transducers and, whenever possible, the primary and secondary transducers are replaced by a single electrical transducer.

Although most of the sensors and transducers are analog type with their output signals being either mechanical or electrical, the role of computers in the modern instrumentation and control activity is making designers of transducers find ways of obtaining digital signals directly from the transducers. Thus transducers that can directly develop digital signals are gaining popularity but, in their absence, transducers whose output signals are ac voltages or pulse trains become the next choice as the number of cycles or pulses during a certain interval of time can be counted. Thus transducers are classified as either analog, frequency generating or digital type. Mechanical transducers, in spite of some basic disadvantages, have an appeal of their own in the sense that if their output signals are displacements, they can be directly converted into digital signals, apart from their suitability for *in-situ* indication or record. For certain process quantities, it is essential to have a combination of mechanical and electrical transducers, so as to derive all the benefits provided by the versatility of the electronic signal-processing circuitry.

As far as the user is concerned, transducers may be classified to represent the quantity they are designed to measure. They may be classified as pressure or force transducers irrespective of the basic principles of operation. However, it may be of some significance, if they are explicitly known as piezoelectric pressure transducers or resistive displacement transducers, thereby conveying the type to which they belong.

A variety of scientific phenomena are utilized for developing the range of transducers for each conceivable physical or chemical quantity. An intelligent choice and combination of them is desired, after a thorough understanding of the system under examination, the range of the amplitude of the measurand, the rate at which it is likely to change, the nature of the output signal desired, and the purpose for which the measurement is undertaken.

Some of the relevant sensors and primary mechanical transducers that develop output signals in the form of displacement, force or air pressure, are briefly described in Chapter 5. Where a single element cannot function as an electrical transducer, a measuring system is constituted so that its output can be converted into a proportionate electrical signal by means of a secondary electrical transducer.

Secondary electrical transducers are described in Chapters 6–8. Electrical transducers may be classified as passive or active type. Passive transducers are those that need to be excited by electrical supply in order to recognise their response to the measurand. Electrical circuit elements which are resistors, capacitors and self and mutual inductors, constitute the group of passive transducers. Active transducers are those that function as energy converters. They draw a certain amount of energy from the medium or system under study and convert it into electrical signal in the form of voltage or current. A thermocouple is an active transducer. It does not need an external source of electrical supply, it itself acts as a source of emf. A closed-loop system can be constituted to serve as a transducer system, having certain decided advantages. Such transducers are known as *feedback transducers* and are briefly covered in Chapter 8.

The basic characteristics that a transducer should possess and the way its performance is rated and specified are discussed in Section 1.5.

1.5   PERFORMANCE CHARACTERISTICS

An instrument may be designed to simply indicate the value of the measurand under static conditions or record and reproduce the variation of the measurand with time truthfully, on a moving chart by means of a pen. The characteristics of the instrument may have to be evaluated so as to judge the suitability of each for the function for which it is designed. Performance characteristics are divided into two groups: (a) static characteristics, and (b) dynamic characteristics. Instruments having satisfactory static characteristics may also be used under dynamic conditions provided they satisfy a few criteria. But certain transducers and instruments are meant for use under dynamic conditions only and are considered unsuitable for static measurements.

1.5.1   Static Characteristics

The static characteristics of a transducer or a measuring system are established by the process of static calibration. By static calibration, the relationship between the output signal and the quantity under study is experimentally determined. The measuring system or transducer responds to the quantity under study but it is also likely to be influenced in its performance by quantities other than the desired signal. In such a case, all efforts should be taken to see that the static calibration is effected under conditions in which all other quantities except the desired input signal are held constant, at agreeable and approved levels during the process of static calibration. Manufacturers and testing laboratories normally furnish the static characteristics of a measuring device or system stating the levels of the other undesirable conditions under which calibration is done. Also, it becomes necessary to repeat the static calibration when each of the disturbing or undesirable condition assumes different levels, with a view to indicating the extent to which the measuring system is likely to be influenced in its performance when the disturbing conditions vary over a known range. Usually, temperature, humidity and vibrations affect the performance of all systems, and it is normal to study the effects of these disturbing inputs on their performance and static characteristic.

The process of static calibration requires the presence of a standard with which the output signal of the measuring device is to be related. If it is an indicating or recording instrument that is under calibration, the indicated value of the pointer on the calibrated scale is checked for its correctness by comparison with a standard source of the quantity under calibration. But, in the case of measuring devices and transducers, the output signal is invariably of different form and dimensions and by the process of calibration, the relationship between the input quantity and output signal has to be established, and this relationship is treated as its static characteristic. It is also known as calibration curve, though in many cases, it is a straight line.

The standards are classified in three ways: absolute, secondary reference, and working standards.

Absolute Standards are those devices designed and constructed as per specifications based on legal international definitions of the various fundamental units of measurement, such as the standard metre. These devices are kept in National Physical Laboratories and the National Bureau of Standards. The term *absolute* signifies the fact that these devices are independent and not relative. They are normally finite and have fixed values.

Secondary Reference Standards are designed and constructed from the absolute standards.

Working Standards are calibrated in terms of secondary standards and are generally meant for on-the-job or field calibration work. They may have fixed but a range of values.

When a measuring system requires to be calibrated, a proper standard whose accuracy is greater than that of the measuring system should be selected. It is better if the working standard is variable in value for calibration at several points of the scale or range. A convention often followed is that the calibration standard should be at least ten times more accurate than the system being calibrated. During the calibration process, it is essential to see that the standard is not disturbed in its value due to the connection or coupling of the measuring system or device with the working standard or due to the technique adopted for comparison of the values of the standard and the measuring device.

The process of static calibration enables the definition of several characteristics of the measuring system. Some of them may be relevant to measuring transducers. The following terms specify the static characteristics of the measuring systems and devices and a better understanding of the terms enables proper selection of the measuring systems for actual application.

Accuracy is defined as the nearness of the indicated value to the true value of the quantity being measured. The measured value invariably differs from the true value because of the effects of disturbing inputs such as temperature, humidity, and because of the performance characteristics of the measuring system itself. The accuracy is, in fact, specified as an uncertainty (or inaccuracy) in the value due to the above factors with declared probability limits. The term, when used in reference to the performance of a transducer, conveys in quantitative terms the extent to which its calibration characteristic is likely to swing in actual operation due to the effects of all disturbances. When the accuracy of an indicating instrument is specified, it indicates the maximum likely departure of the indicated value from the true value. When the process of measurement is based on counting, as in the case of certain digital instruments and counters, the departure from actual to indicated counts is normally zero. If the accuracy of an instrument is stated to be ± 1%, it implies that the maximum departure of the reading from the true value may at the maximum amount to ± 1% of the full-scale indication and that it may be present at any part of the scale, provided that the zero of the instrument is adjusted prior to reading.

Static error is defined as the algebraic difference between the value indicated or conveyed by the output of the measuring system or transducer and the true value of the quantity presented to the input. Error is a positive quantity if the measured value is higher than the true value and is expressed as a percentage of the full scale value of the measuring system.

Correction is defined as the opposite of error, and the true value is obtained by adding algebraically the static error or by applying correction to the indicated value. Thus,

True value = indicated value + correction

Correction is also expressed as a percentage of the full scale value of the measuring instrument.

Static error and correction may also be expressed in the units in which the measured quantity is expressed and as a percentage of the quantity measured.

Uncertainty is expressive of the range of variation of the indicated value from the true value. Uncertainty indicates the probable limits of error which the indicated value may have due to the influence of disturbing inputs. It is bipolar whereas error may be positive or negative depending on whether the indicated value is higher or lower than the true value. Statement of uncertainty signifies the quality of the measuring instrument and hence its accuracy; it is incumbent on the part of every instrumentation engineer to express the uncertainty attendant on each measured value.

Precision defines the degree of refinement with which a measured value is stated. If an input quantity of the same value is applied repeatedly to the instrument under exactly the same environmental conditions, the indicated value is expected to be the same. But in practice, the readings will be clustered around a value in a random manner. The more the number of readings falling very close to the value, the more precise the instrument is. A statistical analysis of such a situation reveals the most probable value as also the probability of a large number of readings falling within a certain range of the most probable value. Simple electrical indicating instruments exhibit such a phenomenon and their readings fall within the range of the detectable change in value. An accurate measuring system should also be precise as there is no point in attempting to obtain high precision from an instrument which is inherently poor in accuracy.

Repeatability is the characteristic of precision instruments. This term is generally applied to quote the standards of performance of measuring transducers. It reflects the closeness of agreement of a group of output signals obtained by the same observer for the same input quantity using the same methods and apparatus under the same operating environment, but over a short time span. Measuring transducers are in continuous use in process control operations and the repeatability of performance of the transducer is more important than the accuracy of the transducer, from considerations of consistency in product quality.

Reproducibility is the closeness with which the same value of the input quantity is measured at different times, and under different conditions of usage of the instrument and by different instruments. The output signals and indications are checked for consistency over prolonged periods and at different locations. Perfect reproducibility ensures interchangeability of instruments and transducers.

Stability defines the ability of a measuring system to maintain its standard of performance over prolonged periods of time. Transducers and instruments of high stability need not be calibrated frequently.

Zero stability defines the ability of an instrument to restore to zero reading after the input quantity has been brought to zero, while other conditions remain the same.

Resolution is the smallest increment of the input quantity to which the measuring system responds. Resolving power or discrimination power is the ability of the system to respond to small changes of the input quantity. If a digital read-out system has a full scale indication of 999, its resolution is 1 or 1 in 999. It does not automatically mean that the accuracy of the instrument is better than 1 in 999. It is desirable that the accuracy matches with the resolving power but it is not always the case with each measuring system.

Sensitivity is defined as the ratio of the change in output signal to a change in the input quantity. It is often referred to as incremental sensitivity or gain as it relates to increments in the signals. From Figure 1.4(a) it is seen that the incremental sensitivity is different for different values of input quantity. But if the static calibration curve is a straight line over the entire range as shown in Figure 1.4(b), the incremental sensitivity is constant over the entire range. In such a case, the overall sensitivity or the static sensitivity becomes the ratio of total change in output signal to the total change in the input quantity. Sensitivity of a measuring system should be high enough to develop a readable or detectable change in output signal for the smallest change in the input quantity.

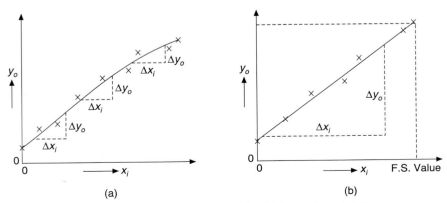

Figure 1.4   (a) Incremental sensitivity; (b) linear characteristics.

Responsiveness denotes the smallest change in the quantity under measurement; which results in an actuating effort required to cause motion of the indicating part of the instrument.

Linearity defines the proportionality between input quantity and output signal. If the sensitivity is constant for all values from zero to full scale value of the measuring system, then the calibration characteristic is linear and is a straight line passing through zero. If it is an indicating or recording instrument, the scale may be made linear. In case there is a zero error, the characteristic assumes the form of equation given by $y_o = mx_i \pm C$, where C is the zero error.

A linear characteristic and two typical nonlinear characteristics are shown in Figure 1.5(a), all starting at the origin but having different terminations for full-scale output values. Nonlinearity in such cases may be expressed in terms of the maximum departure of the characteristic from that of an assumed straight line drawn from zero. The straight line may be best-fit line, accurately determined by the 'least-square' method or an ideal line drawn such that it cuts the nonlinear characteristic as shown in Figure 1.5(b).

But when the basics of performance of a transducer or a measuring instrument yield a square law characteristic, the nonlinearity for such a transducer can also be defined in terms of the departure of the actual characteristic from that of square law.

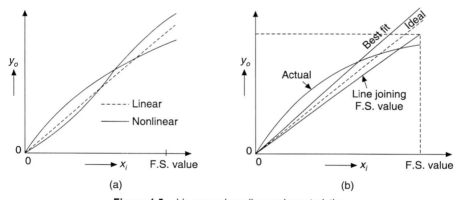

Figure 1.5 Linear and nonlinear characteristics.

Drift is a slow variation in the output signal of a transducer or a measuring system which is not due to any change in the input quantity. It is primarily due to changes in the operating conditions of components inside the measuring system. The drift is noticeable as zero drift and sensitivity drift.

Threshold is defined as the minimum input quantity that is necessary to cause a detectable change in output signal from the zero indication. In a digital readout system, threshold is that input quantity which is necessary to cause a change of one least significant digit of the output signal.

Dead band is defined as the largest change in the input quantity to which the measuring system does not respond. It is due to either static friction (stiction), backlash or hysteresis. Dead band is also known as dead zone or dead space. All elastic mechanical elements used as primary transducers exhibit effects of hysteresis, creep and elastic after-effect to some extent. Pivoted type indicating instruments and recording instruments have the drawback of small zero error due to stiction, backlash and hysteresis. Figure 1.6 shows the characteristics of such systems affected by them.

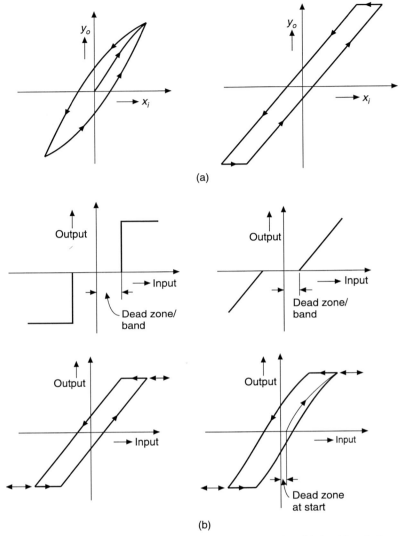

Figure 1.6 Dead band characteristics: (a) Hysteresis characteristics; (b) dead band characteristics.

Span and *range* are the two terms that convey the information about the lower and upper calibration points. The range of indicating instruments is normally from zero to some full-scale value and the span is simply the difference between the full-scale and lower-scale value. But some instruments operate under a bias so that they start reading, for example, voltages from 150 V to 250 V only. The zero of these instruments is suppressed from indication by means of a bias. In such a case, the range is said to be from 150 V to 250 V and the span is 100 V.

Live zero is a term applied to measuring systems whose output signal is not zero for zero input quantity. Some feedback transducer systems are designed to develop output signals of dc current ranging from 4–20 mA when the input quantity changes from zero to the full scale value. For bidirectional variation of the input quantity from zero value, the output signal variation may be effected from 12 mA in either direction over the range of 4–20 mA.

Scale readability varies with the design of the instrument and is partly governed by the sensitivity of the instrument. The term denotes the extent to which the reader is enabled to read the indications. If it is a digital instrument, the reader records the reading as obtained in all the digits, but in the case of analog instruments, his judgement of the reading decides the last digit of the indicated value. The readability of the instrument conveys the degree to which the reader can be precise in recording the reading, but it does not always ensure that the reading is accurate in its value up to the last digit.

Input impedance and output impedance are the two parameters which are invariably used to express the characteristics of an electrical or electronic signal conditioner. Any device or system acting as a source, converter, modifier or receiver of electrical signals may also be characterized by specifying its input and output impedances. The measuring systems of Figures 1.2 and 1.3 are known to consist of several stages, each processing the signal received by it and presenting the same to the next stage in the desired form. Each stage is typified by two ports, one being the input port and the other, the output port. Some of the stages may have additional ports, such as a supply port, to which a source of power is connected to excite the system and other ports to receive or deliver a second and a third signal. For the present consideration, each stage may be considered to have one input port and one output port, each one associated with a corresponding parameter, known as impedance. The terminal stages may be treated as one-port devices functioning either as sources or receivers. An instrumentation system is primarily known to process signals of one physical form or the other, representing the measured data, broadly understood to be information. The flow of information or signals need not and may not always be associated with the flow of power at these ports unless the information-processing stages constitute physical link, or coupling. When two physical devices or systems get coupled to each other, it is necessary to specify the "give and take" characteristics of each, while the signal is transferred from one to the other. It is necessary to imply whether there is any power flow from one to the other along with the signal flow and, if so, to what extent the associated power flow will disturb the state of the signal source and distort the signal under measurement. It is in this context that each two-port network is studied in detail, by electrical and electronic engineers. Although it seems logical to extend all the considerations and theories developed to characterize the four-terminal networks, we shall confine ourselves to show the relevance of such considerations for the present, when applied to the primary measuring elements and such other electrical and non-electrical devices.

When a source of electrical power is connected to a receiver, the latter is considered as a load, and to ensure that the receiver extracts the maximum power from the source, the theorem of maximum power-transfer is applied by electrical engineers and it is shown that the resistance of the load should be equal to the internal resistance of the source. This is known as *impedance matching* and is usually resorted to while matching the power amplifier and its load when the load forms the last stage of a measuring system.

A source of electrical signal may have all the attributes of a source of electrical power except for the fact that one does not wish to disturb the 'state' of the signal source and distort the signal that is intended to be measured or processed. Any measuring system or circuit which is physically coupled to the signal source may be treated as its load, but it is expected that it measures or processes the signal without disturbing the signal source. The state of the signal source should, for all practical purposes, remain the same, whether the measuring system is connected to it or not. Such a situation demands that no power or energy be drawn from the signal source. As far as possible, the designers of electrical and electronic instruments make all efforts to satisfy this condition. But one understands from common experience that a small amount of power transfer takes place, however small it may be, when a physical coupling is made between two systems. When power transfer is involved, there are always two variables or quantities involved, one of which is the quantity under measurement. When the emf e of a voltage source is under measurement, a small current i is drawn by the voltmeter, thereby making the terminal voltage measured as v given by

$$v = e - iR_o$$

where R_o is the resistance of the source, usually called source resistance or output resistance. The measurement would have been perfect if R_o were zero. But the power drawn from the source would be finite and will be given by ei. Hence, if the voltmeter is designed to present an infinite resistance to the source, the value of the current i would have been zero, and the measurement would have been perfect with no power transfer between the source and the voltmeter. The higher the resistance presented by the voltmeter, the more suited the voltmeter is for use with sources of varying values of R_o. When dealing with ac quantities, these resistances are referred to as impedances. Hence it has become customary to generalize the criterion that "the input impedance of all measuring systems should be as high as possible", implying that it should not load the system under examination by drawing power from it. The term *high impedance* should not be taken literally to imply that it is always the ratio of a voltage to current, but should be associated with the principle that the measuring system is endowed with the characteristic property by which it could impede the flow of power and energy from the system when it is coupled to it. An ammeter may be characterized as having a low resistance so that the voltage drop across it becomes very small and hence its power consumption. Designing the ammeter in such a way as to have low resistance has made the measuring system conform to the basic and primary criterion of not involving any significant power transfer in the process of measurement.

It must be appreciated that all sources of signals do not possess the same power handling capacity and hence it is essential to ensure that the physical coupling of a measuring system to the signal source has not robbed some energy from it and that a suitable matching device between them is desired to minimize loading. For example, moving iron voltmeters are used to measure the terminal voltage of 50 Hz alternators, but when used along with electronic low frequency oscillators, the reading will be far removed from the true value.

When the analogy is extended to cover the cases of all electrical and mechanical devices encountered in the instrumentation systems, it becomes necessary to appreciate that there are always two variables involved in the measurement process. If one of them is the variable or quantity under measurement, the other variable becomes an associate variable that enables the knowledge of the extent of power or energy transfer involved. They are termed differently by different authors and are referred to as *flow* or *through variable* (also as pervariable) and *effort* or *across variable* (also as transvariable). The flow variable can be specified and measured at one point in space but effort variable involves two points in space, with one point being considered as reference. The magnitude of the flow variable depends on the extent to which the combination is involved in the energy exchange. The magnitude of the across variable is independent of the amount of exchange. Examples of flow variable are force, momentum, current and charge, whereas velocity, temperature and voltage are across variables. Table 1.1 presents these variables pertaining to the common systems, in two ways: "state" and "rate". The state of a system is representative of the physical condition, as stated earlier, and rate refers to the time-rate of change of the condition. All physical systems, such as thermal, mechanical and fluidic systems, may be modelled as electrical systems; Table 1.1 shows the analogous quantities of these systems. It is worthwhile to note that the analogies result in the following relationships:

$$\dot{x}\,\dot{y} = \text{power}$$

$$\int \dot{x}\,\dot{y}\,dt = \text{energy}$$

It may be noted that when a measuring system is designed for the measurement of x or $\dot{x}$,

Table 1.1 Analogous quantities of thermal, mechanical and fluidic systems

System	Through (flow) variable		Across (effort) variable	
	state (x)	rate ($\dot{x}$)	state (y)	rate ($\dot{y}$)
Electrical	Charge (q)	Current (i)	Flux linkages (ψ)	Voltage (v)
Mechanical (translational)	Linear momentum ($m\,\Delta v$)	Force (f)	Linear displacement	Velocity (v)
Mechanical (angular)	Angular momentum ($J\,\Delta\omega$)	Torque (τ)	Angular displacement (θ)	Angular velocity (ω)
Fluidic	Volume (v)	Volume flow rate (dv/dt)		Pressure (p)
Thermal	Heat (q)	Heat flow rate °C ($(dq/dt)/\Delta T°$)		Temperature ($T°$)

corresponding y or $\dot{y}$, should be minimized so that minimum loading of the signal source takes place. For example, a force transducer should result in the least displacement for being considered as a near ideal force transducer. Likewise, a displacement transducer should involve and need minimum force in order to make an accurate measurement of displacement. A helical spring with high spring-rate, K (= force/displacement), is recommended for force measurements, whereas the one with low K is suited for displacement measurements. Both qualify for being treated as high-impedance devices, when the quantity for which each is considered for application is clearly understood.

The term *output impedance* is normally applicable to voltage signal sources, amplifiers and other signal processing circuits. The measuring system as a whole is treated as a passive system forming a load on signal source and hence the term *output impedance* does not apply to it. But amplifiers and other signal sources that deliver output signals are required to function as ideal voltage sources, possessing very low output impedance. Low output impedance of such voltage sources enables them to preserve their state (in this case, voltage level) in spite of the finite impedance of a measuring system forming its load. In other words, a system acting as a voltage-signal source becomes matched to a measuring system acting as a load, if the output impedance of the source is low and the input impedance of the measuring system is high. To reduce the loading error to 1 per cent, the thumb rule normally adopted is to see that the load impedance is 100 times larger than the source impedance.

If the signal source is a current source, application of the criterion of minimum loading requires reversal of the ratio of impedances.

When designing a multi-stage instrumentation system, the above facts may be kept in mind so as to facilitate easy development of the overall transfer function and tapping of signals from any stage without causing any loading errors.

1.5.2 Dynamic Characteristics

Dynamic characteristics of a measuring system relate to its performance when the measurand is a function of time. Static characteristics are determined when the measurand remains constant in magnitude and does not vary with time. Dynamic response of a measuring system, when subjected to dynamic inputs which are functions of time, depends very much on its own parameters, apart from the nature and complexity of the function. Thus the dynamic response of the measuring system may be considered to consist of two components: one due to its own characteristic parameters and the other due to the nature of the input function.

Though the likelihood of measuring systems being subjected to a variety of measurands of varying description exists, it is customary to specify the performance of measuring systems for the following common input signals (Figure 1.7):

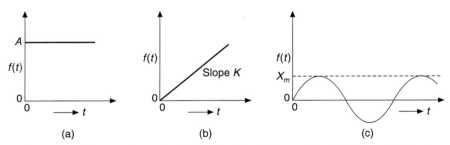

Figure 1.7 Common input signals: (a) Step-input; (b) ramp-input; (c) sinusoidal-input.

1. Step input function with the amplitude, given a step change, ($f(t) = A$ for $t > 0$)
2. Ramp input function with the amplitude changing linearly with time, ($f(t) = Kt$ for $t > 0$)
3. Sinusoidal input function where the amplitude varies sinusoidally with time, ($f(t) = X_m \sin (2\pi ft)$, for $t > 0$), where f is the frequency in Hz.

Measuring devices and systems are normally sluggish in their behaviour and hence their output signals are not faithful reproductions of the input signals. The response may start in most cases as soon as the measuring system is subjected to the input signal, but the output signal takes some time to reach a value which bears a definite and fixed relationship with the input signal. Until then, the response consists of two components: a transient component and a steady-state component. When the transient component reaches a negligible value, the measuring system is said to have gone to steady state. The behaviour of the measuring system under dynamic conditions is mathematically modelled and the solution for the response is sought for the required specific input function. The mathematical models are invariably linear differential equations with constant coefficients, and measuring systems are classified according to the order of the differential equation. Most of the measuring systems belong to either first or second order and hence the dynamic characteristics are specified in such a way as to distinguish one measuring system from the other, for its dynamic performance. Broadly, they are represented by (a) dynamic error, (b) fidelity, (c) bandwidth, and (d) speed of response.

Dynamic error is defined as the algebraic difference between the indicated/recorded value of a measurand and its true value at any instant, when the measurand is a function of time. The dynamic error is a function of time as both the measurand and the output signal of the measuring system are functions of time. This error is zero only in the case of an ideal measuring system of zero order. For measuring systems of higher order, their output signals consist of two components, one pertaining to the transient state and the other to the steady state. Hence dynamic error has two components, viz. transient error and steady-state error.

When the measuring system is subjected to a step-input function (Figure 1.7(a)), the response under steady state is represented by the ratio of the amplitude of the output signal to that of the measurand, as both of them remain constant in magnitude. This ratio is defined as the *static sensitivity* of the system and if it remains constant for a specified range of values of the measurand, the static characteristic becomes a straight line for that range. For all transducer systems responding to static inputs, knowledge of static sensitivity is necessary for estimation of dynamic error. For step-input functions, the transducer system has its dynamic error consisting of only the transient error, as it becomes zero once the system assumes steady state.

For input functions of time as shown in Figures 1.7(b) and 1.7(c), the dynamic error consists of both the transient and steady-state components. For the recording systems, the steady-state error is evaluated from the knowledge of the values of the measurand at any instant and its corresponding recorded signal at that instant. It is a constant in the case of ramp-input function and is a function of time for sinusoidal input function. The ratio of the peak amplitude of the output signal to that of the input signal, known as *dynamic sensitivity,* is also dependent on the frequency of the input signal. It may be nearly equal to the static sensitivity, only for a certain range of frequencies extending from zero. The variation of dynamic sensitivity with frequency for a typical system is shown in Figure 1.8, it is known as gain-frequency (or amplitude-frequency to another scale) characteristic. The steady state error, apart from the error

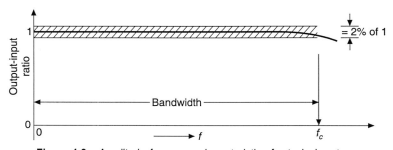

Figure 1.8   Amplitude frequency characteristic of a typical system.

in amplitude, shows up also as phase error, which is the phase difference between the output and input signals. Phase error also depends on the frequency of the input signal, and the variation of phase difference with frequency is known as *phase-frequency characteristic*. The performance of the first and second order physical systems is further explained in Chapter 4.

Fidelity of a measuring system or a transducer refers to its ability to follow instant by instant the variations of the measurand with time. Only a zero-order system possesses excellent fidelity, whereas systems of higher order are not capable of reproducing input signals truthfully at all instants and for all types of time-varying input functions. Hence it has become necessary to establish certain criteria for specifying the fidelity of a measuring system. In this context, the performance of the system under steady state conditions for sinusoidal input functions is chosen as reference for specifying the limits of fidelity. True or excellent fidelity of a system would imply that the waveforms of the output and input signals coincide with each other at all instants under steady state conditions. In such a case, there is neither amplitude error (distortion) nor phase error. However, there is bound to be both types of errors in the performance of measuring systems. Fidelity specification is usually attributed to cover deficiency in amplitude-frequency characteristic and is said to be good if it is flat over a wide range of frequencies extending from zero. Certain systems do not respond to dc input and, in such situations, the flatness is judged from the dynamic sensitivity at some mid-frequency.

Fidelity requirements differ with application. The root mean square (rms) reading moving-iron type indicating meters are required to have the same sensitivity for dc signals and ac signals of frequencies over a small range around 50 Hz. Waveform recorders and cathode ray oscillographs are required to have excellent fidelity with no amplitude or phase distortion for signals of frequencies over a wide range, but not necessarily extending to zero. In all such cases where accurate measurement is the objective, the amplitude-frequency characteristic should remain flat within $\pm$ 2%, with no appreciable phase error.

While specifying the performance of electronic amplifiers and such other electronic equipment meant for entertainment, fidelity specification is relaxed. A greater degree of distortion is tolerated in performance because the human sensory systems cannot detect the distortion at the output end. Hence fidelity is claimed for a wider range of frequencies even though they are quoted as 'hi-fi' (high fidelity) systems. The dynamic sensitivity at zero frequency (or a mid-frequency) is allowed to fall down to 70.7% over the frequency range of its performance.

Bandwidth of a system is the range of frequencies for which its dynamic sensitivity is satisfactory. For measuring systems, the dynamic sensitivity is required to be within $\pm$ 2% of its static sensitivity. Thus the amplitude-frequency characteristic is almost flat as shown in Figure 1.8 right up to a certain frequency from dc, and this range of frequencies is specified as the bandwidth of the measuring system.

For other physical systems, electrical filters and electronic amplifiers, the above criterion is relaxed with the result that their bandwidth specification extends to frequencies at which the dynamic sensitivity is 70.7% of that at zero or the mid-frequency.

Speed of response of a physical system refers to its ability to respond to sudden changes of amplitude of input signal. It is usually specified as the time taken by the system to come close to steady state conditions, for a step-input function. Hence the speed of response is evaluated from the knowledge of the system performance under transient conditions and terms such as time constant, measurement lag, settling time, dead time, and dynamic range are used to convey the response of the variety of systems encountered in practice.

Time constant is associated with the behaviour of a first order system and is defined as the time taken by the system to reach 63.2% of its final output signal amplitude. It is related to the parameters of the system. A system having smaller time constant reaches its final output earlier than the one with larger time constant and hence possesses higher speed of response.

A first-order system having a time constant of T second(s) has its natural frequency represented by ω_n rad/s, which is equal to $1/T$. A coupled system consisting of two first-order systems may also at times be specified by a time constant, when it behaves like an overdamped second-order system so as to convey an idea of its speed of response.

Measurement lag, in general, signifies the time delay in the occurrence of the output signal due to an input signal. It may be specified in different ways, depending on the type of input signal, and it is dependent on the characteristics of the system only. When the input signal is sinusoidal having a frequency of ω_n rad/s, the output signal under steady state conditions lags in phase by $\omega_n T$ rad for a system with a time constant of T seconds, and this phase difference is defined as phase lag of the system,

Certain systems do not respond to input until some time elapses, whatever be the nature of the input signal. Such systems are known as *dead time elements.* But the output signal of the system is a replica of the input signal after T_d seconds as shown in Figure 1.9, and for such systems T_d is defined as the *dead time* of the system.

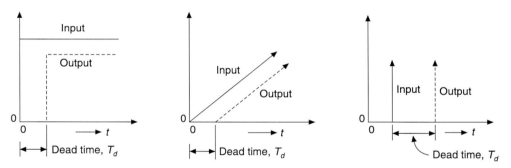

Figure 1.9 Dead time element response.

Settling time is the time taken by a system to be within a close range of its steady-state value. If it is an indicating instrument, one may be interested in knowing the time taken by the pointer to reach a value within ±1% of the final indication, and this time interval is reckoned as its settling time. A smaller settling time indicates higher speed of response. Settling time is also dependent on the parameters of the system and varies with the conditions under which the system operates. The settling time of second-order instruments is affected by the degree of damping provided for the instrument. Figure 1.10 shows the effect of damping on the settling time.

Dynamic range is the range of signals which the measuring system is likely to respond faithfully under dynamic conditions. This is generally expressed as the ratio of the amplitudes of the largest (maximum) signal to the smallest (minimum) signal to which the system is

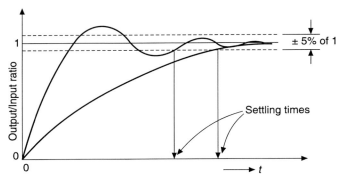

Figure 1.10   Effect of damping on the settling time of indicating instruments.

subjected and the system can handle satisfactorily. The ratio is usually expressed in dB. A dynamic range of 60 dB indicates that the measuring system can handle a range of input signals of amplitudes of 1000 to 1.

Dynamic calibration involves measurement of the dynamic characteristics. A detailed study regarding the behaviour of first- and second-order systems is necessary to appreciate the test procedures for assessment of dynamic characteristics. Chapter 4 shows some of the factors related to the first- and second-order systems.

1.6   ERRORS IN MEASUREMENT

The measurement of a quantity involves the reading of the indicator of an instrument in simple cases or computation of its value from a set of readings of the components and instruments that are put together or assembled to constitute the measuring system. In each measurement, an observer is involved, who is expected to be able to ascertain the extent to which the measured value is close to the true value. In other words, he has to estimate the error in the measured value. No measurement is possible without having some error in the measured value. It is incumbent on the part of the observer to realize the extent of error likely to exist and indicate the same, while furnishing the measured value. It would enable him as well as others to judge whether the measurement is satisfactory or an additional effort is necessary to repeat or improve the measurement so as to minimize the stated error. Error in the measured value is likely to arise due to various reasons, and the observer is not always aware of all the sources of error. He himself becomes a source of error if he has done the measurement carelessly or if he has selected an inappropriate technique for the measurement. If he is a conscious experimenter, he may repeat the measurement not once but many times, depending on the situation. Some errors are likely to arise due to the poor quality and reliability of the instruments and components used while others may be external to him and his instrument. If the measurand remains steady in its value and unvarying with time, the experimenter has larger scope of ascertaining the error. The measurement process is not always confined to steady values and it so happens that the measurand is accessible to him for a short time only, though remaining steady, or its value may be changing with time. The responsibility of identifying the sources of error and fixing or eliminating the errors due to these sources rests with the experimenter. These are classified into three categories: gross errors, systematic errors, and random errors. These are now discussed.

Gross errors These depend largely on the care and vigilance observed by the experimenter and can be minimized to a large extent, if not totally.

Systematic errors These are caused by inherent shortcomings of the instruments and components used in measurement.

Random errors These errors are due to sources and causes that cannot be specified for their effects with certainty on the measured value. They will inevitably remain even if all the systematic and gross errors are eliminated. They are uncontrollable and unavoidable in any measurement and their effect is normally evident by the random character of the readings obtained when measurement is repeated several times. A statistical analysis of the large number of measured values has to be undertaken to obtain the most probable value and estimate the uncertainty attendant on the measured value.

As the gross errors and systematic errors are avoidable and correctable or determinable, they are further elaborated.

1.6.1 Gross Errors

Gross errors are mostly due to lack of knowledge, judgement and care on the part of the experimenter. It is essential to appreciate that, in the process of measuring, the quantity that is being measured undergoes certain changes, though it is difficult to know the extent of these changes. The very act of measuring the quantity alters its value and the experimenter should be able to select the proper measuring technique or instrument that disturbs the measurand to negligible or permissible and acceptable limits. 'Zeroing' the instrument and zero adjustment, if done, before the start of the experiment, reduce the errors in measurement. By making use of the antiparallax mirror, readings can be rendered more precisely. Measuring instruments and methods of measurement for a particular quantity are abundant in literature, but each is recommended for use under certain conditions of operation and with certain assumptions. Observation of the relevant precautions for each measurement and incorporation of proper corrections by computation, in the measurements carried out in different but known operating conditions, render the measured data more accurate and reliable. A meticulous experimenter normally enhances the status of his measurement by repetition.

1.6.2 Systematic Errors

Systematic errors are mostly concerned with the instruments and components used in measurement, which are provided with information related to their static and dynamic characteristics. The significance of differences in the stated and specified characteristics of one instrument with the other should be fully appreciated before they are selected for application. Confirmation of the stated accuracy limits may be done by effecting calibration against standards, before using them. Environmental errors are due to conditions external to the measuring system and all physical systems are affected by temperature, pressure, humidity, magnetic and electric fields or mechanical vibrations. Estimating the extent to which they can degrade the quality of measurement and adopting measures such as air-conditioning, magnetic shielding or some other corrective measure, enable the experimenter guarantee greater accuracy in his measured values.

The dials of variable air capacitors and the scales of deflectional indicating instruments are provided with markings that enable the location of the pointer against the scale precisely. The closeness with which a reading can be obtained depends on the thickness and spacing of the markings. To enable the reading to be obtained with greater precision, some of the instruments are provided with knife-edge pointer and antiparallax mirror and some with a magnifying lens. Selection of instruments and components provided with such additional means and their utilization reduce reading errors.

Wherever possible, the same measurand may be measured by more than one instrument or method in order to ensure that the measurement is done with satisfactory accuracy. Instruments of the same make and specifications are preferred and used by replacing one with the other. Instruments of different make and different specifications may be used to estimate the error due to loading also. It goes without saying that all the above efforts are worthy of attempt and adoption when a measurement is undertaken in a new situation.

1.6.2.1   *Estimation of systematic errors*

It is conventional to record all the digits of the measured value displayed by the digital instrument. If it is analog indication, the reading is estimated to the smallest detectable quantity. The number of significant figures of the measured value is expressive of the preciseness with which the reading is obtained. If the reading is 142.5, it is understood that the value is closer to 142.5 than to 142.4 or 142.6. Another way of expressing the reading is as 142.5 ± 0.05, indicating that the value lies between 142.45 and 142.55.

If a quantity is expressed in three significant figures, it may be written as

$$484 \times 10^2 \quad \text{or} \quad 48.4 \times 10^3 \quad \text{or} \quad 0.0484 \times 10^6$$

but never as 48,400 or 48.40×10^3 since there is a total uncertainty of one unit in the last digit. It is also not correct to write the value as 52.0 if its measured value is found to be 52.

Though the above convention is not strictly observed by many, it is to be understood that the uncertainty is decided by the last significant figure and not by the last of the number of zeroes added after the figure. If two values are recorded with different degrees of preciseness, any mathematical operation carried out on them makes the result as precise as the less precise value; this will be clear from the following example: If $R_1 = 24.4$ ohm and $R_2 = 1.526$ ohm, then $R_1 + R_2$ is to be expressed as 25.9 ohm only, but not as 25.926 ohm. The sum of the values and hence the difference of the values cannot be expressed in more number of digits than the minimum that any of the values has.

The uncertainty attendant on the sum or difference of any two values will be more than the uncertainty of each and will be equal to the sum of the uncertainties, as seen from the following examples: If two values are given as 200 and 800, it is understood that each has an uncertainty of ±0.5, and hence when added or subtracted the result has an uncertainty of ±1. Here the sum is 1000 and is the transition value from 3-digit to 4-digit expression. The number should not be truncated to three digits as the primary numbers are of three digits only.

In multiplication or division, the percentage uncertainty of each value is considered, for determining the significant figures in which the value of the product is presented. Take the value of 51 with a specified uncertainty of ±1%. When 51 is squared, the value is 2601, but it should be written as 26×10^2, specifying that the uncertainty is ±2%, and the value of the product should be represented in the minimum number of digits in which the value is recorded. Similarly, when 51 is multiplied by 10.25 (with an uncertainty of ±0.05%), the product is 522.75, but because of the uncertainties of the original values, it should be represented as 52×10, signifying the extent of uncertainty on the product as ±1%.

It should be remembered that the above conventions and approach are recommended for adoption when dealing with measured values. But if the values are precise numbers with no uncertainty on them, the product and the results of mathematical operations should be presented in all the digits as displayed on calculators, and truncation is applied only when the measured values are under consideration. Students are advised to adopt suitable truncation to present the end-result when computations are done using electronic calculators for solution of problems containing measured data.

In all the above examples, it is emphasized that the end-result of the mathematical operation cannot have and cannot be represented in such a way as to suggest an uncertainty lower than that of the least poorly measured value, used in the operation.

1.6.2.2 Computations with precise values

With the availability of digital instruments possessing high accuracy, it is possible to obtain readings in four or five digits for the measurands. When simple mathematical operations are desired, the end-result is obtained with the desired accuracy by following the simple rules given below, or by using an electronic calculator.

The product ab of two values a and b is expressed as

$$ab = a(c \pm x) = ac \pm ax$$

where $b = (c \pm x)$ and c is so chosen as to become a whole number (integer) and x a fraction. 12.65 may be treated as 12 and 0.65 or as 13 and −0.35.

Similarly, the quotient b/a is expressed as

$$\frac{b}{a} = \frac{c \pm x}{a}$$

or

$$\frac{b}{a} = \frac{a \pm x}{a} = 1 \pm \frac{x}{a}$$

depending on whether b and a are too far apart from each other or too close.

The difference between two fractions is expressed as follows:

$$\frac{a}{b} - \frac{c}{d} = \frac{ad - bc}{bd}$$

When the two fractions are near to unity, these may be written as

$$\frac{a}{b} - \frac{c}{d} = \left(1 \pm \frac{x}{b}\right) - \left(1 \pm \frac{y}{d}\right)$$

where $a = (b \pm x)$ and $c = (d \pm y)$.

Binomial expansion series represented in the form

$$(1 + x)^n = 1 + nx + \frac{n(n-1)x^2}{2} + \ldots \tag{1.1}$$

is helpful in many cases when quick and reasonably accurate end-result is required.

The measurement of a particular quantity necessitates, at times, utilization of readings obtained from more than one instrument or component, the indicated values of which are known to have a certain error. The unknown quantity has to be calculated by using a formula relating it and the readings obtained. The extent of error attendant on the end-result may be tolerable if the relationship or formula used is simple. In calculating the maximum possible systematic error on the end-result, it is necessary to assume that the individual errors of each reading and measured value are all of such polarity as to alter the result in the same direction. It is the most optimistic way of judging the effectiveness of a measuring technique employed. Hence, it is necessary to appreciate that a measuring technique involving less number of readings and measured variables and, of course, with a simpler relationship between them and the measurand, is likely to yield less systematic error on the measurand. When an instrumentation system is designed, a large number of stages are connected in cascade, each effecting a conversion or modification of its input signal. It is necessary therefore to ensure that the overall accuracy between the input quantity and the end-result is within the desired limits.

If the end-result y is given by the sum of three values, u, v and z each of which has its corresponding systematic error given by $\pm \delta u$, $\pm \delta v$ and $\pm \delta z$, then the maximum possible error on y is

$$\delta y = \pm (\delta u + \delta v + \delta z)$$

The percentage error is given as

$$\left(\frac{\delta y}{y}\right) \times 100 = \pm \left(\frac{\delta u + \delta v + \delta z}{y}\right) \times 100 \tag{1.2}$$

or

$$\left(\frac{\delta y}{y}\right) \times 100 = \pm \left(\frac{u}{y}\frac{\delta u}{u} + \frac{v}{y}\frac{\delta v}{v} + \frac{z}{y}\frac{\delta z}{z}\right) \times 100$$

If $y = uv$ or $y = u/v$, it can be proved that the percentage error of y is

$$\left(\frac{\delta y}{y}\right) \times 100 = \pm \left(\frac{\delta u}{u} + \frac{\delta v}{v}\right) \times 100 \tag{1.3}$$

If the end-result involves the products and quotients of a number of values which have percentage systematic errors $\pm a\%$, $\pm b\%$, $\pm c\%$, ..., then the maximum systematic error in the result is $\pm (a + b + c + d + \ldots)\%$.

If $y = u^n$, where n may be positive or negative, integral or fractional, then

$$\left(\frac{\delta y}{y}\right) \times 100 = \pm \left(n\frac{\delta u}{u}\right) \times 100 \tag{1.4}$$

1.6.3   Statistical Analysis of Random Errors

Random errors are also known as residual errors that remain even if all systematic errors are eliminated. It is difficult to locate and isolate the sources of these errors in the measurement, and the experimenter would not know when and why these errors creep into the measured value. Although random errors are not much of a problem in electrical measurements, situations arise in scientific and engineering investigations when measured data display an unaccountable variability (or variance) in the readings. The experimenter-investigator should be conversant with the statistical methods so that he could apply them intelligently for the estimation of the best value of a quantity. Application of the statistical methods is possible only when measurement of the same quantity is repeated many times under the same conditions. It is expected that all systematic errors are removed from the measured values or are assumed to be very small compared to random errors. Even if a certain component is mass-produced by automatic methods, its dimensions or parameters are likely to show random variations in its nominal value when measured, and it is important that the quality assurance engineer be capable of ascertaining the probability of its nominal value lying within certain narrow limits. When measured data is processed by electronic circuits, the end-results displayed are likely to be affected by random noise due to internal and external sources. It is thus essential that the experimenter be able to declare with certainty the probability or the likelihood of the measured value lying within certain narrow limits of the nominal value, although every attention is paid to eliminate the statistical component of error in the measured value.

When a measurement is repeated N times, N readings are obtained as, say $x_1, x_2, x_3, ..., x_N$. A digital instrument made to read the value up to third decimal place, may exhibit randomness in the display of the least significant digit. All the N observations or readings may be grouped into regions of width Δx, which may be taken as the least count of the instrument or the smallest increment in the value that is responded by the instrument (resolution). Each reading or measured value $x_i(i = 1, 2, 3, ..., N)$ is called a *variate,* and the group of N variates is called the *raw data.*

The mean of a set of variates x_i is given by

$$\bar{x} = \frac{(x_1 + x_2 + x_3 + ... + x_N)}{N} = \frac{1}{N} \sum_{i=1}^{N} x_i \tag{1.5}$$

and is simply the arithmetic mean of all readings. The mean is normally taken as the *most probable* value or *the best* estimate of the nominal value, in electrical measurements. The more the number of readings used for mean, the less the random error.

Deviation is the departure of a variate or a given reading from the arithmetic mean and is given by

$$d_1 = x_1 - \bar{x}, \qquad d_i = x_i - \bar{x} \tag{1.6}$$

The deviation may be positive or negative and hence the sum of all deviations is likely to become zero, when the number of variates is large. The mean absolute deviation $\bar{d}$ is defined as the sum of all the absolute values of the deviations (independent of polarity) divided by the number of readings and may be used as an index of the precision of the measurement.

If the deviation d_i of a variate x_i is defined as the difference of x_i from an arbitrary value x instead of $\bar{x}$, it can be proved that the mean of the readings is the most probable value.

The square of the deviation is

$$d_i^2 = (x_i - x)^2 = x_i^2 - 2xx_i + x^2$$

The sum S of the squares of all N deviations is

$$S = \sum_{i=1}^{N} d_i^2 = \sum_{i=1}^{N} x_i^2 - 2x \sum_{i=1}^{N} x_i + Nx^2 \tag{1.7}$$

The value of x that makes S a minimum is obtained by finding dS/dx and equating it to zero. The condition is obtained from

$$-2\sum_{i=1}^{N} x_i + 2Nx = 0$$

giving

$$x = \frac{1}{N} \sum_{i=1}^{N} x_i = \bar{x}$$

A measure of the extent of random error in the measurement may also be given by stating the root-mean-square (rms) deviation which is obtained from

$$D = \left(\frac{1}{N} \sum_{i=1}^{N} d_i^2 \right)^{1/2} \tag{1.8}$$

When N is very large, D is called as the standard deviation σ, and its square σ^2 as the variance or dispersion of x about the most probable value.

When N is not very large, σ is given by

$$\sigma = \left(\frac{1}{N-1} \sum_{i=1}^{N} d_i^2 \right)^{1/2} \tag{1.9}$$

1.6.3.1 *Graphical representation of data*

All the measured values of a particular variable may be divided into groups so as to enable graphical display of the data. The total interval over which the measured values or the variates are scattered is subdivided into smaller intervals of equal width Δx. The smallest increment, which an instrument can respond to and indicate, may be considered as Δx, or it may be of larger width depending on convenience. The grouped data is graphically illustrated in Figure 1.11. This looks like a bar graph with each bar width equal to Δx. The bar heights are equal to the number of measured values lying within each interval Δx and is called the frequency f of the variate. The bar height is usually expressed in normalized form, i.e., the relative frequency, f/N. For relatively small values of N around 20 or 25, the graphical

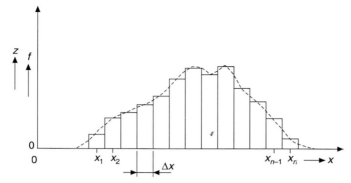

Figure 1.11 Illustration of grouped data.

representation looks like a series of pulses close to each other, and is called a histogram or frequency-distribution graph. It can be approximated to a continuous curve by joining all the top-centres of each pulse. As N becomes larger and Δx is made smaller, the histogram tends towards a smoother curve.

The histogram can also be drawn with the bar height represented by z, where z stands for

$$z = \frac{f/N}{\Delta x} \qquad (1.10)$$

The area of a particular bar is thus numerically equal to the probability that a particular reading will fall within the associated interval. The area of the entire histogram will be unity. Again, if N is very large, and Δx very small, the graph becomes a smooth curve. In the limiting condition, the curve relates x with z (= $p(x)$), with $p(x)$ standing for the probability density distribution function. Thus the probability of a measured value or variate lying between any two values, x_1 and x_2, is given by

$$P(x_1 < x < x_2) = \int_{x_1}^{x_2} p(x)\, dx \qquad (1.11)$$

Besides, the probability of a reading within any chosen value of x is given by

$$F(x) = \int_{-\infty}^{x} p(x)\, dx \qquad (1.12)$$

with $F(x)$ being called the cumulative distribution function.

The commonly used density distribution function is the normal or Gaussian function given by

$$p(x) = \frac{1}{\sigma\sqrt{2\pi}} \exp\left[-(x - \mu)^2/2\sigma^2\right] \qquad (1.13)$$

for values of x stretching to infinity in either direction. μ is normally taken as the mean $\bar{x}$, of a large number of measured values; otherwise, it is the most probable value of x about which the scatter of measured values takes place. The shape of the curve depends entirely on σ as seen from Figure 1.12(a) and $\bar{x}$ (or μ) serves to locate the position of the curve along the x-axis. Figure 1.12(b) shows the curve of $F(x)$ with x. In practice, however, the measured data may not conform totally to the Gaussian distribution as the values of x are not right up to infinite values.

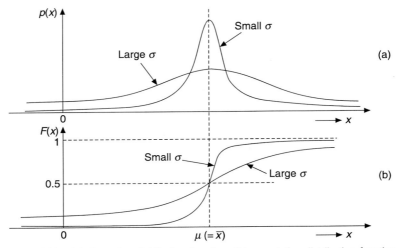

Figure 1.12 (a) Gaussian distribution function; (b) cumulative distribution function.

However, a check on the measured data is necessary to determine whether it can be in close agreement with the normality pattern revealed by Gaussian function. If the experiment is conducted carefully and a large number of measured values are available, the Gaussian distribution function is directly utilized to determine the extent of the random errors present in the measurements, while taking it for granted that $\bar{x}$ may be used in place of μ. Defining a new parameter h equal to $1/\sigma\sqrt{2}$, the function is rewritten in terms of the deviation d as

$$p(d) = \frac{h}{\sqrt{\pi}} \exp\left(-h^2 d^2\right) \qquad (1.14)$$

The parameter h reveals the spread or scatter of the measured values about $\bar{x}$. The larger the value of h, the greater peak the distribution curve has. A larger h implies higher precision and consistency in measurement and hence it is treated as precision index. The shape of $p(d)$ curve is the same as $p(x)$ curve except that there is a shift in the abscissa origin from μ (= $\bar{x}$) to zero. Assuming that all the measured data is in total conformity with the Gaussian distribution function, it can be shown that 68.3 per cent of the measured values lies within a deviation d equal to $\pm\,\sigma$ about the mean value $\bar{x}$. Corresponding values of other deviations and probability percentage are given in Table 1.2.

Table 1.2 Corresponding values of deviation and probability

Probability (%)	Deviation d
50.0	$\pm\, 0.6745\sigma$
68.3	$\pm\, \sigma$
95.4	$\pm\, 2\sigma$
99.7	$\pm\, 3\sigma$

These percentage values are the areas under the curve between the respective d values.

The probable error r is defined as the deviation d equal to $\pm 0.6745\sigma$ within which 50% of the measured values lie scattered about $\bar{x}$. While standard deviation is usually referred to in statistical work, probable error is used to estimate the random error in experimental work.

It is appropriate to understand at this stage that the value of $\bar{x}$ determined from the experimental data may not be true mean value μ, and hence the standard deviation too. It is considered proper to signify the standard deviation computed from experimental data, by s instead of the 'true' σ so that s would stand for sample standard deviation.

In practice, the number of measured values is always finite, about 20 or 25 and in such a case, there is a need to test the data for 'normality'. One common method involves the use of the *chi*-squared (χ^2) statistical test. In brief, the test consists of dividing all the N values into a certain number of groups with each group having values not less than 4 or 5. There should be a minimum of 5 groups, thereby making N not less than 20. The mean of each group is obtained as x_1, x_2, x_3 etc., and it is quite logical to presume that all these mean values are themselves governed by Gaussian distribution having a standard deviation σ' different from the standard deviation σ obtained by considering all the N values as one group. σ' is known as standard error of the mean and is given by $\sigma/\sqrt{N}$. The precision index is thus increased and is proportional to $\sqrt{N}$. For detailed analysis of the χ^2-test procedure and other significance tests for detecting whether differences between sets of observations are due to some real reason or merely due to a chance factor, the reader is advised to refer to textbooks on Statistical Analysis.

As already stated, mass produced components are tested for conformity with the nominal value of the component. If the nominal value of a component is 100, it is expected that when a hundred of them are measured, the mean value should be equal to 100. If the sample standard deviation is 0.25, then it is implied that 68 per cent of all the components produced have their values lying between 99.75 and 100.25. In case the mean happens to be 99.85, with the standard deviation remaining the same, the same percentage of components will have their values lying between 99.60 and 100.10. The departure of the mean value from the nominal value may be attributable to the systematic error of the measuring system or a flaw in the automatic production process.

The probable error of a function of more than one measured value, whose probable errors are known, is evaluated in a different way than the one adopted when dealing with systematic errors. The overall error should be calculated from the knowledge of the probable error of each measured value, and hence the combined error is expressed as follows after obtaining the probable error on the function due to each measured value: If $M = f(u, v, z)$, then the components of the probable error due to u, v and z are

$$\frac{\delta M}{\delta u}\Delta u, \quad \frac{\delta M}{\delta v}\Delta v, \quad \frac{\delta M}{\delta z}\Delta z$$

and the probable error ΔM of M is given by

$$\Delta M = \left[\left(\frac{\delta M}{\delta u}\Delta u\right)^2 + \left(\frac{\Delta M}{\delta v}\Delta v\right)^2 + \left(\frac{\delta M}{\delta z}\Delta z\right)^2\right]^{1/2} \tag{1.15}$$

If Δu represents $a \pm 2s$ limit on u, then ΔM represents $a \pm 2s$ limit on M, and hence 95.4 per cent of the values of M can be expected to fall within these limits.

1.7 CALIBRATION AND STANDARDS

Calibration is an essential process to be undertaken for each instrument and measuring system as frequently as is considered necessary. A reference standard at least ten times more accurate than the instrument under test is normally used. The calibration process is simple. It consists of reading the standard and test instruments simultaneously when the input quantity is held constant at several values over the range of the test instrument. The calibration is better carried out under the stipulated environmental conditions. All industrial grade instruments can be checked for accuracy in the laboratory by using the working standards. All modern electronic instruments that are likely to drift in sensitivity are provided with built-in calibration facility by using suitable circuits along with Zener diodes.

While calibrating, it is customary to take readings both in the ascending and descending order. Calibration thus reveals some of the inherent flaws in the electromechanical instruments and other mechanical transducers involving elastic elements. It is essential to check the linearity between the input and output quantities of many transducers and declare the extent of nonlinearity likely to exist for the range for which it is to be used.

Generally, certification of an instrument manufactured by an industry is undertaken by the National Physical Laboratory and other authorized laboratories where the secondary standards and the working standards are kept. The test instrument is calibrated under several environmental conditions in order to ensure that the grade of performance conforms to its stated specifications. Isolation of systematic errors and random errors is difficult. While calibration in laboratory may be a simple process, statistical approach is necessary at times, as can be seen from the following discussion.

1.7.1 Process of Calibration

The calibration curve of an indicating and recording instrument yields information pertaining to its readings for several input values selected within its full-scale value. The input value is adjusted to an integral value and read on the scale of a working standard instrument. The reading of the working standard is taken as true value and if the corresponding reading of the instrument under calibration is exactly the same as the reading of the working standard, then there is no error. If the situation repeats for all input values, the test instrument is said to be as good as the working standard; otherwise it is the deviation between these readings that one has to confirm. As already stated, even if all the conditions of the environment and the input quantity are held constant, it is quite likely that repeated readings of the two instruments may exhibit a few random errors. In the ideal case, if the readings are equal to each other and have the same indicated value in the least significant digit and in the same number of digits, then the calibration curve between the input and output quantities will be a perfect straight line with a unity slope. In practice, the reading of the test instrument cannot be read to the same number of digits as that of the standard, and each reading is likely to exhibit a certain amount of dispersion around a value which itself has certain deviation from the reading of the standard.

Under these conditions the calibration curve looks like the one shown in Figure 1.13 and is represented as

$$X_o \; (= Y_o) = mX_i + C \tag{1.16}$$

where

X_o = output quantity
X_i = input quantity
m = slope of the calibration curve
C = a constant

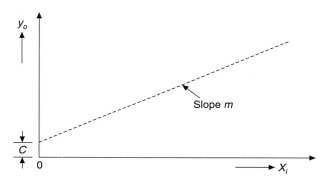

Figure 1.13   Calibration characteristic of an instrument.

In case a transducer is under calibration, Y_o is its output quantity having its dimensions totally different from those of X_i. It is not only essential to determine the values of m and C from the characteristic, but also to define and describe the linearity of the transducer.

The experimenter adjusts the zero of the instrument (or transducer if it has one) and then the input quantity is raised slowly to assume the first integral value where it is controlled to remain constant. The source of input quantity may not always remain steady at this value, in which case readings of the standard and test instruments are taken simultaneously or two observers take the readings at the same instant. The procedure is repeated at other chosen values of input quantity up to the full-scale value and while decreasing it to zero. These values, when plotted, may invariably project a pattern shown in Figure 1.14. The experimenter may be satisfied in most cases to draw a straight line that fits the scattered points by visual judgement or by ensuring that the line has some points above and some below the characteristic. In such a case, it is known as the average calibration curve. Or, the experimenter may choose to define the characteristic by some other criterion.

When high accuracy is attempted in the calibration, it is necessary to repeat the entire calibration as many as 20 to 25 times so that the most probable value for each input quantity is obtained, and these values are plotted to determine the average calibration curve. Even then, if a certain extent of scatter is noticed, it is common to apply the least-squares criterion which minimizes the sum of the squares of the deviations of the X_o (or Y_o) from the input values of X_i from the best-fit line. If the sum is given as

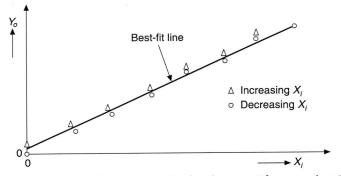

Figure 1.14 Calibration characteristic of an instrument from experiment.

$$S = \sum_{i=1}^{N} [Y_i - (mX_i + C)]^2 \tag{1.17}$$

where,

N = total number of data points

Y_i, X_i = output and input quantities at the ith data point

Then the derivatives of the sum S with respect to m and C are set to zero. Performing these operations, we obtain

$$NC + m \Sigma X_i = \Sigma Y_i$$

$$C \Sigma X_i + m \Sigma X_i^2 = \Sigma X_i Y_i$$

and solving them, we get

$$m = \frac{N \Sigma X_i Y_i - (\Sigma X_i)(\Sigma Y_i)}{N \Sigma X_i^2 - (\Sigma X_i)^2} \tag{1.18}$$

$$C = \frac{(\Sigma Y_i)(\Sigma X_i^2) - (\Sigma X_i Y_i)(\Sigma X_i)}{N \Sigma X_i^2 - (\Sigma X_i^2)} \tag{1.19}$$

The standard deviations of m and C are

$$s_m^2 = \frac{N s_y^2}{N \Sigma X_i^2 - (\Sigma X_i)^2} \tag{1.20}$$

$$s_C^2 = \frac{s_y^2 \Sigma X_i^2}{N \Sigma X_i^2 - (\Sigma Y_i)^2} \tag{1.21}$$

where,

$$s_y^2 = \frac{1}{N} \Sigma (mX_i + C - Y_i)^2$$

The standard deviation of output quantity is s_y. If it is assumed that this standard deviation would be the same for all values of X_i, then s_y can be calculated by using all the data points, without having to repeat any one X_i many times. The alternative approach is to determine the standard deviation of Y_o at each data point and verify whether it is the same for all data points. When N is very large, both approaches may yield the same s_y.

Using the values of m and C, the reading of the test instrument can be calculated for a desired value of X_i. If for a particular case, $X_i = 100$ units, $m = 1.02$, and $C = 0.4$ units, then the test instrument should read 102.4. But if it reads 102 units, then there is a bias or systematic error of -0.4 unit at this reading. Also, from s_y, the probable error on the instrument reading for $2s$ or $3s$ limits can be stated. Thus we are able to isolate the probable and systematic errors by a suitable procedure of calibration. The total inaccuracy of the test instrument is revealed by the calibration process and is defined by the combination of bias (systematic error) and random error (lack of consistency).

When a transducer is calibrated, Y_o and X_i are of different dimensions and it may so happen that C is finite and m be any value other than unity. The overall sensitivity is given by m and the repeatability of the transducer which is an essential characteristic is determined by the value of s_m. The value of C when $X_i = 0$ shows the zero-error and that of s_C, the zero instability. Estimation of the value of m in different environmental conditions is essential to ensure its stability. The calibration should be undertaken for both positive and negative values of X_i for some transducers. The characteristic thus obtained is utilized to declare the static sensitivity, valid for the entire range, which may as well be able to specify the error due to hysteresis. Once the 'best-fit' straight line is drawn, the linearity is then defined in the same way as it is derived for each transducer. For transducers that are considered essentially linear, the specification of nonlinearity is equivalent to a specification of overall inaccuracy.

In practice, the output reading Y_o is obtained, and the corresponding true value and limits of error with which it can be declared, can be obtained, by using the values of m, C and s_y, as follows:

$$X_i = \frac{Y_o - C}{m}$$

$$s_x = \frac{s_y^2}{m^2} \tag{1.22}$$

The method of least squares may also be applied for determining higher order polynomials for fitting the data. It is only required to perform additional differentiation to determine additional constants. For the case of a transducer whose input-output relation is of the form

$$Y = aX^2 + bX + c, \tag{1.23}$$

the sum S is given by

$$S = \sum_{i=1}^{N} [Y_i - (aX_i^2 + bX_i + c)]^2 \tag{1.24}$$

and the derivatives given by the following are set to zero:

$$\frac{\partial S}{\partial a} = 0, \qquad \frac{\partial S}{\partial b} = 0; \qquad \frac{\partial S}{\partial c} = 0$$

The solution for the values of a, b and c is more complex than in the linear case but is easy to evaluate by using calculators and microcomputers.

In all cases of calibration, suitable and intelligent choice of reference standards, fixed and variable, enables the estimation of accuracy and linearity within the desired limits.

1.7.2   Classification of Standards

Standards of mass, length and such other physical quantities are physical devices and systems representing the fundamental unit of the particular quantity. In science and engineering activity, two kinds of units are used: fundamental and derived units. Six quantities, in all, are considered to serve as fundamental units, based on which all other quantities and their units can be defined and derived. Table 1.3 lists the six fundamental quantities, their units and symbols in the SI (Systemé International d'Unites) units which were agreed upon internationally in 1960 for adoption as the standard units in all countries.

Table 1.3   Fundamental SI quantities, their units and symbols

Quantity	*Unit*	*Symbol*
Length	metre	m
Mass	kilogram	kg
Time	second	s
Electric current	ampere	A
Thermodynamic scale	degrees Kelvin	K
Luminous intensity	candela	cd

The derived units are expressed in terms of these six basic and fundamental units by making use of physical laws defining them.

Standards have been developed for all the fundamental units as well as some of the derived mechanical and electrical units. As indicated earlier, they are classified into the following categories:

1. International standards
2. Primary standards
3. Secondary standards
4. Working standards

International standards are those defined and agreed upon internationally. They are maintained at the International Bureau of Weights and Measures and are not accessible outside for calibration of instruments.

Primary standards are those maintained by national standards laboratories in different parts of the world and they are also not accessible outside for calibration. The primary standards established for the fundamental and some derived units are independently calibrated by absolute measurements at each of the national standards laboratories and an average value for the primary standard is obtained with the highest accuracy possible. These are, of course, used for verification and calibration of the secondary standards.

Secondary standards are usually fixed standards for use in industrial laboratories, whereas *working standards* are for day-to-day use in measurement laboratories. Working standards may be lower in accuracy in comparison to secondary standards. The accuracy of secondary standards is maintained by periodic comparison with the primary standards, whereas working standards may be checked against secondary standards.

The above standards for the fundamental and some of the derived quantities are briefly introduced in the following section so as to acquaint the reader with the knowledge of the accuracies attainable when calibration is undertaken against them.

1.7.3 Standards for Calibration

The primary standard of mass is defined as the mass of a cubic decimeter of water at its temperature of maximum density of 4°C, and is maintained at the International Bureau of Weights and Measures near Paris. The primary standard of mass is preserved by the National Bureau of Standards (NBS) to an accuracy of 1 part in 10^8, and the secondary standards to 1 part in 10^6. Working standards of mass are available in a wide range of values to an accuracy of 5 parts in 10^6 (or 5 ppm).

The primary standard of length is one metre long and is the distance between two lines engraved on a platinum-iridium bar preserved in Paris. In 1960, the metre was redefined more accurately and represented by an optical standard. It is defined as equal to 1,650,763.73 wavelengths in vacuum of the orange-red radiation of the krypton 86 atom. The radiation is obtained from a krypton discharge lamp. The accuracy is 1 ppm. Working standards of length are gauge blocks of steel, with two plane parallel surfaces separated by distances having uncertainties of the order of 0.5 µm.

The primary standard of time is one second and is now defined as the interval of time corresponding to 9, 192, 631, 770 cycles of the atomic resonant frequency of cesium-133. The second defined in this form has an accuracy much greater than that defined as 1/86,400 of the mean solar day. An atomic clock with a precision of about 1 µs per day is used as the primary standard of frequency at the NBS. Since frequency is the inverse of time interval, an atomic clock serves as the primary standard for time and frequency and is accurate to 1 part in 10^{11}. By radio-broadcasting signals related to the frequency of the standard, the NBS renders the service of calibration of the secondary and working standards. The working standard of time and frequency is a piezoelectric crystal oscillator of 10 MHz with the crystal kept in a temperature-controlled oven. The drift in frequency is of the order of 3 in 10^7 over a period of one week.

The fundamental unit of electric current is the absolute Ampere. The primary standard is a current balance which weighs the force exerted between two current-carrying coils. The force can be measured accurately and calculated by using a simple relationship involving the geometrical dimensions of the coil.

Equally effective secondary and working standards of current can be set up by using correspondingly accurate voltage and resistance standards.

The absolute value of the ohm in the SI system is defined in terms of the fundamental units of length, mass and time. But the primary standard of resistance of one ohm is a coil of wire of manganin mounted in a double-walled sealed container maintained at constant temperature. A set of 10 one-ohm standards enable cross-check among them. Their stability is found to be within a few parts in 10^7 over several years. The secondary and working standards are likewise made of manganin but made available with values ranging from 10^{-5} to 10^6 ohms. The low-resistance standards are provided with four terminals; two for current and two for potential.

The primary voltage standard is the Weston normal (saturated) cell developing an emf of 1.01858 V (absolute) at 20°C. A number of such cells, kept at 20°C controlled, to within ±0.01°C, have been found to be remarkably stable over a period of 20 years. The drift in emf is only about 1 µV per year. The temperature coefficient of emf is 46 µV/°C. In contrast, the unsaturated type Weston cell has a negligible temperature coefficient of emf and is used as portable secondary and working standard. It is not as stable as the saturated type and its emf may be in the range of 1.018 V–1.020 V. Laboratory working standards are also available with Zener diode as the voltage reference element and are found to be stable with only a drift of 10 ppm in a month. Their small size and accuracy have been responsible for the development of electronic instruments with built-in voltage reference, with consequential improvement in accuracy of indication.

Standards of capacitance are air capacitors whose values of capacitance can be calculated from their dimensions very accurately. NBS maintains primary standards of capacitance, and secondary and working standards are likewise designed with values extending over a wide range. Some of them use mica as the dielectric material as it is found to be a very stable dielectric. Fixed air capacitors are available with accuracies better than ±0.01% and mica capacitors of ±0.1%.

Primary standard of inductance is a standard of mutual inductance, serving as the primary standard for both mutual and self-inductance. It is an air-cored coil system, whose value is checked against other primary standards and has an accuracy of ±0.01%. It is maintained at NBS. Working standards are available for a wide range of values as both fixed and variable type, with accuracies of ±0.1%.

A working standard of magnetic flux is designed for use in laboratories and consists of a stable permanent magnet as the source of constant magnetic flux. It is known as Hibbert's magnetic standard. A coil of known number of turns cuts the flux while falling due to gravitational force, always from the same height, and the induced emf is calculated in terms of the flux linkages, the value of which is guaranteed to an accuracy of ±0.001%.

The thermodynamic temperature is expressed in degrees Kelvin and the triple point of water is fixed as 273.16 K. This is the temperature at which water exists in all states, i.e., ice, liquid water and its vapour. The other fundamental point is fixed at the boiling point of water and the interval is divided into 100 K. The primary standard for temperature uses platinum wire since platinum is a stable metal. The thermometer is used to serve as reference for the above fixed points after its resistance values are determined accurately at these temperatures.

The primary standard for luminous intensity is a black body radiator held at the temperature of solidification of platinum ($\approx$ 2042 K). Its unit is candela which is defined as one-sixtieth of the luminous intensity per cm^2 of the radiator surface. Secondary and working standards of luminous intensity are special tungsten filament lamps, operated at a temperature at which their spectral power distribution in the visible region matches that of the basic standard.

Pressure standards are available as secondary and working standards and their accuracy is determined after checking their values against the fundamental standards of length, mass and time. The basic standards for pressures ranging from medium vacuum to very high pressure are in the form of precision mercury columns and dead-weight piston gauges. Their accuracy varies with the range of pressure for which they are used.

Standards of force depend on standards of mass and acceleration. The acceleration due to gravity can be used as a standard as it is possible to measure it very accurately by means of a pendulum or by a falling body. The numerical value of the acceleration due to gravity can be calculated from the knowledge of altitude above sea level and latitude and hence is used to establish accurately the value of the gravitational force on a known standard mass. Thus the calculable value of g is used to calibrate the force-measuring systems, and the pendulum is the standard for acceleration with an accuracy of one ppm.

Secondary and working standards for other derived quantities, and quantities such as relative humidity depend on development of accurate measuring systems for these quantities. The fundamental quantities on which they depend are determined with high accuracy after comparison with primary or secondary standards. Usually all working standards for these derived quantities have an accuracy of $\pm 0.1\%$.

EXERCISES

1. Indicate the basic stages that constitute a measuring system and define their functions.
2. Distinguish between 'analog indication' and 'digital readout' systems and discuss the merits of each.
3. Distinguish a 'meter' from an 'instrument' and state the primary role of each.
4. Define 'instrumentation' and the importance of instrumentation systems in the modern scientific and technological achievements.
5. Explain the role of electronic circuits in the field of modern instrumentation.
6. What are 'sensors' and 'transducers', and what is the role of each in a measuring system?
7. Classify the transducers according to their characteristics and application.
8. Define 'passive' and 'active' transducers and give an example of each.
9. Distinguish between 'static' and 'dynamic' characteristics of an instrument.
10. Explain the terms 'accuracy' and 'precision' and discuss the need for having them properly matched to each other.
11. Distinguish between 'error' and 'correction' and show how they are usually expressed for an instrument.
12. Define 'total uncertainty' in the reading of an instrument and indicate the factors responsible for this.

13. Define 'repeatability' and 'reproducibility' and discuss why transducers and instruments should possess these characteristics.

14. What is meant by 'stability' of a measuring system? Indicate which class of instruments are required to be more stable.

15. Explain how the 'nonlinearity' of a measuring system is defined and estimated.

16. Define 'drift', 'threshold value' and 'dead-band' of a measuring system and give an example for each.

17. Distinguish between 'range' and 'span' of an instrument.

18. What do the terms 'input impedance' and 'output impedance' convey when related to a measuring system; how should they be chosen?

19. Define the terms 'flow variable' and 'effort variable' and explain their significance in the context of a measuring process.

20. Suggest the flow and effort variables encountered in electrical, mechanical and process measurements.

21. Distinguish between the 'static' and 'dynamic' characteristics of a measuring system and state the relevance of each in a measuring process.

22. Explain the types of test signals used for determination of dynamic characteristics.

23. Define 'dynamic error' and show how it differs with the type of input signal applied to the system.

24. Explain the term 'fidelity' of an instrument and show how it is usually expressed.

25. What does the term 'bandwidth' convey when it is quoted for a measuring instrument?

26. Define the terms 'time constant' and 'settling time' of an instrument and indicate the factors responsible for them.

27. Distinguish the important characteristics of instruments that are totally 'electrical' and totally 'electronic' in nature.

28. Distinguish between 'systematic' and 'random' errors in a measurement and how they are usually minimized.

29. Explain to what extent the experimenter is responsible for introducing errors in a measurement and state some of the ways by which he can reduce them.

30. What is the need for statistical analysis of measured values and when does it arise? Suggest a measurement process when the results are subjected to this analysis.

31. Define 'deviation' and 'standard deviation' of a set of measured values.

32. Define 'probable error' and show how it is extrapolated from the experimental values.

33. Discuss the relevance and process of calibration usually adopted.

34. What are the fundamental SI quantities and how are they defined and expressed?

35. Define the class of standards available for use and calibration process.

36. Write short notes on the standards available for calibration at both the National Physical Laboratories and laboratories of industries and institutions.

SUGGESTED FURTHER READING

Adams, L.F., *Engineering Instrumentation and Control IV*, Hodder and Stoughton, London, 1981.

Carroll, E.C., *Industrial Process Measuring Instruments*, McGraw-Hill, New York, 1962.

Considine, D.M. (Ed.), *Process Instruments and Controls Handbook*, McGraw-Hill, New York, 1957.

Cook, N.H. and Rabinowiez, E., *Physical Measurements and Analysis*, Addison Wesley, Reading (Mass.), 1963.

Doebelin, E.O., *Measurement Systems: Application and Design*, McGraw-Hill, New York, 1966.

Ernest, Frank, *Electrical Measurement Analysis*, Tata McGraw-Hill, New Delhi, 1959.

Fribance, A.E., *Industrial Instrumentation Fundamentals*, McGraw-Hill Kogakusha, Tokyo, 1962.

Holman, J.P., *Experimental Methods for Engineers*, McGraw-Hill Kogakusha, Tokyo, 1966.

Holzbock, W.G., *Instruments for Measurement and Control*, Reinhold Publishing Corp., New York, 1955.

Jones, Barry. E., *Instrumentation: Measurement and Feedback*, Tata McGraw-Hill, New Delhi, 1978.

Lion, K.S., *Instrumentation in Scientific Research*, McGraw-Hill, New York, 1959.

Liptak, B.G. (Ed.), *Instrument Engineer's Handbook*, Vols. I and II, Philadelphia Chilton, Philadelphia, 1969, 1970.

Tuve, G.L. and Domholdt, L.C., *Engineering Experimentation*, McGraw-Hill, New York, 1966.

Wightman, E.J., *Instrumentation in Process Control*, *International Scientific Series*, CRC Press, Columbus, Ohio, 1972.

CHAPTER 2

Signals and Their Representation

2.0 INTRODUCTION

Analysis of physical systems is mostly concerned with the study of the relationship between their input and output signals. A signal may be defined as the variation of any physical quantity with time. Any physical quantity acting as a stimulus to a physical system becomes an input function and is treated as an input signal. The physical system reacts to the stimulus and its response is recognized by a variation in any of its physical characteristics. The physical quantity indicating/denoting its response becomes the output signal.

All signals are generally represented graphically as functions of time, the amplitude of the signal varying with time, as shown in Figure 2.1. To have a better understanding of the nature of the signal, it becomes necessary for the variation of the amplitude with time to be described as a mathematical function or a group of such functions, each defined over a certain specific interval of time.

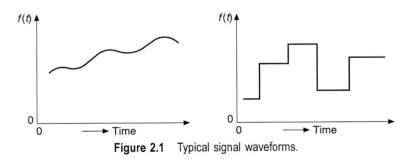

Figure 2.1 Typical signal waveforms.

Signals obtained from physical systems or representing physical phenomena are classified as either deterministic or nondeterministic. Deterministic signals are those that can be represented by an explicit mathematical function of time so that the value of the function at any desired instant can be calculated. Deterministic signals may be periodic or aperiodic. A periodic signal has a waveform that repeats itself at regular intervals of time as shown in Figure 2.2(a). An aperiodic signal occurs only once and exists for a certain duration of time only as illustrated in Figure 2.2(b). Nondeterministic signals cannot be described mathematically and are treated as random signals whose amplitude at any instant cannot be determined or predicted. Random signals are analyzed by application of probability theory and statistical techniques. Figure 2.2(c) represents a typical random signal.

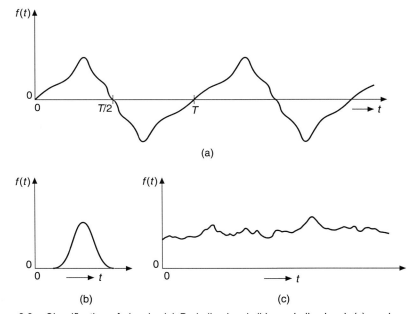

Figure 2.2   Classification of signals: (a) Periodic signal; (b) aperiodic signal; (c) random signal.

In the field of instrumentation, transducers and other physical systems and instruments are subjected to dynamic testing, with a view to assessing their dynamic performance. Also, in actual usage they are designed to measure and respond to input signals of a particular description, while rejecting unwanted signals of random nature. Evaluation of the performance of systems can be done only when the input and output signals are analyzed properly. In this context, it is necessary to recall the basic information relevant to signal representation and description.

Signals may also be broadly classified as either analog or digital. Analog signals are continuous with time and the signal is described by one single value only at any particular instant as shown in Figure 2.3(a). If it assumes values over a range at any one instant, the signal is said to have an instantaneous jump from one value to another. It may assume a new level and remain at that value for a certain amount of time as shown in Figure 2.3(b). Such a signal is known as digital signal. It may assume discrete levels, with jumps from one level to another, at certain intervals of time. Assuming the analog signal to be composed of a series of steps as shown in Figure 2.3(c), with each step at a discrete level, each of which is determined for its amplitude in binary code and a digital signal consisting of a series of pulses having only two levels (0 and 1) spaced at equal intervals of time is brought out to represent its amplitude. The process of conversion is known as analog-to-digital conversion and is accomplished by analog-to-digital converters (ADC). Figure 2.3(d) shows a typical digital signal with pulses representing a binary number, 110101.

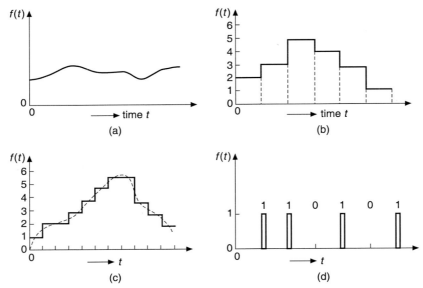

Figure 2.3 Representation of signals: (a) Analog signal; (b) discrete level signal; (c) discretized analog signal; (d) digital signal (binary).

Signals described as mathematical functions of time can be transformed into corresponding signals in frequency domain as frequency is inverse of the time period. Laplace transform and Fourier transform techniques are commonly used in signal analysis; description of commonly occurring signals in time and frequency domains is attempted in Sections 2.1–2.3. The reader may refer to standard literature for details concerning the two techniques. Signals described in frequency domain enable one to analyse systems in a convenient and easy manner.

2.1 LAPLACE AND FOURIER TRANSFORMS

Laplace and Fourier transform techniques are very powerful mathematical tools for analysis of linear systems. The linear systems under test are subjected to certain input signals of standard description and their output signals are analyzed to establish the characteristics of the system.

The Laplace transform technique is an operational method of solving linear differential equations. The differential equation is first transformed into an algebraic equation in complex frequency domain with s as the frequency variable. The Laplace transform of a function $f(t)$ is designated as $F(s)$ and is defined as

$$F(s) = \int_0^\infty f(t) e^{-st} \, dt \tag{2.1}$$

where, $s = \sigma + j\omega$, with σ standing for the real component and ω for the imaginary component. $F(s)$ is also written as $\mathscr{L}[f(t)]$, where, $\mathscr{L}$ stands for the operation carried on $f(t)$. The Laplace transform of the function $f(t)$ exists only if the condition

$$\int_0^\infty |f(t)| e^{-\sigma t} \, dt < \infty \tag{2.2}$$

is satisfied for $\sigma > 0$.

The time function $f(t)$ and its transform are called a transform pair. Table 2.1 lists the Laplace transforms for some commonly used time functions.

Table 2.1 Laplace transform pairs for some commonly used time functions

$f(t)$ [$f(t)$ is finite for $t \geq 0$]	$F(s)$
$f(t)$	$F(s) = \int_0^\infty f(t) e^{-st} \, dt$
$\dfrac{d}{dt} f(t)$	$sF(s) - f(0_-)$
$\dfrac{d^n}{dt^n} f(t)$	$s^n F(s) - \sum_{j=1}^{n} s^{n-j} f^{j-1}(0_-)$
$\int_0^t f(\tau) \, d\tau$	$\dfrac{1}{s} F(s)$
$\int_{0_-}^t \int_{0_-}^t f(\tau) \, d\tau$	$\dfrac{1}{s^2} F(s)$
$f(t - a) u(t - a)$	$e^{-as} F(s)$
$e^{at} f(t)$	$F(s - a)$
$\delta(t)$, unit impulse	1
$u(t)$, unit step	$\dfrac{1}{s}$
t, unit ramp	$\dfrac{1}{s^2}$
$\dfrac{t^n}{n!}$	$\dfrac{1}{s^{n+1}}$
e^{-at}	$\dfrac{1}{s + a}$
$\dfrac{t^{n-1}}{(n-1)!} e^{-at}$	$\dfrac{1}{(s + a)^n}$ (n is an integer)

(Contd.)

Table 2.1   Laplace transform pairs for some commonly used time functions *(Contd.)*

$f(t)$ [$f(t)$ is finite for $t \geq 0$]	$F(s)$
$1 - e^{-at}$	$\dfrac{a}{s(s+a)}$
$\dfrac{1}{b-a}(e^{-at} - e^{-bt})$	$\dfrac{1}{(s+a)(s+b)}$
$\dfrac{1}{b-a}(be^{-bt} - ae^{-at})$	$\dfrac{s}{(s+a)(s+b)}$
$\sin \omega t$	$\dfrac{\omega}{s^2 + \omega^2}$
$\cos \omega t$	$\dfrac{s}{s^2 + \omega^2}$
$\sin(\omega t + \theta)$	$\dfrac{s \sin\theta + \omega \cos\theta}{s^2 + \omega^2}$
$\cos(\omega t + \theta)$	$\dfrac{s \cos\theta - \omega \sin\theta}{s^2 + \omega^2}$
$e^{-at} \sin \omega t$	$\dfrac{\omega}{(s+a)^2 + \omega^2}$
$e^{-at} \cos \omega t$	$\dfrac{s+a}{(s+a)^2 + \omega^2}$
$\sinh at$	$\dfrac{a}{s^2 - a^2}$
$\cosh at$	$\dfrac{s}{s^2 - a^2}$

The inverse Laplace transformation is given by the complex inversion integral

$$f(t) = \frac{1}{2\pi j} \int_{\sigma_1 - j\infty}^{\sigma_1 + j\infty} F(s)\, e^{st}\, ds \tag{2.3}$$

which is a contour integral where the path of integration is along the vertical line $s = \sigma_1$ from $-j\infty$ to $+j\infty$. However, the table of transform pairs may be used to obtain $f(t)$ directly, provided the necessary form of $F(s)$ can be found in the table. The symbol $\mathscr{L}^{-1}$ is used to indicate the inverse Laplace transformation. Then

$$\mathscr{L}^{-1}\{\mathscr{L}[f(t)]\} = \mathscr{L}^{-1}[F(s)] = f(t)$$

Similarly, the Fourier transform of $f(t)$ is given by

$$F(j\omega) = \int_{-\infty}^{\infty} f(t)\, e^{-j\omega t}\, dt \tag{2.4}$$

subject to the sufficient but not necessary condition that

$$\int_{-\infty}^{\infty} |f(t)| \, dt < \infty \tag{2.5}$$

The time function $f(t)$ can be obtained from

$$f(t) = \frac{1}{2\pi} \int_{-\infty}^{\infty} F(j\omega) \, e^{j\omega t} \, d\omega \tag{2.6}$$

Equations (2.4) and (2.6) constitute the Fourier transform pair and are usually written as

$$F(j\omega) = F[f(t)]$$
$$f(t) = F^{-1} \{F[f(t)]\}$$

The Fourier transform is also known as Fourier integral. It is frequently used to describe signals that occur once in some finite time interval and never repeated.

2.2 STANDARD TEST SIGNALS

Physical systems are severely strained when subjected to sudden changes in the input conditions. Their dynamic behaviour may be studied by subjecting them to standard test signals commonly known as an impulse—a step, a ramp, and a parabolic—input function. Another common test signal of great relevance is a sinusoidal signal. Steady-state response to a sinusoidal test signal reveals the nature of the system. It is not necessary to test each system with all the input signals. Depending on the ultimate function for which the system is selected, the test signal may be chosen and tested for its response. Also, if the system function is determined from the mathematical model of the system, the time response can be estimated by making use of Laplace transform techniques.

The step-input function is shown in Figure 2.4(a) and is seen to be a signal that jumps from one level to another instantaneously. It is represented as

$$f(t) = A \, u(t) \tag{2.7}$$

where
$$u(t) = 1 \quad \text{for} \quad t > 0$$
$$\quad\;\; = 0 \quad \text{for} \quad t < 0$$

and A is the change in level (amplitude).

The Laplace transform of the step-input function is

$$F(s) = A/s \tag{2.8}$$

The ramp-input function shown in Figure 2.4(b) is a signal that starts at a value of zero and increases linearly with time. It is represented as

$$f(t) = At \quad \text{for } t > 0 \tag{2.9}$$
$$\quad\;\; = 0 \quad \text{for } t < 0$$

The Laplace transform of ramp-input function is

$$F(s) = A/s^2 \tag{2.10}$$

The parabolic-input function shown in Figure 2.4(c) is represented as

$$f(t) = At^2/2 \quad \text{for } t > 0 \tag{2.11}$$
$$= 0 \qquad \text{for } t < 0$$

The Laplace transform of a parabolic-input function is given by

$$F(s) = A/s^3 \tag{2.12}$$

When the amplitude A of the foregoing input functions is equal to unity, these functions are known as unit-step function, unit-ramp function, etc. Given the unit-step function shown in Figure 2.4(a), the other two functions shown in Figures 2.4(b) and 2.4(c) can be obtained by integration.

The derivative of a step function has a zero value at all times except at the instant at which the jump occurs, i.e. at $t = 0$. At that instant, the derivative assumes infinite value. Such a function is known as *impulse function*.

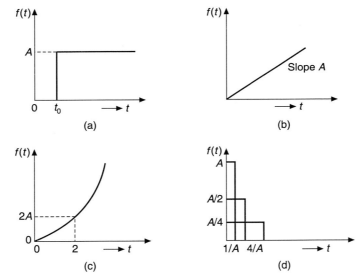

Figure 2.4   Common input signals: (a) Step-input function; (b) ramp-input function; (c) parabolic input function; (d) unit-impulse function.

A unit-impulse function is defined as a signal which has zero value at all times except at $t = 0$, where its magnitude is infinite. It is usually referred to as δ-function and is expressed as

$$\delta(t) = 0 \quad \text{for } t \neq 0 \tag{2.13}$$

$$\int_{-\varepsilon}^{+\varepsilon} \delta(t)\, dt = 1 \tag{2.14}$$

where, ε tends to zero. Since in practice, it is difficult to create an ideal impulse of this expression, it is usually approximated by a pulse of small width of $1/A$ and of height A so that the area of the pulse is unity, as shown in Figure 2.4(d). As the width ε tends to zero, the height tends to infinite value, while maintaining the area at unity.

As already stated, an impulse function is mathematically the derivative of a step function and hence the Laplace transform of a unit-impulse function is

$$\mathscr{L}\,\delta(t) = 1 \tag{2.15}$$

The Laplace transforms for other functions such as exponential, sinusoidal and exponentially varying sinusoidal signals are given in Table 2.1.

2.3 PERIODIC SIGNALS

A periodic function of time can be represented as a series given by

$$g(t) = a_0 + \sum_{n=1}^{\infty} a_n \cos n\omega t + \sum_{n=1}^{\infty} b_n \sin n\omega t \tag{2.16}$$

The function shown in Figure 2.5 is periodic in the interval T as $g(t) = g(t + T)$, and hence T is known as the period of one cycle in seconds. The frequency at which $g(t)$ is periodic is

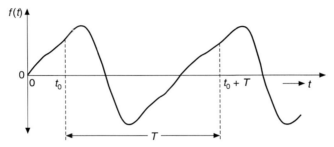

Figure 2.5 A periodic function of time.

indicated by $f(=1/T)$ expressed in Hz (cycles per second). ω is the angular frequency in radians/s and is equal to $2\pi f$. The coefficients a_0, a_n and b_n are known as the Fourier coefficients determined from

$$a_0 = \frac{1}{T} \int_{-T/2}^{T/2} g(t)\,dt$$

$$a_n = \frac{2}{T} \int_{-T/2}^{T/2} g(t) \cos n\omega t\,dt \tag{2.17}$$

$$b_n = \frac{2}{T} \int_{-T/2}^{T/2} g(t) \sin n\omega t\,dt$$

with $n = 1, 2, 3...$.

The series of these trigonometric functions is called *Fourier series* and its representation is known as Fourier series representation. The value of a_0 represents the average value of the function over the period T and is known as the dc component of the signal. The frequency

ω_1 $(= 2\pi f_1)$ corresponding to $n = 1$ is known as the fundamental frequency, and all the higher frequency components for values of $n > 1$ are known as harmonics. a_n and b_n define the amplitudes of the ac components, harmonically related to each other.

The waveform of the periodic function can be analyzed for determining the Fourier coefficients, by choosing the origin of time properly. If it is assumed that $g(t) = g(-t)$, all the '*b*' coefficients will be zero and the function is said to possess even (cosine) symmetry about the time origin. If $g(t) = -g(-t)$, then the '*a*' coefficients will be zero and the function has odd (sine) symmetry. If neither even nor odd symmetry about the time origin exists, then the Fourier series are made up of both the cosine and sine components.

Waveforms of ac supply voltages are usually sinusoidal, and consist of sinusoids of one frequency. The waveform is said to be distorted if it contains components of frequencies which are integer multiples of the fundamental frequency. When physical devices and systems are studied for their response by application of sinusoidal input signals, the output signals may be found distorted, containing components of the fundamental frequency and its harmonics. A measure of the distortion is simply the ratio of the maximum amplitude of the harmonic to that of the fundamental frequency expressed as a percentage. The total harmonic distortion or distortion factor D, due to a large number of harmonics, is given by

$$D = [D_2^2 + D_3^2 + D_4^2 + ...]^{1/2} \times 100 \qquad (2.18)$$

where,

$$D_2 = \frac{\sqrt{a_2^2 + b_2^2}}{\sqrt{a_1^2 + b_1^2}}$$

$$D_3 = \frac{\sqrt{a_3^2 + b_3^2}}{\sqrt{a_1^2 + b_1^2}}$$

For nonlinear systems the distortion in the output signal is such that the amplitudes of the harmonic components progressively decrease with the order of the harmonic n.

A rectangular (square) wave of period T is shown in Figure 2.6(a) possessing cosine symmetry. It can be seen that $a_0 = 0 = b_n$ and the a coefficients are given by

$$a_n = \frac{4A}{n\pi} (-1)^{(n-1)/2}, \quad n = 1, 3, 5 ...$$

and the function $g(t)$ is

$$g(t) = \frac{4A}{\pi} \left(\cos \frac{2\pi t}{T} - \frac{1}{3} \cos \frac{6\pi t}{T} + \frac{1}{5} \cos \frac{10\pi t}{T} + ... \right) \qquad (2.19)$$

A rectangular pulse train of period T possessing cosine symmetry is shown in Figure 2.6(b). The value of the dc component $a_0 = At_0/T$, where t_0 is the duration of the existence of pulse in the time period T. The function is given by

$$g(t) = \frac{At_0}{T} + \sum_{n=1}^{\infty} \frac{2At_0}{T} \frac{\sin (n\pi t_0/T)}{n\pi t_0/T} \left(\cos n \frac{2\pi t}{T} \right) \qquad (2.20)$$

A saw-tooth wave is shown in Figure 2.6(c) and can be represented by

$$g(t) = A \left(\frac{1}{2} - \frac{1}{\pi} \sin \frac{2\pi t}{T} - \frac{1}{2\pi} \sin \frac{4\pi t}{T} - ... \right) \tag{2.21}$$

A triangular wave is shown in Figure 2.6(d) and is given by

$$g(t) = A \left(\frac{1}{2} + \frac{4}{\pi^2} \sum_{n=1,3,5}^{\infty} n^{-2} \cos n \frac{2\pi t}{T} \right) \tag{2.22}$$

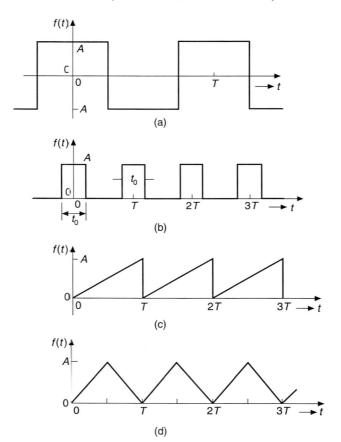

Figure 2.6 Common forms of periodic signals: (a) Rectangular wave; (b) rectangular pulse train; (c) saw-tooth wave; (d) triangular wave.

2.3.1 Complex Form Representation

It is possible to rewrite the Fourier series, making use of the exponential function exp ($j\omega t$) as

$$g(t) = \sum_{n=-\infty}^{\infty} d_n \exp (jn\omega t) \tag{2.23}$$

where

$$j = \sqrt{-1}$$

$$d_n = \frac{1}{2}\sqrt{(a_n^2 + b_n^2)}\, \exp\,(-j\phi_n)$$

$$d_{-n} = \frac{1}{2}\sqrt{(a_n^2 + b_n^2)}\, \exp\,(+j\phi_n)$$

$$\phi_n = \tan^{-1}(b_n/a_n)$$

The exponential representation of the function shows the summation to consist of frequencies having positive and negative values. Although negative frequencies do not carry any physical significance, they have been included for convenience of mathematical representation, enabling better understanding of the processes such as multiplication and modulation of two signals.

It may also be noted that the coefficients d_n and d_{-n} are complex, making $g(t)$ a complex function. But $g(t)$ is always real as d_n and d_{-n} are complex conjugates of each other. The value of d_n is evaluated from

$$d_n = \frac{1}{T}\int_{-T/2}^{T/2} g(t)\exp\,(-jn\omega t)\,dt \tag{2.24}$$

and the average value d_0 of the function is determined separately by substituting $n = 0$ in this equation.

The amplitudes of the positive and negative frequency components are seen to be equal to each other. The coefficient d_n is real when the waveform has symmetry about the time origin; otherwise it is complex.

2.3.2 Line Spectra

The results of the expressions in trigonometric and exponential form may be utilized to describe each frequency component for its amplitude and phase. If the frequencies of the components are distinctly different from each other, they can be represented graphically as the line spectra such that the ordinate at each frequency represents either its amplitude or its phase. The line spectra related to trigonometric representation constitute one-sided spectrum whereas those related to exponential form constitute a two-sided spectrum consisting of both the positive and negative frequency components.

It is observed that the two-sided spectrum represents the coefficient d_n for both the positive and negative frequencies, whereas one-sided spectrum requires both the values of a_n and b_n when the waveform does not exhibit symmetry about the time origin. The phase ϕ_n indicates the instantaneous phase of each frequency component and is dependent on the choice of the time origin.

For the case of a rectangular pulse train shown in Figure 2.6(b), the coefficient d_n is given by

$$d_n = \frac{A}{n\pi}\sin\frac{n\pi t_0}{T} \tag{2.25}$$

and the function $g(t)$ by

$$g(t) = \frac{At_0}{T} + \sum_{n=-\infty}^{\infty} \frac{A}{n\pi} \sin\left(\frac{n\pi t_0}{T}\right) \exp\left(jn\omega t\right) \tag{2.26}$$

The spectral components for both amplitude and phase are shown in Figure 2.7(a) and (b) for the case of $t_0/T = 1/12$. It may be observed that the amplitude of the spectral components for one-sided representation will be twice that of the two-sided representation. The value of ϕ_n is either zero or π.

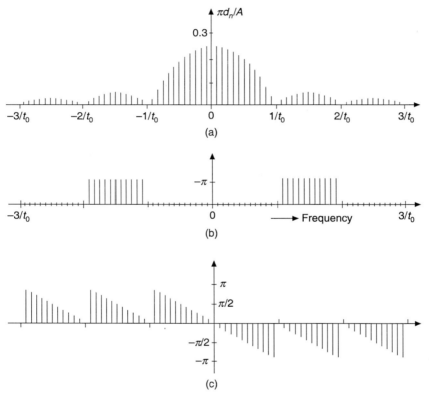

Figure 2.7 Spectra of a periodic rectangular pulse train: (a) Amplitude; (b) phase spectra when time origin coincides with centre of pulse; (c) phase spectra with pulse starting at $t = 0$.

If the time origin is made to coincide with the leading edge of any pulse, the coefficient d_n is given by

$$d_n = \frac{A}{n\pi} \sin\left(\frac{n\pi t_0}{T}\right) \exp\left(-\frac{jn\pi t_0}{T}\right) \tag{2.27}$$

The modulus of d_n is seen to be the same and the amplitude spectra are exactly the same. It is only the phase spectrum that differs as shown in Figure 2.7(c).

2.4 APERIODIC SIGNALS

An aperiodic signal that occurs once in some finite time and is never repeated, may be considered as one with infinite period. Application of this condition to a periodic function results in narrowing the frequency spacing between the spectral components as T becomes larger. At the limiting value of infinity for T, the discrete line spectrum becomes a continuous spectrum. Treating $1/T = \delta f$ and $n\, \delta\omega = \omega$, the expression for d_n becomes

$$d_n = \delta f \int_{-T/2}^{T/2} g(t) \exp(-j\omega t)\, dt \tag{2.28}$$

$$\lim_{T \to \infty} \frac{d_n}{1/T} = \int_{-\infty}^{\infty} g(t) \exp(-j\omega t)\, dt = F(j\omega) \tag{2.29}$$

$F(j\omega)$ is known as the Fourier integral, implying that it is a function of frequency and is complex. It has the dimensions of volts per Hz and is therefore an amplitude spectral density function.

The Fourier transform of a rectangular pulse of amplitude A and duration t_0, shown in Figure 2.8(a), is given by

$$F(j\omega) = At_0 \frac{\sin(\omega t_0/2)}{\omega t_0/2} \text{ V-s} \tag{2.30}$$

The frequency spectrum is shown in Figure 2.8(b).

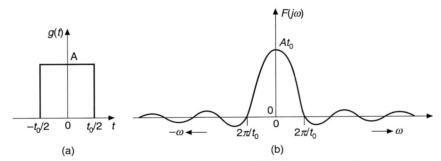

| (a) | (b) |

Figure 2.8 (a) A rectangular pulse; (b) its frequency spectrum.

The unit impulse function $\delta(t)$ or the Dirac delta function is defined as

$$\delta(t) = \lim_{t_0 \to 0,\, A \to \infty} g(t)$$

and $\delta(t) = 0$ for all values of t except $t = 0$ and $\int_{-\infty}^{\infty} \delta(t)\, dt = 1$.

The Fourier transform of the unit impulse function is given by

$$F_\delta(j\omega) = \lim_{At_0 \to 1} At_0 = 1 \tag{2.31}$$

Therefore, the impulse function has a uniform spectral density and the frequency spectrum is flat at unity for $At_0 = 1$, as shown in Figure 2.9.

Similarly, it is possible to develop the Fourier spectra for other forms of aperiodic signals such as triangular and trapezoidal pulses.

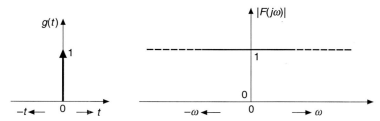

Figure 2.9 A unit impulse function and its frequency spectrum.

All the pulses described so far possess frequency spectra extending over a wide frequency range. There is no geometric form of pulse which is contained within a finite time interval and has a frequency spectrum restricted to a finite narrow band of frequencies. Pulses of narrow spectrum can only be had from pulses that extend over a large duration of time. In this context, it is interesting to note the reciprocal nature of pulses and their spectra. If the waveform of the pulse in time domain is represented by a sin x/x curve, the amplitude spectrum is like a rectangular pulse, with its phase spectrum as zero, as shown in Figure 2.10(a) and 2.10(b). The first zero of the sin x/x pulse occurs at $t = 1/B$, where $2B$ is the width of the amplitude spectrum.

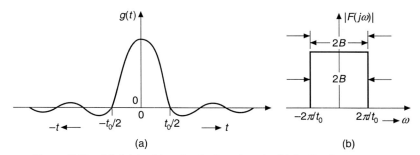

Figure 2.10 (a) A (sin x)/x pulse in time domain; (b) its Fourier spectrum.

2.5 BANDWIDTH

Periodic signals of differing waveforms present discrete amplitude spectra showing the variation of amplitude from one frequency component to another. The amplitude spectrum may be assumed to consist of a frequency range, within which each frequency component significantly contributes to the power of the signal, by virtue of its amplitude. The remaining frequency components may be considered to be of negligible value and hence may be ignored for practical purposes. Thus, when the signal is assumed to consist of only the significant components of frequencies, it is said to be a band-limited signal. The range of these frequencies of the signal is defined as *bandwidth*.

The performance of physical systems is characterized by the way the amplitude and phase of its output signal vary with respect to the input signal. If the response of the system for signals of varying frequencies is described as the ratio of output to input function, it may be represented as

$$\frac{Y_0 \, (j\omega)}{X_i \, (j\omega)} = G(j\omega) = |G(j\omega)| \, \underline{/G(j\omega)} \tag{2.32}$$

A plot of $|G(j\omega)|$ and $\underline{/G(j\omega)}$ vs. ω may yield sufficient information about the characteristics of the system. The system is said to be ideal if the response characteristics are as shown in Figure 2.11(a). Physical systems actually display characteristics such as those shown in Figure 2.11(b) with the value of $|G(j\omega)|$ becoming negligible at some frequencies. Response characteristics of such systems may at least be idealized as shown in Figure 2.12 so as to be treated as a band-limited system. Such a characteristic is treated as that of an ideal low-pass filter having a fixed gain $|G(j\omega)|$ for all frequencies up to ω_1 and zero gain for frequencies beyond ω_1. The frequency ω_1 is known as *cut-off frequency*. The bandwidth of such systems is limited to $2\omega_1 \, (= \Delta\omega)$, whereas for ideal systems it is infinite.

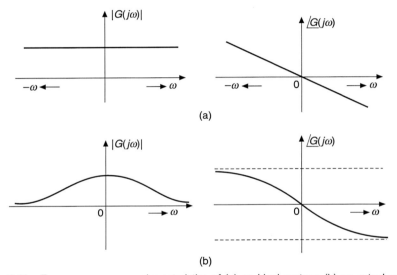

Figure 2.11 Frequency response characteristics of (a) an ideal system; (b) an actual system.

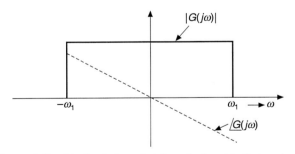

Figure 2.12 Idealized characteristics of a band-limited system.

A rectangular pulse of width t_0, whose Fourier transform is shown in Figure 2.8(b), may be observed to include frequency components with relatively large amplitudes within the frequency range from zero to that where the first zero of $F(j\omega)$ occurs. An ideal filter with its cut-off frequency at ω_1 will allow all these frequency components to pass if $\omega_1 = 2\pi/t_0$. In such a case the product of the pulse width t_0 and the bandwidth $2\omega_1$ is seen to be a constant. In other words, the bandwidth required of systems for satisfactory response or for transmission of the input pulse is inversely proportional to the duration of the pulse being transmitted. The output response in time-domain of an ideal low-pass filter is shown in Figure 2.13, and the distortion introduced may be considered to be tolerable for most purposes, if the system is treated to behave like an ideal low-pass filter.

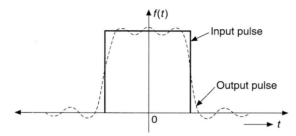

Figure 2.13   Response of an ideal low-pass filter in time-domain.

Although physical systems do not possess the ideal filter characteristics, their cut-off frequency is usually referred to as the frequency at which $|G(j\omega)|$ falls to a value $1/\sqrt{2}$ times that of its value at zero frequency. Such is the case with low-pass filters as shown in Figure 2.14(a). However, for systems and filters having characteristics such as those of Figure. 2.14(b), the peak value

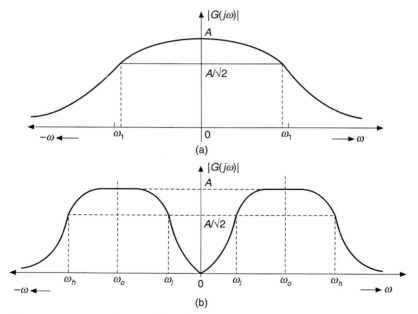

Figure 2.14   Frequency response characteristics of typical systems: (a) A low-pass filter; (b) a band-pass filter.

of $|G(j\omega)|$ occurring at some frequency ω_o is taken with the lower and upper cut-off frequencies, chosen to be those at which the amplitude gain falls to $1/\sqrt{2}$ times the peak value. These cut-off frequencies are otherwise known as half-power points because at these frequencies the power gain of the filter (system) is half of the power gain at ω_0. When physical systems are specified for their dynamic characteristics, the bandwidth of the system is usually indicated. However, the criterion for defining the bandwidth may be decided according to the function for which it is meant. If it is a measuring system, the amplitude response characteristic is required to be flat within ±5% for frequencies within its bandwidth.

2.6 MODULATED SIGNALS

Periodic signals are characterized by their amplitude, frequency and phase with respect to a reference point of time. Likewise, aperiodic signals such as rectangular pulses are characterized by their amplitude, width, and position (or time of occurrence) with respect to the instant of reference. In the field of instrumentation and communication, it becomes necessary to modify one of these characteristics by utilizing another signal. Such a process is known as *modulation*. The signal which is modified or modulated is known as *carrier signal,* and the signal used for modifying the characteristics of the carrier signal is known as *modulating signal.* The resultant signal is said to be a modulated signal. The frequency $\omega_c (2\pi f_c)$ of the carrier signal is usually far higher than the frequency ω_m of the modulating signal. The process of modulation yields several advantages in signal processing and transmission to a far-off location. The modulating signal is generally of greater interest as the function of the carrier signal is only to carry and deliver the modulating signal at the output or receiving end with a little or no loss of integrity. The process of modulation is known as amplitude (*A*), phase (*P*), or frequency (*F*) modulation, depending on which of them is subjected to variation by the modulating signal. If the carrier signal is a pulse, it is referred to as pulse modulation. In all cases of modulation, it is the amplitude of the modulating signal that modifies the characteristics of the carrier signal. It is the modulated carrier signal from which the original modulating signal is to be recovered and reconstructed at the receiving end. It is essential that the amplitude variations of the modulating signal with time are properly reproduced; such a process of reconstruction is known as *demodulation.* The entire process of modulation and demodulation is carried out in the analog domain.

In contrast with analog modulation, *digital modulation* refers to the process of utilizing a series of short duration pulses of constant amplitude to represent the amplitude of the modulating wave. The modulating signal is sampled for its amplitude at regular intervals of time, and during each interval a number of pulses are released in a coded format so as to signify the amplitude of the modulating signal. Such a process constitutes digital modulation and the process of recovering the modulating signal at the output end is *digital demodulation.* Transmission of measured data over a distance by digital modulation and demodulation results in several advantages as compared to the analog process, particularly in respect of preserving the integrity of the modulating signal from the effects of noise.

2.6.1 Amplitude Modulation

The simplest form of amplitude modulation may be obtained by multiplying a carrier signal $e_c(t)$ with the modulating signal $m(t)$ and is reckoned as multiplication-type modulation.

Treating both the signals to be periodic and sinusoidal, the modulated signal can be represented as

$$e_m(t) = e_c(t) \ k \ m(t) = (E_c \cos \omega_c t)(kE_m \cos \omega_m t) \tag{2.33}$$

Rewriting the same, we get

$$e_m(t) = \frac{kE_c E_m}{2} [\cos (\omega_c + \omega_m)t + \cos (\omega_c - \omega_m)t] \tag{2.34}$$

where k is a constant.

The process of multiplication results in the production of two sinusoidal components of frequencies $(\omega_c + \omega_m)$ and $(\omega_c - \omega_m)$, also known as side band frequencies. The value of k can be so adjusted as to make $e_c(t)$ swing from zero to twice its amplitude, as can be seen from Figure 2.15. The two-sided frequency spectra for the modulating and modulated signals are shown in Figure 2.16. It should be noted that the resultant signal consists of only the two side band frequencies with no carrier frequency signal. Hence it is known as double-side band-suppressed carrier (DSBSC) signal. The modulating signal is contained in the two side band frequency signals and it can be recovered from any one of them.

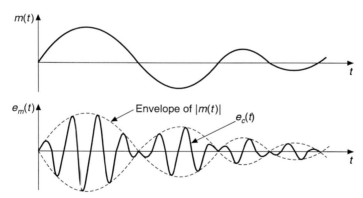

Figure 2.15 Multiplication type modulation signal representation in time domain.

If the modulating signal consists of components of a band of frequencies, the modulation process transposes the base band signal to the region around ω_c, which is higher up in the frequency spectrum, as can be seen from Figure 2.16. It must be noted that the bandwidth of the upper and lower side bands remains the same as that of the baseband modulating signal. It is only necessary to select a carrier signal of a frequency much higher than the highest frequency component of the modulating signal, and the carrier signal is usually sinusoidal in waveform with constant frequency and amplitude.

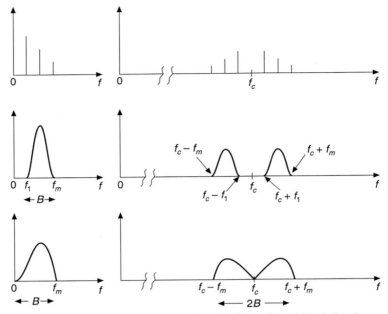

Figure 2.16 Frequency spectra of modulating and modulated signals.

In the field of instrumentation, this type of modulation usually occurs with transducers and transducer readout circuits. The unbalance voltages of Wheatstone bridge networks, excited by ac supply, are modulated signals, with the modulating signal being the resistance variation of one of its arms, due to variation of the physical quantity under measurement. Similarly, output signals of transducers such as linear variable differential transformer, synchros, microsyns etc., are such DSBSC signals. The conversion of dc voltages into ac voltages is a form of such modulation where the carrier is a square wave of a suitable frequency such as 50, 400 or 2000 Hz.

Another form of modulated signal, known as full AM signal, can be obtained by addition of a voltage signal at the carrier frequency to the DSBSC signal, and is given by

$$e_m(t) = E_c \cos \omega_c t + k_1 m(t) E_c \cos \omega_c t$$

It can be rewritten as

$$e_m(t) = E_c(1 + m_a \cos \omega_m t) \cos \omega_c t \tag{2.35}$$

for $m(t) = E_m \cos \omega_m t$. The constant m_a is defined as the 'modulation index' and is controlled to assume values ranging from zero to unity. It is seen that the carrier frequency signal varies in its amplitude from zero to twice its peak value for $m_a = 1$, as shown in Figure 2.17. It is essential to provide phase coherence between the two components of the full AM signal. With the modulation index set to the maximum value of unity, two-thirds of the total power is consumed for the transmission of the carrier signal and only the remaining one-third is available for the upper and lower sidebands. The information signal $m(t)$ is contained in and conveyed by the sidebands only, and hence the full AM signal technique may be treated as an inefficient

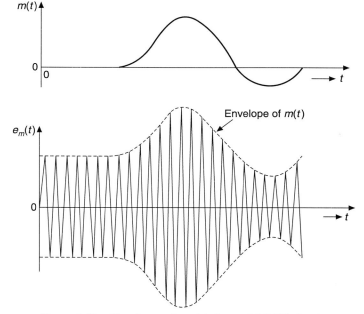

Figure 2.17   Waveforms of modulating and full AM signals.

technique; however, the presence of the carrier signal at the receiving end results in proper demodulation. In case $m(t)$ is the measured data, the amplitude modulation is not adopted since the transmitted signal may be sufficiently curbed in its amplitude due to noise interference, and the signal-to noise ratio may assume unacceptable levels.

2.6.2   Angle Modulation

If the unmodulated carrier signal is represented by $E_c \cos(\omega_c t + \phi_c)$, the angle $(\omega_c t + \phi_c)$ can be subjected to modulation by effecting either the value of ω_c or ϕ_c in accordance with the variation of the amplitude of the modulating signal. Variation in any one of them constitutes variation of the angle contained by $(\omega_c t + \phi_c)$, and hence modulation of this form is known as *angle modulation*. It is phase modulation (PM) if ϕ_c is made to vary instant by instant, in proportion to $m(t)$ as given by

$$\phi_c(t) = k_p\, m(t) \tag{2.36}$$

where k_p is defined as a constant of modulation for phase-modulated signals.

On the contrary, ω_c is made to swing instant by instant, in proportion to $m(t)$ in the case of frequency modulation (FM). However, it may be recalled that the angular frequency $\omega(t)$ at any instant is given by

$$\omega(t) = \frac{d\phi_c(t)}{dt} = \omega_c + k_f\, m(t) \tag{2.37}$$

where, k_f is a constant for frequency modulated signals. It is seen that the carrier signal is of the frequency ω_c in the absence of $m(t)$. Thus it is possible to modulate the angle of a carrier signal, while holding its amplitude E_c at constant value. The total angle $\phi_c(t)$ of the frequency modulated wave is obtained from Eq. (2.37) as

$$\theta_c(t) = \omega_c t + \int_0^t k_f \, m(t) \, dt + \phi_0 \qquad (2.38)$$

where, ϕ_0 is an integration constant representing the phase of the modulated carrier at $t = 0$. ϕ_0 can be treated as zero for frequency modulated signals. From Eq. (2.38) it can be seen that it represents phase modulation where the modulating signal is $\int_0^t m(t) \, dt$ in place of $m(t)$ alone. So a frequency modulated signal can be produced by first integrating the modulating signal $m(t)$ and then using it as the input to a phase modulating circuit. Similarly, a phase modulated signal can be generated by first differentiating the modulating signal and then utilizing it as the input signal to a frequency modulating circuit. Both of them are shown in Figure 2.18.

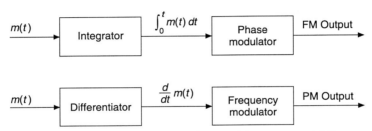

Figure 2.18   Distinction between phase and frequency modulation.

2.6.3   Frequency Modulation

Using Eq. (2.38), the expression for the frequency modulated signal is written as

$$e_m(t)\big|_{FM} = E_c \cos\left[\omega_c t + \int_0^t k_f \, m(t) \, dt + \phi_0 \right]$$

Taking $\phi_0 = 0$ and $m(t) = E_m \cos \omega_m t$, the expression is rewritten as

$$e_m(t)\big|_{FM} = E_c \cos\left(2\pi f_c t + 2\pi \int_0^t k_f E_m \cos \omega_m t \, dt \right)$$

$$= E_c \cos\left(2\pi f_c t + \frac{k_f E_m}{f_m} \sin \omega_m t \right) \qquad (2.39)$$

$$= E_c \cos\left(2\pi f_c t + \frac{\Delta f_c}{f_m} \sin \omega_m t \right) \qquad (2.40)$$

$$= E_c \cos\left(2\pi f_c t + m_f \sin \omega_m t \right) \qquad (2.41)$$

Equations (2.37) and (2.40) show that the frequency of the carrier signal deviates from f_c by $\pm\,\Delta f_c$ and that Δf_c is controlled by the amplitude of the modulating signal. The modulation index m_f of Eq. (2.41) is defined as the ratio of the deviation of carrier frequency to the frequency of the modulating signal and is given in radians. The value of m_f is governed by both the amplitude and frequency of the modulating signal. Treating m_f as the peak phase displacement of the carrier signal due to the modulation process, it is seen that the smaller the value of f_m, the larger the phase displacement.

For modulating signals with f_m much smaller than f_c, the modulated carrier signal can be regarded as a succession of sinusoidal waves of slowly varying period (or frequency) but of constant amplitude as shown in Figure 2.19.

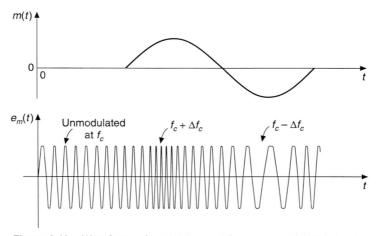

Figure 2.19   Waveforms of modulating and frequency modulated signals.

Expansion of Eq. (2.41) yields

$$e_m(t)\big|_{\text{FM}} = E_c\,[\cos\omega_c t\,\cos(m_f\sin\omega_m t) - \sin\omega_c t\,\sin(m_f\sin\omega_m t)] \tag{2.42}$$

Each of the terms cos (sin) and sin (sin) represents a power series of a power series. The expression for $e_m(t)\big|_{\text{FM}}$ when expanded into a series of sidebands results in

$$e_m(t)\big|_{\text{FM}} = E_c\sum_{n=-\infty}^{+\infty} J_n(m_f)\cos(\omega_c + n\omega_m)t \tag{2.43}$$

where, the functions $J_n(m_f)$ are known as Bessel functions.

It will be seen that there are an infinite number of pairs of sidebands harmonically related to the single modulating frequency ω_m, implying that the bandwidth is very large, almost tending to infinity. The amplitude of the carrier, $J_0(m_f)$, depends on m_f and the amplitudes of the sidebands also vary with m_f. When $m_f \ll 1$, $J_0(m_f)$ tends to unity and $J_1(m_f)$ and $J_2(m_f)$ tend to zero with the result that the carrier signal is only present in the output signal.

Holding ω_m constant and increasing the amplitude of the modulating signal results in the increase of Δf_c and m_f, and so a number of sidebands with significant amplitudes start appearing in the output signal. Considering the sidebands having amplitudes less than 1% of those of the unmodulated carrier, to be of no significance, it can be seen that the number of sidebands N goes on increasing with m_f, as given in Table 2.2. With $m_f < 0.2$, only one pair of sidebands is produced as is the case of full AM. Then, it is known as narrow band frequency modulation. For higher values of m_f, it is called wide band FM.

In practice, the bandwidth of the FM signal is dictated by both the amplitude and frequency of the highest frequency component of the base-band signal. However, the peak frequency deviation of the carrier signal is decided by the amplitude of the highest frequency component only and is given by the product of m_f and the highest frequency of the modulating signal. The bandwidth of the FM signal is N times the frequency of the modulating signal. It is also given by $(N\Delta f_c)/m_f$, as illustrated in Table 2.2.

The frequency spectrum of the FM signal becomes very complicated if the modulating signal consists of more than one frequency component, and the spectrum contains a variety of frequency components. If f_{m1} and f_{m2} are the two frequency components of the modulating signal, then the FM signal consists of sidebands of frequency $f_c + nf_{m1} + mf_{m2}$, where n and m assume any positive or negative integral values.

Table 2.2   Bandwidth requirement for FM signal

Modulation index m_f	Number of significant sidebands, N	Bandwidth as multiple of f_m	Bandwidth as multiple of Δf_c
0.1	2	2	20
0.3	4	4	13.3
0.4	4	4	10
0.5	4	4	8
0.8	6	6	7.5
1.0	6	6	6
1.5	8	8	5.3
2.0	8	8	4
5.0	16	16	3.2
10.0	28	28	2.8
20.0	50	50	2.5
30.0	70	70	2.3

2.6.4   Phase Modulation

The phase modulated (PM) signal is similar to FM signal except for the fact that the modulation index is independent of the frequency of the modulating signal. The phase modulation index m_p is exclusively proportional to the amplitude of the modulating signal, and hence for a chosen value of m_p, the spectrum of a PM signal will be contained within a band of $f_c \pm (N/2)f_m$. If the bandwidth for the PM signal is provided by considering the highest frequency component, it amounts to ineffective utilization of the bandwidth as the sidebands for the lowest frequency will only be around f_c. Hence, in practice, the phase modulation is not favoured. In this respect, FM is preferred for transmission of analog signals. Also, wideband FM systems have the merit of comparative freedom from interference when compared to the narrowband systems.

For a chosen value of m_p, the number of sidebands can be obtained from Table 2.2, taking the value of m_f to be equal to m_p. For nonsinusoidal modulating signals, it is usually seen that the highest frequency component has the lowest amplitude with the result that the number of sidebands for the highest frequency component may result in the same bandwidth as the one decided by large amplitude of the lowest (fundamental) frequency component. However, it may be stated that phase modulation is not widely used in instrumentation systems.

2.7 SAMPLED DATA

In several situations, it may be necessary to obtain the amplitude of a signal at a certain desired instant. For signals that persist for a long time and for some repetitive signals, it may be enough if a set of instantaneous values of the signal are procured at regularly spaced intervals of time. It is necessarily presumed that the signal does not abruptly jump in its amplitude during the intervals between the sampling instants. Under these conditions it is possible to convey all the information contained in the signal by reconstructing it from the set of instantaneous values of the amplitudes. Such a process is known as *sampling process* and the values of the amplitudes are known as *samples*. The question is only regarding the minimum number of such samples required so that the reconstructed signal bears the desired resemblance with the original signal.

It is important in this context to have the knowledge of the frequency spectrum of the signal and its highest frequency. It is understood then that the signal is a band-limited function. If that happens, Shannon's sampling theorem states that the band-limited signal can be completely defined by means of its ordinates at a series of points $1/(2f_{max})$ seconds apart, where f_{max} is the highest frequency of the signal and is usually equal to its bandwidth. The samples are thus obtained at a rate of $2f_{max}$ per second, usually defined as sampling rate. Ideally, the sampling signal is a train of unit impulses of constant amplitude and spaced in time by T which is known as the sampling interval. The sampling process is merely a multiplication of the input signal $x(t)$ by the sampling pulse train $s(t)$. The sampling process and the sampled signal can be seen from Figure 2.20. The sampled signal $z(t)$ presents itself as sampled data and it is seen that $z(t)$ is zero if $x(t)$ is zero. The spectra of the sampling signals and the sampled signal are shown in Figure 2.21. It is observed that the spectra of the input signal repeat themselves as a set of images shifted to higher and lower frequencies. The images are seen to be nonoverlapping and hence to recover the original signal from $z(t)$, it is only necessary to pass the sampled signal through a low-pass filter.

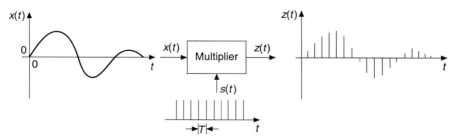

Figure 2.20   Sampling process.

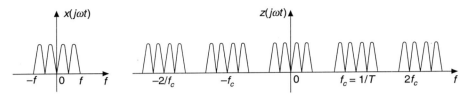

Figure 2.21 Fourier spectra of the sampled signal.

In real practice, the sampling pulses are considered to be of negligible but of finite duration and the sampling rate is made higher than $2f_{max}$.

2.8 PULSE MODULATION

The sampling process described so far may be treated as multiplication type modulation, wherein the input signal $x(t)$ is the modulating signal $m(t)$ and the carrier signal is the sampling signal $s(t)$. The amplitude of each pulse after modulation defines the amplitude of the modulating signal, and hence the sampled data becomes the modulated carrier signal. The sampled data does not contain the carrier signal when $m(t) = 0$, and hence it is equivalent to DSBSC signal. But in pulse amplitude modulation (PAM) systems, it is so arranged that when $m(t) = 0$, the unmodulated carrier appears in the output and hence it is equivalent to full AM. The PAM pulse train becomes the basis for other pulse modulation systems such as pulse duration (or width) modulation (PDM or PWM), pulse position modulation (PPM), or pulse frequency modulation (PFM). Figure 2.22 shows the representation of these signals. The combination of the train of PAM pulses with a coding technique leads to pulse code modulation (PCM).

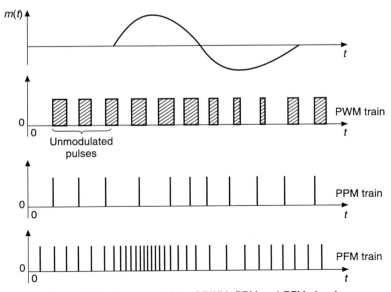

Figure 2.22 Representation of PWM, PPM and PFM signals.

In pulse width modulation system, the width of the pulse t_p is made proportional to the amplitude of the signal sample, whereas in pulse position modulation, the time at which each narrow pulse occurs relative to a mean or reference position, defines the amplitude of the signal sample.

2.9 PULSE CODE MODULATION

The sampled data or the PAM signal covered in the earlier section may be suitable for analog transmission but due to the many advantages of digital transmission techniques, the analog pulses are converted into digital signals prior to transmission. At the receiving end the digital signals are used directly for display and recording or are converted into an analog signal, wherever needed. All attention is paid to reconstruct the original signal as truthfully as possible. Hence the combination of PAM pulses with a coding technique and conversion of the analog signals into digital signals, lead to the pulse code modulation (PCM).

In PCM, the amplitude range of the sampled analog signal is divided into a finite number of discrete levels known as *quanta*. If 32 levels uniquely specify the amplitude range, then it becomes a 5-digit code in the binary system as $2^5 = 32$. In such a case, the samples are said to be simply binary coded. A larger number of discrete levels may be chosen when higher accuracy is desired. The sample amplitude is represented in digital code by a set of five rectangular pulses of same height and duration. Recognition of the levels of these pulses as either near to 0 or near to 1 is all that is needed to reconstruct the amplitude of the analog signal at the receiving end. The conversion of the analog electrical signals into digital signals is accomplished by analog-to-digital converters (ADC). The error due to the actual signal not exactly coinciding with the nearest level of the many levels into which the range is divided, makes up an error in the conversion process and the error in such situations is known as *quantization noise*. The error is treated as noise and will be reduced in effect if the number of levels chosen is larger.

The process of obtaining the PCM signal in a 4-digit code is shown in Figure 2.23. The PCM pulses, representing each sampled amplitude are indicated in Figure 2.24, and the duration of each PCM pulse is decided by the sampling interval and the number of binary digits (bits) used to represent the amplitude. Hence, for higher accuracy, if a larger number of quanta are used, the bandwidth of the channel or link used to transmit these pulses needs to be larger.

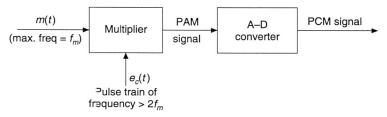

Figure 2.23 Process of quantization.

Where necessary, instead of a binary system, an *m-ary* system with $m = 3$ or 4 may be chosen such that each pulse is characterized by 3 or 4 levels. However, the inherent immunity of the binary coded signal against noise and interference makes it more useful and popular.

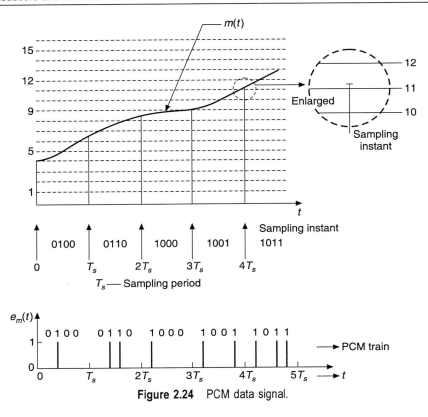

Figure 2.24   PCM data signal.

There are other forms of pulse-code-modulation, such as delta-modulation which makes use of a single-digit code. For further information regarding digital modulation techniques, the reader may refer to standard textbooks on this subject.

EXERCISES

1. Distinguish between periodic and aperiodic signals and give examples for each.

2. What are the characteristics of 'analog' and 'digital' signals? Suggest the sources that produce them.

3. Define 'Laplace' and 'Fourier' transforms and indicate the conditions under which each is applicable.

4. Describe the types of test signals by which physical systems are studied and analyzed for their dynamic behaviour.

5. Define 'distortion' of a periodic signal and how it is estimated.

6. Derive from fundamentals the expressions representing the following waveforms:
   (a) a rectangular wave
   (b) a rectangular pulse train

(c) a saw-tooth wave

(d) a triangular wave

7. What are the line spectra of a signal and how are they represented?

8. Define bandwidth of a signal and explain the way signals are classified according to their bandwidth.

9. Describe the process of modulation and the techniques usually adopted.

10. Distinguish between analog modulation and digital modulation and explain the situations under which each is preferred.

11. Explain the technique of frequency modulation and show that it is the most efficient technique of analog modulation.

12. What are sidebands of a modulated signal and explain their presence in the AM and FM signals.

13. Derive the expression for a frequency-modulated signal and show how the number of side-bands increases with modulation index.

14. Define frequency deviation and modulation index of a FM signal and suggest the recommended (and used) values of each.

15. Distinguish between phase and frequency modulation, defining the modulation index for each case.

16. What is the difference between sampling process and pulse modulation?

17. Explain the techniques of pulse-time modulation and pulse-code modulation and their relative merits.

18. Describe the process of obtaining a PCM signal.

SUGGESTED FURTHER READING

Betts, J.A., *Signal Processing, Modulation and Noise*, ELBS, London, 1975.

Brown, J. and Glazier, E.V.D., *Signal Analysis*, Reinhold Publishing Corp., New York, 1964.

Kuo, Franklin, F., *Network Analysis and Synthesis*, John Wiley, New York, 1962.

Magrab, Edward, B. and Blomquist, Donald, S., *The Measurement of Time-Varying Phenomena: Fundamentals and applications*, Wiley-Interscience, New York, 1971.

van Valkenburg, M.E., *Network Analysis*, 3rd ed., Prentice-Hall of India, New Delhi, 1973.

CHAPTER 3

Electrical Measuring Systems

3.0 INTRODUCTION

Instrumentation systems deal with the measurement of a large variety of physical variables and subsequent processing of the signals into a form suitable for the terminal function. Indication and display is one of the primary objectives of the instrumentation systems and the electrical and electronic indicating instruments and read-out systems constitute the terminal stage. Also, in view of the advantages of electrical signals as detailed in Chapter 1, the primary measuring elements and transducers of the instrumentation system are based on such principles of operation that transform the measurand into an electrical signal. The electrical signal is perceived and realized as either a change in the electrical parameters of the transducer or by the development of emf and current. Whichever be the case, a large variety of electrical and electronic circuits are invariably used to process the output signals of the transducers. In the following sections, it is only proposed to present some of the basic circuits and systems used for the measurement and processing of the electrical outputs of the transducers. Additional electronic circuits are presented in Chapter 9 while other data display and recording systems are described in Chapter 10.

3.1 MEASUREMENT OF CURRENT

Detection and measurement of dc currents ranging from 10^{-5} to 10^{-12} A is necessary when measuring directly the outputs of certain transducers and scientific instruments. Depending on the resistance of the circuit, a measuring system of the required range and internal impedance has to be chosen. The permanent magnet moving coil system offers a solution in most cases, but for circuits developing extremely small currents, special electrometer amplifiers are used.

3.1.1   Permanent Magnet Moving Coil Instruments

Permanent magnet moving coil (pmmc) system is the most widely used of all instruments. The system basically consists of a circular or rectangular coil suspended or pivoted in the magnetic field between the poles of a permanent magnet. It is known as d'Arsonval movement and is referred to as d'Arsonval galvanometer, when provided with a suspended coil. The coil is connected to the external circuit through two helical springs or fine gold ligaments. The current passing through the coil is responsible for torque production, and the coil rotates until the torque developed is equal to the restoring torque due to the control springs.

The relationship between the current and deflection is governed by the pole shape of the magnet and the distribution of the magnetic field in the air gap. If planar poleshoes and a rectangular coil are used as shown in Figure 3.1(a), the angular deflection of the coil will be proportional to the current through it but the deflecting torque is proportional to cos θ_0, where θ_0 is the angle the plane of the coil makes with the direction of the uniform field. Setting the spring so as to exert zero controlling torque for $I = 0$ at $\theta_0 = 0$, the relationship between the deflecting and controlling torque is given by

$$C\theta_0 = BAN\ I \cos\ \theta_0 \tag{3.1}$$

where,

    B = magnetic density, Wb/m^2
    A = effective area of the coil, m^2
    N = number of turns of the coil
    I = current through the coil, A
    C = torque constant, N-m/rad

The static sensitivity $S(= \theta_0/I)$ depends on θ_0 resulting in a scale somewhat linear for small values of θ_0 and becoming crowded for larger values.

For the two cases with radial poles shown in Figure 3.1(b) and 3.1(c), θ_0 is 0° for all positions of the coil and hence scale characteristics are linear but scale length of case (b) is about 160° and that of (c) is nearly 270°. For detection of small currents encountered while balancing Wheatstone bridge networks, optical amplification of the deflection is used by means of a mirror attached to the coil and a lamp-and-scale outfit. High sensitivity is also obtained by concentrating the magnetic field over a narrow region, as shown in Figure 3.1(d). For use with photo-voltaic cells, in exposure meters, the pole pieces and core are arranged, as shown in Figure 3.1(e) so that the magnetic field is nonuniform and the scale characteristic follows logarithmic relationship.

The many advantages of the pmmc system are a direct outcome of the variability in its design features. Of all the electrical instruments, this system consumes least power because the energy stored in the magnetic field is utilized to develop the torque. As a consequence, it has high ratio of torque to weight. It is primarily a current-sensitive instrument, and no torque can be developed unless current passes through the coil. But it has its own merits, in the sense that the system is relatively free from the effects of strong magnetic fields and electric fields. It can be used to measure the sum or the difference of two currents, by using two co-planar coils as shown in Figure 3.1(f). Using two coils with their planes perpendicular to each other or at an angle to each other as shown in Figure 3.1(g), the ratio of two currents I_1 and I_2 can be measured. The torques developed by the currents oppose each other. As the coils are rigidly attached to each other, the angular position assumed by the coil system is dependent on the ratio of currents. For the crossed coil system of Figure 3.1(g), it can be shown that

$$BAN_1I_1 \cos \theta_0 = BAN_2I_2 \cos (90 - \theta_0)$$

and hence

$$\tan \theta_0 = \frac{N_1I_1}{N_2I_2} \qquad (3.2)$$

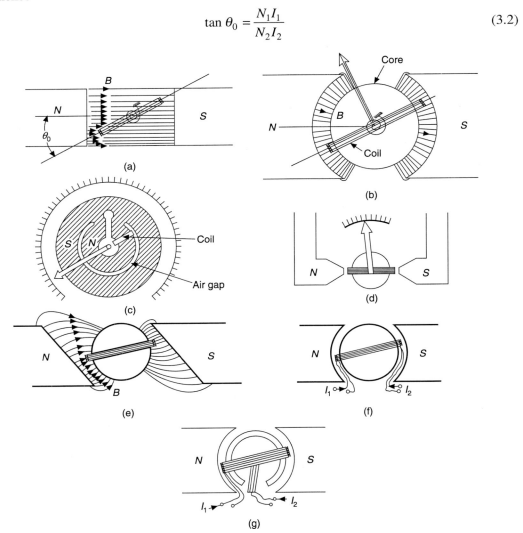

Figure 3.1 Permanent magnet moving-coil systems having (a) uniform field; (b) radial field; (c) circular scale; (d) concentrated field; (e) logarithmic scale; (f) co-planar coil; (g) crossed-coil.

The pmmc system is equivalent to a mass-spring system and hence belongs to the family of second-order systems. A detailed analysis of the behaviour of such systems subjected to time-varying inputs of differing description is presented in Chapter 4. By appropriately choosing the moment of inertia (J) of the moving system, torque constant (C) of the control suspension (or spring) and the damping conditions, it is possible to extend the usefulness of the pmmc system for measurement of other electrical quantities such as charge and magnetic flux-linkages. When the pmmc instrument is designed to indicate dc voltages and currents, it is enough to see that the final indication is attained as fast as possible but with a little or no overshoot. The undamped natural frequency of such voltmeters and ammeters is usually arranged to be about 1 Hz and the damping factor is about 0.95. The pmmc instrument responds to steady quantities, while showing no response to ac quantities of frequency higher than 20 Hz. The coil system and the pointer may be able to follow the instantaneous variation of the ac voltages and currents of a frequency within 2 Hz, but it is difficult for the user to follow the pointer and fix the maximum value. Moreover, the user is normally interested in the root-mean-square (rms) value and there are other instruments that read the rms value, while showing a steady deflection. However, pmmc systems are also designed to function as detectors of ac currents and recording oscillographs.

A vibration galvanometer is a pmmc system which is capable of detecting very small ac currents of a specified frequency. It is normally used to detect the balance conditions of ac bridge networks and ac potentiometers operated at 50 Hz. Hence the undamped natural frequency of the galvanometer is made equal to 50 Hz and the damping is reduced to very low value so that the response amounts to resonance under negligible damping. It is known as vibration galvanometer because of its continuous vibration about its mean value. With the optical amplification provided, the display becomes a band of light extending to the maximum deflection on either side. To satisfy the design criteria, the coil is made lighter, narrower and of a fewer number of turns, and a stiffer suspension fibre is used.

An electromagnetic oscillograph is another pmmc system whose natural frequency is made higher by using a narrow bifilar strip in place of a coil and immersing the same in silicone oil so as to provide liquid damping at a damping factor of 0.7. The oscillograph is intended to plot the waveforms of currents of frequencies up to 40 per cent of its natural frequency. Optical amplification as shown in Figure 3.2 is used by means of a beam of visible or ultraviolet rays and a sensitive paper driven at the desired speed. The sensitive paper, as soon as it is exposed to day light, makes the trace visible. Such reflecting beam oscillographs are generally used for recording of fast transients and waveforms of currents of a frequency range from 0 to 2000 Hz. Wherever necessary, the coils are fed with currents, after suitably amplifying the input voltage/current signals. It is possible to make use of a pen or stylus directly connected to the coil, so as to obtain the trace of the waveforms, but in such cases, the coil should be of a few turns and

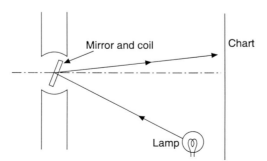

Figure 3.2   Reflecting type prnmc system for recording currents.

the system designed to carry the writing stylus and overcome the friction between the stylus and paper. Such a system is possible for recording faithfully waveforms of currents of frequencies lying within 100 Hz, in which case the undamped natural frequency is about 250 Hz. It is essential, in all these cases, to have the system damped at a damping factor of 0.7 either by electromagnetic means or by oil damping. Other details pertaining to such direct-writing oscillographs may be studied in Chapter 4.

A ballistic galvanometer is one that is designed to possess a much higher natural period of oscillation of about 8-15 seconds. The coil is made heavier and the spring constant lower. It is used primarily as an integrating instrument to measure integral of currents with time that lasts for a duration much smaller than its free period. It is so intended that the current pulse decays to zero before the moving coil system starts moving from its position of rest. The theory concerning the dynamic behaviour of such systems subjected to monopulse inputs may be studied in Chapter 4. The current pulse may be derived from a voltage pulse of induced emf when flux-linkages of a coil change. By noting the first maximum value of the deflection, the amount of charge that passes through the galvanometer is determined. It should be noted that the charge sensitivity of the ballistic galvanometer changes with its damping conditions and hence it is essential to calibrate it under the same damping conditions as those existing under actual test.

A fluxmeter also belongs to the family of pmmc systems and is used to measure magnetic flux-linkages of a coil. The fluxmeter may be considered to be a ballistic galvanometer whose period of oscillation is made very high by making the spring-constant of the suspension practically negligible. The coil is suspended by means of an unspun silk fibre and the connection to the external circuit is made through a pair of fine gold ligaments that exert negligible controlling torque on the moving coil. Without going into the theory, it can be said that the coil system and pointer move from one position to another, when the search coil connected to it experiences a flux-change. As there is no controlling torque, the pointer remains in the new deflected position. Thus, the fluxmeter enables measurement of flux-linkages taking place over a much longer duration of time, such as those in iron-cored circuits, while the difference in the final and initial readings is proportional to the change in flux linkages. The other essential design condition is that the fluxmeter should be overdamped to the extent that at any instant the externally induced emf is equal to the dynamically induced emf of the fluxmeter coil, during its motion in the magnetic field of the permanent magnet.

Apart from all the virtues and versatility of the pmmc system, we should appreciate that by suitable modification of the design, it is customary to generate linear displacements instead of angular displacements. They are treated as force generators and constitute the important class of loudspeakers, in general, and inverse transducers in certain feedback-type measuring systems. The stability of the permanent magnet and the linearity between force (torque) and displacement are the two important characteristics of the pmmc systems, without which attainment of high accuracy and precision in electrical measurements would have been impossible.

Ammeters for measurement of currents larger than 20 mA are provided with shunt resistors such that the current through the moving coil does not exceed 5–25 mA. The shunt is provided normally with four terminals, two for connection to the external circuit and two more for connection to the moving-coil system. The voltage drop across the shunt for the full scale value of each ammeter is usually 60 mV so that the scale of a 12-ohm milliammeter of range 0–5 mA is calibrated to read the total current under measurement.

Measurement of the rms value of ac currents is, however, not possible with the pmmc system since it is given by

$$I_{rms} = \frac{1}{T}\left[\int_0^T i^2(t)\,dt\right]^{1/2}$$

where, T is the period of the ac current. But, the permanent magnet of the pmmc system may be replaced with air-cored coil system so that the current under measurement may be passed through this coil and the moving coil kept in series as shown in Figure 3.3. The average torque developed will be proportional to the rms value of the current. Such a modification led to the development of electrodynamic (or dynamometer) instruments, which can read dc and ac currents of frequencies up to 150 Hz, equally accurately. When the current is limited to 20 mA, the same current passes through the fixed and moving coils and milliammeters of this type are used as *transfer standards* for ensuring the equality between the dc and ac rms

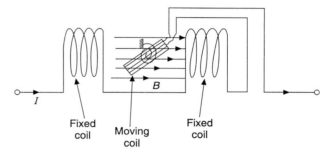

Figure 3.3 Dynamometer type moving coil system.

currents. By providing suitably designed shunts, larger dc currents can be measured. Larger ac currents are measured by using current transformers. The most important application of this electrodynamic system is for measurement of electrical power. Using iron-cored fixed coils instead of the air-cored coils, higher torque can be developed, and such instruments are known as *ferrodynamic* instruments. They are used for measuring not only the ac power but also other electrical quantities such as power factor (single and three-phase) and frequency (25–100 Hz) by addition of another moving coil carried by the same spindle and with currents through them, each differing from the other in phase.

The equality between a dc current and an ac current of a frequency higher than 150 Hz is determined by means of a thermal converter. It consists of a thermocouple and a small heater coil, both enclosed in an evacuated glass bulb. The heater coil is heated by the current under measurement, and the thermo emf is measured by means of a sensitive pmmc system as shown in Figure 3.4. The same thermo emf is read if the rms value of the ac current is equal to the dc current. Such a system is

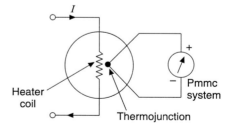

Figure 3.4 Thermocouple type milliammeter.

used as 'Transfer Standard' for ac currents of frequencies extending to MHz range, with suitable modification in the design of the thermal converter. Instruments based on this principle are known as thermal instruments and are very much used for measurement of both dc and ac currents of high frequency (up to 100 MHz).

Measurement of the average value of ac currents is rendered possible by the use of a bridge-rectifier circuit along with a pmmc milliammeter as shown in Figure 3.5. It is essential to realize that these instruments are calibrated to read the rms values of sinusoidally varying currents only. If the waveform is distorted, the reading will be erroneous.

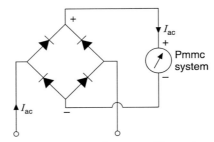

Figure 3.5   Rectifier type milliammeter.

3.1.2   Other Current-Measuring Systems

The other two basic systems that are popularly used for the measurement of electrical currents and other quantities are the moving-iron and induction systems. Both are primarily current sensitive and are restricted for use at power frequency only. Induction-type instruments are meant for use at one frequency (50 Hz) only and they are mostly used for measurement of electrical energy. Moving iron instruments are, however, applicable for use on dc and ac (25–75 Hz) currents and voltages but are not recommended for measurements requiring high accuracy.

Other methods of measuring dc and ac currents are based on the application of electronic amplifiers which amplify the voltage drop across a resistor of suitable value, through which the current under measurement is passed. The amplified voltage may be read on a pmmc system after converting the amplified ac voltage into a dc voltage. Another important device employed is the Hall-converter, which is described in Chapter 7. Instruments using the converter are available for measurement of current and power in both dc and ac circuits. They are not as accurate as the pmmc or the electrodynamic type but possess the merits of small size, and offer scope for measurement of currents over a wider range and frequency.

3.2   MEASUREMENT OF VOLTAGE

The process of voltage measurement primarily consists of determining the potential difference between any two points of either a source or an impedance carrying current through it. One of the terminals is brought to zero potential by connecting it to 'Earth' terminal when the network or the source of voltage is 'floating', and when the measuring system is an electronic voltmeter. Measurement of voltage becomes perfect if the measuring system is voltage sensitive and does not draw any current and power (or energy) from the system under test. All electrical measuring instruments sensitive to current (see Section 3.1) can be used as voltmeters, by adding a resistor of suitable value in series with the milliammeter and calibrating the scale in volts. But it is impracticable to make the measurement without drawing current. Thus all the electrical measuring instruments are restricted for application to situations where the voltages measured are sufficiently large and sources of voltage have low output impedance. Electrostatic voltmeters are fundamentally voltage-sensitive instruments. They are suitable for dc and ac (rms) voltage measurements.

The electrostatic instrument consists of a variable-plate capacitor system, the movable plate of which is subjected to electrostatic force of attraction as shown in Figure 3.6. While measuring dc voltage, the instrument draws no power, but it should be noted that a certain amount of energy is drawn in the initial stage for setting up the electric field between the plates. Later on, no current or power is drawn. While measuring ac voltages, the current is drawn but no power. The energy drawn in the initial stage of dc voltage measurement and the current drawn in the case of ac voltage measurement are so low that the measurement is treated as nearly 'perfect'. The energy stored in the electric field being small, the instrument is not suitable for voltages below 100 V. For voltages higher than 100 V, it can be used as a 'Transfer standard' as the torque produced is the same for dc and ac rms voltages of the same value. It can be used for ac voltages of frequencies up to 50 MHz, but one should not miss noticing the reduction in its input reactance with increasing frequency although its capacitance may be as low as 20 pF.

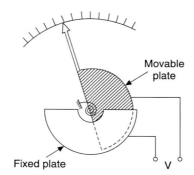

Figure 3.6   Electrostatic moving-vane type system.

In the field of instrumentation, measurement of dc and ac voltages ranging from 10 μV to 10 V is frequently needed and the measuring systems are composed of electrical and electronic circuits and other associated equipment such as rectifiers, potentiometers and galvanometers.

3.2.1   dc Potentiometer Network

Measurement of voltage by dc potentiometer network is widely used for precise and accurate measurement of small emfs obtained from active transducers such as thermocouples. It is a comparison method, in which standard Weston cadmium cell is used as reference of emf. The network is shown in Figure 3.7.

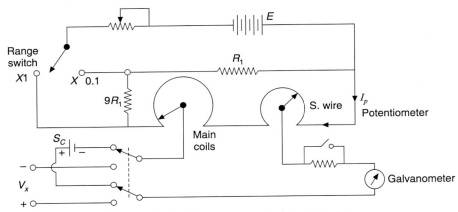

Figure 3.7   Double-range dc potentiometer system.

It has the provision for standardization against the standard cell. The resistance of the potentiometer and the potentiometer current I_p varies with the design. The sensitivity of the galvanometer for the detection of unbalance currents sets the limit for the precision in the measurement while the accuracy is governed by the stability of the standard cell and the potentiometer resistance coils and slide-wire. Commercial versions of the potentiometers are available with varying designs and provisions for range-change, standardization and readout of unknown voltage. The manual balancing method using the galvanometer is replaced by automatic balancing method using a high-gain amplifier and a servomotor. The measurement is rendered faster, enabling recording of slowly varying transients. Figure 3.8 shows the circuit diagram of the self-balancing potentiometer in which a two-phase ac servomotor drives the contactor on the slide-wire and the pen-carriage system for recording. The dc unbalance voltage is converted into a 50 Hz voltage by means of a converter having a vibrating reed that is

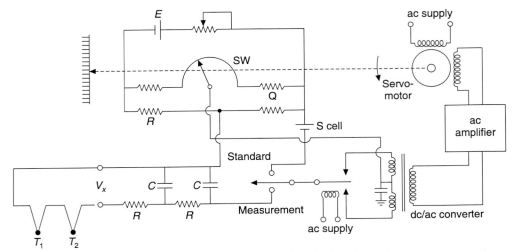

Figure 3.8   Servo-operated automatic potentiometer system.

driven by the 50 Hz supply voltage. The ac amplifier is tuned to 50 Hz and is used to drive the control winding of the servomotor. Depending on the polarity of the unbalance voltage, the motor runs in one direction or the other, but the system is so designed that it runs in such a direction as to reduce the unbalance voltage by the movement of the contactor. The shaft of the motor is mechanically coupled to the pen on the chart and also a pointer on a scale. The chart may be driven at the desired speed. The angular rotation of the shaft is proportional to V_x.

In place of the standard cell, a Zener-diode reference may be used to provide the voltage for standardization. The circuit is designed to measure the thermo emf of a thermocouple. A low-pass filter is used to suppress the unwanted stray ac pickup voltages that may be induced in the long leads of the thermocouple. The cold junction of a thermocouple should be kept at either 0°C or at some fixed temperature. The cold (or reference) junction can be left at room temperature, but provision should be made for compensating the effect of the variation in the room temperature on the indicated emf. All the resistors of the potentiometer circuit except R are of manganin and R is a resistor of wire of a nickel-copper alloy. R is located near the cold junction so that variation in the temperature affects its resistance and hence a correction to the recorded emf automatically. Voltage drop across Q can be made to balance the reference voltage for standardization. Other characteristics of such automatic measuring instruments may be further understood by referring to Chapter 8.

3.2.2 Electronic Voltmeters

Electronic instruments are divided into two categories depending on the way the quantity measured is indicated. Analog type instruments indicate the quantity on a continuous scale by means of a pointer, whereas the digital type instruments display the quantity in numbers. Many digital instruments are none but digital voltmeters with provisions for print-out of the value measured, on a printer or connection to a magnetic tape recorder. The basics of some of these instruments is discussed in the following section. For detailed analysis, the reader is referred to textbooks on electronic instruments.

3.2.2.1 Analog electronic voltmeters

The analog electronic voltmeter consists mainly of three stages: the input voltage attenuator, amplifier, and a pmmc indicating meter. For dc voltmeters, dc amplifiers of the direct-coupled type or the chopper type are used in most cases. Figure 3.9 shows the three stages of a basic dc voltmeter with a FET-input direct-coupled dc amplifier feeding a microammeter. The FET is connected in a source follower mode between the attenuator and the amplifier. The problems

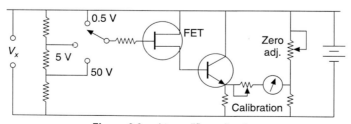

Figure 3.9 dc amplifier-voltmeter.

of drift associated with the dc amplifiers may be overcome by replacing them with the chopper-type dc amplifiers. The chopper converts the dc voltage into ac for amplification by an ac amplifier and then converts back into a dc voltage, proportional to the original input signal. Figure 3.10 shows the circuit of an electronic chopper using four photodiodes (LDRs) instead of synchronous mechanical contactors. The oscillator frequency is about 400 Hz so as to eliminate supply frequency (50 Hz) interference. The light emitted by the neon lamps N_1 and N_2 makes the photoconductive diodes go into states of on/off in a proper sequence so that both the conversion and inversion processes are conducted. Both the input and output signals of the amplifier are of square-wave type and any ripple present in the output is removed by the low-pass filter before it is applied to the pmmc indicator. As the ac amplifier used may be made to possess very high voltage gain, the circuit enables measurement of low dc voltages in the microvolt range. In such cases, the attenuator is not needed. The input impedance of the voltmeter is as high as 10 MΩ.

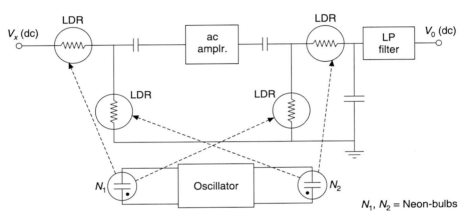

Figure 3.10 Chopper type dc voltmeter.

Differential or balanced amplifiers are particularly useful for the elimination of spurious signals caused by electric stray fields and hence the electronic voltmeters of high accuracy usually have differential amplifiers in the circuit. They are known as *differential voltmeters* which combine the merits of the dc potentiometer network for measurement of voltage with the merits of a high-gain negative feedback dc amplifier. The block diagram of a differential voltmeter (Figure 3.11) shows that the differential amplifier and a microammeter are used to compare a known fraction of the reference voltage with a known and adjustable fraction of the unknown voltage. The unknown dc voltage is applied to the input terminals of the dc amplifier with negative feedback. A fraction of the output voltage of the high gain amplifier is fed back for controlling the closed-loop gain while a suitable fraction is used for comparison against the reference voltage. When the voltage-balance condition is indicated by the difference amplifier and meter, the settings of the range divider and voltage gain of the amplifier section enable the estimation of the unknown voltage. The range of the voltmeter can be as high as 0–1000 V, but the input impedance presented by the input attenuator network cannot be made higher than 1 MΩ.

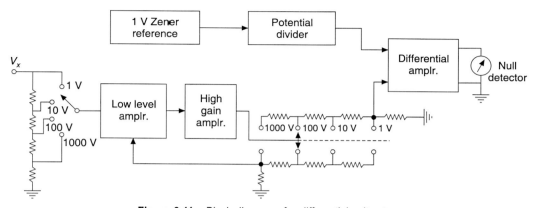

Figure 3.11 Block diagram of a differential voltmeter.

Electronic ac voltmeters can be designed to read either the average value or the rms value. Those average reading voltmeters are basically identical to dc voltmeters except that the unknown ac voltage is rectified either before or after amplification. The scale is calibrated to read the rms value of a sinusoidal wave and hence it is essential to check the voltage for its waveform before using the voltmeter.

An electronic rms-reading voltmeter can be designed, making use of the thermal converter consisting of the thermocouple and a heater coil. A feedback type measuring circuit is shown in Figure 3.12, which uses two thermocouples with matched characteristics and a high gain dc amplifier. Feedback type measuring systems are described in detail in Chapter 8, but it has to be appreciated that the rms value of the heater current of the measuring thermocouple is equal to the feedback dc current of the balancing thermocouple. The indicating meter reads the output voltage of the amplifier under balance conditions and is calibrated in terms of ac rms value of the unknown voltage. The voltmeter is useful for measurement of the nonsinusoidal voltages of frequencies from 10 Hz to 10 MHz. Attention should be given to see that the peak excursions of the waveform do not exceed the dynamic range of the ac amplifier.

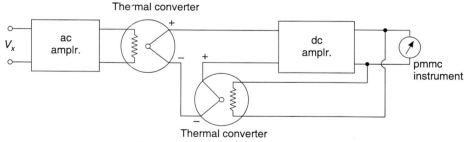

Figure 3.12 rms-reading ac voltmeter.

For measurement of ac voltage of a frequency higher than 10 MHz, the peak-responding voltmeters are generally used. A voltage doubler circuit is shown in Figure 3.13 and the rectified voltage is then fed to a dc electronic voltmeter of suitable range.

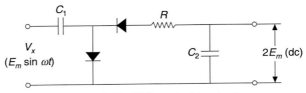

Figure 3.13 Voltage-doubler circuit.

3.2.2.2 Digital electronic voltmeters

The digital voltmeter displays the value of the voltage measured as discrete numerals instead of a pointer reading on a continuous scale. Numerical readout is preferred for various reasons, particularly because it relieves the observer from the strain of reading the value precisely. The measurement is faster and can be brought out in a form suitable for further processing and recording. The digital electronic voltmeters are increasingly becoming popular with the development of integrated circuit (IC) chips at reduced price. Costlier instruments are designed to permit measurements with an accuracy of $\pm 0.005\%$ of full-scale value, and a resolution of 1 ppm. The range extends from 1 V to 1000 V with provision for automatic range selection and overload indication. Additional sophistication is needed for automatic internal calibration and self-checking.

The digital voltmeters are based on either a conversion from analog-to-time domain or analog-to-digital domain. As the final indication is digital in nature, all of them are treated as digital voltmeters.

The self-balancing, servo-operated potentiometer system can be made to provide digital indication, if the servomotor is coupled to a drum-type mechanical indicator, having the digits from zero to nine, imprinted on the periphery of its drum segments. The unknown dc voltage is fed to an input attenuator that provides range-change along with the shifting of a decimal point indicator on the displayed numbers. It is customary to provide overload protection and a filter for rejection of ac noise. These voltmeters are accurate up to $\pm 0.1\%$ of the input range and present an input impedance of 1–10 MΩ. Because of the basics of operation, they are known as continuous balance type digital voltmeters and are popular for measurement of slowly varying voltages with their reading time of about 2 s.

The ramp type digital voltmeter consists of a voltage-to-time converter and a crystal oscillator of 1 MHz frequency. The time interval is measured with the aid of a digital counter that counts the number of pulses that are gated through during the time interval. The gating time interval is proportional to the voltage under measurement, as can be understood from Figure 3.14(a). A ramp voltage generator provides a ramp signal and two comparators provide the necessary 'start' and 'stop' signals when the ramp voltage attains values equal to the unknown voltage and zero. In other words, the time interval T is the time taken by the ramp to rise or fall from zero to a value equal to input voltage. The block diagram is shown in Figure 3.14(b), which includes a sample-rate multivibrator that enables adjustment of the sample rate from a few cycles to about 5 kHz. The sample-rate circuit releases the initiating pulse for the ramp generator to start the next ramp voltage and also a 'reset' pulse to bring the displayed number to zero. The accuracy of conversion is mostly dependent on the ramp linearity and slope stability. The resolution is limited to $\pm 0.05\%$ of full-scale value.

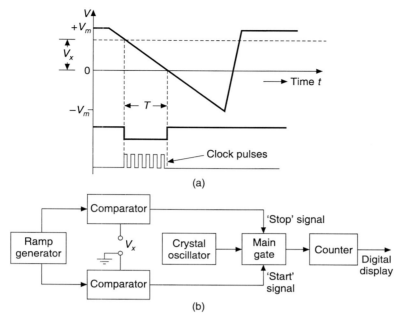

Figure 3.14   (a) Voltage-to-time conversion; (b) block diagram of a ramp type digital voltmeter.

A variation of the ramp type voltmeter is the staircase ramp type digital voltmeter in which the staircase ramp is built up by the count of the counter and hence of the number of pulses released to the counter. A comparator senses the coincidence of the input and staircase voltage and provides a trigger pulse to stop the oscillator. The count reached by them is proportional to the magnitude of the input voltage.

The integrating type digital voltmeter measures the average of the input voltage over a fixed measuring period and uses a voltage-to-frequency converter. The V/F converter is available as an IC chip and it produces an output frequency proportional to the input dc voltage down to and including zero value of the input. The V/F converters have an important role in instrumentation and are explained in Chapter 9. The merit of the integrating type digital voltmeter is its ability to measure accurately in the presence of large noise since the input is integrated. The frequency is normally measured by means of a digital frequency counter although it can also be measured by an analog frequency meter. The accuracy is limited by the stability of the integrator time constant and the stability and accuracy of the threshold level detector of the converter.

In a *dual-slope integration* type digital voltmeter, the integrating amplifier is used to generate two ramps, one due to unknown dc voltage and the other due to a reference voltage, as shown in Figure 3.15. When the integrator output due to V_x crosses the threshold level of the comparator, the crystal oscillator output is gated to the counter. After a predetermined number of counts, N_1 are counted, the input to the integrator is changed to the reference voltage V_s of opposite polarity, and the counter is set to zero. The operation is repeated and counting is continued until the comparator returns to the threshold level. Assuming that the number of counts is N_2 during this period, it can be shown that $V_x/V_s = N_2/N_1$, and the value of V_x is read directly from the counter register. Thus the method enables high accuracy in the measurement

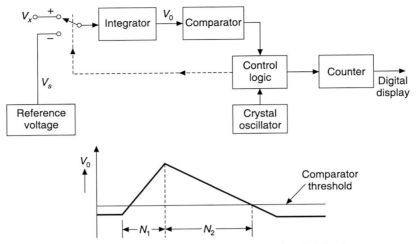

Figure 3.15   Block diagram of dual-slope integrating digital voltmeter.

since it is a comparison method with the reference voltage. It is only necessary that the crystal oscillator and integrator remain stable for the period of each comparison.

All the digital voltmeters described above are relatively slow, providing 12–16 bits in 12.5–200 ms. Digital voltmeters capable of 1000 readings per second and providing 8–12 bits in 1–25 μs use 'successive approximation' technique. Direct conversion from analog domain to digital domain is achieved by the successive approximation converter and it is the most widely used analog-to-digital converter because of its high resolution and high speed. The block diagram of the A/D converter is shown in Figure 3.16. It uses a digital control register with gateable 1 and 0 inputs, a D/A converter with reference supply, a comparator, a control timing loop, and a distribution register which is like a ring counter with a single 1 circulating in it to determine as to which step is in process. The details of functioning are described in Chapter 9, where other A/D converters are also described. The digital output is connected through suitable decoders to the digital display devices, such as LED or LCD seven-segment arrays or Nixie tubes, described in Chapter 10.

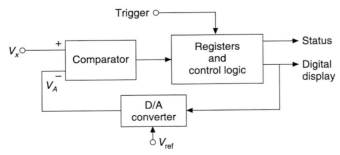

Figure 3.16   Block diagram of a successive approximation type digital voltmeter.

The digital voltmeter and other digital panel meters are normally referred to as 3-digit, 3½-digit, 4-digit, etc. depending on the maximum displayed value. The 3-digit meter displays a maximum value of 999, and therefore it has a resolution of 0.1% of the range. The 3½-digit meter displays a maximum value of 1999 and has a resolution of 1/2000th of the range. Digital multimeters are designed to read dc and ac voltages and currents as well as resistance values, using a basic digital voltmeter of range 0–200 mV and other associated auxiliaries, such as shunts and rectifier circuits.

3.3 MEASUREMENT OF RESISTANCE

Measurement of resistance values extending from a fraction of an ohm to a few megaohms assumes great importance both in the field of electrical measurements and instrumentation. A large variety of resistive transducers exist for the measurement of several non-electrical quantities, and conversion of the resistance changes into proportional electrical voltages becomes an essential requirement. Though several methods for the measurement and direct indication of the resistance values form the subject of electrical measurements, only a few of them are of relevance to the instrumentation engineer. The following methods may be seen for their special features relevant to situations encountered in the measurement of non-electrical quantities.

3.3.1 Indicating Ohmmeters

The simple series type and shunt type ohmmeters are not suitable for instrumentation as they do not possess the desired accuracy and precision. The differential galvanometer (Figure 3.1f) and the crossed-coil pmmc system (Figure 3.1g) offer the means of direct indication using the principle of comparison with a standard resistor. The circuit of the differential galvanometer is shown in Figure 3.17. The two coils of the galvanometer carry currents of opposite polarity and when the deflection is zero, it can be shown that

$$\frac{IR_x}{R_1} = \frac{IR_s}{R_2}$$

$$R_x = R_s \frac{R_1}{R_2} \tag{3.3}$$

where, R_1 and R_2 are the respective circuit resistances shown in Figure 3.17. The measurement needs the values of R_x and R_s to be low enough and that R_1 or R_2 to be variable for bringing the deflection to zero. Small variations in R_x may be directly read off the scale by keeping R_1 and R_2 fixed suitably.

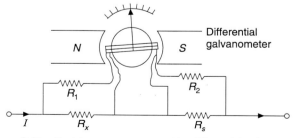

Figure 3.17 Resistance measurement by differential galvanometer.

The crossed-coil type pmmc system offers greater convenience for direct indication of resistance values. If R_1 is the resistance of the coil connected in series with R_x, and R_2 that of the standard resistance circuit (Figure 3.18a), then the ratio of the two currents I_x and I_s is given by

$$\frac{I_x}{I_s} = \frac{R_2 + R_s}{R_1 + R_x} \tag{3.4}$$

and the deflection of the instrument θ_0 depends on the ratio of I_x/I_s. Hence,

$$\theta_0 = f_n(R_x)$$

The scale is graduated in terms of R_x and a suitable form of scale is obtained by shaping pole faces suitably. Insulation testers and Meggers are designed accordingly for measurement of high resistance. The indications are rendered independent of the variations in the battery supply voltage.

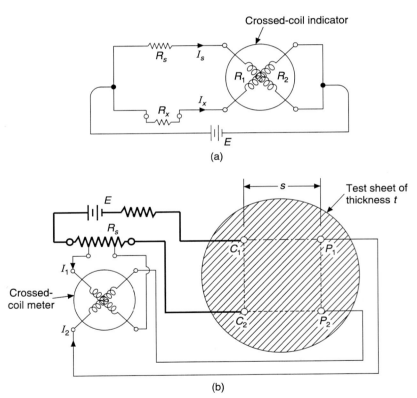

Figure 3.18   (a) Resistance measurement by using a crossed-coil system; (b) location of electrodes on sheet for measurement of resistance of a thick sheet.

An interesting application of the crossed-coil system is shown in Figure 3.18(b), where it is necessary to measure the thickness of a sheet of a good electrically conducting material, which is accessible for measurement from one side only. The test surface is treated as a four-terminal resistor, and a large current is fed into the surface through the current terminals at C_1 and C_2 while the potential difference between P_1 and P_2 is tapped by potential terminals. The four electrodes are provided on a block, as spikes, with all the four located at the corners of a square of side s. The connections of the electrodes to the crossed-coil system are as shown, and the scale is calibrated in terms of the thickness t of the sheet. However, for accurate measurements, the test is repeated at the same current value but with spikes spaced at s, $2s$, $4s$, etc. If θ_1 and θ_2 are the deflections of the instrument with spikes of spacing s and $2s$ respectively, then a calibration curve is provided with values of θ_1/θ_2 and a constant C given by

$$ t = f_n \left(\frac{\theta_1}{\theta_2} \right) s = Cs \tag{3.5} $$

In actual test, only two readings are taken with squares of s and $2s$, and the corresponding value of C is read off the calibration curve to give the thickness t as Cs. The test can as well be conducted by using a source of current and a millivoltmeter. In such a case, the millivoltmeter readings are used in place of θ_1 and θ_2. The method is likely to be accurate up to $\pm 3\%$.

3.3.2 Wheatstone-Bridge Networks

The four-terminal Wheatstone-bridge network (Figure 3.19) is the most popular means of measuring resistance values ranging from 1 Ω to 1 MΩ. They are extremely useful in instrumentation as they can provide either direct indication of the value of an unknown resistor or a voltage proportional to the change in the resistance value from its nominal value. The bridge network excited by dc or ac constitutes an important stage intermediate between the resistive transducer and an amplifier.

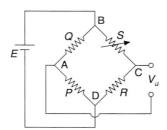

Figure 3.19 Wheatstone bridge network.

The popularity of the resistance bridge networks is due to their capability in providing the desired precision and accuracy. Variable or fixed resistors of manganin can be used in all the arms except the arm of the unknown resistor or transducer.

When a direct indication of the resistance value is desired, the unbalance voltage or current may be measured by a pmmc meter and the scale calibrated accordingly. A centre-zero indicator is normally preferred so that resistance values below and above the nominal value at which the bridge is balanced can be measured directly (see Figure 3.19). At balance, the relationship is given by

$$R_x/S = \frac{P}{Q} = n \qquad (3.6)$$

where,

n = ratio of the bridge

$R = R_x$ = unknown resistance

P, Q, S = other resistors of the bridge

It is conventional to represent the ratio of the bridge n as the ratio of the two resistors in the arms lying on either side of the junction leading to the detector. It can be Q/P as well, because P and Q are the fixed value resistors that enable measurement of R_x in terms of their ratio and a variable resistor S. The nominal value of R_x is the value at which the above balance condition is established. Any deviation of R_x from this nominal value creates unbalance of the bridge, and the unbalance current can be calibrated to indicate the changed value of R_x or its deviation from the nominal value. Knowing the value of the applied voltage E of the bridge, it is possible to calculate the unbalance current. But without going through the development of the relationship, it can be seen that the unbalance voltage and current are proportional to E, and hence the calibration will be liable to error if E is not held constant. In such cases, the crossed-coil system may be used to indicate the resistance value, independent of the fluctuations of the battery supply.

The dc or ac bridge network is very much useful in providing output voltages proportional to the deviation or change of the resistor R_x (or the transducer) from its nominal value, provided the change is very small. In such cases, the bridge is said to be 'slightly unbalanced' as can be seen from the following relationship between the change ΔR and the resulting unbalance voltage V_u. Assuming the voltage source to be ideal, the unbalance voltage V_u between the terminals of A and C is given by

$$V_u = E\left(\frac{R + \Delta R}{S + R + \Delta R}\right) - E\left(\frac{P}{P + Q}\right) \qquad (3.7)$$

Writing $\dfrac{Q}{P} = \dfrac{S}{R} = n$ and $\dfrac{\Delta R}{R} = p$, we get

$$V_u = E\left(\frac{1 + p}{n + 1 + p} - \frac{1}{n + 1}\right)$$

$$= \frac{Epn}{(1 + n)(1 + n + p)} \qquad (3.8)$$

$$\approx \frac{Epn}{(1 + n)^2} \quad \text{if } p \ll n + 1 \qquad (3.9)$$

For small values of p, the unbalance voltage is proportional to the percentage change of R. The linearity in the characteristics for $n = 1$ is within $\pm 1\%$ for changes of R within $\pm 2\%$ of its nominal value. For larger values of n, the linearity improves, but the sensitivity falls as can be seen from Figure 3.20. The sensitivity is maximum for $n = 1$.

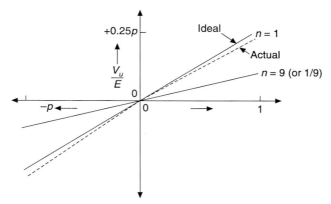

Figure 3.20 Characteristics of a resistance bridge.

Resistive transducers, designed in pairs, may be used in place of S and R so that compensation for temperature changes may be provided while improving linearity and sensitivity. Such transducers are designed to behave as a push–pull pair, i.e. the resistance in one of them increases while in the other it decreases when they are exposed to the quantity under measurement. If p_1 and p_2 correspond to the fractional changes of R and S respectively, the unbalance voltage V_u is obtained as ($R = S$ for push–pull transducers)

$$V_u = \frac{E(p_1 + p_2)}{2(2 + p_1 - p_2)} \tag{3.10}$$

and if $p_1 = p_2 = p$, then

$$V_u = \frac{Ep}{2} \tag{3.11}$$

Thus, the unbalance voltage is found to be double that of the single-transducer bridge, but the primary advantage lies in the linear characteristic between V_u and p. In instrumentation, one comes across the combination of unity-ratio arm bridge networks and push–pull transducers. Push–pull type inductive and capacitive transducers can also be used, but with ac excitation of the bridge network. The advantages of push–pull transducers are discussed in Chapters 6 and 7.

The sensitivity of the bridge can be increased further by increasing E to the extent the resistors of the bridge permit. It is essential to observe that no resistor increases in its resistance due to self-heating.

For a unity-ratio bridge with $E = 2$ V, the unbalance voltage can be as low as 10 μV for a value of 20×10^{-6} for p in the single-ended case and 20 μV for the same value of p in the push–pull case.

When changes in R are required to be indicated on a galvanometer or a sensitive microammeter, the resistance of the bridge network R_{AC} as seen from its terminals A and C, is calculated and used for obtaining the value of the unbalance current I_g. The output (Thevenin) resistance of the bridge network is

$$R_{AC} = \frac{PQ}{P+Q} + \frac{RS}{R+S} = \frac{S+Q}{1+n} \tag{3.12}$$

$$I_g = \frac{Epn}{(1+n)[G(1+n)+(S+Q)]} \tag{3.13}$$

where, G = resistance of the galvanometer.

For small values of p, R_{AC} is considered to remain the same, but in push–pull connection, the value of R_{AC} gets affected for p values larger than 0.05. However, the galvanometer has to be calibrated by actual test in such cases.

When the bridge network forms an intermediate stage of the signal conditioning process, the small unbalance voltage is amplified by suitable amplifiers and, wherever nonlinearity of the bridge requires to be corrected, linearization circuits are used. Considerable amount of attention is necessary while processing the output voltage of the bridge as the amplifiers used respond not only to the unbalance voltages but also to the potentials of the bridge terminals with respect to ground. Therefore, for dc bridge networks, special instrumentation amplifiers are designed, which have the merits of suppressing all spurious input signals and presenting output voltage proportional to V_u with adjustable gain. A brief description of such amplifiers is given in Section 3.5.4.4.

It is possible to excite the bridge from a constant current source instead of a constant voltage source. All the above relations are applicable for voltage sources with negligible internal impedance. Electronic circuits can be designed to function as constant current sources, having very large internal impedance and are preferred with certain transducers.

In Figure 3.21(a), a constant current source is used. For slightly unbalanced bridge, the unbalance voltage V_u can be calculated assuming the bridge voltage across BD terminals to remain constant when R changes by $\pm \Delta R$. Treating the ratio of the bridge as $n = S/R = Q/P$ as already shown and using an additional ratio of P/R $(= Q/S = m)$, the voltage across the bridge can be obtained as

$$V_{BD} = I_S \left[\frac{mR(n+1)}{m+1} \right] \tag{3.14}$$

and using Eq. (3.9) with $V_{BD} = E$, the unbalance voltage V_u is given by

$$V_u = I_S \left(\frac{m}{m+1} \right)\left(\frac{n}{n+1} \right) pR \tag{3.15}$$

Unlike the constant voltage bridge, the constant current source bridge has the unbalance voltage proportional to ΔR $(= pR)$, but not to p as shown in Figure 3.21(b). The sensitivity $(V_u/\Delta R)$ increases with m and n. With $m = n = 10$, and $I_S = 20 \times 10^{-3}$ A, V_u can be 10 μV for a ΔR of 0.6×10^{-3} Ω. If $R = 10$ Ω, V_{BD} is nearly 2 V. The constant current bridge is normally preferred with low-resistance transducers that need to be operated at constant current.

All the resistance bridge networks excited by dc and ac supplies are excellent in speed of response as V_u faithfully reproduces the variations in the resistance of the transducer, as long as the bridge has linear characteristic. At times, bridges with nonlinear characteristic are intentionally used for compensating the nonlinear characteristic of a transducer, so that the overall characteristic is linear over the required range of the measurand.

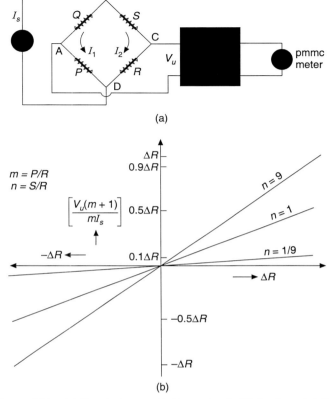

Figure 3.21   (a) Constant-current bridge network; (b) its characteristics.

In actual instrumentation systems, the application of bridge networks requires additional provisions to be incorporated so that V_u does not include any additional components due to interfering inputs. These provisions are as follows:

1. Compensation for the variation of V_u due to variation in E_{BD}
2. Compensation for the errors due to the resistance of the leads of transducer
3. Provision for zero adjustment of the bridge network
4. Provision for calibration of the bridge network for the full-scale value.

Application of bridge networks and operational amplifier circuits has become extensive, as both of them can be operated from the same source of supply. The dc voltage supply required can be sufficiently stabilized and, if necessary, means are incorporated to compensate for voltage drop due to the current in the lead wires of the bridge network. Where a battery supply is used, an adjustable resistor in series with the battery is employed to correct for any deviation in the sensitivity of the bridge, as shown in Figure 3.22.

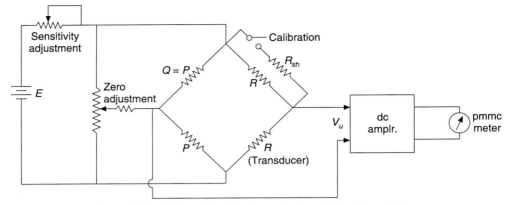

Figure 3.22 Self-contained instrument using a Wheatstone bridge.

Compensation for the resistance and resistance variations of the wires leading to the transducer is essential. The transducer may not always be located near the bridge elements, and the long leads used may change in their resistance values due to temperature gradient along the leads. All the leads must be equal in length and be of the same material and size. They are normally run parallel in a single conduit. The three-lead circuits for a unity-ratio armbridge are shown in Figure 3.23(a) and those for a higher ratio in Figure 3.23(b). When high accuracy in measurement is required, unity-ratio bridge networks are used, and a set of balance points are obtained finally to enable the compensation of the lead resistances in the estimated value.

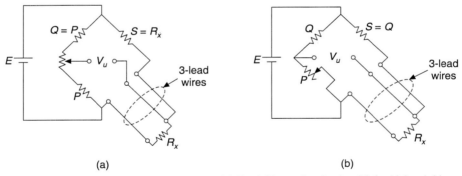

(a)   (b)

Figure 3.23 Leads compensation arrangements: (a) For bridge ratio of unity; (b) for higher bridge ratios.

When single-ended or push–pull transducers are used, zeroing of the bridge network is necessary and adjustments should be provided in such a way that the .contact resistance does not affect the calibration. Figures 3.22 and 3.23 show the provisions for zero adjustment.

When the bridge network is balanced to give zero reading for V_u with the transducer held at its nominal value of resistance, it is equally essential to check that V_u remains the same for the same change in the resistance value of the transducer. This change is simulated by connecting a resistor of suitable value across one of the fixed resistors. If the unbalance voltage thus created is not equal to the desired full-scale value, the variable resistor provided for sensitivity adjustment is altered until it becomes equal (see Figure 3.22).

Figure 3.24 shows an arrangement in which some identical transducers are switched into the bridge network, one after another sequentially. All the transducers may be placed in different locations for monitoring the condition of the entire system. Each transducer is in the bridge-circuit only for a short time. One after another, the transducers are brought into circuit and hence the unbalance voltages from them constitute a series of pulses, spaced apart from each other. The scanning operation may take some time, but each transducer is on only for a much smaller duration. Hence, the applied voltage of the bridge can be increased sufficiently, so as to yield higher sensitivity. The switching arrangement may also be modified such that the bridge network is connected to the supply whenever a transducer is switched on. Such type of pulsed operation reduces the drain on the battery also.

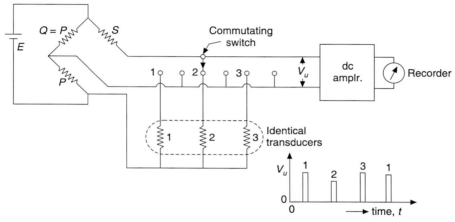

Figure 3.24   Pulsed operation of a Wheatstone bridge.

3.4   MEASUREMENT OF IMPEDANCE

Measurement of impedance involves measurement of the resistive and reactive components of the impedance. The impedance may be inductive or capacitive. It is not correct to assume a coil to be purely inductive, but it is reasonable to assume a loss-free capacitor. A large number of ac Wheatstone bridge networks are available for the measurement of resistive and reactive components of an impedance. but the process of obtaining the balance conditions is not as simple as that of resistance bridge networks. Attainment of balance conditions takes more time and the unbalance voltages cannot be exclusive functions of either the reactance or the resistance component. Hence the usefulness of ac bridge networks in industrial instrumentation is limited in scope, but they are of considerable value in scientific investigations where the speed of measurement is not high. When inductive transducers are considered for use with ac bridge networks, it is assumed that they are high-$Q(= L/R)$ coils so that their resistive components can be neglected in comparison to their reactance components or that they are air-cored coils having their resistive components nearly constant.

Selection of an ac bridge network is based on a few important considerations apart from the choice of frequency, supply voltage, and range of measurement. As pure resistors and capacitors are available, it is advisable that the unknown impedance be expressed in terms of R and C only. The frequency of the supply should not figure in the balance conditions. The balance conditions of the bridge network should yield simple relations that are independent for in-phase and quadrature components. The relationship should not contain higher powers of the values of any component other than unity.

If the transducer is a single self-inductance coil, Maxwell's shunted capacitor bridge of Figure 3.25(a) and Owen's bridge of Figure 3.25(b) satisfy all the above criteria. Moreover, the two bridge networks have no more than one inductor in their arms and hence the likelihood of inductive coupling between the arms is nearly eliminated. The balance equations consist of real and imaginary components which when equated give the relationships between the unknown and known components. For Maxwell's bridge network, L_x and R_x are given by

$$L_x = PSC, \qquad R_x = PS/Q \qquad (3.16)$$

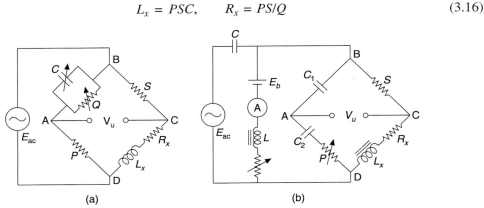

Figure 3.25   ac bridge networks for measurement of (a) self-inductance; (b) incremental self-inductance.

It may be noted that an alternative adjustment of C and Q yields the balancing of the bridge faster, as the factor PS is common to both the conditions. The relationships for calculation of L_x and R_x cannot be any more simpler than those derived.

For the Owen's bridge network of Figure 3.25(b), the balance conditions yield

$$L_x = PSC_1, \qquad R_x = SC_1/C_2 \qquad (3.17)$$

Figure 3.25(b) shows the bridge network excited by both dc and ac so that incremental inductance of the coil can also be determined.

The ratio of the bridge network, as defined earlier, is no longer a real value and it becomes a complex quantity in ac bridge networks. The unbalance voltage consists of two components, one due to variation of L_x from its nominal value and the other due to variation of R_x.

For instrumentation systems, it is desirable that the unbalance voltage is an exclusive function of the change in L_x or the fractional changes in L_x so that it could be related properly to the measurand. It is possible to constitute a coupled coil system so that the total resistance remains the same but the effective self-inductance can be changed by any measurand that affects the coupling coefficient between the primary and secondary coils.

In case the transducer is designed to constitute a push–pull coil system, it can be connected, as shown in Figure 3.26. Two resistors of equal value giving unity ratio are used for the network of Figure 3.26(a), whereas for the network of Figure 3.26(b), the ratio is assumed to be an imaginary quantity given by $j\omega L/R \ (= jn)$. In the latter case, the ac unbalance voltage is given by

$$V_u = \frac{Ep(jn)}{(1 + jn)^2}$$

$$|V_u| = \frac{Epn}{1 + n^2} \tag{3.18}$$

The connection of Figure 3.26(b) results in double the sensitivity than that of Figure 3.26(a), if $\omega L = R$.

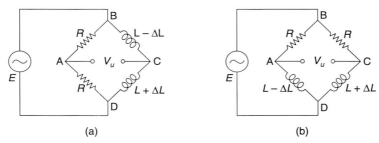

(a)   (b)

Figure 3.26   Methods of connection of a push–pull inductive transducer in an ac bridge.

Capacitive transducers are preferred to inductive transducers, as capacitors can be assumed to be almost perfect and the bridge networks with resistors constituting its ratio arms are convenient in many respects. Push–pull inductive transducers of Figures 3.26(a) and 3.26(b) can be replaced with push–pull capacitive transducers and the unbalance voltages reflect more truthfully the variations in capacitance.

The choice of the frequency of excitation for the ac bridge networks is mostly governed by considerations of the transducer and the measurand, although by choosing a high frequency, inductance coils can be treated to be of high Q, and reactance of capacitive transducers can be brought down to medium values.

A bridge network that has assumed great importance and application is shown in Figure 3.27. It is known as Schering bridge and is used to determine the capacitance and loss

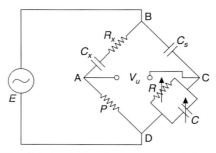

Figure 3.27   Schering bridge for measurement of capacitance.

angle of unknown capacitors. With suitable modifications, the bridge network is used over a wide range of voltages and frequencies, and for capacitors, of low to high values. The ratio of the bridge network is complex and therefore is not of much significance in instrumentation. The unknown capacitor is treated as $R_x - C_x$ series combination, the values of which are given by

$$C_x = C_sR/P, \qquad R_x = PC/C_s \qquad\qquad (3.19)$$

Where C_x is very small in value or when the frequency of excitation is high, it is necessary to shield the system properly. All the components must be guarded from electric fields if bridge is operated at high voltage.

The indication of null or unbalance voltage in the ac bridge networks is usually influenced by the stray magnetic and electric fields, and hence cannot be relied upon unless it is ensured that all such effects are compensated. It is customary to connect an electronic amplifier between the bridge network and the indicator. Earthing one of the detector terminals of the bridge network does not eliminate the effects of stray capacitance. The presence of such sources of error can be detected by reversing the leads from the supply to the bridge network when a different indication will be obtained. Similarly, the connections to any one of the arms may also be checked by reversal. The stray capacitance coupling is most troublesome in high-impedance, high-voltage and high-frequency bridge networks, and connection of one of the terminals of the source of emf or the detector to earth does not offer a solution. The following methods are used to minimize the errors in such cases. A screened and balanced transformer can be used between the bridge network and detector/indicator as shown in Fig. 3.28 when using unity-ratio arms and a push–pull transducer. The transformer is provided with an electrostatic screen between the primary and secondary windings and the screen is connected to the core and earthed. The two halves of the secondary winding are wound in opposite directions and connected in series. When it is connected to the bridge network, stray capacitances between the supply terminal B or D to earth are rendered ineffective. The transformer may also be used in bridge networks with ratio other than unity.

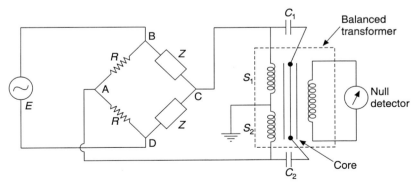

Figure 3.28   Use of balanced transformers for detector.

An ac bridge network having resistive ratio arms can be provided with the 'Wagner earth' connection, and is usually effective as it reduces the potential difference between A, C and earth to zero. It is necessary to balance the bridge for both the positions of a switch connecting *A* and

C alternately to Earth. This method gives good compensation but requires considerable effort and time to satisfy both the balance conditions.

When an ac bridge network is part of an instrumentation system, it is customary to provide it either with the balanced transformer as discussed above or suitably screening the wiring and the components. Balanced transformers may prove useful up to 10 kHz, but for both the audio and radio-frequency bridge networks, screening is effective. Similarly, high voltage Schering bridge network is protected from the intense electric fields by means of electrostatic screening and use of Wagner Earth.

When a single-ended or push–pull type transducer is used in an ac bridge network, the unbalance voltage is taken to be a measure of the unknown non-electrical quantity. If the transducer and bridge network are entrenched in a fixed location and geometrical configuration and are unlikely to be disturbed, the combination is usually calibrated for the specific measurand. However, certain simple measures are taken to see that undue errors do not crop up in indication. The bridged-T network shown in Figure 3.29(a) is one such circuit which has one of the terminals of both the supply and detector/indicator connected to earth. The network is used at radio frequencies for determining L_x and R_x of the coil which is used as a transducer. The twin-T network shown in Figure 3.29(b) is used to determine the difference between the values of C_1 and C_2. C_1 may be fixed and C_2 varied, or C_1 and C_2 may constitute a push–pull capacitive transducer. The detector shows zero when $C_1 = C_2$, and its indication is proportional to the difference in their values. Both the networks are normally used for steady values of the measurands.

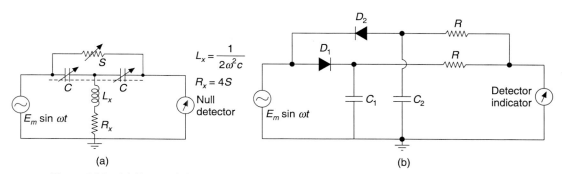

Figure 3.29 (a) T-network for measurement of self-inductance; (b) twin-T network with rectifiers.

In Wheatstone bridge networks, noninductive resistors or loss-free air capacitors are used to make up the ratio arms. It is possible to constitute bridge networks, making use of inductively coupled ratio arms consisting of two coils tightly coupled to each other. The tight coupling close to a coupling factor of unity is achieved by means of a spirally wound tape core or by stacking of annular ring-type thin laminations of low-loss, high-permeability magnetic materials. Two wires taken together are wound on the toroidal core over the entire length, in not more than three layers, so as to minimize the leakage of flux. If the coils are of equal number of turns, and carry equal currents in opposite directions, then the net mmf of the core is zero. In the Blumlein network, the two coils are connected, as shown in Figure 3.30(a) for use with a single ended transducer using unity-ratio, inductively coupled ratio arms. Here, the

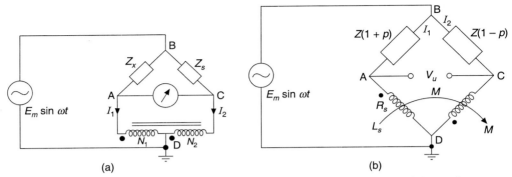

Figure 3.30 Inductively coupled ratio-arm bridge networks: (a) for single-ended transducer; (b) for push–pull transducer.

ratio of the bridge is determined by the ratio of turns of the two coils, and the ratio is precisely known, as the number of turns is exactly counted. When the net mmf of the two coils is zero, there is only resistance drop across AD or CD while under unbalance conditions, each of the arms assumes high reactance drop in addition to the resistive drop. The merits of the above arrangement are:

1. Higher sensitivity near balance as voltage drops across AD and CD are small.
2. With the point D earthed, stray capacitive coupling from points A and C to earth are rendered ineffective at balance.
3. Stray capacitance from B to earth comes across supply, and hence it has no effect on balance.
4. If Z_s and Z_x are widely different, then the balance can be achieved by selecting suitable ratio for N_1/N_2 and finely adjusting the components of Z_s.
5. If $N_1 = N_2$, the bridge network can be used with resistive, inductive or capacitive type push–pull transducers and the unbalance voltage can be related to the value of the measurand.

The bridge network is redrawn to develop the expression for unbalance voltage, with the arms AB and BC making up the push–pull transducer, as shown in Figure 3.30(b). If k is the coefficient of coupling and L_s the self-inductance of each winding, then $M = kL_s$.

Under unbalanced conditions, resistive drop across arms AD and CD may be assumed to be negligible as compared to the reactance drop. Thus, for the slightly unbalanced bridge network, V_u can be derived from the following equations:

$$E = I_1(Z + \Delta Z) + I_1(j\omega L_s) - I_2(j\omega L_s k)$$
$$E = -I_1(j\omega L_s k) + I_2(Z - \Delta Z) + I_2(j\omega L_s)$$
$$V_u = I_1(Z + \Delta Z) - I_2(Z - \Delta Z) \tag{3.20}$$

Simplifying the above equations and treating the value of k as unity, V_u is given by

$$V_u = 4E\frac{\Delta Z}{Z}\frac{j\omega L_s/Z}{\left(1 + p + \dfrac{j\omega L_s}{Z}\right)\left(1 - p + \dfrac{j\omega L_s}{Z}\right) - \dfrac{j^2\omega^2 L_s^2}{Z^2}} \tag{3.21}$$

For comparison with the Wheatstone bridge network, $j\omega L_s/Z$ is treated as the ratio of the bridge, as defined for the conventional bridge, and with push–pull inductive transducers (high Q coils) the ratio is a real number n.

V_u is obtained in terms of p, n and E as

$$V_u = 4Epn \frac{1}{1 + 2n - p^2} \tag{3.22}$$

$$V_u \approx 4Epn/(1 + 2n) \quad \text{if } p^2 << \frac{1}{1 + 2n} \tag{3.23}$$

The sensitivities of the Wheatstone and Blumlein bridge networks are compared for n values ranging from 0.01 to 100 in Figure 3.31(a); Blumlein bridge network has constant sensitivity for values of n larger than 5, while being more sensitive for all values of n. The linearity is higher at higher values of n. Similar observations may be made for the case of resistive transducers by replacing n with jn in relation (3.23).

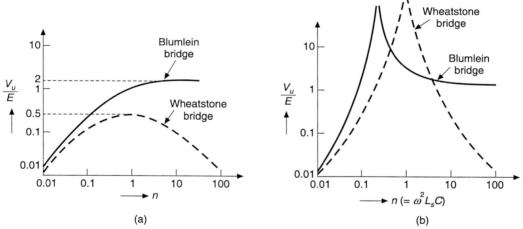

Figure 3.31   Characteristics of Wheatstone and Blumlein bridge networks: (a) For n positive real value; (b) for negative real value.

The bridge network assumes special importance when capacitive transducers are used. The ratio becomes negative and V_u is given by

$$V_u = Ep \frac{4\omega^2 L_s C}{1 - 2\omega^2 L_s C} \tag{3.24}$$

The frequency of the applied voltage E comes into the picture and if ω or the capacitance of transducer assumes such values as to make $\omega^2 L_s C = 1/2$, then the unbalance voltage becomes very large and the bridge behaves as a resonance bridge. A Wheatstone bridge network with the same ratio, goes into resonance when $\omega^2 L_s C = 1$. The relative sensitivities are shown in Figure 3.31(b) for n values ranging from 0.01 to 100. For accurate determination of the value of C, the ratio of 0.5 may be useful, but for stability in sensitivity, a higher value of n is advantageous. Even if slight variations in frequency occur, the bridge sensitivity remains the same.

3.5 ELECTRONIC AMPLIFIERS

All the present day instrumentation systems require electronic amplifiers primarily for amplification of the low-level dc or ac signals obtained from electrical measuring systems and transducers. The input signal may be a voltage or current but the output signal may be required either as voltage or current or power. Amplifiers might also be used as buffer and isolation amplifiers and should be able to respond only to the desired signals in the presence of unwanted interfering signals. A large variety of transistor amplifiers are available for common applications of voltage, current and power amplification with differing input and output impedances. Special amplifiers are designed for amplification of signals arising from sources under certain typical conditions and they may be known as difference amplifiers, chopper amplifiers, electrometer amplifiers, feedback amplifiers etc. which are now discussed.

3.5.1 Difference or Balanced Amplifiers

The basic circuit of a difference input (difference) amplifier is shown in Figure 3.32. It consists of two matched transistors working under same operating conditions. The zero adjustment is provided for obtaining V_o as zero when the potentials of terminals 1 and 2 are brought to zero by shorting them together and earthing. If V_1 is the potential of terminal 1 and V_2 that of terminal 2, then $V_i = (V_1 - V_2)$ and $V_o = A_d(V_1 - V_2)$ assuming that V_1 is at a higher potential than V_2. V_o will appear with reverse polarity if V_2 is at a higher potential than V_1. A_d is known as the 'differential' gain of the difference amplifier. Ideally, the amplifier responds to the difference signal between the input terminals and ignores interference signals that appear in phase at both the terminals. If terminals 1 and 2 are kept floating without being connected to any desired input signal, then the output voltage should also be zero. But due to external stray electric fields, it is likely that V_1 and V_2 may assume the same potential V_{cm} due to their proximity to each other. If the two halves of the amplifier are not perfectly matched and

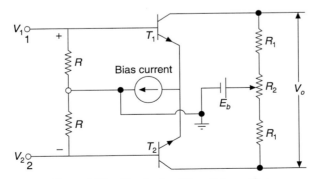

Figure 3.32 Circuit of a differential amplifier.

balanced, a voltage will appear at the output terminals. The signals appearing at 1 and 2 are known as 'common-mode' signals represented by V_{cm} and $V_o = A_{cm}V_{cm}$, where A_{cm} is the *common-mode gain*. This value of V_o due to V_{cm} is known as *common-mode-voltage error*. The ability of the amplifier to reject or minimize that error is referred to as *common-mode rejection* (CMR).

The imperfection of the amplifier causing deviation in its functioning as true 'difference amplifier' is determined by the ratio of A_d/A_{cm}, known as common-mode rejection ratio (CMRR). The difference amplifier should possess as high a ratio as possible, if it cannot be infinity. In other words, CMRR is the ratio of V_{cm} to that value of desired input voltage that would have developed the same V_o as that due to V_{cm}. If long leads are connected to terminals 1 and 2, then the leads are twisted together so as to hold them at the same V_{cm} as far as possible. CMRR is generally expressed in dB as 20 log (A_d/A_{cm}). If for a difference amplifier the CMRR is 100 dB (10^5), the effective error in measurement of V_i with a V_{cm} of 10 V is 0.1 mV $(= 10/10^5)$. If the amplifier is rated for a V_i of 100 mV, then the common-mode error is $\pm 0.1\%$. Considerable attention is required to amplify low level signals developed in high-voltage circuits. Other sources of error arise due to the drift, nonlinearity of the amplifier, and the frequency of the desired input voltage, and so the nominal value of A_d will change. The CMRR is generally specified for supply frequency of 50 Hz and an unbalance of 1 kΩ of input source resistance.

3.5.2 Electrometer Amplifiers

In instrumentation activity, amplification of very low-level dc signal voltages is important. The one-stage amplifier or the direct-coupled two-stage amplifier is not suitable where measurement with high accuracy is desired. The inherent instability due to zero-drift and sensitivity drift renders them unsuitable for application. So, special amplifiers known as *electrometer amplifiers,* are designed by using electrometer valves, operational amplifiers, negative feedback and dc-to-ac converters.

The electrometer valve is designed to possess very low grid current, maximum mutual conductance, and stable anode current for a fixed value of the electrode potential. They are used in matched pairs to constitute a two-stage difference (balanced) amplifier, equivalent to that of Figure 3.32. Making use of the advantages of negative feedback, linearity and constant sensitivity could be achieved.

To amplify dc voltage signals from sources of very high internal impedance, the amplifiers should possess an input impedance 100 to 1000 times larger than that of the source, while showing no zero drift. In this respect, electrometer amplifiers using the electrometer valves are not able to match the requirements. Hence, dc-to-ac converters or choppers using a vibrating-capacitor are used, and such amplifiers are known as vibrating-capacitor electrometer amplifiers. When measuring the emf between the electrodes of a pH measuring system, a high-grade capacitor is alternately connected to the terminals of the electrodes and the input to the ac amplifier, as shown in Figure 3.33(a). The switching operation is carried out by a relay operated from mains supply at 50 Hz. Because there is no direct connection between the measuring circuit or the ac amplifier and the electrodes, it is possible to earth both the ac amplifier and the solution around the electrodes, thereby preventing the effects of interfering signals from stray electric fields.

Another method is to connect a movable plate capacitor to the electrodes directly, and continuously as shown in Fig. 3.33(b) and vibrate the movable plate at a suitable frequency, by means of an electromagnetic drive. The value of the capacitor varies by about 1 per cent and the ac voltage across it is amplified by an ac amplifier. The output voltage is converted back to dc for being read or recorded by an instrument. It is required that the frequency of vibration is much greater than $1/(2\pi CR)$, where R is a high resistance kept in series with C. R can as well be the internal resistance of the emf source. The two capacitors C_1 and C_2, and R_1 block the dc component and allow the ac component only for amplification. The charge on C when the movable plate is at rest is $E_i C (= Q)$. When the plate is made to vibrate about its mean position, the capacitor draws a small amount of charge and energy from the source when the capacitance is increasing, but it is returned to the source during the next half-cycle, and so no power or energy is drawn any more after the initial charging. The energy required to vibrate the plate is drawn from the driving mechanism. The instantaneous current in the circuit $i(t)$ is given by

$$i(t) = \frac{dq(t)}{dt} = \frac{d}{dt}(CE_i) = C\frac{dE_i}{dt} + E_i\frac{dC}{dt}$$

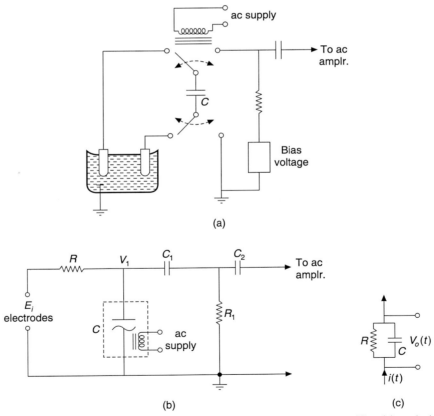

Figure 3.33   (a) Switched-capacitor type amplifier; (b) vibrating capacitor type amplifier; (c) equivalent circuit.

The variation of E_i with time is very small compared with the rate of variation of the capacitance, and hence the first term is negligible.

For a parallel plate with an initial separation between the plates X_0,

$$\frac{\Delta C}{C} = \frac{\Delta X}{X_0} \quad \text{when} \quad \frac{\Delta X}{X_0} < 0.01$$

As ΔX is given by $\Delta X_m \sin \omega_s t$, $i(t)$ is given by

$$i(t) = E_i C \frac{\Delta X_m}{X_0} \omega_s \cos \omega_s t \tag{3.25}$$

The vibrating capacitor acts as a current source with $i(t)$ as the current. From the equivalent circuit shown in Figure 3.33(c), it can be shown that

$$\left| \frac{V_o(j\omega)}{E_i(j\omega)} \right| = \left(\frac{\omega_s CR}{\sqrt{1 + \omega_s^2 C^2 R^2}} \right) \frac{\Delta X_m}{X_0} \tag{3.26}$$

Keeping $\omega_s CR \gg 1$, the amplitude of the ac voltage V across C is made proportional to E_i. If E_i is constant and is from a low internal impedance source, a large value is chosen for R, and the voltage across R is fed to the amplifier, when it has to measure the value of $\Delta X_m/X_0$. In such cases, the vibrating capacitor is functioning as a microphone or as a transducer for the measurement of dynamic force and pressure (see Section 6.3.2). E_i is then referred to as biasing or polarizing emf for the dynamic capacitor microphone. The circuit is thus widely applied for the measurement of pH, dynamic force and pressure and is also referred to as *dynamic condenser amplifier*.

The ac voltage is amplified by high gain amplifiers, without any significant drift and is then synchronously demodulated. Such electrometer amplifiers can have an input impedance of 10^{15} Ω, and a zero drift of about 100 μV over one-day period. The frequency of vibration of the movable plate may be 50 Hz, 400 Hz or 2 kHz, depending on the requirements.

3.5.3 Feedback Amplifiers

The principle of negative feedback between the output and input signals is extensively used to develop several amplifiers, each possessing different operational characteristics. The description and development of such amplifiers may be referred to in standard textbooks, but a brief introduction to those that are particularly useful in instrumentation is considered proper.

The emitter follower is one such amplifier which has high input impedance and very low output impedance. It utilizes 100% negative feedback and hence no voltage gain, and has considerable power gain but no phase inversion. The emitter follower has common ground connection between input and output terminals so that grounding and shielding problems are minimized. An identical amplifier using an operational amplifier is known as voltage follower (voltage gain = 1) with the output voltage following the input voltage faithfully in amplitude and phase. It is used in many situations as a buffer amplifier, between sources of voltage signals and other subsequent stages.

The principle of negative feedback is also applied to develop highly stabilized dc amplifiers having the merits of reasonable dc gain, very low drift and wide bandwidth. The chopper type amplifiers used so far do not possess high bandwidth. If the chopper frequency is 2 kHz, the bandwidth is limited to one-tenth of chopper frequency and, therefore, chopper amplifiers are useful for processing dc voltages varying very slowly with time.

The advantages of negative feedback are combined with the merits of chopper amplifiers, and the combination is known as *chopper stabilized amplifier*. The schematic diagram is shown in Figure 3.34 and the scheme can be used successfully where high input impedance for the amplifier is desired, as is true for pH measurement. Here the high frequency components of the input signal are bypassed from the chopper amplifier and fed to the dc difference amplifier. The high and low frequency components are shown to be fed together to the dc difference amplifier and negative feedback is used to provide uniform gain over a wider range of frequencies. The effect of drift at the input terminals of the dc difference amplifier is reduced by the dc gain of the chopper amplifier. With 100 per cent negative feedback, it is possible to achieve a bandwidth of about 125 kHz, while having very low zero drift as compared to electrometer amplifiers.

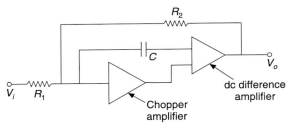

Figure 3.34   Chopper stabilized amplifier.

The negative feedback may also be used exclusively with chopper amplifiers to provide some increase in bandwidth having lower dc gain.

3.5.4   Operational Amplifiers

The operational amplifiers (op-amps) belong to a class of high-gain, dc-coupled differential amplifiers that make an enormous number of applications possible in instrumentation systems. The op-amps constitute one of the most common and important building blocks available in integrated circuit form. A typical op-amp may contain as many as 20 transistors apart from the associated resistors required for interconnection. Because of their high dc-voltage gain (10^4–10^6), op-amps are always used with some external feedback circuits in order to prevent saturation of the output or instability in operation. It is the choice of the feedback circuits that results in the development of committed amplifiers, suitable for a variety of tasks with a variety of desirable performance characteristics. Treating it as equivalent to a basic element such as a resistor or a capacitor, several functional circuits can be configured as RCA circuits, where A stands for the op-amp. A variety of such circuits have indeed enriched the status of the present-day instrumentation systems to the extent that it is assumed by many that the discipline of instrumentation engineering is virtually electronic engineering. It is the special characteristics and near-ideal behaviour of the op-amp that has particularly rendered possible the processing of low-level electrical signals from transducers which are mostly confined to dc and have low frequencies up to 1 kHz. It can be stated that no other electronic element or device approaches its ideal performance to the extent the op-amp does. For a preliminary understanding and appreciation of its potential for application, the op-amp with its idealized characteristics is considered first, whereas the deviation in its performance due to certain imperfections is overcome by limiting its application to such situations where its non-ideal performance is not detrimental to its use.

An op-amp is usually a high-gain dc amplifier with differential input and single-ended output as shown in Figure 3.35(a). The common input point 3 and the output point 4 are usually connected together and grounded. It is common to represent the op-amp by a symbol, as shown in Figure 3.35(b), sometimes without showing the ground terminal. The output voltage v_o is proportional to the difference in potentials of the input terminals marked 1 and 2, and is given by

$$v_o = A_d \, (v_+ - v_-) \tag{3.27}$$

where,

$$A_d = \text{differential voltage gain}$$
$$v_+, \; v_- = \text{potentials at terminals 1 and 2, respectively}$$

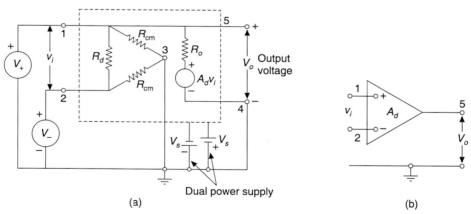

(a)   (b)

Figure 3.35 (a) Op-amp circuit; (b) op-amp representation.

Terminal 2, marked minus, is known as inverting input terminal and terminal 1 as the noninverting input terminal.

The op-amp is idealized by the following characteristics:

1. A_d is infinite.
2. The input impedance is infinite.
3. The output impedance R_o is zero.
4. The bandwidth extends from zero to infinity.
5. v_o is zero when the input voltage v_i is zero.

It may be noted that the notation of marking + for the noninverting input terminal implies the sign of the gain A_d, and not the polarity of the input potential with respect to ground. The op-amp is usually supplied with a symmetrical power supply of +15 V and –15 V with the common or centre-tap connected to ground terminal 4. Two voltage sources may be connected to terminals 1 and 2, as shown in Figure 3.36(a), in which case v_o is proportional to the difference of the two voltages. It can also be used in a single-ended fashion as shown in Figure 3.36(b). In either case, the input voltage v_i needs to be small, as shown in Figure 3.36(c), since A_d is very high and the amplifier saturates for a small value of v_i.

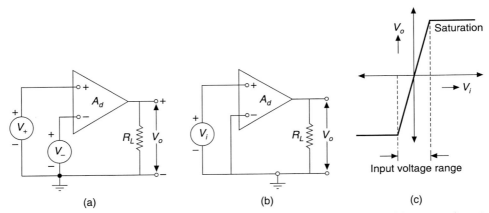

Figure 3.36   (a) Op-amp with two input voltages; (b) op-amp with one input voltage; (c) op-amp characteristic.

3.5.4.1   Basic op-amp configurations

The op-amp is never used in open-loop mode, because of its high gain and associated instability. Several configurations exist for carrying out the desired operation on input signals and they are mostly made up of elements such as resistors and capacitors along with op-amps. All such configurations are based on application of negative feedback. The merits of applying negative feedback to systems that include a high-gain amplifier in the forward path of the closed-loop system are explained in Section 8.1. It is the op-amp that constitutes the high-gain amplifier of the closed-loop configuration and it is not difficult to appreciate the role of such circuits in the field of instrumentation from the three examples. Other applications are presented in Chapter 9.

1. *Inverting amplifier*   The inverting circuit consisting of an op-amp, a feedback impedance, Z_f and a signal source of internal impedance Z_i is shown in Figure 3.37(a). It is seen that a current i_f proportional to v_o is fed back to input side and the currents at the summing point (terminal 2) can be equated as

$$\frac{v_i - v_-}{Z_i} = \frac{v_- - v_o}{Z_f} + i_a$$

where, i_a is the current drawn by op-amp. Writing $v_o = A_d v_-$ since v_+ is zero, and rewriting the equation, v_o is obtained as

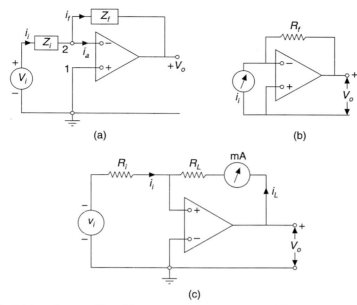

Figure 3.37 (a) Inverting amplifier; (b) current-to-voltage converter; (c) voltage-to-current converter.

$$v_o = -v_i \frac{Z_f}{Z_i} - \frac{v_o}{A_d}\left(1 + \frac{Z_f}{Z_i}\right) + i_a Z_f \tag{3.28}$$

$$v_o = \frac{\left[-v_i \left(\dfrac{Z_f}{Z_i} \right) + i_a Z_f \right]}{\left[1 + \dfrac{1}{A_d}\left(1 + \dfrac{Z_f}{Z_i} \right) \right]}$$

Applying the idealized characteristics 1 and 2, it is seen that Eq. (3.28) simplifies to

$$v_o = -\left(\frac{Z_f}{Z_i} \right) v_i \tag{3.29}$$

It is interesting to note that the relation between input and output voltages is solely governed by impedances in the input and feedback paths. With resistors in place of Z_i and Z_f, the inverting circuit becomes an inverting amplifier whose gain can be greater than or equal to unity, depending on the choice of values for R_i and R_f. Any deviation in the calculated value of gain is due to the presence of i_a and finite A_d. The above relationship also implies that the summing point (terminal 2) is at zero potential (or very nearly at zero potential) and hence it is called virtual earth. The algebraic sum of all currents at virtual earth is thus equal to zero. The configuration has the virtue of its two input terminals at near zero potential and hence there can be no common-mode signal. Usually, signal sources are transducers with a nonzero and possibly variable internal resistance, in which case gain setting by R_f/R_i is not likely to remain constant. It is desirable in such cases to choose a high-valued R_i (0.1 MΩ) which is much larger than the internal resistance of the transducer.

In case the signal source is a constant current source as shown in Figure 3.37(b), it can be seen that

$$v_o = -R_f i_i \qquad (3.30)$$

where i_i is the current from the source. The circuit is used as a current-to-voltage converter and is used with current sources such as photomultiplier tubes.

The inverse operation, viz. voltage-to-current conversion, is shown in Figure 3.37(c) and the load current in R_L is given by

$$i_L = \frac{v_o}{R_L} = \frac{-v_i}{R_i} \qquad (3.31)$$

The circuit is useful for driving the electrolytic cells, indicating meters, motors of chart recorders, etc. The floating load impedance may be the feedback impedance of the op-amp. The load current is limited by the maximum rated output of the amplifier. Wherever larger current is desired, a 'booster' amplifier may be inserted inside the feedback loop.

2. *Noninverting amplifier* The input signal is directly applied between the noninverting input terminal and ground, and the feedback is applied to the other terminal as shown in Figure 3.38(a).

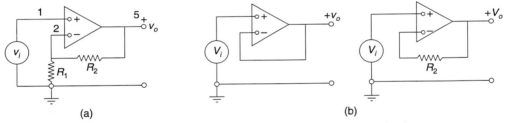

(a)   (b)

Figure 3.38 (a) Noninverting amplifier; (b) voltage follower circuits.

Analyzing the circuit as before and equating the currents at the summing point, it can be shown that

$$\frac{v_i}{R_1} = \frac{v_o - v_i}{R_2}$$

$$v_o = \left[1 + \left(\frac{R_2}{R_1}\right)\right] v_i \qquad (3.32)$$

It should be noted that the effect of the negative feedback is to equalize the potentials of the inverting and noninverting input terminals. There is no phase reversal between v_o and v_i, and terminals 1 and 5 are of the same polarity.

The circuit becomes a voltage follower if R_1 is made infinite or R_2 is made zero, as shown in Figure 3.38(b). The minimum gain is unity and the advantage of the circuit is its high input impedance. The circuit is used as an impedance transformer or isolating buffer. The signal source may be a high impedance source such as a glass pH electrode or a microelectrode inserted into a nerve cell. The voltage follower transforms such a high impedance source into one with nearly zero output impedance, that can drive other subsequent loads without affecting the terminal voltage at the original electrode.

3. *Differential amplifier*   A particular configuration of the inverting amplifier circuit is the differential amplifier using the negative feedback for stabilizing its gain. The circuit is shown in Figure 3.39 and it can be treated as a combination of the inverting and the noninverting amplifier. The output voltage v_o is given by

$$v_o = v_2 \left(\frac{R_4}{R_3 + R_4} \right) \left(1 + \frac{R_2}{R_1} \right) - v_1 \left(\frac{R_2}{R_1} \right) \tag{3.33}$$

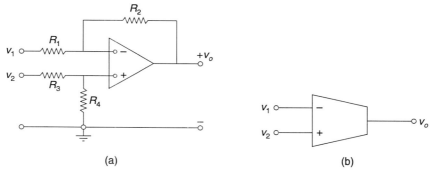

Figure 3.39   (a) Differential amplifier circuit; (b) its representation.

Selecting the resistors to satisfy the condition that $R_1/R_2 = R_3/R_4$ makes the value of $v_o = (v_2 - v_1)R_2/R_1$, giving a true differential amplification.

Any common mode voltage adding to v_1 and v_2 is thus found ineffective. R_2 is generally kept within 100 kΩ to reduce the effects of offset bias currents. The input resistance of the differential amplifier is only $2R_1$ and where higher values are desired an instrumentation amplifier described in Section 3.5.4.4 may be used.

3.5.4.2   Op-amp specifications and limitations

Op-amps have been presented as ideal devices so far but in practice the actual values are neither infinite nor zero. The different types of operational amplifiers available are indicated in Section 3.5.4.4 with each of them designed to satisfy the requirements for a particular application. There are subtle differences in their specifications, and 'manufacturers' data sheets pertaining to them should be consulted before selecting them for each application. A study of the data sheets reveals the fact that they exhibit, to a certain extent, deviation from the ideal characteristics.

Usual specifications include the values of supply voltage (usually ±15 V), rated output (±10 V or ±14 V at ±5 mA or ±10 mA), input voltage range for both series and common-mode signals, input impedance for both differential and common-mode signals. The common-mode rejection ratio is specified for the supply frequency of 50 Hz.

Other specifications pertain to the actual performance and should be studied critically with a view to obtaining the desired performance, after compensating suitably for the limitations in performance of op-amps.

Although the gain of the op-amp is assumed to be infinity, it is normally between 10^4 and 10^8, and the gain is a function of frequency. The bandwidth extends from dc to about 5 or 10 Hz and then falls off at a uniform rate of 20 dB/decade for a frequency-compensated op-amp, as shown in Figure 3.40. The frequency at unity gain is about 1 MHz and the amplifier is said to possess a 1 MHz gain bandwidth product.

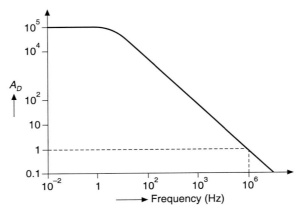

Figure 3.40 Gain-frequency characteristic of an op-amp.

The output voltage of an op-amp is not zero when the input terminals are shorted and grounded. However, it can be brought to zero by applying a small dc voltage (known as input offset voltage) of proper magnitude and sign to either input terminal as shown in Figure 3.41. The null potentiometer is adjusted in each case for compensating the offset voltage. The offset voltage varies with temperature and hence the compensation needs to be checked before using the op-amp. Temperature drift, resulting in reappearance of offset after trimming the op-amp, is usually specified in μV/°C over the operating range of temperature.

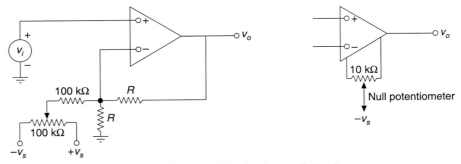

Figure 3.41 Compensation circuits for offset voltages.

Each op-amp has a small dc bias current flowing out of both the input terminals to the ground, and the offset current is the difference in these bias currents. Bias current produces an effect similar to that of offset voltage and it will not be zero even if the offset voltage is zero. Bias current is smaller in case of FET input op-amp and is normally ignored in most cases except while amplifying low-level signals from high impedance sources. Offset current is smaller than bias current and compensation for both offset and bias currents may be made by similar circuits provided externally or by suitably locating an input resistor, as shown in Figure 3.42. Specifications normally include the values of bias currents and the input offset current at a particular temperature as also the variation of each with temperature over the operating range.

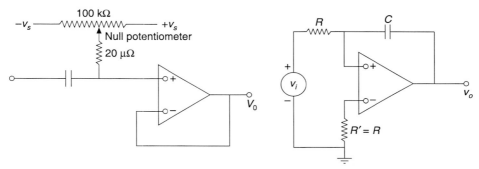

Figure 3.42 Compensation circuits for both offset and bias currents.

Normally, the common-mode operating range is understood to be the range of supply voltages within which the op-amp works satisfactorily. It is nearly equal to the saturation output voltage range.

Other specifications relate to the amplifier drift, input noise level, and slew rate. The *slew rate* is defined as the maximum output-voltage change per unit time. If a square wave is applied to the input terminals, the output voltage will initially be a ramp at the slew rate.

The op-amp with its two input terminals and common ground is treated equivalent to a three-terminal input system having resistances connected, as shown in Figure 3.35. The resistance between the two input terminals is known as *differential input resistance* R_d, whereas the resistance between each input terminal and the ground is known as *common-mode input resistance* R_{cm}. The R_{cm} values are equal to each other but are higher than R_d, by about 100 times. When sources of unequal source resistances are connected to the input terminals, a common-mode error signal of value $(V_{cm}/R_{cm})\,\Delta R_s$ is developed, where ΔR_s is the difference in the source input resistances. When dealing with low-level signals, a check is necessary to see that the common-mode error signal is negligible.

3.5.4.3 Types of op-amps

Most of the op-amps are standard voltage amplifiers built from bipolar junction transistors and are usually known as bipolar differential amplifiers. Other common versions of op-amps are FET differential amplifiers, chopper stabilized amplifiers and varactor bridge or bridge (or parametric) amplifiers. Their operational characteristics differ and each of them has its own special features and application. Most of them are internally compensated for the limitations in their performance and are available for use as building blocks for processing of signals in instrumentation systems. Before selecting them for application, their specifications regarding the values of bias currents, voltage drift, dc gain, and input impedances should be studied from the manufacturers' data sheets.

The bipolar differential amplifier using bipolar junction transistors is the general purpose op-amp that is used over a wide range of temperatures to function as a low-noise amplifier for signals from low-impedance sources.

The FET-input differential amplifier possesses high input impedance because of the addition of a pair of matched field effect transistors to the bipolar differential amplifier. The bias current is reduced but is more sensitive to temperature variations. Such amplifiers are generally preferred for use with high impedance sources and as sample and hold amplifiers.

The chopper-stabilized amplifier described in Section 3.5.3 is mostly drift-free and at the same time provide very high gain of the order of 10^6–10^9. Its main disadvantage is that it can be used either in the inverting or the noninverting mode only, and not as a differential amplifier. It is mostly suitable for use as an integrator and as a buffer amplifier with low-drift and high input impedance.

The varactor bridge amplifier, also known as parametric amplifier, is known for its lowest possible bias currents of the order of 10^{-14}–10^{-12} A. The amplifier utilizes voltage variable capacitors; for details the reader is referred to standard textbooks. The amplifier can withstand high input voltages and can provide extremely high input impedance to serve as buffer amplifier and electrometer amplifier for measurement of very weak currents. They are used as integrators and amplifiers for signals arising from high impedance sources.

3.5.4.4 Instrumentation amplifiers

The instrumentation amplifiers are committed configurations consisting of operational amplifiers, intended primarily for amplification of voltage signals derived from transducer circuits, with accurately adjusted gain values. Any variation in gain affects the accuracy of measurement of the primary quantity. Moreover, the amplifier is required to amplify low-level signals superimposed with common-mode voltages. Thus the major requirements of an instrumentation amplifier apart from the above are:

1. Low drift
2. High input impedance
3. High linearity
4. High CMRR
5. High noise rejection capability.

Thus the instrumentation amplifier is an improved version of the basic differential amplifier with the configuration consisting of three FET-input op-amps connected, as shown in Figure 3.43. High common-mode rejection is realized by connecting two noninverting configurations with op-amps, 1 and 2, and having a common feedback resistor R_1.

The configuration can be analyzed by treating the combination as a linear system and obtaining the expression for v_o by using the superposition theorem. With one terminal applied with the voltage and the other grounded, v_{o1} and v_{o2} can be obtained separately. With v_1 applied to terminal 1 and terminal 2 grounded, v_{o1} and v_{o2} are given by

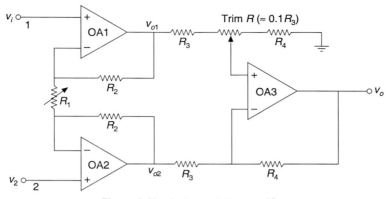

Figure 3.43 Instrumentation amplifier.

$$v_{o1} = v_1 \left(1 + \frac{R_2}{R_1} \right), \qquad v_{o2} = -v_1 \left(\frac{R_2}{R_1} \right)$$

Similarly, when v_2 is applied to terminal 2 and terminal 1 grounded, we have

$$v_{o1} = -v_2 \left(\frac{R_2}{R_1} \right)$$

$$v_{o2} = v_2 \left(1 + \frac{R_2}{R_1} \right)$$

Hence the net effect of the application of v_1 and v_2 is to give $v_{o1} - v_{o2}$ in the form

$$v_{o1} - v_{o2} = \left(1 + \frac{2R_2}{R_1} \right) (v_1 - v_2) \tag{3.34}$$

The second stage is a differential amplifier with a gain of R_4/R_3, and hence the net gain of the configuration is given by

$$A = \left(1 + \frac{2R_2}{R_1} \right) \left(\frac{R_4}{R_3} \right) \tag{3.35}$$

By adjustment of R_1, the required gain values can be obtained.

A trimming resistor R of value equal to 0.1 time R_3 is included to improve the CMRR of the second stage amplifier. The dual-input stage consisting of OA1 and OA2 is seen to provide a differential gain of $(1 + 2R_2/R_1)$ while having a common mode gain of unity only. Thus there is an overall improvement in the CMRR of the configuration by a factor of $(1 + 2R_2/R_1)$. Typical values of CMRR range from 80 to 110 dB. The dedicated versions for use with dc Wheatstone bridge networks possess high CMRR values of 140 dB for 50 Hz voltages.

Instrumentation amplifiers are finding increasing application in the amplification of the output signals obtained from thermocouples, strain gauge bridges and biological electrodes.

When the Wheatstone bridge is connected to the differential amplifier, as shown in Figure 3.44, the equivalent voltages and resistances from the inverting and noninverting input terminals of the amplifier to the ground will be unequal in values. The equivalent voltages are equal only when the bridge is perfectly balanced. Even then, the resistance values are unlikely to equal each other, in which case the difference will be considered as source unbalance resistance, and there will be an error in v_o due to the common-mode signals at the input terminals of the amplifier. The higher the value of E_{BD}, the larger will be the error. Hence, for such situations where very low-level voltages present along with high common-mode voltages have to be amplified, the instrumentation amplifier is used.

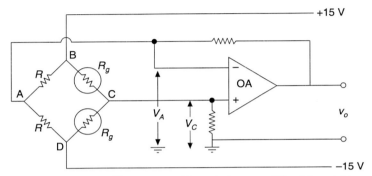

Figure 3.44 Differential amplifier connected to a strain gauge Wheatstone bridge network.

For applications that call for very high-gain common-mode voltages or isolation impedances with galvanic isolation, isolation amplifiers with transformer or optical coupling are recommended for use.

3.5.5 Isolation Amplifiers

The isolation amplifier, sometimes called isolator, is an amplifier circuit with its input circuit galvanically isolated from its power supply and output circuits. It is specifically recommended for use when accurate amplification and measurement of dc and low-frequency voltage signals are required with high CMRR, and high common-mode voltages of the order of a few kV are present along with the desired signals. They are invariably recommended for application in electrical environments of the kind associated with medical equipment used for patient monitoring and instrumentation systems requiring high-voltage bridge circuits, etc. In an isolation amplifier, there should be no possibility for dc current to flow between the input, output and the power supply stages. There may be isolation between any two stages or between all the three stages, depending on the requirements. The isolation results in lowering the drift also.

For low drift with time and temperature, the chopper type op-amps offer the solution when dc bridge outputs are amplified. In such cases, the bridge excitation is obtained from a floating source such as a battery, or an isolated dc-to-dc converter or an ac supply coupled via transformer. The output of the bridge, though grounded, may be treated as floating with respect to the excitation. The schematic diagram is shown in Figure 3.45. However, it is essential to

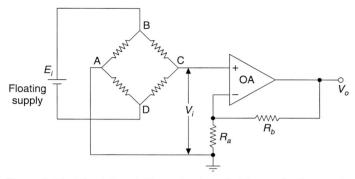

Figure 3.45   Wheatstone bridge network excited from a floating supply.

choose a highly stable floating power supply when high precision and accuracy are desired in the measurement.

The simplest means adopted for isolation is through transformer coupling. The isolation amplifier shown in Fig. 3.46 employs transformer coupling after the input signal is converted to an ac signal. Similarly, the dc output signal is again converted to an ac signal for effecting the feedback. Thus the amplifier can provide a common-mode input resistance R_{cm} of about 10^{12} Ω and a CMRR of 110 dB at 50 Hz with a 5 Ω source unbalance resistance. The dc power supply for the input and output stages may be obtained by converting the primary dc supply into an ac supply of suitable frequency and converting the same back to dc after obtaining the necessary isolation through transformers. It may make the system look unwieldy, but when very high common-mode rejection with low drift is desired, isolation by transformer coupling is adopted. There is every chance for a small capacitative coupling between the transformer windings.

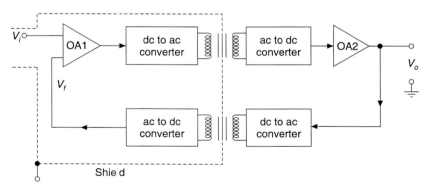

Figure 3.46   Isolation amplifier using transformer coupling.

Another commonly adopted technique is through electro-optical coupling by means of light emitting diodes (LEDs) and photodiodes. The photodiode is mounted inside a package along with the LED so that the electrical signals converted into light signals in the LED are received by the photodiode. Thus electrical isolation is achieved by optical coupling. The combination is known as *optocoupler*. The scheme of connections is shown in Figure 3.47, where a voltage-to-current amplifier is used to drive the LED and a current-to-voltage converter

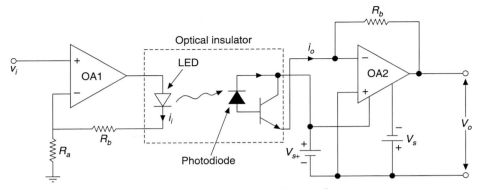

Figure 3.47   Isolation by optical coupling.

develops the output voltage. The current gain of i_o/i_i is normally between 0.1 and 0.5. The bandwidth of the LED opto-coupler extends up to 1 MHz. The circuit is useful for positive voltage signals only. The overall voltage gain (= v_o/v_i) may not be constant as it is likely to vary slightly with temperature. Power supplies for both the driver and receiver stages should be separate, thereby permitting complete isolation.

Opto-couplers are normally recommended for use in environment having a lot of electrical noise and mostly when digital signals are to be transmitted and processed.

3.5.6   Charge Amplifiers

The measurement of charge deposited on the plates of a capacitor can be done by finding the potential difference across them, treating its capacitance as constant. The charge may leak through its own equivalent shunting resistance and hence the measurement should be effected before any fall takes place. Also, the measurement becomes erroneous due to the loading effect of any measuring system and the leads used for connection. A voltage amplifier may be used for amplifying the voltage across the capacitor, provided it has considerably larger input resistance and smaller input capacitance in comparison to those of the charged capacitor. Even then, the varying lengths and quality of the connecting leads may considerably attenuate the measured values of charge.

Piezoelectric crystals develop charges across their opposite faces, when subjected to mechanical stress, and the charge leaks away through its equivalent shunt resistance, even though the mechanical stress is held constant. Thus for measurement of very slowly varying stresses, if not steady forces, an amplifier suitably designed to develop output voltage signals truly proportional to the charge as also the stress is essential. In other words, the bandwidth of the amplifier should extend down to very low frequencies, say 0.01 Hz.

The piezoelectric crystal may be treated as a charge generator with the charge $q(t)$ given by

$$q(t) = K_f F_i(t) = K_d \frac{\Delta x(t)}{X_o}, \qquad i(t) = \frac{dq(t)}{dt} \tag{3.36}$$

where, K_f and K_d are constants, $F_i(t)$ is the force applied and $\Delta x(t)/X_o$ is the mechanical strain of the crystal along the direction of $F_i(t)$. The voltage across the crystal may be obtained by finding the voltage drop across the parallel combination of the crystal capacitance C_i and its shunting resistance R_{sh}, due to the current $i(t)$ as the current from a constant current generator. If R_{sh} is assumed to be infinite, then the true voltage across C becomes

$$v_i(t) = \frac{K_d}{C_i} \frac{\Delta x(t)}{X_o} \tag{3.37}$$

The measurement of this voltage is rendered possible by making use of a charge amplifier, whose output voltage is independent of the effect of the impedances of the connecting cable and the operational amplifier that constitutes the measuring circuit.

A similar problem may be seen to arise when measuring the voltage drop across R of the circuit shown in Figure 3.33(b), where one plate of the capacitor is subjected to vibration at some desired frequency. The system may be considered to serve as a means for vibration studies where the amplitude of vibration $\Delta x(t)$ is very small as compared to the initial spacing X_o of the plates. The measuring system is desired to develop output voltages proportional to $\Delta x(t)$ to enable vibration measurements at low frequencies. The charge amplifier may be used to measure the voltage across R as given in Eq. (3.26), without any attenuation due to leads and the amplifier.

The circuit of the charge amplifier is shown in Figure 3.48(a), where C_i may be treated as the capacitance of the capacitive transducer or the piezoelectric transducer and R_{sh} the equivalent shunting resistance of the crystal (= R of the parallel circuit shown in Figure 3.33(b)). The operational amplifier is assumed to be of gain A. A capacitor C_f is connected in the feedback path.

Adding the currents at the inverting input terminal of the op-amp results in the expression (for sinusoidal input voltage)

$$(v_i - v_-)\, j\omega C_i = (v_- - v_o)\, j\omega C_f + \frac{(v_-)(1 + j\omega C_1 R_{eff})}{R_{eff}} \tag{3.38}$$

where,

$C_1 = C_c + C_a$
C_c = capacitance of the cable
C_a = input capacitance of the op-amp
R_{eff} = equivalent resistance of the input resistance R_a of the amplifier, equivalent shunt resistance of the cable R_c, and R_{sh}, all in parallel

R_c decreases and C_c increases with the length of the cable and their values depend on the type of cable used. Dealing with cases of very short cable lengths and high R_a op-amps and substituting $V_0 = -VA$, it can be shown that

$$\frac{v_o(j\omega)}{v_i(j\omega)} = \frac{-j\omega C_i R_{sh} A}{[1 + j\omega R_{sh}(AC_f + C_1)]}$$

The value of AC_f is very large in comparison to C_1, and hence the expression simplifies to

$$\frac{v_o(j\omega)}{v_i(j\omega)} = \frac{-j\omega C_i R_{sh} A}{1 + j\omega R_{sh} AC_f} \tag{3.39}$$

If C_f and A are so chosen as to make $\omega R_{sh} AC_f \gg 1$, the input-output voltage relationship is given by

$$\frac{v_o(j\omega)}{v_i(j\omega)} \approx \frac{-C_i}{C_f} = A_i \qquad (3.40)$$

This is the ideal voltage gain of the charge amplifier at very high frequencies, i.e. when $\omega \gg 1/(R_{sh}AC_f)$. Otherwise, the gain is given by

$$\left| \frac{v_o(j\omega)}{v_i(j\omega)} \right| = (A_i \omega C_f AR_{eff}) (1 + \omega^2 C_f^2 A^2 R_{eff}^2)^{-1/2} \qquad (3.41)$$

The value of R_{eff} can be replaced by R_c if it is much smaller than R_a and R_{sh}. Under steady state conditions the gain is zero. For values of $\omega C_f AR_{eff}$ equal to 10, the amplitude response will be within -5% of A_i and by proper choice of C_f and A, the response can be extended to very low frequencies. For a typical case, A may be taken as 10^4, R_c as 10^4 Ω, and C_f as 10000 pF, in which case $AC_f R_{eff}$ is equal to 1 s. The frequency at which error of -5% occurs is thus about 0.5 Hz. Any increase in the value of C_f may result in extending the bandwidth to still lower frequency, but the sensitivity of the system as given by Eq. (3.40) becomes lower. Hence, it is advisable to choose an op-amp having higher gain than 10^4.

In practice, the amplifier gets saturated because of bias currents through the op-amp charging the capacitor C_f. The capacitor does not also possess infinite resistance. To overcome this problem, an alternative path is provided across C_f by means of a resistor R_f of high value as shown in Figure 3.48(b).

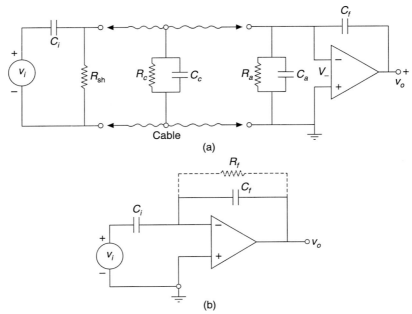

Figure 3.48 (a) Charge amplifier circuit; (b) simplified charge amplifier circuit.

The effect of R_f in the performance of the charge amplifier may be deduced from the relation

$$\frac{v_o(j\omega)}{v_i(j\omega)} = \frac{-C_i}{C_f} \frac{j\omega R_f C_f}{1 + j\omega C_f R_f} \tag{3.42}$$

Comparing this equation with Eq. (3.41), it is clear that R_f should be larger than $R_{eff}A$, so as to render the circuit behave as a charge amplifier. For enabling static measurements of force, quartz crystals may be used and the charge amplifier designed for use along with them may require R_f between 10^{10} Ω and 10^{14} Ω and C_f between 100 pF and 10^5 pF.

The high frequency response of charge amplifier when used with piezoelectric crystals is limited by the natural frequency (mechanical) of the crystal material. The ideal gain A_i is, in fact, affected by variation in the value of C_i due to changes in the dielectric constant of the material with temperature.

Thus the charge amplifier configuration has rendered measurement of charge possible, without showing any effect of the connecting cable on the accuracy of measurement. The measurement is free from the loading effects, while at the same time extending the bandwidth right down to very low frequencies, amounting to static measurements.

The charge amplifier is equally useful for measurement of amplitudes of vibration at low frequencies.

3.5.7 Power Amplifiers

In some cases, amplifiers with sufficient power output in response to the input voltage signal are required. Normally, the amplifiers draw negligible power from the source of input signal and they are designed to have sufficient voltage or current gain. An amplifier designed to have voltage gain of less than 1 may have substantial power gain.

The class B push–pull transistor power amplifier is most commonly used. Power transistors are capable of handling currents up to 10 A at collector voltages of about 100 V. The main advantages of class B operation are high power efficiency and low distortion. Of the three possible symmetrical connections, the common-emitter connection shown in Fig. 3.49 has larger power gain than the others. The common-base connection is preferred when low power

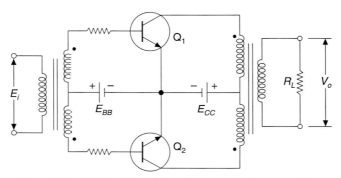

Figure 3.49 Push–pull amplifier of common-emitter connection.

gain with low distortion is required. The common-collector configuration yields larger power output with reasonable power gain and low distortion. The common-emitter connection has greater distortion and is normally used with loads that do not respond to the harmonics in the output signal.

An alternative circuit for a transformer-coupled amplifier is shown in Figure 3.50 in which two operational amplifiers are used to drive the transistors. The transistor Q_1 conducts on the positive signal cycles and Q_2 on the negative. The peak-to-peak output voltage V_o is about $0.9V_s n$, where n is the turns ratio between the total primary and the secondary windings.

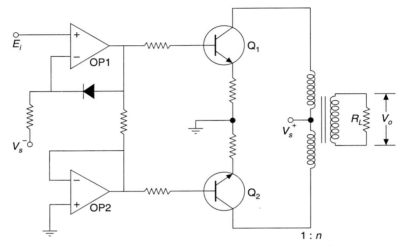

Figure 3.50   Push–pull amplifier with transformer coupling.

Push–pull transistor amplifiers can also be adopted for dc applications. The simple circuit shown in Figure 3.51(a) employs a matched pair of npn and pnp junction transistors. Each of the pair possesses characteristics which are the negative image of the other. Matched pairs are available for power outputs of about 10 W.

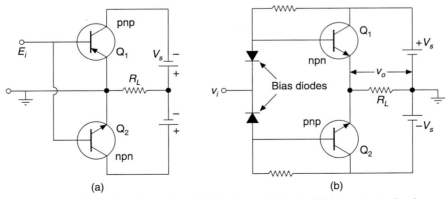

Figure 3.51   Push–pull amplifier: (a) For dc input signals; (b) for ac input signals.

By using two bias diodes as shown in Figure 3.51(b), the amplifier can be used as an ac amplifier. Bias is adjusted so that both the transistors are slightly conducting when the input signal is zero.

In ac servomechanisms, the frequency of the input signal to the amplifier is the carrier frequency of 50 Hz or 400 Hz. The amplifier may be designed to have maximum power gain at frequencies close to the carrier frequency. The load impedance is inductive, and a capacitor is used across it for improving the power factor. The amplifiers designed for such loads are known as *servo amplifiers*. The control winding of a two-phase servomotor forms the load impedance in many applications. The induction motor develops output torque reacting to the fundamental frequency component of current only. The harmonics are thus automatically rejected, but they may saturate the amplifier and result in making the control winding hotter.

3.6   MEASUREMENT OF PHASE ANGLE

The measurement of the phase angle and power factor of load impedance on single-phase circuits is accomplished in several ways. Power factor meters of the dynamometer type directly indicate the power factor and phase angle of the load. The meter consists of two light coils with their axes at right angles to each other and mounted on the spindle carrying the pointer. The fixed coils carry the load current, as shown in Figure 3.52 and develop the magnetic field in which the moving coils work. The two crossed coils carry equal currents but at quadrature to each other and develop torques opposing each other. The moving system deflects until the torques balance each other and the angular displacement is a measure of the power factor and phase angle. There is no control spring attached to the moving system. The scale is calibrated such that it reads both leading and lagging phase angle and it reads unity power factor at the middle of the scale. With slight modification in the arrangement of the moving coils and connections, it may be designed to read the power factor of three-phase balanced loads. However, these are designed for use at one frequency which is the power frequency.

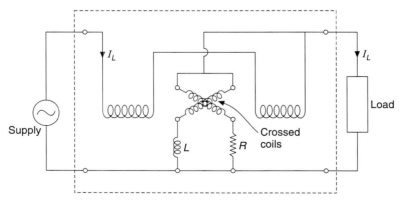

Figure 3.52   Phase angle meter for single phase.

The phase difference between any two voltages at frequencies above 10 Hz can be measured by using a cathode ray oscillograph. The voltages after sufficient amplification are applied to the X and Y plates of the oscillograph, and the phase angle is estimated from the Lissajous pattern obtained on the screen. The measurement cannot be very accurate. At frequencies below 5 Hz, an X-Y plotter may be used.

At frequencies well below 5 Hz, the phase angle can be determined accurately by measuring the time-interval t_0 between the zero crossings of the two signals of the same frequency. If T_0 is the time period of the signals, then the phase angle ϕ in degrees is given by

$$\phi = \frac{360 t_0}{T_0}$$

Electronic phase meters are designed to indicate the phase angle directly by a pointer on a scale and the block diagram of the meter is shown in Figure 3.53(a). The two signals are suitably amplified and shaped so that their zero-crossings can be detected accurately as shown in Figure 3.53(b). A flip-flop circuit is used to obtain a rectangular pulse of duration t_0 during the time period T_0. The positive zero crossing of V_1 sets the flip-flop and that of V_2 resets it. Thus a series of rectangular pulses at intervals of T_0 are obtained, so that the average value of the pulse train may be used for indication by a permanent magnet moving coil meter.

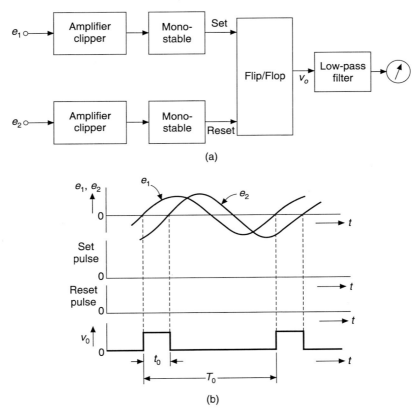

Figure 3.53 (a) Block diagram of an electronic phase angle meter; (b) principle of operation.

A digital readout of the phase angle can also be obtained by using a digital time interval meter, described in Section 3.8.

3.6.1 Phase-Sensitive Detectors

A phase-sensitive detector is used for detection and measurement of the amplitude and phase of a periodic signal with respect to another reference signal of the same frequency. It is also known as lock-in amplifier and synchronous demodulator. In instrumentation, a large number of situations exist where the desired information resides in both the amplitude and phase of a signal. For example, the unbalance voltage of an ac bridge network has its magnitude and phase dependent on the components of the unknown impedance. The excitation voltage of the bridge serves as the reference signal.

The basic scheme of operation of a phase-sensitive detector (p.s.d.) is shown in Figure 3.54(a). The reference signal is assumed to be a square wave driving the switch towards the contactor 1 during the positive half-cycle and to contactor 2 during the negative half-cycle. The sinusoidal test signal is amplified to assume reasonable amplitude and then applied to the switching system. The reference and test signals are of the same frequency and hence the output of the adder consists of full-wave rectified signal, if the signals are in phase. Then the average value is read on a pmmc system and is maximum. When the test and reference signals are in phase, the operation is similar to an ac-to-dc converter, but when there is a phase difference between them, the average output voltage and the meter reading become functions of the phase angle θ of the test voltage with respect to the reference voltage. When $\theta = 180°$, the reading reaches a negative maximum, and when $\theta = 90°$, the reading is zero as can be seen from Figure 3.54(b). The low-pass filter removes the ac ripple components and presents ripple-free dc output voltage for indication. The reading can be shown to be $(2/\pi)V_{i\,max} \cos \theta$.

In case the frequency of the test signal voltage is not exactly the same as that of the reference signal, the output of the adder in each half-cycle becomes erratic, changing from positive to negative values, thus finally becoming zero when averaged over many cycles. The ability of the phase-sensitive detector to reject signals such as random noise which differs in frequency from that of the reference signal, depends on the averaging time of the smoothing filter. The longer the averaging time, the greater is its rejection capability. Thus the p.s.d. is used to process signals that are 'buried' in noise and present output voltages with high signal-to-noise ratio.

For use with test signals of low to moderate frequencies, the circuit of a p.s.d. with an FET acting as a single-pole, double-throw switch is shown in Figure 3.55. The output response is limited by the capacitor C in the low-pass filters. The op-amp 2 provides the output voltage with respect to ground.

Most of the phase-sensitive detectors may be designed with the additional provision for shifting the phase of the reference signal until a maximum or a zero reading is obtained. Thus the circuit also provides the means for searching components of the test signal, having the frequency of the reference signal, but with a different phase angle. It is sufficient that the noise components do not constitute so large an amplitude in comparison to the component being searched, as to cause saturation of the amplifier at the input end.

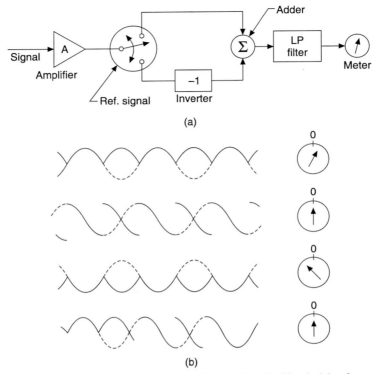

Figure 3.54 (a) Basic scheme of a phase-sensitive detector (p.s.d.); (b) principle of operation of p.s.d.

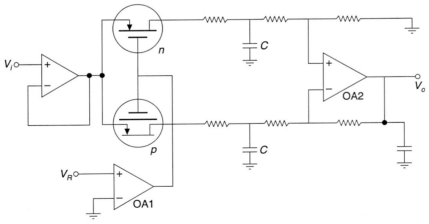

Figure 3.55 Circuit of a phase-sensitive detector.

The low-pass filter of the p.s.d. can be replaced by an integrator that integrates the output of the adder for a definite length of time. Such an arrangement has the advantage that the final output of the integrator is a definite value related to the amplitude and phase of the component of the test signal having the reference frequency, and a digital voltmeter can be used for indication.

3.7 FREQUENCY MEASUREMENT

Frequency can be measured in a variety of ways: These can be simple indicating instruments for use at power frequencies or can be based on other techniques such as resonance for measurement of high frequencies.

The schematic diagram of a frequency meter of the electrodynamometer type is shown in Figure 3.56. The inductor L_1, and the capacitor C_1 form a resonant circuit which is tuned to a frequency slightly lower than the lowest frequency read on the scale, whereas L_2 and C_2 are tuned to a frequency slightly higher than the top-point of the scale. Usually the range of the frequency meter is either 45–55 Hz (or 25–75 Hz) with 50 Hz read at the middle of the scale. The moving coil carries the sum of the two currents in the two tuned circuits and lies in the magnetic field developed by the field coils carrying these currents. The torque developed by the moving coil is balanced by a control spring.

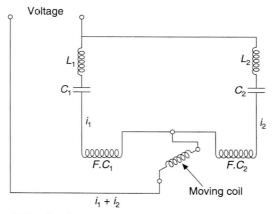

Figure 3.56   An electrodynamometer type frequency meter circuit.

Frequency can be most conveniently and accurately measured by counting the number of cycles of the unknown signal for a precisely controlled time interval. The instrument designed for frequency measurement mainly consists of an electronic counter and a time standard. The block diagram of the instrument is shown in Figure 3.57. The input signal is amplified to a suitable level and passed to a Schmitt trigger so that it is converted into a square wave. The square wave is differentiated and clipped resulting in a series of pulses separated by the period of the original input signal. The time standard is, in fact, a frequency standard and is a piezoelectric crystal oscillator generating a voltage whose frequency is very stable. The crystal is kept in a temperature controlled oven. The oscillator frequency is usually chosen at 1 MHz or 10 MHz. The time-base output is shaped by a Schmitt trigger so that positive spikes spaced at 1 μs interval are presented to a number of decade dividers. With six decades it is possible to obtain time intervals from 1 μs to 1 s. The output pulse from the time-base selector switch is passed through a Schmitt trigger to the gate control flip-flop. The gate flip-flop assumes a state such that an 'ENABLE' signal is applied to the main gate. As this is usually an AND gate, the input signal pulses are passed on for counting and display by the decimal counting units.

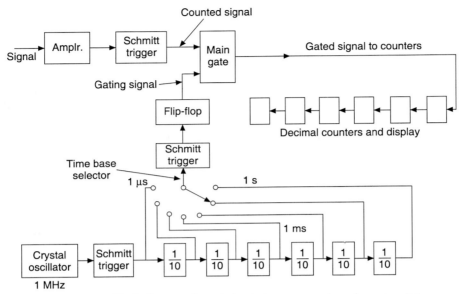

Figure 3.57 Block diagram of an electronic frequency meter using a counter.

The counting goes on until the second pulse from the decade divider assembly arrives at the control flip-flop. The gate control assumes the other state which removes the 'enable' signal from the main gate. The main gate closes and no further pulses are admitted to the decimal counting units. Thus the display corresponds to the number of input pulses received during a precise interval selected by the time-base. The instrument is more suited for the measurement of high frequency; for low frequency signals, the count will be smaller and hence it is likely to result in larger error at lower frequencies. Therefore, it is desirable to measure the period of the signal when the frequency is lower than 1 kHz.

The building blocks used in Figure 3.57 are rearranged so that the counted signal and the gating signal are reversed. The arrangement shown in Figure 3.58 has the gating signal provided

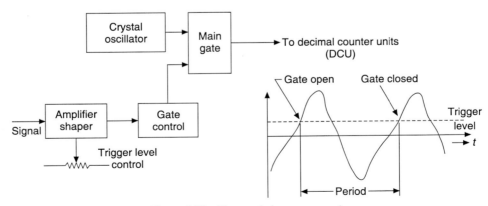

Figure 3.58 Time-period measurement.

by the unknown input signal, which now controls the opening and closing of the main gate. The input signal is properly shaped and the trigger level control is so adjusted that the triggering occurs on the steepest portion of the input signal. The precisely spaced pulses from the crystal oscillator are counted for one period of the unknown frequency and displayed by the decimal counting units.

The accuracy of measurement of the time period may be enhanced by keeping the main gate open for more than one period of the unknown signal. This is known as *multiple-period average mode* of operation.

3.8 TIME-INTERVAL MEASUREMENTS

There are many occasions in instrumentation when the measurement of the time interval between two events is needed. The basic building blocks used for frequency measurement can be interconnected in a slightly different manner as shown in Figure 3.59(a) to enable the measurements of the time interval. The events under investigation should be made to generate two electrical pulses, so that the first pulse is used to open the gate and the second to close it. Means are provided for controlling the trigger level as well as selection of the slope of each input signal.

Figure 3.59(b) shows the application of the time interval meter for measurement of the width of a rectangular pulse. A positive slope is selected for the 'start' signal and negative slope for the 'stop' signal.

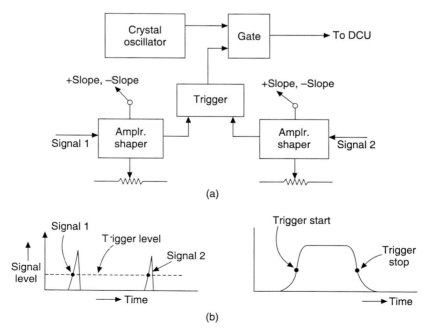

Figure 3.59   (a) Block diagram of a time-interval meter; (b) pulse-width measurement.

When a phase angle between two sinusoidal signals of the same frequency is desired, the above scheme can be employed. In such a case, it is necessary to bring the amplitudes of both the input signals to the same value and adjust the triggering levels to be the same. This method enables measurement of phase angle with high resolution, say of $0.1°$.

EXERCISES

1. Explain the reasons for the popularity of a permanent-magnet moving-coil system and show how some variations in design enable the measurements in different situations.

2. Show that the pmmc system is a current-operated system; also show how the same is made suitable for measurements of (i) voltage, (ii) electrical charge, and (iii) magnetic flux linkages.

3. Explain how a vibration galvanometer functions as an ac current detector.

4. Discuss the merits of a dc-potentiometer system and indicate the provisions made in the commercial versions to enhance its versatility.

5. Discuss briefly the constructional details of a self-balancing potentiometer.

6. List the important characteristics that distinguish the performance of the electrical and electronic type voltmeters.

7. Draw the circuit of an electronic differential voltmeter and explain its functioning. What are its merits when compared to other simpler versions?

8. What is a thermal converter? Explain how it can be used to make up an accurate electronic rms-reading voltmeter.

9. Discuss the basic differences in the performance of analog and digital voltmeters and indicate the merits of digital indication.

10. Explain briefly the techniques employed for the design of digital voltmeters.

11. Explain, by means of circuits, how resistance can be measured and indicated on a scale in analog form. Discuss the scale form of each.

12. Discuss the reasons for the popularity of Wheatstone bridge networks for the measurement of resistance and explain how precise and accurate measurements are possible by using this network.

13. Show that in a slightly unbalanced Wheatstone bridge network, the unbalance voltage is proportional to the deviation in unknown resistance from its nominal value. Discuss the range of linearity.

14. Show how the sensitivity and linearity of a Wheatstone bridge network can be improved.

15. Distinguish the relative merits of exciting the bridge from (i) a constant voltage source, and (ii) a constant current source.

16. Contrast the performance of dc and ac bridge networks and explain the conditions under which impedance measurements can be done easily and accurately.

17. Draw the circuits of two popular ac bridge networks that are used for the measurement of impedance of coils and capacitors.

18. Describe a transformer-coupled ratio-arm bridge network and contrast its distinguishing features in performance and application when compared to the conventional Wheatstone bridge network.

19. Show that a unity-ratio transformer-coupled bridge has considerable appeal for application in instrumentation. Obtain the expression for the sensitivity of the bridge and discuss the range of linearity and the effect of frequency when used for measurements of inductive and capacitive impedances.

20. What is an electrometer amplifier and what is its special application?

21. Draw the circuit of a vibrating-plate capacitor amplifier and show how it enables measurement of emfs from very high impedance sources.

22. Explain how the application of negative feedback in amplifiers affects their performance.

23. What are the important characteristics of an operational amplifier? Why has it become a part of all electronic circuitry nowadays?

24. Draw the circuits of (i) an inverting amplifier and (ii) a non-inverting amplifier, and discuss the situations under which each is preferred for use.

25. What is a differential amplifier and how is it configured? Derive the relationship between the input and output voltages.

26. What is an instrumentation amplifier and what is its special role in instrumentation? Draw the scheme of connections and derive the expression for its gain.

27. What is an isolation amplifier? Discuss its application in instrumentation and briefly explain the types of amplifiers available.

28. Draw the circuit of a charge amplifier and show how it enables measurement of electrical charge. Obtain the relationship between charge and output, and discuss its limitations, if any.

29. Explain how electrical and electronic phasemeters are designed for measurement of phase angle.

30. Show how the electrical dynamometer type system can be successfully designed to measure (i) electrical power, (ii) power factor, and (iii) frequency of ac voltage.

31. Explain briefly the techniques adopted for measurement of time interval, showing how it is indicated.

SUGGESTED FURTHER READING

Boros, A., *Electrical Measurements in Engineering*, Elsevier, Amsterdam, 1985.

Buckingham, H. and Price, E.M., *Principles of Electrical Measurements*, English Universities Press, London, 1955.

Cooper, William, D. and Helfrick, Albert, D., *Electronic Instrumentation and Measurement Techniques*, 3rd ed., Prentice-Hall of India, New Delhi, 1970.

Ernest Frank, *Electrical Measurement Analysis*, Tata McGraw-Hill, New Delhi, 1959.

Gregory, B.A., *Introduction to Electrical Instrumentation and Measurement Systems*, ELBS and Macmillan, London, 1981.

Harris, F.K., *Electrical Measurements*, John Wiley, New York, 1959.

Kinnard, I.F., *Applied Electrical Measurements*, John Wiley, New York, 1958.

Kurt, S. Lion, *Elements of Electrical and Electronic Instrumentation*, McGraw-Hill Kogakusha, Tokyo, 1975.

Oliver, B.M. and Cage, J.M. (Eds.), *Electronic Measurements and Instrumentation*, McGraw-Hill, New York, 1971.

4

Dynamics of Instrument Systems

4.0   INTRODUCTION

An instrument system is invariably a combination of some physical elements connected together to accomplish a desired objective. A variety of mechanical and electrical components are used to make up the transducer and instrument systems for indication and recording of physical variables. It is important to realize the performance of these systems under static and dynamic conditions from the knowledge of the behaviour of each component of the system.

For purposes of analysis, certain assumptions are usually made concerning the behaviour of each element, and a simpler mathematical model is developed. The physical laws governing the behaviour of each element are utilized to obtain the mathematical model of the physical system. The development of such models usually results in the formation of standard differential equations with constant coefficients and hence these equations are considered as linear models. The composite physical system is then characterized as linear time-invariant and is represented by means of a relationship between its input (excitation) and output (response) variables. For the common simple transducer and instrument systems, it is customary to relate its output variable with a single input variable and study its performance when excited by input functions, treated as standard input signals.

The transfer function approach is usually adopted for the study of physical systems and the transfer function of a linear time-invariant system is defined as the ratio of the Laplace transform of the output variable to the Laplace transform of the input variable, under the assumption that the initial values of both variables are zero and the system wakes up to respond to the excitation function from the instant of its application. The transfer function (TF) is denoted by $G(s)$ and is written in the form

$$G(s) = \frac{Y_o(s)}{X_i(s)} = \frac{R(s)}{E(s)} \tag{4.1}$$

where, $Y_o(s)$ and $R(s)$ represent the output (or response) variable and $X_i(s)$ and $E(s)$ represent the input (or excitation) variable. $G(s)$ is also defined as the system function. The highest power of s (= the complex frequency) in the denominator of the transfer function determines the order of the physical system. Most of the transducer and instrument systems belong to either first or second order and the evaluation of their response in time domain for standard input signals is very much desired for classifying their response as satisfactory or unsatisfactory.

The transfer function is used to estimate the transient and steady state response of the system when subjected to input signals of varying descriptions. However, the response of the instrument systems for step-input and sinusoidal input functions is what is mostly desired. Performance criteria are generally specified for the response of the system for both the static and dynamics inputs and any lack of conformity with the specifications is revealed by actual testing of the systems. Those systems that may be found to make up the resonance or found to belong to third order and above, have to be dynamically compensated for satisfactory performance as a stable system, or are limited for use on such input functions having a frequency content far removed from the region, leading to undesirable operation.

In Sections 4.1–4.6 a few basic elements and some instrument systems consisting of such elements are briefly described and their transfer functions derived. Their performance is revealed from consulting the normalized performance of systems of first and second order, and is indicated in Section 4.1.

4.1 GENERALIZED PERFORMANCE OF SYSTEMS

Most of the simple physical systems, when modelled, result in ordinary linear differential equations with constant coefficients generally represented as

$$a_n \frac{d^n y_o(t)}{dt^n} + a_{n-1} \frac{d^{n-1} y_o(t)}{dt^{n-1}} + \cdots + a_1 \frac{dy_o(t)}{dt} + a_0 y_o(t)$$

$$= b_m \frac{d^m x_i(t)}{dt^m} + b_{m-1} \frac{d^{m-1} x_i(t)}{dt^{m-1}} + \cdots + b_0 x_i(t) \tag{4.2}$$

where $y_o(t)$ is the output variable and $x_i(t)$ is the input variable, and a's and b's are constants representing the physical parameters of the system. Both $y_o(t)$ and $x_i(t)$ are functions of time. The value of n defines the order of the system. The above representation relates $y_o(t)$ with $x_i(t)$ in time domain.

When the relationship between $y_o(t)$ and $x_i(t)$ is transformed into the complex frequency domain, it is represented by the transfer function $G(s)$ and is given by

$$G(s) = \frac{Y_o(s)}{X_i(s)} = \frac{N(s)}{D(s)} = \frac{b_m s^m + b_{m-1} s^{m-1} \cdots b_1 s + b_0}{a_n s^n + a_{n-1} s^{n-1} \cdots a_1 s + a_0} \tag{4.3}$$

where $s = \sigma + j\omega$.

Assuming the initial values of $y_o(t)$ and $x_i(t)$ at $t = 0$ to be zero, the complete solution for $y_o(t)$ can be obtained. The solution consists of two parts: one due to the forcing function $x_i(t)$ and the other due to its natural characteristics. The solution is usually represented as

$$y_o(t) = y_o(t)\big|_{\text{cf}} + y_o(t)\big|_{\text{pi}} \qquad (4.4)$$

where $y_o(t)\big|_{\text{cf}}$ is known as the complementary function part of the solution. $y_o(t)\big|_{\text{cf}}$ is obtained by equating the denominator $D(s)$ to zero and solving it. $D(s)$ is also known as the characteristic equation and is used to determine the transient component of the response which invariably dies with time. What remains in time after the transient dies, is known as particular integral part $y_o(t)\big|_{\text{pi}}$, which reflects the behaviour of the system due to the forcing function. This part of the solution is treated as the steady state response.

Most of the physical systems are studied for their response when subjected to forcing functions, usually described as standard input functions such as step-input, ramp-input etc. The Laplace transforms of such input functions are given in Section 2.2 and are useful in determining the response of the system for these input functions, in time domain.

One of the most important input functions is the sinusoidal input, and the response of the system is usually desired under steady state conditions. As the system is assumed to be linear, the steady state response is characterized by the relationship between the input variable and the output variable in terms of the ratio of their amplitudes and the phase difference between them. The steady state response for sinusoidal input is obtained from the transfer function by writing $j\omega$ for s and evaluating the relationship at different frequencies ranging from zero to infinity. The ratio becomes

$$\frac{Y_o(j\omega)}{X_i(j\omega)} = |M| \angle\phi \qquad (4.5)$$

where $|M|$ represents the ratio of the amplitudes and $\angle\phi$ the phase difference between the input and output variables.

The values of $|M|$ and ϕ vary with the frequency of the forcing function and when plotted against frequency, they constitute the frequency-response curves of the system.

For other input functions, the Laplace transform of the input function is obtained and used to evaluate the response $y_o(s)$, from which the complete solution for $y_o(t)$ in time domain is arrived at.

4.1.1   Zero-order Systems

Systems having a constant value for the transfer function $Y_o(s)/X_i(s)$ are classified as zero-order systems and a constant value, K signifies the static sensitivity of the system. Such systems have the output variable closely following the input variable at all instants of time without showing any time or phase lag. In other words, at all instants of time and for all values of $x_i(t)$, the amplitude of $y_o(t)$ is K times the amplitude of $x_i(t)$. Zero-order systems do not result in any dynamic error.

It must be noted, however, that no physical system is so ideal as to be called a zero-order system. It is only some of the elements of the system that may display the characteristics of a zero-order system. Depending on the conditions of usage, certain systems are considered to function as zero-order systems.

4.1.2 First-order Systems

First-order systems are those that are characterized by a transfer function represented as

$$G(s) = \frac{Y_o(s)}{X_i(s)} = \frac{b_0}{a_1 s + a_0}$$

and rewritten as

$$G(s) = \frac{K}{1 + s\tau} \tag{4.6}$$

where, $K = b_0/a_0$, the static sensitivity and $\tau = a_1/a_0$, the time constant of the system in seconds.

Systems characterized by Eq. (4.6) are studied for their response when subjected to standard input signals.

In the case of a unit step input function, $X_i(s) = 1/s$, and hence $Y_o(s)$ is given by

$$Y_o(s) = \frac{1}{s} \frac{K}{1 + s\tau}$$

giving

$$y_o(t) = K(1 - e^{-t/\tau}) \tag{4.7}$$

Equation (4.7) reveals the fact that $y_o(t)$ assumes a final value of K slowly with time. The fastness with which the system responds or in other words, the speed of response, is dependent on the value of τ. The smaller the value of τ, the higher the speed of response. It is essential to check whether a particular measuring system is fast enough in its indication of the final value, and hence a useful criterion is adopted for the time taken to settle down to a value within some prescribed limits (usually ± 5%) of its final value, as shown in Figure 4.1(a).

The dynamic performance may be represented with the help of curves drawn in a normalized fashion between $y_o(t)/K$ and t/τ. Figure 4.1(b) shows the response of a system that is subjected to the unit step input at $t = 0$, and when the input is switched off at $t = T$.

It should be observed that by the time $t = \tau$, the value of $y_o(t)$ assumes 63.2% of its final value and it falls by that amount from its steady state value in the same time as shown in Figure 4.1(b). Accordingly, it is a usual practice to define the time constant of a first-order system, as stated in Section 1.5.2, as the time taken to reach 63.2% of its final value, when subjected to a step-input function.

If the input function is of the unit-ramp type, then the input–output relationship is given by

$$Y_o(s) = \frac{1}{s^2} \frac{K}{1 + s\tau}$$

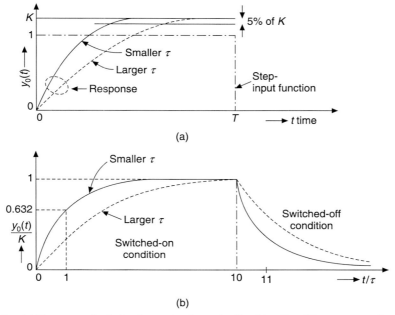

(a)

(b)

Figure 4.1   (a) Response of a first-order system for a step-input function; (b) response under switched-on and switched-off conditions.

and when solved, $y_o(t)$ is given by

$$y_o(t) = K[e^{-t/\tau} + t - \tau] \qquad (4.8)$$

The response of the system for unit ramp input is shown in Figure 4.2. If the system is ideal, it should result in an output signal $y_o(t) = Kt$, but there is a deviation from this value due to its time constant. Hence the dynamic error is given by

$$\text{Dynamic error} = + K(\tau e^{-t/\tau}) - K\tau \qquad (4.9)$$

The first term of the net dynamic error dies with time and hence it constitutes transient error, whereas the second term $K\tau$ becomes the steady state error. Under steady state conditions, the amplitude of output attains the true value after τ seconds only.

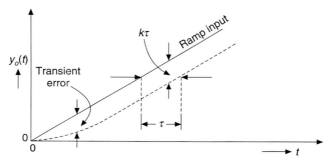

Figure 4.2   Response of first-order system for a ramp-input.

Similarly, the response for a unit-impulse function is represented by

$$Y_o(s) = \frac{K}{1 + s\tau}$$

$$y_o(t) = \frac{K}{\tau} e^{-t/\tau} \tag{4.10}$$

If the strength of the impulse is A units, the response becomes A times the one given by Eq. (4.10). The response shown in Figure 4.3(a) is for an impulse of strength of A units but of duration T seconds, and that in Figure 4.3(b) is for an ideal impulse.

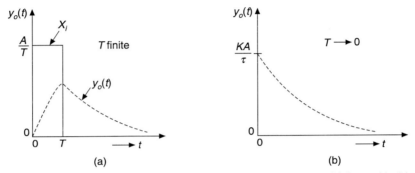

Figure 4.3   Response of first-order system: (a) For a prolonged impulse-input; (b) for an ideal impulse-input.

For sinusoidal input functions, the frequency response is determined from the relation

$$\frac{Y_o(j\omega)}{X_i(j\omega)} = \frac{K}{1 + j\omega\tau}$$

$$= \left| \frac{K}{1 + \omega^2\tau^2} \right| \angle \tan^{-1} - \omega\tau = |M| \angle \phi \tag{4.11}$$

At zero frequency, i.e., under dc excitation, the value of $|M|$ becomes equal to K with $\phi = 0$. Treating the natural frequency of the system, ω_n, as given by $1/\tau$, the frequency response curves relating $|M|$ and $\angle \phi$ with ω/ω_n ($= \omega\tau$) are shown in Figure 4.4.

4.1.3   Second-order Systems

Second-order systems are characterized by a transfer function given by

$$G(s) = \frac{Y_o(s)}{X_i(s)} = \frac{b_0}{a_2 s^2 + a_1 s + a_0}$$

which can be rewritten as

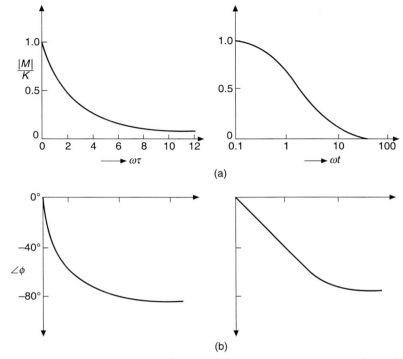

Figure 4.4 Frequency response characteristics of a first-order system: (a) For amplitude; (b) for phase.

$$G(s) = \frac{K}{s^2\left(\dfrac{a_2}{a_0}\right) + s\left(\dfrac{a_1}{a_0}\right) + 1} \tag{4.12}$$

where, K = static sensitivity (= b_0/a_0). The undamped natural frequency ω_n of the second-order system becomes $\sqrt{(a_0/a_2)}$ and the ratio a_1/a_0 signifies the damping conditions of the system. Representing the ratio of the damping coefficient under any damping condition to the coefficient that would result in critical damping conditions, by a dimensionless factor ξ, commonly known as damping factor (or damping ratio), it can be shown that

$$\xi = \frac{a_1}{2\sqrt{a_0 a_2}} \tag{4.13}$$

$$2\xi/\omega_n = a_1/a_0 \tag{4.14}$$

Equation (4.12) can be rewritten as

$$\frac{Y_o(s)}{X_i(s)} = \frac{K}{\left(\dfrac{s^2}{\omega_n^2} + \dfrac{2\xi s}{\omega_n} + 1\right)} \tag{4.15}$$

and is used for finding $Y_o(s)$ for the different input functions.

For a unit step input function, $Y_o(s)$ is given by

$$Y_o(s) = \frac{K}{s\left(\dfrac{s^2}{\omega_n^2} + \dfrac{2\xi s}{\omega_n} + 1\right)}$$

whose solution yields $y_o(t)$ for the different damping conditions given by

$$\frac{y_o(t)}{K} = \left[1 - \frac{\exp(-\omega_n \xi t)}{\sqrt{1-\xi^2}} \sin\left\{\omega_n\sqrt{(1-\xi^2)}t + \sin^{-1}\sqrt{1-\xi^2}\right\}\right] \quad \text{for } \xi < 1 \qquad (4.16)$$

$$\frac{y_o(t)}{K} = [1 - (1 + \omega_n t)\exp(-\omega_n t)] \quad \text{for } \xi = 1 \qquad (4.17)$$

$$\frac{y_o(t)}{K} = \left[1 - \frac{\exp(-\omega_n \xi t)}{\sqrt{\xi^2-1}} \sinh\left[\omega_n\sqrt{(\xi^2-1)}t + \sinh^{-1}\sqrt{\xi^2-1}\right]\right] \quad \text{for } \xi > 1 \qquad (4.18)$$

A plot of $y_o(t)/K$ against $\omega_n t$ is shown in Figure 4.5 and is useful for determining the fastness of response of a system. The way $y_o(t)$ reaches the final value from its initial value of zero depends on the damping conditions, i.e., on the value of the damping factor ξ. If $\xi = 0$, the system is undamped and Eq. (4.16) makes $y_o(t)/K$ equal to $(1 - \cos \omega_n t)$. The system continues to oscillate around its final value of unity with the peaks touching the values of 0 and 2. As the value of ξ increases, the amplitude of oscillations starts diminishing progressively with time and the system is said to be underdamped. As it reaches the final value after prolonged oscillations, the settling time is considerably large. The settling time becomes less as ξ increases, and when $\xi = 1$, the oscillations will totally cease and $y_o(t)$ is given by Eq. (4.17). It is seen from Figure 4.5 that as the time taken to reach the final value is the least, and under such damping conditions, the system is said to be critically damped. However, any further

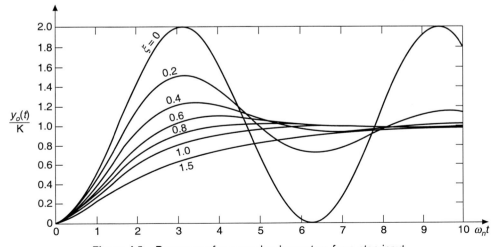

Figure 4.5   Response of a second-order system for a step-input.

increase in the value of ξ delays the assumption of the final value, and thus increases the settling time. When $\xi > 1$, the system is overdamped. Under these conditions, the response is governed by Eq. (4.18). Physical systems that are likely to be subjected to step input functions are thus required to be critically damped and all the second-order type indicating instruments are designed to work under damping factors nearing unity. When working at ξ of 0.7 or 0.8, the settling time may become larger than that of critical damping, but the oscillatory response, with $y_o(t)/K$ overshooting the final value, ensures positive response of the system without being stuck anywhere in its movement. At a damping factor of 0.68, the indicated value will be within $\pm 5\%$ of the final value within the time of about $0.28/\omega_n$.

The first peak value of $y_o(t)/K$ for slightly underdamped conditions is obtained from Eq. (4.16) and is used to determine the first overshoot which is given by

$$M_p = \exp\left[-\pi\xi / \sqrt{1 - \xi^2}\right] \qquad (4.19)$$

M_p is usually expressed as a percentage of the final value. Experimental determination of the percentage overshoot enables the estimation of the damping factor of the system at which it is working.

The response of the second-order system for unit ramp input is obtained from

$$Y_o(s) = \frac{K}{s^2 \left(\dfrac{s^2}{\omega_n^2} + \dfrac{2\xi s}{\omega_n} + 1\right)} \qquad (4.20)$$

whose solution for underdamped conditions is given by

$$\frac{y_o(t)}{K} = \left[t - \frac{2\xi}{\omega_n} + \frac{\exp(-\omega_n \xi t)}{\omega_n \sqrt{1 - \xi^2}} \sin\left(\omega_n \sqrt{(1 - \xi^2)}\, t + \phi\right)\right]$$

where,

$$\tan \phi = \frac{2\xi\sqrt{1 - \xi^2}}{2\xi^2 - 1} \qquad (4.21)$$

The general performance of the system is shown in Figure 4.6. It is seen that there is a steady state error of $2\xi K/\omega_n$. The steady state error decreases as ω_n increases and is proportional to ξ. Under steady state conditions, there is a time lag of $2\xi/\omega_n$ in the indication of the true value. For a given ω_n if ξ is reduced, oscillations persist for a longer time, but the steady state time lag and steady state error become less.

It is quite realistic to assume that electrical and electronic instruments are subjected to step and ramp-input excitations but other physical instruments, designed for measurement of pressure or temperature are unlikely to experience step changes of input quantities. Hence the input is considered to change from the initial value in a ramp fashion until it becomes constant. Such a change is treated as terminated ramp input function and is represented in Figure 4.7 assuming that

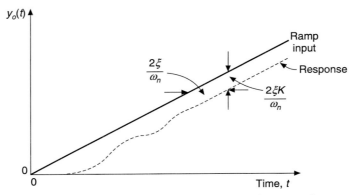

Figure 4.6   Response of a second-order system for a ramp-input.

$$x_i(t) = \frac{t}{T} \quad \text{for} \ \ 0 \le t \le T \tag{4.22}$$

$$= 1 \quad \text{for} \ \ T \le t < \infty$$

$$\frac{dy_o(t)}{dt} = 0 = y_o(t) \quad \text{at} \ \ t = 0$$

The solution for the case of a second-order lightly damped system can be obtained by treating the input as a sum of two ramp inputs, one at $t = 0$ and the other at $t = T$. It can be proved that a lightly damped mechanical system results in satisfactory response with considerably attenuated oscillations under transient conditions, provided ω_n of the system is sufficiently large compared to $1/T$, where T is the period at the end of which the input function assumes steady value. The validity of these observations is understood when the configuration of a pressure measuring system is studied.

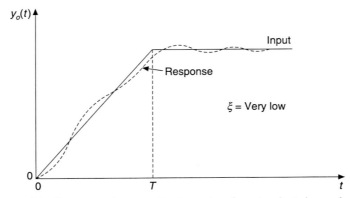

Figure 4.7   Response of a second-order system for a terminated ramp-input.

The frequency response of the system is, as already seen, obtained from its transfer function and is given by

$$\frac{Y_o(j\omega)}{X_i(j\omega)} = \frac{K}{\left[-\left(\dfrac{\omega}{\omega_n}\right)^2 + \dfrac{2j\omega\xi}{\omega_n} + 1\right]} \tag{4.23}$$

Writing $\omega/\omega_n = \eta$, the ratio of the frequency of the forcing function to its natural frequency, the response is expressed as

$$\frac{Y_o(j\omega)}{X_i(j\omega)} = \frac{K}{[(1-\eta^2)^2 + 4\xi^2\eta^2]^{1/2}} \angle\phi = |M| \angle\phi \tag{4.24}$$

where

$$\phi = \tan^{-1} \frac{2\xi\eta}{\sqrt{1-\eta^2}}$$

The variation of $|M|$ and $\angle\phi$ with ξ is shown in Figure 4.8 and the curves constitute the frequency response curves of the system under steady state conditions. A study of these curves reveals the fact that the magnitude of $|M|$ remains constant at K for frequencies up to $0.4\omega_n$ provided ξ is kept at 0.707. At this damping factor, the peak of $|M|$ disappears and hence the bandwidth of the system is considered to be maximum. It is important that these conditions be observed and fulfilled in design so that the system can be used as a recording instrument for input signals ranging in frequency from zero to that given by $0.4\omega_n$. It is equally essential that the recording of these sinusoidal signals is done with no error in phase or, in other words, the dc component of the signal, the fundamental and its harmonics are recorded truthfully both in amplitude and phase. There should neither be amplitude distortion nor phase distortion. Such a requirement is fulfilled, if the system has a linear relationship between ϕ and η for the range of frequencies lying within $0.4\omega_n$. An examination of the frequency response curves for ϕ shows that this condition is almost satisfied. It has already been stated that a choice of ξ of 0.7 is equally satisfactory for an indicating instrument, if a slightly larger settling time is tolerated. Thus instruments meant for the double function of indication and recording are designed to operate at a value of ξ equal to 0.7.

It is equally important to understand how second-order systems respond to impulse input functions. The impulse response in time domain reflects its natural behaviour and is used to relate the impulse response with the impulse strength. The input functions can be treated as impulses or pulses of such duration at the end of which only the system begins to respond. It is essential to assume that the system is provided with certain energy within the duration of the pulse T, and its response begins only after the pulse ceases to exist.

Thus it is necessary to evaluate the initial conditions such as $y_o(t)$ and $\dot{y}_o(t)$ at $t = 0_+$ so as to be incorporated in the solution obtained for its natural behaviour. As the impulse is no more present, the system reaches its final value which is the same as that with which it starts. The basic Eq. (4.2) may be integrated to give

$$a_2 \frac{dy_o(t)}{dt} + a_1 y_o(t) + a_0 \int_0^T y_o(t)\, dt = b_0 \int_0^T x_i(t)\, dt \tag{4.25}$$

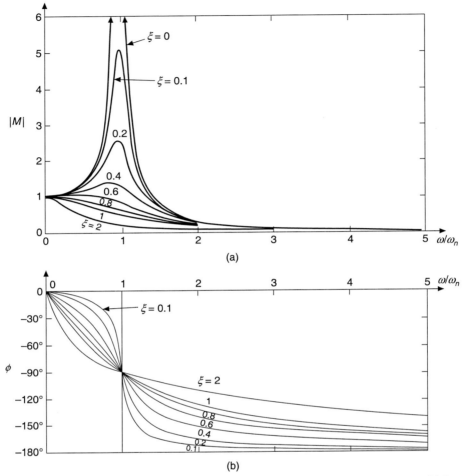

Figure 4.8   Frequency response characteristics of a second-order system: (a) For amplitude; (b) for phase.

As the duration of the pulse is very short, it is assumed, that $y_o(t)$ at $t = T$ is zero and that the system has an initial rate of change given by $dy_o(t)/dt$. When these conditions are applied, Eq. (4.25) yields the initial velocity as

$$\frac{dy_o(t)}{dt}\bigg|_{t=T} = \frac{b_0}{a_2}\int_0^T x_i(t)\, dt = KA\omega_n^2 \qquad (4.26)$$

where A is the strength of pulse. The response of the system for unit impulse input is given by

$$Y_o(s) = \frac{1}{\left(\dfrac{s^2}{\omega_n^2} + \dfrac{2\xi s}{\omega_n} + 1\right)}$$

from which the value of $y_o(t)$ is obtained in a normalized manner as

$$\frac{y_o(t)}{KA\omega_n^2} = \frac{1}{\omega_n\sqrt{1-\xi^2}} \exp(-\xi\omega_n t) \sin(\omega_n\sqrt{1-\xi^2}\,t) \quad \text{for } \xi < 1 \tag{4.27}$$

Similarly, for overdamped and critically damped conditions, $y_o(t)$ is given by

$$\frac{y_o(t)}{KA\omega_n^2} = t \exp(-\omega_n t) \quad \text{for } \xi = 1 \tag{4.28}$$

$$\frac{y_o(t)}{KA\omega_n^2} = \frac{1}{\omega_n\sqrt{\xi^2-1}} \exp(-\omega_n\xi t) \sinh\left\{\omega_n\sqrt{(\xi^2-1)}\,t\right\} \tag{4.29}$$

The response curves are shown in Figure 4.9. From the above equations, the value of the first peak for $y_o(t)$ can be evaluated and related to the strength of the pulse or impulse, for measurement of the same. It is quite important to realize that the duration of the pulse is much smaller than the natural time period, $2\pi/\omega_n$. The value of $y_o(t)$ when $t = \pi/\omega_n$ represents the first peak and is seen to be a function of ξ. The smaller the value of ξ, the larger is the value of the first peak.

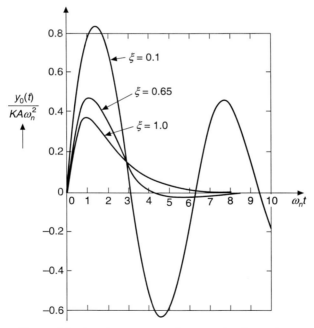

Figure 4.9 Impulse response of a second-order system.

The above working conditions are satisfied when measurements of electrical charge and magnetic flux linkages are done by permanent magnet moving coil systems which are so designed that they possess as large a time period as possible.

4.1.4 Higher Order Systems

Higher order physical systems may be studied for their performance when they are subjected to standard input signals and the response becomes further complicated. Unless they are properly damped, they may become unstable in performance. Usually, instrument systems are designed to operate under conditions that keep the system stable, and dynamic compensation is adopted to obtain the desired performance and limit its working at frequencies far removed from the region of resonance or unstable operation.

4.2 ELECTRICAL NETWORKS

Electrical networks, configured from the combination of the three basic circuit elements—resistor, capacitor and a self-inductor—enable formation of analogous systems portraying the behaviour of the various other physical systems. They can be modelled to represent even the complicated physical systems by making use of their mathematical description. The simple networks are illustrated in this section to enable the reader to perceive the analogous quantities between the electrical quantities and other physical quantities, pertaining to a physical system. Electrical networks, in combination with ideal voltage amplifiers, can be designed to obtain the desired transfer function between the output and input voltages. Such networks are known as *active networks*, whereas *passive networks* are those that are made up of the passive elements, R, L and C.

Each one of these basic elements is characterized by the relation between the current flowing through the element and the voltage developed across it. Such relationship serves as the basis for the formulation of network equations.

The resistor is basically a zero-order element and its resistance R is given by

$$\frac{V_R(s)}{I(s)} = R \tag{4.30}$$

treating $V_R(s)$ as the output variable and $I(s)$ as the input variable.

The basic relationship governing the current $i(t)$ passing through a self-inductor of L henries and the voltage $v_L(t)$ developed across it is given by

$$v_L = L\frac{di(t)}{dt}$$

The Laplace transform of this equation is

$$V_L(s) = LsI(s) - i(0\_) \tag{4.31}$$

Assuming that the initial current at $t = 0$ is zero, the transfer function becomes

$$\frac{V_L(s)}{I(s)} = sL$$

Similarly, the basic relationship of a capacitor of C farads is given by

$$v_c(t) = \frac{q(t)}{C} = \frac{\int_{0^-}^{\tau} i(t)\,dt}{C}$$

thereby yielding

$$V_c(s) = \frac{I(s)}{sC} + v_c(0_-) \tag{4.32}$$

Treating the initial voltage $v_c(0_-)$ of the capacitor as zero, the transfer function becomes

$$\frac{V_c(s)}{I(s)} = \frac{1}{sC}$$

The ideal voltage amplifier is usually associated with a voltage gain A and is given by

$$\frac{V_o(s)}{V_i(s)} = A \tag{4.33}$$

The ideal voltage amplifier is thus assumed to be a zero-order system for most of its applications in the field of instrumentation. Active networks are formed by a combination of the ideal amplifier with the passive elements R and C, and they serve a variety of applications in signal processing.

A first-order RC network is shown in Figure 4.10, and the transfer function between $V_o(s)$ and $V_i(s)$ can be obtained as

$$\frac{V_o(s)}{V_i(s)} = \frac{1/sC}{R + \dfrac{1}{sC}} = \frac{1}{1 + sCR} = \frac{1}{1 + s\tau} \tag{4.33}$$

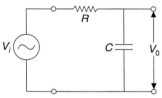

Figure 4.10 A simple RC network.

The static sensitivity of the network is seen to be unity and its time constant τ is equal to the product of R and C. When the network is excited by sinusoidal signals, it serves as a low-pass filter as seen from Figure 4.4.

Figure 4.11 shows two such networks in cascade, the first one consisting of R_1 and C_1 and the second consisting of R_2 and C_2. Without going into the detailed steps, the transfer function of the network is given by

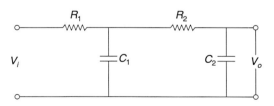

Figure 4.11 Two simple RC networks in cascade.

$$\frac{V_o(s)}{V_i(s)} = \frac{1}{s^2 C_1 C_2 R_1 R_2 + s(R_1 C_1 + R_2 C_2 + R_1 C_2) + 1} \tag{4.34}$$

When two networks are connected together with an isolating amplifier of unity voltage gain, it is seen that the transfer function simplifies to

$$\frac{V_o(s)}{V_i(s)} = \frac{1}{s^2 C_1 C_2 R_1 R_2 + s(R_1 C_1 + R_2 C_2) + 1} \qquad (4.35)$$

The additional term sR_1C_2 of Eq. (4.34) is due to loading of the second network on the first network and is known as the coupling term. In case the two networks are characterized by their time constants $\tau_1(= R_1C_1)$ and $\tau_2(= R_2C_2)$, then Eq. (4.35) is given by

$$\frac{V_o(s)}{V_i(s)} = \frac{1}{s^2 \tau_1 \tau_2 + s(\tau_1 + \tau_2) + 1} \qquad (4.36)$$

The combination is thus similar to a second-order system having its undamped natural frequency of $\sqrt{1/\tau_1\tau_2}$ and the damping factor governed by τ_1 and τ_2.

If the two networks are made similar in all respects, i.e. $\tau_1 = \tau_2$, then Eq. (4.36) represents the behaviour of a critically damped second-order system, whereas Eq. (4.34) represents the behaviour of an overdamped system with $\xi = 1.5$.

The observations made above are sufficient to show that cascading of two first-order systems does not result in an underdamped second-order system. The coupling term adds to the slowness of response of the system and when the second stage does not constitute any load on the first, the combination is likely to become critically damped.

The three basic circuit elements can be connected together in different ways to constitute second-order systems. Figure 4.12 shows the network in which the transfer function relating the voltage across the capacitor with the input voltage is drawn up and is given by

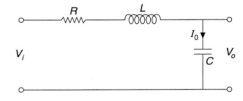

Figure 4.12 A simple RLC network.

$$\frac{V_o(s)}{V_i(s)} = \frac{1}{s^2 LC + sCR + 1} \qquad (4.37)$$

The undamped natural frequency is given by $1/\sqrt{LC}$, and the damping conditions are governed by R.

The R-L-C series network of Figure 4.12 may as well be seen to provide a different transfer function if $I(s)$ is considered as the output variable, and in such a case

$$\frac{I_o(s)}{V_i(s)} = \frac{sC}{s^2 LC + sCR + 1} \qquad (4.38)$$

When the network is excited by a sinusoidal voltage $V_i(j\omega)$, then the ratio becomes

$$\frac{I_o(j\omega)}{V_i(j\omega)} = \frac{j\omega C}{[-\omega^2 LC + j\omega CR + 1]} \qquad (4.39)$$

showing that the series network goes into resonance at $\omega = 1/\sqrt{LC} = \omega_n$ and under resonance, the output current is decided and controlled in value only by R. The value of $I_o(j\omega)$ assumes a maximum value at resonance and is zero on dc and at $\omega = $ infinity. Figure 4.13 shows the resonance curve.

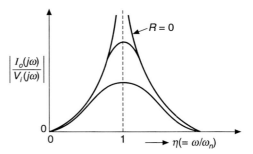

Figure 4.13 Resonance curve of an RLC network.

The three circuit elements may be connected in parallel to each other to constitute a parallel resonant circuit. Considering the voltage across the combination as the output variable and the current through the combination as the input variable, we may obtain an expression similar to that of Eq. (4.39). Then the resonance curve shows that the current goes through a minimum at $\omega = 1/\sqrt{LC} = \omega_n$.

It is possible to draw up networks in a variety of ways with or without the active element, the ideal voltage amplifier, and arrive at different transfer functions between the input and output variables.

An ideal transformer is another electrical system that is considered as a zero-order element, when used in networks. The ratio of the output voltage on no load to the input voltage is totally governed by turns ratio, and hence its transfer function is represented by

$$\frac{V_o(s)}{V_i(s)} = \frac{V_{\text{sec}}(s)}{V_{\text{pri}}(s)} = \frac{N_s}{N_p} = T$$

4.3 MECHANICAL SYSTEMS

All physical elements have mass, and the mass of each is said to constitute an inertial element. When a force F is applied to the mass, a reaction force is produced, which is equal to the product of the mass M and its acceleration and acts in a direction opposite to that of acceleration. The forces are represented by

$$F(t) = M \frac{d^2 x(t)}{dt^2}$$

giving

$$\frac{X(s)}{F(s)} = \frac{1}{s^2 M} \tag{4.40}$$

Similarly, when a body is rotating, a reaction torque T is produced, which is equal to the product of its moment of inertia J and its angular acceleration. This torque acts in a direction opposite to that of angular acceleration and is given by

$$T(t) = J \frac{d^2 \theta(t)}{dt^2}$$

giving

$$\frac{\theta(s)}{T(s)} = \frac{1}{s^2 J} \qquad (4.41)$$

The linear and torsional springs are used to provide restoring force and torque respectively. When they are strained in the linear range, the restoring force and torque are given by

$$F(s) = KX(s) \quad \text{or} \quad T(s) = K\,\theta(s) \qquad (4.42)$$

where, K represents the stiffness of the spring.

The transfer function of the spring element shows that it belongs to a zero-order system.

There is friction whenever surfaces of mechanical systems come in sliding or moving contact. The friction is mainly of two types: Coulomb friction or static friction (stiction) and viscous friction. The static friction force is a constant and is the minimum force required to move the body from its state of rest. The viscous friction force, on the other hand, comes into existence when there is a relative velocity between the sliding surfaces, and is given by

$$F(t) = B\frac{dx(t)}{dt}$$

where, B represents the damping coefficient, yielding thereby

$$\frac{X(s)}{F(s)} = \frac{1}{sB} \qquad (4.43)$$

The viscous friction force acts in a direction opposite to that of velocity.

There are many situations in which viscous friction is intentionally brought into play, and the arrangements provided are usually known as dash pots. Figure 4.14 shows an arrangement in which a piston moves in a cylinder filled with oil. The flow of oil from one side to the other through the narrow clearance between piston and cylinder offers resistance to the movement of the piston. The viscous friction force is known as damping force. Similar arrangements provide damping torque under rotary motion, the torque being given by

$$T(s) = Bs\,\theta(s) \qquad (4.44)$$

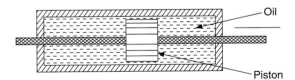

Figure 4.14   Dashpot system for damping.

Most of the mechanical systems are combinations of the above three elements and are generally referred to as mass-spring-dashpot systems.

Figure 4.15 shows a mass M attached to a spring and a dashpot. The other end of the spring and the cylinder of the dashpot are rigidly connected to a reference surface. The force equation can be written as

$$F(s) - Bs\,X_m(s) - KX_m(s) = Ms^2 X_m(s)$$

where $F(s)$ is the externally applied force to move the mass M against the spring and damping forces.

Rearranging the terms, the transfer function relating the output displacement of mass M and the input force may be obtained as

$$\frac{X_m(s)}{F(s)} = \frac{1}{Ms^2 + Bs + K} \qquad (4.45)$$

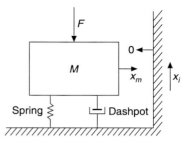

Figure 4.15 A spring-mass system.

The configuration of mass M with respect to the housing as shown in Figure 4.15 may be modified to resemble a mass inside a lift subjected to motion along the direction of force F. The housing along with its spring-mass-dashpot system constitutes a siesmic system in which the housing is rigidly connected to the structure under movement. The relative displacement $x_r(t)$ between the mass and the housing at any instant may be related with the absolute movements of the structure, $x_i(t)$, and the proof mass, $x_m(t)$. The movement of the lift upwards develops reaction force equal to $M d^2 x_m(t)/dt^2$, and this force is opposed by spring and dashpot. So,

$$M \frac{d^2 x_m(t)}{dt^2} = B \frac{dx_r(t)}{dt} + K\,x_r(t) \qquad (4.46)$$

But

$$x_m(t) = x_i(t) - x_r(t)$$

So,

$$M \frac{d^2 x_i(t)}{dt^2} = M \frac{d^2 x_r(t)}{dt^2} + B \frac{dx_r(t)}{dt} + K\,x_r(t) \qquad (4.47)$$

Taking all the initial conditions as zero, the transfer function becomes

$$\frac{X_r(s)}{X_i(s)} = \frac{Ms^2}{Ms^2 + Bs + K}$$

$$\frac{X_r(s)}{s^2 X_i(s)} = \frac{1/\omega_n^2}{\dfrac{s^2}{\omega_n^2} + \dfrac{2\xi s}{\omega_n} + 1} \qquad (4.48)$$

The transfer function enables measurement of $s^2 X_i(s)$, which is the acceleration along the direction shown. The static sensitivity between the relative displacement and acceleration is $1/\omega_n^2$. The frequency response curves resemble those given in Figure 4.8. The seismic system is useful for measurement of absolute acceleration, absolute amplitude of vibration etc.

Similarly, for a rotational system, the angular displacement and the input torque are related to give similar transfer function as

$$\frac{\theta(s)}{T(s)} = \frac{1}{Js^2 + Bs + K} \qquad (4.49)$$

All the analog indicating instruments are provided with pointers which are in turn connected to the torque developing systems. As is evident from the transfer function, these systems belong to the class of second-order systems.

Gear trains are used in instrument servomechanisms primarily to bring down the total angular displacement to within 300° for indication over a circular scale. Proper gear ratio also matches the motor to load mechanically. Usually, the servomotor driving the load through gear train runs at higher speed but low torque, and hence the gear train provides the necessary torque amplification and speed reduction. The gear trains of mechanical systems are equivalent to transformers of electrical systems in this respect. In instrument servomechanisms, the load on the motor is mostly inertial.

The primary gear with N_1 teeth is coupled to the secondary gear with N_2 teeth, as shown in Figure 4.16. When the motor drives the primary gear by an angle θ_1, the inertial load turns through an angle θ_2. If r_1 be the radius of the primary gear and r_2 be that of secondary, then $\theta_1 r_1 = \theta_2 r_2$, and hence it can be written as

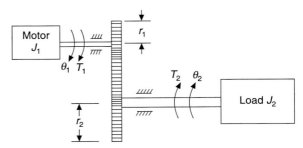

Figure 4.16   A gear train assembly.

$$\frac{\theta_2}{\theta_1} = \frac{r_1}{r_2} = \frac{N_1}{N_2} \qquad (4.50)$$

where, N_1 and N_2 are the number of teeth of the primary and secondary gears respectively.

Assuming that there is no loss of power in the gears, the work done by primary gear is equated to that of the secondary gear, and hence $\theta_1 T_1 = \theta_2 T_2$, resulting in

$$\frac{T_1}{T_2} = \frac{\theta_2}{\theta_1} = \frac{N_1}{N_2} \qquad (4.51)$$

where, T_1 is the load torque on the primary gear due to the rest of the gear train and T_2 the torque transmitted to the secondary gear.

In practice, a speed reduction by a factor of 10 is adopted by properly choosing N_1 and N_2, and the sturdiness of the gears is dependent on the torques transmitted. In instruments and instrument servomechanisms, where torques are usually much smaller, nylon gear trains are used, sometimes resulting in speed reduction by a factor of 1000 in three stages.

Considering the load torque on the secondary gear to be entirely inertial, it can be shown that

$$T_2 = J_2 \frac{d^2\theta_2}{dt^2}$$

Hence

$$T_M = J_1 \frac{d^2\theta_1}{dt^2} + \frac{N_1}{N_2}\left(J_2 \frac{d^2\theta_2}{dt^2}\right) \qquad (4.52)$$

where,

T_M = torque developed by the drive motor
J_1 = moment of inertia of motor and primary gear
J_2 = moment of inertia of load member and secondary gear

T_M can now be rewritten as

$$T_M = \left[J_1 + J_2 \left(\frac{N_1}{N_2} \right)^2 \right] \frac{d^2\theta_1}{dt^2}$$

Thus the equivalent moment of inertia of gear train referred to motor shaft is

$$J_{1eq} = J_1 + J_2 \left(\frac{N_1}{N_2} \right)^2 \tag{4.53}$$

Similarly, the equivalent viscous friction of gear train is given by

$$B_{1eq} = B_1 + B_2 \left(\frac{N_1}{N_2} \right)^2$$

where, B_1 and B_2 are the coefficients of viscous friction attendant on the primary and secondary shafts respectively.

In electrical systems, a transformer with a primary winding of N_1 turns and a secondary winding of N_2 turns is useful in matching the load impedance connected to the secondary winding suitably so as to result in maximum transfer of power from primary to the secondary. The equivalent impedance $\dot{Z}_{1eq}$ presented to source of supply is given by

$$\dot{Z}_{1eq} = \dot{Z}_1 + \dot{Z}_2 \left(\frac{N_1}{N_2} \right)^2 \tag{4.54}$$

where,

$\dot{Z}_2$ = total impedance of secondary winding and load impedance

$\dot{Z}_1$ = impedance of primary winding

It is important to realize that, while choosing the gear trains, the gear ratio is so chosen as to satisfy the criterion of transfer of maximum power from the driving shaft to the driven shaft by making $J_1 = J_2(N_1/N_2)^2$ in situations where the load on the secondary gear is almost inertial.

Levers transmit translational motion and force in the same manner as the gears transmit rotational motion and torque. Considering the levers to be ideal in the sense that they are massless and frictionless, the input and output quantities (see Fig. 4.17) are related to l_1 and l_2 as

$$\frac{x_2}{x_1} = \frac{l_2}{l_1}$$

and
$$\frac{f_2}{f_1} = \frac{l_1}{l_2} \text{ (for balance of forces)} \tag{4.55}$$

(see Figure 4.17).

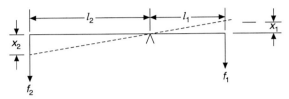

Figure 4.17   A simple lever assembly.

4.4 ELECTROMECHANICAL SYSTEMS

Electromechanical systems constitute a large section of electrical measuring systems and other instrumentation systems. Most of the electrical indicating instruments used for the measurement of voltage, current or power are electromechanical systems whose output variable is an angular displacement. Other devices such as tachogenerators, servomotors or servovalves may be treated as electromechanical systems whose outputs may be brought out either in analog, frequency or in digital domain. The primary function of all such electromechanical devices is to convert either a mechanical variable into an electrical one or vice versa. The electrical variables are voltages and currents, whereas the mechanical variables are angular displacement, angular velocity, force or torque. The transfer functions relating some of them are simple and direct, allowing them to be treated as zero-order systems, while others are slightly involved, rendering them as either first or second-order systems.

Among these systems, the tachogenerators are based on induced emfs due to the movement of conductors in magnetic fields, and hence the developed emf $E_o(s)$ is proportional to speed $\Omega(s)$ as given by

$$\frac{E_o(s)}{\Omega(s)} = K_g \tag{4.56}$$

where, K_g is known as either a generator constant or a back emf constant.

Similarly, the torques $Q_o(s)$ developed by the moving coil of a permanent magnet moving coil system is related to its current $I(s)$ as

$$\frac{Q_o(s)}{I(s)} = BAN = K_t \text{ (ref Eq. (3.1))} \tag{4.57}$$

where, K_t is called the torque constant.

The force motor (like the above torque motor) consisting of a pot type permanent magnet in whose field a cylindrical coil carries current, is represented in a similar way, by a transfer function as

$$\frac{F_o(s)}{I(s)} = K_f \tag{4.58}$$

where, K_f is the force constant.

The torque developed by a dc motor is, likewise, considered to have its torque developed proportional to its armature current $I_a(s)$, and K_t of Eq. (4.57) is used as its transfer function.

The above devices are very much used in measuring systems and feedback instrumentation systems, and their application as zero-order systems has resulted in development of effective feedback systems. However, current-to-torque or current-to-force conversion, using Eqs. (4.57) and (4.58), is not possible unless the systems are excited from constant current sources. Otherwise, it is necessary to treat the current as derived from a voltage source of finite output resistance and, in such a case, the transfer function cannot be as simple as shown in Eqs. (4.57) and (4.58). The circuit depicted in Figure 4.18 represents the case of a moving coil of a pmmc system or the armature of a constant field (permanent magnet or otherwise) dc motor. The coil is shown to be of a resistance R and self-inductance L. Whenever the coil is in motion, there is dynamically induced emf in the coil proportional to the angular velocity and hence the equation relating the applied voltage $e(t)$ and the opposing voltages is given by

$$e(t) = Ri(t) + L\frac{di(t)}{dt} + K_t \frac{d\theta(t)}{dt} \tag{4.59}$$

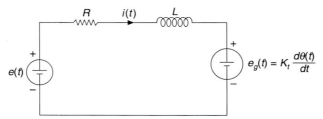

Figure 4.18   Equivalent circuit of an electromechanical system.

Taking the Laplace transform of the variables, the equation becomes

$$E(s) = (R + sL)\, I(s) + K_t s\, \theta(s)$$

and hence $I(s)$ is given by

$$I(s) = \frac{E(s) - K_t s\, \theta(s)}{R + sL} \tag{4.60}$$

Treating L as negligible and ineffective as is done in the case of pmmc instruments, the torque developed is given by

$$Q_o(s) = \frac{K_t[E(s) - K_t s\, \theta(s)]}{R} \tag{4.61}$$

with the load torque consisting of inertial, viscous damping and control spring type, the dynamic behaviour is characterized by the equation relating the output angular displacement $\theta(s)$ with $E(s)$, as

$$K_t \frac{E(s)}{R} = \left[Js^2 + \left(B + \frac{K_t^2}{R} \right) s + K \right] \theta(s) \tag{4.62}$$

$$\frac{\theta(s)}{E(s)} = \frac{K_t / R}{Js^2 + B's + K} \tag{4.63}$$

where, $B' = B + (K_t^2/R)$.

The term K_t^2/R is the viscous damping coefficient added to the existing air (viscous) damping and is the result of interaction between currents due to back emf and the magnetic field. The term K_t^2/R is usually referred to as electromagnetic or eddy-current damping coefficient. For a pmmc system, it increases with reduction of R.

In the case of a constant field dc motor excited from a voltage source, the transfer function between output angular velocity and applied voltage becomes, after writing $K = 0$ and $B = 0$,

$$\frac{s\theta(s)}{E(s)} = \frac{\Omega_o(s)}{E(s)} = \frac{1/K_t}{\left[\dfrac{JR}{K_t^2} s + 1 \right]} \tag{4.64}$$

The value of JR/K_t^2 defines the motor mechanical time constant and for a particular motor, it is proportional to the total armature circuit resistance. To reduce the time constant, it is desirable to excite the motor from a voltage source of negligible resistance, so that R is mostly governed by the resistance of the motor armature.

Equation (4.61) shows how the torque is related to applied voltage and the angular velocity of a dc motor under steady state conditions. Figure 4.19 shows the torque speed characteristics for different applied voltages. For a specified applied voltage, the torque speed characteristic has negative slope of K_t/R, and it is essential for a motor to come to rest when the applied voltage is made zero when used as a servomotor. Such a torque-speed characteristic ensures positive viscous damping as seen from Eq. (4.62).

It is for the above reason that the two-phase induction motor, while used as a servomotor in position and speed control schemes, is so designed that torque-speed characteristics are linear and of negative slope as shown in Figure 4.19.

Equation (4.63) relating the output angular displacement and applied voltage enables estimation of the behaviour of the permanent magnet moving-coil instruments, and a family of such instruments constitute the range of the galvanometers, as described in Section 3.1.1. The choice of suitable values of K and J enables the system to possess the desired undamped natural frequency. The number of turns N and the area of the coil A also influence the value of J, apart from effecting K_t and R, and thereby the electromagnetic damping coefficient. It is interesting to realize how the variation in the design brought about by the proper choice of the above has rendered the pmmc system suitable for accurate measurements of voltage, current, charge, flux linkages as also for the detection and recording of ac currents.

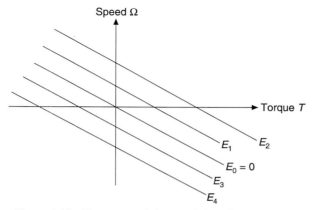

Figure 4.19 Torque-speed characteristics of a servomotor.

The electrical time constant L/R, as seen from Eq. (4.60), is generally considered to be much smaller compared to the mechanical time constant JR/K_t^2, and hence, while determining the dynamic behaviour of electromechanical systems, the assumption that L is negligible does not amount to much error in the analysis.

The force motor assembly consisting of a cylindrical moving coil and a permanent magnet is shown in Figure 4.20 and is used in place of a dashpot to develop equivalent viscous damping by electromagnetic means. The pot magnet assembly produces a steady magnetic field of flux density B in the annular air gap of depth b. The moving coil is replaced by a thin copper cup of mean diameter $2R$, and of wall thickness d. If the cup is made to move with a velocity $v(t)$, in a direction perpendicular to the flux lines, an emf $e(t)$ is induced in the cup and is given by

$$e(t) = 2\pi R B \, v(t)$$

The resistance of the cup lying in the magnetic field is given by $R_c = 2\pi R \rho / bd$, where ρ is the specific resistivity of cup material and the corresponding current $i(t)$ produced develops the damping force given by

$$F(t) = 2\pi R B \, i(t)$$

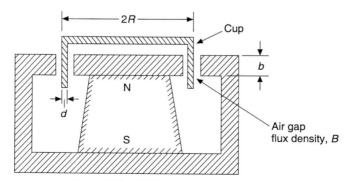

Figure 4.20 Eddy-current damping system.

Thus the damping coefficient B' can be shown to be

$$B' = \frac{F(t)}{v(t)} = 2\pi R \frac{B^2 bd}{\rho} \qquad (4.65)$$

The system is thus very useful for developing accurate and predictable viscous damping effect.

4.5 THERMAL SYSTEMS

Thermal systems are, in fact, mechanical systems that are subjected to the processes of heat flow. Heat is a form of energy and it can be transferred from one point to another through material media and stored as internal energy.

Heat transfer takes place by conduction from the point at a higher temperature to the point at a lower temperature, through the solid medium separating the two points. The heat flow rate may be considered as an analog of current (through-variable) and the temperature as electrical potential (across-variable).

The thermal resistance R of the path of heat transfer is given by

$$R = l/kA$$

where

k = thermal conductivity of the material (solid) medium separating the two points
A = area of cross-section through which heat transfer takes place
l = length of the path of heat transfer (= thickness of solid medium)

The value of k/l is usually referred to as the heat transfer coefficient h, and is used to represent the resistance of the interface (or film) between a fluid and a solid.

Writing $q(t)$ as the quantity of heat, the rate of heat transfer is represented by

$$\frac{dq(t)}{dt} = \frac{T_1 - T_2}{R} = hA\Delta T(t)$$

where, $\Delta T(t)$ is the temperature difference between the two points 1 and 2.

The transfer function relating the heat flow rate and the temperature difference is thus given by

$$\frac{\Delta Q(s)}{\Delta T(s)} = hA \text{ (= thermal conductance)} = \frac{1}{R} \qquad (4.66)$$

If a body receives heat energy, it may be stored in it as internal energy. In case it does not entail any work done and if there is no change of state of the body, then the heat energy received raises the temperature of the body. The ratio of the net amount of heat received and stored in the body in a certain time interval to the rise in its temperature is defined as the thermal storage capacity or thermal capacitance C which is given by

$$\frac{\Delta Q(s)}{\Delta T(s)} = C = mc \qquad (4.67)$$

where,

> m = mass of the body
> c = specific heat of the material of the body

A simple thermometer system may be considered as a combination of a thermal capacitance and a thermal resistance. The thermometer bulb of mass m and of specific heat c receives the heat and stores it, the effect of which is noticed as a rise in its temperature. Figure 4.21 shows the bulb surrounded by the fluid whose temperature T_1 is under measurement. If T_2 is the temperature of the bulb, the instant by instant variations of T_1 and T_2 are represented by an equation based on the theorem of conservation of energy, and the relationship is given by

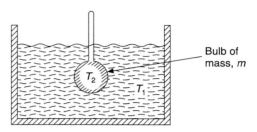

Figure 4.21 A simple thermometer system.

$$hA[T_1(t) - T_2(t)] = mc\frac{d}{dt}T_2(t) \tag{4.68}$$

Taking the Laplace transform and rewriting, the transfer function relating $T_1(s)$ and $T_2(s)$ is obtained as

$$\frac{T_2(s)}{T_1(s)} = \frac{1}{1 + s(mc/hA)} \tag{4.69}$$

Here, the value of mc/hA is the time constant of the simple thermometer, making it a simple first-order system. The time constant of a thermometer is not truly a constant as the heat transfer coefficient h varies with the properties of the fluid surrounding the bulb.

In case the thermometer is protected by means of a jacket, the combination becomes equivalent to cascading of two first-order systems, thereby making the system slower and overdamped as explained in Section 4.2.

The foregoing analysis is based on a few assumptions, one of them being that the heat transfer coefficient h cannot truly reflect the entire conduction and convection processes of heat transfer. Strictly speaking, the heat transfer takes place at the interface of the fluid and the glass bulb at the first instance and then the transfer from the inner surface of the glass bulb to mercury inside the bulb. However, it should be noted that the preciseness with which the dynamic behaviour of thermal and other fluidic systems is estimated cannot be high for reasons that can be well understood from the above example.

4.6 FLUIDIC SYSTEMS

Fluidic systems are those mechanical systems in which the flow of liquids and gases (or vapours) from one point to another constitutes variations in the pressures they build up in the containers or the connecting tubes. Several pneumatic and hydraulic instruments are based on the flow of gases (air) and liquids. The simplifying assumption that liquids are incompressible makes the analysis of hydraulic systems far simpler, but the compressibility of air makes the analysis of pneumatic systems far more complex.

However, for the sake of understanding of the dynamic behaviour of some simple systems and the equivalence with the electrical systems for modelling purposes, a few devices are considered for analysis. The fluid flow-rate may be considered as an analog of electrical current while the pressure at a point in a fluid is analogous to electrical potential. Hence, it is possible to define the pneumatic or hydraulic resistance of a connecting pipe through which the fluid flows as well as the pneumatic or hydraulic capacitance of a storage vessel. The parallelism adopted in defining these parameters for a thermal system in Section 4.5 through Eqs. (4.66) and (4.67) may be attempted for fluidic systems as well.

In each case, the theorem of conservation of mass of the fluid under flow may have to be applied to derive the necessary relationships. Wherever the relationship between the pressure and flow rate or between any other two variables is nonlinear, the slope of the characteristic between the variables in question is taken at the operating point and is used to relate the increments of these two variables for determining the transfer function. Such an approach leads to linearization of the model of the system, and the transfer function derived is known as *quasi-linear transfer function.*

For the sake of automatic control, a large variety of fluidic devices are used in fluidic systems in which the flow of fluids is controlled and utilized to generate output signals of desired description. In the field of instrumentation, a few elastic elements, such as diaphragms, bellows and Bourdon tubes are used for measurement of pressure. Due to application of pressure, the elastic element or transducer, being flexible, gets distorted, resulting in output displacements conveyed to a pointer or a secondary transducer. Thus the element may be considered equivalent to a spring-mass system with practically no damping if surrounded by air cushions of large volume on either side. But in actual operation, its natural characteristics are considerably affected by the presence of a fluid medium and the size and configuration of the chambers holding the medium on either side. For example, a diaphragm is connected as shown in Figure 4.22(a) such that on one side it has the fluid contained in the pressure vessel, whereas on the other side it has a small chamber totally enclosed with air at some desired pressure. The natural frequency of the diaphragm and the damping factor under which it is operating will be considerably modified if there is a liquid medium on both sides.

When it is not possible to connect the diaphragm in flush with the walls of the vessel, a connecting tube or capillary is used as shown in Figure 4.22(b). The cavity or chamber at the diaphragm side of the connecting tube may have a volume V, which may be different from the volume of the connecting tube. It is important to realize that any change in pressure occurring at the mouth of the capillary takes time to be transmitted to the other end, and hence the indicated value of pressure does not reflect the true pressure under dynamic conditions. The dynamic behaviour of such a pressure measuring system cannot be estimated correctly, as there are several factors to be considered for its analysis. However, a few assumptions always enable a simplified approach for predicting the performance. The complexity or the simplicity of analysis is revealed by a study of the following aspects relevant to pressure measuring systems.

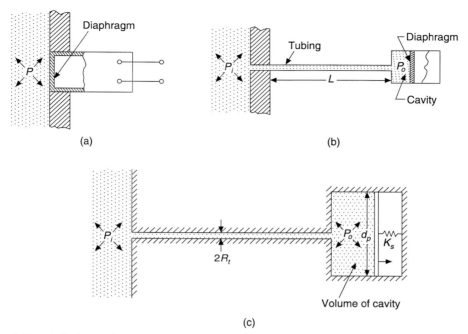

Figure 4.22   A diaphragm for pressure measurement: (a) In flush with walls of tank; (b) connection through a capillary tube and equivalent of the diaphragm; (c) to a spring-load system.

1. When a small change in pressure about its mean value occurs in front of the capillary tubing, the pressure wave travels through the tubing to the other end at the velocity of sound in the medium contained in the nonrigid tube. The velocity of sound, v, for such a case is given by

$$v = \left[\frac{E_f}{\rho(1 + 2R_t E_f / tE_t)} \right]^{1/2} \tag{4.70}$$

where,

$\quad E_f$ = bulk modulus of the fluid
$\quad \rho$ = mass density of the fluid
$\quad E_t$ = modulus of elasticity of tube material
$\quad R_t$ = inner radius of the tube
$\quad t$ = thickness of tube walls

The time taken for the pressure wave to travel the length of the tubing is known as the *transportation time* (or *dead time*) and is represented as τ_{dt}. In such a case, the pressure changes $\Delta P_i(s)$ and $\Delta P_o(s)$ are related by

$$\frac{\Delta P_o(s)}{\Delta P_i(s)} = \exp(-s\tau_{dt}) \tag{4.71}$$

Hydraulic and pneumatic lines of considerable length are used for transmission of pressure, and in such cases, the above transportation lag needs to be considered. However, for pressure measurement, most of the time, the pressure transducer is connected by a small length of capillary, and hence the variation of pressure along the tube due to the wave motion is very small. Also, for sufficiently low frequency fluctuations of pressure, and for small amplitudes of ΔP_i about a high mean value, it can be assumed that the particles of the fluid move together as a lump along the tube at a certain average velocity. The inertia effects of the moving fluid and other moving parts become prominent at higher frequencies only.

2. The flow of fluid in the capillary tube is taken as laminar as the pressure changes are very slow. But, at higher frequencies of fluctuations, the flow may be considered to be of either transition or turbulent region. Under these circumstances, necessary modifications for the relationship between the pressure drop cross the tubing and the average velocity of flow have to be incorporated.

For laminar flow conditions, the resistance offered to the flow due to the viscosity of the fluid and the viscous damping coefficient due to the presence of the connecting tube are obtained in the following manner: From the basic definition of dynamic viscosity, μ is given by

$$\mu = \text{velocity gradient} = \text{shearing stress}$$

and the velocity gradient of the flow at a distance of r from the centre line is given by

$$\frac{dv}{dr} = \frac{\pi r^2 \Delta P}{2\pi r L \mu} \tag{4.72}$$

where,

ΔP = pressure difference at the two ends of the tube
L = length of the tube
v = velocity of flow at a distance r from the centre line

The velocity of the fluid at $r = R_t$ is zero and, using this condition, v is obtained as

$$v = \frac{\Delta P}{4\mu L}(R_t^2 - r^2) \tag{4.73}$$

where, R_t is the inner radius. The velocity at the centre line, v_o, is thus given by

$$v_o = \Delta P \frac{R_t^2}{4\mu L}$$

The average velocity v_{av} is taken as the mean of the velocity at $r = R_t$ and that at the centre line and hence is given by

$$v_{\text{av}} = \frac{\Delta P R_t^2}{8\mu L} = \frac{v_o}{2} \tag{4.74}$$

The volume flow rate is $\pi R_t^2 v_{\text{av}}$. Substituting the value of v_{av}, the total volume of fluid that flows out is given by

$$q(t) = \int_0^\tau \frac{\pi R_t^4 \Delta P}{8\mu L} \, dt$$

thereby giving the transfer function between the volume flow rate and the pressure difference ΔP as

$$\frac{sQ(s)}{\Delta P(s)} = \frac{\pi R_t^4}{8\mu L} = \frac{1}{R_h} \tag{4.75}$$

where, R_h is considered as the hydraulic resistance of the capillary tube.

From Eq. (4.73), the velocity of the fluid contained in an annular element of thickness dr at radius r is known and hence the kinetic energy of this element of fluid is given by

$$d(\text{K.E.}) = \frac{1}{2} (2\pi r \, dr \, L\rho) v^2 \tag{4.76}$$

Writing $v = v_o (1 - r^2/R_t^2)$ and integrating Eq. (4.76), the kinetic energy of the entire fluid in motion in the capillary tube is obtained as

$$\text{K.E.} = \frac{\pi \rho L v_o^2 R_t^2}{6} = \frac{4\pi \rho L v_{\text{av}}^2 R_t^2}{6} \tag{4.77}$$

From the kinetic energy, it is possible to determine the equivalent rigid mass that would possess the same K.E. as the fluid.

3. Any chamber or cavity that stores the fluid may be considered to be an equivalent of hydraulic capacitance. Consider the cavity of volume V shown in Figure 4.22(b) as a chamber with its capacitance C_h defined by the ratio of the incremental volume ΔV of fluid added to the increment in its pressure ΔP. Then C_h is given by

$$C_h = \frac{\Delta V}{\Delta P} = \frac{\Delta Q(s)}{\Delta P(s)} \tag{4.78}$$

As liquids are incompressible, C_h can also be defined as $\rho \Delta V/\Delta P$. In Figure 4.22(b), the diaphragm may be considered to be infinitely stiff, in which case the cavity volume remains the same.

Alternatively, the flexible diaphragm may be considered to yield to the pressure change and result in a corresponding change in cavity volume. The displacement generated by the diaphragm may be treated as equivalent to that of a spring-loaded piston as shown in Figure 4.22(c) or to that of a bellows which is free to expand on one side.

An analysis of the case of the rigid chamber with the connecting tube indicates that it behaves as a first-order system, whereas the flexible diaphragm behaves as a second-order system, by taking into account the inertia of moving parts and fluid.

Considering the cavity of Figure 4.22(b) to remain the same in its volume in spite of the additional volume of fluid that enters the cavity, it can be said that the pressure of the fluid in the cavity should have undergone a change with the addition.

Writing $\Delta P(s)$ of Eq. (4.75) as $[P_i(s) - P_o(s)]$, the incremental volume of fluid added is given by

$$\Delta Q(s) = \frac{P_i(s) - P_o(s)}{sR_h}$$

which would have resulted in its pressure $P_o(s)$, and hence

$$\Delta Q(s) = C_h P_o(s) = \frac{P_i(s) - P_o(s)}{sR_h}$$

Rewriting the terms, it is seen that

$$\frac{P_o(s)}{P_i(s)} = \frac{1}{1 + sC_h R_h} \tag{4.79}$$

The product $C_h R_h$ is thus the time constant of the system, where R_h is given by Eq. (4.75).

It is necessary to estimate the value of C_h either from the knowledge of the way in which the diaphragm deflects under pressure or by means of an experiment.

Considering the diaphragm to be equivalent to a bellows of spring rate K_s, it can be shown that the increase in volume of the bellows when subjected to a pressure difference is given by

$$\Delta V = \left(\frac{\pi R_b^2 \Delta P}{K_s}\right) \pi R_b^2 = C_h \Delta P \tag{4.80}$$

thereby giving

$$C_h = \frac{\pi^2 R_b^4}{K_s} \tag{4.81}$$

where, R_b is the mean radius of bellows. The time constant $C_h R_h$ of the bellows connected by means of a tubing to a pressure vessel is equal to

$$\tau_h = C_h R_h = \frac{8\pi \mu L}{K_s}\left(\frac{R_b}{R_t}\right)^4 \tag{4.82}$$

It is seen that the time constant is smaller if the connecting tube is smaller in length and larger in internal diameter.

The spring rate of bellows is usually less than that of diaphragms, and hence with less flexible elastic elements, an additional volume of fluid needs to flow into the chamber and, at times, if R_t is smaller than R_b, the fluid flows into the chamber through the capillary at a much higher velocity. Hence the K.E. of the moving fluid in the capillary may be considered to contribute an equivalent additional mass to the moving parts of the transducer and thus lower its natural frequency.

If the piston or the end plate of bellows is moving at a velocity of dx/dt, then the fluid flow rate into the cavity should be such that

$$\pi R_t^2 v_{av} = \pi R_b^2 \frac{dx}{dt}$$

or

$$v_{av} = \left(\frac{R_b}{R_t}\right)^2 \frac{dx}{dt} \tag{4.83}$$

The K.E. of the moving fluid inside the connecting tube is given in Eq. (4.77) from which the equivalent mass M_e, which should be added to the mass M of the piston or the equivalent moving part of bellows, may be obtained as

$$\frac{M_e}{2}\left(\frac{dx}{dt}\right)^2 = \frac{4\pi\rho L v_{av}^2 R_t^2}{6}$$

and substituting the value of v_{av} from Eq. (4.83), we get

$$\frac{M_e}{2}\left(\frac{dx}{dt}\right)^2 = \frac{2\pi\rho L R_b^4}{6R_t^2}\left(\frac{dx}{dt}\right)^2 \tag{4.84}$$

$$M_e = \frac{4\pi\rho L R_b^4}{3R_t^2} \tag{4.85}$$

From Eq. (4.74), the pressure drop across the connecting tube is given by

$$\Delta P = \frac{8\mu L}{R_t^2} v_{av}$$

and hence the viscous force on the piston due to this pressure drop becomes the damping force $B(dx/dt)$. Therefore,

$$\pi R_b^2 \frac{8\mu L}{R_t^2} v_{av} = B\frac{dx}{dt}$$

Substituting the value of v_{av}, B is given by

$$B = 8\pi\mu L\left(\frac{R_b}{R_t}\right)^4 \tag{4.86}$$

Thus the equation of the system is written as

$$(M + M_e)\frac{d^2x}{dt^2} + B\frac{dx}{dt} + K_s x = \Delta P \pi R_b^2 \tag{4.87}$$

The analysis reveals the effect of the moving fluid on the undamped natural frequency of the transducer and the viscous damping it presents. The density and viscosity of the fluid can sufficiently influence the dynamic performance of the transducer.

4.6.1   Liquid Manometer Systems

Liquid filled manometers are used for pressure measurements, and the application of the basic relationships derived so far makes the manometer a second-order system.

The manometer shown in Figure 4.23 is of a U-tube configuration, the two limbs of which are connected to two sources of pressure. If the sources are gas-filled chambers, the inertial and viscous forces arising from gas flow may be ignored in comparison to those due to flow of manometric liquid.

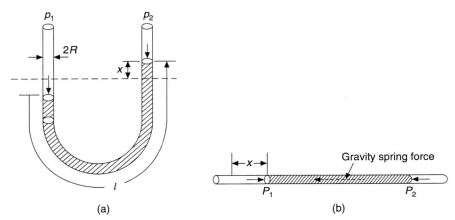

(a)   (b)

Figure 4.23   (a) A liquid-filled manometer system; (b) its equivalent.

When $P_1 = P_2$, the levels of the manometric liquid are the same, and the displacement of each liquid level is treated as zero. When P_1 differs from P_2, the liquid levels suffer a displacement of x but in opposite directions. The net difference of $2x$ in levels sets up a restoring force on the liquid and is given by $2\pi R^2 \rho g x$. The restoring force is equivalent to that of a spring.

The difference in pressure $\Delta P(P_1 \sim P_2)$ sets up the input force, bringing about the difference in levels. This input force is given by $\pi R^2 \Delta P$, where R is the inner radius of the tube.

Under dynamic conditions, the inertial and viscous forces come into existence. Taking v_{av} as represented by $\dot{x}$, the viscous force, using Eq. (4.74), is given by $\pi R^2 (8\mu L/R^2)\dot{x}$. From Eq. (4.77), the equivalent mass of fluid in motion is obtained and the inertial force is thus given by $(4/3)\,\pi \rho L R^2 \ddot{x}$.

Balancing the above forces working on the manaometric liquid, the dynamic behaviour of the system is given by

$$\pi R^2 \Delta P = \pi R^2 \left(\frac{4}{3}\rho L\ddot{x} + \frac{8\mu L}{R^2}\dot{x} + 2\rho g x \right) \tag{4.88}$$

The manometer system is thus seen to belong to a second-order system with its undamped natural frequency ω_n and the damping factor ξ given by

$$\omega_n = \sqrt{3g/2L} \tag{4.89}$$

$$\xi = \frac{2\mu}{R^2 \rho} \sqrt{\frac{3L}{2g}} \tag{4.90}$$

In all cases of manometer systems, the value of ξ works out to be far less than 1 and hence the system is usually very much underdamped.

4.6.2 Pneumatic Systems

The pressure measuring system shown in Figure 4.22 is an example of a pneumatic system when the pressure of a gas vessel is under measurement. The essential difference between the pneumatic and hydraulic systems is that gases possess high compressibility and hence a small volume of gas ΔV admitted into the cavity of volume V raises the pressure of the cavity by ΔP without any increase in the volume of the chamber. The gas in the cavity thus undergoes compression under adiabatic conditions and the adiabatic bulk modulus E_a of the gas is defined as

$$E_a = \frac{\Delta P}{\Delta V/V} = v_s^2 \rho \tag{4.91}$$

E_a is also equal to $v_s^2 \rho$, where v_s is the velocity of sound in the gas and ρ the mass density of the gas. Hence the gas in the cavity of volume V serves as a cushion providing the equivalent of spring-effect.

Other assumptions made for the analysis of liquid systems in Section 4.6.1 are applicable to pneumatic systems as well. Thus, assuming that ΔP_i is the small change in the input pressure, it is desired that the change in the pressure of the cavity or the output pressure change, ΔP_o, be related to ΔP_i which is considered to be relatively much smaller in value than P_i. Its rate of variation with time is sufficiently low so as to make the assumption that a lump of gas travels through the tube slowly. The density and pressure of the gas inside the tube are assumed to be constant because the length of the tube is invariably much smaller than the wavelength of the pressure wave. Also, the equivalent of mass in motion is taken as the mass contained in the tube.

With the above assumptions, the equation for the dynamic equilibrium is set up. When the input pressure rises by ΔP_i from its steady value of P_i, the effect of it is to create a force of $\pi R^2 \Delta P_i(t)$ at the inlet of the capillary tubing.

The effect of this force on the mass of gas contained in the tube at that instant should be studied. The mass of the gas contained in the tube is $\pi R^2 L \rho$, and the entire lump of gas is assumed to suffer a push, the effect of which is to release into the cavity at the other end a mass of gas of volume ΔV given by $\pi R^2 \Delta x(t)$. $\Delta x(t)$ is treated as the net displacement of the lump of gas due to the force created at the input end. As the cavity is assumed to be provided with rigid walls, the effect of the addition of gas of volume ΔV is only to raise its pressure by $\Delta P_o(t)$. From Eq. (4.91), $\Delta P_o(t)$ is given by

$$\Delta P_o(t) = \frac{E_a \pi R^2 \Delta x(t)}{V} \tag{4.92}$$

The inertial force required to make the mass of gas go into motion is $\pi R^2 L \rho (d/dt) \, \Delta x(t)$. The viscous force to be overcome due to its movement with an average velocity of $(d/dt) \, \Delta x(t)$ is

$$8\pi\mu L \frac{d}{dt} \Delta x(t)$$

The balance of the input force and the inertial and viscous forces is $\pi R^2 \Delta P_o(t)$, and the equation relating these forces under dynamic conditions becomes

$$\pi R^2 \Delta P_i(t) - \pi R^2 L \rho \frac{d^2}{dt^2} \Delta x(t) - 8\pi L \mu \frac{d}{dt} \Delta x(t) = \pi R^2 \Delta P_o(t) \tag{4.93}$$

Replacing the derivatives of $\Delta x(t)$ with the derivatives of $\Delta P_o(t)$, from Eq. (4.92), $\Delta P_i(t)$ is given by

$$\Delta P_i(t) = \frac{L\rho V}{\pi R^2 E_a} \frac{d^2}{dt^2} \Delta P_o(t) + \frac{8\mu L V}{\pi R^4 E_a} \frac{d}{dt} \Delta P_o(t) + \Delta P_o(t) \tag{4.94}$$

Thus the connecting tube and the rigid chamber at the receiving end behave as a second-order system with the undamped natural frequency ω_n and damping factor ξ being given by

$$\omega_n = R\sqrt{\frac{E_a \pi}{L\rho V}} \tag{4.95}$$

$$\xi = \frac{4\mu}{R^3} \sqrt{\frac{LV}{\pi\rho E_a}} \tag{4.96}$$

The values of E_a and ρ used in Eqs. (4.95) and (4.96) are valid when the pressure of the gas is P_o and, when P_o deviates from the assumed value, ρ and E_a differ and hence ω_n and ξ cannot be truly taken as constants of the system.

It is clear from the foregoing analysis that the mass of gas contained in the connecting tube is assumed to be incompressible, and its adiabatic bulk modulus is used in respect of the gas contained in the cavity. This is permissible as long as the volume of the connecting tube is much smaller than the volume of cavity. When the volumes are comparable, the above analysis fails and the values of ω_n and ξ get sufficiently modified.

In case the inertial effects of the gas in the tube are ignored, then the system simplifies to a first-order system, with the transfer function relating the increments in the input and output pressures being given by

$$\frac{\Delta P_o(s)}{\Delta P_i(s)} = \frac{1}{1 + s\tau_{pn}} \tag{4.97}$$

where,

$$\tau_{pn} = \frac{8\mu L V}{\pi R^4 E_a}$$

As the pneumatic resistance R_{pn} is given by $8\mu L/\pi R^4$, the pneumatic capacitance $C_{pn} \, (= \Delta V/\Delta P)$ becomes $V/E_a (= V/\rho v_s^2)$.

Taking the connecting tube as equivalent to an electrical transmission line characterized by self-inductance and capacitance of the line, the equivalent of pneumatic self-inductance L_{pn} of the connecting tube is obtained from

$$\omega_n^2 = \frac{1}{L_{pn}C_{pn}} = \frac{R^2 \pi E_a}{L \rho V}$$

and is given by $L_{pn} = L\rho/\pi R^2$.

4.6.3   A Flapper Nozzle System

The case of a stiff-walled chamber of constant volume connected to a pressure source by means of a capillary tube is presented in Section 4.6, and its time constant is derived (Eq. (4.79)) for a hydraulic system. Similarly, it can be shown that the pneumatic system also behaves as a first-order system under certain assumptions, although its time constant is not really a constant but is dependent on other conditions such as temperature and pressure. Both the systems are equivalent to an RC network shown in Figure 4.10.

A similar system is shown in Figure 2.24(a), with the difference that the chamber of volume V is connected to a constant pressure source and the fluid is drawn from the chamber through an outlet (tap). The volume rate of flow of the fluid from the chamber may be varied by adjusting the area of opening of the outlet. Thus the pressure of the fluid in the chamber is governed by the nature and area of opening provided by the inlet and outlet ports.

In the case of a hydraulic system, the chamber can be an open tank, the liquid level of which signifies the pressure of the liquid in the tank, as seen in Figure 4.24(b).

In both the above cases, it is necessary to observe that the load-side restrictor is so adjusted for its opening that the rate at which fluid enters the chamber is always larger than that flowing out; otherwise, the chamber gets emptied finally.

The equivalent electrical network is shown in Figure 4.24(c). If the connecting tube between the constant pressure source and the chamber is a capillary tube, it can be assumed that the supply-side restrictor R_i has a constant value. But the load-side restrictor R_o is variable and may be equivalent to a nonlinear resistor, if it is not a capillary tube. From the equivalent network, the transfer function of the system may be obtained, provided the proper values of R_i, R_o and C are available.

The transfer function of the electrical system is given by

$$\frac{V_o(s)}{V_i(s)} = \frac{R_o}{R_i + R_o} \frac{1}{\left(1 + sC\dfrac{R_o R_i}{R_o + R_i}\right)} \tag{4.98}$$

As supply pressure $V_i(s)$ is held constant, $V_o(s)$ is governed by R_o, and the effective time constant of the system is determined by both R_i and R_o.

Based on the above arrangement, a transducer, popularly known as flapper-nozzle valve, is designed and used for the measurement of minute displacements. Such transducers are designed for use on both hydraulic and pneumatic power supplies. The hydraulic system is designed suitably to withstand high pressures. The pneumatic system is operated at an air

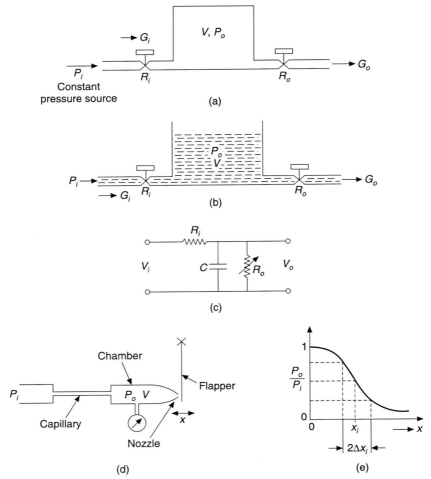

Figure 4.24   (a) A closed storage tank; (b) an open tank; (c) equivalent network; (d) flapper-nozzle value; (e) its characteristics.

pressure of about 13.79 N/cm² (20 psig). As discussed in Section 4.6.2, the development of the transfer function for the pneumatic flapper-nozzle system is more involved and subject to a number of assumptions. The linearization technique is used to derive the incremental transfer function so as to reveal the factors on which the sensitivity and the time constant depend.

The flapper-nozzle valve shown in Figure 4.24(d), consists of a chamber of small volume, connected to a constant pressure source on one side through a capillary tube and on the other side vented to atmosphere, through a nozzle. A flapper is used in front of the nozzle so as to regulate the rate of flow of fluid bleeding out through the nozzle into the atmosphere. When the flapper is tightly held against the nozzle, no fluid leaks out and hence the pressure in the chamber ultimately reaches the valve of supply pressure as governed by Eqs. (4.79) and (4.97). If the flapper is held too far away, it has no effect on the outlet flow rate, and the flow rate from the nozzle is entirely governed by its size and configuration. In such a case, it may so happen that whatever fluid that flows in may flow out and the pressure of the fluid in the chamber finally becomes equal to atmospheric pressure.

As the system is designed for measurement of displacements, the supply-side restrictor and the nozzle are so designed and adjusted that the pressure of the fluid in the chamber (output pressure, P_o) never falls to the atmospheric pressure, even with the flapper held far away from the nozzle. Thus the output pressure P_o is varied over a range as shown in Figure 4.24(e) by the movement of the flapper in front of the nozzle. A small movement of the flapper can result in a large change of P_o, and hence it is treated as a displacement-to-pressure converter. The output pressure of the nozzle-back chamber is measured with a pressure-measuring system.

In the case of a hydraulic valve, the fluid is assumed to be incompressible and the total volume of the chamber and the pressure measuring system (such as bellows) undergoes variation with P_o and position of the flapper. Thus the system is characterized as a first-order system with its time constant derived as per Eq. (4.82).

In the case of a pneumatic valve, the compressibility of gas contributes significantly to its sluggishness.

For deriving the transfer function of the system, the principle of conservation of mass is applied to the volume V of the chamber by considering that the difference between the mass entering the chamber and that leaving the chamber adds to the storage of some mass in the chamber, thereby altering its pressure.

Under steady state conditions, the inlet mass flow rate G_i is equal to the outlet mass flow rate G_o, with the output pressure held at P_o. Assuming that the temperature of the gas remains constant throughout the flow process through the restrictors, the effect of a small displacement ΔX_i of the flapper from its mean position on the output pressure P_o can be evaluated from the knowledge of the values of R_i and R_o. As the flow of compressible gases through restrictions is complex, R_i and R_o can be determined experimentally. R_i is estimated from the curve shown in Figure 4.25(a) and is given by

$$\frac{1}{R_i} = \frac{dG_i}{dP_o}$$

Figure 4.25 Estimation of parameters from experimental curves: (a) Supply-side restrictor; (b) load-side restrictor.

Similarly, R_o is determined from the experimental curve shown in Figure 4.25(b) and is given by

$$\frac{1}{R_o} = \frac{dG_o}{dP_o}$$

The outlet mass flow rate G_o is also dependent on the position of the flapper and is expressed by

$$K = \frac{dG_o}{dX_i}$$

When the flapper is displaced by ΔX_i from its mean position, P_o changes by ΔP_o in a time interval Δt, due to the addition of mass in the chamber.

Expressing the mass of gas added to the chamber as ΔM, it can be shown that

$$\Delta M = \left[\frac{\Delta P_o}{R_i} - \left(\frac{\Delta P_o}{R_o} + K\Delta X_i\right)\right]\Delta t \tag{4.99}$$

The gas is assumed to be perfect, obeying the gas law

$$P_o V = M R_g T_o$$

where,

T_o = temperature of the gas
R_g = gas constant
M = mass of the gas contained in the chamber at the pressure P_o
P_o = pressure of the gas in the chamber of volume V under steady state

As V and T_o are assumed to remain constant, the increase in pressure is given by

$$\Delta P_o = \frac{R_g T_o}{V} \Delta M$$

Using this value of ΔM in Eq. (4.99), the incremental transfer function is given by

$$\frac{\Delta P_o(s)}{\Delta X_i(s)} = \frac{-K}{\left[\left(\dfrac{1}{R_o} - \dfrac{1}{R_i}\right) + s\dfrac{V}{R_g T_o}\right]} \tag{4.100}$$

$$= \frac{-K R_{eq}}{\left(1 + s\dfrac{R_{eq}V}{R_g T_o}\right)} = \frac{-K R_{eq}}{1 + s\tau} \tag{4.101}$$

where,

$$\frac{1}{R_{eq}} = \left(\frac{1}{R_o} - \frac{1}{R_i}\right)$$

$$\tau = \frac{VR_{eq}}{R_g T_o}, \text{ the time constant}$$

The nozzle back chamber is usually of small volume (about 1–5 cc) to hold the time constant at a low value. The time constant becomes more, if the pressure-measuring system is flexible. R_{eq} is variable because of the nonlinearity of R_o. Thus the time constant is dependent on the operating conditions, and for large signals, it assumes a different value.

4.7 FILTERING AND DYNAMIC COMPENSATION

Most of the physical systems may be characterized as frequency-selective in their behaviour since they do not possess the same sensitivity for signals of differing frequencies. The physical systems whose transfer functions are represented by Eqs. (4.11) and (4.24) and whose characteristics are shown in Figures 4.4 and 4.8 have some common features. They possess maximum sensitivity for dc signals and the sensitivity decreases with frequency. When they are used as measuring systems, they are said to be flat in their sensitivity up to a frequency f_c at which the ratio of the dynamic sensitivity to static sensitivity is within a specified value, say, 0.98 or 0.95. The range of frequencies from zero (dc) to f_c constitute the bandwidth, when used as a measuring system, and f_c is termed as the *cut-off frequency.*

The frequency-selective behaviour of such physical systems may be utilized for other purposes such as filtering and dynamic compensation in the field of instrumentation. Filtering is the process of allowing signals of a certain frequency range to pass through while rejecting the rest. The range of frequencies to be passed through is known as *pass-band* and that of the rejected is *reject-band.* The transition from passband to reject-band, if it occurs abruptly at f_c, is called an *ideal* filter and the characteristics of such filters will be similar to 'brick-wall' as shown in Figure 4.26 by solid lines.

The dashed curves are, in fact, characteristics corresponding to practical filters. When dealing with electronic amplifiers and such other equipment, it has come to be understood that the frequency at which the sensitivity falls to $1/\sqrt{2}$ times the maximum value is the cut-off frequency (also called corner frequency). Figure 4.26 shows characteristics of four common types of filters classified broadly as

1. Low-pass
2. High-pass
3. Band-pass
4. Band-reject.

An observation of the characteristics of these filters suggests that low-pass filter is the filter that passes off signals of frequencies ranging from zero to the cut-off frequency f_c, whereas the high-pass filter is that which rejects this band and allows signals ranging from f_c to infinity. Band-pass and band-reject filters are those that allow or reject a band of frequencies, somewhere within zero and infinity. Low-pass and high-pass filters are simple to construct, whereas band-pass and band-reject filters can be slightly complex. A variety of frequency response characteristics may be considered and physical systems designed to realize them. The

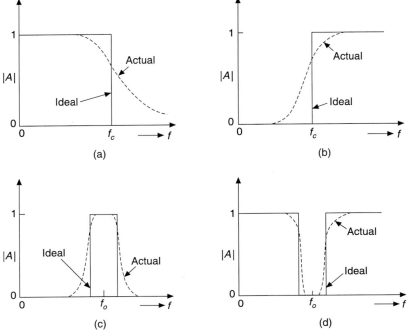

Figure 4.26 Characteristics of ideal and practical filters: (a) Low-pass; (b) high-pass; (c) band-pass; (d) band-reject.

frequency response characteristics of one physical system can be modified by cascading it with another physical system so as to result in a desired form of characteristic and, in such situations, it is understood that the second physical system is used to effect dynamic compensation. Physical systems using mechanical, pneumatic or electrical components may be constructed to perform the task of filtering or dynamic compensation. However, in instrumentation, much of the signal processing is done on electrical signals and hence electrical networks are discussed in Section 4.7.1.

4.7.1   RC Networks as Filters

The simple RC network of Figure 4.10 may be treated as a low-pass filter with its cut-off frequency equal to its natural frequency ω_n defined by $1/\tau$. The gain of the filter, when expressed in dB, is equal to zero at zero frequency (dc) and for frequencies beyond f_c, the slope of the characteristic is -6 dB per octave or -20 dB per decade. If the characteristic is plotted between gain in dB and frequency (on logarithmic scale), the characteristic may be approximated to the form shown in Figure 4.27 such that it is flat up to f_c, and beyond, it has a slope of -6 dB per octave. The characteristics thus approximated and obtained are known as "Bode plots" and are used for quick comparison of filter performance and design purposes. The approximated characteristic has an error of $+3$ dB at the cut-off frequency, and errors at other neighbouring frequencies may be computed and used, where necessary. Low-pass and other filters are thus characterized by

1. pass-band frequencies,
2. sensitivity (or gain in dB) at cut-off frequency (or frequencies), and
3. slope of the characteristic (rate of attenuation) in the transition region.

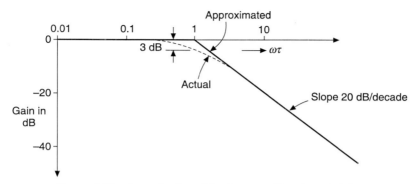

Figure 4.27 Approximation of the low-pass filter characteristic.

The simple RC filter network of Figure 4.10 is the simplest low-pass filter and is known as the first-order type, being characterized by a single time constant. If two such networks are cascaded together, as shown in Figure 4.11, it can be seen that the filter characteristic is improved in its slope, although the pass band is reduced. While designing filters, attention is paid, in most of the cases, to obtain the gain-frequency characteristic, in the desired form so that the unwanted signals having frequencies other than those of wanted signals are contained or suppressed to the desired extent. The phase-frequency response characteristic is usually not accorded much importance in the design of filters. Several methods of approximating the desired gain-frequency characteristic are developed and corresponding filters are designed. They invariably become higher-order filters, as the power of *s* in the denominator of the filter transfer function becomes higher than 2. The filters thus designed are broadly classified as *passive* or *active*. Passive filters are those that are constructed by using the three electrical elements *R*, *L* and *C*, whereas the active filters consist of mostly resistors, capacitors and high gain op-amps. In active, filters, the use of *L* as an element is totally restricted, since it is not a pure element and since it is a source of noise in the network in which it is used.

The positions of *R* and C when interchanged result in the simple high-pass filter shown in Figure 4.28(a). Its transfer function may be shown to be

$$\frac{V_o(s)}{V_i(s)} = \frac{s\tau}{1+s\tau} \tag{4.103}$$

with the result that the frequency response characteristic is flat for frequencies above f_c and has the same slope of -6 dB per octave for frequencies lower than f_c as shown in Figure 4.28(b).

It is possible to make use of one low-pass filter and one high-pass filter, each having a different cut-off frequency from the other, so that the combination has the desired characteristics of a band-pass or band-reject filter. Figure 4.29(a) shows that the cut-off frequency of the low-pass filter is higher than the cut-off frequency of the high-pass filter, thereby yielding a band-pass filter, with the band-pass frequencies lying between their cut-off frequencies. Figure 4.29(b) shows how the band-reject characteristic is similarly obtained by cascading one low-pass and one high-pass filter.

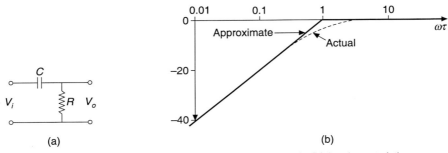

(a) (b)

Figure 4.28 (a) A simple RC high-pass filter network; (b) its characteristic.

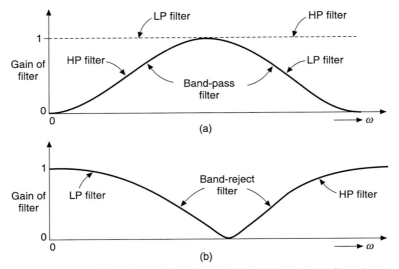

Figure 4.29 (a) Band-pass filter characteristic; (b) band-reject filter characteristic.

There are situations where a signal of a certain desired frequency alone is desired to be suppressed (or rejected totally) or passed through. Tuned LCR circuits of the series or parallel type may be used to pass signals of one frequency and attenuate other frequency components to a large extent, by having them sharply tuned. Similarly, bridged-T and twin-T networks can be used as filters that suppress signals of one frequency to a large extent. Such filter networks are known as narrow band-pass or narrow band-reject filters.

The bridged-T filter is shown in Figure 4.30(a) and its transfer function may be obtained as

$$\frac{V_o(s)}{V_i(s)} = \frac{R_1 R_2 C_1 C_2 s^2 + C_1(R_1 + R_2)s + 1}{R_1 R_2 C_1 C_2 s^2 + [C_1(R_1 + R_2) + R_1 C_2]s + 1} \tag{4.104}$$

The network yields a minimum value for signals of a frequency ω equal to $\omega_n = \dfrac{1}{\sqrt{R_1 R_2 C_1 C_2}}$,

as shown in Figure 4.30(b). The slope of the characteristic and the value of the function at $\omega = \omega_n$

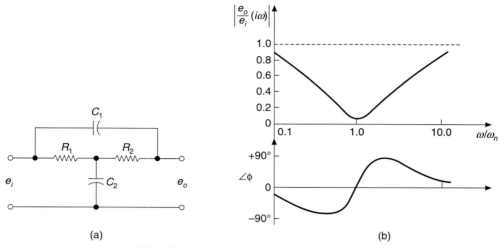

(a) (b)

Figure 4.30 (a) A bridged-T filter network; (b) its characteristic.

depend on the component values. The network does not have much scope of application for obvious reasons.

The twin-T filter is shown in Figure 4.31(a) and it is usually favoured when a particular frequency component is required to be totally suppressed. The transfer function is more complex, showing that it forms a third-order system, with its numerator and denominator being cubic polynomials.

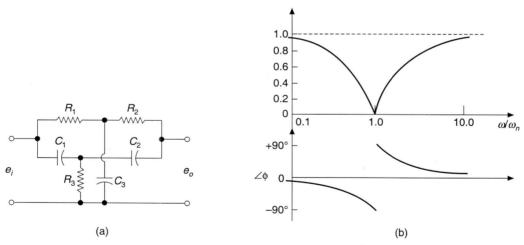

(a) (b)

Figure 4.31 (a) A twin-T filter network; (b) its characteristic.

The filter is usually simplified by making $R_1 = R_2$ and $C_1 = C_2$ so that it becomes a symmetrical network. The response characteristic is shown in Figure 4.31(b), showing that a sharp null can be obtained at the frequency of rejection. Due to the shape of the characteristic, the filter is called *notch filter*.

For common applications, $C_1 = C_2$ and $R_1 = R_2$, and a sharp null at the rejection frequency is obtained by making $C_3 = 2C_1$ and $R_3 = R_1/2$. The rejection frequency is given by

$$\omega_n = \frac{1}{R_1 C_1} \qquad (4.105)$$

4.7.2 Dynamic Compensation

Widely differing gain-frequency characteristics can be obtained by using resistors and capacitors, and such networks are used for dynamic compensation of system behaviour. Three such simple networks are shown in Figure 4.32.

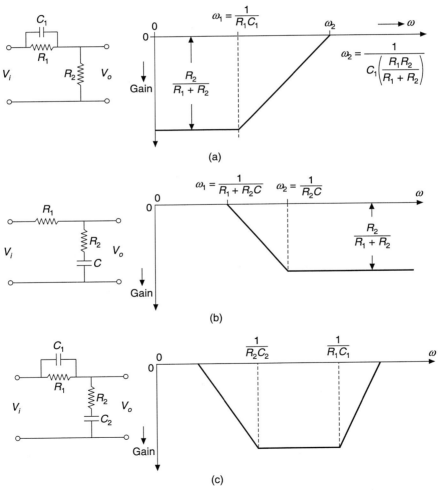

Figure 4.32 RC networks for dynamic compensation: (a) Lead network; (b) lag network; (c) lead lag network and their response characteristics.

The network shown in Figure 4.32(a) is usually known as *lead network,* whose transfer function can be obtained as

$$G_2(s) = \frac{V_o(s)}{V_i(s)} = \frac{R_2}{R_1 + R_2} \frac{1 + sC_1R_1}{1 + s\left(\dfrac{R_2}{R_1 + R_2}\right)C_1R_1}$$

$$= \alpha \frac{1 + s\tau_1}{1 + s\alpha\tau_1} \qquad (4.106)$$

where,

$$\alpha = \frac{R_2}{R_1 + R_2}$$

If $V_i(s)$ is the output signal of a first-order physical system (or a transducer system) having its transfer function given by

$$G_1(s) = \frac{V_i(s)}{X(s)} = \frac{K}{1 + s\tau_1}$$

then, it is possible to cascade the physical system and the network such that the overall transfer function $G(s)$ becomes equal to $G_1(s)G_2(s)$. Hence,

$$G(s) = \frac{K\alpha}{1 + s\alpha\tau_1} \qquad (4.107)$$

The combination is now seen to possess a smaller time constant although the static sensitivity has become as much less as the time constant. By a proper choice of R_1 and R_2, the value of α may be obtained and any loss of sensitivity may be compensated by using amplifiers at a later stage. Such a process of compensating the sluggishness of a physical system and speeding its dynamic performance is known as *dynamic compensation.*

The network shown in Figure 4.32(b) is referred to as lag network and has its transfer function given by

$$G_2(s) = \frac{V_o(s)}{V_i(s)} = \frac{1 + sR_2C_2}{1 + sC_2(R_1 + R_2)} \qquad (4.108)$$

The network of Figure 4.32(c) is a combination of the above two networks effected in such a way as to result in a lag-lead network. The transfer function can be obtained as

$$G_2(s) = \frac{V_o(s)}{V_i(s)} = \frac{(1 + sC_1R_1)(1 + sC_2R_2)}{[1 + sC_2(R_1 + R_2)]\left[1 + sC_1R_2\left(\dfrac{R_2}{R_1 + R_2}\right)\right]} \qquad ((4.109)$$

The corresponding asymptotic gain-frequency characteristic of each network is presented alongside the network.

All the above networks are shown to be composed of resistors and capacitors only as introduction of an inductive element causes problems by virtue of the magnetic field it creates as well as by its induced emfs due to external fields. RC networks, however complicated they happen to be, cannot yield output voltage larger than the input voltage and hence the voltage gain varies between zero and unity. Also, the transfer functions derived are based on the assumption that they are excited by ideal voltage sources and the output terminals are virtually under open-circuit conditions. Where higher gains are desired, they are used in combination with high-gain amplifiers (or op-amps), and a variety of filters can be configured in several ways so as to yield the desired characteristics for magnitude and phase. Some simple active filters are described in Section 9.5.

EXERCISES

1. What is meant by order of a system and what does it signify?

2. What are the characteristics of a zero-order system and how does it react to standard test input signals?

3. Derive the transfer function of a first-order system and identify the constants by which it is characterized.

4. Define *time constant* of a first-order system and explain how its dynamic performance is affected when it is subjected to a step-input function.

5. Define the *dynamic error* of a first-order system and derive the expressions for the same when it is subjected to standard test input signals.

6. Obtain the expression relating the dynamic and static sensitivities of a first-order system and show that the ratio varies with frequency.

7. Discuss the merit of testing a first-order system by applying a unit-impulse input function and show how the output varies with time.

8. Define the *undamped natural frequency* of a second-order system and show how its natural frequency is affected by damping.

9. Define the *damping factor* of a second-order system and show how its response varies with time when subjected to a step-input function under different damping conditions.

10. Define the *settling time* of physical systems and show how it varies with damping factor for second-order systems.

11. What is meant by *critical damping* conditions of a system? Explain why certain physical systems are operated under critical damping conditions.

12. Discuss the performance of a second-order system when subjected to a (i) step input, and (ii) sinusoidal input function. Suggest suitable values for the damping factor at which the system is preferable to operate in each case.

13. What is a *terminated ramp input* function? Indicate the relevance of analyzing the performance of systems for such input functions.

14. Show the response curves for a second-order physical system subjected to an impulse test input function.

15. Discuss the nature of problems faced when the order of a physical system is larger than two.

16. Discuss the performance of two first-order systems when cascaded. Will the combination become an underdamped system?

17. Draw the equivalence between the R-L-C electrical networks and the mechanical mass-spring-dashpot systems and show how they are connected together to constitute second-order systems.

18. Identify zero-order elements used in electrical and mechanical systems.

19. Explain the need for using gear trains in mechanical systems and show how the load member connected to the gear train is reflected to the driving side.

20. Explain the presence of electromagnetic damping in the case of electromechanical systems and the factors on which it depends.

21. Obtain the transfer function relating the speed and applied voltage for the case of a dc motor and identify its time constant.

22. Show that a motor having negative slope between torque and speed has positive damping conditions.

23. Define *thermal resistance* and *thermal capacitance* of a thermal system and derive the time constant of a thermal system.

24. Explain why the time constant of a thermometer system is not really a constant.

25. Draw the analogy between electrical network elements and the parameters of a fluid-flow system.

26. Show that a capillary tube is equivalent to a resistor in electrical circuits and obtain the expression for the 'hydraulic resistance' of a capillary tube.

27. Define *hydraulic capacitance* and show how it can be evaluated for the simple element of a bellows.

28. Show that liquid-filled manometers behave as underdamped second-order systems.

29. Discuss the essential differences in the behaviour of hydraulic and pneumatic systems.

30. Show that a long connecting tube of a chamber used for storage of gases makes the entire system behave like a second-order system.

31. Describe a flapper-nozzle system and indicate its basic principles of operation and application.

32. Obtain the transfer function of a flapper-nozzle system operated on air supply and show how its time constant is minimized by design.

33. Discuss the role of filters in electrical systems and show how they are classified.

34. Explain what is meant by dynamic compensation and its importance when dealing with transducers.

35. Draw the schemes of lead and lag networks and derive their transfer functions. Show how their amplitude-frequency response characteristics look like.

36. Draw the lead-lag network and show how its amplitude-frequency response characteristic looks like.

37. Distinguish between active and passive filters and their relative merits in performance and configuration.

38. Explain why self-inductors are not used in filter networks?

SUGGESTED FURTHER READING

Boros, A., *Electrical Measurements in Engineering*, Elsevier, Amsterdam, 1985.

Cannon, Robert, H., Jr., *Dynamics of Physical Systems*, McGraw-Hill, Kogakusha, Tokyo, 1967.

Doebelin, E.O., *Measurement Systems: Application and Design*, McGraw-Hill, New York, 1966.

Gibson, John, E. and Tutuer, Frank, B., *Control System Components*, McGraw-Hill, Kogakusha, Tokyo, 1959.

Harris, F.K., *Electrical Measurements*, John Wiley, New York, 1959.

Holman, J.P., *Experimental Methods for Engineers*, McGraw-Hill, Kogakusha, Tokyo, 1966.

Jones, Barry. E., *Instrumentation, Measurement and Feedback*, Tata McGraw-Hill, New Delhi, 1978.

Macfarlane, A.G.J., *Engineering System Analysis*, Harrap, 1964.

Nagrath, I.J. and Gopal. M., *Control System Engineering* (2nd ed.), Wiley Eastern, New Delhi, 1982.

Van Valkenburg, M.E., *Network Analysis*, 3rd ed., Prentice-Hall of India, New Delhi, 1983.

CHAPTER 5

Mechanical Transducers

5.0 INTRODUCTION

Industrial instrumentation is concerned mostly with the measurement of a large number of mechanical quantities such as temperature, pressure, flow-rate of fluids, and density. In vehicular transportation systems, measurement of other quantities such as speed, velocity, acceleration, attitude etc. is essential. Accuracy of control of the above quantities cannot be achieved unless they are measured with a high degree of accuracy. Under steady state conditions, measurement of these quantities is carried out, since long, by mechanical instruments. In most of these instruments, it is possible to identify a primary sensing element or a sensor that comes in contact with the system and responds to the condition of the system under measurement. At times, the primary sensing element in combination with a few elements, may constitute a primary measuring system. The response of such primary devices is utilized for indication and record. Their output signals may be mechanical in nature and they are in most cases, either displacement, change of pressure or force. These output signals may be further processed for purposes of indication and recording.

Mechanical transducers are those primary sensing elements that respond to changes in the physical condition of a system and deliver output signals related to the measurand, but of a different form and nature. For example, a bimetallic element reacts to changes in temperature and its response is noticeable in the form of mechanical displacement. It is seen as a direct effect, without any other intermediate signal being available for processing. Mechanical transducer systems may also be constituted in such a way as to render the output signal in the desired form and at the desired level for further processing. In such a case, the transducer system consists of additional stages further to the sensing element for processing the output signal of the sensing element. All of them are broadly classified as primary transducers as they respond first to the nonelectrical quantity under measurement.

The mechanical transducers are distinguished from the electrical transducers since their output signals are also mechanical by nature. Electrical transducers respond to nonelectrical quantities but develop output signals, which are electrical by nature. Resistance thermometer element, for example, is an electrical transducer as a change in temperature is recognized as a change in the resistance of the element. For each measurand, it is possible to suggest both mechanical and electrical transducers, but it is always desirable to minimize the number of stages of conversion with an eye on reduction of total error in measurement.

In this chapter, a brief review of the primary transducers and transducer systems is presented and it may be noted that they are all mechanical by nature, i.e. their output signals are linear or angular displacements, mechanical strain or pressure. They are presented with a view to complete the literature, although they may not offer always the best means for measurement. An intelligent selection of a primary transducer needs to be made along with the proper choice of a secondary transducer that delivers output signal (which is electrical by nature) and that satisfies the demands of the instrumentation system. It may so happen that, in certain cases, it is not necessary to have any mechanical transducer for a certain nonelectrical quantity, as is the case of temperature measurement by means of a thermocouple. In environments such as those of chemical processes and petroleum refineries, mechanical transducers are preferred due to the existence of fire hazards with electrical systems. In such cases, pneumatic transducers with air-pressure variations as output signals are preferred and used, although they may be slow in operation. Some of the mechanical transducers and transducer systems presented in this chapter may continue to dominate the present industrial scene for one reason or another and may be of some help in effecting modernization of the instrumentation systems to meet the present day needs, without entirely rejecting the existing primary transducer systems. Table 5.1 presents the range of primary mechanical transducers and transducer systems, available for the common mechanical quantities, and the nature of output signals developed by them.

5.1 BASICS OF TEMPERATURE MEASUREMENT

The temperature of a body is one of the fundamental parameters by which its degree of hotness or coldness is identified. When a substance acquires or releases heat, it is reflected through some changes in its physical dimensions and state, which constitute the primary means by which temperature measurement is carried out.

The substances employed may be solids, liquids or gases and the instruments developed are widely known as bimetal-strip, liquid-in-glass, vapour-pressure and gas thermometers. While measuring temperature, they happen to develop output signals such as mechanical displacement or pressure, and these can be further converted into electrical signals. These basic mechanical instruments may be treated as the primary mechanical transducer systems.

5.1.1 Absolute Thermodynamic or Kelvin Scale

Lord Kelvin defined a scale of temperature in terms of the mechanical work which may be obtained from a reversible heat engine working between two temperatures, and which does not depend on the properties of a particular substance. If Q_1 is the heat given up by the source at temperature T_1 and Q_2, the heat received by sink at temperature T_2, for the heat engine working on the ideal Carnot cycle, $Q_2/Q_1 = T_2/T_1$ (applying the second law of thermodynamics). Temperature scale defined on this basis is known as *thermodynamic temperature scale*. The unit of thermodynamic temperature, so defined, is the Kelvin (K). The Kelvin is defined as 1/273.16 of the temperature of the triple-point of water which is +273.16 K. The thermodynamic temperatures are realized by using the gas law $PV = mRT$, obeyed by ideal gases, where P is the absolute pressure, V is the volume of gas of mass m, R is the gas constant of the gas, and T is the temperature on the absolute or thermodynamic (or Kelvin) scale. Constant volume gas thermometers, using hydrogen for temperatures up to 500°C and nitrogen from 500°C to 1500°C, are constructed and thermodynamic temperature values are established and compared with values in the practical temperature scale.

5.1.1.1 *The International Practical Temperature Scale*

The gas thermometer using gas laws, though useful as the standard of reference, is complex and cumbersome to operate. Hence for standardization of thermometers in laboratories, 11 basic fixed points assigned as standards of temperature are established and thereby an International practical temperature scale (IPTS) is provided.

Table 5.1 Mechanical transducers and transducer systems

Mechanical quantity to be measured	Description of transducer/ Transducer system	Output Signals			
		Displacement	Force/ Torque	Strain	Pressure
1. Temperature	Bimetallic element	√	√		
	Fluid expansion systems	√	√		
2. Pressure	U-tube and bell type manometers	√			
	Ring balance manometer	√			
	Metallic diaphragms	√		√	
	Capsules and bellows	√			
	Bourdon tubes	√			
	Membranes	√			
3. Force (weight)	Spring balance	√		√	
	Cantilever	√		√	
	Diaphragms	√		√	
	Pneumatic and hydraulic load cells				√
	Column and proving ring load cells	√		√	

(Contd.)

Table 5.1 Mechanical transducers and transducer systems (*Contd.*)

Mechanical quantity to be measured	Description of transducer/ Transducer system	Output Signals			
		Displacement	Force/ Torque	Strain	Pressure
4. Torque	Torsion bar	√		√	
	Flat spiral springs	√			
	Dynamometer		√	√	
	Gyroscope	√			
5. Density of liquids	Hydrometer	√			
	Air bubbler system				√
	U-tube weighing system		√		
6. Liquid level	Float elements	√	√	√	
	Manometer system	√			
	Diaphragms	√		√	
	Container weight		√		√
7. Viscosity	Capillary tube				√
	Concentric cylinder system		√		
8. Flow rate of fluids	Pitot static tube				√
	Flow-obstruction elements			√	√
	Rotating vane system	√			
	Rotameter float system	√			
9. Displacement	Flapper nozzle system				√
10. Absolute displacement, velocity and acceleration	Seismic system	√			
11. Vehicle attitude	Gyroscope	√	√		

√ Signifies the availability of the output signal for the particular transducer or transducer system.

The boiling and freezing points are with reference to the standard pressure of 101.325 kN/m^2 (= 760 mm of Hg). This scale conforms closely to the thermodynamic scale and is easily reproducible.

The IPTS defines the Celsius temperature (θ) as

$$\theta = T - T_0$$

where T is the thermodynamic temperature in Kelvin units and $T_0 = 273.16$ K.

The methods by which the intervals between the fixed points are accurately established for equivalence are identified and utilized for standardization and calibration. Figure 5.1 indicates the fixed points of the IPTS (1968).

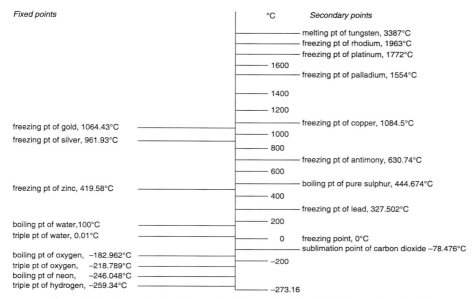

Figure 5.1   Some fixed points on the International Practical Temperature Scale.

5.1.2   Bimetallic Element

A bimetallic element consists of strips of two metals of differing coefficients of thermal expansion, bonded or welded together to form a single piece. At the temperature at which they are bonded, it is a flat and straight strip, but when it is heated to a higher temperature, it starts curling or bending towards the side of metal with lower coefficient, as shown in Figure 5.2(a). The layer with higher coefficient α_1 is called the active side, while the other with lower coefficient is known as the passive side.

The element is fixed at one end and kept in the environment whose temperature is to be measured. A rise in temperature by $T°C$ causes the element to bend into an arc of a circle. The angle ϕ between the original plate axis and the tangent drawn from the tip to the deflection line is given by

$$\phi = \frac{6E_1E_2t_1t_2\,(t_1+t_2)(\alpha_1-\alpha_2)}{4E_1E_2t_1t_2\,(t_1+t_2)^2+(E_1t_1^2-E_2t_2^2)}\,L\Delta T \tag{5.1}$$

$$= k_T L\Delta T \text{ (radians)} \tag{5.2}$$

where,

E_1, E_2 = Young's moduli of the materials of layers 1 and 2, respectively

t_1, t_2 = thicknesses of layers 1 and 2

α_1, α_2 = coefficients of thermal expansion of the materials of layers 1 and 2, respectively

L = length of the original element

ΔT = rise in temperature

k_T = bimetallic element sensitivity in radians per m/°C

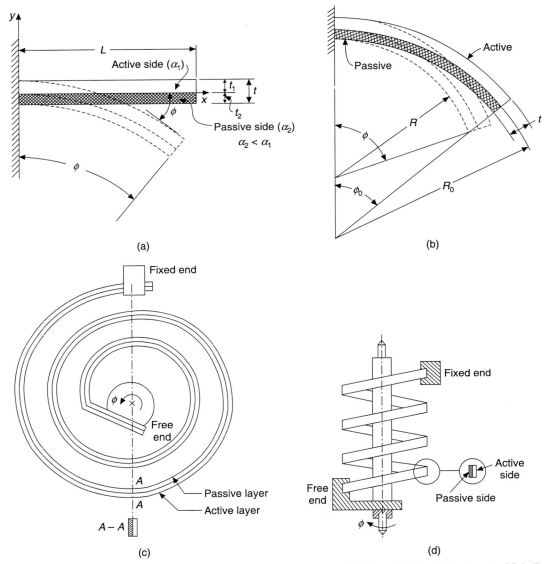

Figure 5.2 (a) Flat bimetallic element; (b) curved bimetallic element; (c) flat-spiral bimetallic element; (d) helical bimetallic element.

The sensitivity assumes a maximum value when the term $(E_1 t_1^2 - E_2 t_2^2)$ becomes zero, or when

$$\frac{t_1}{t_2} = \sqrt{\frac{E_2}{E_1}} \tag{5.3}$$

Under these conditions, k_T is given by

$$k_T = \frac{3}{2} \frac{\alpha_1 - \alpha_2}{t_1 + t_2} = \frac{3}{2} \frac{\alpha_1 - \alpha_2}{t} \tag{5.4}$$

where t is the thickness of the element $(t_1 + t_2)$.

The deflection (d) of the tip of the element, from its original position, can be shown to be given by

$$d = k_T \frac{L^2}{2} \Delta T \tag{5.5}$$

It must be noted that Eq. (5.5) is true as long as the element is not under any load along its length up to the tip.

If the original shape of the bimetallic element is circular, as shown in Figure 5.2(b), the deformation is measured in terms of the difference $\Delta \phi_T$, which is $(\phi - \phi_0)$ and is given by

$$\Delta \phi_T = k_T L \, \Delta T \tag{5.6}$$

To obtain higher sensitivity, the curved bimetallic element is extended to form (i) a flat spiral (Figure 5.2c), and (ii) a helical spiral (Figure 5.2d). In each case, the thickness of the element t is kept much smaller than the radius of curvature. They produce an appreciable angular deflection of the free end, about 270°. The long bimetallic ribbon in the case of helical spiral is wound without break into several compensated helices, arranged coaxially one within the other so as to compensate for lateral displacements of the spindle connected to its free end. In fact, the spindle, when carrying a disc known as shaft angle encoder, enables the measurement of temperature directly in digital fashion. Both the flat and helical spirals develop angular deflections proportional to temperature.

All the elements are brought out in various sizes and ranges by choosing proper materials. They are designed for temperatures up to 600°C. The passive element is, in most cases, of Invar, and the active element is of stainless steel or brass. The thickness of active layer is made higher because of the lower value of Young's modulus of the active side material. Changes in the value of sensitivity k_T occur because of variation of thermal coefficient with temperature and usage. However, the bimetallic temperature sensors are accurate up to ±1% and their repeatability and linearity are high enough for use in industry.

5.1.3 Fluid Expansion Systems

Thermometers based on the expansion of liquids and gases have been in use since long both in laboratories and the industry. Liquid-in-glass thermometers consist mainly of a glass bulb containing the liquid and a stem of glass tubing with fine capillary bore. The level of liquid in the glass tube is indicative of the temperature under measurement. The level changes in the stem can be converted into an electrical signal by using a suitable transducer. For industrial application, the glass bulb and the capillary are replaced by a steel bulb and capillary. The change in volume of mercury or the filling liquid is sensed by a Bourdon tube element

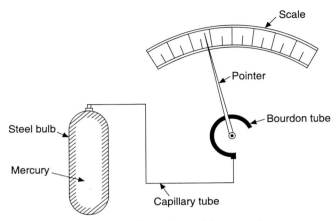

Figure 5.3   Mercury-in-steel thermometer.

connected to the capillary as shown in Figure 5.3. The bulb, the capillary and the Bourdon tube are completely filled with the liquid, at a high pressure. When the liquid in the bulb expands, the Bourdon tube uncurls and indicates the temperature. The deformation or the displacement signal of the Bourdon tube element may form the means for further signal conversion.

Similarly, the output-displacement signals of the Bourdon tube element connected to the capillary tube of the vapour pressure thermometer can be utilized for further transduction in instrumentation systems.

Table 5.2 presents the liquids normally used for the above two types of thermometers and their range.

Table 5.2   Thermometer liquids and ranges

Type of thermometer	Liquid	Range
1. Liquid-in-steel bulb thermometers	Mercury	$-40°C$ to $+650°C$
	Xylene	$-40°C$ to $+400°C$
	Alcohol	$-45°C$ to $+150°C$
	Ether	$+20°C$ to $+90°C$
2. Liquid-in-glass thermometers	Mercury	$-35°C$ to $+500°C$
	Alcohol	$-80°C$ to $+70°C$
	Toluene	$-80°C$ to $+100°C$
	Pentune	$-200°C$ to $+30°C$
	Creosote	$-5°C$ to $+200°C$
3. Vapour pressure thermometers	Water	$+120°C$ to $+220°C$
	Toluene	$+150°C$ to $+250°C$
	Ethyl Chloride	$+30°C$ to $+100°C$
	Methyl Chloride	$0°C$ to $+50°C$
	Argon	Very low temps
	Sulphur dioxide	$+30°C$ to $+120°C$

5.2 BASICS OF PRESSURE MEASUREMENT

Measurement of pressure assumes great importance at all stages of scientific and engineering measurements. The techniques available for measurement may be divided into three main types: those meant for pressures (i) above the atmospheric pressure and relative to it (known as gauge pressure), (ii) lower than the atmospheric pressure and extending to very low values, designated as negative gauge pressure or vacuum, and (iii) very high pressure expressed in atmospheres. Absolute pressure is with respect to absolute vacuum and is equal to the force per unit area exerted by a fluid on the walls of its container. It is indicated by the subscript 'abs', as for example, P_{abs} (or P_a). Differential pressure is the net force per unit area exerted by the fluids present on either side of the wall separating them. Gauge pressure is indicated by the subscript g, as for example, P_g. All pressure gauges are normally calibrated to read pressures higher than atmospheric pressure and read zero, when test pressure is exactly equal to standard atmospheric pressure.

Atmospheric pressure is to be expressed in N/m^2, while in practice it is referred to the height of the mercury column of a barometer. Standard atmospheric pressure is the pressure due to a column of mercury, 76 cm high at 0°C at sea level. It is equal to 101.325 kN/m^2.

In meteorology, pressure is expressed in *bars,* one bar being a pressure of 10^5 N/m^2 (10^6 dynes/cm^2). A smaller unit is the microbar. Very low pressure and vacuum are generally expressed in units of *micron,* one micron being equal to the pressure exerted by a column of mercury of 1 μm, while a *torr* refers to a column of mercury of 1 mm.

Conventional mechanical instruments are based on two distinct methods:

1. Direct measurement by comparison with the pressure due to a column of liquid of known density.

2. Indirect measurement of the pressure, by measuring the effect of application of pressure over a known area of an elastic element, such as the flat diaphragm or Bourdon tube.

In most of the above methods, the levels of liquid columns and the deformation and strains developed in the elastic elements can be easily measured by using the secondary electrical transducers for displacement and strain measurements. In the following sections, all those primary mechanical transducers and systems that enable pressure measurements in conjunction with the secondary electrical transducers are briefly presented. Table 5.3 lists some of the elastic materials and their properties.

5.2.1 Manometers

1. The single-tube manometer is in fact, a *piezometer* connected to a pipe at the point where the static pressure of the fluid is desired. The connection is shown in Figure 5.4(a). The gauge pressure P_g at A is given by ρgh, where ρ is the mass density of the fluid in the pipe and g, the acceleration due to gravity. The piezometer utilizes the fluid, under measurement and hence is unsuitable for gases.

2. The twin-tube *manometer* employs tubes of equal cross-section, and is connected together to form a U-tube, as shown in Figure 5.4(b). It is used for measurement of gas pressure or liquid pressure, provided the liquid does not mix or react chemically with the

Table 5.3   Properties of elastic materials

Material and composition	Young's modulus, E (Nm^{-2}) × 10^{-9}	Modulus of rigidity, G (Nm^{-2}) × 10^{-9}	Coefficient of linear expansion, α (m/m/°C) × 10^6	Thermal coefficient of E $\left(\frac{\Delta E}{E}\middle/°C\right)$ × 10^5	Thermal coefficient of G $\left(\frac{\Delta G}{G}\middle/°C\right)$ × 10^5	Remarks
1. Spring steel (Fe, C, Mn)	210	80	11.8	−24	−24	Used for springs and membranes
2. German silver (Cu, Zn, Ni)	110	38	16.2	−35	−37	Corrosion resistant, used for springs, helical and flat.
3. Copper rolled	105	40	16.7			Used for soft springs and ligaments
4. Duralumin rolled	68	27	22.6			Used for proving rings
5. Phosphor bronze (Cu, Sn)	105	45	17.8	−36	−40	Corrosion resistant, good electrical conductor, used for instrument springs, Bourdon tubes and diaphragms
6. Aluminium rolled	95		22.9			Used for proving rings.
7. Beryllium copper (Cu, Be)	110–130	40–50	16.6	−35	−33	Low hysteresis, good electrical conductor, used for instrument springs, Bourdon tubes and diaphragms
8. Invar (Ni, Fe)	147	56	1.08	+4.8		Low thermal expansion used for instrument springs
9. Elinvar (Ni, Cr, Si, W, Mn, Fe, C)	148	57	0.6	−0.66	−0.72	Low thermal expansion used for instrument springs
10. Isoelastic (Ni, Cr, Mn, Fe)	180	64	7.2	−3.6 to 2.7		Stable properties
11. Ni-span C (Ni, C, Mn)	165–185	70	8.1	−1 to +1		Stable properties
12. Constantan (Cu Ni)	156	60	17			Used for hard springs
13. Platinum			8.9			Stable properties
14. Stainless steel			11.0			Used for diaphragms, bellows and membranes
15. Rubber	0.008					Used for membranes
16. Bronze			17.5			Used for bellows

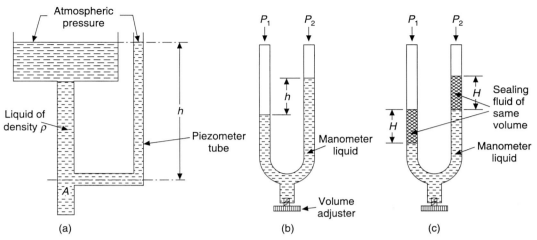

Figure 5.4 (a) Piezometer tube; (b) U-tube manometers for direct use; (c) U-tube manometer with sealing fluid.

manometric liquid. The manometric liquid is mercury in most cases or water for low gauge pressures. The pressure difference $P_1 - P_2$, is, as before, given by $\rho g h$, where ρ is the mass density of the manometric liquid.

When it is necessary to measure pressure of a liquid that must not come in contact with the manometer liquid, a sealing liquid is used as a buffer, in each limb, on top of it as shown in Figure 5.4(c). The sealing liquid must be lighter than the manometric liquid but heavier than the fluid under test. Equal quantity of the sealing liquid is added to each limb of the manometer so that the two amounts stand balanced.

3. The well- or cistern-type manometer is widely used because of its convenience in requiring reading of only one level of the manometric liquid, instead of the two levels of the U-tube type. This system shown in Figure 5.5(a) is favoured because one secondary transducer located on the narrow-section tube serves the purpose of pressure measurement. The level of the liquid in the well having its cross-section A, much larger than that of the other tube a, remains practically the same, unaffected by pressure variation in P_1. The difference in pressure, ΔP, is given by

$$P_1 - P_2 = \Delta P = h\rho g\left(1 + \frac{a}{A}\right)$$

$$\approx h\rho g \quad \text{when } a << A \tag{5.7}$$

To increase sensitivity in the measurement, the narrow tube may be kept tilted at an angle β with the vertical axis. With the length of the column measured as shown in Figure 5.5(b), the difference in pressure, ΔP, can be shown to be given by

$$\Delta P = R\rho g\left(1 + \frac{a}{A}\right)\cos \beta \tag{5.8}$$

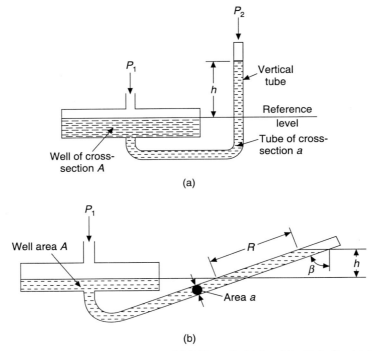

Figure 5.5   Well-type manometers: (a) With vertical tube; (b) with inclined tube.

Table 5.4 indicates some of the manometric liquids and their uses.

Table 5.4   Manometric liquids

Liquid	*Sp. gravity at 20°C*	*Remarks*
1. Water	1.0	Useful for small pressure differential only. There is problem of evaporation.
2. Transformer oil	0.864	Useful for small pressure differential. No problem of evaporation exists.
3. Dibutylphthalate	1.047	Does not mix with water.
4. Carbon tetrachloride	1.605	Problem of high evaporation rate exists.
5. Tetrabromoethane	2.964	Evaporates slowly.
6. Mercury	13.56	Does not mix with other liquids, or evaporate. Does not wet the sides of the tube. Used for high pressure differential.

5.2.2   Ring-balance Manometer

The ring-balance type manometer is slightly different in construction and is widely used for the measurement of low differential pressures arising in gas-flow metering. It mainly consists of a hollow annular ring of circular section, made of either a light alloy of aluminium or of a light transparent material such as polythene. The ring is supported at its centre on a knife-edge groove or any such friction-free system as seen from Figure 5.6. Two pressure-measuring chambers are formed by a partition at its upper end, and connection with the test pressures is established at the two inlet ports, through flexible tubing shown on either side of the partition. The manometer liquid, filling partly the lower portion of the ring, serves as the lower partition. When $P_1 = P_2$, the ring is balanced such that the control weight is in its lowest position. When there is a pressure difference, it sets up a rotating moment and the ring assumes a new position where the rotating moment is balanced by the controlling moment set up by the control weight. If θ is the angle through which the ring has turned, then the controlling moment and the rotating moment are given by

$$\text{Controlling moment} = WR_2 \sin \theta$$
$$\text{Rotating moment} = AR_1 \, \Delta P$$

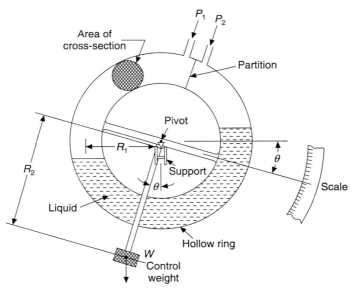

Figure 5.6   Ring balance manometer.

Therefore, ΔP is given by

$$\Delta P = \frac{WR_2 \sin \theta}{AR_1} \tag{5.9}$$

where,

W = counter weight (= Mg), N
R_1 = mean radius of the ring, m
R_2 = length of the arm carrying W, m
A = area of cross-section of ring, m^2

The calibration of the instrument is independent of the liquid used, but it should be noted that P_1 should not be so high as to drive the manometric liquid so much to the bottom of the ring as to let the test gas bubble through to the low pressure side.

5.2.3 Bell-type Manometer

One of the versions of the Bell-type manometer, which allows the use of an electrical displacement transducer, is shown in Figure 5.7. It consists mainly of an inverted bell of thin material, kept in position by means of a spring at its top and held inside a large container. The liquid inside the container covers the open end of the bell. The lower pressure is led into the bell through the tubing and the higher pressure acts on the outside of the bell. The upward and downward forces, when balanced, decide the position of the ball.

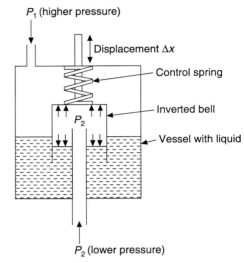

$$(P_1 - P_2)A = \Delta PA = K \Delta x \qquad (5.10)$$

where A is the area of the upper face of bell in m^2, the spring constant in N/m and ΔxK the change in position of bell as indicated by spring in m. The size of the system becomes large with A which needs to be large for measurement of small values of ΔP.

Figure 5.7 Inverted bell-type manometer.

5.2.4 Thin Plate Diaphragms

The most widely used pressure-sensitive primary transducer is the elastic element—the thin plate diaphragm. It is primarily a circular plate made from metal alloys such as bronze, phosphor bronze, beryllium copper and stainless steel. The thickness of the plate and its size depend on the range of pressure to be measured. Normally, the ratio of the diameter to thickness may vary from 25 to 100 with the thickness ranging from 0.1 mm to 5.0 mm. The plate is kept clamped between the flanges or, in certain cases, welded around its circumference. The flat diaphragm shown in Figure 5.8 has been very popular in instrumentation systems, where the diaphragm is subjected to forces from the test pressure only and none else, such as those from the linkages found in indicating-type pressure gauges. The small displacements due to the deformation of the diaphragm under pressure are picked up by noncontact type electrical secondary transducers, or the surface strains of the diaphragm are sensed and measured by the tiny electrical strain gauges of negligible mass. Some of the diaphragms allow integration of strain gauges on their surface. The high popularity of the thin plate diaphragm is also due to its applicability to measurement of dynamic pressures at frequencies up to 10 kHz. Independence from the effects of temperature on the characteristics is possible by use of special alloys, known as Ni-Span-C (a nickel iron alloy). To make the diaphragms more stable and rugged, they are precision machined from thicker plates, so as to minimize the effects of

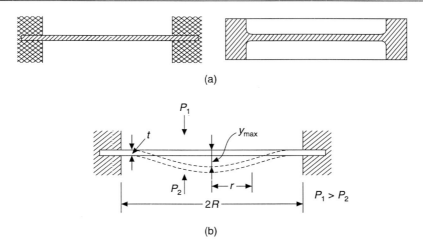

(a)

(b)

Figure 5.8   Flat-plate diaphragms: (a) Under no load; (b) under load.

clamping at the rim on their characteristics. At small pressures, the deformation of the diaphragm and the consequent displacement at the centre along with the surface strains are mainly due to bending only. The following relations hold good for cases where the deflection of the diaphragm at its centre, y_{max}, is about one-third of its thickness. With increased pressure, tensile stresses in addition to bending deformations appear, and the relationship between the deflection y_{max} at centre and the pressure difference ΔP across the diaphragm becomes nonlinear, as seen from the equation

$$\Delta P = \frac{16Et^4}{3R^4(1-\mu^2)}\left[\left(\frac{y_m}{t}\right) + 0.488\left(\frac{y_m}{t}\right)^3\right] \tag{5.11}$$

where,

  ΔP = pressure difference, N/m^2
  R = radius of the diaphragm, m
  t = thickness of the diaphragm, m
  E = Young's modulus of the material, N/m^2
  μ = Poisson's ratio

For values of $y_m t$ up to 0.3, linearity can be within satisfactory limits and Eq. (5.11) simplifies to

$$y_m = \frac{3(1-\mu^2)R^4}{16Et^3}\Delta P \tag{5.12}$$

The deflection y at any radius r from the centre is given by

$$y = \frac{3(1-\mu^2)}{16Et^3}(R^2 - r^2)^2\,\Delta P \tag{5.13}$$

The relationships lose validity when the diaphragm displacements are utilized to result in indications by a pointer on a scale. They are applicable when used in conjunction with noncontact type secondary electrical transducers. Circumferential and radial strains at several points on the surface of the diaphragm can be measured by resistance strain gauges. Nonmetallic diaphragms made from materials such as teflon, leather, neoprene and rubberized fabric are also used for measuring low pressures, but in such cases, they are backed by soft metallic springs.

5.2.5 Membranes

A membrane is a very thin diaphragm, mostly used for very low pressure measurements. It is made from nonmetallic materials such as polythene and leather and, in certain cases, from metallic foils. Polythene film diaphragms are coated with copper or gold to serve as the movable plate of a capacitor. They are generally kept in radial tension, and their stiffness to bending forces is negligible. Hence, the membranes deflect under pressure, the surface of which becomes part of a spherical surface as shown in Figure 5.9. The deflection of the membrane at its centre is given by

$$Y_{\max} = \frac{R^2}{4S} \Delta P \tag{5.14}$$

where, R is the radius of the membrane in m and S, the radial tension in N/m.

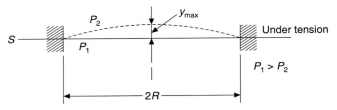

Figure 5.9 Thin membrane under stress.

5.2.6 Corrugated Diaphragms and Capsules

If thin diaphragms are provided with corrugations as shown in Figure 5.10(a), the effect will be greater flexibility and sensitivity to pressure variations. The shape of the radial cross-section of the diaphragm is called its profile and the pressure-deflection characteristic is influenced by the profile. Corrugation results in stress more due to bending than by tension. The deflection at the centre will be as much as 2% of the diameter of the diaphragm, and the characteristic can also be made nonlinear by choice of the profile. A *capsule* is made of two identical corrugated diaphragms, so as to form a leak-proof chamber as shown in Figure 5.10(b), and is referred to as an *aneroid*. The fluid under measurement is admitted to the chamber. While one diaphragm is rigidly held, the other deflects and results in double the displacement of a single diaphragm. For increasing the sensitivity further, two or more capsules are connected in series and the net displacement is proportional to the number of capsules.

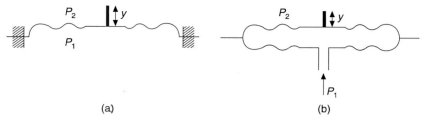

Figure 5.10   (a) A corrugated diaphragm; (b) capsule.

The flexibility of each diaphragm increases with the number of corrugations and decreases with their depth. But greater flexibility renders them unsuitable for situations where mechanical vibrations and acceleration act on the system. The same alloys used for flat diaphragm are used for capsules. The central part of the diaphragm carries, in all cases, a round disc which serves on one side to communicate the displacement, while on the other side it enables the mounting of the capsule on its support, allowing admission of the pressure to be measured. Aneroids have their chambers evacuated to pressures as low as 0.1 mm Hg and sealed so as to be used for measurement of atmospheric pressure. Those intended for measuring the differential pressure arising in fluid-flow measurements are called *open capsules*. The capsule is normally float-mounted such that the rim of the capsule is free to move radially, when developing displacements along its axis.

The characteristic of a corrugated diaphragm is described by a cubic equation given by

$$\Delta P = A y_m + B(y_m)^3 \tag{5.15}$$

where,

$$A = \frac{Et^3}{R^4} a, \qquad B = \frac{Et}{R^4} b$$

with a and b entirely influenced by the profile of the diaphragm and its material.

5.2.7   Bellows Element

The bellows element is a thin-walled cylindrical cup-like structure with a number of folds for its cylindrical surface along the axis of the cup as shown in Figure 5.11(a). The folds are known as *corrugations* or *convolutions*. The formation of the convolutions for a thin-walled seamless tubing or cup is a complex process, and hence welded bellows are manufactured by welding separately stamped annular diaphragms with a large hole at the centre as shown in Figure 5.11(b). Precision bellows are normally of special metallic alloys used for generating much larger displacements with pressure than the other elastic elements. They are also made from plastic material and rubber for use at much lower pressures, and do not require much precision.

Bellows are axially stretched or compressed by forces developed due to the difference in pressure between the inside and outside of the element. Bellows differing widely in their sizes are available for a variety of applications in instrumentation. The diameter of the bellows ranges from 7 mm to 150 mm, while the wall thickness ranges from 0.08 mm to 0.3 mm. The merits

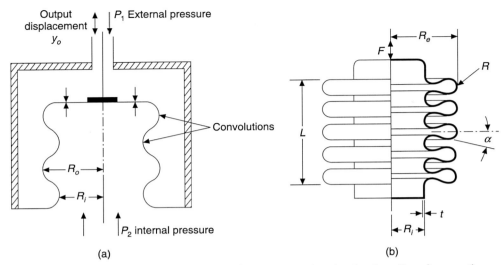

Figure 5.11 (a) A simple bellows element; (b) cut-away section showing formation of corrugations.

of a bellows element are its low spring rate, low flexural rigidity, and a near-linear characteristic between pressure and displacement. But the limitation in their application to measurement is due to their lack of zero stability. To some extent, this difficulty is minimized by using it in conjunction with a high quality helical spring at the cost of its spring rate. The sensitivity of the bellows depends on various parameters such as the fold angle α, ratio m of the radius of curvature of the fold, R to internal radius R_i, and the ratio, c of the external radius R_e to R_i. However, the axial deflection y_m for the case with $\alpha = 0$ is given by

$$y_m = \pi R_m^2 \, \Delta P \, \frac{1 - \mu^2}{Et} \, \frac{n}{A_0 + B_0 \dfrac{t^2}{R_i^2}} \tag{5.16}$$

where,

n = No. of convolutions

R_m = mean radius $\dfrac{R_e + R_i}{2}$ m

P = pressure difference N/m^2

μ = Poisson's ratio of the material

E = Young's modulus N/m^2

t = thickness of walls m

R_i = internal radius m

A_0, B_0 = constants depending on m and c

The effective area of the bellows for the calculation of the force is obtained by considering the mean radius. Bellows element is used as output transducer also, in which the pressure of a fluid is converted into a force, conveyed by the spindle attached to the bellows. Bellows element is normally preferred to be used in compression mode for obtaining better linearity. Hence the pressure external to the element is kept higher than the one inside.

5.2.8 Bourdon Tube Elements

Bourdon tubes are the most favoured mechanical elements for the measurement of pressure and constitute an important group of primary transducers for converting pressure into linear and angular displacements. The simplest and the best form of the Bourdon tube element consists of a thin-walled tube of an oval or flat-sided cross-section, bent into a circular arc of a central angle ranging from 200° to 270° (hence known as C-type Bourdon tube) as shown in Figure 5.12. One end of the tube is attached to a rigid base, enabling communication with the source of pressure, while the other end is closed and sealed, having freedom of movement. Due to the pressure difference between the inside and outside of the tube, the cross-section of the tube is deformed and the tube bends to acquire a circular cross-section, the resulting effect being the displacement of the free tip. The initial coiling angle ψ_0 decreases to ψ as shown, and the difference of these two angles, $\Delta\psi$, and the consequent displacement of the free tip y are linearly related to the pressure difference ΔP. The net displacement is about 3-4 mm for a tube of 10 cm diameter, and $\Delta\psi$ is about 10°. Thin-walled tubes having a ratio of t/b less than 0.6, where t is the thickness of walls and b, half-breadth of the tube (half of the minor axis, if elliptical cross-section) are used for pressures up to 35 kN/m², and thick-walled tubes having higher ratios above 0.6 are used for high pressure measurement. Expressions for $\Delta\psi$ and y for thin-walled tubes of elliptic and flat-sided cross-section are linearly related to ΔP as follows:

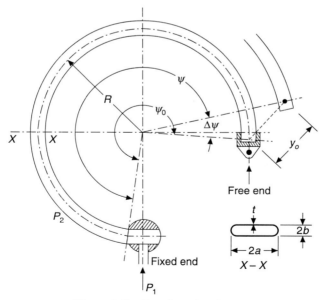

Figure 5.12 Bourdon tube element.

$$\frac{\Delta \psi}{\psi_0} = \Delta P \left[\frac{1-\mu^2}{Ebt} R_0^2 \right] \left(1 - \frac{b^2}{a^2} \right) \left(\frac{\alpha}{\beta + \chi} \right) \tag{5.17}$$

$$y_{\text{radial}} = \frac{\Delta \psi}{\psi_0} R_0 (1 - \cos \psi_0) \tag{5.18}$$

$$y_{\text{tangential}} = \frac{\Delta \psi}{\psi_0} R_0 (\psi_0 - \sin \psi_0) \tag{5.19}$$

$$y = \sqrt{(y_{\text{rad}}^2 + y_{\text{tan}}^2)} \tag{5.20}$$

where

ΔP = pressure difference, N/m^2

R_0 = radius of the arc, m

μ = Poisson's ratio of the material

E = Young's modulus of the material, N/m^2

a = semi-major axis of the tube, m

b = semi-minor axis of the tube, m

t = thickness of walls of tube, m

χ = $R_0 t/a^2$, a dimensionless constant

α, β = dimensionless constants obtained from Table 5.5

$\Delta \psi, \psi_0$ = angles, radians

For the case of a Bourdon element, with $\psi = 270°$, y is given by

$$y = 5.8 R_0 \frac{d\psi}{\psi_0} \tag{5.21}$$

In the case of thermometers, the volume variation ΔV, which is the increase of the volume of the liquid due to increase in temperature, has to be related to the displacement y given by

$$y = \frac{\Delta V \alpha [(\psi_0 - \sin \psi_0)^2 + (1 - \cos \psi_0)^2]^{1/2}}{12ab\psi_0 n(1 - b^2/a^2)} \tag{5.22}$$

and the change in tube volume, ΔV, in terms of the difference in pressure, ΔP, is given by

$$\Delta V = 12 \Delta P \frac{1-\mu^2}{E} \frac{R_0^3 a \psi_0}{t} \left(1 - \frac{b^2}{a^2} \right) \left(\frac{\eta}{\beta + \chi^2} \right) \tag{5.23}$$

where η is a constant depending on the type and size of tube cross-section.

Table 5.5   Constants pertaining to Bourdon tube

Tube cross-section	Elliptical			Flat sided		
a/b	α	β	η	α	β	η
1	0.750	0.083	0.197	0.637	0.096	0.149
1.5	0.636	0.062	0.149	0.549	0.110	0.151
2	0.566	0.053	0.142	0.548	0.115	0.144
3	0.493	0.045	0.121	0.480	0.121	0.131
4	0.452	0.044	0.111	0.437	0.121	0.122
5	0.430	0.043	0.105	0.408	0.121	0.115
6	0.416	0.042	0.102	0.388	0.121	0.110
7	0.406	0.042	0.100	0.372	0.120	0.107
8	0.400	0.042	0.098	0.360	0.119	0.105
9	0.395	0.042	0.097	0.350	0.119	0.103
10	0.390	0.042	0.095	0.343	0.118	0.101

Though difficult to manufacture, tubes of elliptic cross-section possess higher sensitivity. Thick-walled tubes are chosen for high pressure measurements, while a dumb-bell shape is used with a view to minimizing its volume.

5.3   BASICS OF FORCE MEASUREMENT

Force and weight measurements are common in industries, and a good number of mechanical instruments are available for the measurement of weight. Conventional weighing techniques are based mostly on beam-balance principle which is used to compare known and unknown weights. Force is a vector quantity, and hence when a measurement is taken, it is essential to observe that the direction of application of the force and the direction of the force to which the measuring system is expected to respond, coincide. Otherwise, the measurement is likely to be incorrect. Sources generating forces vary widely in their nature and communication of forces to the measuring system requires suitable mechanisms. Alternatively, primary measuring elements should be so designed as to become an integral part of the force generator. Section 5.2 has dealt with the primary transducers suitable for measurement of uniformly distributed forces due to fluid pressure. The weight of a body is the force with which it is attracted towards the centre of the earth due to earth's gravity, and it can be applied to a primary transducer as a concentrated force or load acting at a point. Primary devices that enable such measurements are known as *load cells*. In practice, such primary devices are expected to enable measurements of both static and dynamic forces.

In the following sections, primary mechanical transducers, mostly elastic elements, such as springs, diaphragms, cantilevers and proving rings are briefly described. Also, hydraulic and pneumatic type load cells, developing output signals of mechanical nature, such as displacement and air pressure, are briefly explained. All the force-sensing primary transducers are easily calibrated under steady state conditions, by using standard weights.

5.3.1 Helical Spiral Springs

Helical spiral springs are used in spring balances for measurement of weights of bodies. The weight under measurement is balanced by the restoring force of the spring. The helical spiral spring is, in fact, a common element in most of the instrumentation systems for reasons of its compactness and linearity in its characteristic. The attached weight sets up the force along its axis. With one end rigidly held, the elongation of the spring along its axis due to the application of force serves as the displacement signal for further processing or for indication purposes. The helix angle α is kept within 6°, so that the wire is treated as a straight round bar twisted around its axis by the moment FR, where F is the force applied and R the mean radius of the helix as shown in Figure 5.13. The axial displacement y is expressed, in terms of the force F, as

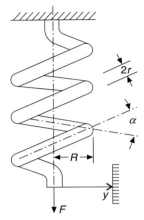

Figure 5.13 Helical spiral spring.

$$y = \frac{4FnR^3}{Gr^4} \qquad (5.24)$$

where

F = force applied, N
R = mean radius of the helical coil, m
r = radius of wire, m
G = modulus of rigidity, N/m^{-2}
n = No. of coils of the spring
y = axial displacement of the spring

The value of n is taken normally less than the actual number by 2 and the ratio of R/r is kept about 7 or 8. The maximum stress in wire cross-section is given by

$$\sigma_{\max} = \frac{2FRK}{\pi r^3} \qquad (5.25)$$

where K is a constant depending on the ratio R/r and is approximately 1.2 for the ratio of 7 or 8. The springs are of square section, when better elastic strength is desired. The helical coil is with half the turns clockwise and the other half anti-clockwise.

5.3.2 Cantilever Beams

A cantilever beam, as shown in Figure 5.14, with its one end rigidly fixed in the support and the other end being subjected to a concentrated force F, forms a suitable means for measurement of weight and force. The bending moment proportional to the force results in a displacement $y_{\max}$ at the tip, while the top and bottom surface will be under tensile and compressional strains respectively.

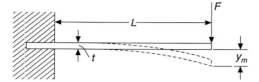

Figure 5.14 Cantilever beam.

The expressions for the displacement y_{max} and the mechanical strain ε_{max} are

$$y_{max} = \frac{4FL^3}{Ebt^3} \tag{5.26}$$

$$\varepsilon_{max} = \frac{6FL}{Ebt^2} \tag{5.27}$$

where,

F = force, N
L, b = length and breadth, m of the beam respectively
t = thickness of the beam, m
E = Young's modulus of the material, N/m^2

The dimensions of the beam are determined by the magnitude of the force and the proportionality limit. For small forces, leaf springs fixed at one end are used.

5.3.3 Beams Held at Both Ends

For a beam or strip supported freely at both ends and loaded at the centre by the force as shown in Figure 5.15, the relationships between the force F and the output signals are

$$y_m = \frac{FL^3}{4Ebt^3} \tag{5.28}$$

$$\varepsilon_{max} = \frac{3FL}{2Ebt^2} \tag{5.29}$$

where, F, L, t, b and E are as in Section 5.3.2. It should be noted that the maximum strain occurs at the cross-section at which the force is applied.

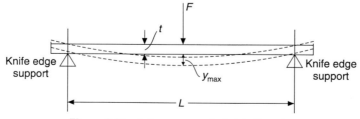

Figure 5.15 A beam supported at both ends.

For the beam held rigidly inside the support, as shown in Figure 5.16, the deformation will be as shown and the deflection and the strain will be reduced to one-fourth of the values given in Eqs. (5.28) and (5.29).

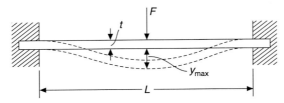

Figure 5.16 A beam rigidly held at both ends.

5.3.4 Diaphragm Elements

Circular plate diaphragms serve as plate springs and can be used for the measurement of forces. The displacement of the diaphragm at its centre when supported freely at its periphery of radius R, as shown in Figure 5.17, is given by

$$y_m = \frac{3+\mu}{1-\mu} \frac{3FR^2(1-\mu^2)}{4\pi Et^3} \tag{5.30}$$

and when it is rigidly fixed in the supporting structure, y_m becomes

$$y_m = \frac{3FR^2(1-\mu^2)}{4\pi Et^3} \tag{5.31}$$

where,

R = radius of diaphragm, m
t = thickness of the diaphragm, m
μ = Poisson's ratio
E = Young's modulus of the material, N/m^2

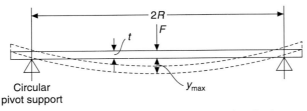

Figure 5.17 A diaphragm supported around perimeter.

Diaphragm elements are used for measurement of small forces, and when they are of reasonable size, can be used alongwith strain gauges.

5.3.5 Column-type Load Cells

Load cells are primarily intended for measurement of weight of bodies such as slowly moving vehicles. They are as well designed for various applications where concentrated forces can be conveyed to the load cells through mechanical linkages. Those cells meant for weighing are provided with platforms onto which vehicles are admitted at slow speed or are provided with supports for hanging the body to be weighed. A column type load cell primarily consists of either slender rod, robust column of rectangular or circular cross-section, or even the squat cylinder shown in Figure 5.18. These columns of regular configurations enable the measurement of deformations with reasonable accuracy, though ultimately they are calibrated against standard weights. It is essential to recognize the importance of transmitting the force uniformly over the entire cross-section A of this load cell; it is also essential to recognize that there is no other force working on the column apart from the one under measurement, acting along the axis of the column. Due to the stress F/A, the surface of the column undergoes compressional strain ε_a along its axis and tensile strain along its circumference ε_t. These strains are measured conveniently by the resistance type strain gauges, by locating them suitably on the outside surfaces. The strains are given by

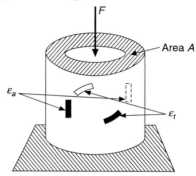

Figure 5.18   A column-type load cell.

$$\varepsilon_a = \frac{F}{AE} = \varepsilon_1 = \varepsilon_3 \qquad (5.32)$$

$$\varepsilon_t = \frac{-\mu F}{AE} = \varepsilon_2 = \varepsilon_4 \qquad (5.33)$$

The size of the column permits the attachment or bonding of the strain gauges as shown, and so utilization of the axial displacement is not considered any more efficient. The column is held rigidly in vertical position on a vibration-free mounting and the force is transmitted through a piston or any other means, to the entire sectional area of the column. They are designed to measure up to 2000 T.

Distributing the four strain gauges around the periphery such that the tensile and compressional strain gauges are alternated, it is possible to achieve both temperature compensation and immunity to bending stresses due to forces inadvertently applied at an angle to the axis of the column. Some-times, the column is held vertically by guard plates so as to increase its stiffness in the radial direction.

5.3.6 Proving Ring-type Load Cell

For smaller weights, column type load cells cannot be used, as their size prohibits the location of strain gauges. More strain per unit applied force is possible by using configurations employing bending stresses, such as the cantilever beam. Another configuration having all the advantages and compactness and ease of application is the proving ring.

The proving ring is also known as ring dynamometer which consists of circular ring as shown in Figure 5.19(a) having radial thickness t which is very small as compared to its mean radius R. When a weight is attached or a force is applied, the circular ring becomes elliptical and the circumferential stresses developed result in tensile or compressional strains on the inner and outer surfaces of the ring. The axial width or depth b of the ring may permit location of four strain gauges, two for compressional strain and two for tensile strain. For the arrangement of strain gauges as illustrated in Figure 5.19(a), the strains measured are due to bending stresses only and all strains are equal in magnitude since t is much smaller than R, and are given by

$$\varepsilon = \frac{1.08FR}{Ebt^2} \tag{5.34}$$

where,

F = force applied, N
b = axial width of ring, m
t = thickness of the ring, m
R = mean radius of the ring, m
E = Young's modulus of the material of the ring, N/m^2

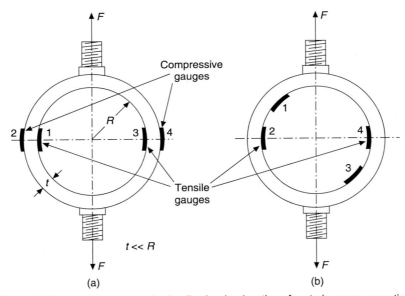

Figure 5.19   Proving ring-type load cells showing locations for strain gauge mounting.

Strain gauges 1 and 3 indicate tensile strain while gauges 2 and 4 indicate compressional strain. Normally, in practice, the stiffness due to the addition of the loading hooks or the supporting lobes and abutments reduces the strain values. While manufacturing, some proving rings are machined from a solid block of the material, in such a way as to make the supporting lobes become an integral part of the ring.

Alternatively, when the ring size does not lend itself to use with strain gauges, the deformation of the ring along the direction of force is measured in terms of the displacement, which is in fact the difference between the major axis of the ellipse and the diameter of ring, both measured internally. The displacement is given by

$$y = \frac{1.79FR^3}{Ebt^3} \tag{5.35}$$

The displacement measurement is done by either dial gauges or secondary electrical transducers. When strain gauges are located as shown in Figure 5.19(b), a little more sensitivity is obtained because of the additional stresses at the points where gauges 1 and 3 are located.

Other configurations such as the one shown in Figure 5.20(a) are available. In fact, any supporting link holding the weight under measurement gets stressed and the associated strains can be measured by strain gauges located suitably at points where the maximum strain occurs. Such devices need only to be calibrated often by means of standard weights. Some typical elements are shown in Figure 5.20(b).

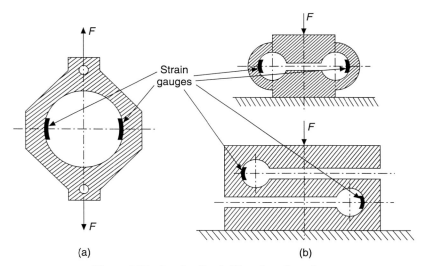

(a)   (b)

Figure 5.20   Load cells of different configurations.

Proving rings can be of any metal or alloy, and for higher sensitivity, aluminium is used. They are designed for loads down to 50 kg, while for lower values, it is difficult to satisfy the requirement of having very low thickness-to-diameter ratio.

5.4   BASICS OF TORQUE MEASUREMENT

Application of a force along the axis of a cylindrical rod results in an increase in its length along the direction of force, whereas application of a force tangential to the surface of the rod results in rotation of the rod about its axis. The moment of the force F is FR if R is the radius of the rod, and when the forces are acting on the rod tangential to the surface and at the two points, diametrically opposite to each other as shown in Figure 5.21. The moment constitutes a twisting moment or a torque on the rod equal in magnitude to *2FR*. If the rod is not free to rotate and its one end is rigidly held fixed, the other end undergoes the twisting moment. The deformation set in is recognized by an angular displacement of the rod about its axis of rotation. It can also be sensed as surface strain. A line on the surface of the rod originally parallel to the axis of rotation suffers an angular shift which becomes zero, once the applied torque becomes zero.

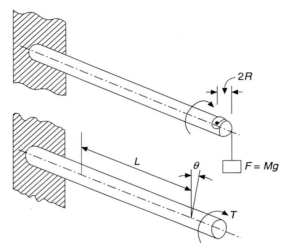

Figure 5.21   A solid-section rod under static torque.

Torque measurement involves, fundamentally, measurement of force and the length of the arm perpendicular to the force and extending from it to the axis of rotation. The torque applied to bodies held rigidly or free may be measured directly by measuring the force by means of force transducers. However, in practice, torque measurement is often undertaken to estimate the shaft power of machines serving as sources and sinks. Dynamometers are known to be the torque-measuring systems for such machines, based on the measurement of force. Another common requirement of torque measurement is in respect of measuring the shaft power conveyed by a source machine to a sink machine under dynamic conditions. The rotating shaft connecting the two machines serves as the primary transducer or a torsion bar, or a torsion tube is inserted for sensing the torque transmitted by the machine to its load. The output signals of the torsion bar are picked up by novel transducer systems, consisting of either strain gauges or other electrical transducers. Systems based on these principles are referred to as torquemeters.

5.4.1   Torsion Bar

A torsion bar is simply a metallic rod of circular or rectangular cross-section used for measuring the torque transmitted by a machine to its load. It is connected between the driving source and the load such that it is in perfect alignment with their axes of rotation. The torque transmitted by the source to the load strains the torsion bar and measurements of the angular displacement or the surface strain are used to determine the torque. The angular displacement of a hollow cylinder as shown in Figure 5.22(a) is given by

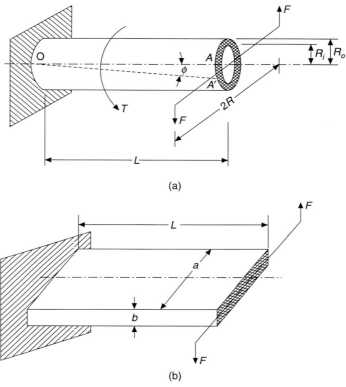

(a)

(b)

Figure 5.22　(a) A hollow cylinder as torsion bar; (b) torsion bar of rectangular section.

$$\phi = \frac{2LT}{\pi G(R_o^4 - R_i^4)} \tag{5.36}$$

where,

ϕ = angular displacement, radians
L = active length of torsion bar, m
T = torque transmitted, N-m
R_o, R_i = outer and inner radii of the tube, m
G = modulus of rigidity of bar material, N/m^2

ϕ refers to the angular displacement of point A relative to point O. Several electrical methods of measuring this displacement are available and presented in Chapters 6 and 7.

When torsion bars of circular cross-section are used, Eq. (5.36) simplifies to

$$\phi = \frac{2LT}{\pi GR_o^4} \tag{5.37}$$

If the torsion bar is flat or is of rectangular cross-section, as shown in Figure 5.22(b), the angular displacement, is given by

$$\phi = \frac{TL}{Gab^3 \left[\dfrac{16}{3} - 3.36\left(\dfrac{b}{a}\right)\left\{1 - \left(\dfrac{b^4}{12a^4}\right)\right\}\right]} \qquad (5.38)$$

where, a and b refer to the dimensions of the section shown.

Alternatively, the torque is obtained from measurements of the surface shear strain, by locating strain gauges precisely at 45° to the shaft axis as shown in Figure 5.23. The merits of this method are independence from ambient temperature variations and insensitivity to bending or axial strains. The surface shear strain of each gauge is given by

$$\varepsilon_{45°} = \pm \frac{2T}{\pi GR_o^3} \qquad (5.39)$$

Torsion bars of square cross-section are preferable to those of circular cross-section, for mechanical reasons.

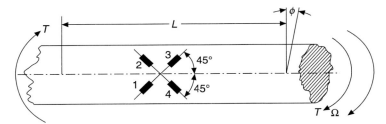

Figure 5.23   Dynamic torque measurement by strain gauges.

5.4.2   Flat Spiral Spring

If a flat spiral spring of n turns is rigidly connected to a spindle at its innermost turn, and the outermost turn is subjected to a force as shown in Figure 5.24, the torque developed is equal to FR, and the consequent angle of rotation of the spindle ϕ, is a measure of the torque. The angle ϕ is given by

$$\phi = \frac{12\pi TRn}{Ewt^3} \qquad (5.40)$$

where

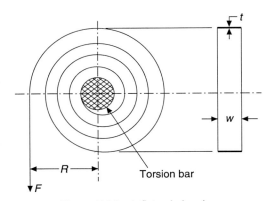

Figure 5.24   A flat spiral spring.

w = width of the strip, m
t = thickness of the strip, m
E = Young's modulus of the material of the spring, N/m^2

Small torques developed in instrument mechanisms can be measured by using this method.

5.5   BASICS OF DENSITY MEASUREMENT

Measurement of density of solids, liquids and gases is very important. This is more so when they are continuously flowing from one site to another. It is very difficult to carry out the measurement of the density of solids under these conditions, but techniques are available for the measurement of liquids and gases. Mass density is defined as the mass per unit volume, whereas weight density is the weight-force per unit volume. Specific gravity is known as the ratio of the density of the substance to the density of a water at 4°C. Specific volumes of all substances vary with temperature while those of compressible fluids are influenced by pressure also. It is necessary to indicate the pressure and temperature at which the density is measured and, wherever necessary, corrections are to be applied for obtaining the density at the standard temperature and pressure.

Apart from all conventional techniques available for the determination of the density of fluids in the laboratory, certain techniques suitable for use with some of the mechanical transducers already described or other electrical transducers are presented in the Section 5.5.1.

It is advisable to have the variations in density converted into either displacement, pressure or force, for further processing these for indication and other purposes. Load cells can be used to weigh the container having a fixed volume of liquid when the container is provided with a continuous flow of liquid into it after suitably sampling from the main flow and an overflow device as shown in Figure 5.25. Other techniques employ floats such as the hydrometer and other systems, resulting in output signals of pressure force or displacement.

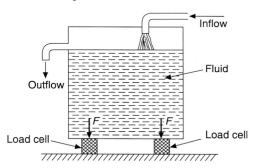

Figure 5.25   Density measurement by load cells.

5.5.1   Hydrometer System

The simplest system for measuring the density of liquids is the hydrometer which consists of a glass float that is weighted at the bottom with mercury or lead balls to make it float upright, as shown in Figure 5.26(a). The float has a hollow stem inside—a graduated scale—which serves to indicate the density of the liquid. When floating in the liquid, the scale is read at the position of the surface of the liquid. If the level of the liquid in the container is held constant by using an overflow tube, the vertical displacements of the float and stem can be used for indication of the density. A short steel needle attached to the stem will serve as the moving iron needle of the linear variable differentia] transformer that translates the displacements into an electrical signal. The system is calibrated by using liquids of known densities.

The buoyant force on the float acting along the axis of the hydrometer stem can be used to load a cantilever type load cell if the force is of reasonable magnitude.

Alternatively, three floats, each differing in its volume and density from the others, are connected together to form a single assembly while having the freedom to rotate, as shown in Figure 5.26(b). The buoyant forces working on all the three decide the angular position of the assembly and by suitably designing the sizes of the floats, it is possible to make the angular deflection proportional to the density of the liquid.

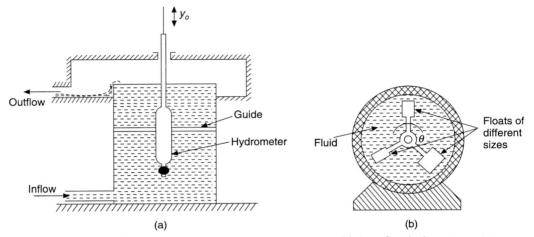

Figure 5.26   (a) Hydrometer system for density measurement; (b) three-float hydrometer system.

5.5.2 Air Bubbler System

In these air bubbler systems, air pressure, built up due to admission of air through a tube into the container of a liquid, is measured. One of the systems is shown in Figure 5.27(a), in which air is supplied past a regulating valve so that it just escapes from the bottom end of the tube as bubbles. Then the pressure in the tube equals the pressure due to the head of liquid, h, above the lower end of the tube, and is given by

$$\Delta P = h\rho g \tag{5.41}$$

where

P_o = gauge pressure in tube, N/m^2
h = depth of the liquid column, m
ρ = density of the liquid, kg/m^3
g = acceleration due to gravity, m/s^2

Measurement of pressure differential ΔP, by any transducer is used to indicate the density of the liquid.

Alternatively, two tubes kept at different depths may be used along with a manometer for the density measurement as shown in Figure 5.27(b). In such a case,

$$H\rho g = \Delta h \rho_m g \tag{5.42}$$

or

$$\rho = \rho_m \frac{\Delta h}{H}$$

where,

ρ, ρ_m = densities of the liquid and manometric liquid respectively

Δh = difference in levels of manometer

H = difference in the depths of the tubes from the liquid surface

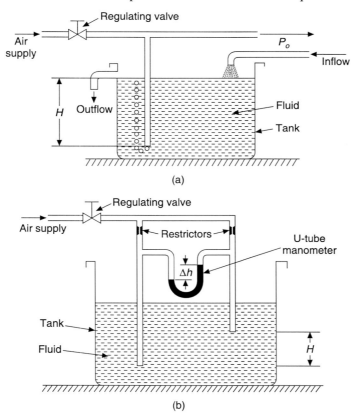

(a)

(b)

Figure 5.27 Fluid density measurement: (a) By air-bubbler system; (b) by U-tube manometer system.

When gas density has to be measured, the gas is pumped continuously by means of a blower run at constant speed. The pressure drop across the blower is proportional to the density.

5.5.3 U-tube Weighing System

The liquid whose density is under measurement is made to flow through a U-tube kept horizontally on flexures located at the open ends of the tube as shown in Figure 5.28. The weight of the tube and its liquids content are measured by any weight or force measuring transducer, whether pneumatic or electrical. Normally, the weight of the tube and its contents are measured by the null-balance principle using feedback techniques, which will be described in Chapter 8.

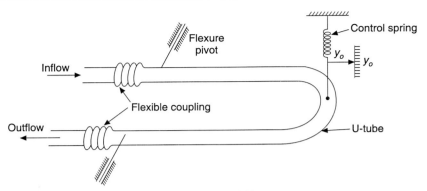

Figure 5.28 U-tube weighing system.

5.6 BASICS OF LIQUID-LEVEL MEASUREMENT

Measurement of the level of liquids in closed or open containers is essential in many industrial processes. The on-line measurement of the level of solid or granular material presents considerable difficulties, whereas for the liquid level measurement, a large number of techniques are available. Those involving mechanical elements are now described: developing output signals such as linear or angular displacement, pressure or force, which render the direct indication or further conversion to electrical signals by secondary transducers. However, the nature and properties of the liquid determine the technique to be employed. Direct measurement is done by means of dip-stick- or sight-glass and hook-type level gauges, which are unsuitable for on-line measurement. Other commonly used systems are now given, of which the float level gauge is popular.

5.6.1 Float Element

For open or vented containers, a suitably designed float enables the development of output signals in the form of linear or angular displacements as shown in Figure 5.29. Direct calibration

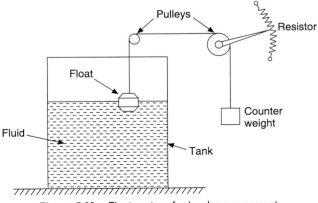

Figure 5.29 Float-system for level measurement.

of these displacements is done in level gauges, whereas secondary transducers of electrical or mechanical nature enable the conversion of these displacements into a signal suitable for recording, remote indication or control. The indications are independent of the uniformity of density of the fluid. The weight of the float and the counter weight are so chosen that the float remains half-immersed in the liquid. Slight variations in the density of the liquid may not result in much error in indication, when floats are made lighter.

Alternatively, a float whose weight is always greater than the weight of the liquid displaced (otherwise known as displacer) is used to measure this level, as shown in Figure 5.30. In this case, the vertical movement of the float is almost prevented, by being attached rigidly to a shaft or spindle, which serves as a means of transmitting the net downward force to a force transducer such as a cantilever beam or leaf spring. Considering the float to be of area of cross-section A, and of mass m, the net downward force on the float, F, is given by

$$F = mg - \rho g h A$$

giving

$$h = (mg - F)/(\rho g A)$$

$$\Delta h = \left(\frac{-1}{\rho g A} \right) \Delta F = K \Delta F \tag{5.43}$$

The change in level is proportional to the change in force, and calibration can be suitably accomplished. The liquid should be of uniform density all through.

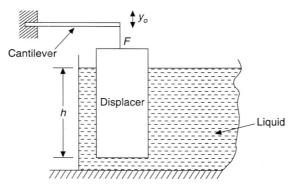

Figure 5.30   Displacer system for level measurement.

5.6.2   Level-to-Pressure Converters

The pressure due to a column of liquid of constant density ρ is $\rho g h$, and this pressure measurement is carried out by any pressure transducer such as diaphragm or mercury manometer as shown in Figures 5.31(a) and 5.31(b).

For the manometer shown in Figure 5.31(b), it is seen that $\rho g h = (\rho_m h_m - \rho y)g$, where ρ_m is the density of mercury.

If the mercury levels in the two legs of the manometer are adjusted initially to be at the same height as the bottom of the container, then $h_m = 2y$. Hence,

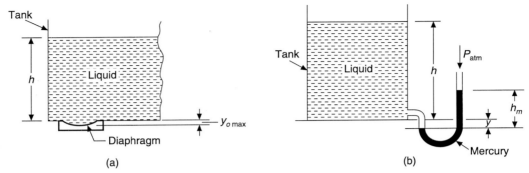

Figure 5.31   Level measurement: (a) By diaphragm; (b) by manometer.

$$h = h_m \left(\frac{\rho_m}{\rho} - \frac{1}{2} \right)$$

$$\Delta h = K \, \Delta h_m \qquad\qquad (5.44)$$

The mercury level can be measured by electrical transducers.

For the case with closed containers, the connection of manometer is modified as shown in Figure 5.32. A condensing chamber is provided at the top of the second tube of the manometer and is filled with the same liquid as in the container. The location of the condensing chamber is such that the level of the liquid in it is always above the level that can be attained in the container. The pressure P_t at the top of the liquid may be due to pressure of the vapour of the liquid or it may be intentionally held at some high value. If the level of the liquid in the container is brought to be the same as the one in the condenser, then both the levels of mercury in the two legs will assume the same value and will be at a depth of y_0 from the bottom of the container.

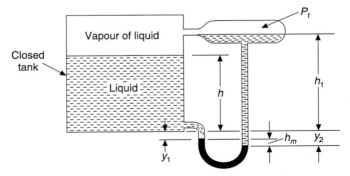

Figure 5.32   Level measurement in closed tanks.

For any other level h of the liquid in the container, the levels of mercury in the legs will be as shown in Figure 5.32. Hence, it can be shown that

$$(h + y_1)\rho g + \rho_m h_m g = (h_1 + y_2)\rho g \qquad\qquad (5.45)$$

Since

$$y_1 = y_0 - \frac{h_m}{2}$$

$$y_2 = y_0 + \frac{h_m}{2}$$

by simplifying Eq. (5.45), h is obtained as

$$h = h_1 - h_m \left(\frac{\rho_m}{\rho} - 1 \right) \tag{5.46}$$

and $\Delta h = K\Delta h_m$ as h_1 is held constant. Alternatively, any other transducer such as the diaphragm type can be used in place of the manometer for sensing the pressure difference between the top and bottom of the closed container, while retaining the condenser.

5.6.3 Level-to-Force Converter

If the liquid is of constant density, then the weight of the closed container is obtained by means of load cells. Using containers of uniform cross-section, the level of the liquid can be ascertained from the weight. The container is free from any of the attachments described in Section 5.6.2.

5.7 BASICS OF VISCOSITY MEASUREMENT

Viscosity is the property of flowing fluids and it quantifies the resistance to flow set-up within the fluid. It is entirely due to the internal friction between the moving particles of the fluid. Viscosity is responsible for most of the dissipation of energy in transportation of liquids and gases through pipelines. The effectiveness of lubricating oils depends, among other factors, on the viscosity of the oils. On-line measurement of the viscosity of oils is a common need in industry and for fluid-flow studies, a knowledge of the viscosity of the liquid or gas is essential.

Viscosity can be understood as that property which determines the magnitude of the resistance of the fluid to a shearing force. Fluid flow is said to be *laminar* or *viscous* if the flow is taking place in such a way that adjacent layers of the fluid move slowly in one direction only. If adjacent layers are moving with differing velocities, the constant interchange of molecules and momentum creates resistance to the relative motion of the layers. A shear force is the force component tangential to the surface of the layers and this force divided by the area of the surface is the shear stress over that area. If the adjacent layers of fluid of infinitesimal thickness are separated by a small distance dx, and their velocities are v and $v + dv$, then a velocity gradient dv/dx exists between these layers. This velocity gradient is the result of application of the shear stress and is considered to be the rate of deformation. Newtonian fluids are those for which the shear stress is proportional to the velocity gradient. Newton's equation for viscosity is given by

$$\tau = \mu \frac{dv}{dx}$$

where τ is the shear stress and the proportionality factor μ is called the coefficient of *absolute* or *dynamic viscosity* of the fluid. *Fluidity* is a term used to describe the property of fluids and is given by $1/\mu$. Kinematic viscosity v is equal to the absolute viscosity divided by the mass density ρ. Dynamic viscosity has the units of N s/m^2 in SI system and dyne-s/cm^2 (called Poise) in the cgs system. Kinematic viscosity has the units of m^2/s in the SI system and of cm^2/s (called Stoke) in the cgs system. Water at 20°C has an absolute viscosity of 1.008 centi-poises.

Many lubricating oils and most of the gases and their liquids are Newtonian, whereas thick liquids like honey, rubber suspensions, synthetic oils and thermosetting plastics are nonNewtonian (also called pseudoplastic) in nature. Temperature has the greatest influence on viscosity. The viscosity of liquids decreases with increase in temperature, whereas the viscosity of gases increases with increase in temperature. Viscosity may be understood to refer to the physical property of clean fluids; if these are contaminated or contain any particulate matter such as wood pulp, they are treated differently.

Laboratory methods of measuring the viscosity of liquids are time-consuming and are not suitable for on-line measurements. They are either the falling-sphere or Saybolt viscometers. Techniques based on the principles of these instruments are developed for on-line measurements in industry employing some of the pressure or torque-sensing transducers. In the following sections, only two such systems for liquids are presented.

It is important to note that the flow should be laminar or streamlined when the viscosity of the fluid is under measurement. The particles of the fluid in the capillary tube or pipe of constant diameter should be moving parallel to the walls of the tube, and there should not be any movement in the radial direction or across the stream. The flow pattern changes its character from laminar type once the velocity of the fluid crosses a critical value. The critical velocity is dependent on the pipe diameter d, dynamic viscosity μ, and mass density ρ of the fluid. The flow pattern is seen to be laminar once the value of $d\rho v/\mu$ is below 2000, and the value of $d\rho v/\mu$ is known as the Reynolds number (Re) which is dimensionless. For Re values above 4000, the flow is turbulent, while for the intermediate values it is in transitional phase. Measurement of the mass flow rate ($= \pi \rho v d^2/4$) through a capillary tube may be used to estimate the critical velocity of a fluid. Viscosity measurements are carried out on fluids after ascertaining that the flow pattern is laminar.

5.7.1   Viscosity-to-Pressure Converter

The most common method of viscosity measurement is by measuring the pressure drop in laminar flow through a capillary tube. They are, in general, known as orifice-type viscometers, and are available as Redwood (UK) and Saybolt (USA) viscometers. When the flow through the capillary is established as shown in Figure 5.33, it is necessary to calculate the volume flow-rate by measuring the volume of liquid collected in a certain time interval. The pressure difference across the capillary ΔP is taken as $h\rho g$, and the dynamic viscosity is calculated from the Hagen-Poissullle Equation

$$\frac{\pi d^4 \Delta P}{128 \dot{Q} L} \tag{5.47}$$

where,

$\dot{Q}$ = volume flow rate of fluid
L = length of the capillary tube

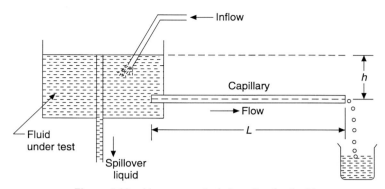

Figure 5.33   Measurement of viscosity of a liquid.

For accurate measurements, ΔP may be taken by measuring $P_1 - P_2$, where P_2 is the pressure at a point far enough downstream from the entrance of the tube in cases where L is less than $125d$.

For continuous viscosity measurements of a fluid, it is essential that a constant flow rate is maintained through the capillary tube by means of a constant flow pump and the pressure drop across the tube is continuously measured by a pressure transducer, while the temperature is held constant.

When the fluid is a compressible gas, the temperature of the gas and the gas constant come into the picture and the measurement of dynamic viscosity becomes more complex. The dynamic viscosity of the gas at the absolute temperature T is obtained by measurement of the mass flow rate ω and is given by

$$\mu = \frac{\pi R^4 (P_1^2 - P_2^2)}{16 \omega R_g T} \tag{5.48}$$

where, P_1 and P_2 are the pressures at the two ends of the capillary tube, R_g is the gas constant, and T, the temperature of fluid.

5.7.2   Viscosity-to-Torque Converter

The development of torque proportional to the viscosity of a liquid at varying velocities is possible in the method under discussion by using two vertical concentric cylinders, one rotating and the other inner one kept static, separated by a small gap. In Figure 5.34 is shown the case of the inner cylinder kept stationary, with the result that the torque experienced by it due to the rotation of the outer cylinder is measured by a torque transducer or angular displacement transducer. In the Mac-Michael viscometer, it is a cylindrical bob that is suspended by a torsion wire that allows the measurement of torque. It is essential, however, to keep the radius of the inner cylinder much larger than the gap between the outer cup and the inner cylinder. The dynamic viscosity μ is given by

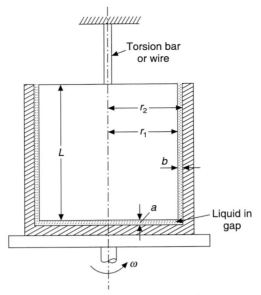

Figure 5.34   Rotating cylinder type viscometer.

$$\mu = \frac{T}{\pi\omega r_1^2 \left[\dfrac{r_1^2}{2a} + \dfrac{2Lr_2}{b} \right]} \tag{5.49}$$

where,

T = torque experienced by inner cylinder

ω = angular velocity of outer cup

r_1, r_2, a and b = the dimensions of the cup and cylinder as shown in Figure 5.34.

Shearle's method uses the concentric cylinders for measuring the high values of viscosity.

5.7.3 Viscosity-to-Displacement Converter

A viscometer for the measurement of the viscosity of a fluid utilizes a float specially designed such that its position varies with the viscosity of the fluid, while the flow rate of the fluid is held constant. In Figure 5.35 floats 1 and 2 are kept inside a tapered glass tube. Float 1 is designed to be insensitive to viscosity and its position determines the flow-rate only. Float 2 is sensitive to viscosity and its position is measured by a displacement transducer. The throttling valve sets up a constant pressure difference across the meter and the constant flow rate is ascertained by the position of float 1.

The system enables the measurement of viscosity of both liquid and gas. It has to be calibrated for each fluid and is suitable for continuous monitoring of the viscosity of the fluid.

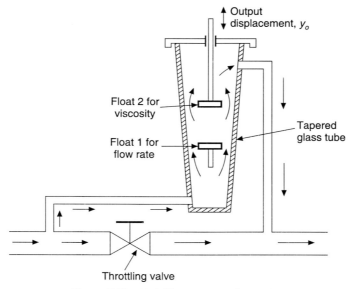

Figure 5.35 Variable area type viscometer.

5.8 BASICS OF FLOW MEASUREMENT

In all the production processes, the measurement of the rate of flow of solids, liquids and gases assumes great importance. Equally important is the knowledge of the total quantity that has gone past a section during a certain interval of time. Thus, meters are broadly classified as flow-rate meters (also known as flow meters or flow gauges) and quantity meters. Quantity meters may directly read the total quantity, or the output signals of the flow meter may be integrated over the required interval of time to indicate the total quantity. Flow meters may be calibrated to read either the volume flow rate or the mass flow rate. Flow meters for liquids normally read the volume flow rate; those for gases may indicate either the volume flow rate or the mass flow rate.

The techniques available for flow measurements vary widely due to (a) the nature and properties of the medium, (b) the conditions under which flow measurement has to be carried out, (c) the range of the flow rate, and (d) the accuracy desired.

Quantity meters and mass flow meters for liquids are presented in Chapter 8 while flow meters for liquids and gases utilizing the primary mechanical measuring elements for converting the flow rate into a suitable mechanical output signal are presented in Sections 5.8.1–5.8.6.

In this context, it is relevant to recall the fundamentals concerning the flow of fluids. Fluid flow is considered to be *steady* if the fluid velocity at any given point is constant in time. If the velocity varies erratically from point to point as well as from time to time, it is said to be *nonsteady* or *turbulent* It is known as *rotational* if the element or particles of the fluid at each point has some angular velocity about that point. Liquids are treated as incompressible and hence flow of liquids is considered as *incompressible* and that of gases as *compressible*. If the paths of all particles of the fluid in a pipeline of unvarying cross-section are exactly parallel to the walls of the pipe, then the flow is described as *streamlined, laminar* or *viscous*. For values of Re (Reynolds number = $d\rho v/\mu$) < 2000, the flow pattern is laminar, which is true for viscous liquids flowing through a capillary tube. For values of Re > 4000, the flow is turbulent. The swirls and eddies produced in turbulent flow may become so violent as to make the paths of the particles helical, with the particles moving in spirals along the pipe. *Pulsating* flow occurs when a single-acting reciprocating pump delivers the fluid through a pipe, in which case the velocity and the pressure of the fluid vary rapidly with time. To render the flow steady, straightening vanes are introduced to suppress the helical flow and a cushioning chamber of large capacity is inserted in the line to suppress the pulsations of velocity and pressure.

Bernoulli's theorem is applied to obtain the total energy of the fluid at each of the two points M and N separated by a distance and in a tapered pipeline as shown in Figure 5.36. It is assumed that (a) the walls of pipeline are smooth and frictionless, (b) viscous forces are negligible, (c) there is no storage or leakage of fluid in the region between the two points, and (d) the temperature of the fluid is the same at the two points. If a unit mass of the fluid is considered, the total energy at section 1 through M is the sum of velocity energy (= $1/2\ V_1^2$), pressure energy (= P_1/ρ_1), and the potential energy (= gZ_1), and is equal to the total energy at section 2 through N, as per the law of conservation of energy. Representing the values of the variables, with subscript 2 at the section through N, we have the energy balance equation given by

$$\frac{V_1^2}{2} + \frac{P_1}{\rho_1} + gZ_1 = \frac{V_2^2}{2} + \frac{P_2}{\rho_2} + gZ_2 \tag{5.50}$$

where,

P_1, P_2 = pressures
V_1, V_2 = velocities of flow
ρ_1, ρ_2 = mass densities of the fluid
Z_1, Z_2 = elevations of the points from a reference plane

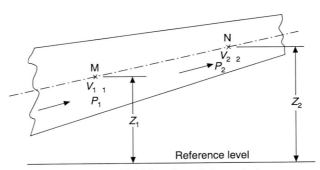

Figure 5.36   Fluid flow through tapered pipe.

For liquids which are assumed to be incompressible, $\rho_1 = \rho_2$, and for the flow through horizontal pipeline the equation simplifies to

$$P_1 - P_2 = \Delta P = \frac{\rho}{2}(V_2^2 - V_1^2) \tag{5.51}$$

where, ΔP is the differential pressure. Equating the volume flow rates at the two points M and N, it is seen that

$$\dot{Q} = \dot{Q}_1 = A_1 V_1 = \dot{Q}_2 = A_2 V_2$$

giving

$$V_1 = V_2(A_2/A_1) = V_2 m \tag{5.52}$$

where, m is the ratio of areas of cross-sections A_2 and A_1 of the pipe at N and M respectively. Rewriting Eq. (5.51), V_2 is given by

$$V_2 = \frac{\sqrt{(2\Delta P / \rho)}}{\sqrt{(1 - m^2)}} = E\sqrt{(2\Delta P / \rho)} \tag{5.53}$$

where, E is known as the *velocity of approach factor,* and is equal to $1/\sqrt{(1 - m^2)}$. Measurement of the differential pressure P and the knowledge of A_2, m and ρ enable the determination of volume flow rate $\dot{Q}$.

In deriving Eqs. (5.51)–(5.53), the velocity at M is assumed to be the same at all points of the section normal to flow and passing through M. In fact, the velocity of the particles of the fluid at the walls of the pipe is zero, while that at the centre of the pipe is maximum. The velocity distribution or the profile depends on the Reynolds number. Figure 5.37 shows the velocity profiles which are symmetrical about the centre-lines for the two cases (i) the laminar flow (Re < 2000) and (ii) turbulent flow (Re > 4000). There will be considerable deviation from this symmetry if the section under consideration is within a distance equal to six times the pipe diameter from the location of a bend, valve or any other obstruction.

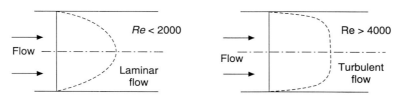

Figure 5.37   Velocity profiles for fluid flow.

In practice, one comes across situations in which fluid flow takes place through orifices and obstructions of different sizes and geometrical configurations. To measure the velocity and the volume flow rate of liquids and gases, orifice plates and other types of obstructions are inserted in the flow so as to develop a reasonable value of pressure difference across them. It is essential to realize how the flow patterns exist in such cases. Figure 5.38 indicates the flow pattern of a jet of liquid gushing out from a hole made in the vertical wall of a tank. The jet is seen to contract in its area of cross-section as it leaves the hole. At the section a–a the jet has minimum cross-section and maximum velocity. This section at which the flow lines become parallel for the first time after emerging from the hole is known as *vena contracta,* and the reduction in cross-section of the flow is denoted by a factor known as *coefficient of contraction,* C_c, given by

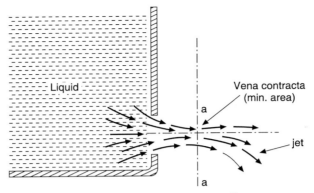

Figure 5.38 Liquid jet from an orifice.

$$C_c = \frac{\text{area of jet at vena contracta}}{\text{area of the orifice or hole}}$$

C_c varies with the size and shape of the orifice as well as the nature of its edges. For a small sharp-edged orifice, it is about 0.65. Similarly, another factor known as *coefficient of velocity* C_v is used to reveal the effect of the orifice on the velocity and is given by

$$C_v = \frac{\text{actual velocity at vena contracta}}{\text{theoretical value of velocity}}$$

The product of C_c and C_v is defined as the coefficient of discharge C_d given by

$$C_d = \frac{\text{actual discharge}}{\text{theoretical discharge}}$$

The values of C_d are experimentally determined and furnished by the manufacturers, for the different types and sizes of obstructions manufactured by them. The values of C_d and the values of theoretical discharge enable the determination of actual discharge. It must be again emphasized that C_d is not a constant for an obstruction and that it is also dependent on the Reynolds number.

When dealing with the flow of gases through pipelines and obstructions, the above equations have to be modified, taking into consideration the effects of temperature and pressure on behaviour of gases. The gas laws are obeyed by most gases under conditions remote from critical temperatures and pressures. When gas flow takes place through an orifice, change of pressure occurs so suddenly that the gas cannot absorb any heat energy from its surroundings. When it expands due to reduction in pressure, it does work by using its own heat energy with the consequent fall in temperature. Using the gas laws applicable for such an adiabatic expansion of a gas, an *expansion factor Y*, is used to obtain actual mass flow rate M, as given by mass flow rate (for compressible flow) $\dot{M}$ which is equal to Y times the mass flow rate (for incompressible flow). The factor Y is dependent on the ratio of specific heats k of the gas, the ratio β of orifice diameter to the pipe diameter, and the ratio of downstream to upstream pressure P_2/P_1, and is given by

$$Y = \left[\left(\frac{P_2}{P_1} \right)^{2/k} \frac{k}{k-1} \frac{1 - \left(\frac{P_2}{P_1} \right)^{(k-1)/k}}{1 - \left(\frac{P_2}{P_1} \right)} \frac{1 - \beta^4}{1 - \beta^4 \left(\frac{P_2}{P_1} \right)^{2/k}} \right]^{1/2} \tag{5.54}$$

Table 5.6 contains the values of Y for a range of β, k and P_2/P_1.

Table 5.6 Some expansion factors for compressible flow, Y

P_2/P_1	k	Ratio of diameters $\beta = \dfrac{d(orifice)}{D(pipe)}$			
		0.30	0.40	0.50	0.60
0.95	1.40	0.973	0.972	0.971	0.968
	1.30	0.971	0.970	0.968	0.966
	1.20	0.968	0.967	0.966	0.962
0.90	1.40	0.944	0.943	0.941	0.935
	1.30	0.940	0.939	0.935	0.931
	1.20	0.935	0.933	0.931	0.925
0.85	1.40	0.915	0.914	0.910	0.902
	1.30	0.910	0.907	0.904	0.896
	1.20	0.902	0.900	0.895	0.887
0.80	1.40	0.886	0.884	0.880	0.868
	1.30	0.876	0.873	0.869	0.857
	1.20	0.866	0.864	0.860	0.848

If the gas contains water vapour, an additional factor known as *moisture correction factor* N is used to obtain the correct mass flow rate. N is equal to unity for a dry gas. The net correction for the mass flow rate is worked out by incorporating the values of upstream temperature and pressure of the gas along with the specific gravity of dry gas relative to air. Values of N are available in literature.

The techniques based on the principles explained above lead to two distinct types of fluid flow meters: (i) those enabling volume and mass flow rate measurement utilizing the pressure differential across an obstruction, and (ii) those in which direct measurement of velocity of flow is obtained. Some of the primary mechanical transducers and systems that enable further transduction of their output signals into electrical signals are described in the following sections. There are some other commonly used mechanical flow and quantity meters which are described in detail in standard textbooks; some others of recent origin based on novel physical phenomena are presented in Chapters 7 and 8.

5.8.1 Pitot-static Tube

When a fluid is flowing in a horizontal pipeline in a streamlined fashion, it is known that the centre-line of the pipe coincides with the velocity vector and that the velocity varies from zero at the walls of the pipeline to the maximum at the centre. At any cross-section of the pipe, if the velocity at any point is required, a measuring element that does not alter the velocity profile has to be inserted into the tube. In other words, the sensing element introduced should be infinitely small in size. A Pitot tube having a small opening facing the direction of fluid flow is kept in the pipe as shown in Figure 5.39(a) to measure the velocity of the stream at the mouth of the tube. The fluid impinging on the open end will be brought to rest and the kinetic energy of the moving particles will be converted into pressure energy. Hence the pressure built up in the tube, known as total pressure or stagnation pressure, will be higher than the *free-stream pressure* or the *static pressure* that would have been present in the absence of the tube. This excess pressure head is measured as Δh and this increase in pressure is known as *impact pressure*. With V_2 becoming zero in Eq. (5.51), the excess pressure $(P_2 - P_1)$ is given by $(P_2 - P_1) = \Delta P = \rho g \Delta h$, and the velocity of the stream at the mouth of the pitot tube is given by

$$V_1 = \sqrt{2g\,\Delta h} \tag{5.55}$$

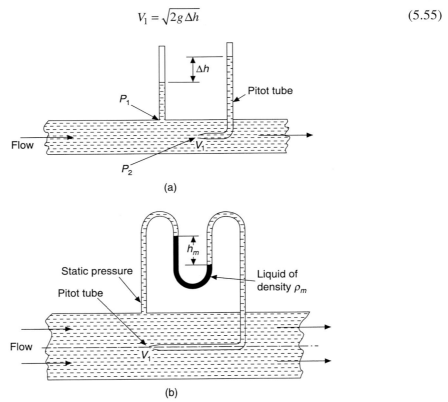

Figure 5.39 Static and Pitot tubes for velocity-of-flow measurement: (a) Static tube; (b) Pitot tube.

In fact, tapping the wall of the pipe and connecting a tube manometer as shown in Figure 5.39(b) enables the measurement of the difference between the free-stream pressure P_1, and the stagnation pressure P_2 as $\Delta P = h_m \rho_m g$. The value of V_1 is given by

$$V_1 = \sqrt{2h_m \rho_m g / \rho} \tag{5.56}$$

where,

h_m = difference in levels of the manometer
ρ_m = mass density of the manometric liquid
ρ = mass density of fluid in motion

It is important to measure the static pressure in the vicinity of the mouth of the Pitot tube for accurate measurements of the local velocity, and so combinations known as Pitot-static tubes, as shown in Figure 5.40, are designed for use with a U-tube, inclined-tube manometer or any other differential pressure transducer. They are available in different sizes, and a correction factor for each is provided by the manufacturers for calculation of the actual velocity. The correction factor varies with the geometry and type of tube-tip and the provision to sense the static pressure. Pitot-static tubes are very much used for investigating the point-to-point variations of the velocity of a fluid flow. It is essential to recall that the tube axis should coincide with velocity vector. The Pitot tubes are used for both liquid and gas flow velocities and for measurements of the velocity of aircraft relative to air. An angle of misalignment of the tube up to 5° from the direction of flow has negligible effect on the velocity values obtained. It is suitable for fluids which are clean without any particulate or suspended matter. For gases the expression for velocity gets modified to

$$V_1 = \left[\frac{2P_{\text{stat}}}{\rho} \frac{k}{k-1} \left\{ \left(\frac{P_{\text{stag}}}{P_{\text{stat}}} \right)^{(k-1)/k} - 1 \right\} \right]^{1/2} \tag{5.57}$$

Pitot-static tube lends its application to velocity measurements in open channels also. Velocity of the fluids should be somewhat higher for developing appreciable stagnation pressure.

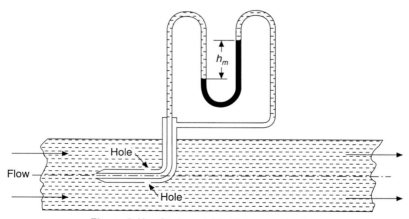

Figure 5.40   Static and Pitot tube combination.

5.8.2   Flow-Obstruction Elements

While the Pitot-static tube measures the velocity at a point in a stream, with practically no disturbance to the flow pattern, the obstruction elements are used for the measurement of the mean flow velocity as also the flow rate, with considerable disturbance to the flow pattern and consequent pressure loss due to obstruction. There are four commonly used versions of the obstruction element: these are known as venturi, flow nozzle, Dahl tube, and orifice-plate. Their geometrical configurations and other details are shown in Figure 5.41. The upstream pressure P_1 and the downstream pressure P_2 are obtained by suitable tappings provided on either side of the obstruction and the pressure differential $\Delta P(P_1 - P_2)$ is used to obtain the theoretical

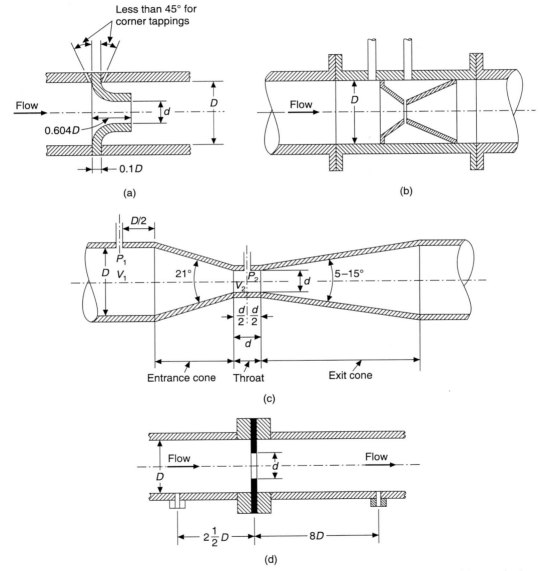

Figure 5.41   Obstruction elements for fluid flow rate: (a) Flow-nozzle; (b) Dahl flow-tube; (c) venturi tube; (d) orifice plate.

value of the mean flow velocity V_2 by using Eq. (5.53). The actual volume flow rate is calculated from the mean flow velocity V_2, the area of opening offered by obstruction is $A_2 (= \pi d^2/4)$, and is given by

$$\dot{Q} = C_d E A_2 V_2 = K A_2 V_2 \tag{5.58}$$

where,

K = flow coefficient (= $C_d E$)
C_d = discharge coefficient
E = velocity of approach factor

The discharge coefficients and the flow coefficients are determined experimentally for each size and version of obstruction element and provided by the manufacturers. It should be remembered that these coefficients vary with the flow conditions as determined by the Reynolds number, and the accuracy of the flow rate determination depends on the application of the correct coefficient.

There are generally three accepted positions for the upstream and downstream taps, and the differential pressure measured depends to some extent on the location of these taps. The flow coefficients furnished by the manufacturers take into account the effects of tap location. Flange taps are normally used with orifice plates and form an integral part. They are located 2.54 cm away from the faces of the orifice plate. Corner taps are also used for orifice plates as shown in Figure 5.41 with the holes opening into the corners formed by pipe wall and the orifice plate. Another combination (vena contracta taps) employs upstream tap at a distance of D (pipe diameter) from the upstream face of the obstruction, and the downstream tap at $D/2$ away from the downstream face. These locations amount to those of vena contracta yielding maximum pressure drop. The third combination (pipe taps) employs upstream tap at $2.5D$ away from the obstruction and downstream tap at $8D$ away. These taps provide a pressure differential equal to the total loss of pressure.

The flow nozzle shown in Figure 5.41(a) is a one-piece obstruction element which can be welded into the pipe for the measurement of high flow velocity of water or steam at high pressure.

The Dahl tube given in Figure 5.41(b) consists of a short length of parallel lead-in pipe followed by the converging upstream cone and the diverging upstream cone. A circumferential gap is kept between the two cones and the downstream pressure is tapped at the location of the gap. Strangely, Dahl tube produces larger pressure differential with lower pressure loss as compared to the venturi.

The venturi consists of three parts: entrance cone, throat, and exit cone as shown in Figure 5.41(c), and for the location of the taps, it has a total loss of pressure equal to 10–20% of the pressure differential across it. Though it is considerably long and costly, it offers high accuracy, low pressure loss, and suitability for flow rate of fluids containing small quantities of fine particulate matter.

The orifice plate in Figure 5.41(d) is the simplest and cheapest element, with its coefficient of discharge being the lowest at about 0.6. It is a thin metallic disc arranged concentric with the pipe in most cases. It may be eccentrically located when intended for use with fluids containing small traces of particulate matter. The segmental orifice plate has a hole that is partly circular located below its centre. The three types are shown in Figure 5.42. The size of the opening is designed so as to produce approximately the maximum pressure difference at the maximum rate of flow. The edge of the orifice plate on the upstream face should be sharp as rounding or burring considerably affects the flow rate. The dimensions of one of the sharp-edged versions are shown in Figure 5.42. Depending on the nature of the fluid and its flow rate, they may be made of nickel, gun metal, ebonite or plastics.

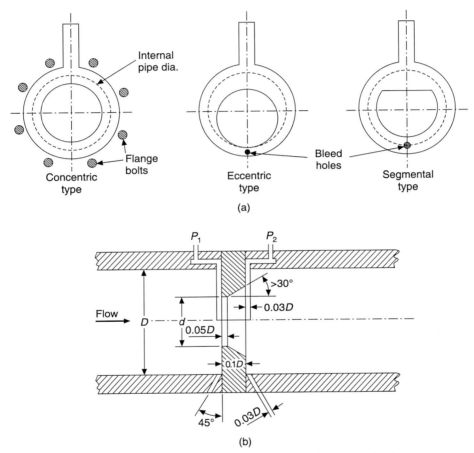

Figure 5.42   (a) Various types of orifice plates; (b) details of a sharp-edged orifice plate.

All the measurements involving obstruction elements have their useful range restricted from one-third to the full scale value of the velocity because of the relationship between velocity and the differential pressure.

5.8.3   Centrifugal Force Element

Another novel element that develops a pressure difference due to the flow of liquid in a pipe, but without being inserted into the flow causing obstruction, is the centrifugal element. It is, in fact, not an additional fixture to the pipeline, but any long horizontally laidout pipeline would serve the purpose if at a suitable location it could be given a right angle bend in such a way as to form a smooth arc of a circle of radius R in the horizontal plane. The centrifugal forces created by the change of direction of flow of the liquid inside the pipe are utilized and the difference in pressures on the outside and inside of the bend $(P_1 - P_2)$ is measured by a pressure transducer. Pressure taps are located in the inner and outer circumferential walls of the bend, as shown in Figure 5.43, on the centre-lines of the two legs of the bend and in the same radius plane. The difference in pressure depends on the velocity of flow and the density of liquid. The mass flow rate M is given by

$$M = C_d A \sqrt{\rho(P_1 - P_2)} \qquad (5.59)$$

where,

C_d = discharge coefficient
A = area of pipe cross-section
ρ = mass density of the liquid

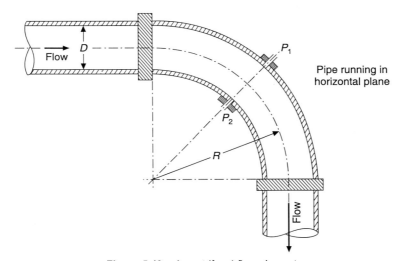

Figure 5.43   A centrifugal flow-element.

To develop reasonable difference in pressure, it is necessary to have a high value for the flow velocity. $P_1 - P_2$ referred in Eq. (5.59) is very small, which is an advantage by itself. It is essential that the pipe should run horizontally for at least a distance of 25 pipe diameters on either side of the bend.

5.8.4   Static Vane Elements

If a light, flat rectangular vane suspended freely about its upper edge is kept in a horizontal stream of liquid, it will be deflected from its vertical position, as shown in Figure 5.44 due to the change in momentum of liquid impinging normally on the flat surface of the vane. The angle of deflection depends on the force which is a function of the velocity of flow. For each liquid

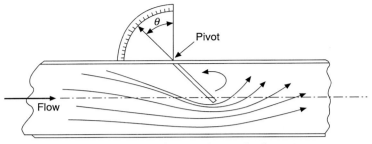

Figure 5.44   A static vane element for flow rate.

of density ρ, the relationship between the angle of deflection, θ, and the velocity of flow have to be experimentally determined and calibration charts provided. The relationship is far from linear.

The static-vane type element may be used for open and closed channels. When used for closed pipes, the vane element may be made to work as a cantilever, with its one edge rigidly fixed in the walls, while the other edge is at the centre of the pipe. The deflection of the cantilever spring or the strain on its surface becomes a measure of the velocity. It must be noted that these elements cause considerable disturbance and pressure loss in the flow.

5.8.5   Rotating-vane Systems

The force that produces deflection of the static vane in the stream may be utilized to produce continuous rotation, if a set of vanes are attached radially around a pivoted spindle as shown in Figure 5.45. It will be seen that there is one vane in the stream at any one time, in which case the rate of rotation of the spindle is proportional to the mean flow velocity. The total number of revolutions may be used to indicate the total volume of liquid that passed through the system. If the density is constant, the readings may be used for mass flow rate and the total mass of liquid that flows through over the desired interval of time. They can be used for flow in either direction and calibrated for both horizontal and vertical flows. The merits of the rotating-vane system are smaller pressure loss as compared to the static-vane element and suitability for use with secondary transducers for direct digital indication of flow rate and quantity flown.

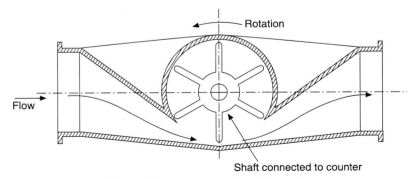

Figure 5.45   Rotary-vane type flow meter.

Considerable design modifications of the basic rotating-vane flow meter resulted, in course of time, in the development of turbine meters having rotors as shown in Figure 5.46(a) and (b) with helical vanes, or aerofoil-shaped vanes. Primary consideration is given to reduce the effects of friction and effect considerable improvement in the accuracy and linearity of the relationship between the angular velocity and flow rate. These flow meters are available for wide ranges of flow rate with negligible pressure drop and high repeatability.

The common turbine-type flow meter shown in Figure 5.46(b) has an axially mounted freely rotating turbine rotor, having its axis coinciding with the centre-line of the pipe and the direction of flow of the fluid. The fluid in motion impinges on the rotor blades, resulting in the development of a torque on each blade or wing of the rotor due to the shape and curvature of the blades. The rotor goes into rotation with an angular velocity proportional to the fluid velocity. Under steady state conditions, the volume flow rate is proportional to the angular velocity. The rotor is supported by ball or sleeve bearings on a shaft which in turn is held rigidly inside the meter. The rotor speed is nowadays measured by means of an electromagnetic

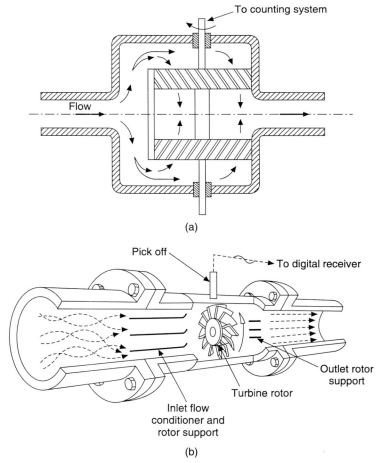

Figure 5.46 Turbine flowmeters: (a) Vertical shaft type; (b) horizontal shaft type.

transducer and associated digital readout instead of the conventional mechanical counter. The number of blades is normally three or four and the entire system is designed to stand the stream pressure and for introduction into the pipeline.

5.8.6 Rotameter-float System

The rotameter-float system employs a float kept in a fluid stream so that its position is a measure of the velocity of the fluid. The fluid whose velocity has to be measured is passed through a tapered glass tube as shown in Figure 5.47(a). The float obstructs the flow and the fluid flows through the annular clearance between the float and the inner wall of the tube. When the forces working on the float due to the upward flow of the fluid are balanced by the weight and buoyancy forces, the float will come to rest. The clearance between the bottom of the float and the tapered glass varies in area with the position of the float, and the flowing fluid through this clearance sets up a differential pressure across the top and bottom surfaces of the float. It can be seen that the float always assumes a position for each velocity holding the differential

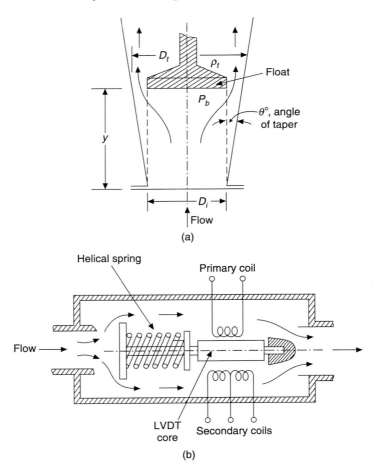

Figure 5.47   Rotameter-float type flowmeters: (a) for vertical flow; (b) for horizontal flow.

pressure constant. Hence the instruments based on this principle are known as constant head variable area type flow meters. They are also known as flow rotors. The position of the float is calibrated in terms of fluid velocity or it is converted into electrical signals by means of secondary transducers.

The upward force on the float is given by $A_f(P_b - P_t)$, where A_f is the area of the bottom surface of the float exposed to the flow and P_b and P_t are the pressures at the bottom and top respectively. The downward force is $V_f(\rho_f - \rho)$ where V_f is the volume of the float and ρ_f and ρ are mass densities of the float and fluid respectively. Neglecting viscous forces working on the float, the upward force can be expressed as

$$A_f(P_b - P_t) = V_f(\rho_f - \rho)g \qquad (5.60)$$

or

$$(P_b - P_t) = \Delta P = \frac{V_f g(\rho_f - \rho)}{A_f}$$

yielding the volume flow rate $\dot{Q}$ given by

$$\dot{Q} = AC_d \sqrt{\frac{2V_f g(\rho_f - \rho)}{A_f \rho}} \qquad (5.61)$$

where A is the area through which flow takes place and C_d is the discharge coefficient.

Assuming that the angle of taper of tube is small and that the diameters of the float and the tube at the inlet end D_i, it can be shown that

$$A = \pi D_i y \tan \theta = ky \qquad (5.62)$$

where $k = $ a constant equal to $\pi D_i \tan \theta$. If the density of the liquid is constant, the volume and mass flow rates are each proportional to the displacement of the float, y. The ratio of the mass flow rate $\dot{M}$ to the displacement is known as meter constant and is effected by 0.5% for a change in density of the liquid by 10%. The mass flow rate may, however, be made totally independent of density, by making the density of the float double that of the fluid. The float has to be made hollow or of solid light plastic material. To make the float independent of viscosity drag, the length of the float is made small while ensuring that the Reynolds number of the flow is not greater than 2000.

The tube is made of borosilicate. The guiding spindle or shaft carrying the float at the lower end passes through the upper part of the tube, with the upper end serving as an index on the scale. Or, alternatively, a secondary electrical transducer for displacements converts y into a proportionate electrical signal. The case so far presented pertains to flow in the vertical direction through the tube. If the flow is horizontal and the tube is in the horizontal position, the float is backed by a spring as shown in Figure 5.47(b). The spring force is proportional to the displacement of the float and is not a constant. Hence the flow rates are nonlinearly related to displacement. The secondary transducer and its associated circuitry may be made to correct for this nonlinearity and present a signal proportional to the velocity.

The rotameter flow system is versatile in that it can be designed to measure the flow rates of liquids widely ranging in their viscosities and volume flow rates as low as 0.1 cc/min.

5.9 DISPLACEMENT-TO-PRESSURE TRANSDUCER

The need to convert small displacements and variations in clearances between any two surfaces into corresponding air pressure variations arises in many situations in industries. The output signal can as well be a variation in hydraulic pressure where fast action is desired. The flapper-nozzle valve briefly described in Section 4.6.3 is shown in Figure 5.48. The outlet port consists of a nozzle of diameter d_n, kept in close proximity of the flapper which is freely suspended from a pivot. The nozzle-back chamber is fed from a constant pressure source through an orifice restrictor or a capillary tube. The position of the flapper with respect to the nozzle determines the pressure of the fluid in the chamber. The pressure of the fluid in chamber, P_o, is the output signal of the system and hence the valve is treated as a displacement-to-pressure transducer.

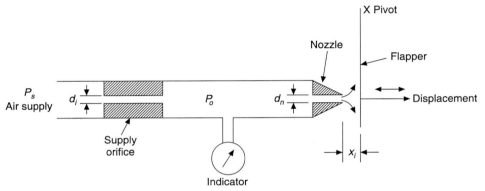

Figure 5.48   A flapper-nozzle displacement transducer.

As the supply pressure is held constant, the ratio of P_o/P_s may vary from zero to unity when the clearance between the nozzle and flapper, x, varies from a large value to zero, as shown in Figure 5.49. The displacement-pressure characteristic is nonlinear at the extremes, but sufficient linearity exists over a certain range of x. Hence the operating point is chosen at the mid-point of the linear range and a mean position for the flapper is determined for each transducer.

Under steady state conditions, the mass of the fluid entering the chamber is equal to the mass bleeding through the nozzle. In the case of a hydraulic valve, the liquid is assumed to be incompressible and hence the volume flow rates may be considered for obtaining the relationship between the input displacement and the output pressure.

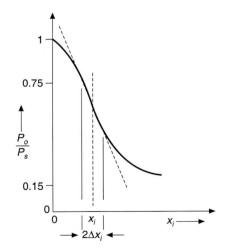

Figure 5.49   Displacement-pressure characteristic of a flapper-nozzle valve.

The volume flow rate of the liquid through the supply side orifice is given by

$$\dot{Q}_i = C_1 \frac{\pi d_i^2}{4} \sqrt{2(P_s - P_o)}$$

where

d_i = diameter of the orifice
C_1 = discharge coefficient

The volume flow rate through the nozzle is governed by the surface area of the cylindrical opening between the flapper and nozzle, and is therefore given by

$$\dot{Q}_o = C_2 \pi d_n x_i \sqrt{2P_o}$$

where

d_n = diameter of nozzle
C_2 = discharge coefficient

The outside (atmospheric) pressure is considered negligible in its effect on $\dot{Q}_o$ in comparison to that of P_o.

Under equilibrium conditions, $\dot{Q}_i = \dot{Q}_o$, thereby resulting in

$$\frac{P_o}{P_s} = \frac{1}{1 + 16\dfrac{(C_2 d_n x_i)^2}{C_1 d_i^2}} \tag{5.71}$$

Assuming $C_1 = C_2$, it can be shown that the displacement-pressure characteristic has maximum slope when x_i is about $0.14(d_i^2/d_n)$. However, the operating point for each case is so chosen that for equal displacements of the flapper to either side of the mean position equal changes in output pressure from the mean value are obtained. The ratio of P_o/P_s swings normally between 0.15 and 0.75. Commercial flapper-nozzle valves are available for use on both hydraulic and pneumatic power supplies.

Analysis of pneumatic flapper-nozzle valves may be carried out on lines similar to the analysis presented in Section 4.6.3, using the perfect gas equation and assuming conservation of mass flow. They are operated at an air pressure of about 13.79 N/cm² (20 psig) and are designed to provide output pressure variation between 3 and 15 psig, with its mean value at 9 psig.

The range of displacement of the flapper required to change the output pressure over the above useful range is called *throttling range,* and in most cases, it is as low as 0.1 mm. It is this high sensitivity of detection of small displacements that has made the transducer assume great importance in mechanical instrumentation, especially in situations where fire hazards exist.

The volume of the chamber is kept around 5 cc to reduce the time constant of the transducer which is usually about 100 ms.

The pneumatic (or hydraulic) reaction force on the flapper, due to the jet of the fluid bleeding out may be compensated by converting the arrangement into a push–pull system. This system consists of two similar transducers with one flapper located centrally between the two nozzles and subjected to displacements normal to the direction of streaming fluid.

Supply of air must be at constant pressure and must be dust-free and filtered. The flapper is usually actuated by sensing elements, such as a bimetallic strip, and the output pressure is utilized for indication. It is the basic element of many pneumatic and hydraulic instrumentation systems. It must be noted that it is highly susceptible to vibrations.

5.10 SEISMIC DISPLACEMENT TRANSDUCER

The seismic system mainly constitutes a mass-spring-dashpot combination housed inside a chamber. The housing chamber is rigidly connected to the body of the vehicle or a platform that is in motion. The vehicle or platform may be subjected to vibration, velocity or acceleration, and the seismic system enables the measurement of these quantities. The seismic system consists of a proof mass M supported by means of a spring and provided with a damping arrangement of the dashpot type as shown in Figure 5.50. It is a system having a single degree-of-freedom movement, and the mass is constrained to move only along the axis perpendicular to the platform. The absolute motion of the platform is represented by $x_i(t)$ and that of the mass by $x_m(t)$. When the platform is at rest, the mass assumes an initial position with respect to the housing frame. When the platform is in motion, in the upward direction, the mass tends to remain fixed in its spatial position with the result that it moves downwards. At any instant, the relative displacement between the housing frame and the mass is given by $x_o(t)$, where

$$x_o(t) = x_i(t) - x_m(t)$$

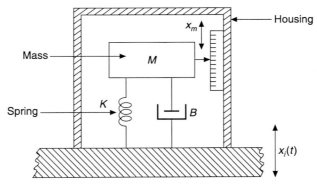

Figure 5.50 Basic seismic system.

Applying Newton's law, it is seen that

$$Kx_o(t) + B\dot{x}_o(t) = M\ddot{x}_m(t)$$

and replacing $x_m(t)$, the equation of the system becomes

$$M\ddot{x}_i(t) = M\ddot{x}_o(t) + B\dot{x}_o(t) + Kx_o(t)$$

The transfer function of the system is

$$\frac{X_o(s)}{X_i(s)} = \frac{s^2 / \omega_n^2}{(s^2 / \omega_n^2) + (2\xi s / \omega_n) + 1} \tag{5.72}$$

where

ω_n = the undamped natural frequency = $\sqrt{K/M}$

$$\xi = \frac{B}{2\sqrt{K/M}}$$

The above equation suggests a means of measuring $x_i(t)$ by measuring the relative displacement $x_o(t)$, and any relative displacement transducer can be used to measure $x_o(t)$ and provide electrical signals proportional to $x_o(t)$. For measurement of velocity and acceleration of vehicles in motion, either on land or in space, no reference is available, and in such cases, the seismic system enables measurement of $x_i(t)$ and its derivatives through the measurement of $x_o(t)$ by means of a suitable relative displacement transducer. Hence the seismic system is generally known as absolute displacement transducer (or pick up) in contrast to the relative displacement transducer that requires a fixed reference point or platform on which the fixed part is located and the moving part is connected to the vibrating structure. Even where a reference platform is available, the seismic system is preferred due to convenience in its application.

Expressing the ratio of the forcing frequency ω to ω_n as η, the frequency response of the system is given by

$$\frac{X_o(j\omega)}{X_i(j\omega)} = \frac{-\eta^2}{1 - \eta^2 + 2j\eta\xi} = |M| \angle\phi \tag{5.73}$$

Figure 5.51 represents the amplitude and phase response. For values of ω much larger than ω_n, it is seen that the ratio of $X_o(j\omega)/X_i(j\omega)$ is nearly equal to unity and the phase difference is zero. At a damping factor of about 0.7, the system has no peak in its amplitude characteristic. The bandwidth of the system will be larger if it is designed to possess a lower value of ω_n, which means that the spring must be softer and the mass larger. The size of M is dependent on the mass of the structure to which the seismic system is coupled, and it is essential to see that the frequency and amplitude of vibrations of the structure or platform are not altered due to its addition. The seismic system, thus, renders a measurement possible, which otherwise is very difficult to undertake. Vibration studies of several vehicle platforms and machinery are undertaken by using such seismic vibration transducers. It goes without saying that $x_o(t)$ is measured by a transducer that has the same bandwidth as that of the seismic system.

Figure 5.52 shows the seismic system for measurement of torsional vibration amplitudes.

Another important application of the seismic system is for measurement of absolute acceleration of vehicles in motion. The performance of the system can be studied by rewriting Eq. (5.72) as

$$\frac{X_o(s)}{s^2 X_i(s)} = \frac{1 / \omega_n^2}{s^2 / \omega_n^2 + 2\xi s / \omega_n + 1} \tag{5.74}$$

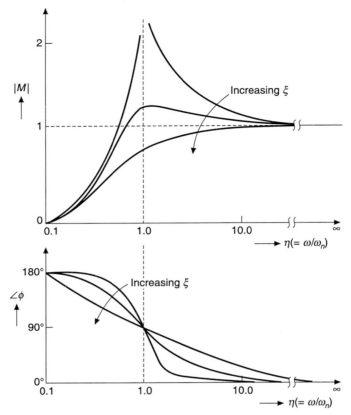

Figure 5.51   Amplitude and phase characteristics of a seismic displacement transducer.

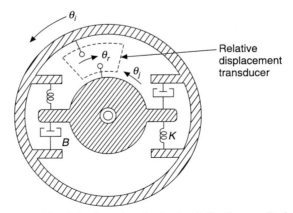

Figure 5.52   Seismic system for torsional vibration amplitudes.

where, $s^2 X_i(s) = A_i(s)$, the acceleration of the vehicle. The frequency response is given by

$$\frac{X_o(j\omega)}{A_i(j\omega)} = \frac{1/\omega_n^2}{1 - \eta^2 + 2j\xi\eta} = |M| \angle\phi \tag{5.75}$$

Figure 5.53 represents the amplitude and phase response and it is seen that the response is flat up to a frequency given by $0.4\omega_n$ if the damping factor is chosen as 0.707. The sensitivity is given by $1/\omega_n^2$, indicating that an effort to increase the bandwidth reduces the sensitivity. The design criterion for accelerometers is to have larger bandwidth, and hence a harder spring and lower mass are desirable.

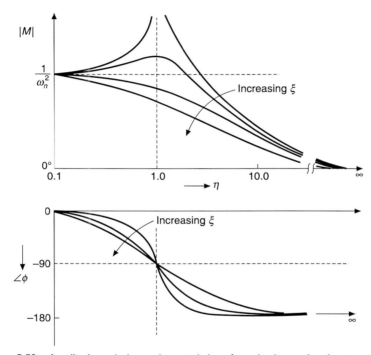

Figure 5.53 Amplitude and phase characteristics of a seismic acceleration transducer.

The seismic vibration transducer and the seismic acceleration transducer have both yielded signals proportional to their corresponding inputs of a certain frequency range. To obtain the values of absolute velocity of the vehicle, it is necessary to differentiate the output signals of the absolute displacement transducer or to integrate the output signals of the absolute acceleration transducer. In fact, integration of the signals obtained from accelerometers is preferred to differentiation.

Equation (5.72), when modified to give the transfer function of the system between $X_o(s)$ and the velocity input $V_i(s)$, becomes

$$\frac{X_o(s)}{sX_i(s)} = \frac{X_o(s)}{V_i(s)} = \frac{s/\omega_n^2}{\dfrac{s^2}{\omega_n^2} + 2\xi s/\omega_n + 1} \tag{5.76}$$

When the frequency response plot is obtained, it is seen from Figure 5.54 that amplitude is zero at both zero and infinite frequencies, and around resonance it is determined by the damping ratio ξ. Hence, it is difficult to obtain flat amplitude response for any required and reasonable range of frequencies. Thus the system does not present itself as a suitable means for measurement of vibratory velocities.

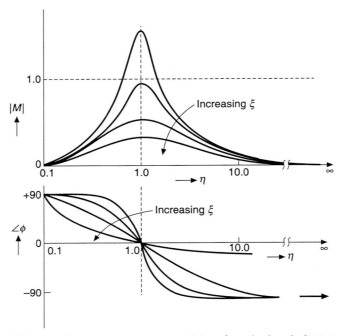

Figure 5.54   Amplitude and phase characteristics of a seismic velocity transducer.

Another possibility exists for obtaining a transfer function of the form given by Eq. (5.72). This possibility is by replacing the relative displacement transducer with a relative velocity transducer, whose transfer function is given by

$$E_o(s) = K_1 s X_o(s)$$

Replacing $X_o(s)$ by $E_o(s)/K_1 s$ in Eq. (5.72), the transfer function becomes

$$\frac{E_o(s)}{sX_i(s)} = \frac{K_1 s^2 / \omega_n^2}{(s^2 / \omega_n^2) + (2s\xi / \omega_n) + 1} \tag{5.77}$$

where $E_o(s)$ is the output signal of the relative velocity transducer. The frequency response plots resemble the curves of Figure 5.51 except that the sensitivity of the combination is K_1 instead of unity. Thus, the basic seismic system has been found to be extremely important in vibration studies and acceleration and velocity measurements of vehicles in motion. For each application, a suitable secondary transducer of desired bandwidth has to be selected so as to obtain accurate measurements. A variety of combinations of the basic seismic system and a secondary transducer exist and they are known as piezoelectric accelerometer or strain gauge accelerometer.

5.11   BASICS OF A GYROSCOPE

A free gyroscope, in its simplest form, consists of a wheel mounted on an axle and supported on a gimbol system such that the axis of the wheel may turn in any direction. In practical gyroscopes the wheel is replaced by the rotor of a polyphase induction motor, designed to operate at speeds as high as 20,000 rpm. The wheel is brought up to a high speed of rotation and left to itself to spin. The axis of rotation is called the spin axis. The other two axes perpendicular to the spin axis are used as input and output axes. The centre of mass of the rotor or wheel coincides with the centre of the supporting frame.

A free gyroscope with its supporting gimbol system is shown in Fig. 5.55. The spin axis is shown to coincide with the z-axis. The angular momentum H_s of the rotor is $J_s\Omega_s$, where J_s is its moment of inertia and Ω_s is its angular velocity. H_s is a vector whose direction is along the positive direction of z-axis in accordance with the convention of representation of the

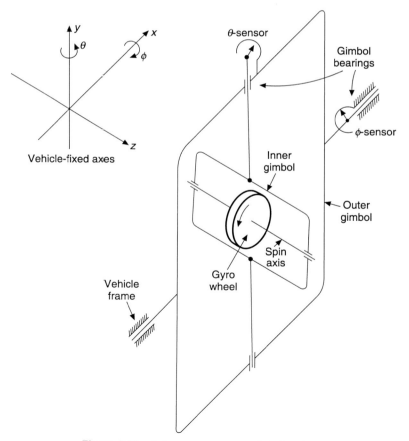

Figure 5.55   A free gyroscope on gimbol system.

direction of the angular momentum along the direction of advance of a right-hand screw while the direction of rotation is the same as the direction in which the screw is turned.

The inner gimbol supports the gyrowheel and is free to rotate about the *y*-axis. Similarly, the outer gimbol supports the inner gimbol and is free to rotate about the *x*-axis. The entire assembly is mounted inside a vehicle in motion such that the spin axis coincides with one of the three mutually perpendicular axes specified for the vehicle.

If the gimbol bearings are frictionless, the angular momentum will not change unless and until acted upon by external forces. Thus the spin axis which is initially set for the gyrorotor will be preserved with a high degree of fixity and hence remains fixed in space serving as a reference axis. Thus the first and foremost property of the free gyro is that its rotor behaves as if it is stiff and rigid in its state and unaffected by any disturbing torques.

If a small torque is applied about the *x*-axis in the positive direction in such a way as to make the rear flat side of the gyro wheel come to the top in the horizontal plane, the angular momentum should be disturbed. This is because of Newton's law which says that torque is the time rate of change of angular momentum. Since the torque is a vector pointing along the positive *x*-axis, the change in angular momentum must also point along the same direction. If the torque τ_x is applied for a small time Δt, then

$$\tau_x \, \Delta t = \Delta H_s$$
$$= J_s \Delta \Omega_s \quad \text{since } J_s \text{ is a scalar quantity}$$

Figure 5.56 shows the net effect of the applied torque. Since the torque is a vector pointing along the positive *x*-axis, the change in angular velocity must also point along the same direction. Hence the spin axis of the gyroscope suffers an angular displacement $\Delta \theta_y$ about the *y*-axis in the *x-z* plane and in the positive direction. If the angle $\Delta \theta_y$ is assumed to be small due to the torque impulse ($\tau_x \, dt$) being small, then from Figure 5.56 it is seen that

$$d\theta_y = \frac{\tau_x dt}{H_s} \tag{5.78}$$

or

$$\frac{d\theta_y}{dt} = \Omega_y = \frac{\tau_x}{H_s} \tag{5.79}$$

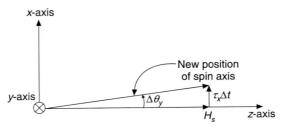

Figure 5.56   Effect of torque impulse on spin-axis.

Similarly, it can be shown that the spin axis suffers an angular displacement $\Delta \phi_x$ in the negative direction for torque impulse input along the positive direction of the *y*-axis, resulting in

$$\Omega_x = \frac{-\tau_y}{H_s} \tag{5.80}$$

$d\theta_y$ and $d\phi_x$ are known as angles of precession and Ω_y and Ω_x are velocities of precession.

The second important property of the free gyroscope is thus seen to be that, if a torque is applied about an axis transverse to the spin axis, the rotor turns about a third axis which is in quadrature to both. This tilting movement is known as *precession*. On removing the torque the precession disappears. From Eqs. (5.79) and (5.80), it may be noted that the applied torque results in a velocity rather than an acceleration and that the moment of inertia about the spin axis comes into picture rather than the moment of inertia about the axis of the applied torque.

The above observations concerning the properties of the free gyroscope suggest the possibility of utilizing the gyroscope for measurement of angular displacements and torques applied about one axis in terms of angular displacements about another axis. In the usual nomenclature of gyroscopes, the support axis of the gyro gimbol which is established by construction at right angles to the spin axis (SA) is defined as the output axis, OA. For the system shown in Figure 5.55, the y-axis is the OA, and the third axis (the x-axis) becomes the input axis, IA. The spin axis coincides with spin reference axis (SRA) when $d\theta_y$ is zero and coincidence is reflected by the null indication of zero output of the electrical position sensor of the output axis. When $d\theta_y$ is zero, the input axis is perpendicular to both OA and SRA (or SA).

So far the gyroscope has been assumed to be operating in ideal conditions with no gimbol-bearing friction. A constant friction torque about any axis acts as an input torque τ_f and results in an equivalent precession velocity ($= \tau_f/H_s$), and hence becomes a major source of error. It can be minimized by choosing a high value for H_s. Instead of increasing the mass and size of the gyro-rotor, its speed of rotation is usually brought up to a high value of about 20,000 rpm. Also, an unbalance of the gimbols results in gravitational forces and torques that tend to precess the gyro. The effect of all such imperfections is to cause drift of the spin axis of the gyro and means should be provided externally for correcting the drift at regular intervals of time. High quality gyroscopes are built with drift rates of less than 1° in 12 hours.

A better understanding of the behaviour of a gyroscope can only be obtained by considering all the forces at work when the spin axis and the angular momentum vector of the spinning rotor are shifted from the space reference axis. Such a situation arises due to the presence of small input torques and hence of small angular displacements of θ and ϕ of the OA and IA respectively. When considering $d\theta/dt$ about the y-axis, the moment of inertia J_y of everything that rotates when the inner gimbol turns in its bearing, should be taken into account. In the same way the moment of inertia J_x of all that rotates about the x-axis should be considered when the outer gimbol turns in its bearing.

Assuming that J_x and J_y as seen from the space-fixed x and y axes remain unaffected in their values due to the small angular displacements of ϕ and θ, the external applied torques τ_x and τ_y may be equated to the corresponding rates of change of angular momentum along the x and y axes. Resolving the angular momentum of the spinning rotor as shown in Figure 5.57, the torque equations can be obtained as

$$\tau_x = \frac{d}{dt}\left(H_s \sin\theta + J_x \frac{d\phi}{dt}\right) \tag{5.81}$$

$$\tau_y = \frac{d}{dt}\left(-H_s\cos\theta\sin\phi + J_y\frac{d\theta}{dt}\right)$$ (5.82)

Although it is possible to use a single free gyroscope to measure the two angles θ and ϕ, it is desirable to limit its usefulness to measure one angle only. As defined earlier, the y-axis is the OA and the x-axis is the IA for the arrangement shown in Figure 5.57. If the rotation of the gyro about the OA is restrained by means of a control spring and viscous damping, Eq. (5.82) will be modified as

$$\tau_y - B\frac{d\theta}{dt} - K_s\theta = \frac{d}{dt}\left(-H_s\cos\theta\sin\phi + J_y\frac{d\theta}{dt}\right)$$ (5.83)

where K_s is the stiffness of the control spring and B the coefficient of viscous damping.

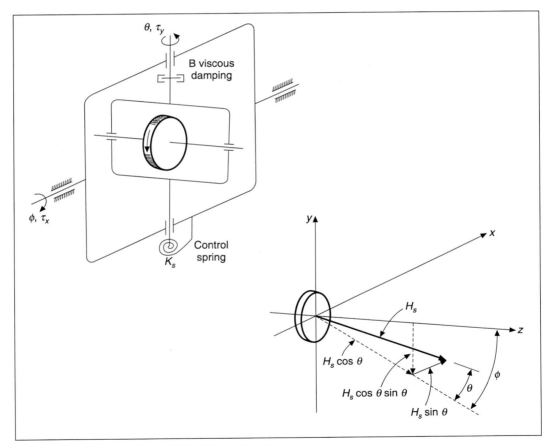

Figure 5.57   Effect of input torques on angular momentum.

Since θ and ϕ are defined as small angles only, Eqs. (5.81) and (5.83) can be simplified as

$$\tau_x = H_s \frac{d\theta}{dt} + J_x \frac{d^2\phi}{dt^2} \tag{5.84}$$

$$\tau_y - B\frac{d\theta}{dt} - K_s\theta = -H_s\frac{d\phi}{dt} + J_y\frac{d^2\theta}{dt^2} \tag{5.85}$$

These equations define the inter-relationship between the two torques τ_x and τ_y to the two angles θ and ϕ. Each of the two angles depends on both τ_x and τ_y as can be seen from the transfer functions obtained from the Eqs. (5.84) and (5.85):

$$\phi(s) = \frac{(J_y s^2 + Bs + K_s)\tau_x - H_s s\tau_y}{s^2[J_x J_y s^2 + BJ_x s + (H_s^2 + J_x K_s)]} \tag{5.86}$$

$$\theta(s) = \frac{J_x s\tau_y + H_s\tau_x}{s[J_x J_y s^2 + BJ_x s + (H_s^2 + J_x K_s)]} \tag{5.87}$$

For the single-axis gyro, $\tau_y = 0$, and hence for an input displacement of ϕ backed by a torque τ_y about the x-axis (IA), the output angle θ is obtained from Eqs. (5.86) and (5.87) as

$$\frac{\theta(s)}{\phi(s)} = \frac{H_s s}{J_y s^2 + Bs + K_s} \tag{5.88}$$

Equation (5.88) defines the transfer function of a single-axis restrained gyro, useful for measuring angular displacements conveyed to the gyroscope about the input axis. For an air borne vehicle, three such single-axis gyros are required to define completely the three axes of motion. Two free gyros may also be used to define the same, but their cross-coupling or interaction problems render the system complex for analysis and application. High performance systems employ three single-axis constrained gyros. The transfer function given by Eq. (5.88) suggests two possible methods of utilizing the gyro as (a) rate gyro and (b) rate integrating gyro.

5.11.1 The Rate Gyro

The rate gyro is used to measure absolute angular velocity, $s\phi(s)$ in terms of $\theta(s)$, and the TF relating them is obtained from Eq. (5.88) as

$$\frac{\theta(s)}{s\phi(s)} = \frac{H_s / K_s}{(s^2 / \omega_n^2) + (2s\xi / \omega_n) + 1} \tag{5.89}$$

where

$$\omega_n = \sqrt{K_s / J_y} \quad \text{rad/s} = \text{natural frequency}$$

$$\xi = B/2\sqrt{J_y K_s} = \text{damping factor}$$

The rate gyro is made more sensitive by having a high value for the angular momentum, H_s, and a soft spring of lower K_s. The natural frequency ω_n is around 10 to 100 Hz and a reasonable amount of damping is necessary to prevent oscillatory response.

When an aircraft is in motion, the vehicle axes are fixed as roll, pitch and yaw-axes as shown in Figure 5.58. To measure all three components of angular velocity along these directions, three rate gyros are mounted in the vehicle.

5.11.2   Integrating Gyro

If the control spring of the constrained gyro is removed, K_s becomes zero and the TF relating $\theta(s)$ and $\phi(s)$ becomes

$$\frac{\theta(s)}{\phi(s)} = \frac{H_s}{J_y s + B} = \frac{H_s / B}{1 + s\tau} \tag{5.90}$$

where $\tau = J_y/B$ = time constant in seconds.

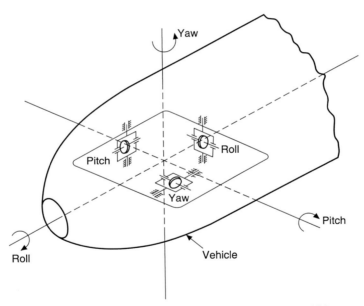

Figure 5.58   Mounting of three rate gyros in air-borne vehicle.

It is seen that the output angle θ is a direct measure of the input angle ϕ, and the integrating gyro behaves like a first-order system with a time constant τ. For small input displacements of ϕ, the output signal θ is proportional to ϕ instead of the rate of change of ϕ. Hence the restrained rate gyro with no control spring becomes the integrating gyro. High sensitivity for the integrating gyro is again obtained by having a high value for H_s and a lower coefficient of damping B. The rate integrating gyro is the basis of highly accurate inertial navigation systems where it is used as a reference to maintain the stable platforms in a fixed attitude within a vehicle while the vehicle moves arbitrarily. Sensitive and accurate position sensors, such as microsyn, are used to detect θ, and their output signals are utilized to drive servomechanisms that maintain the platform in a fixed angular orientation. Three accelerometers are mounted on the stable platform measuring the acceleration of the vehicle in the three specified directions. The outputs of the three accelerometers give an accurate measure of vehicle motion. The accelerometers and the gyroscopes form the most fundamental components for all inertial navigation and guidance systems. Gyroscopes are used to establish a spatial reference for the accelerometer and as a source of vehicle attitude information.

EXERCISES

1. Explain how bimetallic elements enable measurement of temperature and show how they are shaped for use.

2. Discuss the factors on which the sensitivity of bimetallic elements depend and how their sensitivity is maximized.

3. If a flat bimetallic element of length of 4 cm is to be made up of Invar and brass, determine the thickness of each strip to be used and the sensitivity of the element for a temperature rise of 1°C.

4. Briefly discuss the merits of using bimetallic elements in place of fluid expansion thermometers, when measurements of temperature in industry are involved.

5. Indicate the range of mechanical measuring systems commonly available for measurement of pressure directly and indirectly. State the units in which pressure is usually expressed when measured directly and indirectly.

6. Describe the constructional features of a ring-balance manometer and show how the deflection of the pointer is related to pressure.

7. Explain why metallic plate diaphragms are popular in industry for pressure measurements. State the factors on which the range of pressure to be measured by each element is decided.

8. In a particular case, phosphor bronze diaphragm of thickness 0.2 mm and of diameter 4 cm is used for measurement of pressure difference across it. Taking its Young's modulus as 105×10^9 N/m^2 and Poisson's ratio as 0.3, determine the range of pressure for which it can be used so as to have almost linear relationship between the deflection at the centre and the pressure difference. Estimate the extent of nonlinearity in the relationship, if the deflection at the centre is 0.08 mm (use Eq. (5.11)).

9. Distinguish the basic differences in the construction and performance of thin plate diaphragms and membranes and indicate the materials used for each of them.

10. Describe the constructional features of a capsule and a bellows and show how they are useful for pressure measurements.

11. Describe the reasons for the popularity of the Bourdon tube element for pressure measurement and show how the elements are constructed.

12. Suggest the range of mechanical transducers commonly used for force measurements.

13. What is a load cell and how is it made for use in industry?

14. Using Eq. (5.24), design a helical spring for use in a spring balance that enables a linear scale length of 8 cm and useful for weighing 10 kg. Suggest the material used.

15. Explain how cantilever beams can be used for force measurement. Discuss the nature of output signals.

16. Show how thin plate diaphragms also enable measurement of force. Show how the force is applied to the diaphragms.

17. Describe the constructional features of (a) column-type and (b) proving-ring type load cells, and show how their output signals enable further processing for indication.

18. What is a torsion bar and how is it used for torsion measurements?

19. Explain how the density of liquids contained in vessels under static conditions can be measured using (a) hydrometer (b) air bubbler system.

20. Explain the techniques by which levels of liquids in vessels can be converted into mechanical signals of displacement, pressure and force for measurement and indication of the level suitably.

21. Describe the arrangement necessary for the measurement of liquid level in closed containers and explain the need for the same.

22. Define 'absolute (dynamic) viscosity' and 'kinematic viscosity' of a fluid and indicate the units in which they are expressed.

23. Distinguish between 'laminar' and 'turbulent' flow of fluids and show how they can be distinguished and their effects noticed.

24. Explain the technique of measuring dynamic viscosity by means of a capillary tube.

25. Explain the concentric-cylinder technique of measuring the dynamic viscosity of liquids. Derive the relationship between the torque developed and viscosity.

26. Explain the nature of problems faced while measuring volume and mass flow rate of fluids.

27. Applying Bernoulli's theorem, obtain the pressure difference between a pair of points separated by a distance in a tapered pipeline kept horizontal and having a liquid flowing through it. From this, derive the velocity of flow at a point in downstream in terms of the pressure difference.

28. Define 'velocity of approach factor' and 'coefficient of velocity' and 'coefficient of discharge' when dealing with the flow of liquids.

29. Explain why gas flow measurements are complicated and what measures are taken to get somewhat accurate measurements.

30. Describe a Pitot-static tube and explain its usefulness for flow measurements.

31. Indicate the types of obstruction elements used for fluid-flow measurements and comment on their relative merits for application.

32. Describe how flow measurements of liquids may be undertaken by using (i) a static vane, and (ii) a rotating vane system. Discuss the nature of output signals obtainable from each.

33. Describe the constructional features of a rotameter and comment on its applicability for flow measurements.

34. Describe a flapper-nozzle valve and show how it can be used as (a) a displacement-to-pressure converter and (b) a displacement amplifier. Indicate the nature of power supplies on which it can be used.

35. Describe a basic seismic system and derive the basic relationship between its input and output displacements.

36. Explain how a basic siesmic system is exceptionally suitable for measurements of (a) absolute displacements and (b) absolute accelerations. Discuss its limitations, if any, while measuring the same.

37. Describe the constructional features of a free gyroscope and explain its performance.

38. What are the important performance features of the gyroscope, which lend themselves for application for measurements of angular displacements and torques?

39. For a single-axis gyro, derive from fundamentals, the relationship between the input and output angular displacements when an input torque is applied to it.

40. Describe (a) the rate gyro and (b) the integrating gyro, and suggest their application in inertial navigation systems.

SUGGESTED FURTHER READING

Adams, L.F., *Engineering Measurements and Instrumentation*, Hodder and Stoughton, London, 1975.

Adams, L.F., *Engineering Instrumentation and Control IV*, Hodder and Stoughton, London, 1981.

Ambrosius, E.E., Fellows, R.D. and Brickman, A.D., *Mechanical Measurements and Instrumentation*, Ronald Press, New York, 1966.

Beckwith, T.G. and Buck, N.L., *Mechanical Measurements*, Addison-Wesley, Reading (Mass.), 1961.

Boros, A., *Electrical Measurements in Engineering*, Elsevier, Amsterdam, 1985.

Carroll, G.C., *Industrial Process Measuring Instruments*, McGraw-Hill, New York, 1962.

Doebelin, E.O., *Measurement Systems: Application and Design*, McGraw-Hill, New York, 1966.

Fribance, A.E., *Industrial Instrumentation Fundamentals*, McGraw-Hill, Kogakusha, Tokyo, 1962.

Gibson, J.E. and Tuteur, F.B., *Control System Components*, McGraw-Hill, New York, 1958.

Holman, J.P., *Experimental Methods for Engineers*, McGraw-Hill, Kogakusha, Tokyo, 1966.

Kirk, F.W. and Rimboi, Nicholas, R., *Instrumentation* (3rd ed.); American Technical Publishers Inc., Chicago, 1975.

Liptak, B.G. (Ed.), *Instrument Engineers Handbook*, Vols. I & II, Philadelphia Chilton, Philadelphia, 1969, 1970.

Neubert, H.K.P., *Instrument Transducers*, Clarendon Press, Oxford, 1975.

Noltingk (Ed.), *Jones' Instrument Technology*, Vol. I, Butterworth Scientific Publications, London, 1956.

O'Higgins, *Basic Instrumentation*, McGraw-Hill, New York, 1966.

Patranobis, D., *Principles of Industrial Instrumentation*, Tata McGraw-Hill, New Delhi, 1976.

Wightman, E.J., Instrumentation in Process Control, International Scientific Series, CRC Press, Columbus, Ohio, 1972.

CHAPTER 6

Passive Electrical Transducers

6.0 INTRODUCTION

Electrical circuits consist of combinations of the three passive elements: resistor, inductor and capacitor. The primary parameters that describe them are respectively resistance, self or mutual inductance and capacitance. Any change in the parameter of the element can be recognized only when the element is made 'live' by electric energization or excitation, otherwise the element is in 'dead' state. Hence transducers that are based on the variation of the parameters due to application of any external stimulus are known as *passive transducers*. In this chapter, resistive, inductive and capacitive transducers are presented along with the several possibilities available for making use of them for measurement of physical and chemical variables. Wherever possible, sections are subdivided in such a way as to identify the element of the transducer and the measurand, such as strain-gauge flow transducer and capacitive strain transducer.

Basic characteristics of each transducer, its limitations and, where necessary, relevant signal processing circuitry are presented. Additional insight is provided for transducers that are more powerful and popular, so as to acquaint the reader with the developments in transducer technology. Though the criteria for the design of transducers have been enumerated, details concerning actual designs are not given. Published literature and other books may be consulted for such information.

6.1 RESISTIVE TRANSDUCERS

Applicability of resistive elements for the measurement of several nonelectrical quantities has been exploited to a great extent, resulting in the development of resistive transducers, each playing the role of either a primary or a secondary transducer. The primary factors that make resistive elements popular and powerful as transducers are versatility in regard to their size and resistance value, high speed of response, availability of a large variety of electrical circuitry yielding high resolution in measurements, and suitability for use in circuits with dc and ac supplies. The dc resistance R of a conductor of length l_x, having a uniform area of cross-section a and with specific resistivity ρ of its material is given by

$$R = \frac{\rho l_x}{a}$$

The resistor may as well be a column of a liquid or gas, contained in a chamber with a pair of electrodes. Change in the value of the resistive element can be brought about by subjecting the element to external stimulus that effects either dimensions of the element or its resistivity. Table 6.1 indicates the nonelectrical quantities that can be measured by the resistance transducer.

Table 6.1 Resistive transducers and the range of measurands

Effect	*As a primary transducer*	*Through the use of secondary transducers*
1. Change in length l_x	Linear and angular displacements, thickness	Temperature, pressure, weight, force, torque, level, density, fluid flow rate, viscosity, velocity, acceleration and attitude of vehicles in space
2. Change in length and area of cross-section	High pressure strain	All the above quantities except velocity and attitude
3. Change in resistivity	Temperature	Thermal conductivity, gas composition, fluid flow rate, wind velocity and direction, low pressure and vacuum and infra-red radiation
	Visible radiation	Proximity, intensity of visible radiation, and displacement
	Magnetic field Humidity Moisture content	Proximity and displacement

The fundamental relationship between the resistance and length of the conductor may be utilized for continuous monitoring of the thickness of a sheet of insulating material under production, while its resistivity and composition are held constant. The dimensional changes can be brought about by subjecting the resistive elements, to pressure, force of torque, directly or by means of some primary transducers. Resistive strain gauges enable measurement of strain of mechanical members on to which they are bonded. By a suitable choice of the primary mechanical transducers, it is possible to extend the scope to the measurement of density, level and flow rate of liquids. Composite transducer systems utilizing the resistive strain gauge for the measurement of the above quantities are either set up for each situation or are commercially manufactured.

It is, however, to be noted that the dynamic response of the transducer system becomes controlled and limited by the primary transducer, though the resistive element or strain gauge by itself has very good dynamic response. It is equally important to recognize the influence of temperature on the dimensions, and hence, all transducer systems utilizing dimensional changes should be worked at the temperature at which calibration is effected. Also, correction factors should be made available for use if steps are not taken to compensate for changes in temperature.

The resistivity of the material medium constituting the path of current in the resistor varies with the temperature and composition of the medium. Resistance thermometers are known, since long, to be exceptionally good for temperature measurements and bolometers for thermal (infra-red) radiation. In fact, they are primary transducers for these quantities. The resistivity of carbon and some other semiconductor materials is affected by pressure. The granules of these materials, when subjected to pressure, get closer and the resistivity and resistance falls. Systems known as *Carbon-piles* are used in voltage regulators. Resistivity variations of hygroscopic materials such as lithium chloride, due to absorption of moisture, are used as hygrometers. Moisture content of materials in production, such as paper, is continuously monitored by measuring the variation in resistivity. In such cases, an electrode system, for holding the material or medium, equivalent to that of a parallel-plate capacitor, is necessary. Conductivity cells of this nature are used to measure the concentration of electrolytic solutions and extended suitably to cover the measurement of flow-rate of certain liquids. Suitably designed electrode systems are also used to measure the density of liquids and pressure of gaseous atmosphere ranging from 10^{-1} to 10^{-12} mm Hg.

Other applications of resistive elements as primary transducers include the measurement of magnetic field, voltage and intensity of optical radiation.

It is observed that resistivity variation can occur due to more than one influencing factor, and hence care must be taken to see that all other interfering quantities are controlled and held constant. Temperature should be held constant while composition measurements are done.

While in all the above measurements the resistive element is kept near about the ambient temperature, techniques in which the element is brought to a temperature far above its surroundings are developed for continuous monitoring of the flow-rate of fluids (hot-wire anemometers) and composition of binary gas mixtures (thermal conductivity gauges). It is obvious from the above that the dynamic response of the resistive element is dependent on the ultimate form of transducer system, with its protective casing etc., the characteristics of the medium and the system under investigation.

In the following sections, common resistive transducers, which have become powerful in many respects, in the field of instrumentation are presented.

6.1.1 Resistance Thermometers

Resistance thermometers are primary electrical transducers enabling measurement of temperature changes in terms of resistance changes. The resistive element is usually made of a solid material, a metal, metallic alloy or a semiconductor compound. The resistivity of metals increases with temperature, while that of semiconductors and insulators generally decreases. Wire wound elements employ considerable length of wire, and if free to expand, the length also increases with increase in temperature. Hence as temperature changes, the change in resistance will be due to changes in both length and resistivity. Materials used for resistance thermometers have temperature coefficient of resistivity much larger than the coefficient of thermal expansion. The temperature coefficient of resistance, α, is given by

$$\alpha = \frac{1}{\Delta T} \frac{\Delta\rho}{\rho_0} = \frac{1}{\Delta T} \frac{\Delta R}{R_0}$$

where

ΔT = change in temperature, °C

$\Delta\rho/\rho_0$ = fractional change in resistivity

$\Delta R/R_0$ = fractional change in resistance

ρ_0, R_0 = resistivity and resistance respectively at 0°C.

The resistance R_T at any other temperature $T°C$ is given by

$$R_T = R_0(1 + \alpha\Delta T) = R_0(1 + \alpha T) \tag{6.1}$$

Each metal or metallic alloy obeys the above relationship over a range of temperatures and each resistance transducer is generally limited to measure temperature within this range. Nonlinearity sets in at higher temperatures and in such cases the relationship is modified as

$$R_T = R_0(1 + \alpha_1 T + \alpha_2 T^2 + \dots + \alpha_n T^n) \tag{6.2}$$

where α_1, α_2, ..., α_n are constants applicable for each metal.

Platinum, nickel, copper and tungsten are the commonly used metals for the wire wound elements. Resistivity of these metals and others are given in Table 6.2 showing the range of temperature of each. The temperature-resistance characteristic of platinum is linear and reproducible over a wide range of temperatures. It is available in pure form and is relatively unaffected by environmental conditions. Platinum elements are used for precision thermometry and a resolution of ±0.0001°C is possible when used along with suitable measuring circuits. Nickel has a high temperature coefficient of 0.0068 while platinum, copper and tungsten have a coefficient around 0.004. Platinum is highly stable when compared to other metals and alloys and hence is used as the basis of the International Temperature Scale from −190°C to +660°C. Alloys of metals do not possess any particular advantages for temperature measurements. Phosphor bronze is favoured for low temperature thermometry below 7 K.

Table 6.2 Properties of some metals and metallic alloys

Material	Density (g/cm³)	Temperature coefficient of resistance × 10³	Resistivity (μΩ m)	Thermal conductivity (cal/cm sec°C) × 10³	Gauge factor	Temperature coefficient of expansion × 10⁶	Remarks
Advance			0.45		2.1	14.4	Used for strain gauges
Aluminium	2.7	4.5		500		25.5	
Brass	8.4					18.9	
Bronze	8.7			200			

(Contd.)

Table 6.2 Properties of some metals and metallic alloys (*Contd.*)

Material	Density (g/cm³)	Temperature coefficient of resistance × 10³	Resistivity (μΩ m)	Thermal conductivity (cal/cm s °C) × 10³	Gauge factor	Temperature coefficient of expansion × 10⁶	Remarks
Carbon		−0.7					
Constantan		±0.04	0.48		2.1	17.0	Used for resistors
Copper	8.9	4.3		930		16.7	For coils and temperature measurement, −200°C to +250°C
Duralumin	2.79					22.6	
Ferry			0.4		2.2	12.5	Used for strain gauges
Gold	19.31	4.0		750			
Isoelastic		0.175				4.0	
Karma		0.02	1.25		2.1	10.0	Used for strain gauges
Manganin	8.5	±0.02					
Nichrome V		0.1	1.0		2.5	13.2	Used for strain gauges
Nickel	8.9	6.8	0.065		−12	12.8	For temperature measurement, −100°C to + 350°C
Platinum	21.48	3.92	0.1	1000	4.8	8.9	For strain gauges and temperature measurement, − 250°C to + 700°C; stable metal
Silver	10.5	4.1					
Stainless steel	7.9			110		11	
Titanium						9	
Tungsten	19.34	4.8					Used for hot wire filaments

The wire-resistance thermometer, as shown in Figure 6.1, usually consists of a coil wound on a mica or ceramic former, which also serves as a supporting mount for the coil. The coil is wound in bifilar form so as to make it noninductive. Such coils are available in different sizes and with different resistance values ranging from 10 ohms to 25,000 ohms. The open type element is brought in direct contact with the fluid whose temperature is to be measured. In such cases, the dynamic response depends on the size and quantity of resistive material of the element and the characteristics of the fluid. The fluid should be at rest and measurement will be in error if it is in motion. If the fluid is corrosive, the element will be spoiled, and if the liquid is reasonably conductive temperature measurement becomes erroneous. Hence the elements are normally enclosed in a protective tube of pyrex glass, porcelain, quartz or nickel, depending on the range of temperatures and the nature of the fluid. The tube is evacuated and sealed or filled with air or any other inert gas and kept around atmospheric pressure or in some cases at a higher pressure. As explained earlier, protective tubes lower the speed of response considerably. To obtain satisfactory degree of repeatability, say ±0.1°C, for the industrial grade thermometers, the resistive elements are designed such that the wire remains annealed and strain-free for the operating temperature range. Such strain-free designs are produced by allowing the wire to expand or contract freely as temperature changes. Two common forms of the elements on formers are shown in Figure 6.1.

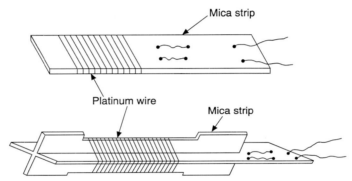

Figure 6.1 Resistance thermometer coils on mica formers.

Thin film resistive elements of platinum are available for surface temperature measurements. They are much smaller in size, having high speed of response relative to the wire type. Platinum is vacuum deposited to form a thin film of 0.1μ thick on an insulating substrate which in turn is bonded to a plastic backing. Glass encapsulated platinum films, deposited on a ceramic substrate, are available in sizes of 10 × 3 × 1 mm. The resistance-temperature characteristic of such film type thermometers differs considerably in slope from that of pure platinum. This deviation is attributed to the process of deposition of the film of platinum on the substrate. However, accuracy and repeatability of the order ±0.01°C is obtained, while having the merits of suitability for measurements of surface temperature of structures point by point, and high speed of response. They are very much used in aero-space industry. They render measurements of temperature in small and low thermal capacity systems possible. Films of nickel, gold, silver and germanium are also available but they are restricted to measurements of temperature ranging from 20°C to 200°C. Care must be taken to see that no strain is imparted to the element, while mounting or cementing the element on the surface of structures and also during the period of measurement.

Resistance of the elements falls within the range of 10–1000 ohms and hence for its measurement Wheatstone bridge circuits are exceptionally suitable, when provided with compensation for the leads of the resistance thermometer. The three-lead and four-lead connections are satisfactory in most cases. For accurate measurements with platinum resistance thermometry, the Müeller bridge configuration is employed. For laboratory work, null-balance type bridge circuits with dc excitation are satisfactory, but for continuous monitoring of process temperatures required in industry, deflectional-type bridge circuits or servo-operated, self-balancing bridge circuits are used. Bridge circuits with d.c. excitation are common, with filters for noise suppression. Excitation of ac has the merit of eliminating the effects of thermo-emfs developed in the leads of the thermometer and other components of the bridge. Special electronic circuits and amplifiers for use with bridge circuits are described in Section 3.5.

It is known that for deflectional type bridge circuits, the output voltage of the bridge is not linearly related to the fractional change in resistance for values greater than 0.05. Most of the resistance thermometer elements do not possess linearity for the entire operating range. But the important requirement of the present-day instrumentation systems is that the output voltage signals should be linearly related to the temperature over the operating range. It is possible to compensate the nonlinearity of the resistance-temperature characteristic by a suitably designed bridge having bridge ratio higher than unity. Where the transducer characteristic is sufficiently linear, bridge output nonlinearity is offset by means of electronic circuits, using operational amplifiers. The current through the resistance thermometer element should be of a magnitude that does not cause any self-heating error. A larger current may yield higher sensitivity, but at the cost of accuracy. In no case should its temperature rise beyond that of its surroundings, although it is possible to apply correction due to self-heating. The bridge output voltage is proportional to its excitation voltage. Depending on the rate of change of temperature and the speed of response of the thermometer system, temperature can be sampled by bringing in the resistive element or the entire bridge circuit into excitation and keeping it energized electrically for a much smaller duration, during which time the output voltage of the bridge is recorded. Such a pulsed excitation of the element or bridge yields much higher output voltages without causing much self-heating error. I^2Rt remains the same when current is increased by 10 times while the duration for which it is applied is reduced by 100 times. Pulsed excitation allows time sharing of the bridge among several resistance thermometers as shown in Figure 6.2, yielding much larger output voltages, as is necessitated in many data acquisition systems.

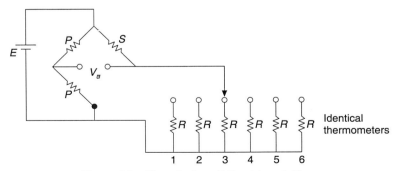

Figure 6.2   Time-sharing of Wheatstone bridge.

To obtain the average temperature of a furnace or the average surface temperature of a metallic structure, a number of resistance thermometers may be located at several points and all of them may be connected in series so as to constitute one arm of the Wheatstone bridge as shown in Figure 6.3(a). Similarly, it is possible to obtain the temperature gradient inside a system or on the surface by locating two matched resistive elements either of the wire type or the film type, at two points separated from each other. They are connected in the adjacent arms of a unity-ratio arm bridge as shown in Figure 6.3(b). Temperature difference of 0.05°C is easily detected by such arrangement.

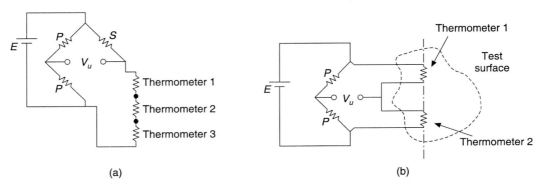

(a)   (b)

Figure 6.3 Measurement (a) of average temperature; (b) of differential temperature.

6.1.1.1 Thermistors

Resistance thermometer elements made from a class of materials known as semiconductor compounds are called *thermistors*. These compounds are sintered mixtures of sulphides, silicates, oxides and selenides of metals such as nickel, manganese, cobalt, copper, zinc, iron, aluminium and uranium. Thermistors are thermally sensitive resistors and, depending on their composition, the resistivity may have its value anywhere from 10^{-1} to 10^9 Ω-cm. Resistance values of thermistors at 20°C may lie in the range of 100 Ω to 1 MΩ. The materials are very brittle and hence elements in wire form cannot be obtained. The elements are in the form of beads, rods and discs of different sizes. The beads and discs are available with glass envelope encapsulation. The beads may be as small as 0.1 mm in diameter and may have their resistance as high as 10 kΩ.

Most of the semiconductor materials used for thermometry possess high resistivity and high negative temperature coefficient of resistance. There are some compound mixtures having positive coefficient. The resistance-temperature characteristic is far from linear and is governed by the relation

$$R_T = R_0 \exp\left[\beta\left(\frac{1}{T} - \frac{1}{T_0}\right)\right] \tag{6.3}$$

where R_T and R_0 are resistance values of the thermistor at absolute temperatures T K and T_0 K respectively and β a constant of the material expressed in degrees Kelvin (absolute). The typical value of β at 25°C is 4000, for commonly used materials for temperature measurements. The temperature coefficient of resistance obtained from Eq. (6.3) is

$$\alpha_T = \left(\frac{dR_T}{dT}\right)\bigg/R_T = -\frac{\beta}{T^2} \tag{6.4}$$

The resistance-temperature characteristics of the thermistor materials are shown in Figure 6.4, along with those of some metals. It is seen that the thermistors are limited in their application as they can be used only for the range –100°C to +300°C. Some thermistors containing sintered aluminium oxide are used up to about 1000°C. Thermistors are quite useful for compensating electrical circuits for changes in ambient temperature.

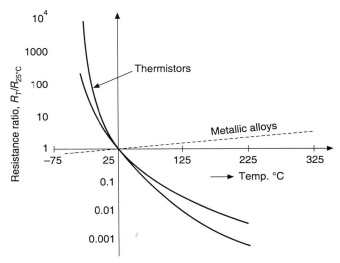

Figure 6.4 Resistance temperature characteristics of thermistor materials and metal alloys.

The merits of thermistors over the wire-resistance elements for temperature measurements lie in their high values of temperature coefficient, relatively small size (making localized temperature measurement possible), low thermal capacity, high speed of response, and large resistance value. Time constant varies between 1 s to 20 s, depending on the type of protecting tube and, when directly used, it may be 100 ms. Stability and reproducibility are generally not satisfactory and they have to be frequently calibrated. Improvement in stability is brought about by exposing them to a temperature slightly above that to which they might be used. High resolution in temperature measurements under dynamic conditions is possible with thermistors since its resistance varies to the extent of 80 Ω for a change of 1°C. Very small temperature gradients can be measured by means of thermistors. When used in bridge circuits, no lead compensation is necessary. Wheatstone bridge circuits with suitable bridge ratios are used, but it is difficult to compensate for the nonlinearity of the thermistor characteristic. The high speed of response, small size and low cost favour the thermistor to become a matching component to go with the integrated electronic circuits, which are also light, compact and powerful to process the resistance changes into electrical signals of desired range. These circuits are described in Section 6.1.3.

6.1.1.2 Semiconductor Temperature Sensors

Semiconductor temperature sensors are those elements which are based on the temperature sensitivity of semiconductors such as germanium and silicon, either in pure or in doped form. Silicon has a positive temperature coefficient, about 0.7%/°C and linearity to within ±0.5% in its operating range of −65°C to 200°C. Silicon sensors are available in rod form with resistance values ranging from 10 Ω to 10 kΩ. Silicon crystals doped with boron exhibit positive temperature coefficient in the range −50°C to +250°C but have negative coefficient below −50°C, and a nonlinear resistance temperature characteristic.

Similarly, germanium crystals doped with arsenic, gallium or antimony are used for low and cryogenic temperatures in the range of 1 K to 35 K. The resistance-temperature characteristic enables achievement of very high resolution of the order of ±0.0001 K and high sensitivity. The resistance of 7000 Ω at 2 K falls to 6 Ω at 60 K. Extreme caution is observed to ensure that there is no self-heating error when the semiconductors are used for cryogenic temperature measurements.

6.1.1.3 Errors in Temperature Measurements

While measuring the temperature of a body or a fluid, it is assumed that the transfer of heat energy from the body to the thermometer element takes place by conduction as they are brought in physical contact with each other and that the amount so transferred is negligible as compared to the thermal capacity of the body, resulting in no alteration in its actual temperature. Static errors in temperature measurements can be minimized by a proper selection of the resistance thermometer or any other thermometer system for its size and thermal capacity. Of course, dynamic errors will be considerably reduced along with static errors, if thermometer elements of the thermistor type or the platinum foil or film type are chosen. Although the element may be sufficiently fast in response, the cascading effect of a protective sheath or well of porcelain with the element will be to make the combination slower in response by 10 to 20 times in most cases. In industrial measurements, the element is inserted into furnaces by means of a long protecting tube and the whole assembly may be rigidly fixed to the walls of the furnace. In such cases, heat transfer by conduction takes place along the walls of the tube to the furnace walls, and there will be a drop in temperature of the tip where the element is located. When the element is used directly without the protecting sheath, a small amount of heat may be conducted away from the element through the connecting leads. Errors in measurement due to the above reasons are known as *conduction errors*. In practice, it is necessary to consider errors due to convection and radiation also. The laws governing the rate of transfer of heat $\dot{Q}$ by the three modes are as follows:

(i) *Fourier's law for conduction*

$$\dot{Q} = -kA \frac{\partial T}{\partial x} \tag{6.5}$$

where,

k = thermal conductivity

A = area through which heat is transferred

$\dfrac{\partial T}{\partial x}$ = temperature gradient

(ii) *Newton's law of cooling for convection*

$$\dot{Q} = hA(T_1 - T_e) \tag{6.6}$$

where,

      h = convection heat transfer coefficient
      A = surface area of the body exposed to the environment
      T_1 = temperature of the hot body
      T_e = temperature of the environment immediately surrounding the hot body

(iii) *Stefan-Boltzmann's law for radiation*

$$\dot{Q}_{1-2} = K\sigma\varepsilon A(T_1^4 - T_2^4) \tag{6.7}$$

where,

      $\dot{Q}_{1-2}$ = heat transferred from the hot body at T_1 to the hot body at T_2
      A = surface area of the hot body at T_1
  T_1, T_2 = absolute temperatures of the hot body and other at T_2 $(T_2 < T_1)$
      σ = Stefan's constant
      ε = emissivity of the surface of the hotter body
      K = a constant related to the geometrical configurations of the bodies

In the case of coal-fired or gas-fired furnaces, the thermometer element is held inside, where the hot gases are in motion and the heat transfer by convection is prominent in raising the temperature of the thermometer. Apart from the properties of the gases, the velocity of gases is also responsible and hence the temperature measured will not be accurate and will be more than the actual temperature of the gases.

Heat transfer by radiation also takes place between the thermometer system and the walls of the furnace and other incandescent solids in the furnace. For obtaining accurate values of temperature, it is essential to apply corrections due to the causes mentioned above or to take proper precautions to minimize the errors from each. Temperature measuring probes consisting of the element and the sheath are designed in different ways to suit the conditions of each application so as to minimize the errors and, wherever possible, enable the calculation of errors. Error due to radiation effects becomes large where the temperatures of the probe and the walls of the furnace differ considerably. In such cases, radiation shields are employed to protect the probe from the sight of the surroundings. If the shield is close to fluid temperature, heat transfer from probe will be very much minimized. Simple shields consist of metallic or ceramic sleeves, as shown in Figure 6.5(a), around the probe in such a way as not to arrest the flow of the fluid. Emissivities of probe and shield surfaces and the geometrical configurations of the probes also contribute to minimization of the radiation error. A bright shiny probe surface reduces the radiation error to some extent when compared with the one covered with soot. Figure 6.5(b) shows the radiation shield which arrests the flowing fluid so that the temperature measured amounts to the temperature of the fluid under static conditions.

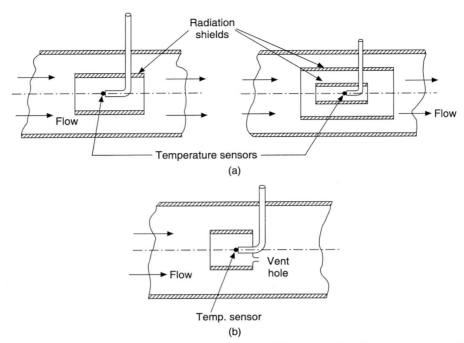

Figure 6.5 Radiation shields for temperature sensors: (a) Single and multiple type open shields; (b) single shield arresting the flow.

Assuming that no radiation shields are used and there is no conduction error, it is possible to calculate the fluid temperature from the following energy balance equation:

$$hA(T_f - T_{th}) = \sigma A\varepsilon(T_{th}^4 - T_w^4) \qquad (6.8)$$

where, σ = Stefan's constant
h = heat transfer coefficient due to convection from the gas to the thermometer probe
A = surface area of the probe
ε = emissivity of the probe surface
T_f, T_{th}, T_w = temperatures of the fluid, thermometer and the wall of the furnace respectively

Values of h for each application are rather difficult to obtain. However, the equation helps in estimating the error which is likely to result during the measurement.

6.1.1.4 Temperature Measurement of Fluids in Motion

Temperature of a fluid is measured by immersing the thermometer element or probe into the fluid which is held steady and at rest. There is no relative motion between the fluid and the thermometer. If the actual temperature of the fluid in motion is desired, it is clear that there should be no relative motion between them which, in other words, means that the thermometer should be moving with the fluid with the same velocity. As this is impossible, the fluid in motion is brought almost to rest in the vicinity of the temperature probe kept inside an enclosure known as stagnation chamber. Due to the velocity suddenly becoming zero, the kinetic energy of the flowing mass is converted to thermal energy, resulting in an increase in temperature of the fluid trapped in the stagnation chamber. Assuming that there is no heat transfer of any other kind from the chamber, and treating the fluid to have been brought to rest adiabatically, it can be shown that the increase in the temperature of the fluid around the thermometer is given by

$$\Delta T = (T_1 - T_2) = \frac{V^2}{2C_p} \tag{6.9}$$

where,

T_1 = stagnation temperature of the fluid in the stagnation chamber
T_2 = free-stream or static temperature
V = free-stream velocity
C_p = specific heat of the fluid at constant pressure

In fact, the equation assumes prominence where gas flows at high velocity, e.g. air velocity around an aircraft. Equation (6.9) is modified for such cases and is rewritten as

$$\frac{T_1}{T_2} = 1 + \frac{v-1}{2}M^2 \tag{6.10}$$

where,

M = Mach number, being the ratio of V to acoustic velocity

v = ratio of specific heats $\left(= \dfrac{C_p}{C_v}\right)$ of the gas

T_1, T_2 = absolute temperatures

For air, T_1/T_2 (= $1 + 0.2M^2$), resulting in an error of only +1% for air velocity around 70 m/s.

For estimation of errors at higher velocities, Eq. (6.10) may be used. In practice, however, the stagnation chamber is provided with a ventilating hole as shown in Figure 6.6, with the ratio of A/a between 10 and 20. The probe design varies with the application, and no probe is really free from errors due to the small heat losses from conduction and radiation. Even though the conduction and radiation losses are zero, the temperature indicated will still be lower than the stagnation temperature as calculated from Eq. (6.10). Each probe is tested for its deviation in its performance and is expressed in terms of a recovery factor R, defined by

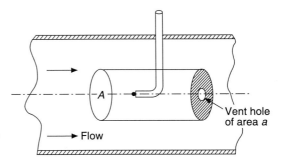

Figure 6.6   Stagnation chamber for temperature measurement.

$$R = \frac{T_{th} - T_2}{T_1 - T_2} \tag{6.11}$$

where T_{th} is the stagnation temperature indicated by thermometer. R is generally determined by experiment and is usually between 0.75 and 0.99. In Figure 6.7 is shown a probe that has a recovery factor of 0.99 to 1. The thermometer element may be kept open and bare for fast response. If conduction and radiation losses are present in a particular application, additional correction or a correction factor defined in the same way as the recovery factor should be applied, implying that the process is not exclusively adiabatic. For such cases, the correction factor will never reach the value of unity, even if the specially designed probe of unity recovery factor is used. The correction factor is determined in actual conditions of usage, whereas the recovery factor of a probe can be determined in laboratory using standard apparatus.

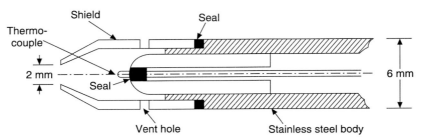

Figure 6.7   A high-speed temperature probe.

6.1.1.5 *Thermal Radiation Detectors*

So far, in all measurements of temperature it was insisted that physical contact between the thermometer system and the body/fluid under test be established. At temperatures above 1400°C, it is not possible to establish physical contact due to the practical limitations of the materials employed for the thermometer elements. Hence there is a necessity for evolving methods by which the temperature of the hot body can be measured from a distance, without getting into physical contact with the body. The inaccessibility of the hot body, its size, and its thermal capacity also sometimes do not permit any physical contact to be made. In industry, the surface temperature of a moving in-got of hot metal or the temperature of a body kept inside a furnace may be desired. Hence instruments based on the measurement of radiant energy emitted by the hot body are developed and used to estimate the temperature of the body, and they are popularly known as *radiation pyrometers*.

Radiation takes place independently of the medium surrounding the source and is entirely electromagnetic by nature. Radiation is classified into several regions or bands, depending on its characteristics and wavelengths, although its velocity of propagation remains the same. All hot bodies emit visible radiation in the band of wavelengths between 0.3µ and 0.72µ ($\mu = 10^{-6}$ m) and infrared (thermal) radiation, in the range 0.72µ–1000µ. The energy emitted at each wavelength varies with wavelength and the temperature of the hot body. If the hot body is black, it is a near perfect radiator, in the sense that it radiates more energy at any one temperature than any other body at the same temperature. A black body also absorbs all radiation falling on it without transmitting or reflecting any. The absorptive power and the emissive power of a black body are treated equal to unity. When, almost all wavelengths emitted by a hot body are considered, the emissive power is called the *total emissive power* and similarly the total absorptive power.

The total energy E_t radiated in watts/cm² of surface, by a perfectly hot body is given by the Stefan-Boltzmann's law:

$$E_t = \sigma T^4 \tag{6.12}$$

where,

T = temperature of the hot body, K

σ = Stefan's constant (= 5.672×10^{-12} W/cm²-deg⁴)

Pyrometers based on the application of this relationship are known as *total radiation pyrometers*. Most of the energy radiated by hot bodies having temperatures above 250°C falls within the range of wavelengths between 0.72μ and 10μ as shown in Figure 6.8. Hence the thermal radiation within this band of wavelengths requires to be focussed on to a detector which should function as a black body absorbing all the thermal radiation incident on it. If the detector is at a temperature T_d, then the net radiant energy absorbed by the detector is given by

$$E_{\text{net}} = \sigma(T^4 - T_d^4) \tag{6.13}$$

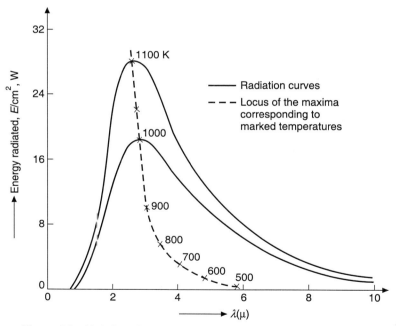

Figure 6.8   Variation of radiant energy with wavelength of a black body.

If $T \gg T_d$, E_{net} will be as given by Eq. (6.12). The increase in the temperature of the detector will be a measure of this energy received, and hence the temperature of the hot body. To render the radiation detector sensitive and fast in response, it is made as small as possible in size and low in thermal capacity. It is enclosed in an evacuated cell so as to minimize convection loss to the surroundings and mounted on insulating blocks to reduce conduction losses.

The radiation detector can be a thin strip of blackened metal foil of platinum and is usually referred to as bolometer. Change in the resistance of this foil is indicative of the temperature of the hot body. Thermistor discs of surface area of 1 mm^2 enclosed in a cell, having a suitable window for admission of radiation, are also used as detectors. Energy levels of the order of 3×10^{-8} W can be detected and the time constant of these detectors is about 3 ms.

6.1.2 Hot-wire Resistance Transducers

The process of heat transfer by conduction is seen to be the basis for measurements of temperature of systems, solid or fluid, by bringing the resistive element in physical contact with the system. Care is taken to see that heat transfer between the system under test and the resistive element by convection and radiation is negligible by preventing the flow of fluid and suitably designing the jacket enclosing the element. Electrical circuitry employed to sense the resistance changes has been suitably energized to ensure that there is no self-heating error due to the current flowing through the element.

In this section, resistive elements heated by currents, and kept at temperatures well above those of their surroundings are employed to enable measurement of certain quantities related to the state and properties of the medium around the element, through the measurement of its resistance. The medium whether a liquid or a gas, may be in motion or at rest. By properly controlling the operating conditions of the measuring system and through suitable design, heat transfer either by conduction or convection only is allowed to take place. When thermal equilibrium between the hot wire and its surrounding medium is established, the temperature of the resistive element as also its resistance becomes a function of the variable under measurement. It is shown that it is possible, by using these principles, to measure liquid level, flow rate, and velocity of liquids and gases (hot-wire anemometry), low pressure and vacuum (Pirani gauge), thermal conductivity of gases, and to determine the composition of gas mixtures (thermal conductivity gauges).

The resistive element employed for the above measurements is usually a fine wire of platinum or tungsten, which is mounted inside a metallic or glass enclosure as shown in Figure 6.9. The wire is heated to a temperature of 300°C to 400°C, which is far in excess of that of the surrounding medium and enclosure. When thermal equilibrium conditions are attained, the heat generated in the wire will be balanced by the heat loss by conduction, convection and radiation. Assuming that the heat loss due to conduction through the lead wires and the supporting mounts and radiation from the surface of the hot-wire filament are negligible compared to the convection loss, the energy-balance equation can be written as

$$I^2 R_w = JhA(T_w - T_e) \tag{6.14}$$

where,

I = current through the filament, amp

R_w = resistance of the filament at the temperature of T_w, ohms

T_w = temperature of the filament, °C

T_e = temperature of environment, °C

h = heat transfer coefficient of the film of medium immediately surrounding the filament, cal/cm²-sec-°C

A = surface area of the filament exposed to the medium, cm²

J = mechanical equivalent of heat (= 4.186 J/cal)

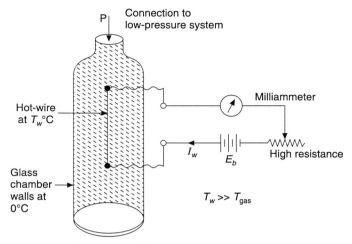

Figure 6.9 Hot-wire resistance transducer.

By virtue of T_w being made much higher than T_e, which is taken as the temperature of the surrounding medium, and by holding T_e constant, it can be seen that any change in h would disturb the thermal equilibrium attained earlier, and thus necessitate a change in R_w for a constant current which, in other words, is a change in T_w. The new value of R_w is thus a measure of the factors influencing a change in h. The heat transfer co-efficient h is

$$h = \frac{k}{l}\bigg|_{l \to 0} + f_n(v, \rho, s) \quad \text{[see Section 4.5]} \tag{6.15}$$

where,

k = thermal conductivity
ρ = density
s = specific heat
v = velocity

all pertaining to the medium. Variation in any one of these properties or in the velocity of the fluid can be related to the variation in the resistance of the filament, if all the other factors remain fixed. When the medium is gaseous, density varies with pressure and hence measurement of the pressure of the gas is possible. Measurement of the velocity of a fluid or gas is possible, if the fluid or gas remains clean without any suspended matter and constant in its composition. The design of the measuring system and the size and temperature of the hot wire filament help minimize certain errors in measurement, while compensation for errors due to certain interfering variables is also possible. It should be realized that expressions relating the variables of interest serve as guidelines for the development and design, and in practice, each measuring system, based on these techniques, requires calibration. While stability and reproducibility are always important, for long term measurements, the merits of these systems lie in their application potential for on-line measurements, their size and speed of response. Other relevant basic characteristics and design considerations, pertaining to each measuring system are given in the remaining part of this Section (6.1.2).

6.1.2.1 Hot-wire Level Transducer

The level of a liquid in a container is measured by means of a fine wire or thin ribbon of platinum kept vertically inside the column of liquid as shown in Figure 6.10. Constant current is passed through the wire such that it is at about 200°C in the absence of the liquid. When it is partly immersed in the fluid, heat transfer takes place from the exposed portion of the wire to the mixture of air and vapours of the liquid (if the container is closed) and from its immersed portion to the liquid. The heat developed in the hot wire should in no case be so large as to raise the temperature of the liquid from its initial temperature. The total resistance of the exposed portion and the immersed portion is a measure of the liquid level. Calibration is required for each liquid and each container. In most cases, the resistance-level characteristic may not be linear. The properties of the fluid should be known first and it must be ascertained that introduction of the hot wire will not result in any reaction. This method is successfully applied for determining the level of liquefied gases.

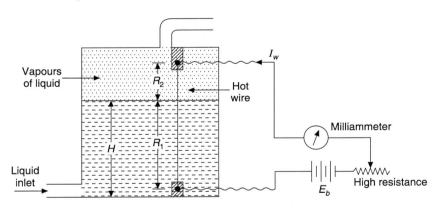

Figure 6.10   Hot-wire type liquid-level gauge.

6.1.2.2 Hot-wire Pressure Transducer

Hot-wire pressure transducer, commonly known as Pirani vacuum gauge, is used for the measurement of pressure of gases, in the range 1 mm Hg–1 μm Hg.

When a gas is admitted into the chamber containing the hot-wire filament, heat transfer takes place between the molecules of the gas and the hot wire. The gas is brought to rest, and hence with velocity being zero, h becomes equal to k; the thermal conductivity of the gas is given by Eq. (6.15). The molecules of the gas are in a state of continuous vibration and collisions among themselves and the walls of the chamber are frequent. Whenever a molecule comes into contact with the hot wire, during its random motion, heat is collected from the hot wire, with the result that its kinetic energy increases. After a series of perfectly elastic collisions with the other molecules, the kinetic energy as well as the heat energy is finally transferred to the walls of the chamber. When the pressure of the gas in the chamber is below 10 mm Hg, the number of molecules inside the chamber becomes less, with the consequent increase in the mean free path of the molecules becoming finally much larger than the spacing between the filament and the walls of the chamber. As the pressure becomes lower, the number of molecules participating in collisions becomes less and hence the rate at which heat transfer from the hot wire takes place becomes smaller. For the same heating current, the temperature of the filament goes up at lower pressures and hence its resistance. As the process of heat transfer takes place, due to the gas molecules going into a state of vibration about their mean positions, the heat transfer coefficient is equal to the thermal conductivity of the gas. Thermal conductivity of gases at such low pressures is known as molecular thermal conductivity, as distinguished from the one at higher pressures above this range. The thermal conductivity of gases at normal pressures is independent of pressure, and different gases have different thermal conductivities. The thermal conductivity of a gas is about 5 to 10 times the molecular thermal conductivity.

As already stated, radiation loss and conduction loss through the leads should be minimized by proper design of the system. Low emissivity metals such as platinum or tungsten are used for the filament. The resistance of the filament is about 15 ohms and increases by 10% for the change of pressure from 1 mm Hg to 1 μm Hg. The heating current is chosen to be between 10 mA and 100 mA, resulting in a temperature of 100°C to 400°C for the hot wire. In certain designs, four-coiled tungsten filaments such as those in electric bulbs are kept in parallel and supported inside a glass chamber, cylindrical or spherical in shape. The temperature of the glass chamber should be held constant at the value at which the calibration of the gauge is done. As already seen, the mean free path of the gas at low pressures decides the size and shape of the chamber. Hence the response time is as much as 60 s in most cases.

The rate at which heat is lost from the filament due to conduction in the low pressure gas is proportional to the pressure and the difference in temperatures of the filament and chamber. The resistance-pressure characteristic is nonlinear, as shown in Figure 6.11(a).

Unity-ratio Wheatstone bridge circuits, as shown in Figure 6.11(b), are normally used, yielding output voltages linearly related to pressure. The reference chamber is identical in all respects to the test chamber except that it is totally free from any gas and sealed. Such an arrangement is known to compensate for errors due to conduction through leads and radiation. Changes in the excitation voltage of the bridge and the ambient temperature changes are also compensated. Characteristic curves relating the unbalance voltage with the pressure are obtained for each gas, while the entire system is often checked for calibration using a standard gas or air.

Higher output voltages are obtained by using thermistors in place of the wire and coiled filaments. In some gauges, the temperature of the filament is directly measured by a thermocouple or a thermopile, incorporating all the provisions for compensation of the errors, by using the reference chamber.

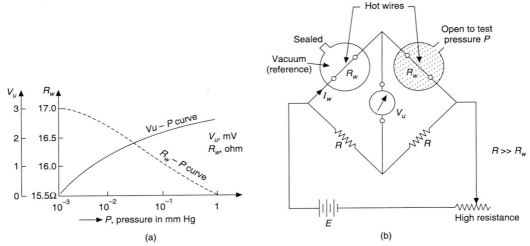

Figure 6.11 (a) Typical characteristics of a Pirani gauge; (b) Wheatstone bridge with test and reference gauges.

6.1.2.3 Thermal Conductivity Gauge

The hot-wire resistance transducer described so far for the measurement of the low pressure of gases is seen to measure the molecular thermal conductivity of the gas at low pressure. In fact, with some modifications in design, it can be made to measure the thermal conductivity of gases at normal pressure. The thermal conductivity of the gas, k, is in fact a function of its specific heat at constant volume C_v and its viscosity μ and is given by

$$k = \alpha \mu C_v \tag{6.16}$$

where, α is a constant depending on the atomicity of the gas.

The hot-wire thermal conductivity gauge differs in its size and shape from the Pirani gauge. The gas at normal pressure possesses a very large number of molecules vibrating about their mean positions and colliding with the neighbouring molecules. The kinetic energy, as also the heat collected from the hot wire, are finally transferred to the walls. The mean free path of each molecule is very small and the spacing between the filament and the walls always allows such a large number of collisions as to render the thermal conductivity measurable, independent of pressure. The volume of the chamber is small but the thermal capacity of the walls of the chamber has to be larger as the rate of heat transfer is higher at normal pressures. Usually, the chamber is cylindrical in shape, with the hot wire kept centrally along its axis, and the walls are made of copper or brass and sufficiently thick, making the thermal capacity of the chamber very large. Thus the heat conducted by the gas does not contribute to any appreciable increase in the temperature of the chamber. Table 6.3 presents the thermal conductivities of some gases.

Table 6.3 Thermal conductivity of some gases at atmospheric pressure

Gas	Thermal conductivity (cal/cm-sec°C) $\times 10^7$
Air	583
Oxygen	591
Nitrogen	581
Argon	424
Helium	3,480
Hydrogen	4,160
Chlorine	187
Carbon monoxide	560
Carbon dioxide	352
Ammonia	522
Methane	721
Ethane	436
Acetylene	453
Acetone	237
Chloroform	158

It is observed that hydrogen has a large value of thermal conductivity. Whatever be the nature of the gas, the same number of molecules are contained in the cell volume, at a given temperature and pressure. The mean kinetic energy of each molecule is proportional to its absolute temperature. At any temperature, molecules of a less dense gas have lower mass for each molecule and hence higher mean velocity. Thus the heat transfer is faster for Hydrogen. While measuring the thermal conductivities of two different gases, it is necessary that they be admitted into the cell at the same temperature and pressure. The thermal conductivity of a gas increases with temperature, and the value of k_T at a temperature of T °C is given by

$$k_T = k_0 \frac{b + 273}{b + T} \left(\frac{T}{273} \right)^{3/2} \tag{6.17}$$

where,

k_0 = thermal conductivity at 0°C
T = absolute temperature in °C
b = a constant of the gas tested

This relationship holds good for the temperatures ranging from −80°C to +100°C. The temperature coefficient of thermal conductivity is thus a function of temperature and is not the same for all gases. Care has to be taken to minimize the effect of temperature on the measurements.

Although Wheatstone bridge circuits using a test cell and a reference cell do not preclude their application to the measurement of absolute thermal conductivity of a gas at normal temperature and at pressures above 10 cm of Hg, they are commonly used for on-line analysis for composition of gas mixtures, through the comparison of the thermal conductivities of the gas mixture and a reference gas of known composition. The reference cell contains the standard gas of known composition at normal pressure and is generally sealed. The test gas is admitted into the cell continuously at the same normal pressure but at a very small rate of, say, 20 cc per minute. The error due to convection losses at this flow rate is expected to be tolerable for industrial measurements, provided the pressure is regulated at normal value. Convection losses are observed to be less in cells with the wire mounted vertically. Both the test and standard gases may be aspirated and passed through the cells at the same but very low flow rate, or a small fraction of the test gas from the main line is obtained through a by-pass, as shown in Figure 6.12. For accurate measurements, it is necessary to hold both the test and reference gases

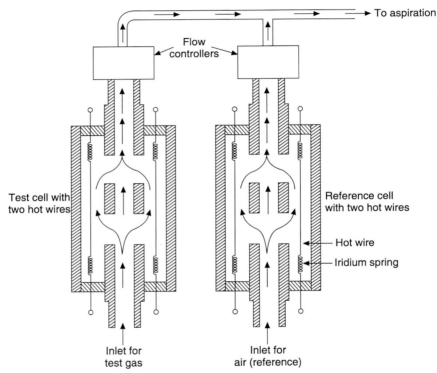

Figure 6.12   Gas analysis system consisting of test and reference thermal conductivity cells.

at the same pressure, temperature, humidity and low flow rate by means of suitable regulating systems. Unbalance voltage of the bridge is used to relate the difference in the conductivities of the test and reference gas with the composition of the test gas. Wherever there is need to protect the wire from any effects of the test gas, such as its flow rate, a porous plug is used for molecular diffusion of the gas into the cell. To compensate for the loss in speed of response, the volume of the cell is brought down to 0.5 cc. One particular design, known as kathorometer cell, is shown in Figure 6.13. The platinum wire is fused with glass, bent into U-shape, and is sealed into an outer glass sheath. A metal tube encloses the glass sheath, with the intervening space being filled with Wood's metal (melting point 65°C). The reference cell is identical in design except that the glass tube is sealed at both ends with the standard gas inside. Otherwise,

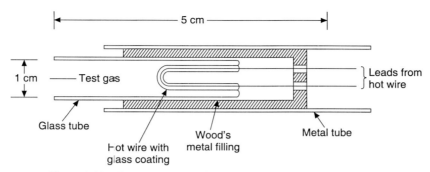

Figure 6.13   Kathorometer cell for thermal conductivity measurement.

both the cells may be closed by means of porous plugs. Four cells are arranged in a metal block, of which two form test cells and the other two reference cells. They are connected in the Wheatstone bridge circuit such that each pair of cells occupy the two opposite arms. The unbalance voltage of the bridge is calibrated in terms of the thermal conductivity of the composition of the test gas.

6.1.2.4   *Gas Analysis for Composition*

The absolute thermal conductivity of air (20% oxygen and 80% nitrogen) is about three times that of chlorine and one-seventh of hydrogen. As oxygen and nitrogen have thermal conductivities very near to each other, air is taken as reference to compare the values of other gases. Hence, it is difficult to determine the composition if the two constituents of the mixture have gases of nearly the same thermal conductivity. It is far simpler and direct to estimate the percentage of one constituent of a binary gas mixture (only two gases) if one of them has a much larger or a much smaller thermal conductivity than the other. For example, estimation of carbon dioxide content in the air of a fruit-storage chamber is one such application. It should be noted that the relationship between the conductivity of the gas mixture and the concentration of the constituent gases depends on the molecular and physical properties of the gases and the laws of forces between the molecules during collision. Hence for each gas mixture, a standard gas/gas mixture (or known proportions) is chosen and calibration curves are provided between the composition and unbalance voltage of the bridge circuit.

6.1.2.5   *Hot-wire Anemometer*

An important application of the hot-wire resistance transducer is for the measurement of flow rate of fluids. Flow rate of nonconducting liquids in open channels and closed pipes and of gases in closed pipes can be measured very conveniently by suitably locating the hot-wire filament. The hot-wire filament as suggested in Section 6.1.2 is usually a fine wire of platinum or tungsten, and is mounted in the flow channel, by means of a support as shown in Figure 6.14(a). The diameter and the length of the wire depend on the size of the pipe and the maximum flow rate which has to be measured. The diameter of the wire may vary from 5 μm to 300 μm, and the length is about half the diameter of the pipe. The filament is centrally located inside the pipe such that the axis of the wire is normal to the direction of fluid flow.

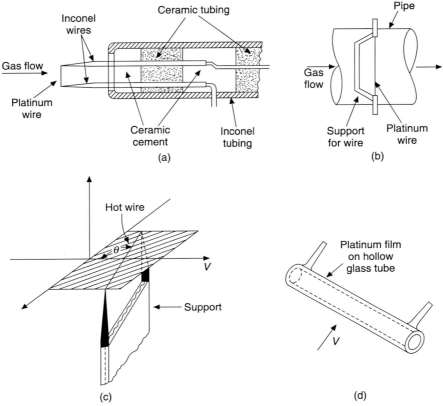

Figure 6.14 (a) Hot-wire anemometer probe; (b) bare hot-wire in pipe; (c) effect of θ; (d) film type anemometer.

The fluid should be free from any particulate matter as, otherwise, due to the impact of the particles, the wire may snap. When used for the measurement of flow rate of liquids and of gases at high flow rates, the filament should be rugged enough to withstand the force on it due to the flow. The liquid whose flow rate is under measurement should be nonconducting so that the resistance of the filament remains unaffected due to the surrounding medium.

The more important application of the hot-wire filament is in hot-wire anemometry. The hot-wire filament is mounted on a probe as shown in Figure 6.14(b) and is exposed to wind. In such a situation, the measurement is affected by the wind velocity. The direction of flow of the wind should be known so as to orient the filament properly. The wind may have an average velocity above which fluctuating components of a high frequency may be present. In wind tunnel experiments, it is necessary to carry out wind turbulence measurements. Special probes are designed for such applications.

A single wire probe responds essentially to the component of velocity perpendicular to it, if the angle between wire and velocity vector is between 25° and 90°, as shown in Figure 6.14(c). For $\theta = 0$, the heat loss by convection is about half that for 90°.

The actual flow direction may be estimated by means of a single wire probe, if the flow angle is roughly known. The wire is set at about 50° from the flow direction and its resistance is measured. The probe is then rotated in the opposite direction until an angle is found at which the resistance measured is the same as before. The bisector of the angle between the two locations is thus taken as the true flow direction.

Special probes for a quick estimation of the true flow direction are designed such that they carry two matched wire filaments, having their axes at 90° to each other. When the two filaments indicate equal resistance values, the flow direction is assumed to lie at 45° to either of the filament axes.

Where sufficient ruggedness for the filament against the onslaught of any dust particles in the wind is desired, hot-film elements are used. The film of platinum is deposited on a glass or ceramic base as shown in Figure 6.14(d).

Equation (6.14) provides the basis for the measurement of the fluid rate by means of the hot wire. The loss of heat from the hot wire occurs due to conduction and convection as the wire is in contact with the flowing fluid. The wire temperature is affected by the flow rate of the fluid and, therefore, its resistance can be a measure of its velocity, provided its temperature and density remain constant.

% Using Eqs. (6.14) and (6.15), the energy-balance equation can be rewritten as

$$I^2 R_w = JA(k + K\sqrt{V})(T_w - T_f) \tag{6.18}$$

where,

K = a constant
V = velocity of the fluid
T_f = temperature of the fluid

All other quantities are as indicated in Eq. (6.14).

The current in the hot filament is chosen so as to make T_w much higher than T_f. Usually, T_w may be around 400°C. However, it is important that T_w be so chosen that the filament will not burn out in case the flow rate becomes zero. The maximum flow rate which is desired to be measured may also be the criterion for the choice of T_w. As already stated, the heat loss due to radiation and conduction through the leads of the wire may be minimized by suitable design and choice of T_w.

Hot-wire anemometers are commonly available in two forms: (i) the constant-temperature type, and (ii) the constant current type.

In the constant-temperature mode of operation, the current through the wire requires to be adjusted such that the hot wire assumes the same temperature. As the flow rate V increases, a higher value of current is required to establish the thermal equilibrium. In such a case, the relationship between the current I and the flow rate V is obtained from Eq. (6.18) as

$$I^2 = K_1 + K_2\sqrt{V} \tag{6.19}$$

where, K_1 and K_2 are constants.

The adjustment of current I may be made manually or by automatic means.

In the constant current mode of operation the current in the hot wire is held at a constant value and the resistance variations due to flow rate variations are measured.

Suitable measuring techniques are chosen for each mode operation, depending on whether the flow rate is steady of fluctuating. These are now described.

(i) *Constant-temperature anemometer.* For the measurement of steady velocities, the constant temperature mode of operation using a Wheatstone bridge is often used. The measuring circuit is shown in Figure 6.15(a), in which a galvanometer is used to detect the balance conditions. The bridge is connected to a battery voltage source in series with an adjustable resistor of a very large value (say, 2000 Ω). The equivalent resistance of the bridge is of the order of 1–20 Ω. The voltage drop across the hot wire is measured by a high resistance millivoltmeter whenever the bridge is brought to balance condition. R_1 and R_w are very low in resistance (about 1 Ω) and the resistors R providing the unity ratio of the bridge are sufficiently high in resistance so that most of the current flows through R_w and R_1.

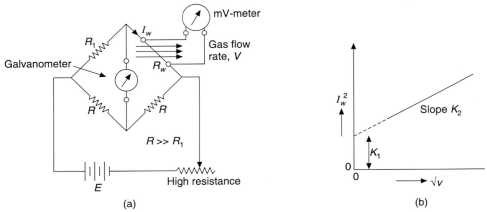

(a)   (b)

Figure 6.15   (a) Constant-resistance type anemometer; (b) calibration curve.

The measuring circuit is calibrated first by exposing the hot wire to known velocities over a desired range and using the same liquid/gas for which it is ultimately used. The pressure and temperature of the fluid should be maintained at the same values during calibration and usage later. The velocities of the fluid are measured by additional techniques and should be known with reasonable accuracy.

Initial adjustment of the measuring circuit is done by creating a low velocity (V_1) of flow across the hot wire. The hot-wire current I_w is brought up to a value high enough for adequate sensitivity but low enough to prevent wire burn-out at the low initial velocity setting. Allowing enough time for attainment of thermal equilibrium, the bridge is balanced by adjusting R_1 (or any of the other resistors). The value of I_w is obtained from the reading of the millivoltmeter. For each subsequent value of velocity, the corresponding I_w for balancing the bridge is obtained.

A plot of I_w^2 vs. $\sqrt{V}$ is shown in Figure 6.15(b) and is used as the calibration curve for the specified medium of flow at the specified temperature and pressure. Adjustment of the current I_w to attain balance conditions cannot be done if the velocity of the fluid is fluctuating.

(ii) *Constant-current anemometer.* In the constant-current mode of operation, the current through the hot wire is kept constant at a suitable value. The measuring circuit shown in Figure 6.15(a) is modified so as to function as a constant current anemometer. The circuit is redrawn in Figure 6.16(a), in which a high resistance millivoltmeter is used to measure the unbalance voltage of the bridge. The operating value of I_w is chosen and set at a proper value, taking the same precautions regarding the burn-out of the wire. The unbalance voltage V_u is plotted against the flow rate V. Before using the measuring circuit in actual measurements, a check is necessary for the value of I_w, which should be the same as that specified during calibration. The calibration curve is shown in Figure 6.16(b), and the characteristic is seen to be far from linear.

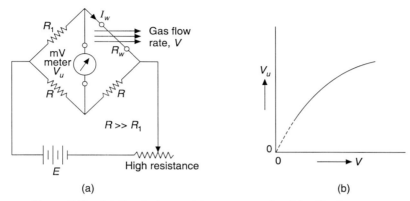

(a)   (b)

Figure 6.16 (a) Constant-current type anemometer; (b) calibration curve.

The range of velocities for which the constant-current type is used is necessarily low because of the likelihood of the wire burn-out when the flow stops. Any attempt to safeguard against this danger would mean choice of a lower I_w for the upper limit of velocity or a lower value of velocity for the upper limit with a satisfactory value of I_w.

The measuring circuit of the constant-current anemometer can be used for the measurement of steady velocities as well as the rapidly fluctuating components such as the turbulent components superimposed on an average velocity.

The frequency of the fluctuating components may be very low in the case of liquids; however, for wind turbulence measurements, it may be as high as 100 kHz. In such a case, the thermal inertia of the hot-wire filament becomes applicable, and hence dynamic measurement of velocities becomes impracticable. The time constant of the hot wire may be reduced by a proper choice of its size and configuration but cannot be brought down to such a low value as to make the measuring system flat in response up to a frequency of 100 kHz. As can be seen from the following analysis, the time constant is dependent on other factors also, such as the temperature coefficient of resistance of the material of wire, its operating conditions of velocity and heater current. The time constant of the hot-wire filament may be derived for both the modes of operation by using the basic Eq. (6.14). As the variables involved are nonlinearly related, the transfer function is obtained by incremental analysis by considering increments of I, R_w and T_w at an operating point at which thermal equilibrium is already established.

In the case of constant-current anemometer, the incremental values of ΔR_w and ΔV are considered as I is held constant. After the thermal equilibrium is attained, the increase in velocity from V to $V + \Delta V$ results in a change of resistance of the wire from R_w to $R_w + \Delta R_w$, and the energy-balance equation under these conditions is given by

$$[I^2(R_w + \Delta R_w) - JA\{K_1 + K_2\sqrt{(V + \Delta V)}\}(T_w + \Delta T_w - T_f)]\,\Delta t = Mc\,\Delta T_w \tag{6.20}$$

where,

M = mass of the hot wire

c = specific heat of the material of the wire

ΔT_w = increase in temperature of hot wire

Δt = interval of time

Under thermal equilibrium at the operating point,

$$I^2 R_w = JA(K_1 + K_2\sqrt{V})(T_w - T_f)$$

The resistance of the wire and its temperature are related by $R_w = R_0(1 + \alpha T_w)$, and hence

$$T_w = (R_w - R_0)/R_0 a, \qquad \Delta T_w = \Delta R_w/R_0\alpha \tag{6.21}$$

where R_0 = resistance of wire at 0°C.

Writing $K_2\sqrt{(V + \Delta V)}$ as $K_2\sqrt{V}(1 + (\Delta V/2V))$, the energy balance under dynamic conditions (neglecting T_f since $T_w \gg T_f$) can be written as

$$I^2\Delta R_w - JA\left\{(K_1 + K_2\sqrt{V})\frac{\Delta R_w}{R_0\alpha} + \frac{K_2\Delta V}{2\sqrt{V}}\frac{R_w - R_0}{R_0\alpha}\right\} = \frac{Mc\,\Delta R_w}{R_0\alpha\,\Delta t}$$

The transfer function (TF) relating to the increments may be obtained from the above equation as

$$\frac{\Delta R_w(s)}{\Delta V(s)} = \frac{K_v}{1 + s\tau} \tag{6.22}$$

where

$$K_v = \frac{-JAK_2(R_w - R_0)}{2\sqrt{V}\{(K_1 + K_2\sqrt{V})JA - I^2 R_0\alpha\}}$$

$$\tau = \frac{Mc}{\{JA(K_1 + K_2\sqrt{V}) - I^2 R_0\alpha\}}$$

Similarly, the constant of the hot wire for the case of constant-temperature mode of operation can be derived. The wire temperature and its resistance are affected by the increments of both the current and velocity. The energy-balance equation is given by

$$(I + \Delta I)^2(R_w + \Delta R_w) - JA\{K_1 + K_2\sqrt{(V + \Delta V)}\}(T_w + \Delta T_w - T_f) = Mc\frac{\Delta T_w}{\Delta t}$$

As the system is treated to be linear for increments of ΔI and ΔV at the operating point, the principle of superposition may be used after obtaining the transfer function for each input separately. $\Delta R_w(s)/\Delta V(s)$ is already given in Eq. (6.22).

Thus, to obtain $\Delta R_w(s)/\Delta I(s)$, the energy-balance equation can be rewritten as

$$(I + \Delta I)^2 \, (R_w + \Delta R_w) - JA \, \{(K_1 + K_2 \sqrt{V}) \, (T_w + \Delta T_w)\} = Mc \, \frac{\Delta T_w}{\Delta t}$$

The TF relating ΔR_w and ΔI can be obtained as

$$\frac{\Delta R_w(s)}{\Delta I(s)} = \frac{K_i}{1 + s\tau} \tag{6.23}$$

where,

$$K_i = \frac{2IR_w R_0 \alpha}{[JA(K_1 + K_2 \sqrt{V}) - I^2 R_0 \alpha]}$$

$$\tau = \frac{Mc}{[JA(K_1 + K_2 \sqrt{V}) - I^2 R_0 \alpha]}$$

Thus the TF relating ΔR_w with the two incremental inputs of ΔV and ΔI is given by the sum of the TFs given in Eqs. (6.22) and (6.23).

It is observed that the time constant is the same for both the incremental inputs of velocity and current. In each case, the time constant is governed by the operating conditions around which the increments are considered. The time constant ceases to be valid when the operating point is changed. The time constant also varies if the increments in velocity and hence the amplitudes of the fluctuating components become large.

In practice, the time constant is about 1 ms, and hence the measuring systems do not possess flatness in response beyond 150 Hz. Hence, additional techniques of dynamic compensation of the measuring system have to be applied.

Dynamic compensation by means of lead networks as explained in Section 4.7.2 may be used for the measuring system of Figure 6.16(a). The unbalance voltage of the bridge may be amplified by a dc amplifier and applied to the compensating network. The output voltage indicates both the average and fluctuating components of velocity.

If the fluctuating components are only desired for display on an oscillograph, the measuring system can be modified, as shown in Figure 6.17. The ac amplifier used removes the dc component and hence signals related to the fluctuating components are only allowed to pass through. If passive networks are only used for dynamic compensation, two such networks may be connected in cascade with a buffer amplifier inserted between them. They may be designed to possess flat response right up to a frequency of 100 kHz.

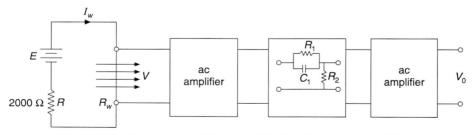

Figure 6.17　Measurement of flow-rate of fluids under dynamic conditions.

Dynamic compensation of a measuring system may also be effected by using negative feedback. The Wheatstone bridge of the constant-temperature anemometer, shown in Figure 6.15, may be modified to function as a feedback system and become self-balancing by altering the current in the hot wire whenever it is affected by a change in velocity. The circuit shown in Figure 6.18 uses a high gain amplifier which amplifies the unbalance voltage and converts it into a proportional current which in turn is used to excite the bridge. The system is so designed that the output voltage across R_0 is proportional to the amplitude of the velocity component. The application of negative feedback enables the reduction of the time constant by a factor of 100, and the system is made suitable for display and recording of both the steady and fluctuating components. They are designed to possess flatness in response up to 50 kHz. The detailed analysis of the feedback type anemometer is presented in Section 8.5.

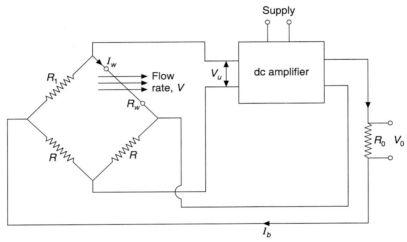

Figure 6.18   Self-balancing constant resistance type anemometer.

It may be further noted that when the dynamic compensation is effected by the lead networks, it is necessary to adjust the parameters of the network, if the medium of flow is changed.

The hot-wire technique of measuring the velocity of flow has assumed great significance since the measurement can be done without causing any disturbance to the already existing environment. It can be used for measurement of very low velocities; it can also be used when the volume rate of flow is low. The hot wire can be located in pipes of small size without causing any pressure drop in the fluid stream. Of course, it can only measure the average velocity of flow in the pipe. As already stated it is unsuitable for velocity measurement if the fluid is a conducting liquid. Its application is mainly for gas flow and wind velocity measurements and in the laboratory for flow velocity measurements of non-conducting liquids and gases.

6.1.3   Resistive Displacement Transducers

The simplest transducer for converting linear (translatory) or angular (rotary) displacements into a change of resistance is the resistive element provided with a movable contactor. The change in resistance is brought out by only a change in the length or portion of the resistor from one end to the point of contact. When measuring the relative motion between a moving body and a fixed body, the moving body is mechanically coupled to the contactor and the relative displacement is measured. Although the variable resistor of this kind can be treated as the primary transducer for mechanical displacements, its main application in instrumentation is to function as a secondary electrical transducer converting the output displacement signals of the primary mechanical transducers such as the Bourdon tube and cantilever elements into corresponding resistance changes. When used as a secondary transducer, its design is mostly influenced by the fact that the mechanical loading of the primary transducer, due to its coupling with the resistive transducer, is negligible. In other words, the force required to move the contactor of the variable resistor is limited to values specified by the output characteristics of the primary mechanical transducer. The variable resistor will, in fact, become an ideal displacement transducer, if the force required to move the contactor is zero. Thus there is a wide range of resistive displacement transducers, generally known as resistance potentiometers (resistance pots), with considerably differing force requirements. As already stated, when transducer systems, employing a primary mechanical transducer and a variable resistor are built, it is essential that the mechanical input characteristics of the latter are matched with the output characteristics of the former.

The transducer of resistance R_p is usually energized by dc or ac supplies, and electrical output signals are obtained by using simple electrical circuits shown in Figure 6.19. Circuit (a) provides dc output currents of range 4–20 mA or any other desired range, whereas others provide voltage outputs. Circuit (c) is the usual variable potential divider, also known as single-ended potentiometer circuit in contrast with that of (d), a push-pull potentiometer circuit. Circuits (b) and (d) develop bipolar outputs for bidirectional motion about the central point.

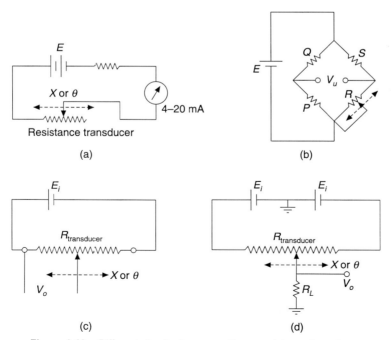

Figure 6.19   Different circuits for connecting a resistance transducer.

Potentiometer circuits of (c) and (d) provide very high sensitivity, in volts/degree of angular displacement, or volts/cm, provided the power rating of the potentiometer permits application of larger E_i. In most applications potentiometer circuits are favoured.

Other important factors for consideration regarding resistance potentiometers are linearity, resolution, loading (electrical) errors, and noise level in output voltages. Resistance values range from 1 Ω to 1 MΩ, depending on whether they are of wire-wound type or film or conducting plastic type. Theoretically, stretched wire or a film-type potentiometer possesses infinite resolution, whereas the wire-wound type has its resolution limited by its number of turns. The latter has the additional advantage of its range of angular displacement extended to 50 or 60 revolutions, and is known as *helical potentiometers* (helipots). The resistance wire is wound on an insulated copper mandrel or any other flexible mandrel of insulating material and then bent into a circle for a single-turn potentiometer or a helix to form a multiturn helical potentiometer.

Linearity between resistance and displacement generally depends on the uniformity of the wire diameter and the uniformity of the winding. Considering the fact that the output signals are voltages, it is essential to take into account the effect of electrical loading due to the voltmeter coming in parallel with the tapped portion of the potentiometer. If the voltmeter is electronic in nature, and has high input impedance, the loading effect will be negligible, with the ratio of V_o/E_i of Figure 6.19(c) being the same as R_x/R_p. For a linear transducer, R_x/R_p is the same as the fractional value x, which is the ratio of the displacement given to the contactor, to its full-scale value. Hence under no-load conditions,

$$x = \frac{R_x}{R_p} = \frac{V_o}{E_i}$$

For the same position of contactor, the output voltage will be lower if R_L, the resistance of voltmeter forming the load, is finite and this new value of V_o', if taken to represent the displacement, is given by

$$\frac{V_o'}{E_i} = x'$$

With the true value being x, the error is given by

$$\text{error} = x' - x$$

Representing R_p/R_L by m, the percentage error can be obtained as

$$\frac{100(x' - x)}{x'} = [-mx(1 - x)]100 \tag{6.24}$$

It is seen that error is negative and becomes zero at $x = 0$ and $x = 1$. For intermediate values of x, this error contributes to the deviation of the inherently linear characteristic of the transducer, as seen from Figure 6.20. By keeping m within 0.05 or 0.1, linearity may be brought to satisfactory limits. The nonlinearity of the characteristic for finite values of R_L can be compensated by designing the potentiometer with an inherently nonlinear characteristic or by introducing a nonlinear variable resistance in series with R_L.

The resolution of a wire-wound potentiometer is decided by the displacement required for the contactor to move from one turn to the next turn. It may be possible to construct potentiometers with 100–200 turns per cm length of the former in which case the resolution is limited to 0.1 mm or 0.05 mm and the characteristic between output voltage and displacement

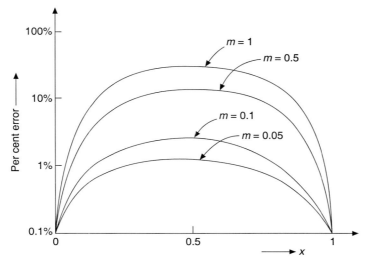

Figure 6.20　Error due to loading of a potentiometer type resistance transducer.

becomes a stepped curve instead of being continuously straight. To decrease the size of the steps, a finer wire may be used, resulting in more number of turns per cm, but the contactor, by virtue of its area of contact, may bridge and short two or three turns at a time in its traverse, resulting in minor jumps in output voltage, due to current fluctuations already set in. Thus linearity is considerably influenced by resolution. Higher resolution and better linearity are possible in the case of film potentiometers, but they may become sources of Johnson noise and random noise due to their high resistance values. In addition to the effects of the above unwanted voltages affecting the resolution and linearity, additional sources of noise are due to the thermo emf and contact emf between the different materials used for wiper, wire, etc. The resistance between wire and contactor undergoes variation in movement, especially due to dust particles, and generates 'contact noise'. The contactor may bounce in certain cases of wire-wound potentiometers, resulting in loss of contact for a short time. Rapid motion of wiper on the potentiometer also brings in noise of high frequency in the output voltage. Rubbing action between dissimilar metals also produces noise voltages. The pressure of the contactor is a critical factor in the design of each potentiometer, as the stability of the characteristic as well as its useful life are affected by wear-out of the potentiometer wire.

While mechanical considerations favour low-force or low-torque potentiometers with long life, electrical considerations for the design favour matching of the properties of the materials used for the wire and wiper. Wire materials should possess low temperature coefficient of resistance, low contact emf with wiper material, and low thermo emf with other materials such as copper, used for connections. For copper and nichrome wires, wiper material of gold-silver is preferred, whereas for manganese wire, an alloy of nickel-silver may be used.

Depending on the size, force or torque required for movement of the contactor, and the type of the transducer, each resistance potentiometer transducer has its specifications which include the maximum rate of traverse of the wiper, the range of frequencies of input displacements, and the maximum number of to-and-fro motions that the potentiometer can be subjected to before its performance deteriorates.

6.1.4 Resistive Strain Transducer

When regular bodies are subjected to stress, strains developed can be measured electrically and relationships between the forces responsible for the stress and the resulting electrical output signals can be established. The resistive strain gauge element is one such device that plays the role of a secondary transducer in sensing the tensile or compressive strain in a particular-direction at a point on the surface of a body or structure. If the modulus of elasticity (Young's modulus) E of the material of the body is known, the stress can be calculated and utilized to identify the magnitude of quantities affecting the stress. Strain-gauge pressure transducers and strain-gauge accelerometers are such transducer systems employing the strain gauges as the secondary electrical transducer along with a suitable primary mechanical transducer for converting the basic quantity under measurement into stress. In certain cases, the strain-gauge element may be stressed directly, allowing the strain measured to be related to the applied stress and hence the force applied.

When a wire of length l' and of cross-section a is subjected to longitudinal stress σ, its length increases from l' to $(l' + \Delta l')$ due to tensile strain along its axis and its cross-sectional area decreases from a to $(a - \Delta a)$ due to compressive strain at right angles to its axis and along its circumference. The ratio $\Delta l'/l'$ is known as the axial unit strain occurring along the direction of the applied stress and the ratio $\Delta d/d$, in this case, is the lateral or transverse unit strain. They are related to each other as

$$\frac{\Delta d}{d} = -\mu \frac{\Delta l'}{l'} \tag{6.25}$$

where,

μ = Poisson's ratio (0.3 for most of the metals working within elastic limits)
d = diameter of the wire

These dimensional changes affect the resistance value of the resistor as can be seen from the following derivation. The resistance of a wire of uniform cross-section a is given by

$$R = \frac{\rho l}{a}$$

where, ρ = resistivity.

Taking logarithms for both sides, we get

$$\ln R = \ln l - \ln a + \ln \rho$$

Differentiating, we obtain

$$\frac{dR}{R} = \frac{dl}{l} - \frac{da}{a} + \frac{d\rho}{\rho}$$

But

$$\frac{da}{a} = -2\mu \frac{dl}{l}$$

Hence,

$$\frac{dR}{R} = \frac{dl}{l}(1 + 2\mu) + \frac{d\rho}{\rho}$$

$$\frac{dR/R}{dl/l} = G = 1 + 2\mu + \frac{d\rho/\rho}{dl/l} \tag{6.26}$$

Here, G stands for the '*gauge factor*', also known as 'strain-sensitivity factor', and dl/l is the axial unit strain ε. For all wires drawn from metals and metallic alloys, Poisson's ratio μ is taken as 0.3 and $(d\rho/\rho)/(dl/l)$ is almost zero. Most of the metals and metallic alloys exhibit an average value of 2 for the gauge factor. Table 6.2 shows the range of values for the gauge factor, for some metals and metallic alloys. Ferry, advance and constantan are used commonly up to temperatures below 400°C, whereas karma and nichrome V are applicable at temperatures up to 1000°C. Iso-elastic and nichrome V are suited to measurements of dynamic strain.

6.1.4.1 Unbonded Strain Gauges

The resistance-wire strain gauges are available in two forms: unbonded and the bonded ones. In the unbonded type, a fine wire of 25µ diameter is kept stretched between two insulating posts held rigidly on two different platforms that can move relative to each other under external forces. While such a system forms the simplest assembly of Figure 6.21(a), actual transducer systems employing four saphire posts and holding four equal lengths of tungsten-platinum wire of 5µ diameter, are common as the four wires can form a Wheatstone bridge circuit providing temperature compensation. The four posts, as shown in Figure 6.21(b), are mounted on a star-spring structure. When the centre of the star spring is subjected to the force under measurement,

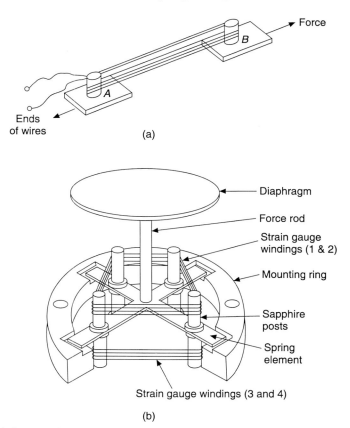

Figure 6.21   Unbonded type strain gauges: (a) Simple assembly; (b) a transducer system for pressure measurement.

the star-spring flexes with each pair of the strain gauge elements on opposite sides going into strains of opposite polarity. The resulting unbalance voltage of the bridge circuit is proportional to the force and hence the pressure on the diaphragm. The whole assembly is encased in a housing with provisions for admission of pressure and for electrical connections. They are available for pressure ranges from as low as 0–0.15 to 0–660 bars.

6.1.4.2   Bonded Strain Gauges

In unbonded strain gauges, the wires of the elements are exposed to air or the environment. The housing normally protects the assembly and is designed particularly to minimize the effects of vibrations. These unbonded strain gauges are designed for particular applications and are not suitable for measurements of local strains. To make them suitable for application, bonded strain gauges are developed in widely differing sizes. The resistive element may be in the form of wire, foil or film of the material. The most common type is the wire-resistance gauge, with the wire looped back and forth as shown in Figure 6.22(a) to form a flat grid in one plane. The number of loops is limited to 3 or 4 with the loops so closely spaced as to minimize the length of the wire lying along the transverse direction *OB* in comparison to that along the longitudinal direction *OA*. Such an approach in construction renders the strain gauge sufficiently insensitive to strains along the transverse direction and hence its 'cross-sensitivity' becomes very low. Single-wire types with thick copper wires welded at the ends, as shown in Figure 6.22(b), may be preferred for use where cross-sensitivity has to be zero. Alternatively, the wire may be wound on a thin insulating strip. The wire may also be wound in a helix on a hollow cylindrical paper tube and then the combination is flattened as shown in Figure 6.22(c). Both of these wrap-around type strain gauges are thicker and higher in resistance. Foils of thickness from 10μ to 15μ may also be used as seen from Figure 6.22(d) in place of wires of circular section. The

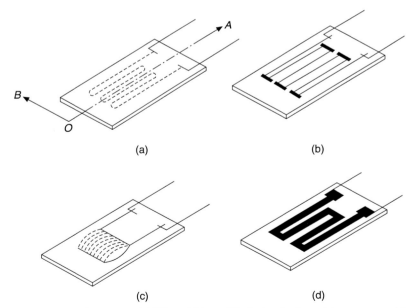

Figure 6.22   Bonded type strain gauges: (a) Wire-grid type; (b) single wire type; (c) wrap around type; (d) foil-type.

photo-chemical process of manufacturing them has brought down the gauge length to 3 mm. Thin-film gauges are also developed by vacuum deposition of films on thin insulating base, having gauge factors slightly higher than those of the above types.

All the above resistance elements are cemented securely by nitrocellulose cement or phenolic resin, to two sheets of flexible insulating material, one on each side of the element. Once cemented, the combination becomes an integral component and the element cannot be retrieved. The bonded strain gauges, thus constructed, are available from manufacturers along with the data concerning gauge factors.

For accurate measurement of the surface strain of any structure, it is necessary that the bonded strain gauge become integral with the test surface. For each type of gauge and test surface, proper cementing agents should be used. The area where the strain gauge is to be mounted is first polished by emery cloth or polishing wheel and then all traces of greasy film are removed by cotton swabs dipped in acetone. Similarly, one of the surfaces of the strain gauge is also cleaned for mounting on the structure. For mounting paper-backed gauges, nitrocellulose cement is generously applied to the surface of the structure as well as the gauge surface, and then the gauge is mounted on the surface, such that the gauge length (axis) is along the direction in which the strain has to be measured. Excess cement is squeezed out by application of gentle pressure. It is held in position until it is fully dried up. After gluing the bakelite-backed gauges with phenolic cement (araldite), they are held in position but with greater pressure and then slowly heated up to about 170°C. Foil-type and film-type gauges with epoxy plastic film backing are glued to the test surface by means of epoxy cement, after cleaning them with trichlorethylene. It may be heated at 60°C or 70°C for three hours for drying. The process of cementing thus enables the strain gauge element to become integral with the test surface. It is equally important to see that the gap between the test surface and the gauge element is negligible enough, in comparison with the depth of the neutral surface of the structure from the test surface. In this respect, foil type and film-type gauges will be closer to the test surface, whereas the wrap-around type will be farthest. Also, they possess the merit of having higher surface area/cross-section ratio and higher resistance, which are desirable for faithful transmission of strain from the test surface to gauge element and slightly higher sensitivity. The foil and film type are available as transferable type, permitting transfer of the gauge from the temporary backing to the test surface coated with the epoxy cement. Improper bonding of the gauge to the surface becomes a source of errors due to hysteresis, creep and zero drift, apart from lowering the stability for long term measurements.

6.1.4.3 Sources of Error

Bonded strain gauges have assumed great importance for application in situations involving both static and dynamic measurements of strain. The size of the gauge permits point-to-point investigation of strains occurring on surfaces of structures. Metallic wire strain gauges have proved their long term stability. Their application potential has considerably increased with the availability of a range of good quality cements for bonding the strain gauges to the surfaces of structures. The gauge factors of the wire- and foil-type gauges are seen to be sufficiently constant for some of the materials and are normally furnished by manufacturers after considerable testing and experimentation. To enhance the accuracy and reliability in the measurements of strain, sources of error other than those due to mounting are seen to be due to

1. lead connections and measuring circuits;
2. moisture and humidity effects; and
3. effects of ambient temperature.

Gauge element and the leads from it should be properly soldered or welded as any variation in the contact resistance between them is thought to be due to strain, more so if the variation occurs during the period of measurement. Repeated cycles of strain may cause fatigue of the joint and contact may be lost finally. Hence dual-lead gauges are provided for gauges which are used for long-term measurements. Figure 6.23 shows the connection of an intermediate wire between grid and the lead wire, which improves the fatigue endurance of the gauge. The gauge may be located far away from the measuring circuitry, in which case compensation for the variation in the resistance of the long lead from gauge should be provided, as is done in the case of thermometer circuitry. The gauge may be provided with four leads, two for potential and two for current, when it is used for measurements of dynamic strain. Special care has to be taken when bridge circuits are employed for measurement of strain.

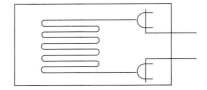

Figure 6.23   Dual-lead type strain gauge.

Moisture absorption by paper-backing and the cement used for bonding can change the resistance of the element and the insulation resistance between the gauge element and the test surface that serves as ground. The volume change of the cement may also strain the element and cause errors in measurement. For long term stability of the gauges, adhesives and cements used should be nonhygroscopic and should not become brittle in course of time. Once lodged on to the structure, there is no possibility of isolating the actual strains and the interfering strains due to moisture and humidity.

The values of strain normally encountered in engineering measurements range from 10^{-6} to 10^{-3}. Taking the gauge factor of 2, and the gauge resistance as 120 Ω, the change in resistance of the gauge due to a strain value of 10^{-6} is as low as 0.24×10^{-3} Ω. It is essential that there should not be any other interfering input that affects the resistance of the gauge element. It may be recalled that the resistivity of the material changes with the temperature of the element and hence its resistance. The strain gauge material and the material of the structure on which it is mounted have their coefficients of linear expansion, which may not always be the same. Thus temperature of the surface and hence of the strain-gauge element has considerable influence on the resistance of the element due to the above-mentioned reasons. Sometimes, it is possible that resistance changes due to temperature effects may swamp the resistance change due to strain. For accurate measurement of strain, it is essential that the temperature effects be either fully compensated or are taken care of by suitable correction after calculation. Once a strain gauge is bonded to the surface of a structure, and is used in a Wheatstone bridge circuit, the unbalance voltage developed will be treated as if it were due to the strain of the structure, if all other factors that contribute to the development of unbalance voltage are not compensated fully.

6.1.4.4 Temperature Compensation

It is obvious that the temperature of the strain gauge and the body on which it is cemented should remain at the same temperature at which they are cemented together, throughout the period of strain measurement. Then any increase in resistance of the gauge can be considered to be due to strain of the structure transmitted to the strain gauge. In practice, however, it is difficult to hold the entire system under test, which may be large or small in size, at one temperature, throughout the period of measurement.

The solution for such situations is always found, as in earlier cases, in having two identical gauges, one a dummy gauge and the other an active gauge, in a unity-ratio Wheatstone bridge circuit, as shown in Figure 6.24(a). The dummy gauge is cemented to the surface of an identical

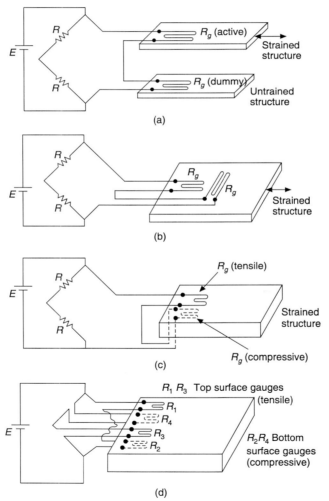

Figure 6.24 Wheatstone bridge circuits for temperature compensation using: (a) One active and one dummy strain gauge; (b) two active gauges on the same side; (c) two active gauges on opposite sides; (d) four active gauges.

structure, kept free from any mechanical stress. When both structures are under stress-free conditions and at the same temperature, the bridge is balanced. An equal rise in temperatures of both the structures raises equally the resistance values of both the active and dummy gauge, and the bridge still remains balanced. When the test structure is strained, the resistance of the active gauge only goes up and the unbalance voltage V_u is proportional to the strain. As the fractional change p (= $\Delta R/R$) is far less than unity, V_u is given by

$$V_u = \frac{Ep}{4} = \frac{EG\varepsilon}{4} \tag{6.27}$$

In case the test surface is wide enough to admit another strain gauge, by the side of the active gauge, as shown in Figure 6.24(b), temperature compensation is as good as in the above case provided the second gauge is mounted so as to be sensitive to strain occurring at 90° to the other strain. If the strain in the transverse direction is zero, the second gauge has its resistance effected by temperature only and the effect of temperature is the same for both gauges, as they are proximate to each other. The above relationship, as given in Eq. (6.27) for unbalance voltage, still holds good.

In most cases, the transverse strain is of opposite sign and is equal to $\mu\varepsilon$, where μ is Poisson's ratio of the material of the structure. Accordingly, unbalance voltage will be modified as

$$V_u = \frac{EG\varepsilon}{4}(1 + \mu) \tag{6.28}$$

The sensitivity can be doubled while having temperature compensation by using two active gauges mounted on the test structure but located in such a way that the two gauges are strained equally but in opposite sense. Such an arrangement is possible only with basic structures such as a cantilever which may have the strain gauges mounted on opposite sides of the cantilever as shown in Figure 6.24(c). However, it is essential that both the top and bottom surfaces of the cantilever be always at the same temperature, and such a condition is possible only with cantilevers and others that are physically small in size. Any difference in the temperatures of the two surfaces will cause error in measurement. The unbalance voltage V_u, when fully compensated for temperature, will be twice as much as the voltage given by Eq. (6.27) if both gauges measure axial strain.

The arrangement of Figure 6.24(c) may suggest automatically another possibility as given in Figure 6.24(d) for temperature compensation while doubling the value of V_u obtained. Two active gauges mounted side by side on the top surface are under tensile strain, and occupy the diagonally opposite arms of the Wheatstone bridge, while the other two under compressive strain occupy the remaining two arms. It may be noted, in this context, that all primary mechanical transducers may not always enable the location of the strain gauges, in the said manner. Provided that the entire structure is at the same temperature, the arrangement of Figure 6.24(d) has the merit of not requiring the resistors for the ratio arms. As the bridge is symmetrical, the thermo emfs do not figure in V_u.

Investigations of strain in girders and steel structures require measurements over long stretches of time. A large number of strain gauges may be located on the structures at several points, and the strain at each point may be sampled at regular intervals or continuously recorded. Strain gauges used in such situations should be stable and the gauge factors should remain the same. The gauge elements undergo cyclic changes of strain as well as temperature, and may break under fatigue. Each element is tested for its life-span in terms of the number of cycles of strain it can stand, without yielding to fatigue.

Even though temperature compensation is adopted by the methods suggested, it is, at times, necessary to compute the equivalent strain that shows up due to temperature effects. The combined effect of temperature on the resistance of the element is given by

$$\frac{\Delta R}{R} = [(\alpha_s - \alpha_g)G + \alpha_t]\Delta T \qquad (6.29)$$

where,

α_s, α_g = thermal coefficients of expansion of the materials of test surface and gauge respectively

α_t = thermal coefficient of resistance of gauge material

ΔT = increase in the temperature of the gauge and test surface

At the start of the investigations, the temperature of the test surface is measured, and what is required is that the temperature should remain the same all through the period of measurement. Any increase in temperature by $\Delta T°C$ would show up as an increase in the resistance, causing spurious indication of strain, even though the structure remains unstrained. When the bridge circuit is constituted with a single active gauge, the unbalance voltage due to the thermal effects must be computed and correction applied to the net voltage indicated.

If the test material has larger thermal coefficient of expansion than that of the gauge element, the element will be under tensile strain and the combined effect will be larger. For each material of the test structure, it is possible to select a gauge material having the same coefficient of expansion. Temperature compensation may also be done by using negative coefficient materials in series with the element. In this context, it is worthwhile to recall the merit of transducer systems using four unbonded wires of equal length as the four arms of a Wheatstone bridge.

6.1.4.5 *Circuitry for Strain Gauge Instrumentation*

In all the above circuits, the Wheatstone bridge circuit is excited by dc voltage. The unbalance voltages are very small and need amplification by suitable electronic amplifiers. Normally, the stability of the circuit is judged by its zero drift. The bridge circuit may as well be excited by an ac supply as all the four arms are almost resistive. The dynamic strains of the test structures may be, at times, of a frequency range up to 10 kHz, in which case the frequency of ac excitation voltage should be at least ten times higher. Electronic amplifiers handling signals arising in such situations are presented in Section 3.5.4.4. An instrumentation system enabling static and dynamic measurements is shown in Figure 6.25 along with the additional provisions for (a) zero adjustment, (b) sensitivity adjustment, and (c) on the spot calibration facility. The high-valued shunt resistor, R_{sh}, when coming across the active gauge, affects its value which can be calculated. As the gauge factor of the gauge is known, the calculated equivalent strain

given by $\dfrac{R_g}{[G(R_g + R_{sh})]}$ is compared with the indicated value, and the calibration verified.

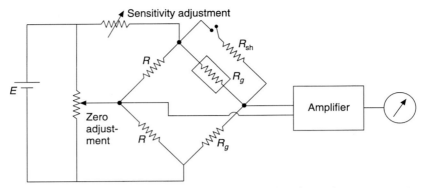

Figure 6.25   Wheatstone bridge with additional provisions for strain measurement.

Dynamic strains occurring over short durations of time and transient strains may often be measured by a simple circuit using a single resistance gauge. The voltage across the strain gauge element may consist of two components: (a) a dc component due to static strain and (b) an ac component due to dynamic strain. It should be understood that the period over which dynamic/transient strain is being measured is so small that no variation in the value of E_i or in the temperature of the element is expected to occur. The dc component may be blocked by a capacitor, and only the ac components are passed on for amplification and record on an oscillograph. Temperature compensation is not considered essential. Normally the ac component is very small—of the order of 5–50 μV—and hence strain gauges of higher gauge factor, though of higher resistivity, are favoured for such situations. Metallic alloy gauges of iso-elastic with gauge factor of 3.6 or semiconductor strain gauges with gauge factors as high as 100 can be used to provide higher output voltages.

6.1.4.6   Semiconductor Strain Gauges

Most of the metals and metallic alloys selected for use as materials for strain gauges so far have yielded gauge factors of about 2 and their resistance variations are seen to be entirely due to dimensional changes of the wire. The resistivity of the material is seen to remain constant and insensitive to strain. The strain sensitivity of the gauge or its gauge factor could be improved if the term $(d\rho/\rho)/\varepsilon$ is finite and positive. Semiconductor materials such as silicon and germanium have yielded considerable sensitivity of resistivity to strain, while opening up new vistas of developments in instrumentation, through integration of the strain sensors with diffused electronic components for purposes of signal processing. It is quite obvious that enough attention should be paid to hold the temperature of the semiconductor strain gauge constant during measurements of strain, unless it is proved that resistivity change due to temperature is negligible in comparison to that due to strain. The gauge factor for the semiconductor strain gauges is given by

$$G = \frac{dR/R}{dl/l} = 1 + 2\mu + \frac{d\rho/\rho}{dl/l}$$

or

$$G = 1 + 2\mu + \pi E \tag{6.30}$$

where,

π = piezo-resistive coefficient

E = Young's modulus of the material

The resistivity change due to strain is known as piezo-resistive effect and guage factors of the order of 100 to 150 have become possible because of this effect. Contribution to the gauge factor due to dimensional changes is thus seen as negligible. Hence it is worthwhile to recall all the findings with regard to the behaviour of pure crystals of silicon and germanium when doped with some impurities and the effects of doping on the resistivity and gauge factor.

6.1.4.7 Temperature Effects on Semiconductor Gauges

The resistivities of both silicon and germanium are very high in pure state but are considerably lower when doped with atoms of phosphorus or boron. The resistivity of the material and hence the gauge resistance falls with increased degree of doping. The case of *n*-type silicon doped with phosphorus, as seen from Figure 6.26(a), reveals that its resistivity and the thermal coefficient of resistance can be lowered to a large extent under high degree of doping. Figure 6.26(b) shows the variation of gauge factor with temperature for several degrees of doping. The gauge factor falls with temperature under all degrees of doping, and high degree of doping results in constancy of gauge factor over the temperature range of –50°C to +150°C. It is clear from the above observations that the doped semiconductor strain gauges provide a lot of versatility, which is impossible in the case of wire-wound strain gauges. High gauge factors of 150–200 are desirable for measurements of dynamic strain and temperature stability is essential for static strain measurements even though gauge factor is low. A wide range of *p*-type and *n*-type semiconductor strain gauges with different degrees of doping is available. For reasons of integration and other advantages, silicon semiconductor strain gauges have become popular for a variety of applications.

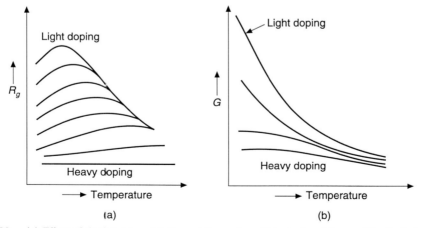

Figure 6.26   (a) Effect of doping on resistivity and thermal coefficient of resistance; (b) effect of doping on gauge factor.

6.1.4.8 Linearity

Thin filaments of about 150 μ thickness are cut from single-crystal doped silicon with crystal orientations specified as (111) or (100) according to the Miller index. The longitudinal piezo-resistive coefficient, along the length of the filament, is very much dependent on the direction of current through the filament, with respect to the crystallographic axes. The width of the filament being very small, the current is passed along its length. Unfortunately, the semiconductor strain gauges, whether of silicon or germanium, have highly nonlinear relationship between the gauge factor and strain. The relationship, in general, may be written in the form

$$G = (dR/R)/\varepsilon = C_1 + C_2\varepsilon + C_3\varepsilon^2 + ... \tag{6.31}$$

where, C_1, C_2 and C_3 are constants valid for each material.

At a room temperature of 20°C and for strain values lower than 10^{-4}, the p (111) type silicon has its characteristics given by

$$G = 175 + 72,625\varepsilon \quad \text{for light doping}$$

$$G = 119.5 + 4000\varepsilon \quad \text{for heavy doping}$$

For strain values higher than 10^{-4}, the nonlinear term becomes much higher in each case.

Similarly, the n-type silicon and germanium both display similar nonlinearity in the relationships, but their gauge factors are negative.

6.1.4.9 Linearization and Temperature Compensation

Semiconductor strain gauges are normally single filaments of lengths varying from 2 mm to 10 mm. The electrodes are formed by vapour deposition, and ohmic electrical contacts are made with fine gold wires attached by means of a thermocompression method. The strain gauges are then brought to their nominal resistance values of 120 Ω or 350 Ω by electrolytic etching. The filaments are bonded in between two films of insulating material of bakelite, phenolic or epoxy origin. Some of the ways of providing the gold wires and lead-ribbons are shown in Figure 6.27. Two filaments of either the same material or one of p-type and the other of n-type may be provided on the same backing so as to be suitably connected in Wheatstone bridge circuits.

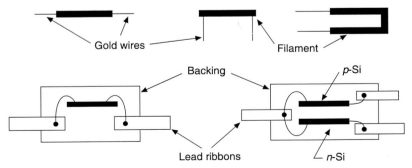

Figure 6.27 Semiconductor strain gauges and their lead-wire assembly.

Mounting these strain gauges on the surface of structures is done in the same way as in the case of metallic wire strain gauges. Film-type strain gauges are formed of the *p*-type silicon, by thermal evaporation in vacuum, directly on a substrate of silicon dioxide on a silicon diaphragm or any other structure, so as to constitute a composite transducer system for measurement of pressure. In all such applications, stability and long life are ascertained by repeated calibration and checking.

If a single strain-gauge element is chosen to form one arm of the Wheatstone bridge circuit, the fractional change in its resistance *p*, due to strain of the order of 10^{-3}, is in the range of 0.2–0.25. For values of *p* in this range, the output voltage of the bridge is also nonlinear. As the two nonlinearities are of opposite nature as shown in Figure 6.28, it is possible to obtain a linear relationship between bridge output voltage and strain, by suitably designing the ratio of the bridge. It is necessary to control the temperature of the environment in which the measuring system is located; otherwise, temperature effects will show up as apparent strain and lead to errors in measurement.

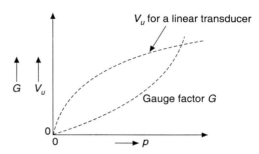

Figure 6.28 Wheatstone bridge and semiconductor strain gauge combination for linear characteristic.

One or two pairs of semiconductor strain gauges can be used in push-pull fashion to result in higher sensitivity and better linearity. For the same tensile or compressive strain at a point, a push–pull pair can be formed by one *p*-type and one *n*-type semiconductor gauge. Using push–pull pairs does not ensure temperature compensation, as is observed in the case of metallic gauges.

Temperature compensation to some extent is possible, through the use of constant current bridge supplies or by insertion of temperature-dependent resistors in series with constant voltage supplies.

Zeroing or biasing a semiconductor strain-gauge bridge is a tricky affair. It is found that it can be done in a much easier way by balancing the output voltage of the strain-gauge bridge with the output voltage developed from a conventional resistance bridge excited from the same source as shown in Figure 6.29.

6.1.4.10 Application of Strain Gauges

As already seen, the metallic strain gauges are sufficiently stable, even though the gauge factor is low, and hence they can be relied upon for measurements of strain over long periods. These strain gauges can be used in large numbers for stress analysis work, each one being a replica of the other. They are also used for building complete transducer systems for the measurement of several physical quantities as explained in Sections 6.1.4.13–6.1.4.18. The major advantages of semiconductor strain gauges are their higher strain sensitivity and smaller size. Miniaturization of strain gauge transducers is possible through the use of semiconductor strain gauges, with corresponding increase in sensitivity of the transducer system. Each of the semiconductor strain gauges requires matched circuits for temperature compensation and

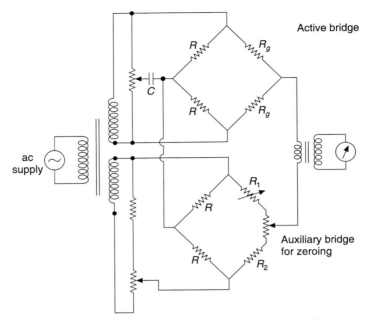

Figure 6.29 Zeroing of a semiconductor strain gauge bridge.

linearization, and hence these are unsuitable for stress analysis investigations. They are particularly suitable for dynamic and transient strain measurements occurring over short duration.

The semiconductor gauges possess higher fatigue life and higher elastic strain as compared to metallic gauges. Semiconductor materials are photosensitive and hence the gauges should be protected from strong light.

6.1.4.11 *Calibration of Strain Gauges*

The user is furnished with the gauge factor of a strain gauge by its manufacturer. When the strain gauges are mass produced, each of them is expected to have the same resistance and gauge factor. The rigour of production techniques may enable them to have the same resistance and the same gauge factor, but certainly within practical limits. Gauges may be selected by sampling, and their resistance values may be controlled to be within ±0.4%. But the most important parameter, viz. the gauge factor, cannot be provided to the user for any bonded strain gauge, as it cannot be removed from the surface after effecting the calibration. By cementing it to a simple regular shaped member such as a cantilever, strain can be accurately determined from the theory or measured by other direct means. Hence no calibrated strain gauge is available for use by the experimenter, and whatever is furnished, is from the knowledge of the calibration effected on a similar gauge. It is this peculiar difficulty with strain gauges that compounds the errors in strain measurements attendent on the technique of locating and mounting the gauge on the test member. Each of the unbonded strain gauges can be provided with the necessary calibration results, but the bonded gauges are provided with gauge factors with uncertainties of ±1%, and all strain measurements using bonded gauges cannot be more accurate than ±1%.

6.1.4.12 Experimental Stress Analysis

It has so far been understood that the size of the bonded strain gauge enables the measurements of strain at a point. To a large extent, this is true when semiconductor strain gauge is used, as its size is much smaller than that of the metallic gauge. But, in fact, the element or filament is finite in its active length, which enables the measurement of strain along the direction of its length or, in other words, along the gauge axis. The mid-point of each gauge is generally marked on it by the manufacturer, and hence the average strain indicated by the gauge will be numerically equal to the strain at the mid-point, if it is a uniaxial strain and if the rate of change of strain with distance from the mid-point is constant. Otherwise, the strain measured will not be correct. But in case the test surface is subjected to biaxial stresses, then strain gauges with zero cross sensitivity are preferred for use and, in this respect, semiconductor strain gauges are more suitable than metallic gauges. In fact, semiconductor gauges are used for testing members of very small size. It has been mentioned that the practical difficulties of proper circuits for use with semiconductor gauges, apart from the range of uncertainty on their gauge factors, prohibit their application to experimental stress analysis. Metallic strain gauges are invariably looked upon as the best and reliable means of estimating the stresses at various points on the surfaces of huge structures. In most cases the directions of stress are not known. Extensive studies of experimental stress analysis are carried out on structures related to air-craft, space-craft, ship-body, skeleton of bridges, etc. In this context, a few more aspects concerning the application of strain gauges are relevant.

Assuming that each strain gauge measures the strain at its mid-point, it may be interesting to know the effect of the elevation of the strain gauge element, from the actual test surface, as seen from Figure 6.30. The nature of the surface and the depth of the test surface from the neutral plane of structure, where there is no strain, can produce sufficient error in the measured values of strain and hence in the estimation of stresses. The test structure may not always provide uniaxial stress and may not be a body of regular geometrical configuration. Although there are special gauge-patterns developed for such studies, each gauge

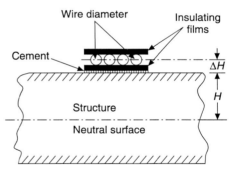

Figure 6.30   Elevated strain gauge mounting on the test surface.

should be mounted closest to the test surface. The thickness of the cements used for cementing the bonded gauge to the surface, the thickness of the film encasing the gauge and the thickness, or diameter of the gauge element, all contribute to errors in the strain measured, as the strain becomes larger with increasing separation of test-point from the neutral plane of the structure. Undoubtedly, therefore, metallic strain gauges are considered to be satisfactory for experimental stress analysis of structures, sufficiently huge in size and large in thickness.

6.1.4.13 Stress-strain Relationships

So far we have seen that the stress in a particular direction is obtained by multiplying Young's modulus with the strain measured along that direction by aligning the gauge axis accordingly.

It is true only for cases of uniaxial stress. In reality, the test point at which the gauge is located is a free surface, where the situation is *biaxial* as there will be lateral strain in addition to the strain along the direction of stress. Poisson's ratio is utilized to identify the lateral strain. It must be realized that the approach is simplified by treating the situation as an extension of the rules governing the uniaxial stress system.

In practice, however, the strain at a point on a surface is dependent on the way the structural member is subjected to forces, and a triaxial strain condition may arise if a couple is applied to a body in addition to forces. As the surface is free, there is no stress normal to the surface and the system is treated as a biaxial system. Taking the case of a regular body, such as the cylindrical pressure vessel shown in Figure 6.31(a), and working out the effects of stresses along the x- and y-axes, the surface strains along these axes at a point P can be obtained as

$$\varepsilon_x = (\sigma_x - \mu\sigma_y)/E$$

$$\varepsilon_y = (\sigma_y - \mu\sigma_x)/E \tag{6.32}$$

where,

σ_x = longitudinal stress along the axis of the cylinder
σ_y = hoopstress along the circumference
E = Young's modulus

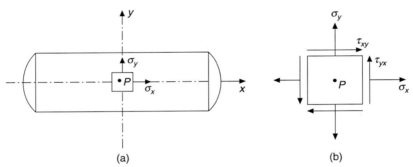

(a)   (b)

Figure 6.31   A cylindrical vessel under pressure; (b) an element on surface under shear stresses.

Solving these equations, σ_x and σ_y can be obtained as

$$\sigma_x = \frac{E(\varepsilon_x + \mu\varepsilon_y)}{1 - \mu^2}$$

$$\sigma_y = \frac{E(\varepsilon_y + \mu\varepsilon_x)}{1 - \mu^2} \tag{6.33}$$

The calculations are aided by the fact that the two principal axes along which the stresses occur are known in this case.

In structural analysis, one does not always know these principal directions, and hence to enable computation of the third unknown quantity θ_1, the angle made by one of the principal directions with the x- or y-axis. The complete stress-strain analysis is possible only with the addition of one more strain gauge making a total of three gauges for such situations. The case is better explained by the example of the cylindrical vessel which is applied with a torque about its axis, in addition to its internal pressure. The surface element at P will be under shear stress as shown in Figure 6.31(b), where τ_{xy} is the shear stress caused by the torque and τ_{yx} is the balancing complementary shear stress. The element is still under biaxial stresses, but the principal stresses have suffered an angular shift θ in the direction of torque. The principal strains also occur along these directions, the one along σ_1 being maximum and the other minimum. The shear strains are maximum along the two directions bisecting the directions of principal strains.

To enable stress analysis of structures, relatively easily, special strain gauge arrangements, known as strain gauge *rosettes* and incorporating two or three grids, are developed (see Figure 6.32).

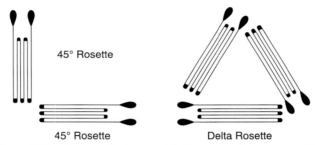

Figure 6.32 Rectangular type and Delta type strain gauge rosettes.

Where, the directions of the two principal stresses are known, a rectangular rosette with two elements at right angles to each other is used. For the completely unknown biaxial stress condition, delta rosette or the T-delta rosette with four elements is used. The fourth gauge of T-delta rosette is used to check the results obtained from the other elements. Strains measured by each element are used finally to get the stress pattern at the point of location of the rosette.

When the above rosettes are used, stress-strain relationships for structural analysis are involved; for a detailed analysis, one may refer to text books on the subject. Simpler relationships valid for rosettes forming an integral part of the transducer systems are given in the next section.

6.1.4.14 *Strain-Gauge Transducer Systems*

Resistive strain gauge elements may be considered to act as primary transducers in stress analysis but, in fact, as explained in Chapter 5, they are used as secondary transducers in conjunction with a wide range of elastic transducers, for measurement of nonelectrical quantities. The combination is normally identified as strain-gauge transducer systems and, depending on the quantity for which they are meant, they may be known as strain-gauge accelerometer or strain-gauge pressure transducer etc. It may be recalled that the strain-gauge transducer systems are very powerful and popular in application, as they have the following inherent advantages of the strain gauges:

1. Negligible loading effect on the primary transducer
2. Formation of a zero-order system resulting in flat frequency response up to 100 kHz
3. Formation of push–pull systems for temperature compensation
4. Very small size permitting corresponding reduction in the size of the primary elastic transducer
5. Wide range for selection in respect of size, gauge factors and resistance of elements.

Some of the transducer systems commercially available, for the measurement of several nonelectrical quantities, are now described, suggesting the enormity of the scope of application of the strain gauges.

6.1.4.15 Strain-Gauge Force Transducer

The primary quantity to which the strain gauge element responds to is, in fact, the force which, when applied to the elastic member, results in strain. The cantilever used in Figure 5.14 is the primary elastic transducer of the force measuring system. When a known mass is attached to the cantilever, the unbalance voltage can be calibrated in terms of either force or weight, thereby forming a simple and convenient means of weight measurement. If the centre of the strain gauge is located at a distance of L_1 from the free end, the strain can be calculated from

$$\varepsilon = \frac{6FL_1}{Ebt^2}$$

where F is the force applied. The system also enables the determination of the gauge factors of strain gauges, once the Young's modulus of the material of the cantilever, E, and the dimensions of the cantilever-length (L), breadth (b), and thickness (t)—are known accurately. The frequency range of dynamic forces, which can be measured is, in fact, limited by the natural frequency of the cantilever which may be equally high. But, in actual conditions, the mass of the force-generating system and the stiffness of the coupled system decide the natural frequency which varies from one to another. It has already been explained that ideal force transducers are those that result in no displacement when force is applied, and so .the cantilevers with high stiffness (larger thickness) are preferred. Higher stiffness yields higher natural frequency for the transducer.

Equally popular but for larger forces and weights, are the column-type and ring-type load cells described in Sections 5.3.5 and 5.3.6. Proving-ring type has the merit of higher sensitivity, smaller size and lower natural frequency, when compared to the column-type cell. The columns range in size from cylinders of 7 cm in diameter and 10 cm in length to those of 30 cm in diameter to 60 cm in length. Load cell designs of the above-mentioned types enable measurements of weights and forces ranging from 0.5 N to 10^4 N. Four strain gauges are located on the column and proving-ring type load cells as shown in Figure 6.33(a) and 6.33(b). To improve the accuracy of measurements, temperature compensation for the variation of Young's modulus with temperature is necessary and is effected by connecting a temperature-dependent resistor R_T in series with R_{se} as shown in Figure 6.33(c). Variations in the temperature of the load cell affect the value of R_T, thereby altering the voltage applied to the bridge suitably.

6.1.4.16 *Strain Gauge Flow Transducer*

If the cantilever beam with the strain gauge mounted is held in a stream, as shown in Figure 6.34, it acts as a restrained static vane. The cantilever beam is subjected to a drag force, and the net effect of the force can be used as a measure of the fluid velocity. Being a maximum at the tip of the cantilever, the force decreases towards the fixed end and hence a coefficient is used to express the force F as

$$F = C_d \frac{A\rho V^2}{2} \tag{6.34}$$

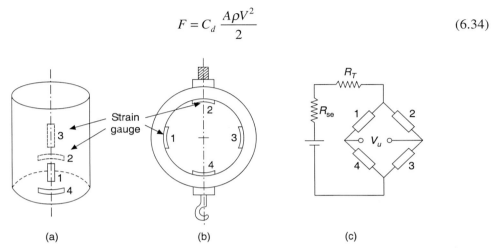

Figure 6.33 Location of strain gauges on: (a) The column type; (b) the proving ring type load cells; (c) Wheatstone bridge with compensating resistors.

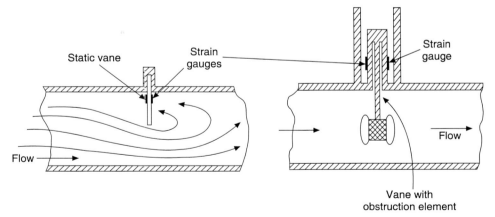

Figure 6.34 Strain gauge flow transducer.

where,

C_d = drag coefficient
ρ = mass density of the fluid
V = velocity of the fluid
A = cross-sectional area of the pipe

Any other suitably shaped body may also be used, provided the disturbance created by its insertion is within permissible limits. A hollow tube with gauges mounted on the outside is used with the intention of providing isolation of the gauges from the effects due to the impact of fluid. The measuring system has to be calibrated for its characteristic between flow rate and unbalance voltage of the bridge. For dynamic flow studies, the characteristic needs to be linearized. In certain cases, the orifice plate is provided with many holes and the force on it is sensed by strain gauge elements. The natural frequency is found to be about 1.5 kHz and the damping factor is dependent on the fluid.

6.1.4.17 Strain Gauge Accelerometers

The cantilever element and the solid or hollow column element, provided with four strain gauges mounted suitably on their surfaces, serve as suitable means for measurement of linear accelerations. The force generated on a mass M, when subjected to acceleration $\ddot{y}$, is $M\ddot{y}$. If the mass is suitably, located on the primary transducers, the strain developed due to the force can be sensed by the strain gauges. The unbalance voltage of the bridge circuits will be linearly related to the acceleration. Figure 6.35 shows the strain gauge accelerometer, with the proof mass supported at the free end of the cantilever. The housing of the accelerometer is filled with silicone liquid, for the damping of the system, and it is rigidly fixed to the structure undergoing motion, along the vertical direction. Under constant acceleration, the longitudinal strain $\varepsilon_{\max}$ at the fixed end is known (Eq. (5.27)) and thus the static sensitivity between strain and acceleration becomes

$$\frac{\varepsilon_{\max}}{\ddot{y}} = \frac{6ML}{Ebt^2} \tag{6.35}$$

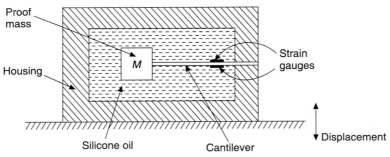

Figure 6.35 Strain gauge accelerometer.

Neglecting the contribution of the cantilever to the mass of the system, the undamped natural frequency of the cantilever-mass system ω_n is given by $\sqrt{(K/M)}$, where K is the spring rate of the cantilever.

From Eq. (5.26), the spring rate K is obtained as

$$K = \frac{\text{force}}{\text{displacement}} = \frac{Ebt^3}{4L^3} \tag{6.36}$$

giving thereby

$$\omega_n^2 = \sqrt{\left(\frac{Ebt^3}{4ML^3}\right)} \tag{6.37}$$

From the above relations, it can be observed that higher sensitivity is possible at the cost of natural frequency. A wide range of accelerometers are available with differing sensitivities and bandwidths. As stated earlier, column type elements provide higher natural frequency and hence accelerometers using them possess higher bandwidths. The mass of the column and proof mass are sufficiently high as to render them unsuitable for use in vehicles where size and weight are at a premium.

The sensitivity and natural frequency are calculated in the same manner as before, and are given by

$$\text{Sensitivity} = \frac{\varepsilon_{\text{axial}}}{\ddot{y}} = \frac{M}{AE} \tag{6.38}$$

$$\omega_n = \sqrt{\frac{AE}{ML}} \tag{6.39}$$

where,

A = cross-sectional area of the column
L = length of the column
E = Young's modulus of the material of the column

Natural frequencies of the order of a few kilohertz are obtained easily by a proper choice of M, L and A. The column type accelerometers are not provided with oil-damping, and hence the damping factor is low and less than 0.1. Accelerometers are available for measurement of linear accelerations up to 2500 m/s^2, i.e., 250 g. As the duration of acceleration of any vehicle is normally short, semiconductor strain gauges may be preferred, although they need frequent calibration.

6.1.4.18 Strain Gauge Torque Transducer

In Section 5.4.1, torsion bars of different cross-sections are shown to enable static and dynamic torque measurements. Cylindrical shafts connecting the driving and driven machines are used as primary transducers, and strain gauges mounted on the surface of the shaft, as shown in Figure 6.36, are used to measure the shear strain given by Eq. (5.39).

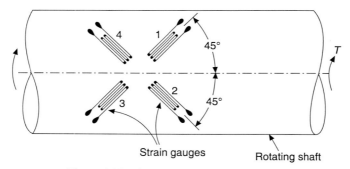

Figure 6.36 Strain gauge torque transducer.

When a hollow shaft is used, the strain is given by

$$\varepsilon_{45°} = \pm\frac{TR_o}{\pi G(R_o^4 - R_i^4)} \qquad (6.40)$$

where,

T = torque transmitted
G = modulus of rigidity
R_o, R_i = outer and inner radii of the hollow shaft

Two gauges are in tensile mode and the other two in the compressive mode. All the four gauges are available in the form of a rosette. Slip rings are used for connecting the supply voltage to the bridge and the unbalance voltage to the indicator. The contact resistance between the brushes and slip rings is a problem. Hence a noncontacting rotary transformer is devised for application in such situations. The shaft material is normally ferromagnetic in nature and hence the shaft is used as the core. Two rotary transformers are provided on the shaft, as shown in Figure 6.37, such that the induction of voltages takes place between the coaxially wound coils of each pair. The frequency of excitation voltages may be chosen to be within the range of 400 Hz to 10 kHz. Completely secure output signals free of noise can be obtained by application of telemetry techniques, as detailed in Chapter 11.

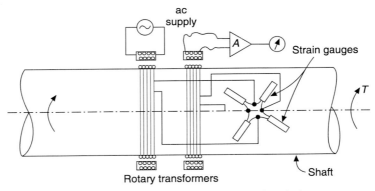

Figure 6.37 Torque transducer for rotating shafts.

Commercial strain gauge torque-meters are available for 1–10^3 N-m torque, and for 4000 rpm–24000 rpm speeds. The natural frequency of the system is mostly determined by the stiffness of the torsion bar and the inertia of the loads connected at either end of the bar. Damping is again very low for obvious reasons.

6.1.4.19 *Strain Gauge Pressure Transducer*

For pressure measurements, a thin plate diaphragm is shown to be the most commonly used primary transducer, providing output signals of linear displacements. It is explained in detail in Section 5.2.4 and is considered to be equally suitable for use with strain gauges. The net force acting on the diaphragm of area A is ΔPA, where ΔP is the pressure difference across the diaphragm. This force can be made to act at the free end of a cantilever, as shown in

Figure 6.38. Strain gauges mounted on the cantilever and connected to form a Wheatstone bridge circuit provide the output signals of voltage proportional to pressure. Even though the number of signal conversions is four, the linearity is considered to be high. Both the diaphragm and the cantilever contribute to the total stiffness of the system, and the cantilever stiffness should be higher than that of the diaphragm to minimize the effect of hysteresis of the diaphragm at low ranges.

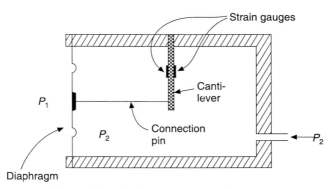

Figure 6.38 Strain gauge pressure transducer.

Other alternative designs for the above system are available, in which a double cantilever is used along with either a corrugated diaphragm, a capsule or a bellows. The system, although sound in principle and operation, can be simplified in practice by considering the diaphragm itself to be provided with four strain gauges, located at such points where compressive and tensile strains exist. For the case of a thin plate flat diaphragm, clamped at the edges and subjected to a uniform pressure difference ΔP, the tangential stress s_t and radial stress s_r at any point on the low pressure side are given by

$$s_t = \frac{3R^2\mu}{8t^2}\left[\left(\frac{1}{\mu}+1\right)-\left(\frac{1}{\mu}+3\right)\left(\frac{r}{R}\right)^2\right]\Delta P$$

$$s_r = \frac{3R^2\mu}{8t^2}\left[\left(\frac{1}{\mu}+1\right)-\left(\frac{3}{\mu}+1\right)\left(\frac{r}{R}\right)^2\right]\Delta P$$

(6.41)

where,

μ = Poisson's ratio
R = radius of the diaphragm
t = thickness of the diaphragm
r = distance of the point from the centre of diaphragm

The stresses are valid for small pressures only when the centre deflection is less than one-third of the thickness of the diaphragm. A plot of the stresses, showing their variation with r shown in Figure 6.39, reveals the existence of points where tensile and compressive strains are available. Gauges 1 and 3 are located as close to the edge as possible to sense the radial strain, whereas gauges 2 and 4 are located almost at the centre to sense the tangential strain which is

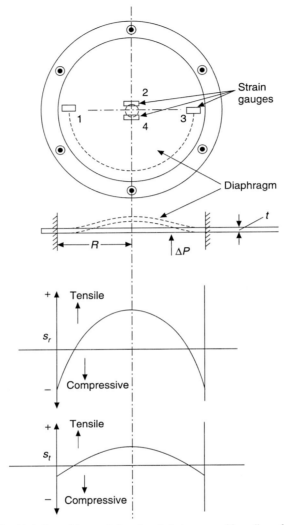

Figure 6.39   Variation of tangential and radial stresses with radius of diaphragm.

maximum at the centre. It must be noted that the net strain, as indicated by any of the strain gauges, has to be computed by using the relations given in Eq. (6.32) as the surface of the diaphragm is under biaxial stress. The strain gauges, connected in push-pull fashion in the bridge circuit provide the temperature compensation and the unbalanced voltage V_u can be computed once the net strain of each gauge is computed. However, each diaphragm, with its strain gauges mounted on it, has to be calibrated.

Strain gauge rosettes are available for use with diaphragms, in various sizes to cover the entire surface area of the diaphragm. The radial elements of the rosette, shown in Figure 6.40 sense the radial strain and the spiral about the centre senses the tangential strain. The strain values become lower for thinner diaphragms ultimately becoming insignificant for detection by

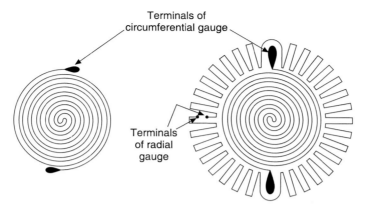

Figure 6.40   Strain gauge rosettes for use with circular diaphragms.

strain gauges. Each diaphragm has its rating for the maximum pressure difference to be applied, so as to be within specified linearity limits. Metallic strain gauges are used with diaphragms whose R/t ratio is in the range of 20–200, and semiconductor strain gauges may extend the range further because of their high sensitivity.

Integrated circuit technology has in fact provided the scope of integrating the semiconductor strain gauges on a silicon diaphragm, along with the additional temperature-dependent resistors for temperature compensation and other operational amplifier circuitry for amplification of voltages. The entire system is commercially made available in different sizes and, when mass produced, their accuracy is stated to be within ± 2% of the specifications. The advantages, are easy replaceability, small size and larger output voltages, required for application to data acquisition systems.

The undamped natural frequency f_0 of the diaphragm is given by

$$f_0 = \frac{2.56t}{\pi R^2} \left[\frac{E}{3\rho(1 - \mu^2)} \right]^{1/2} \text{ Hz} \tag{6.42}$$

where,

    ρ = mass density of the material of the diaphragm, kg/m^3
    E = Young's modulus, N/m^2
    μ = Poisson's ratio,
    t = thickness of diaphragm, m
    R = radius of diaphragm, m

When liquid pressure is under measurement, f_0 becomes lower because of the inertial effect of the fluid. However, it is possible to have diaphragms having natural frequencies ranging up to 5 kHz.

Another interesting elastic member for use with strain gauges is the strain tube and it is referred here to suggest the scope for development of primary transducers for pressure measurement. The strain tube is hollow, as shown in Figure 6.41(a), and is provided with strain gauge wire woven on the surface to constitute the elements for sensing the longitudinal and lateral strains developed when the catenary diaphragm communicates the effect of pressure and strains the tube. The system is said to possess a very high value of natural frequency of about 45 kHz.

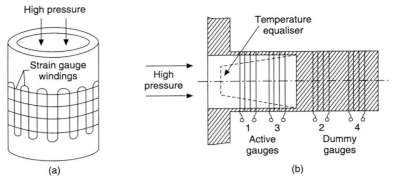

Figure 6.41   Strain tubes for pressure measurements.

The strain tube may as well be applied with the pressure directly as shown in Figure 6.41(b). The tube has a hollow portion, which is provided with two coils wound over its outer surface. These coils are glued to the surface and serve as strain gauge elements undergoing circumferential tensile strain. The other portion is solid and has two coils wound on it to serve as two dummy strain gauges. The inside solid core is intended to provide equalization of temperature, bringing up quickly the temperature of the solid portion to that of the gas under test. For thin tubes whose thickness of walls, t, is very small compared to R_i, the strain of the active element is given by

$$\varepsilon = \frac{\Delta P R_i}{tE}\left(1 - \frac{\mu}{2}\right) \tag{6.43}$$

where, R_i = the internal radius of the tube.

The volume of the tube is small and hence does not load the source of pressure under investigation. The dynamic pressure changes occurring in small internal combustion engines can be studied by attaching this transducer to the walls of the cylinder of the engine. They are designed for pressures ranging up to 10 MN/m^2. They possess the advantages of having higher overload ratio, as compared to the silicon diaphragm gauges.

6.1.5   Resistive Pressure Transducers

A change in the resistance of an electrical conductor can be brought about by subjecting it to an increase in the external pressure exerted from all directions. Very high pressure is needed to effect a perceptible change in the resistance. The bulk modulus of the material of the conductor and the geometrical shape of the conductor may be used to determine the changes in the dimensions and therefore the change in resistance. If these dimensional changes due to the bulk compression effect are considered, it should always result in an increase in the resistance of the conductor. But the experimental studies reveal a decrease in the resistance in the case of conductors of aluminium, copper, iron, carbon and constantan and an increase in the case of conductors of antimony, bismuth, lithium and manganin. In the case of cesium, the resistance falls first and then increases. It is considered that the resistivity of the material is also affected by the application of pressure. The net change in the resistance of the conductor is given by

$$(R_p - R_o) = \Delta R = R_o \beta \Delta P \tag{6.44}$$

where,

R_o, R_p = resistance values at normal pressure, P_o and at $(P_o + \Delta P)$ respectively
ΔP = increase in the pressure from the normal pressure P_o
β = pressure coefficient of resistance of the material of the conductor

It is this direct effect of pressure on the resistance of an electrical conductor that distinguishes the resistive pressure transducer from the pressure transducer that uses strain-gauges mounted on an elastic element. The conductor of the resistive pressure transducer plays the role of a primary transducer, whereas the resistive strain gauge element is a secondary electrical transducer, always needing a primary mechanical transducer for the measurement of a nonelectrical quantity.

In this context, two simple devices, the carbon microphone and the carbon pile, may be considered for the principles of their operation. Pressure variations due to sound waves are conveyed through a diaphragm to the granules of carbon, kept packed in the chamber of the microphone. Due to the application of pressure, the gap between the granules decreases and the contact area between them increases, resulting in the decrease of resistance of the conductor consisting of the granules. Similarly, the carbon pile consists of a large number of thin discs of carbon granules, held between a pair of electrodes. The resistor thus formed has its resistance lowered when a force is applied to the electrodes or when the area of the electrode is exposed to the pressure under measurement. In fact, the pile of these discs is spring loaded, so as to be under an initial pressure. The nonlinear relationship between the resistance and pressure changes prohibits these two devices for consideration as transducers for measurement. However, carbon-compound resistors of 100 ohms have been used for measurement of pressures up to about 30 kN/m^2 with satisfactory linearity and negligible hysteresis. The resistor is kept suspended by fine wires inside a cavity filled with grease, so as to serve as an insulator around the resistor and as a medium for communication of the pressure to the resistor. The electrical noise in the resistor and the instability of the resistor do not favour their application for high pressure measurement.

For measurement of much higher pressures of the order of 10^4 to 10^5 atm, fine wires of manganin or an alloy of gold with chromium (2.1%) having a resistance of 100 ohms are used. The fractional change in the resistance of the conductor due to change in dimensions and resistivity is given by Eq. (6.26), as

$$\frac{\Delta R}{R} = \frac{\Delta L}{L} - 2\frac{\Delta d}{d} + \frac{\Delta \rho}{\rho} \tag{6.45}$$

The conductor is a fine wire whose ends cannot be subjected to the external pressure as they have to provide connection to the external leads. Hence stress on the end faces, $\sigma_z = 0$. The cylindrical surface is under biaxial stress, making σ_x and σ_y equal to $-\Delta P$.

The corresponding strains are given by

$$\varepsilon_x = \varepsilon_y = -\frac{\Delta P}{E}(1 - \mu) = \frac{\Delta d}{d}$$

$$\varepsilon_z = \frac{\Delta L}{L} = \frac{2\mu \Delta P}{E} \tag{6.46}$$

Using the values of $\Delta d/d$ and $\Delta L/L$ in Eq. (6.45), $\Delta R/R$ is given as

$$\frac{\Delta R}{R} = \frac{2\,\Delta P\,\mu}{E} + \frac{2\,\Delta P}{E}(1-\mu) + \frac{\Delta\rho}{\rho}$$

$$\beta = \frac{\Delta R/R}{\Delta P} = \frac{2}{E} + \frac{\Delta\rho/\rho}{\Delta P} \qquad\qquad (6.47)$$

For these materials having β negative, it is obvious that resistivity is lowered with an increase in pressure. The values of β for bismuth, rubidium and antimony are very high as compared to manganin and gold alloy but the characteristic is highly nonlinear. Table 6.4 presents the values of β for some of the metals and alloys.

Table 6.4 Pressure coefficient of resistance of some metals and alloys

Material	Pressure coefficient in Ω/Ω per kg/cm^2
Gold alloy	0.91×10^{-7}
Manganin	2.34×10^{-7}

Gold alloy and manganin have linear characteristics while at the same time possess very low temperature coefficient of resistance. Even though gold alloy is costly, it has much lower temperature coefficient of resistance than manganin, thereby avoiding the need for temperature compensation. Pressure coefficient of resistance is almost independent of temperature.

High (hydrostatic) pressure gauges are based on this principle and have been constructed for pressures ranging up to 20×10^3 kg/cm^2. They are known as Bridgman gauges for high pressure measurements and are used for pressure ranges beyond the capacity of Bourdon tube gauges and strain gauge pressure transducers. The constructional details of a gauge are shown in Figure 6.42, and it may be observed that the conductor, in the form of a loosely wound coil, is enclosed in a flexible kerosene-filled bellows which serves the purpose of transmitting the external pressure to the conductor. The resistance change is measured by a Wheatstone bridge

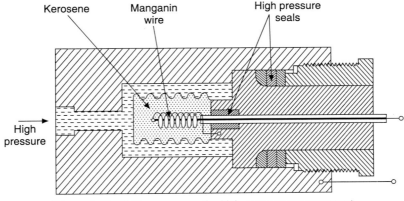

Figure 6.42 Bridgman gauge for high pressure measurement.

circuit. Although the conductor is extremely fast in response, the transducer system as a whole is limited in its bandwidth due to the presence and size of the fluid chamber in which the wire is kept.

6.1.6 Resistive Moisture Transducers

Measurement of water content in substances is very important as it may considerably affect their properties, quality, composition, rate of decay or decomposition etc. The substances may be solids, solutions or gases. Gases and air contain water in vaporous state. Aqueous solutions contain certain quantity of salts, and solids such as grain, tobacco and wood contain water during their formation.

In textile and paper industries, on-line measurement of the moisture content in the textile fibres and paper is essential. It is also important to maintain a high level of humidity in the atmosphere of textile industries. Also, colour printing, drying of wood, leather and lacquers and successful storage of meat, fruits and eggs requires careful humidity control at the desired levels.

Moisture content in a solid substance is ascertained in the laboratory by obtaining the weights of the sample before and after drying and expressing the difference as a percentage of the weight of the substance after drying.

The state of humidity of gases or air is usually expressed in terms of absolute humidity or relative humidity.

The *absolute humidity* of a gas is defined as the mass of water vapour present in a unit volume of the gas and may be expressed in grams per cubic metre.

The *relative humidity* (RH) of gas is expressed in the following ways:

$$\text{RH} = \frac{\text{(mass of water vapour present in a given volume of gas)}}{\begin{array}{c}\text{(mass of water vapour necessary to saturate the same}\\ \text{volume of gas at the same temperature}\end{array}}$$

$$= \frac{\text{(pressure exerted by the vapour)}}{\text{(saturation vapour pressure at the given temperature)}}$$

$$= \frac{\text{(saturation vapour pressure at the dew point)}}{\text{(saturation vapour pressure at the given temperature)}}$$

The instruments used to measure the humidity of a gas are called *hygrometers*. In the laboratory, the humidity of a gas is determined by absorbing the water vapour contained in a known volume of the gas by passing it through a weighed drying tube holding a suitable dehydrating agent.

The first and classical method of determining the relative humidity of air is by the wet and dry bulb thermometer system known as *psychrometer*. Continuous measurement of humidity is nowadays mostly carried out by measurement of the dew-point. When humid air is cooled slowly and continuously, a temperature reaches at which the water vapour present in the air is sufficient to saturate the air. Any further cooling will only cause the vapour to be deposited on the surfaces of objects in its vicinity, in the form of dew. The temperature at which this dew formation occurs is defined as the *dew-point* for the sample of air.

Aqueous solutions are formed when salts, acids or bases are dissolved in water. The addition of these substances to water results in an instantaneous formation of atoms or groups of atoms which carry electrical charge. These charged atoms or groups are known as ions and are responsible for conduction of electricity through the solution. Such substances are known as electrolytes, and the solution as electrolytic solution.

The concentration of a solution may be expressed as percentage by weight, meaning that a 10 per cent solution has 10 grams of the substance (solute) dissolved in 100 grams of water.

Another method of expressing the concentration is in terms of the solution of 'normal concentration' or the 'normal solution' (N). A solution has normal concentration if 1 gram-equivalent weight of the solute is dissolved in 1 litre of solution. 0.1 N solution has only one-tenth of a gram-equivalent weight, dissolved in one litre of solution. For example, sodium chloride has a gram-equivalent weight of 58.5 (23 + 35.5) grams, and hence 1 litre of normal sodium chloride solution has 58.5 grams of the salt in one litre of water.

Changes in the concentration of a solution affect the electrical resistivity and the conductivity, but there is no simple relationship between conductivity and the total concentration of the solute except at very low concentrations (Figure 6.43). Measurement of conductivity of the solution does not yield information about the types of ions present in the solution. In most of the situations of chemical process control, the type of solute is the same and the concentration only needs to be measured continuously. Similarly, water purity is measured by measuring its conductivity.

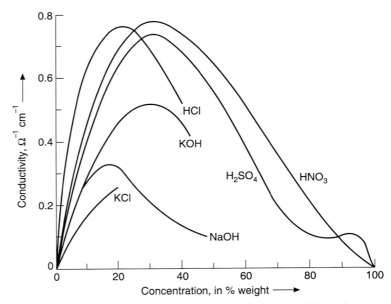

Figure 6.43 Characteristics of solutions between electrical conductivity and concentration.

For all the above applications, suitable transducer systems with electrodes are required to hold the sample of substances and enable measurements of resistance and conductivity. For each application, there is a cell and a calibration curve relating the input and output quantities.

6.1.6.1 Moisture Content of Solids

It is known that the electrical resistivity of materials such as wood, grain, flour and a large variety of substances in powder form decreases when the moisture content of the material increases. The majority of the techniques used for measuring the moisture content of such materials involve measurement of electrical resistance. The material is held in close contact with a pair of electrodes of suitable size. All samples of powder materials, or grain or textiles should be brought to a high degree of compression between the electrodes. A variety of electrode systems exist, each suitable for a certain range of materials. Figure 6.44 shows the details of the electrode system and the provision made for compaction of the powder material. After filling the cup with a measured quantity of the sample, the top cap holding the spring-loaded piston is screwed down fully so as to develop the required pressure on the material. Actual moisture content of the sample is determined by weighing it after it is fully dried. For each material and its electrode system, the characteristic between resistance and moisture content is established by calibration.

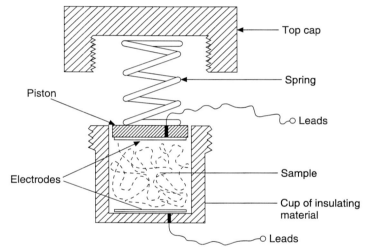

Figure 6.44   Test cell for moisture measurements of granular substances.

6.1.6.2 Concentration of Electrolytic Solutions

Measurement of the electrical conductivity of solutions enables one to estimate electrolytic impurities in water as low as 0.1 part per million (ppm). Transducers with properly disposed electrodes make continuous monitoring of water purity and air pollution possible. Objectionable gases such as sulphur dioxide, hydrogen sulphide and ammonia are absorbed in suitable solutions, and the variation in the conductivity of the solution is taken as the index of air pollution.

When an electrolyte is dissolved in water, some of its molecules will split up into ions (atoms charged positively or negatively), and this process is known as *dissociation*. Molecules of the solute will be dissociating and at the same time ions will be recombining. Finally, a state of equilibrium will be established between the ions and the undissociated solute. The degree of dissociation, defined as the fraction of the solute dissociated, depends on the substance and the dilution of the solution. The greater the dilution, the higher is the degree of dissociation, but at the same time the number of ions in a unit volume falls. Although a higher degree of dissociation results in an increase of the specific conductivity K, the net effect is a decrease in K. Figure 6.43 shows the variation of K with concentration (inverse of dilution) for some common electrolytes.

To measure the influence of the degree of dissociation alone, another quantity, equivalent conductivity λ (as distinct from the specific conductivity K) is used. It is the conductivity of a solution, 1 cc of which contains one gram-equivalent weight of the solute, and is given by

$$\lambda = K\phi \qquad (6.48)$$

where,

   ϕ = the number of cc which contains one gram-equivalent of the solute at the same dilution.

For highly diluted solutions, K is nearly proportional to the molecular concentration, c (= $1/\phi$), as is borne out by the curves of Figure 6.43. It is found that for concentrations below 5%, the equivalent conductivity is constant, and hence measurement of the specific conductivity of a solution is the most convenient technique for estimation of dissolved ionized solute in a dilute solution.

The equivalent conductivity of a solution at infinite dilution is the sum of the equivalent conductivities of the anions and cations formed (Kohlrausch's law). Ionic conductivity increases with temperature, and the temperature coefficient of the ionic conductivity is different for ions of different atoms and groups of atoms. So the equivalent conductivity also increases with temperature. For certain electrolytic solutions, it increases first, becomes maximum, and then falls with increase of temperature. The dielectric constant and viscosity of the solvent are also affected by temperature and hence the conductivity of the solution. Therefore, it is essential to provide temperature compensation by addition of suitable means to the measuring circuits. For many solutions, at 18°C, the conductivity increases by 2% per 1°C rise of temperature.

The specific conductivity of pure water is found to be 0.043×10^{-6} mhos, whereas that of ordinary water is about 4×10^{-6} mhos at 18°C. The presence of one gram-equivalent of potassium chloride in 50 litres of solution increases the specific conductivity of ordinary water by about 1000 times. The effectiveness of the technique, as seen by this example, has led to the design of a large variety of transducer systems (also known as conductivity cells) suitable for use in laboratory and industry.

The ratio of the spacing between the electrodes to the area of the electrodes is known as cell constant and cells are available with cell constants ranging from 0.1 to 100. For each solution, a cell is chosen with a cell constant that brings the cell resistance within the range 10–10000 ohms. To minimize the errors due to polarization, the area of the electrodes may be made larger, and the frequency of excitation of the ac bridge circuit is kept high at about 1000 Hz. The material of the electrodes is either gold or platinum provided with a coating of platinum black. Three typical cells are shown in Figure 6.45. The cell of (a) is for immersion in the solution, whereas that of (b) is for on-line measurement with the solution flowing through the cell. The cell(c) is provided with four terminals, two each for current and potential.

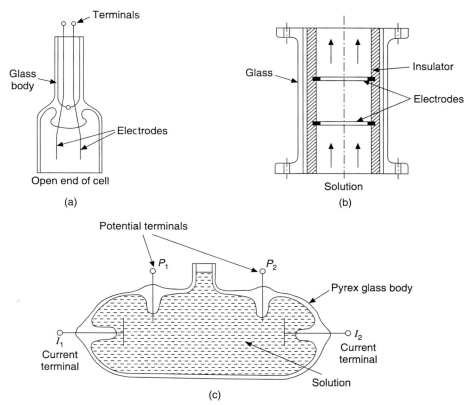

Figure 6.45 Electrode arrangement in typical conductivity cells: (a) For immersion; (b) for on-line measurement; (c) for filling in with solution.

Alternating current Wheatstone bridge circuits are used to measure the conductance of the solution. In case of a single cell, the circuit of Figure 6.46(a), is used. A resistance thermometer (or a thermistor) of suitable value and temperature coefficient of resistance is used as R_4 to provide temperature compensation. It is placed in the solution in close proximity of the cell. The circuit shown in the figure employs a reference cell filled with a standard (of known concentration) sample of the same solution as that of the test cell. The reference cell is in sealed condition, but is placed in the test solution in close proximity of the test cell for providing the necessary temperature compensation. The reference cell may as well contain a suitable and stable reference solution which has the same temperature coefficient of conductivity as that of the test solution. However, it must be noted that the unbalance voltage of the bridge circuits may not be linearly related to the conductivity as the fractional change in the resistance of the test cell is considerable in most cases.

The ac potentiometer circuit shown in Figure 6.46(b) measures the voltage drop developed across a noninductive standard resistor R_s due to the current I_c. By keeping the value of R_s much smaller than the resistance of the cell, the voltage drop across R_s is made proportional to the conductivity of the solution. The Wheatstone bridge circuit consisting of a resistance

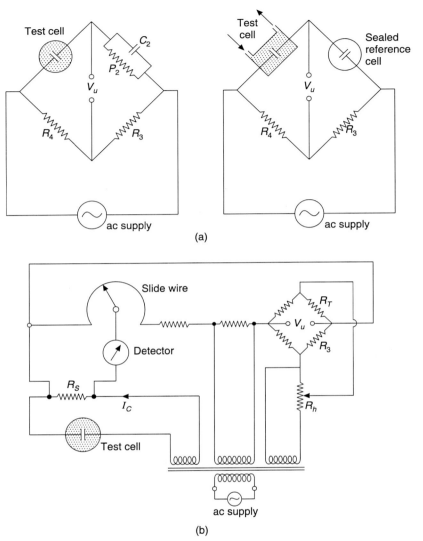

Figure 6.46   Measuring circuits for conductivity of solutions: (a) ac Wheatstone bridges; (b) ac potentiometer.

thermometer R_T and three manganin resistors provides the temperature compensation. The unbalance voltage V_u of the bridge is used to boost the voltage applied to the potentiometer circuit. At 18°C, V_u is made zero by adjusting R_3. For a certain desired change of temperature, the value of V_u can be adjusted by varying R_h so as to facilitate the temperature compensation when different solutions are tested.

6.1.6.3   *Relative Humidity of Gases*

Certain substances have the property of absorbing or giving off moisture very rapidly in such a way that the quantity of moisture they contain depends on the partial pressure of the water vapour present in the atmosphere surrounding the substance. Such substances are known as *hygroscopic substances*. Some salts like lithium chloride, zinc chloride and calcium chloride are hygroscopic by nature. Electrical conductivity of these hygroscopic salts is dependent on their moisture content and hence electrical hygrometers are developed using this property of the hygroscopic salts. These hygrometers enable continuous measurement of relative humidity.

The resistive element consists of a tubular wick made of glass fibres impregnated with lithium chloride, and mounted over a thin-walled metal tube electrically insulated from the wick. Two parallel wires of silver or platinum are helically wound around the wick, leaving sufficient spacing between the two wires all through. The two wires serve as electrodes, and the coating of lithium chloride on the wick provides the path for the current flowing from one wire to the other, as shown in Figure 6.47. The resistance of the path between the two wires is measured by simple circuits by using the ac source of emf. The relationship between humidity and resistance is, however, nonlinear and each element covers only a small range of the order of 10 per cent of the relative humidity. At least 10 elements are necessary to cover the range of 5–99% of the relative humidity of atmosphere. Time taken for the establishment of equilibrium with the surrounding atmosphere is about 3–10 s. Accuracy of measurement by these resistive elements may be within ±1.5%, but resolution may be within ±0.15% of relative humidity.

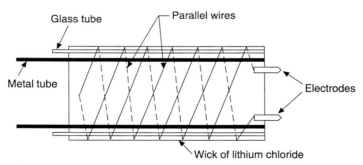

Figure 6.47   Lithium chloride cell for relative humidity measurement.

6.1.7   Resistive Magnetic-flux Transducers

Permanent magnets are used in many devices as sources of stored fixed energy. In transducers, small variations in the magnetic flux available across an air gap can be brought about by changes in the quantity to be measured. For example, the thickness gauge of Figure 6.48 consists of a permanent magnet, with its magnetic path split into two parallel paths, one of which is the sheet of the magnetic material. The other is the shunt path with an air gap, across which changes in magnetic field occur due to variations in the thickness of the sheet. Similarly, in dc machines, actual flux-density at several points in the air gap between the stator and rotor has to be measured by a suitable transducer. It is desired that the transducer be so thin as to permit introduction into narrow air gaps and be of such size as to distinguish variations in flux density from one point to another.

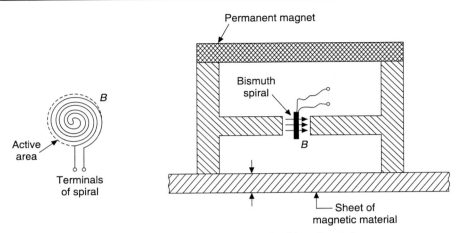

Figure 6.48 Thickness gauge using bismuth spiral.

The resistivity of certain substances changes with the intensity of magnetic field to which it is exposed. These substances are known as *magneto-resistive* materials. Some metals exhibit this property at low temperatures, but bismuth shows considerable effect even at room temperatures. Indium antimonide is another magneto-resistive material. The material is drawn in the form of thin flat strip so that it can be wound into a flat bifilar spiral of 0.5–3 cm in diameter and cemented between two sheets of insulating material. The change in resistance, due to variation in the net magnetic flux perpendicular to the plane of the spiral, is measured by means of dc Wheatstone bridge networks. The resistivity of bismuth doubles when exposed to a flux density of about 2 Wb/m^2 and the percentage change of resistance is nearly proportional to flux density over the entire range except at the lower end.

The magneto-resistance effect is considerable in the case of indium antimonide and indium arsenide even at room temperatures, and hence they are preferred for application in several transducers. Even though indium antimonide is more sensitive, indium arsenide is commonly used in transducers as its resistivity is 100 times more than that of antimonide. But the changes in resistivity with the changes in magnetic field or flux density are not linearly related. Calibration curves are provided for each spiral showing the variation of resistance of spiral with the intensity of magnetic field. The change in resistance is proportional to the square of the change in intensity at low values. For accurate measurements of steady magnetic fields, temperature compensation by using a dummy spiral or by any other means is essential. Self-heating errors should be avoided by holding the value of current through the spiral at a very low value.

Production processes enable manufacture of extremely small-sized filaments of the above and they are throwing open tremendous opportunity for application in modern instrumentation systems. One such application is the measurement of small displacements conveyed to the strip, along the direction of a magnetic field, produced by a permanent magnet. Although the size and sensitivity of the magneto-resistive transducer offers scope for several applications, the problems of induced emfs in the spiral under dynamic conditions limit its application to almost steady state measurements.

6.1.8   Resistive Optical Radiation Transducers

The electrical resistivity of certain materials changes when exposed to light, and they are known as photoresistive (or photoconductive) materials. All semiconductors are photoresistive in nature and their resistivity falls when light is incident on them. The effect is noticeable in selenium, silicon and germanium. Also, compounds such as lead sulphide, activated polycrystalline cadmium sulphide and cadmium sulpho selenide exhibit this property and they are used as basic materials for the construction of photoconductive cells. Each material has its own spectral response characteristic as seen from Figure 6.49, and none of them possess uniform sensitivity for all wavelengths of the visible region of the electromagnetic radiation. The resistivity is very high in dark and it may change by three decades for a change of light intensity by three decades.

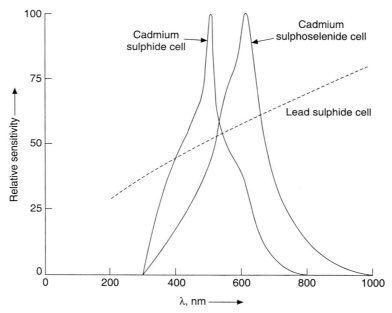

Figure 6.49   Spectral response of photoconductive cells.

The photoconductive cell usually consists of a deposit of a photoconductive material on a ceramic substrate. Electrodes of tin or indium are embedded on the photoconductive material by evaporation, in an interdigitating pattern. The electrodes make ohmic contact with the deposit and enable measurement of the resistance of the cell. The cells may be constructed with active area of the cell ranging from 1 mm^2 to 35 mm^2. The cell resistance depends on the properties of the photoconductive layer, the dimensions of the cell, and the electrode configuration.

The measurement of the resistance of the cell can be made by using dc bridge and potentiometer circuits. A dc voltage is applied across the series combination of the cell and a small resistor whose value is kept much smaller than the lowest possible value of the cell under maximum illumination conditions. The voltage drop across the small resistor is measured on a dc potentiometer. When the radiation falling on the cell is chopped mechanically, the voltage drop is amplified by ac amplifiers.

Temperature has practically no influence on the performance and sensitivity of the photoconductive cell as long as the illumination level is not low. All semiconductor cells have improved sensitivity with lowering of temperature, and hence temperature compensation is needed if their higher sensitivities are utilized at temperatures below $-100°C$. The time constant of the cells ranges from 0.05 ms to 2 ms and is dependent on the cell size and intensity of illumination. Apart from the photoemissive and photovoltaic cells, the photoconductive cells are used in a variety of applications involving illumination measurements and detection of optical radiation in spectrophotometers.

6.2 INDUCTIVE TRANSDUCERS

Inductive transducers are those in which the self-inductance of a coil or the mutual inductance of a pair of coils is altered in value due to a variation in the value of the quantity under measurement. Transducers based on the above principle are found to possess considerable sensitivity and scope of application, especially for displacement and thickness measurements.

Some of the general observations regarding the operation and performance of the inductors may be recalled in this context for an appreciation of the limitations that render them somewhat less popular in instrumentation systems. Detection of changes in inductance can be done only by using the inductors in ac circuits. If the coils are air-cored, they act as a source of alternating magnetic fields that would interfere with all the neighbouring circuits while the coils themselves link with the external magnetic fields. The problem becomes more acute when the coils are used in high frequency circuits. Proper electric and magnetic shielding becomes essential. If the coils are iron-cored, the magnetic flux is contained in the path of iron and interference effects are minimized. The frequency of ac supplies used in the circuits should be at least 10 times higher than the frequency of the input quantity. Iron losses of the cores add to the resistance of the copper of coils and vary with the frequency of excitation. Inductance variation may not always be brought about without variation in resistance. But one looks for output signals to be exclusively functions of changes in inductance.

Measurement of displacements, by changing the tapping point of a solenoidal or a toroidal coil, as shown in Figure 6.50(a), may be practicable, but the output voltages are governed by both the self-inductance and resistance of the tapped portion. Hence they are unsuitable for displacement measurements though, in this way, they bear resemblance to the resistance potentiometers.

When displacements are measured in the way shown in Figure 6.50(b), the movable core (armature) is subjected to the electromagnetic force of attraction, which cannot be treated to be insignificant in most cases. The displacement source has to overcome this force for outward displacement of the armature from the fixed core.

The output signal is an induced emf of the secondary coil if a coil system is used as a mutual inductance-type or transformer-type transducer as shown in Figures 6.51(a) and 6.51(b). Under steady-state conditions, the induced emfs are of the same frequency as that of excitation voltage, and will not be exactly in phase opposition with the primary applied voltages. Under dynamic input conditions, the induced emfs are amplitude modulated, by the input displacement

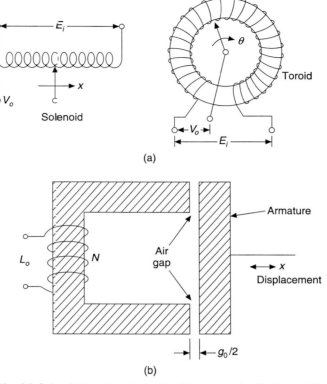

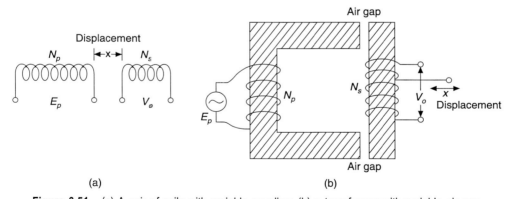

Figure 6.50 (a) Solenoidal and toroidal coils; (b) iron-cored coil with variable air gap.

Figure 6.51 (a) A pair of coils with variable coupling; (b) a transformer with variable air gap.

signals. Such is the case with the unbalance voltages of the ac Wheatstone bridge circuits having a self-inductance type transducer in one of its arms. Enough care needs to be exercised, while processing the output signals, as a faithful reproduction of the input signals is always desired.

It is also necessary to bear in mind that the relative permeability of the magnetic materials is not a constant and that it varies with the intensity of magnetization. Whether the coil system of the transducer is partly iron-cored or almost iron-cored, it is necessary to hold the magnetization currents to such low values as to operate the core under its initial magnetization conditions. Such a situation would not only permit the assumption of the initial permeability as a constant for design purposes, but also minimize the electromagnetic force of attraction on the movable armature.

To surmount some of the problems raised above, inductive transducers are constructed using push–pull coil systems. It can be seen from the following sections that single-coil transducer systems are used only in certain particular situations, such as thickness measurement of sheets.

The coils that are commonly used for inductance transducers are primarily of two types: (a) coils that are wound on insulating coil formers (or self-supporting coils) for use with ferromagnetic cores, and (b) coils that are equivalent to long solenoids with one or two layers of winding or short solenoids of a large number of layers of winding. Closely wound toroidal coils, although equivalent to a long solenoid, have the merit of independence from the effects of external magnetic fields but are not suited to displacement measurements. Simple configurations of the types shown in Figure 6.50(b) enable the variation of self-inductance to occur due to the displacements of the movable core or armature.

The self-inductance of each configuration is worked out from the basic relation, $L = N^2/R$, where R is the reluctance of the magnetic path. For the iron-cored coil with a small air gap g_0, R is given by

$$R = \frac{l_m - g_0}{\mu_0 \mu_r A} + \frac{g_0}{\mu_0 A}$$

where,

l_m = total length of magnetic path
A = area of cross-section of core
μ_0 = absolute permeability of air
μ_r = relative permeability of iron

which, if rewritten, has the form

$$R = \frac{l_m}{\mu_0 A} \frac{1 + (g_0/l_m)(\mu_r - 1)}{\mu_r}$$

As the relative permeability μ_r is higher than unity, and as the core is normally operated at very low magnetization level, the above expression may be simplified to

$$R = \frac{l_m}{\mu_0 \mu_{\mathrm{eff}} A} = \frac{l_m}{\mu_0 A} \left[\frac{1 + (g\mu_i/l_m)}{\mu_i} \right] \tag{6.49}$$

where,

μ_{eff} = effective permeability of the composite path
μ_i = initial relative permeability of the core material.

In the case of the long solenoid shown in Figure 6.52(a), variation in the self-inductance of the coil is effected by the movement of the core inside the coil. The movable core or plunger, as it is called, may be cylindrical in shape or tapered. The magnetic field strength H at any point P on the axis is given by

$$H = \frac{NI}{2l_c} \left[\frac{l_c + 2x}{[4r^2 + (l_c + 2x)^2]^{1/2}} + \frac{l_c - 2x}{[4r^2 + (l_c - 2x)^2]^{1/2}} \right] \tag{6.50}$$

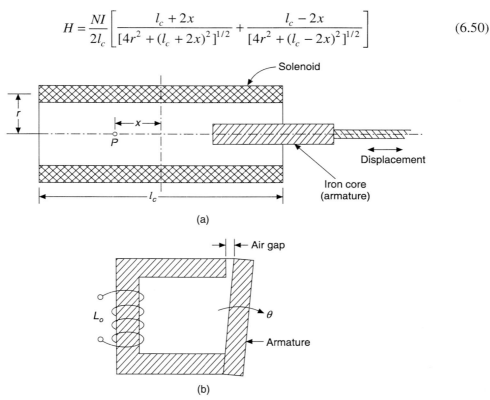

Figure 6.52 (a) A solenoidal coil with a plunger-core for linear displacements; (b) an iron-cored coil for angular displacements.

where,

x = distance between P and centre of coil,
r = radius of the coil,
l_c = length of the coil,
N = number of turns of coil
I = current in the coil

For reasons already explained, these single-ended coil systems are not favoured for application to measurements of displacements. The push–pull systems are treated in Section 6.2.2.

In the above configurations a change in self-inductance is caused by variation of g alone. Transducers based on the variation of g, as also the reluctance of the magnetic path, are known as *variable reluctance* transducers. Reluctance variation can as well be brought about by variation of M_r, and in such cases, they are known as *variable permeability transducers*. Composition of magnetic materials may be monitored by variable permeability transducers.

When a pair of coils is used, the mutual inductance between them is decided by the coefficient of coupling k, and is given by $M = k/\sqrt{L_1 L_2}$, where L_1 and L_2 are the self-inductance values of the coils. If the coils are connected in series, the effective self-inductance of the combination is given by $L = L_1 + L_2 \pm 2M$, depending on the way they are connected. The coefficient of coupling can be varied in several ways and the corresponding changes in L or M are used to provide output signals. Based on the variation of coupling, various configurations emerge, of which some have become very popular for measurements of linear and angular displacements.

A third alternative to the above is offered by a completely air-cored coil system and a short-circuited ring or sleeve. They are known as *proximity transducers*. The effect of the eddy currents flowing in the short-circuit single turn secondary, on the self-inductance of the primary coil forms the basis of this transducer. Although they are not highly sensitive, they have a linear characteristic over a large range of input displacements.

For the core and armature, the magnetic material that possesses high permeability and low loss is preferred. The inductive coil should be of high Q-value and the design values of the core, coil and air gap should yield high Q for the material selected. Some of the soft magnetic materials normally used for inductive transducers are presented in Table 6.5 along with their properties.

Table 6.5 Properties of some soft magnetic materials for transducer application

Property	Materials		
	Mu metal	*Radiometal*	*Hyrho radiometal*
Initial permeability (μ_i)	60,000	6,000	4,000
Maximum permeability (μ_r)	2,40,000	40,000	65,000
Saturation flux density, (Wb/m^2)	0.77	1.6	1.4
Remnance (Wb/m^2)	0.45	1.0	1.0
Coercivity (A/m)	1.0	10	10
Hysteresis loss (J/m^3/Hz)	3.2	40	45
Resistivity ($\mu\Omega$-m)	0.60	0.45	0.75
Density (g/cm^3)	8.8	8.3	8.3
Young's modulus (N/m^2)	185×10^9	170×10^9	170×10^9
Tensile strength (N/m^2)	540×10^6	430×10^6	430×10^6

When inductive transducers are designed for operation at voltages of frequency of the order of 1 MHz, ceramic magnetic materials are preferred for the core and armature. These are known, in general, as *ferrites* (Fe_2O_4) and are further classified as Manganese–zinc ferrites (Ferrox cube A), Nickel–zinc ferrites (Ferrox cube B), and Magnesium–manganese ferrites (Ferrox cube D). All these materials are hard and brittle as porcelain. They can be cast into different shapes and sizes, depending on their application. They are available in the form of rods, toroids, pots and E-, I- or U-shaped cores. Compared to the properties of soft magnetic materials, they have much lower incremental permeability but much higher resistivity. Eddy current losses and hysteresis losses are extremely low even at high frequencies. Normally, Manganese–zinc ferrites are used for cores of inductive transducers. Variation in composition can alter the properties of the ferrites and the grades of Ferrox cube A material available and their properties are presented in Table 6.6.

Table 6.6 Properties of some ferrite materials (Manganese–zinc ferrites)

Property	Grade		
	A_1	A_4	A_{13}
Initial permeability (μ_i)	700	1200	1850
Saturation flux density (Wb/m^2)	0.36	0.36	0.34
Resistivity ($\mu\Omega$-m)	0.80	0.20	1.00
Density (g/cm^3)	4.70	4.80	4.60

6.2.1 Inductive Thickness Transducers

In industry, the measurement of the thickness of rolled sheets or mass-produced objects is a common requirement. The material of the test sheet or object may be magnetic (iron or steel), nonmagnetic and conducting (Aluminium or copper) or nonmagnetic and nonconducting (bakelite or paint). Inductive transducers meant for such purposes are known as inductive thickness gauges. As the thickness is of primary interest, it is important that the properties of the materials, such as permeability and resistivity, should remain constant. Each gauge is suitably designed for use with the test object and calibrated by making use of reference sheets or slabs of known thickness but of the same material of the test object.

Variable reluctance-type inductance transducers prove handy for most of the applications. An E-, U- or I-shaped yoke of high permeability material is provided with one coil for the self-inductance type and a pair of coils for the mutual inductance type. The magnetic path is completed through the test piece of magnetic material, as shown in Fig. 6.53. The yoke is usually laminated to limit the eddy currents produced when the coil is excited by alternating current. The attraction force of the yoke on the armature and the weight of the yoke may help in reducing the air gap between the yoke and the test piece. However, the surfaces of the test piece and the yoke are kept smooth for a closer contact. If the reluctance of the yoke is made negligible as compared to that of the test piece, the self-inductance L of the coil is proportional to the thickness t of the test piece and is given by

$$L = \frac{N^2 \mu_0 \mu_r bt}{l} \tag{6.51}$$

where, b and l are the width and length respectively of the test piece and μ_r is the relative permeability of the material.

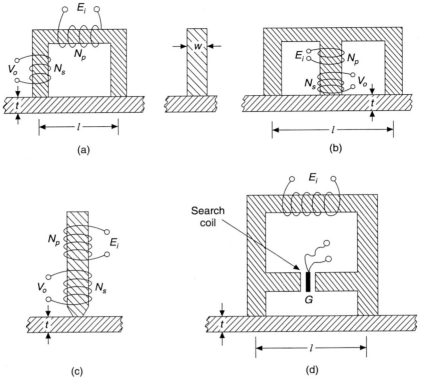

Figure 6.53 Different arrangements for measurement of thickness of metallic and magnetic sheets.

The thickness of sheets of magnetic material as well as insulating material may be obtained by any of the arrangements shown in Figures 6.53(a) and 6.53(b). In the case of insulating material, the sheet is kept between the yoke, and a magnetic material backing of known thickness. The reluctance of the path is almost governed by the thickness of insulating sheet. Measurement of thickness of test pieces ranging from 25 μm to 2.5 mm is possible by the above methods with an accuracy of 2–5%.

The primary coil of the system shown in Figure 6.53(c) is excited from a relatively high frequency source as the reluctance variation with the thickness of the sample will be very small. However, it is possible to measure variations in the thickness of conducting material sheets. The induced emf of the secondary coil may be used for direct indication and calibration.

An alternative is shown in Figure 6.53(d), where the test object of magnetic material forms a low reluctance shunt path for the magnetic flux across the gap G. The induced emfs of the search coil serve as the output signals of the transducer. The primary coil is excited from a constant voltage source of suitable frequency.

6.2.2 Inductive Displacement Transducers

A simple form of the inductive displacement transducer, as shown in Figure 6.50(b), consists of a U-shaped laminated ferromagnetic core and an I-shaped armature subjected to input displacements. Reluctance variation due to a change in air gap causes variation in the self-inductance of the coil.

Using the relationship of Eq. (6.49) for the effective permeability of the composite path of iron and air, the self-inductance L_0 of the coil for an air gap of g_0 is obtained as

$$L_0 = \frac{N^2 \mu_0 A}{l_m} \left(\frac{\mu_i}{1 + g_0 \mu_i / l_m} \right) = K \frac{1}{g_0 + \dfrac{l_m}{\mu_i}} \tag{6.52}$$

where $K = N^2 \mu_0 A$. When the gap decreases by Δg, the value of L_0 increases by ΔL, giving

$$L_0 + \Delta L = K \left(\frac{1}{(g_0 - \Delta g + l_m / \mu_i)} \right)$$

and the fractional change in L_0

$$\frac{\Delta L}{L_0} = \frac{\Delta g}{g_0} \frac{1}{\left(1 + \dfrac{l_m}{\mu_i g_0} - \dfrac{\Delta g}{g_0} \right)}$$

When rewritten, this equation becomes

$$\frac{\Delta L}{L_0} = \frac{\Delta g}{g_0} \frac{1}{1 + \dfrac{l_m}{\mu_i g_0}} \frac{1}{\left(1 - \dfrac{\Delta g / g_0}{1 + \dfrac{l_m}{\mu_i g_0}} \right)} \tag{6.53}$$

$$\frac{\Delta L / L_0}{\Delta g / g_0} = S \frac{1}{1 - S \Delta g / g_0} \tag{6.54}$$

where, S = the incremental sensitivity of fractional changes which, when $S\Delta g/g_0$ is $<< 1$, is given by $1/(1 + l_m/\mu_i g)$. For small values of $S(\Delta g/g_0)$, the expression

$$\frac{\Delta L / L_0}{\Delta g / g_0}$$

can be expanded to give

$$\frac{\Delta L / L_0}{\Delta g / g_0} = S \left[1 + \frac{S\Delta g}{g_0} + S^2 \left(\frac{\Delta g}{g} \right)^2 + \dots \right] \tag{6.55}$$

Similarly, for increase in g_0, the decrease in L_0 is given by

$$\frac{\Delta L / L_0}{\Delta g / g_0} = S \left[1 - \frac{S\Delta g}{g_0} + S^2 \left(\frac{\Delta g}{g} \right)^2 - \dots \right] \tag{6.56}$$

For a choice of l_m/g_0 equal to 50, a ferrite core with an initial permeability of 1000 provides a value of 0.95 for S, whereas a mu-metal core with an initial permeability of 50,000 offers a value of nearly unity. As S cannot be much different from unity, satisfactory linearity between fractional changes in inductance and fractional changes in gap at any chosen g_0 can be found only for values of $\Delta g/g_0$ within ±0.01.

However, it is worthwhile to notice the large sensitivity between incremental changes of L_0 and g_0 from Eq. (6.54) which gives $\Delta L/\Delta g$ for decreasing values of g_0 as

$$\frac{\Delta L}{\Delta g} = \frac{L_0}{g_0} S \left[1 + \frac{\Delta g S}{g_0} + \left(\frac{\Delta g S}{g_0} \right)^2 + \dots \right] \tag{6.57}$$

Characteristic curves (*a* and *b*) showing the variation of L_0 with g_0 (Figure 6.54) indicate that the ratio of L_0/g_0 is not a constant and that it decreases with increasing value of g_0.

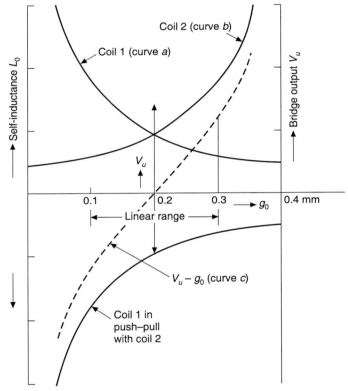

Figure 6.54   Characteristic curves showing the variation of self-inductance with air gap for a single-ended and push–pull transducer.

Some of the basic problems associated with the above single-ended system are overcome by using two identical cores with identical coils so as to constitute a push–pull transducer. The armature is subjected to displacements in the same direction as that of the electromagnetic force, and the net force on the armature is zero when it is lying symmetrically between the two cores. The armature is said to be in the null (or neutral) position, and in this position both the coils have the same self-inductance. Under these conditions, all the ac bridge circuits shown in Figure 6.55 will be under balance, and hence V_u will be zero. Any movement of the core from the null position increases the self-inductance of one coil and decreases the self-inductance of the other, resulting in a linear characteristic between V_u and g over a limited range around g_0 as seen from curve (c) of Figure 6.54.

The push–pull arrangement further provides the following advantages:

1. The system is immune from the effects of external magnetic fields.
2. The force required to move the armature from the null position is considerably reduced (thereby increasing the mechanical input impedance).
3. The force required to overcome the electromagnetic pull on the armature is the same for movements in either direction from the null position.
4. There is very little magnetic flux that interferes with the neighbouring circuits.

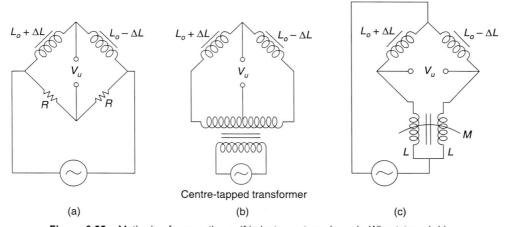

(a)   (b)   (c)

Figure 6.55   Methods of connecting self-inductance transducer in Wheatstone bridge.

Sensing magnetic flux variations due to reluctance variations can as well be done by measuring the emfs induced in an additional coil suitably located in the transducer. All the advantages of the push–pull self-inductance transducer can be retained while the bridge circuits for processing the parameter changes can be replaced by a high impedance voltmeter to measure the magnitude of the induced emf. They are better known as *variable reluctance mutual-inductance* (or *transformer) transducers*. Some of the more common configurations are shown in Figure 6.56. In systems (a) and (b), the armature is subjected to linear displacements along the direction of electromagnetic pull while that of (c) is subjected to angular displacements. In the arrangement of (a), the two identical coils on the cores are connected in series and excited in such a way that the net induced emf in the secondary winding on the armature is zero, when the armature is in the null position. The arrangement of (b) has the primary winding on the armature, while the two secondary windings on the limbs are connected in phase opposition so

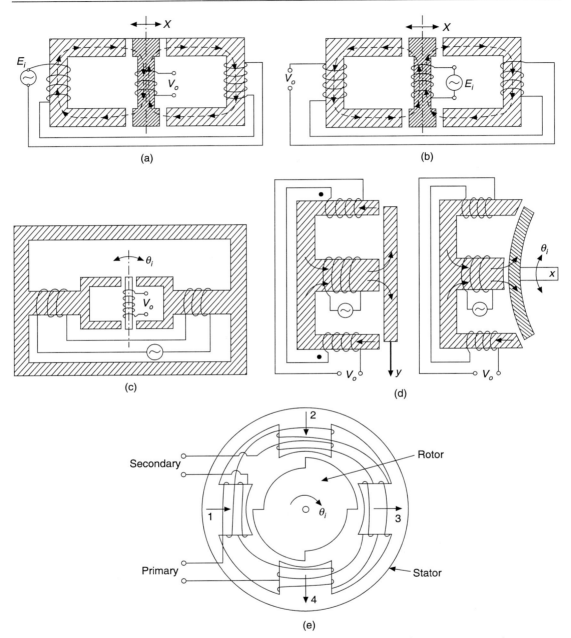

Figure 6.56 Common configurations of the variable reluctance transformer type transducer: (a), (b) for linear displacement (c) for angular displacement; (d) E-pick-offs; (e) microsyn.

that net output voltage is zero when the armature is in the null position. In system (c), the armature has no flux through its cross-section, when it is in null position, but when it is subjected to an angular displacement, a low reluctance path through the length of the armature is formed, and the flux in the armature induces emf in the secondary.

The configurations of (d) are commonly known as E-pick offs. The armature is subjected to displacements transverse to the direction of electromagnetic pull, the force required being much smaller than for case (a) or (b). The primary winding on the centre limb is excited by ac supply and the two secondary windings on the outer limbs are connected in series so as to have their induced emfs in phase opposition. When the armature is in the null position, the net emf is zero.

The transducer of (e) is also a member of the family of transformer transducers, but is generally known as microsyn, or magnesyn, or telegon, under their patented names. The device consists of a stator and a rotor (armature), both laminated. The rotor does not carry any winding and it is shaped like the figure of eight looking like a two-pole rotor. It serves the function of an armature in varying the reluctance of the stator magnetic circuits. The stator has four poles, and each is provided with a primary and a secondary winding. All the primary coils are connected in series and excited from an ac supply. The polarity of the windings is such that the adjacent poles 1 and 2 are of the same polarity and the other poles, 3 and 4, are of opposite polarity. The secondary windings are also in series but connected in such a way that the emfs induced in the windings on poles 1 and 3 are in phase opposition to the emfs induced in the windings on poles 2 and 4. The null position of the rotor is shown in the figure and, in this position, flux in all the four poles is the same, and hence net output voltage is zero. The reluctance of the magnetic path of the pole-pair 2–4 is equal to that of the pole-pair 1–3. Any angular displacement of the rotor in the anticlockwise direction increases the reluctance of the pair 1–3 and decreases that of 2–4. The corresponding effect on induced emfs results in a net output voltage. The output voltage is proportional to the angular displacement over a range of ±7° from the null position.

All the above transformer transducers have certain common performance characteristics. The output voltages are proportional to the input displacements over a small range but they are very sensitive. The output voltage is of the same frequency as that of the supply voltage. If an rms reading, ac meter is used to read the output voltage, it is not possible to recognize the polarity of the input displacements. The characteristic looks like a V-curve. The output voltage is neither in phase nor in phase opposition to the applied voltage due to the finiteness of the leakage impedance of the primary winding and the iron loss of the core. Hence the output voltage, when seen in relation to the primary applied voltage, consists of an additional component, known as quadrature voltage. When the transducer is used to measure positional errors in servo-mechanisms for position control, it is essential to eliminate the quadrature voltage. A variable capacitor and a high resistance potentiometer are connected, as shown in Figure 6.57, and adjusted until output voltage is either exactly in phase or in phase opposition to applied voltage.

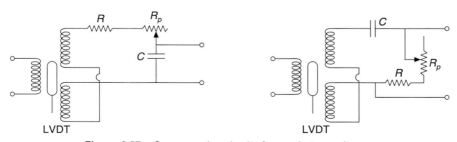

Figure 6.57 Compensation circuits for quadrature voltages.

Choosing cores of suitable materials and lamination thickness, the transducers can be used on ac supplies of frequencies ranging from 50 Hz to 10 kHz. The bandwidth of the transducer extends up to one-tenth of the excitation frequency. Under dynamic conditions, the output voltages are amplitude-modulated signals of excitation (carrier) frequency, the modulating signals being the dynamic input displacements. The output voltages have to be suitably processed for faithful reproduction, record and display of input displacements.

6.2.2.1 Induction Potentiometers

Two coils coupled to each other, such that the orientation of one of them with respect to the other determines the induced emf in one of them, may be used for measurement of angular deflections over a range of $\pm$ 90°. The two coils shown in Figure 6.58(a) constitute an equivalent of a transformer with variable coupling between primary and secondary. The mutual inductance M is maximum when the coils are coaxial, and zero when they are in quadrature. If θ_i is the angle between the coil axes, the mutual inductance and the induced emf in the secondary coils are given by

$$M = M_{\max} \cos \theta_i$$

$$e_0 = (KE_m \sin \omega_{ex}t) \cos \theta_i \qquad (6.58)$$

where

$$K = \text{a constant}$$

$$E_m \sin \omega_{ex}t = \text{excitation voltage of frequency } \omega_{ex}.$$

Although the above system can be considered to function as a variable self-inductance potentiometer, with the effective self-inductance given by

$$L_{eff} = L_1 + L_2 \pm 2M_m \cos \theta_i$$

provision of a closed magnetic circuit with iron core yields some of the advantages claimed in Section 6.2.1. Figure 6.58(b) shows such an arrangement, with the two coils mounted, one on the stator and the other on the rotor. The rotor is usually dumbbell shaped or of any other suitable shape, which, as far as possible, provides uniform gap over the entire periphery. The coils may be concentrated or distributed over the periphery. The concentrated coil system gives an output voltage which is proportional to θ_i over a very small range around the null point as seen from Eq. (6.58), whereas provision of distributed windings results in the extension of the linear range to $\pm$ 90°. The devices of this kind belong to the class of induction potentiometers, under the patent names of linvar, indpot, etc. They are normally designed for use at excitation frequencies of 50 Hz or 400 Hz, providing sensitivities of the order of 1 volt/degree of rotation. The devices are available in different sizes ranging from 10 mm to 75 mm in diameter. The need for provision of a pair of slip rings and brushes to deliver the output signal makes the induction potentiometer less popular as compared to microsyn, for which the range of measurement is limited to $\pm$ 5°.

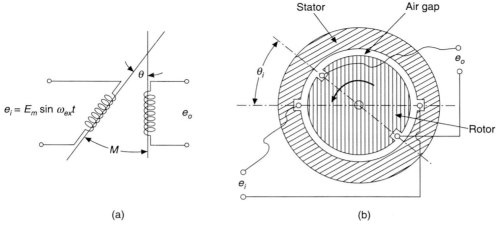

Figure 6.58 (a) Coupled-coils for angular displacement; (b) rotary induction potentiometer.

6.2.2.2 Synchro Pairs

All the above displacement transducers consist of two members, the relative motion between them being treated as the input displacement. In fact, the stator is held on a rigid reference base, while the movable armature or rotor is actuated by the moving object. All the displacement transducers are, in fact, devices that measure relative displacement between two bodies, objects or shafts that are so proximate to each other as to enable the mechanical coupling between them and the transducer. But in servomechanisms, there is a need to measure the relative angular displacement between two shafts that are located at a distance from each other. At times, it is necessary to measure the sum or difference of two angular positions. To enable such measurements, devices similar to induction potentiometers are developed and used in pairs. They are commonly known as synchros, selsyns or autosyns, but are invariably used in pairs.

A synchro primarily belongs to the class of transformer-type displacement transducers that develop induced emfs related to the angular position of its rotor, when excited by ac supply. Synchros are always used in pairs and are normally known as either synchro-transmitter-control transformer pair or synchro-transmitter-receiver pair. The function of the transmitter remaining the same in both pairs, the output of the control transformer is an induced emf, whereas the output of the receiver is a mechanical displacement.

The basic synchro unit is a single-phase device, although its construction resembles that of a miniature three-phase generator. The stator and rotor are of laminated silicon or of any other high grade steel. The stator is provided with a set of three identical coils of the concentric type, which are so located that their axes are mutually at 120° apart. The stator is provided with slots in which windings are distributed and the slots are often skewed to minimize slot effects. The rotor is provided with an excitation winding to function as a two-pole rotor and slip rings are used for connection to ac supply.

There are three different types of rotors as shown in Figure 6.59(a). The salient *pole type rotors* are invariably used for the synchro-transmitters (or generators) and receivers (also known as repeaters or motors), whereas the *umbrella type* and the *cylindrical type* are used in synchro-control transformers. The cylindrical rotor has the merit of possessing uniform reluctance all over the air gap.

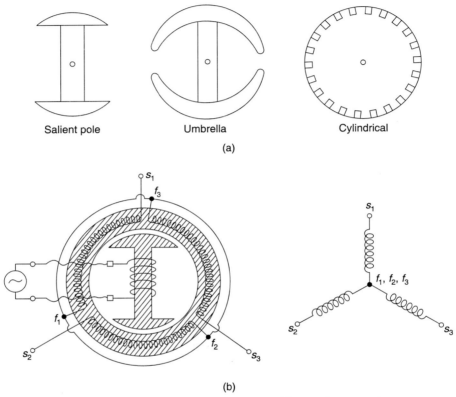

(a)

(b)

Figure 6.59   (a) Shapes of synchro-rotors; (b) a synchro generator.

The role of the synchro-transmitter in any one of the pairs can be understood by referring to Figure 6.59(b). The sinusoidal voltage applied to the rotor winding sets up alternating flux along its axis and distributed nearly sinusoidally in the gap around the periphery. The stator windings are star-connected, and each has an induced emf proportional to the cosine of the angle between the rotor axis and the stator winding axis, as in the case of induction potentiometer. The induced emfs in the three windings reach a peak value of $E_p = KE_r \sin \omega_{ext} t$ when their axes coincide with the rotor axis. K being the ratio of transformation.

When the rotor axis is at an angle of θ with the stator coil 2, the three emfs are given by

$$E_{1n} = E_p \cos (\theta + 120°)$$
$$E_{2n} = E_p \cos \theta \qquad\qquad (6.59)$$
$$E_{3n} = E_p \cos (\theta + 240°)$$

The three terminal emfs are obtained from Eqs. (6.59) as

$$E_{12} = \sqrt{3}E_p \sin(\theta + 240°)$$

$$E_{23} = \sqrt{3}E_p \sin(\theta + 120°) \tag{6.60}$$

$$E_{31} = \sqrt{3}E_p \sin\theta$$

It is observed from the above that E_{31} becomes zero when E_{2n} is at peak value. Under these conditions, the synchro is said to be in its *electrical zero* position and the rotor axis, coinciding with the axis of stator coil 2, serves as the reference axis in a plane perpendicular to the axis of rotation of rotor. The three emfs of Eq. (6.60) are in time phase and their magnitudes uniquely define the position of rotor with respect to the reference axis. Each one varies sinusoidally, with θ reaching a maximum at different positions of the rotor.

A synchro-control transformer is connected to the synchro-transmitter, as shown in Figure 6.60, and the pair is used as an error detector in automatic position control scheme. The error is taken as zero when the pole axes of the two rotors are exactly at 90° to each other, and error is the angle of deviation from this quadrature of the rotor axes. When the stator coils are connected as shown, circulating currents flow through both the sets of stator coils and an alternating flux pattern is set up in the control transformer. If the stator coils of control transformer have exactly the same orientation with reference axis as those of transmitter, the alternating flux pattern set up in the air gap of the synchro-control-transformer will have the same spatial orientation as that of the transmitter rotor. The induced emf in the rotor winding of the control transformer is zero, as its pole axis is at quadrature with the flux pattern. Under these conditions, the synchro pair is said to be zeroed and any deviation of the rotor of the control transformer from this alignment results in an induced emf which is proportional to the deviation if it is very very small. If θ and α are the angular orientations of the transmitter and control transformer rotors respectively, as shown in Figure 6.60, then the pair is in zeroed condition provided $(\theta - \alpha) = 90°$. If $(\theta - \alpha)$ becomes $90° \pm \delta$, there will be an induced emf in the output winding which is given by

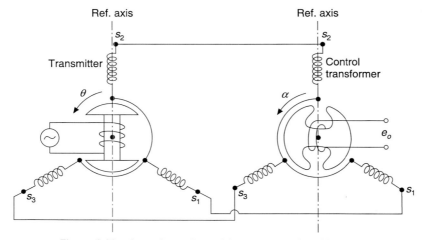

Figure 6.60   A synchro pair used for detection of position-error.

$$e_o(t) = K'(E_r \sin \omega_{ex}t) \cos (\theta - \alpha)$$

$$= \pm K'(E_r \sin \omega_{ex}t) \sin \delta \qquad (6.61)$$

where,

K' = a constant of transformation

δ = angle of deviation from quadrature

When the above pair is used in position-control schemes, the synchro pair is zeroed first, so that the output signal voltage is zero. Any deviation from this condition constitutes an error angle of δ, resulting in the development of an error signal proportional to δ, as long as δ is small. The output voltage signal is, as is the case with all ac excited systems, an amplitude modulated signal.

It may be recalled that the transmitter, serving as a generator, should be of low impedance windings for its stator. The control transformer has its stator windings of high impedance so that more than one control transformer can form the load of the transmitter.

The synchro-transmitter-receiver pair is used as a mechanical follower system, with the receiver (repeater) rotor following and reproducing the angular position of the transmitter rotor. The two synchros are located at a distance from each other, but electrically connected, as shown in Figure 6.61. The rotors of both the transmitter and the repeater are excited by the same ac supply. The ac flux pattern due to currents in the stator windings of repeater draws and aligns its rotor along its axis at all times and for all positions assumed by the transmitter rotor. The magnetic attraction between the rotor and the flux pattern set-up is responsible for the development of torque on the rotor until both of them are in alignment. The torque is actually a sinusoidal function of the angular difference between the two rotor positions, but for small differences, it is nearly proportional to the angular difference. When used for remote indication of position, the load on the shaft of repeater is only a pointer or a disc, and only a small torque is required to overcome the friction in the bearings. It is this need to overcome the friction that causes a small deviation from exact correspondence between the rotor positions. Any additional load torque on the shaft of the repeater synchro can only be borne by departing further from

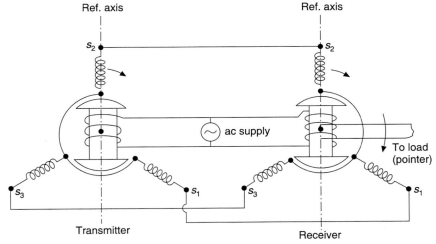

Figure 6.61 A synchro pair used as a mechanical follower.

positional correspondence. Accuracy of synchro-repeater systems depends on the *torque gradient* expressed as torque developed per unit angle of error. As the distance between transmitter and receiver increases, the resistance of the connecting cable increases, resulting in loss of accuracy.

Since the receiver is electrically identical with the transmitter, the rotor of transmitter experiences the same torque as the rotor of the receiver. Hence, the system may be considered to act as an electrical torque transmitter. The rotor of the receiver synchro is also of salient pole type and is usually provided with a vibration damper, to damp out oscillations under transient conditions. The system by itself has no damping.

When a control transformer is required to respond to the difference or sum of two angular positions of shafts, a differential synchro is inserted between the synchro-transmitter and the control transformer. The differential synchro is similar to the synchro-transmitter in all respects except that the rotor has three windings distributed in the slots of the laminated cylindrical rotor. They are star-connected with the terminals connected to slip rings for external connection. Figure 6.62 shows the manner in which the differential synchro is connected. The entire system is in zeroed condition, as explained earlier, when the rotor of synchro transmitter, all stator windings 2, and the rotor winding 2 of the differential synchro are aligned along the reference axis, and the rotor of the control transformer is at quadrature with reference axis. The output signal is zero in this condition. If the rotors are displaced from the reference axis in the positive direction, then the output signal is proportional to $\sin (\theta \sim \phi)$, where θ and ϕ are their orientations with respect to the reference axis. The system is useful in generating command signals from two different locations for the control of the position of a shaft, in automatic position control schemes. For displacements of the rotor of the differential synchro in the negative direction, it will respond to the sum of the angles. It is possible to use more than one differential synchro when more than one additional input angle has to be inserted into the system.

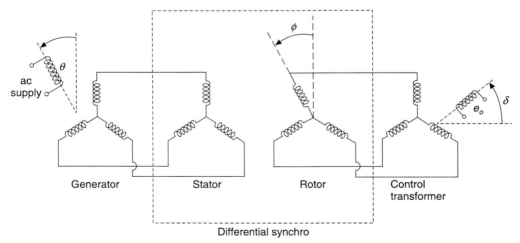

Figure 6.62   Differential synchro used in a synchro pair.

6.2.3 Movable Core-Type Inductive Transducers

Movable core-type inductive transducers consist of a coil system, along whose axis an iron core moves. The reluctance of the magnetic path is mostly due to air gap, and the position of the core inside the coil system determines the self-inductance of each coil. The inductance of the coil depends on the length of that part of the core which lies inside the coil. When plotted, the magnetic field strength along the axis, as given by Eq. (6.50), shows that the change in self-inductance will be considerable, only if the core is about half-way through the coil. Tapered cores of about the length of the coil are used for measurement of liquid level, with the core acting as the balancing weight for the float. Although the alternating magnetic field distribution around the coil poses no problem in such cases, it is necessary to provide a slotted ferromagnetic sleeve surrounding the coils of such transducers to act as magnetic shield.

The push–pull type self-inductance transducer consists of two identical coils wound on the same former, but each occupying one-half of the length from the centre, as shown in Figure 6.63. While in operation, the coils carry equal currents in such a direction as to produce a null point at the centre of the transducer. If the core is symmetrically located inside the coil system, it does not experience any magnetic pull and each coil has the same value for self-inductance. Any movement of the core from the null position disturbs the equality in the values and results in unbalance voltage V_u, when the transducer is connected in any of the bridge circuits shown in Figure 6.55.

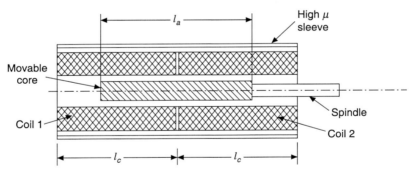

Figure 6.63   Push–pull type self-inductance transducer.

Such variable inductance transducers are available in different sizes and ranges, enabling measurement of linear displacements of a frequency range up to 2 kHz. The core is slotted longitudinally to reduce core loss at high frequencies, as shown in Figure 6.63. A similar transducer, with the addition of another coil acting as the primary coil, has all the merits of the above and is known as *linear variable differential transformer*. The mutual inductance between the primary coil and each of the secondary coils is the same when the core is in the null position. The secondaries are connected in series opposition so that their induced emfs oppose each other, giving an output signal proportional to displacement. Other details are presented in Section 6.2.3.1. Due to the small size of the core and the negligible force required for the movement of the core from the null position, the above transducers constitute secondary transducers for use with Bourdon tubes and diaphragms (pressure measurements), hydrometers (density measurements), and cantilevers and proving rings (load cells). Their high sensitivity and high resolution make them exceptionally suitable for many such applications. The frequency of the ac supplies used for these transducers should be at least 10 times higher than the frequency of the displacement signal. Relative amplitudes of vibration occurring at frequencies up to 2 kHz can be measured, provided the coil system can be held rigidly on a stationary platform serving as reference. The dynamic response is limited due to the mass of the core and the electromagnetic spring forces acting on it. The output signals are processed by phase sensitive detectors for indication and record.

6.2.3.1 Linear Variable Differential Transformer

The linear variable differential transformer is basically a mutual-inductance type transducer with variable coupling between the primary and the two secondary coils. It is equivalent to E pick-off in its operation except that the reluctance of the magnetic path is mostly due to the air path. It consists of a primary coil, uniformly wound over a certain length of the transducer, and two identical secondary coils symmetrically wound on either side of the primary coil and away from the centre, as shown in Figure 6.64. The iron core is free to move inside the coils in either direction from the null (central) position. When the primary coil is excited by ac supply, the induced emfs of the secondaries are equal to each other, with the core lying in the null position. The secondaries are connected in series but in phase opposition so that the resultant output voltage is zero. Displacement of the core in either direction from the null position results in output voltages proportional to displacement but of opposite polarity. The output voltage, as read by an ac rms voltmeter, is shown in Figure 6.65, and it is observed that there is a residual voltage at the null position; due to stray capacitance coupling between primary and secondaries, and the characteristic is linear over a limited range.

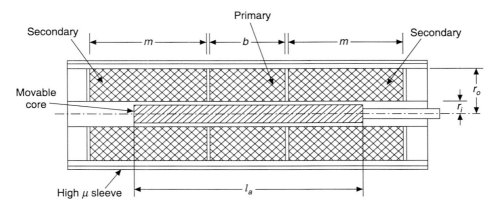

Figure 6.64 Linear variable differential transformer.

The range of displacement for which the transducer is linear within certain limits is determined by the dimensions of the transducer. The movable core is slightly more than three times the length of the primary winding and lies symmetrically with respect to the primary in its null position. The primary coil is of length $b,$ whereas each secondary is of length m. The net induced emf (E_o) of the secondary coils is given by

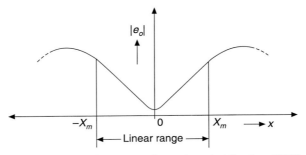

Figure 6.65   Displacement-voltage characteristic of an LVDT.

$$E_o(j\omega) = j\omega I_p \left[\frac{4\pi N_p N_s \mu_0 bx}{3ml_n(r_o/r_i)} \left(1 - \frac{x^2}{2b^2} \right) \right] \tag{6.62}$$

where,

ω = supply frequency, rad/s
I_p = primary current
N_p, N_s = No. of turns of primary and secondary windings
r_o, r_i = outer and inner radii of the coil system
x = displacement of the core from null position
μ_0 = absolute permeability of space $(4\pi \times 10^{-7}$ H/m)

The nonlinearity term, $x^2/2b^2$ in Eq. (6.62), is dependent on the length of the primary winding b, and for a desired range of x_{max} and error due to nonlinearity ε, the length of the primary winding is given by

$$b = x_{max}/\sqrt{2\varepsilon} \tag{6.63}$$

The length of the secondary winding is

$$m = b + x_{max}$$

The length of the core and the length of the secondary are kept a little more to accommodate the small spacing between the primary and each secondary winding. The ratio of r_i/l_a is about 0.05 and the ratio of r_o/r_i varies between 2 and 8. The number of secondary turns should be as large as possible to produce larger sensitivity. Since it is likely to be connected to high impedance loads and amplifiers, the secondary windings can be of finer wire.

The small residual output signal at the null position may be reduced by using either a balanced and centre-tapped excitation source as shown in Figure 6.66(a) or by using a high resistance potentiometer and an adjustable capacitor, as illustrated in Figure 6.66(b). As is the case with similar transducers, the output signal is an amplitude modulated sine wave, and demodulation is effected by using phase sensitive circuits. If the induced emf of each secondary is available separately, a simple demodulation circuit using two full wave rectifiers and a cascaded R-C filter, is used as shown in Figure 6.67. Faithful reproduction of the input displacement signals is possible in this case by suitably matching the impedances at each stage as indicated in the circuit.

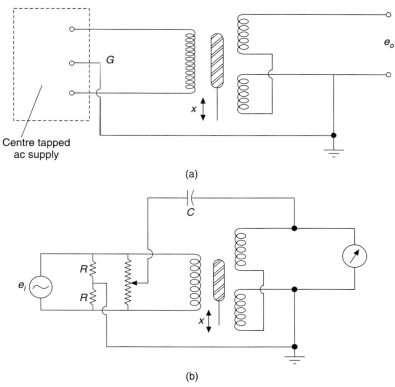

(a)

(b)

Figure 6.66   Circuits for compensation of null voltages by using: (a) A centre-tapped transformer; (b) a high resistance potentiometer.

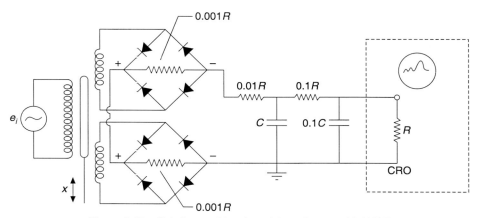

Figure 6.67   Polarity sensitive demodulator for use with LVDT.

Being continuously variable, there are no resolution problems with the LVDTs which are available in different ranges, the lowest being as low as 3.5 nm with an accuracy of 0.5%. Sensitivity as high as 10–40 mV per μm is not difficult to obtain at higher frequency of excitation. The mass of the core may vary from 0.1 g to 5 g, and the maximum force due to magnetic pull may be about 100 dynes. LVDTs can be designed to respond to angular displacements as well for full scale displacements of ±40°, but their application is not common.

6.2.4 Eddy Current Type Inductive Transducer

This transducer consists basically of a coil that produces magnetic field at right angles to the surface of an electrically conducting plate. When the coil carries alternating current, the eddy currents produced in the conducting surface act like a shorted secondary and hence the effective self-inductance of the primary coil is reduced. The conducting sheath is normally of nonmagnetic material but it can also be of magnetic material. The transducer can be used as a proximity meter to sense the approach of a metallic object near the coil and calibrated to indicate the closeness for each situation. The effective change in the self-inductance of the coil depends on the separation, thickness and electrical conductivity of the material of the surface, on coil dimensions and frequency of the ac excitation. The merit of the system lies in the fact that there is no mechanical link between the coil and surface. The geometrical orientation of the surface of the approaching object should be in conformity with the design and calibration. For nonmagnetic surfaces, the excitation frequency may be in the radio frequency range.

For regularly shaped objects like a thin-walled circular ring or sleeve (hollow tube) of copper or any other conducting material, it is possible to work out approximately the variation in the self-inductance of the coil. The ring or sleeve acts as a single-turn secondary. The effective inductance L_1' of a coil coupled with a secondary is given by

$$L_1' = L_1(1 - k^2)$$

where,

L_1 = self-inductance of primary coil

k = coefficient of coupling between primary and secondary

For the case of a sleeve of length l_2 and radius r_2 over a primary coil of length l_1 and radius r_1 as shown in Figure 6.68(a), the coefficient of coupling is

$$k = \frac{r_1}{r_2}\left(\frac{l_2}{l_1}\right)^{1/2}$$

and so the effective inductance is given by

$$L_1' = L_1\left[1 - \left(\frac{r_1}{r_2}\right)^2 \left(\frac{l_2}{l_1}\right)\right] \tag{6.64}$$

For a displacement of the sleeve by Δl_2, the change in inductance is

$$\Delta L_1' = -L_1 k^2 \frac{\Delta l_2}{l_2}$$

For convenience, the push–pull coil system of Figure 6.68(b) is used and the unbalance voltage V_u of the bridge circuit is calibrated in terms of Δl_2 for the system under consideration. The sensitivity is low but the range can be large. The main objection for the wide use of the transducer is due to extensive spread of its magnetic field in and around the coil.

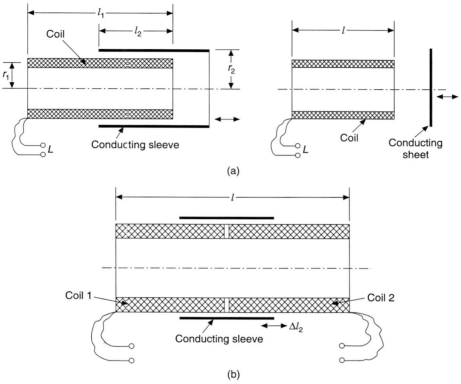

Figure 6.68 Eddy-current type inductive transducers: (a) Single-ended type; (b) push–pull type.

6.3 CAPACITIVE TRANSDUCERS

Capacitive transducers are, in fact, primary transducers for measurement of displacements; they are also known as proximity transducers in the sense that they measure the nearness of an object without any mechanical coupling between them. The only invisible coupling is through the electrostatic forces of attraction between the object and one plate of the capacitor which, of course, is so low that the transducer can be treated as an ideal displacement transducer. It is a nonloading, noncontact, and noninvasive type transducer for displacement measurements.

The parallel plate capacitor is widely used in many applications. The capacitance is given by

$$C_o = \frac{\varepsilon_o \varepsilon_r A}{t}$$

where,

ε_o = absolute permittivity of free space
ε_r = dielectric constant of the medium between the plates (relative permittivity)
A = area of the plates
t = separation between the plates

Table 6.7 presents properties of some dielectric materials.

Table 6.7 Properties of some dielectric materials

Material	Relative-permittivity (ε_r)	Resistivity (Ωm)	Density (g/cm^3)	Dielectric strength (kV/mm)
Bakelite	4.4		1.2	10–40
Glass	5–8	10^9–10^{12}	2.2–4.0	20–30
Marble	8–10	10^8	2.7	6–10
Mica	7–8		2.8	50–200
Paraffin	2.2	3×10^{16}	0.4–0.9	20–30
Polystyrene	2.2	$5 \times 10^{13} - 5 \times 10^{15}$	1.05–1.65	25–50
Porcelain	6.5	3×10^{12}	2.4	20
Rubber	2.6–3	4×10^{11}	1.7–2.0	15–25
Araldite	3.3			
Silica	3.8			
Quartz	4.5		2.65	
Silk	4.5			
Silicon oil	2.7			
Water	80		1.0	
Kerosene	2.0		0.8	
Glycerine	56.2		1.26	
Carbon tetrachloride	2.2			
Ethyl alcohol	25.0			

In proximity meters and other systems for measurement of displacement and amplitude of vibration, t is varied, resulting in variation of capacitance. Displacements conveyed to the movable plate may be so arranged as to either vary A, the area common to the plates, or alter the extent of penetration of a dielectric material inside the plates, as is observed in the case of the movable iron inductance transducers. The capacitance of such transducers is, in certain cases, as low as 1 pF and, with the help of suitable electrical circuitry, it is possible to detect very small variations in capacitance and measure displacements of the order of 25 nm.

A variety of transducers are developed, based on the above principles for the measurements of pressure, strain and thickness of insulating materials in the form of thin sheets. On-line measurements are possible, and the transducers are available for measurements of displacements of a frequency range up to 20 kHz. The capacitance microphone is one example of its many applications.

Capacitive transducers with cylindrical electrodes are also popular for measurements of pressure of fluids and level of fluids and granular materials.

The variation effect of dielecric constant due to variation in composition, absorption of moisture and other effects is used for composition measurements. The dielectric constant of certain insulators and semiconductors varies with temperature, and it is possible to make use of this effect for measurement of temperature.

The basic nonlinearity in the relationship between the capacitance and the quantity under measurement can be overcome in most cases by employing either push–pull systems or other associated electrical circuits.

The variation in the capacitance is, in most cases, so small that it is difficult to isolate the cable capacitance and stray capacitance effects. Electrodes have to be rigidly supported on high grade insulating materials so that capacitance variation due to effects of moisture, pressure or vibration on the supports does not interfere with the capacitance variation due to quantity under measurement. Ceramic materials are generally preferred to plastic and other organic materials. The movable plate is generally grounded but parasitic potentials may be induced on the ungrounded plates. Other metallic parts such as diaphragms, electrodes and transducer housing also require a high degree of form stability and mechanical rigidity. Nickel iron alloys are preferred for these parts. The plates should be rhodium or gold plated so as to render them anticorrosive. The air gap should be protected from humidity and dust. While measuring pressure of fluids, the diaphragm materials should be of a material compatible with the properties of the fluid.

Electrostatic screening of the leads of the transducer is essential throughout, without a break from the transducer of the circuit. Any change in the capacitance of the cable appears as a change in the capacitance of the transducer and hence special cables are used, which have carbon coating on either side of the dielectric separating the core and screen.

All capacitive transducers have an initial capacitance C_0, and it is likely that the value may drift slowly with time due to reasons explained above. It is necessary that all signal processing circuitry should have means of zeroing the readout and facilitate the use of the calibrated scale for measuring the unknown quantity. Percentage changes of capacitance lower than 0.1% are detected and measured by using resonant circuits which convert the change of capacitance into a change of resonant frequency.

6.3.1 Capacitive Thickness Transducer

Thickness of thin insulating sheets such as glass mica, paper and bakelite can be measured by capacitive transducer. The plates can be so arranged that the sheet, which is under continuous production, may be passed in between the plates without touching them, as shown in Figure 6.69. As the thickness is continuously monitored, the dielctric constant of the material should remain constant. The total separation between the plates, $a,$ is larger than the thickness t of the sheet, but while measuring the thickness in the laboratory, no air gap is necessary.

The capacitance of the system is

$$C_0 = \frac{\varepsilon_0 A}{a - t + (t / \varepsilon)} \tag{6.65}$$

where ε = the dielectric constant of the material of the sheet.

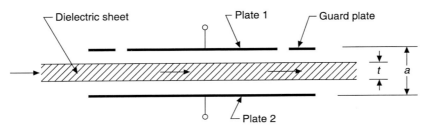

Figure 6.69   Capacitive thickness transducer.

An increase in the thickness of the sheet by Δt causes an increase of the capacitance by ΔC, and hence

$$C_0 + \Delta C = \frac{\varepsilon_0 A}{[a - (t + \Delta t)(1 - 1/\varepsilon)]}$$

Solving for $\Delta C/C_0$, we get

$$\frac{\Delta C}{C_0} = \frac{\Delta t}{t} N \frac{1}{1 - N(\Delta t/t)} \tag{6.66}$$

where,

$$N = \frac{\varepsilon - 1}{1 + \varepsilon(a - t)/t}$$

For $N(\Delta t/t) \ll 1$, the fractional change in capacitance is proportional to the fractional change in its thickness. The sensitivity factor N, relating these fractional changes, is as much as the dielectric constant of the material when the air gap to thickness ratio falls below 0.01, and the sensitivity falls as the ratio becomes larger. The value of N is also dependent on the dielectric constant of the material. Any increase of N by reducing the air gap to thickness ratio decreases the range of linearity.

A unity-ratio arm bridge circuit may be used with test and reference capacitors with the latter having the sheet of the same material but of thickness t_0 between its plates. The unbalance voltage V_u may be calibrated in terms of $\Delta t/t_0$. Very small variations in t_0 can be detected and indicated.

6.3.2   Capacitive Displacement Transducers

As explained in Section 6.3, the capacitive displacement transducer is fundamentally a proximity transducer, in the sense that the movable plate or electrode may be the conducting surface of any object in the vicinity of the fixed plate. For convenience, the parallel plate capacitor is used, with one of its plates being subjected to movement by a mechanical primary transducer. If the transducer has a solid insulating material of dielectric constant ε, as shown in Figure 6.70(a), the capacitance C_0 is given by

$$C_0 = \frac{\varepsilon_0 A}{x_0 + t/\varepsilon} \tag{6.67}$$

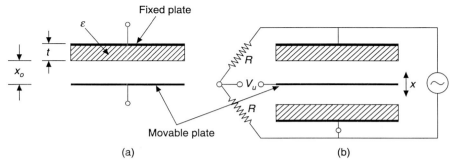

Figure 6.70   (a) Single-ended capacitive displacement transducer; (b) push–pull type transducer in a unity-ratio Wheatstone bridge.

If the air gap is decreased by Δx, the capacitance increases by C which is given by

$$C_0 + \Delta C = \frac{\varepsilon_0 A}{x_0 - \Delta x + t/\varepsilon}$$

The fractional change in capacitance is

$$\frac{\Delta C}{C_0} = \frac{\Delta x}{x_0 + t} \frac{N}{1 - N\Delta x/(x_0 + t)} \tag{6.68}$$

where,

$$N = \frac{1 + t/x_0}{1 + t/x_0 \varepsilon}, \quad \text{the sensitivity factor}$$

For small values of $N\Delta x/(x_0 + t)$, percentage changes of capacitance are proportional to percentage changes of total separation between the plates. If the solid dielectric material is removed, N becomes equal to unity, but its presence improves sensitivity. As seen in Section 6.3.1, any improvement in sensitivity results in nonlinearity, and the range of displacements for which the system is linear is reduced. However, electrical circuits that respond to very minute changes in C_0 are available, and the improvement in the sensitivity by using a dielectric is not normally carried out. Also, there are no solid dielectrics which are stable enough, possessing high values of ε. Even if quartz is used, improvement is only by 10 times.

The range of linearity can be extended, to a large extent, by using another fixed electrode, thus converting it into a push–pull transducer, as shown in Figure 6.70(b). Unity-ratio arm Wheatstone bridge or Blumlein bridge circuit may be used for measuring $\Delta C/C_0$, and the latter can result in very high sensitivity if the bridge is so designed that $\omega_{ex}^2 LC_0 = 1/2$. Even for single-ended transducers, oscillators having the tank circuit consisting of L and the capacitive transducer are highly suitable, as the small changes in the frequency of oscillation due to change in capacitance are easily detected.

Other forms of capacitive transducers useful for measurement of large linear displacements are shown in Figures 6.71(a) and 6.71(b). The electrodes are fixed in position, but the solid dielectric moves into the space between the electrodes, thereby altering the capacitance of the transducer. The mass of the moving system and the electrostatic force on the dielectric limit the dynamic response of the transducer to low frequencies.

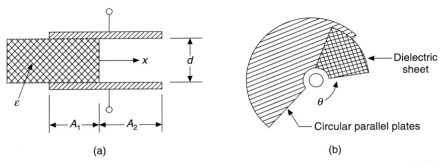

(a)   (b)

Figure 6.71 Capacitive transducers using movable dielectric core: (a) for linear displacement; (b) for angular displacement.

6.3.2.1 Proximity Transducer

In certain applications, the proximity of an object with respect to the fixed plate of the transducer is desired. Electrical circuits that develop output voltages proportional to the separation between the plates are available. The circuit shown in Figure 6.72 uses an operational amplifier of high gain, giving output signal e_o proportional to x_o. The moving object is provided with a plane conducting surface, if it does not behave like one. The object is earthed and the fixed plate is so designed as to have much smaller area than the movable surface and is provided with a guard ring as shown in Figure 6.72. The output signal e_o is given by

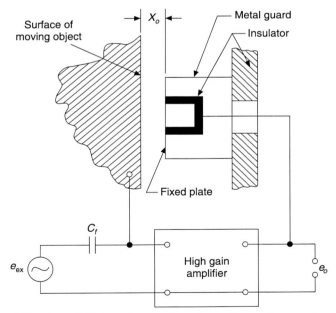

Figure 6.72 A proximity transducer system along with signal processing circuit.

$$e_o = -\frac{C_f}{C_x} e_i = -\frac{C_f x_0}{\varepsilon_0 A} E_m \sin \omega_{ex} t \qquad (6.69)$$

where,

C_f = capacitance of the standard capacitor

$E_m \sin \omega_{ex} t$ = sinusoidal applied voltage

If the object is vibrating, the amplitude of vibrations can be measured, and in such a case, e_o is an amplitude-modulated wave. Demodulation is effected by circuits explained earlier and it is necessary that the frequency of vibration be less than one-tenth of the frequency of ac supply used. Normally, the supply frequency can be about 50 kHz, thereby enabling measurement of displacements of a frequency up to 5 kHz. Commercial instruments are developed with probes having diameters ranging from 1 mm to 25 mm for full-scale displacements of 25 μm to 12.5 mm respectively. Amplitudes of vibration as low as 25 nm can be measured with the smallest probe from the above.

The most significant merit of the proximity detector is that it does not constitute any load on the object, as the electrostatic force of attraction is so low.

A proximity meter of a different electrode arrangement, which is free from the effects of earth capacitances, is shown in Figure 6.73. *A* and *B* are the live electrodes between which an earthed screen *S* is present. An earthed object with a plane metallic surface is in the proximity of the two plates *A* and *B*, and the fringe capacitance *C* between them varies with the distance of the object, *x*, from them. Measuring the capacitance between *A* and *B*, proximity or the distance *x* of the object can be measured. The capacitance–distance characteristic is approximately linear even for large values of *x*.

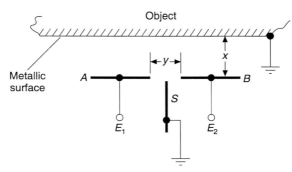

Figure 6.73 A proximity transducer using two plates and an earthed object.

The value of *C* is very small but with suitable circuitry the small variations can be measured.

6.3.2.2 *Capacitive Strain Transducer*

A strain gauge based on the principle of capacitance variation with plate separation is developed, making use of two arched metal strips to support the electrodes of the capacitor, as shown in Figure 6.74(a). When the structure is strained, there is a change in the differential height of the arches as well as the gap between the electrodes. The height variation of each arch strip is calculated from

$$x^2 - x_0^2 = -\frac{15}{64} \varepsilon L (\varepsilon L + 2w_0)$$ (6.70)

where,

ε = strain
x = height of arch under strain
x_0 = initial height of arch under no strain
w_0 = unstrained width of arch
L = gauge length

The gauge factor $\left(= \dfrac{\Delta C / C_0}{\varepsilon} \right)$ is about 100 and the gauge is used for measurements of strain up to ± 5000 μ at temperatures as high as 600°C.

An alternative arrangement is shown in Figure 6.74(b) in which the bowing of the arched metallic parts due to strain changes the gap between the electrodes. The flexible insulating strips and electrodes are cemented to the arched parts. The capacitance between the two live electrodes gives a measure of the strain. The strain–capacitance characteristic is not linear. Using a polyimide film of insulation, the strain gauge is used up to 300°C. The capacitance of the gauge is about 1 pF for a gauge length of 2 cm, and has a gauge factor of about a few hundreds.

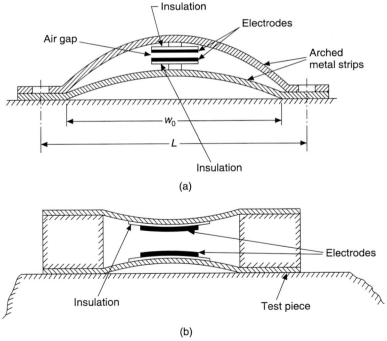

Figure 6.74   Capacitive strain transducers using (a) plate separation; (b) gap changing by arching.

6.3.2.3 *Variable Area Capacitive Transducer*

A transducer for measurement of angular displacements and angular vibrations is shown in Figure 6.75(a). It consists of a set of flat serrated plates, and any small relative movement between them in the plane of the plates causes a change in capacitance due to the change in the area common to the serrations. Neglecting the effect of fringing, the capacitance C_0 of the transducer, when all the teeth are in alignment, is given by

$$C_0 = \frac{\varepsilon_0 nbl}{d} \tag{6.71}$$

where,

l = length of each tooth
b = width of tooth along the axis of rotation
d = gap between the teeth
n = number of pairs of teeth

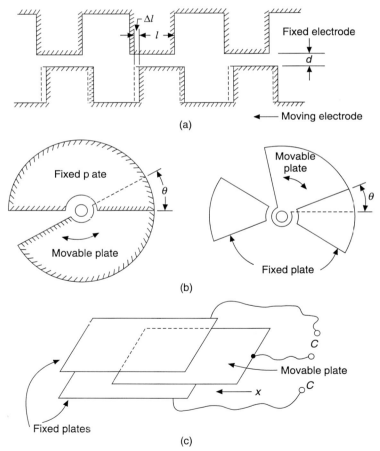

(a)

(b)

(c)

Figure 6.75 Capacitive transducers for angular displacement using (a) Serrated plates, (b) semicircular plates, and (c) parallel plates for linear displacement.

Any reduction in l due to small angular displacement results in a change of capacitance given by

$$\Delta C = \frac{\varepsilon_0 nb\,\Delta l}{d}$$

and the fractional change in capacitance is

$$\frac{\Delta C}{C_0} = \frac{\Delta l}{l}$$

The above relations hold good for cases with $l \gg d$, and for small angular displacements only.

A simpler form of capacitive transducer for angular displacements is the capacitor with semicircular plates of the variable area type, as shown in Figure 6.75(b). The capacitance is proportional to the common area and hence θ. A range of 180° can be obtained with an accuracy of $\pm$ 0.1% and transducers are available in differential type and multiple plate type. In the rotary type transducers, the movable plate moves in a direction normal to the direction of electrostatic forces. An equivalent form of transducer for measurement of large linear displacements is shown in Figure 6.75(c).

6.3.2.4 *Capacitive Tachometers*

Capacitive transducers meant for measurement of angular displacement can as well be used for measurement of angular velocity when the total angular displacement is less than 180°. They are extremely suitable for situations in which contact cannot be established and the inertia of the electromagnetic tachometer is prohibitively large. The capacitive tachometer is suitable for low angular velocity. If the change in capacitance is proportional to change in angular displacement, as is the case with circular plate type, the angular velocity is proportional to dC/dt. If the capacitor is charged from a source of voltage E_b, the current flowing in the capacitor is $E_b(dC/dt)$ and the voltage drop across a small resistor due to this current can be the output signal.

For measurement of angular velocities involving many revolutions per second, the capacitor is charged during one-half of the revolution by means of a contactor on the shaft or slip rings, and discharged through a milliammeter during the other half of the revolution. The average discharging current is proportional to the speed. With the provision of four slip rings to be contacted during one revolution, the tachometer is rendered polarity sensitive.

6.3.2.5 *Capacitive Pressure Transducers*

For pressure measurements, it is convenient if the metallic diaphragm or the membrane forms the movable plate of the transducer as shown in Figure 6.76. The capacitance change due to the deformation of the clamped diaphragm can be calculated using Eq. (5.13) for the deflection y at any radius r from the centre of the diaphragm.

Considering an annular element of width dr at a distance r from the centre, its capacitance can be given by

$$dC = \frac{\varepsilon_0 2\pi r\,dr}{d - y}$$

where,

d = initial spacing between the electrodes without any pressure difference ΔP

y = displacement of the annular element from its initial position

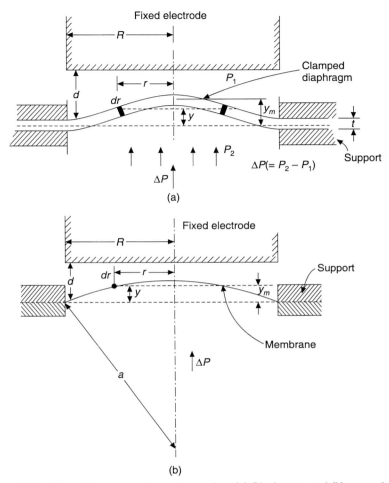

Figure 6.76 Capacitive pressure transducer using: (a) Diaphragm, and (b) a membrane.

For small deflections with $y/d \ll 1$, $1/(d-y)$ can be written as

$$\frac{1}{d-y} = \frac{1}{d}\left(1 + \frac{y}{d}\right)$$

and the total capacitance due to the deflected diaphragm is given by

$$C_0 + \Delta C = \int_0^R dC = \frac{2\pi\varepsilon_0}{d}\int_0^R \left(1 + \frac{y}{d}\right) r\, dr \tag{6.72}$$

Substituting y with $\dfrac{3}{16}\Delta P\left(\dfrac{1-\mu^2}{Et^3}\right)(R^2 - r^2)^2$, the fractional change in capacitance can be given by

$$\frac{\Delta C}{C_0} = \frac{(1-\mu^2)R^4 \Delta P}{16E\,dt^3}$$ (6.73)

As is the case with all parallel plate capacitors, y_m/d is kept within 0.01 for linearity between $\Delta C/C_0$ and ΔP. It is possible to extend the linearity range by converting it into a push–pull system.

A similar expression for the case of the stretched membrane under deflection can be derived as follows: For small deflections, the relation between the deflection y at any point away from the centre is given by

$$y = \frac{\Delta P}{4}\frac{R^2 - r^2}{S}$$ (6.74)

As the deflection forms part of the surface of a hollow sphere of radius a, it can be seen that

$$(y_m - y)\,[2a - (y_m - y)] = r^2$$

$$y_m(2a - y_m) = R^2$$

Eliminating $2a$ from the above equations, we get

$$\frac{R^2}{y_m} = \frac{r^2}{y_m - y} - y$$

$$\frac{y}{y_m} = 1 - \frac{r^2}{R^2} + \frac{y_m^2}{R^2}\left(\frac{y}{y_m} - \frac{y^2}{y_m^2}\right)$$

Hence,

$$\frac{y}{y_m} \approx \left(1 - \frac{r^2}{R^2}\right) \text{ assuming } \frac{y_m^2}{R^2} \ll 1$$ (6.75)

Considering an annular element of width dr at a distance r from the centre, the partial capacitance of the element and therefore the total capacitance due to the entire membrane is given by

$$C_0 + \Delta C = 2\pi\varepsilon_0 \int_0^R \frac{r\,dr}{d - y}$$ (6.76)

Using the expression for y from Eq. (6.75), $\Delta C/C_0$ is obtained as

$$\frac{\Delta C}{C_0} = \frac{1}{2}\frac{y_m}{d} + \frac{1}{3}\left(\frac{y_m}{d}\right)^2 + \frac{1}{4}\left(\frac{y_m}{d}\right)^3$$ (6.77)

For small deflections with $\dfrac{y_m}{d} \ll 1$,

$$\frac{\Delta C}{C_0} \approx \frac{1}{2}\left(\frac{y_m}{d}\right) = \frac{\Delta P R^2}{8Sd} \quad \text{(see Eq. 5.14)}$$ (6.78)

From the above assumptions, sufficient linearity may be achieved by making $y_m/d \approx 0.01$ and $y_m/R \approx 0.01$. For a membrane of 1 cm radius and 1 cm separation, the capacitance is about 0.3 pF, and the electrode arrangement with the connecting cable will considerably alter this value. Special circuitry to detect and process the small capacitance changes is necessary. Wherever possible, push–pull system may be employed to improve linearity.

The differential transducer of the above type, used to measure small differential pressures in spectrophotometers is shown in Fig. 6.77. A thin stainless steel diaphragm is clamped between two glass slabs, to serve as the movable plate. The spherical depressions of a maximum depth of 25 μm are formed in the glass slabs and used as fixed plates when gold coated. Very small differential pressures of the order of 0.001 μbar are measured by such arrangement.

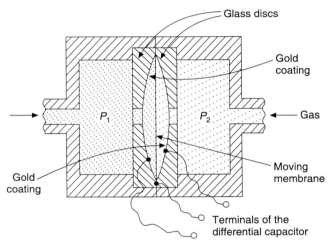

Figure 6.77   Differential capacitive transducer for differential pressure.

The dielectric constants of gases and liquids vary with pressure. When the pressure of such fluids is measured, the cylindrical electrode arrangement shown in Figure 6.78 is used so

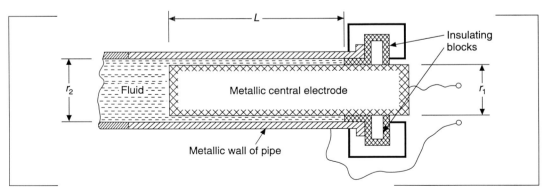

Figure 6.78   Cylindrical capacitive transducer for pressure measurements.

that the pressure is continuously measured under flowing conditions. The walls of the metallic pipe are used as the outer electrode, and a solid cylindrical rod running along the pipe serves as the inner electrode.

The capacitance C_0 of the transducer is given by

$$C_0 = \frac{2\pi\varepsilon_0\varepsilon_r L}{\ln (r_2 / r_1)} \tag{6.79}$$

where,

$\quad L$ = length of the central electrode
$\quad r_2$ = inner radius of the pipe
$\quad r_1$ = radius of the central electrode
$\quad \varepsilon_r$ = dielectric constant of the fluid

The dielectric constant of air at 19°C changes from 1.0006 at 1 atm to 1.0548 at 100 atm. The dielectric constant of benzene changes by 0.5% in the range of 1 to 100 atm. Variation of ε_r with pressure is observed in certain solid substances such as the crystal of sodium chloride, potassium chloride and potassium bromide. But dielectric constant does not change linearly with pressure in all these cases. Hence the capacitance-pressure characteristic is nonlinear. The transducer may be used for indication of pressure under static or slowly varying conditions.

It is considered essential to use bridge circuits so that a standard capacitor having similar electrode arrangement but with the fluid under static and standard pressure is kept as reference for comparison in one arm of the bridge circuit.

6.3.2.6 Capacitive Microphone

The capacitive microphone is a modified version of the capacitive pressure transducer in which the vibratory displacements of the membrane due to pressure variations result in variation of the capacitance. Sound waves are, in fact, pressure waves or fluctuations of the pressure of air around the normal pressure of 1 atm. Capacitive microphone is used to reproduce the pressure fluctuations faithfully. The range of frequencies involved is from 20 Hz to 15 kHz, and the transducer is designed to be immune from the effects of any variation in the atmospheric pressure. Atmospheric pressure changes very slowly with time but the change may be considerably larger than the amplitude of the fluctuating component. The capacitive microphone membrane should be sensitive to the fluctuating components only and special provision must be made to equalize the pressure on either side of the membrane so as to prevent it from bursting due to the atmospheric pressure variations.

The constructional features of the capacitive microphone are shown in Figure 6.79. It consists of a very thin metallic membrane kept under radial tension. The capillary tube or hole connecting the chamber behind the membrane to atmosphere is intended to equalize the pressure on either side of the diaphragm. The capillary path and the volume of the microphone chamber act as a low pass filter allowing the slow variations of the atmospheric pressure but preventing the pressure fluctuations of sound waves from appearance on the backside of the membrane. Thus at any instant, the membrane is subjected to pressure fluctuations on the front side only, and the deformation of the membrane is as given in Eq. (6.74). The fixed plate of the capacitor is held rigidly inside the chamber and is provided with holes so as to allow the

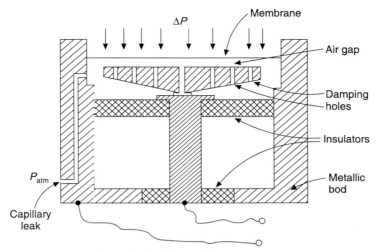

Figure 6.79   A capacitive microphone.

air to pass freely from the space between the electrodes to the other side of the chamber. These holes are known as *damping holes*. The clearance between the electrodes is very small and hence the presence of air between them resists the deflection of the membrane if the damping holes are not provided. Also, the damping effect of the holes helps in controlling the amplitude of vibration of the membrane at frequencies near about its undamped natural frequency. Normally, the capacitive microphone is designed to have flat frequency response from 10 Hz to 20 kHz. In this frequency range, the response of the diaphragm is equivalent to that of a spring mass-dashpot system. The diameter of the membrane varies from 5 mm to 25 mm and the spacing between the electrodes is kept about 20 μm. The volume of the microphone chamber and the length of the capillary should be very small so as to render the pneumatic time constant very small, as explained in Section 4.6.3.

The capacitance variations develop output voltages across the high resistance R. A high impedance amplifier amplifies the voltages. The polarizing or biasing voltage E_b is kept high enough, as permissible with the air gap between electrodes. The sensitivity is in the range of 1–5 mV/μbar.

Based on the above design principles, capacitive pressure transducers having bandwidth up to 100 kHz can be developed. They are used to measure the noise level of machinery and for analysis and diagnosis of vibration problems of machinery.

The human ear is more sensitive to frequencies around 1 kHz and it can detect pressure fluctuations at this frequency, having an rms value of 0.0002 μbar. This value of pressure is accepted as standard of reference for measurement of sound levels. The sound pressure level is defined in dB and is given by

$$\text{Sound pressure level} = 20 \log \frac{p}{0.0002} \text{ dB} \tag{6.80}$$

where, p = rms sound pressure in μbar. Hence, the average threshold of hearing for human beings is reckoned as 0 dB, and higher sound levels are expressed as positive numbers. Capacitive microphones are used in sound level meters.

6.3.2.7 *Capacitive Level Transducer*

The capacitor with the concentric cylindrical electrodes can be used to measure the level of liquids that are nonconducting and insulating. The electrode assembly is kept inside the container, as shown in Figure 6.80(a), with the provision of holes at the lower end of the outer cylinder for admission of liquid. The capacitance C, due to the two columns of liquid and its vapour, is given by

$$C = 2\pi\varepsilon_0 \frac{\varepsilon_1 h_1 + \varepsilon_2 h_2}{\ln (r_2 / r_1)} \tag{6.81}$$

where, ε_1 and h_1 denote the dielectric constant and height of the liquid and ε_2 and h_2 refer to the remaining space filled with the vapour of the liquid. ε_2 can be taken as unity. The capacitance level characteristic is linear except at the extremes where the initial and stray capacitance becomes prominent. The outer electrode is normally earthed. Compensation for the variation of ε_1 with temperature may be necessary for most of the liquids.

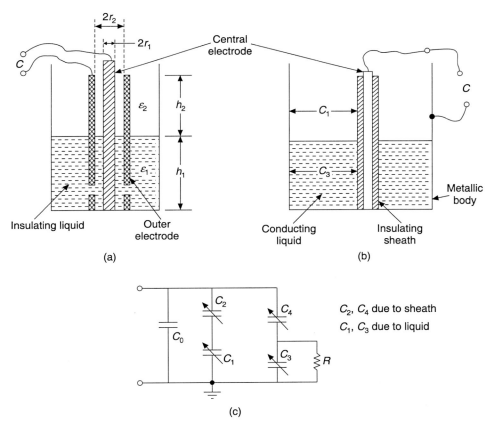

Figure 6.80   A capacitive level transducer for use in: (a) Insulating and conducting liquids; (b) conducting liquid only; (c) equivalent circuit.

For conducting liquids, the central electrode is provided with an insulated capping, as shown in Figure 6.80(b), so as to prevent short circuiting of the capacitance by the liquid. The equivalent circuit of the system is shown in Figure 6.80(c), and the net capacitance is related to the level of the liquid.

In both the above systems, the outer electrode may be the metallic tank itself, if it is cylindrical in shape, or the system as a whole may be connected to the container through a pipe.

The technique is well suited to measurements of levels of liquid metals at high temperatures and liquefied gases at very low temperatures. Levels of powdered or granular solids can also be measured as long as they are dry and have constant composition.

6.3.3 Capacitive Moisture Transducer

The dielectric constant of pure water is about 80 and that of most insulating materials, solids or liquids, is less than 10, and so it is possible to measure the moisture content of these materials by measuring the dielectric constant of the moist solid or solution of the substance in water. The technique can be extended for application to other combinations, if the variation in the dielectric constant is due to variation of the proportion of one substance in the mixture. The equivalent series or shunt resistance of the capacitor, representing the dielectric losses of the sample, may also be used to indicate the moisture content. Two identical capacitors, one holding the test sample and the other the dry sample, may be used in an ac bridge circuit, and the equivalent loss resistance as well as the capacitance may be measured by balancing the bridge. As the capacitance values increase with moisture and equivalent shunt resistance falls, the arm with dry sample may be shunted by a variable capacitor and resistor as shown in Figure 6.81, and their values may be calibrated against the moisture content. Otherwise, the unbalance voltage may be directly used for calibration. One particular advantage of solids is that no additional means are necessary for them to compact the test material between the electrodes for good contact as is the case with resistive moisture transducers.

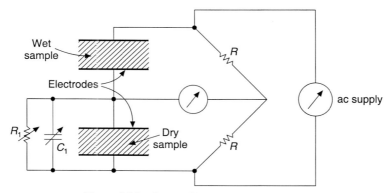

Figure 6.81   A capacitive moisture transducer.

The choice of the frequency of supply voltage for the bridge circuit or of any other technique is critical as the electrical conductivity and dielectric constant of water change considerably at frequencies above 10 kHz. A preliminary understanding of the behaviour of the moist material is necessary before the transducer is calibrated. Also, the test and reference sample have to be maintained at the same temperature. The technique is suitable and reliable for solid materials that absorb moisture all through the substance. Certain materials allow moisture to settle on the surface only but do not absorb, and in such cases, the method is unsuitable and measurement of moisture will not be correct.

Salinity of water due to addition of contaminants and salts can be measured by noting the change in dielectric constant. Any contamination of oil in water can also be estimated. A section of the pipe through which the water or any other solution passes, may be used with two plane electrodes placed in it. If the diameter of the pipe is not large, a set of electrodes may be attached to the inside surface of the pipe, with certain length of the pipe being made of an insulating material.

Dry air at 45°C has a dielectric constant of 1.000247, and when saturated with moisture, it is 1.000593. Hence the variation in capacitance of any electrode arrangement will be so small that it is very difficult to measure accurately.

6.3.3.1 Capacitive Hygrometer

A more practical form of hygrometer employs the arrangement shown in Figure 6.82(a). The central part of the transducer is an aluminium rod acting as one electrode. The rod is oxidized over part of its length over which is provided a thin layer of graphite or of an evaporated metal.

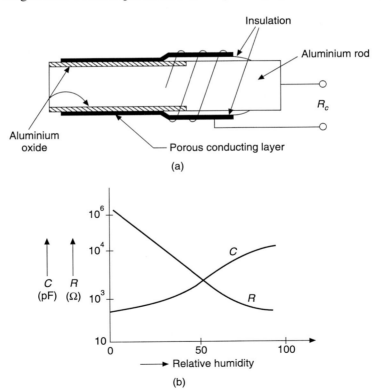

(a)

(b)

Figure 6.82 (a) A capacitive hygrometer; (b) characteristic curves showing the effect of humidity on R and C.

Moisture is absorbed through this thin porous layer, by the aluminium oxide, and the equivalent capacitance between the outer metallic layer and aluminium rod undergoes variation because of the amount of moisture absorbed. When equilibrium is reached with the moist atmosphere, the resistance and capacitance of the capacitor are measured. The variation of both components is shown in Figure 6.82(b) and can be used as a measure of the relative humidity. To some extent, the resistance variation is linear, but capacitance variation is nonlinear.

EXERCISES

1. Briefly suggest the reasons for the popularity of resistors for use as transducers. Indicate the physical quantities that can be measured by bringing dimensional changes in a wire-wound resistor.

2. Suggest the commonly used metals and alloys for making up resistance coils for measurement of temperature. When do you prefer platinum for the resistance thermometer coil?

3. Describe foil and film type resistors and their application as transducers.

4. What is meant by leads-compensation, and how is it applied in resistance thermometry?

5. What are thermistors and what physical quantity is precisely measured by them?

6. Show how the resistance-temperature characteristic of a thermistor looks like and comment on its suitability for temperature measurements.

7. Discuss the relative merits and demerits of thermistors in comparison to wire-wound resistance coils.

8. Discuss (a) the reasons for which a jacket is provided for a resistance thermometer element and (b) the effect on its performance.

9. Explain why there is large scope for errors in temperature measurement by resistance thermometers. Show how these errors can be minimized while making a measurement.

10. Explain the nature of problems faced in measurement of temperature of a fluid in motion. Describe how the probes are designed for use in such situations.

11. Explain how the temperature of hot bodies can be measured without physical contact and from a distance.

12. Describe a bolometer, what does it measure?

13. Show that a hot-wire resistor can be used for the measurement of fluid level and fluid-flow rate.

14. Explain how vacuum and pressure of gases can be measured by using a hot-wire resistor. Discuss the nature of the problems faced in measurement.

15. Describe the construction and operation of a thermal conductivity gauge and show that it can be used for analysis of binary gas mixtures.

16. Explain the basic principles involved in hot-wire anemometry and show how gas flow rates in closed pipes can be measured satisfactorily?

17. Distinguish between the operational features of the constant-temperature and constant-current type anemometers.

18. From the basic relations governing the hot-wire anemometer, obtain the relationship for its incremental sensitivity and time constant when operated under (a) the constant-temperature mode (b) the constant-current mode.

19. Explain the techniques adopted for increasing the bandwidth of the basic hot-wire anemometer.

20. Show that a wire-wound resistance potentiometer can be used for measurement of linear and angular displacements. Comment on its linearity and resolution and the factors affecting these characteristics.

21. Explain the 'loading effect' on the accuracy of a resistance-potentiometer transducer when used for measurement of displacements.

22. Define the *gauge factor* of a resistance strain-gauge and obtain the expression for the same in terms of other constants.

23. Distinguish between bonded and unbonded strain gauges and comment on their suitability for measurement of physical quantities.

24. What are the important precautions to be taken while using metallic wire strain gauges?

25. Why are resistance strain gauges used in pairs?

26. Define the term *Piezo-resistive coefficient* and explain why semiconductor strain gauges have high values for the gauge factor.

27. What are the differences in the construction and application of the wire type and semiconductor type strain-gauges?

28. Discuss the field of application of strain gauges and the situations in which they are most suited for use.

29. What are strain gauge rosettes and how are they used in experimental stress analysis?

30. Describe the primary transducers to be used along with strain gauges for measurement of force, liquid flow rate, and acceleration. In each case, explain how the gauges are placed on the primary transducers.

31. Explain how the torque transmitted by a motor to its load is measured by means of strain gauges. Show how these are mounted on the shaft connecting them.

32. Show how the thin plate diaphragm is used alongwith strain gauges for measurement of pressure.

33. Describe the principles of operation and constructional features of a Bridgman gauge and derive the relationship between the input and output quantities.

34. What are conductivity cells and how do they enable measurement of the concentration of electrolytic solutions?

35. Draw the electrical measuring circuits used alongwith the conductivity cells and show how the measurement is carried out. What are the precautions to be observed in the measurement?

36. Explain how resistive transducers can be designed and used for the measurement of the relative humidity of air and gases. What is the basic limitation of the arrangement?

37. What is magneto-resistance effect and what are the materials that exhibit this phenomenon? Suggest one important application in instrumentation.

38. Define photo-resistive effect and mention the materials that exhibit this effect. Discuss its application potential in instrumentation.

39. Explain why inductive transducers are mostly iron-cored. Suggest the important considerations to be applied while designing and using them.

40. Explain the techniques by which thickness of (a) magnetic and (b) nonmagnetic metallic sheets can be measured. Show the arrangement of coils and specimen sheet used.

41. Describe how a self-inductive transducer can be designed by means of U and I stampings for measurement of displacements. Derive the relationship between fractional changes in self-inductance and displacements conveyed to the movable armature.

42. Describe the push–pull type self-inductive transducer and discuss the merits of this over the single coil transducer.

43. What are transformer-type transducers and how are they configured for measurement of linear and angular displacements?

44. Describe the constructional and operational features of (a) E-pick-off and (b) microsyn, and show why they are popular for measurement of linear and angular displacements.

45. Describe the construction and operation of an induction potentiometer. Discuss its limitations.

46. Explain the basic principles of operation of synchros and discuss the need for variations in their design.

47. Discuss the use of a synchro transmitter-control transformer pair and how the combination is used in position control systems.

48. Describe how a synchro pair can be used for conveying angular displacements to a distant point.

49. Describe the constructional features of a linear variable differential transformer and comment on its merits in comparison to a push–pull self-inductive transducer.

50. Discuss the factors limiting the bandwidth and sensitivity of a linear variable differential transformer.

51. Comment on the linearity of the linear variable differential transformer and show how the linear range is estimated from design constants.

52. Explain how eddy-current phenomenon can be utilized to make up a proximity meter.

53. Show that a parallel plate capacitor serves as the most suitable transducer for measurement of linear and angular displacements.

54. Discuss the problems encountered while measuring small displacements by capacitive transducers.

55. Show how output voltage signals proportional to linear displacements imparted to the movable plate of a parallel-plate capacitor, may be obtained.

56. Show how a capacitive transducer can be used to monitor the thickness of an insulating sheet in motion, without making physical contact; comment on the linearity and sensitivity of the system.

57. Describe the techniques of measuring the proximity of a metallic or nonmetallic object by means of a capacitive system; comment on the linearity of the systems suggested.

58. Show how a capacitive strain transducer can be designed for measuring the surface strains of large structures.

59. Describe the design features of capacitive pressure transducers and obtain expressions relating the capacitance variation with pressure for the cases of (a) thin plate diaphragm and (b) a metal-coated membrane.

60. Explain how a capacitive transducer can be used as a microphone and show what are the additional considerations applied while designing the same.

61. Explain how moisture content in grain, tobacco, wood and granulated powder can be measured by means of a capacitive transducer.

62. Is it possible to measure relative humidity by means of a capacitive transducer? Show how the capacitance variation is measured.

63. Show that a capacitive transducer can be used to provide the means for measuring angular velocity when the total angle of rotation is limited to 100°.

SUGGESTED FURTHER READING

Adams, L.F., *Engineering Measurements and Instrumentation*, Hodder and Stoughton, London, 1975.

Boros, A., *Electrical Measurements in Engineering*, Elsevier, Amsterdam, 1985.

Doebelin, E.O., *Measurement Systems: Application and Design*, McGraw-Hill, New York, 1966.

Gibson, J.E. and Tuteur, F.B., *Control System Components*, McGraw-Hill, New York, 1958.

Holman, J.P., *Experimental Methods for Engineers*, McGraw-Hill Kogakusha, Tokyo, 1966.

Kirk, F.W. and Rimboi, N.R., *Instrumentation* (3rd ed.), American Technical Publishers Inc., Chicago, 1975.

Liptak, B.G. (Ed.), *Instrument Engineers Handbook*, Vols. I and II, Philadelphia Chilton, Philadelphia, 1969, 1970.

Neubert, H.K.P., *Instrument Transducers*, Clarendon Press, Oxford, 1975.

Noltingk (Ed.), *Jones' Instrument Technology*, Vol I: Mechanical Measurements, Butterworth, Scientific Publications, London, 1985.

Patranobis, D., *Principles of Industrial Instrumentation*, Tata McGraw-Hill, New Delhi, 1976.

Sachse, Herbert, B., *Semiconducting Temperature Sensors and Their Applications*, John Wiley, New York, 1975.

Wightman, E.J., *Instrumentation in Process Control*, CRC Press, Columbus, Ohio, 1978.

7

Active Electrical Transducers

7.0 INTRODUCTION

In Chapter 6, Electrical transducers based on passive electrical elements are described. Elements that provide electrical signals in the form of voltages or currents, dc or ac, are treated as active transducers, in the sense that they function as sources of energy, though very small, and do not need excitation from an external source of electrical power. Thermocouples and piezoelectric crystals are the known popular electrical transducers. However, transducer systems which are combinations of a few physical elements and can function in a way similar to the above are included in this chapter. Transducers that develop output signals in *time domain* and *digital domain* such as the frequency-generating transducers and shaft-angle encoders are also covered in this chapter.

Electrode systems that develop potentials when coming in contact with solutions and bio-tissues are briefly explained for completeness of information pertaining to active transducers. Electrochemistry and analytical instrumentation are the two subjects that provide considerable information related to the science and technology of electrode systems. However, sufficient details pertaining to each transducer, its scope of application and limitations are spelt out while describing the techniques used for the development of digital transducer systems.

7.1 THERMOELECTRIC TRANSDUCERS

A thermocouple is a thermoelectric device that converts thermal energy into electrical energy. The thermocouple is used as a primary transducer for measurement of temperature, converting temperature changes directly into emf. Three phenomena which govern the behaviour of a thermocouple are the *Seebeck effect*, the *Peltier effect*, and the *Thompson effect*.

7.1.1 Thermoelectric Phenomena

If two wires of different metals A and B are joined together to form two junctions and if the two junctions are at different temperatures, an electric current will flow round the circuit. This is the *Seebeck effect.* Seebeck arranged 35 metals in order of their thermoelectric properties. The current flows across the hot junction from the former to the latter metal of the following series:

Bi—Ni—Co—Pd—Pt—
U—Cu—Mn—Ti—Hg—Pb—
Sn—Cr—Mo—Ph—Ir—Au—Ag—
Zn—W—Cd—Fe—As—Sb—Te.

If metal A is of copper and metal B of iron, then the current flows from copper to iron at the hot junction and from iron to copper at the cold (reference) junction as shown in Figure 7.1(a). If copper wire is cut, an emf will appear across the open circuit. The Seebeck emf depends on the difference in the temperatures of the two junctions.

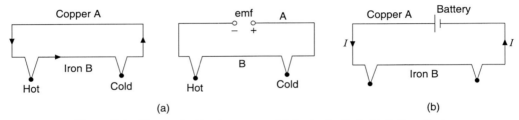

Figure 7.1   Thermoelectric phenomena: (a) Seebeck effect; (b) Peltier effect.

The reverse phenomenon is the *Peltier effect.* An external source of emf is connected, as shown in Figure 7.1(b), and a current is forced through the junctions. It is observed that heat is absorbed when the current flows across the iron-copper junction from copper to iron and liberated if the flow of current is reversed. The amount of heat liberated or absorbed is proportional to the quantity of electricity that crosses the junction, and the amount of heat liberated or absorbed when one ampere passes for a second is called the *Peltier coefficient.*

Another reversible heat-flow effect, called the *Thompson effect;* involves the contribution of emf to Seebeck emf in the wire of the same metal, if a temperature difference existed within that particular conductor. When a current flows through a copper conductor having a thermal gradient (and hence a heat flow) along its length, heat is liberated at any point where the current is in the same direction as the heat flow while heat is absorbed at any point where these are opposite. In iron, on the other hand, heat is absorbed at any point when the current flows in the direction of heat flow while heat is liberated when the current flows in the direction opposite to the flow of heat.

It will be seen, therefore, that the Seebeck effect is a combination of both the Peltier and Thompson effects and will vary according to the difference of temperature between the two junctions and with the metals chosen for the couple. It is essential that no current is allowed to flow in the thermoelectric circuit, if the Seebeck emf is under measurement. Seebeck emf measured by a high resistance voltmeter (or dc potentiometer) is different from the one measured by a conventional low impedance millivoltmeter. Potentiometric voltage measurements yield high accuracy in measurement of Seebeck emf. The total Seebeck emf produced is thus partly due to Peltier effect and partly due to Thompson effect. The Peltier emfs are assumed proportional to the temperature of the junction while Thompson emfs are proportional to the difference between the squares of the junction temperatures. Thompson emf is very low as compared to Peltier emf. Temperature measurements based on the thermoelectric phenomena rely entirely on empirical calibration and application of the thermoelectric laws established from a lot of experience.

The emf produced by any thermocouple with the junctions at any two temperatures may be obtained from a thermoelectric diagram. Lead does not show any measurable Thompson effect. Hence the emf acting on a couple formed of a metal and lead for a change of temperature of one junction is represented on this diagram. The characteristic showing the emf per degree against the temperature of the hot junction is known as the *thermoelectric line*; Figure 7.2 shows such lines for some common metals. The ordinate is taken as positive when for a small increase of temperature, the current flows from lead to the metal at the hot junction. These lines enable computation of emf acting around a thermoelectric circuit formed of any two metals having their junctions at any two temperatures.

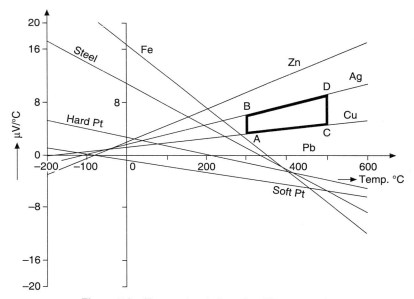

Figure 7.2   Thermoelectric lines for different metals.

Thermoelectric lines for certain metals are of negative slope. If a silver-copper thermocouple has its junctions maintained at 300°C and 500°C respectively, then the emf acting around the circuit is represented by the area of the trapezium ABCD, i.e., [(AB + CD)/2] × (200). But if copper and iron are used with the cold junction at 0°C, the emf goes on increasing with temperature of the hot junction up to 275°C where the two thermoelectric lines intersect. With further increase in temperature, the emf starts falling and becomes zero at 550°C. Later, the polarity of emf is reversed and its value goes on increasing with increase in temperature. The temperature at which the thermoelectric lines intersect is known as *neutral temperature*. While choosing materials to form a thermocouple, it is important to observe that the neutral temperature does not fall within the range of temperature to be measured.

To measure the thermo-emf in a thermoelectric circuit, introduction of a millivoltmeter or some voltage measuring system is necessary, which amounts to addition of conductors and formation of junctions. The following laws are established to help understand the effects of additional junctions.

(*i*) *Law of intermediate metals.* A third homogeneous metal C is introduced into one of the conductors as shown in Figure 7.3(a). The net emf of the circuit remains the same as long as the new junctions formed are at the same temperature. Also, it is at times necessary to insert the third metal C between wires of A and B by opening the junction at T_1 as shown in Figure 7.3(a). In such a case, the net emf again remains the same as long as the newly formed junctions are at T_1 only.

(*ii*) *Law of intermediate temperatures.* The third conductor of metal C introduced into the circuits of Figure 7.3(a) may assume any temperature without affecting the net emf as long as its two junctions with other metals are at the same temperature.

If a certain thermocouple produces an emf E_1 when its junctions are at T_1 and T_2, and it produces E_2 when at T_2 and T_3, it will produce $E_1 + E_2$ when the junctions assume temperatures of T_1 and T_3 as shown in Figure 7.3(b).

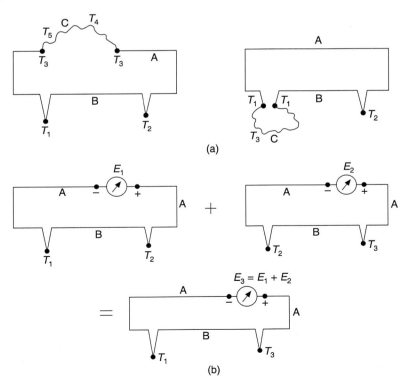

Figure 7.3 Illustration of thermoelectric laws: (a) Law of intermediate metals; (b) law of intermediate temperatures.

Applying the above laws, it can be seen that:

1. Long lead wires connecting the junctions may be exposed to an unknown temperature environment, without affecting the net thermo emf.

2. A voltage measuring system may be connected as shown in Figure 7.3, while observing that the new junctions formed are at the same temperature.

3. Thermocouple junctions may be formed by soldering, welding or brazing.

4. If the cold junction is kept at a temperature other than 0°C, it is possible to apply correction by referring to standard tables of emfs, provided the temperature of the cold junction is accurately known.

7.1.2　Common Thermocouple Systems

While many metals and alloys exhibit the thermoelectric effect, only a small number of them are widely applied for temperature measurements in industry. The materials chosen should not change in composition with time under prolonged usage due to the effect of its environment and temperature. The calibration characteristic should remain constant with usage. Generally, all industrial thermocouples are protected from contamination and mechanical strains, which may affect their calibration.

Of all the metals used for thermocouples, platinum is stable and platinum/platinum–rhodium thermocouple is the primary standard for temperatures between 630.5°C and 1063°C. Its sensitivity is only about 6 μV/°C, and is used up to 1500°C. It is most stable even in oxidizing atmospheres. Constantan (Ni 40%, Cu 60%) is another alloy that is used with copper, iron or chromel (Ni 90%, Cr 10%). Copper/constantan thermocouple has the maximum sensitivity of 60 μV/°C and is useful for the range from –200°C to +400°C. Iron/constantan is the most widely used thermocouple for industrial applications for the range of temperatures from –150°C to +1000°C. It can be used in oxidizing environment up to 750°C and reducing environment up to 1000°C. Chromel/alumel (Ni 94%, Mn 3%, Al 2%, Si 1%) is another useful thermocouple for temperatures from –200°C to +1300°C, and is preferred for use in nonreducing environment at temperatures between 700°C and 1300°C. The temperature-emf characteristics of the above common thermocouples are shown in Figure 7.4. In all the above cases, the polarity of the material mentioned first is positive for temperatures greater than the reference junction temperature.

For temperatures above 1500°C, such as those encountered in nuclear reactors and jet and rocket engines, special alloys are developed using Boron, Tungsten and Molybdenum. Boron/graphite thermocouple has a sensitivity of about 40 μV/°C and can be used up to 2500°C. Tungsten/molybdenum is used for temperatures between +1250°C and 2500°C. Rhodium/iridium is useful up to 2000°C. Alloys of tungsten and rhenium are applicable up to 2500°C.

For the reference junction, an ice bath is commonly used. For the most accurate measurements, reference junction is kept in a triple-point-of-water apparatus, whose temperature is 0.01 ± 0.0005°C. The quality of wire produced from batch to batch may vary slightly with the result that the emf produced may not exactly conform to the values made available from standard tables. For accuracy, it is better to calibrate each thermocouple formed and then use it for measurement of temperature.

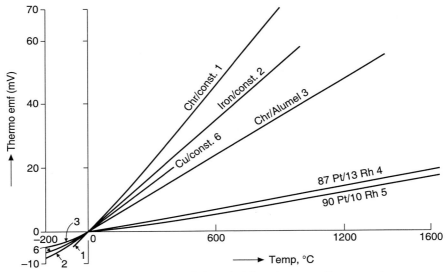

Figure 7.4 Temperature-emf characteristics of common thermocouples.

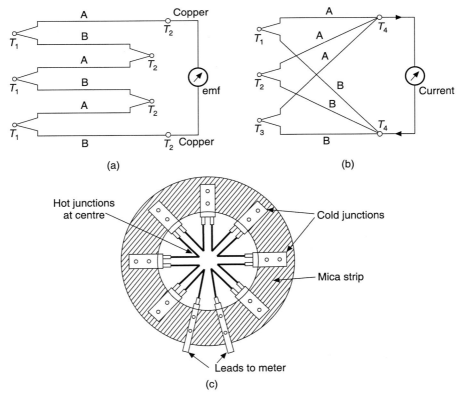

Figure 7.5 (a) Thermocouples in series; (b) thermocouples in parallel; (c) a thermopile.

In order to obtain a higher output emf, two or more thermocouples may be connected in series, as shown in Figure 7.5(a), and for measurement of average temperature, a parallel connection of Figure 7.5(b) may be used. To detect infrared (thermal) radiation, the series arrangement using as many as 25 thermocouples of fine silver and bismuth wires or chromel and constantan wires, is used. The series arrangement is known as thermopile and is shown in Figure 7.5(c). The expanded view shows the blackened junction pairs surrounded by an annular ring of mica which serves as both electrical and thermal insulation. The thermopile measures the temperature difference between the hot junctions where the radiation is brought to focus and the ambient temperature surrounding the detector. Provision for compensation of ambient temperature variation is essential.

The smallness of the size and the low thermal capacity of the bare thermocouple, as is the case with thermistor, make it extremely suitable for measurement of rapidly changing internal temperature of hot bodies and surface temperatures of structures from point to point. The thermal contact of the hot junction with the body is usually established by means of a contacting plate or contacting probe, as shown in Figures 7.6(a) and 7.6(b). They are designed to have very low time constant of the order of 1–10 μs. The system shown in Figure 7.6(a) has two ribbons of different metals separated by thin mica insulation. The end surface of the probe is drawn on a file or emery cloth, so that fine particles of each metal cross the insulation and flow to the other side, thereby forming numerous microscopic hot-weld thermojunctions. Subsequent abrasive action results in the formation of new junctions as the surface wears away. The system

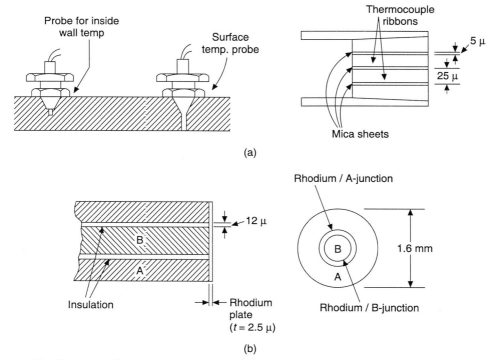

Figure 7.6 Thermocouple probes for surface temperature measurements using: (a) Two metals, and (b) three metals.

shown in Figure 7.6(b) is formed by depositing a thin layer of rhodium over the end of a coaxial pair of metals A and B. The thermo emf between A and B is unaffected by the presence of rhodium plate as its junctions with A and B are at the same temperature.

In *suction type* thermocouple system, the thermocouple is exposed to the hot gas after it is extracted from the hot furnace and made to flow continuously past the hot junction. The thermocouple is provided with a radiation shield so that it does not radiate more heat to the cooler walls of the gases, than that it receives by radiation.

The *immersion type* thermocouple system is intended to measure the temperature of hot liquids and gases by immersing it into the medium of the test fluids. For temperatures up to 500°C, they are directly used if a rapid indication is desired; otherwise, a protecting tube of mild steel of sufficient length is used. The thermoelectric characteristic of the thermocouple wires changes due to oxidation in the test fluids, and the wires may be corroded in certain fluids. Mechanical protection of the couple is equally essential. The protecting tube or sheath is either metallic or of some insulating refractory material such as fire clay, porcelain or fused silica. For prolonged exposures, a sheath of nickel-chromium is used and it is normally formed by boring from a solid rod. Refractory sheaths are more impervious to gases at higher temperatures than metal sheaths, but they are more liable to fracture if heated or cooled suddenly or unequally. Where necessary, the refractory sheath is protected by an outer sheath of fire clay, which is more resistant to thermal shock. The sheaths are provided with a terminal block on an insulating head up to which the leads of the thermocouple are brought out. All through the length of the sheath, the leads from the hot junction are insulated suitably by beads of enamel, porcelain or other refractory material. The hot junction is located at the tip of sheath and the sheath is either filled with inert gas or evacuated and sealed. All other considerations applicable to probes of resistance thermometers also apply to thermocouple probes to reduce errors due to radiation and conduction.

For quick measurement of the temperature of the hot molten steel, the thermojunction is protected by a vitreous silica sheath, thick enough to withstand the mechanical stresses due to sudden immersion, but thin enough to attain the actual temperature in about 15 seconds. The sheath is about 12 cm long, 7 mm in diameter, and 1 mm in wall thickness. A new silica sheath is used for each immersion. The sheath is connected to another protecting tube long enough for use and it is above the level of the hot metal. New hot junctions are formed after a few immersions, by cutting off the immersed portion of wires and by twisting the leads of wires drawn from spools provided at the cold end of the protecting sheath. For highly accurate measurements, the platinum/platinum–rhodium thermocouples are used.

In most cases, it is necessary to use long leads from the terminal head for connection to the millivoltmeter or recording instrument. The temperature of the head may vary even though both the terminals may be at the same temperature at any one time. As the terminal head serves as a cold junction, any variation in its temperature alters the reading of temperature. If the leads are of the same materials as the thermocouple, then the cold junction is transferred to the terminals of the indicator, where the temperature is likely to remain constant. If the thermocouple wires are expensive, the same effect is obtained by using *compensating leads* made of other materials that have the same thermoelectric characteristics but are less costly. Such a compensation is effective for a narrow range of temperature, 0-100°C only. For platinum/platinum–rhodium thermocouples, compensating leads of copper/copper–nickel alloy are used, with copper wire being connected to platinum–rhodium wire.

Wherever compensation for variation of cold junction temperature is required, a Wheatstone bridge circuit as shown in Figure 7.7(a) is used with one of its arms having a resistor of nickel wire or a thermistor and located near the cold junction. The permanent magnet moving-coil type millivoltmeter is provided in certain cases with each thermocouple, and in such cases, automatic compensation for temperature of cold junction is provided inside the instrument, by means of a bimetallic strip along with the control spring, as shown in Figure 7.7(b). The bimetallic strip is at the temperature of the cold junction, as the thermocouple leads are connected to the terminals of the instrument, and the pointer motion is properly controlled by the strip.

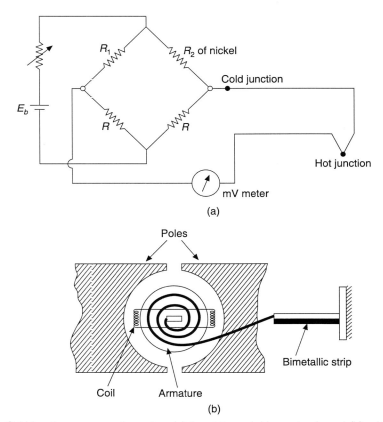

(a)

(b)

Figure 7.7   Cold junction compensation using: (a) A resistance bridge network, and (b) a bimetallic strip.

The mass of metal forming the hot junction is so small that the dynamic response of the thermocouple alone is quite satisfactory for most situations. But the protecting sheath and the separation between the walls of the sheath and hot junction lower the speed of response considerably. As already shown, the system behaves like an overdamped second-order system. Fortunately, there are electrical circuits that may be used for speeding up the response of the system, and hence there is no need to compromise the design of the sheaths in this respect. Circuits for dynamic compensation are presented in Section 4.7.2.

7.2 PIEZOELECTRIC TRANSDUCERS

Piezoelectric transducers are converters of mechanical energy into electrical energy and are based on the direct piezoelectric effect observed in certain nonmetallic and insulating dielectric compounds. Electrical charge is developed on the surface of the crystals, when they are under mechanical strain due to application of stress. Due to their high mechanical rigidity they are treated as near-ideal transducers for measurement of force and thereby pressure, acceleration, torque strain and amplitudes of vibration. They owe their popularity to their small size, high natural frequency, linearity, high sensitivity, wide measuring range, and polarity sensitivity. The commonly used materials are stable enough for all applications at temperatures up to 200°C. The small capacitance of the transducer and its high insulation resistance cause some problems for measurement of charge developed and the consequent voltage across the faces. The charge leaks away through its insulation resistance, and hence special amplifiers such as charge amplifiers are used to measure the charge. The transducer by itself is unsuitable for measurement of steady quantities due to the leakage of charge. However, it is possible to measure almost steady quantities, by means of suitably designed charge amplifiers.

The *anisotropic effect* noticed in p-n junctions of semiconductor diodes and transistors is allied to the piezoelectric phenomenon. The application of localized stress on the upper surface of a semiconductor junction results in a change of current across the junction. Such devices are known as piezoelectric transistors and are used for measurement of small pressure and force.

Conversion of electrical energy into mechanical energy is possible by using the same device. The application of electric potential between the surfaces of a crystal results in a change of its physical dimensions. This is the inverse (reverse) effect and is also known as *electrostriction*. The effect is widely applied for generation of ultrasonic waves. Generating (transmitting) and receiving (detection) ultrasonic systems are useful for detection and location of underwater objects, noninvasive measurements, and nondestructive testing.

7.2.1 Piezoelectric Phenomenon

In 1880, the Curie brothers discovered the direct piezoelectric effect in certain solid crystalline dielectric substances such as quartz. Under mechanical stress, the crystal is strained and the consequent charge developed is proportional to strain. Quartz is silicon dioxide (SiO_2) and is available as a natural substance. The atoms are arranged in the crystal as shown in Figure 7.8(a), forming a hexagon in the plane of paper while the optical axis (z-axis) is perpendicular to the xy-plane. For the three Si atoms, the six oxygen atoms are lumped in pairs, thereby forming a hexagonal crystal. The x and y axes are referred to as electrical and mechanical axis respectively. Under stress-free conditions, all charges are balanced, but when a force is applied along the x-axis, the balance is disturbed and electrical charge is developed on the two faces A and B, as shown in Figure 7.8(b). This is known as longitudinal effect. A force along the y-axis also distorts the arrangement of atoms, and charges are developed on the two faces A and B, as shown in Figure 7.8(c) and this is referred to as *transverse effect*. Due to the symmetry along the optical axis, no effects are noticed when force is applied along the z-axis.

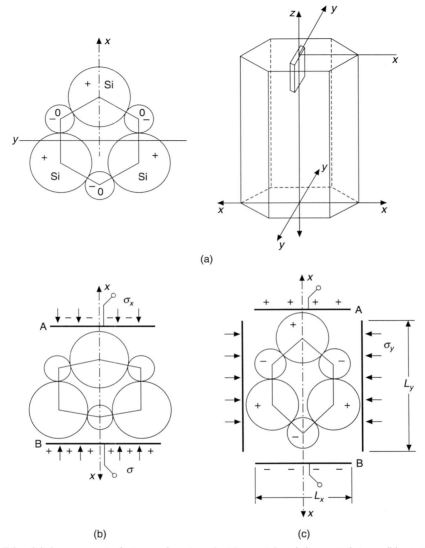

Figure 7.8   (a) Arrangement of atoms of a piezoelectric crystal and the crystal axes, (b) crystal under longitudinal effect; (c) crystal under transverse effect.

The characteristic features of the longitudinal effect are that the charge generated is independent of the area of the crystal and its thickness in the x-direction. The charge developed on a given area of the crystal face is proportional to the area affected by the pressure and thus proportional to the total force applied normal to the surface.

However, when a force is applied in the transverse (y) direction, the charge generated on A and B depends on the lengths (L_x, L_y) of the faces in the x and y directions.

Application of shear stress τ about any of the three axes may also yield charge on the faces perpendicular to the x-axis. The charge sensitivity or the piezoelectric d-coefficient is the charge developed per unit force. The net piezoelectric effect is represented by the vector of electric polarization $\overline{P}$ as

$$\overline{P} = P_{xx} + P_{yy} + P_{zz} \qquad (7.1)$$

where, x, y and z refer to the conventional orthogonal system related to the crystal axes and P_{xx} indicates the net effect on the face perpendicular to the x-axis due to application of axial stresses σ and shear stresses τ to the crystal. P_{xx} is related to these stresses as

$$P_{xx} = d_{11}\sigma_{xx} + d_{12}\sigma_{yy} + d_{13}\sigma_{zz} + d_{14}\tau_{yz} + d_{15}\tau_{zx} + d_{16}\tau_{xy} \qquad (7.2)$$

where, σ_{xx}, σ_{yy} and σ_{zz} are the axial stresses along x, y and z directions and τ_{yz}, τ_{zx} and τ_{xy} are the shear stresses along the yz, zx and xy planes. According to convention, the first subscript of the d-coefficient signifies the direction of electrical effect and the second, that of the mechanical effect. For example, d_{14} refers to charge development on a face perpendicular to the x-axis due to torque applied about the x-axis as shown in Figure 7.9(a). Similarly, P_{yy} and P_{zz} may be expressed in relation to the charges developed on other faces with the applied stresses.

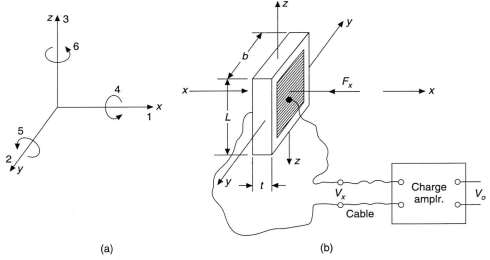

(a)   (b)

Figure 7.9 Piezoelectric crystal wafer used as force transducer.

For quartz, all d-coefficients for P_{zz} are zero. P_{xx} and P_{yy} have their coefficients given by

$$\begin{aligned}
P_{xx} &= d_{11}\sigma_{xx} + d_{12}\sigma_{yy} + d_{14}\tau_{yz} \\
P_{yy} &= d_{25}\tau_{zx} + d_{26}\tau_{xy} \\
d_{11} &= -d_{12} \\
d_{25} &= -d_{14} \\
d_{26} &= -2d_{11}
\end{aligned} \qquad (7.3)$$

The array of the piezoelectric coefficients differs from one material to another and some coefficients are of zero value.

Wafers or slices of thickness t and area A $(= Lb)$ are cut normal to the x-axis of the quartz crystal as shown in Figure 7.9(b). For a force F_x applied along the x-direction, the charge developed is

$$Q_x = d_{11}F_x \qquad (7.4)$$

As quartz is an insulator, the surfaces on which charge accumulates are provided with metallic electrodes or coated with gold to serve as electrodes for connection to external measuring circuits. Assuming that the area of the coated surface is almost the same as that of the face, the capacitance of the transducer is given by

$$C_x = \frac{\varepsilon_0 \varepsilon_r Lb}{t}$$

and the voltage across the electrodes

$$V_x = \frac{Q_x}{C_x} = \frac{d_{11}F_x}{C_x} = \frac{d_{11}\sigma_{xx}t}{\varepsilon_0 \varepsilon_r} = g_{11}\sigma_{xx}t \qquad (7.5)$$

$$g_{11} = \frac{V_x}{t\sigma_{xx}} = \frac{d_{11}}{\varepsilon_0 \varepsilon_r} \qquad (7.6)$$

The piezoelectric g-coefficient is defined as the voltage sensitivity and is expressed as the electric field ($= V_x/t$) produced per unit stress. This method of using the piezoelectric crystal for measurement of force or pressure is known as thickness expander (TE) mode.

If a force F_y is applied along the y-direction, the charge and electric field developed in the x-direction become

$$Q_x = d_{12}\frac{L}{t}F_y \qquad (7.7)$$

$$V_x = \frac{d_{12}}{\varepsilon_0 \varepsilon_r}\sigma_{yy}t \qquad (7.8)$$

$$g_{12} = \frac{V_x/t}{\sigma_{yy}} = \frac{d_{12}}{\varepsilon_0 \varepsilon_r} \qquad (7.9)$$

This method of application of force is known as length-expander (LE) mode and is useful for measurement of small forces by bringing up the ratio of L/t to suitable values.

There is another set of coefficients known as h-coefficients defining the voltage sensitivity as the electric field developed per unit strain. They are obtained by multiplying the g-coefficients by the Young's modulus Y of the material, valid for the appropriate crystal orientation.

$$h_{11} = g_{11}Y = \frac{V_x/t}{\Delta t/t} \qquad (7.10)$$

Other modes of utilizing the piezoelectric effect are known as (a) volume expander (VE), (b) thickness shear (TS), and (c) face shear (FS), as seen from Figure 7.10.

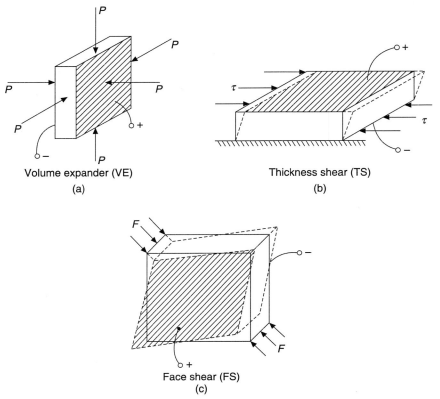

Volume expander (VE)
(a)

Thickness shear (TS)
(b)

Face shear (FS)
(c)

Figure 7.10   Modes of utilizing piezoelectric effect.

7.2.2   Piezoelectric Materials

The materials exhibiting the piezoelectric phenomenon are divided into two groups: natural and synthetic. The natural group consists of quartz, Rochelle salt and tourmaline, whereas the synthetic group consists of ammonium dihydrogen phosphate (ADP). Lithium sulphate (LS), and Dipotassium tartarate (DKT). The basic properties of these materials are given in Table 7.1. Depending on the crystal structure, discs or wafers are cut and used for measurement of force in one or the other of the modes described. The d-coefficient applicable for each mode is given for each material.

Quartz has no useful shear mode. It is the most stable material and artificially grown quartz is normally preferred as it is purer than the natural quartz. Its d-coefficient and relative permittivity may remain unaffected by temperature. Only the volume resistivity falls sharply at higher temperatures beyond 100°C.

Tourmaline is the only material exhibiting a large sensitivity in the volume expander mode related to the z-axis and its net d-coefficient d_h (hydrostatic coefficient) is given by

$$d_h = (d_{33} + 2d_{31})$$

where $d_{31} = 0.31d_{33}$. Its useful temperature range is large (up to 1000°C).

Table 7.1 Properties of piezoelectric materials

Material	Crystal cut and mode	ε_r	d-coefficient (C/N) $\times 10^{12}$	ρ (Ω-m) vol. resistivity	Density (kg/m³)	Young's modulus (N/m²) $Y \times 10^{-6}$	Temperature range	Remarks
1. Quartz	0°x-TE	4.5	2.3	10^{12}	2.65	80	550°C	Stable, unaffected by humidity
	0°x-I.F.	4.5	2.3	10^{12}	2.65	80	550°C	Safe stress—98 $\times 10^6$ N/m²
2. Tourmaline	0°z-TE	6.6	1.9	10^{11}	3.10	160	1000°C	Unaffected by humidity
	0°z-VE	6.6	2.4	10^{11}	3.10	—	1000°C	for hydraulic pressure
3. Rochelle salt	0°x-FS	350	550	10^{10}	1.77	—	45°C	Twister bimorph Sensitive to
	45°x-LE	350	275	10^{10}	1.77	19.3	45°C	Bender bimorph humidity low
	0°y-FS	9.4	54	10^{10}	1.77	—	45°C	Twister bimorph stress of 14.7
	45°y-LE	9.4	27	10^{10}	1.77	10.7	45°C	Bender bimorph $\times 10^6$ N/m².
4. ADP (Ammonium dihydrogen phosphate)	0°z-FS	15.3	48	10^{8}	1.80	—	125°C	Twister bimorph Low stress
	45°z-LE	15.3	24	10^{8}	1.80	19.3	125°C	Bender bimorph of 20 $\times$ 10^6 N/m²
5. Lithium sulphate	0°y-TE	10.3	16	10^{10}	2.06	46	75°C	Hydraulic pressure
	0°y-VE	10.3	13.5	10^{10}	2.06	—	75°C	

Rochelle salt is the material that is being produced on industrial scale for producing gramophone pick-ups and crystal microphones. It has the highest relative permittivity among the natural group and has high shear sensitivity of d_{14}. For accurate measurements of force, it is unsuitable due to the large effect of temperature on its relative permittivity even at room temperatures.

ADP crystals possess the lowest resistivity which is also temperature dependent. With temperature compensation, they are used in acceleration and pressure transducers. The crystals are cut so as to respond to face shear (FS), and length expanders (LE) are cut from FS plates in a diagonal direction.

Lithium sulphate is highly sensitive when used in VE mode, but it can also be used in TE mode.

7.2.2.1 Ferroelectric Materials

There are certain polycrystalline ceramic compounds which exhibit the property of retaining electric polarization when exposed to intense electric fields. These materials are known as ferroelectric materials (equivalent to ferromagnetic materials), and after polarization, their behaviour is similar to the piezoelectric materials. Three such common substances which are popularly used for piezoelectric transducers are Barium titanate ($BaTiO_3$), lead zirconate–titanate, and lead metaniobate. The crystals are kept under strong dc electric field and heated up slowly up to a temperature higher than their Curie temperatures. Then they are allowed to cool slowly in the presence of the electric field. After cooling and removal of the electric field, it is seen that they retain certain electric polarization and exhibit piezoelectric properties. They possess very high values of relative permittivity and high sensitivity. Positive polarization is along the z-axis, and the array of d-coefficients is as follows:

$$
\begin{array}{cccccc}
0 & 0 & 0 & 0 & d_{15} & 0 \\
0 & 0 & 0 & d_{24} & 0 & 0 \\
d_{31} & d_{32} & d_{33} & 0 & 0 & 0 \\
\end{array}
\tag{7.11}
$$
$$d_{15} = d_{24}, \quad d_{31} = d_{32}$$

Some of the properties of the above materials are given in Table 7.2. It is seen that they are suitable for use in all the modes. The main advantage of these materials is that they are free from the limitations imposed by the crystal structure, and hence they can be moulded to any size and shape. The direction of polarization can be set at will during the process of production, and depending on the shape and size of the transducer desired. The electrodes are usually fired-on coatings of silver or palladium, to which connecting wires are soldered on prior to polarization. Even though the d-coefficients and ε_r are slightly affected by temperature, their ratio is nearly constant, thereby rendering g-coefficients nearly constant with temperature. The versatility in respect of their piezoelectric properties, and the several ways in which they can be brought about in size and shape have opened up avenues for measurement of all nonelectrical quantities based on force.

Table 7.2 Properties of piezoelectric ceramic materials (TE Mode)

Material	ε_r	d_{33} (C/N) $\times 10^{12}$	ρ (Ω-m)	Density (kg/m^3)	Young's modulus Y (N/m^2) $\times 10^{-6}$	Curie temperature	Remarks
1. Lead zirconate titanate-5A	1750	356	10^{14}	7.6	59	285°C	High charge output
2. Lead zirconate titanate-5H	3400	593	10^{11}	7.5	48	193°C	Very high charge output
3. Lead zirconate titanate-7H	425	150	10^{9}	7.6	72	350°C	Very high voltage output
4. Lead metaniobate	250	80	10^{9}	5.8	47	400°C	High Curie temperature
5. Barium titanate	1500				86		

7.2.2.2 Piezoelectric Semiconductors

In 1961, it was discovered that a localized stress on the upper surface of the p-n junction of a semiconductor diode caused a very large reversible current change in the current across the junction. The phenomenon is due to the anisotropic stress effect in p-n junctions, and devices utilizing this effect are known as piezoelectric diodes and transistors. The variation of current across the junction of a Germanium diode for forward and reverse voltages is shown in Figure 7.11(a), and it will be observed that considerable change in the magnitude of the current results from application of a few grams of localized force. Moreover, the change is reversible. The behaviour of a silicon n-p-n planar transistor is shown in Figure 7.11(b). The force is applied to the surface by means of a pointed stylus. The current gain of the transistor decreases with increase of force, and the capacitance between base and collector changes in a similar fashion. These changes enable circuitry to deliver analog and digital outputs. Without any amplification, analog output voltages of the order of 20% of the supply voltage can be obtained with supply voltages ranging from 1 V to 50 V. They are useful for measurement of small forces of the order of 10^{-3} N with a linearity of ±1%. They are highly sensitive and can withstand large overloads. The mechanical resonant frequency is around 150 kHz. The characteristics are affected by temperature and hence they are used in environments where the temperature is held constant. These devices can be integrated with other electronic circuits so as to deliver outputs in the desired form.

7.2.3 Piezoelectric Force Transducer

Piezoelectric crystal or element primarily responds to force-input and possesses all the desired characteristics of an ideal force transducer. The element can be directly stressed by application of force at one point of the surface. Multiple forces can also be applied at more than one point of the surface, and summed by using one single crystal.

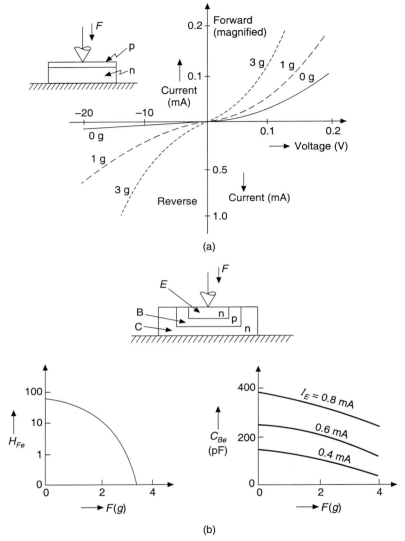

Figure 7.11 (a) Piezoelectric semiconductor diode and its characteristics; (b) piezoelectric semiconductor transistor and its characteristics.

To increase the charge sensitivity, more than one element can be used to form a transducer system and such combinations are known as *bimorphs* or *multimorphs* (or piezopile), depending on whether they are of two elements or more. The series and parallel connected bimorphs are shown in Figure 7.12(a). A multimorph of four elements, which develops four times the charge of a single element, is shown in Figure 7.12(b). It should be noted that the four elements are mechanically in series but electrically in parallel and hence the net capacitance of the transducer increases correspondingly. The series electrical connection may as well be employed if higher voltage sensitivity is desired, provided the reduced capacitance is a satisfactory match to the

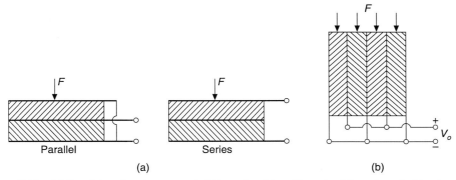

Figure 7.12   (a) Parallel and series connected bimorphs; (b) multimorph of four piezoelectric elements.

signal processing circuits. When bimorphs are made up of ceramic elements, the direction of polarization of the two elements should be noted, and then connected so as to develop charges and voltages under stress as shown in Figure 7.13(a). Bender-type bimorphs of all the piezoelectric materials can be made from single elements of either thickness expander mode or length expander mode.

A twister bimorph is shown in Figure 7.13(b), with the force applied at *A*, while the remaining three corners *B, C* and *D* are held rigidly. These are made up of face shear elements of ADP and Rochelle salt cemented together to form the three-point twister with the diagonal on the top surface expanding under force. If the four corners can be subjected to concentrated forces as shown in the four-point twister of Figure 7.13(b), the expanding diagonals will be perpendicular to each other and on opposite sides of the bimorph.

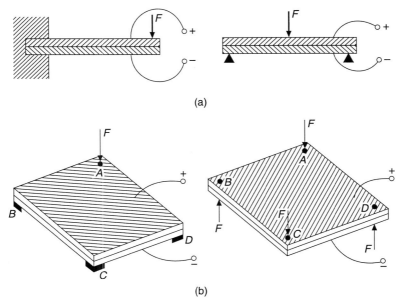

Figure 7.13   (a) Bender type bimorphs; (b) twister type bimorphs.

7.2.4   Piezoelectric Strain Transducer

Any piezoelectric element of the length expander mode, if cemented to the surface of the structure under stress, the strain in the structure is transmitted to the element and a voltage proportional to strain is directly available from the transducer. The output is obtained by using the *h*-coefficient given by

$$V_o = het \qquad (7.12)$$

where, e = strain, m/m

t = thickness of the element, m

The sensitivity of the transducer is very high. With elements of 1 mm thickness, the sensitivity of quartz element is as high as 4.5 V/microstrain and that of the ceramic element, about 0.8 V/microstrain. The advantages of the ceramic element, though of lower sensitivity, are its high capacitance and smaller value for lower cut-off frequency. However, they are not usually preferred for measurement of strain under static conditions. Piezo-resistive strain transducers, though known to be suited for transient strain measurements, are not as sensitive as the piezoelectric type. If accuracy and stability are of primary interest, metallic alloy resistive strain gauges are chosen, especially when static strain is monitored over a long period of time.

7.2.5   Piezoelectric Torque Transducer

A cantilever-type bender bimorph can be used as a twister bimorph for the measurement of torque as shown in Figure 7.14. The twisting moment may be due to a small force transmitted through a lever or may be obtained directly by connecting it to a driving shaft/spindle as obtained in instrument mechanisms. The sensitivity is high and is therefore very much useful for measurement of small driving torques under dynamic conditions. These elements are of Rochelle salt for use at room temperature and of ADP for temperatures up to 100°C.

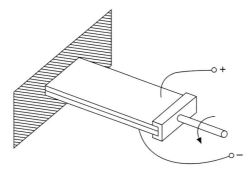

Figure 7.14   A cantilever type twister bimorph.

7.2.6   Piezoelectric Pressure Transducers

Piezoelectric transducers are more suitable for pressure measurements under dynamic conditions only and are often used as microphones, hydrophones, and engine pressure indicators.

In the case of the piezoelectric microphone, the diaphragm and the bimorph are connected together by means of a fine needle (spindle) as shown in Figure 7.15. The natural frequency of the diaphragm, the bimorph, and the associated system should be made higher than the highest frequency to be responded to (10 kHz normally). When used in sound level meters, it is essential for microphone to have flat frequency response up to 10 kHz.

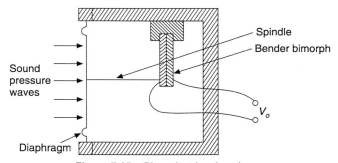

Figure 7.15   Piezoelectric microphone.

Large pressure variations occurring at frequencies up to 20 kHz in internal combustion engines are measured by using multimorphs (piezopile) of quartz elements. The surfaces of the elements, connecting electrode surfaces in between and the diaphragm or load plate at the extremes, should be optically flat, and no air should be trapped in between as it would reduce the natural frequency of the system. The transducer is prestressed so as to enable pressure fluctuations about a mean value to be measured. The prestressing is produced by a thin-walled tube under tension, as shown in Figure 7.16(a). A very thin diaphragm of flexible material is used for sealing. The preload may also be developed by a stiff diaphragm, as shown in Figure 7.16(b). The net force F_1 to which the piezopile responds is given by

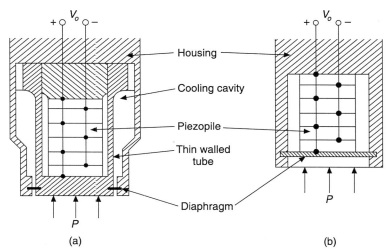

Figure 7.16   Piezoelectric pressure transducers prestressed by (a) a thin-walled tube; (b) a thick diaphragm.

$$\frac{F_1}{F} = \frac{k_1}{k_1 + k_2}$$

where,

F = the total force acting on the transducer
k_1 = spring-rate of piezopile
k_2 = spring-rate of the preloading tube or diaphragm

The cooling cavity is intended to suppress the effects of external temperature variations, and when the temperature of the engine is high, water cooling should be employed. As the piezopile has a very high natural frequency of the order of 500 kHz, it is possible to correspondingly design the entire transducer system in such a way as to obtain flat frequency response over a frequency range of 30 Hz–50 kHz.

Hollow spheres and hollow cylinders of Barium titanate or any other ceramic material enable measurement of air-blast pressures and underwater pressures transients. The ideal blast-pressure transducer should be nondirectional, and hence the hollow sphere is the best choice. The dimensions of the transducer must be small compared with the wavelength of the blast wave. However, a small hollow cylinder shown in Figure 7.17 is used in most cases. The outer and inner surfaces are metallized and used as electrodes. The walls are polarized in a radial direction. The tube cavity may be sealed against the external pressure and the blast pressure is applied to the outer surfaces. The cylinder responds to the pressure P_e in all the three modes as shown in Figure 7.17. For a thin-walled hollow tube, the open circuit voltage generated by the radial stress and tangential stress is given by

$$V_o = P_e \left(g_{33}b \frac{a-b}{a+b} - g_{31}b \right) \tag{7.13}$$

where,

g_{33}, g_{31} = the g-coefficients of the material
b = outer radius
a = inner radius

Since g_{33} and g_{31} are of opposite signs and $(a - b)$ is negative, V_o goes through zero at some critical ratio of outside-to-inside diameter. A proper choice of the ratio of the thickness of the wall to outer diameter is necessary: for lower pressures it is held below 0.2 and for higher pressures well above 0.35, thereby avoiding the critical region.

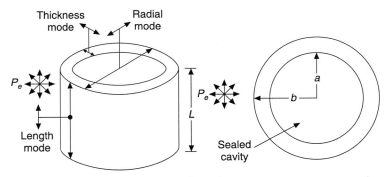

Figure 7.17   Pressure transducer for underwater pressure measurement.

7.2.7   Piezoelectric Acceleration Transducer

The design features of an acceleration transducer follow the pattern of a force transducer except for the addition of a proof mass for developing force under acceleration inputs. The single crystal or the piezopile is prestressed by screwing down the cap on the hemispherical spring shown in Figure 7.18(a). If it is ceramic material element, prestressing renders the system more linear in its characteristic, as shown in Figure 7.18(b). The transverse sensitivity of the system must be minimized by means of strong guide springs or a diaphragm. An acceleration transducer, compensated for transverse acceleration inputs, is shown in Figure 7.18(c). The two independent seismic systems are so connected as to yield additive output for acceleration in the desired direction. However, for transverse acceleration, the output signals are made to cancel each other by so orienting their *y*-axes.

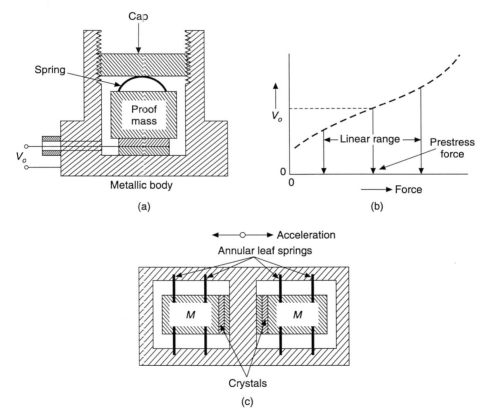

Figure 7.18   (a) Piezoelectric acceleration transducer; (b) its input–output characteristics; (c) compensation for transverse acceleration.

If the slight departure in the linearity and stability of the ceramic piezoelectric materials are tolerated, their application in acceleration measurements yields light-weight accelerometers (without needing the proof mass) of higher sensitivity and larger bandwidth. The bender bimorph can be used as a cantilever element with acceleration inputs applied normal to the plane of the bimorph. The self (though distributed) mass of the bimorph itself acts as the proof mass of the accelerometer. A novel design employs a piezoelectric disc bonded to a thin metallic disc as shown in Figure 7.19, and it is supported at its centre by a stand.

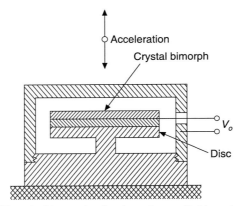

Figure 7.19   Piezoelectric acceleration transducer with bimorph as proof mass.

The natural frequency of the piezoelements is so high and far removed from the range of frequencies normally required of accelerometers or, in fact, all the piezoelectric transducers, since there is no need for additional damping in the system. By choosing quartz as the material for the pile and through the use of charge amplifiers, the lower cut-off frequency is pushed to as low a value as possible, whereas in the case of ceramic material elements the high value of ε_r results in much larger bandwidths extending towards much lower frequency. Accelerometers employing ceramic material elements have become so light and compact as to suit the needs of air-borne equipment and inertial navigation systems.

7.3 MAGNETOSTRICTIVE TRANSDUCERS

Magnetostrictive transducers are similar to piezoelectric transducers and are based on the application of the magnetostriction phenomenon. They are converters of mechanical energy into magnetic energy and are also known as magnetoelastic transducers. The phenomenon is reversible and the devices developed convert energy from one form to another. The natural frequency of the transducers can be as high as 10 kHz and are very much used as transmitters (senders) and receivers in vibration and acoustic studies. The transducers possess very high mechanical input impedance and are suitable for measurement of force and hence acceleration and pressure. They can measure large forces, both static and dynamic. They are rugged in constructional features and, when used as active transducers, the output impedance is low. Nickel and nickel-alloys are mostly used. It is the basic nonlinearity in the B-H characteristic which is responsible for its limited scope of application, especially when high accuracy is desired. Each transducer needs calibration before use.

7.3.1 Magnetostriction Phenomenon

Certain ferromagnetic materials are considerably affected in their magnetic properties when they are mechanically stressed. This phenomenon is known as magnetostriction (Villari effect) and is particularly significant in nickel and nickel–iron alloys. The noise in transformer is due to the reverse effect. The shape and size of the B-H characteristic and the B-H loop is sufficiently altered when the material is subjected to tensile, compressive or shear stress. The B-H characteristics of nickel and nickel–iron (Ni, 68%) alloy are presented in Figure 7.20, showing the effect of increasing tensile stress σ on the materials. Similarly, the magnetization

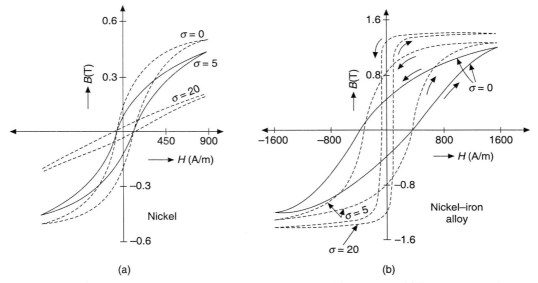

Figure 7.20 B-H characteristics under different stress values: (a) For nickel; (b) for nickel–iron alloy.

characteristic is affected and it is observed that the permeability increases with increase in tensile stress in the case of nickel–iron alloys and decreases in the case of pure nickel. The change in the shape of the B-H loop alters the remnance B_r of the material. When B_r and permeability decrease with increase in stress, it is known as negative magnetostriction; otherwise, it is called positive magnetostriction. Also, the percentage of nickel in the nickel–iron alloy has considerable influence on the characteristics. The materials are sensitive to the polarity of stress and hence the transducers enable measurement of alternating forces. Some ferrite materials such as 'Ferroxcube B' exhibit magnetostriction of considerable degree but due to their brittleness, they are not used. Even the magnetostrictive transducers employing nickel or nickel–iron alloys would have competed with the piezoelectric transducers, if the effects of stress on permeability and remnance are linear without showing any hysteresis. The transducers need mechanical bias, so as to make them suitable for application of cyclic stress. Figure 7.21(a) shows the variation of B with stress at different values of H, and Figure 7.21(b) shows the effect of superposition of cyclic torsion on tensile stress for the case of a nickel sample.

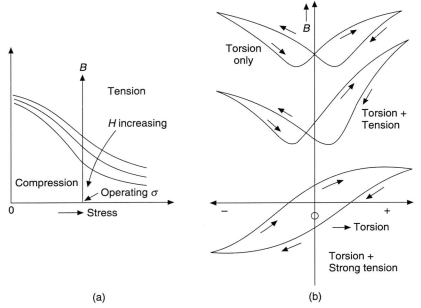

(a)   (b)

Figure 7.21   Characteristics of a nickel sample: (a) For *H* variation; (b) for superposed cyclic torsion.

7.3.2   Magnetostrictive Force Transducer

The self-inductance of an iron-cored coil changes if the core characteristic is changed due to application of force. It is the mechanical strain that affects the orientation of the magnetic domains, and hence the change in the value of effective permeability. The magnetic path should be continuous with no air gap present. The core may be laminated. The laminations are stacked to form the core, and a coil is provided to enable measurement of its self-inductance. The coil current is so adjusted as to make the self-inductance maximum and make it most sensitive to stress. One of the simple configurations commonly employed is shown in Figure 7.22. The arrangement allows measurement of large static forces and 10–20% change in self-inductance is observed with nickel and nickel–iron alloy transducers. For reasons already explained, it is difficult to attain high accuracy in measurement of force and weight. Though for reasons of ruggedness it is used for measurement of large force, the accuracy is limited to 2–5% of full scale. Although it is possible to measure dynamic forces by this method, the following method is preferred for various reasons. The core can be polarized by means of dc mmf so that the operating point of flux density is held at some convenient value below saturation. Application of stress σ results in a change of B by $\pm\Delta B$, depending on the material. The sensitivity of the transducer is defined as the ratio of ΔB to σ and is given by

$$S = \frac{\Delta B}{\sigma}\bigg|_{B = B_o} \tag{7.14}$$

where, B_o = operating point of flux density.

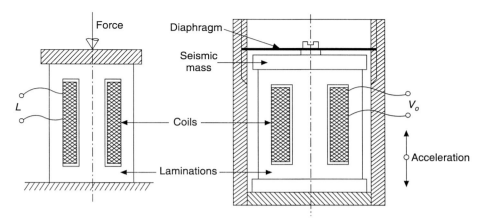

Fig. 7.22   Magnetostrictive force transducer.   **Fig. 7.23**   Magnetostrictive acceleration transducer.

For small sinusoidally varying σ, corresponding variations of ΔB are assumed to be sinusoidal. If a coil is provided on the core, the induced emf will be proportional to σ and sinusoidal. The sensitivity is observed to be maximum in the case of nickel–iron (Ni 68%) alloy when B_o is adjusted to $1/\sqrt{3}$ of saturation flux density. It is approximately equal to 3×10^{-8} T/N.

The operating flux density B_o may be chosen as the remnant flux density B_r for reasons of simplicity and stability. The sensitivity may be lower but it is preferred since bias winding is not needed. The fall in sensitivity can be made up by providing more turns in the pick-up coil, utilizing the window space of the bias winding. The emf induced in the winding is given by

$$e(t) = SAN \frac{d\sigma(t)}{dt} \tag{7.15}$$

where, A is the area of coil and N the number of turns.

Transient forces and stresses can be measured by integrating $e(t)$ before it is displayed on the oscillograph. Though the system becomes equivalent to a permanent magnet, the core needs to be laminated. Influence of strong external magnetic fields may alter the sensitivity sufficiently unless it is further shielded.

7.3.3   Magnetostrictive Acceleration Transducer

To extend the application of the transducer for measurement of acceleration, addition of proof mass is required. The mass of the core itself serves as proof mass to some extent and additional mass is provided by a brass cylinder of at least an equal mass, as shown in Figure 7.23. To prevent the transducer from responding to transverse accelerations, the brass cylinder is guided by a flexible diaphragm. The induced emf of the coil is integrated in such a way as to extend the bandwidth of the system towards the lower frequencies. As compared to piezoelectric accelerometers, these transducers are of larger size and mass and are lower in accuracy. While measuring acceleration, the variation in the earth's magnetic field affects the sensitivity. Laminations and coil should be rigidly held in position so as not to be affected under high accelerations.

7.3.4 Magnetostrictive Torsion Transducer

Magnetostrictive torsion transducer consists of a nickel wire of 0.5–1 mm diameter kept stretched between the poles of a permanent magnet and having a small stylus rigidly attached to it at the mid-point. The wire is prestressed by twisting it, before being installed into the position. Two pick-up coils of fine wire are wound round the wire on either side of the mid-point, as shown in Figure 7.24. Any displacement of the stylus to one side or the other increases the torsion on one side and decreases it by an equal amount on the other side. This results in an increase of magnetic flux in one-half and a decrease in the other half. The corresponding induced emfs are in phase opposition and are processed by suitable networks as in the case of linear variable differential transformer. It is used as phonograph pick-up and is designed to have flat frequency response over 150 Hz–15 kHz frequency range. Due to the nonlinearity and hysteresis in the performance, it is normally limited for use when time-varying torsions of small amplitude are to be measured.

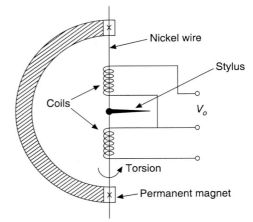

Figure 7.24 Magnetostrictive torsion transducer.

7.4 HALL-EFFECT TRANSDUCERS

The Hall effect was discovered in 1879 and belongs to one of the galvanomagnetic phenomena in which the interaction between the magnetic field and moving electrical charges results in the development of forces that alter the motion of the charge. The Hall effect is observed in all metals, but it is very much prominent in semiconductor materials. A thin strip of bismuth or n-type germanium is subjected to magnetic field B normal to its surface as shown in Figure 7.25, while it carries a current I along the length of the strip, but normal to B. The magnetic field exerts a force (known as *Lorentz force*) on the electrons moving at a velocity v, with the result that some of them drift towards the edges of the strip. The edge surfaces act like charged electrodes and the potential difference measured between P and Q is known as Hall

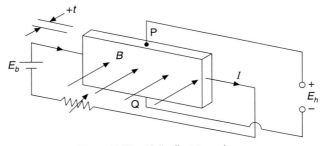

Figure 7.25 Hall effect transducer.

potential E_H which increases with increase of B and I. The build-up of the charge on the edge surfaces will, in turn, develop an electric field (Hall field) of such a polarity that counteracts the collection of charges on the surfaces. The force on the electrons due to Hall field and the Lorentz force balance each other finally. The time required to reach this equilibrium is about 10^{-14} s.

If e is the charge of the electron, then the Lorentz force Bev and the force due to Hall field are equal to each other. Hence

$$Bev = eE_H/b \qquad (7.16)$$

$$E_H = Bbv \text{ (volts)}$$

where,

B = the magnetic flux density, T
v = velocity, m/s
b = width, m

The electrons and the free charge carriers assume a velocity along the length of the strip, which is proportional to electric field along the direction of motion. If the mobility of the charge carriers is represented by χ, then v is given by

$$v = \chi E_b/L \qquad (7.17)$$

and using $E_b = I\rho L/bt$, v is given by $\chi I\rho/bt$. Hence,

$$E_H = \rho\chi BI/t = K_H BI/t \qquad (7.18)$$

where,

K_H = Hall constant $(= \chi\rho)$
t = thickness of the strip, m
L = length of the strip, m

The Hall coefficient depends on the number of free charge carriers per unit volume. The coefficient for metals and semiconductors is given by

$$K_H = \frac{K}{e} \frac{(n\chi_n^2 - p\chi_p^2)}{(n\chi_n + p\chi_p)} \qquad (7.19)$$

where,

K = a positive constant $(K = 1$ for metals and $3\pi/8$ for semiconductors)
n, p = No. of free electrons and holes per m^3
χ_n, χ_p = mobility of the electrons and holes respectively
e = charge of the electron

Table 7.3 provides the values of the Hall coefficients for n-type conductors.

As the number of free charge carriers in semiconductors is far less than in metals, semiconductors possess a very large value for Hall coefficients. However, semiconductors are highly sensitive to temperature. The current passage through the element creates noise and the current should be limited to very low values. The advantages are that semiconductors are extremely fast and require very little space.

Table 7.3 Hall coefficients of *n*-type materials

Material	K_H Vm/AT
Copper	-5.3×10^{-11}
Silver	-9×10^{-11}
Bismuth	-5×10^{-7}
Silicon	-10^{-2}
Germanium	-3.5×10^{-2}
Indium antimonide	-6×10^{-4}
Indium arsenide	-9×10^{-3}
Mercury selenide	-7.36×10^{-6}
Mercury telluride	-1.47×10^{-6}

7.4.1 Applications of Hall Transducers

The Hall effect transducer is primarily suited for the measurement of magnetic fields. It is preferred for mapping of magnetic field around conductors carrying large currents under different flow conditions, with and without the presence of metallic objects in their vicinity. This transducer has made it possible to measure dc and ac currents in conductors without interruption of the circuit and without making any electrical contact with the conductor. The range of current that can be measured in this manner can be as low as 0–1 mA or as high as 0–1 kA. Wattmeters of different ranges are available for measurement of dc and ac power at frequencies up to 10 kHz. *B* is proportional to load current and *I* to load voltage.

In the field of instrumentation, the Hall element is highly valued for its speed of response in detection of changes in the magnetic field to which it is exposed. The advantages are its small size and high sensitivity. It is used as a proximity detector as it does not require to establish a mechanical link with the test object. It is used to measure the change in the strength or direction of the magnetic field due to the displacement or nearness of the test object. In other words, it can be used as a displacement transducer, provided the test object can distort the magnetic field due to its motion. Otherwise, the test object has to be mechanically coupled to the Hall effect element, already lying in a steady magnetic field of constant strength. Its application as angular displacement transducer and proximity detector can be understood from Figure 7.26. As the element can respond to quick changes in the field, it is equally applicable for measurement of amplitudes of vibration of objects and count the number of fast moving objects across the magnetic field.

In all the above applications, the current through the element should be held constant at about 5–20 mA dc by using constant current sources. The value of E_H for the case of an *n*-type germanium element, carrying a current of 10 mA is 1.4 mV when exposed to a magnetic field of 0.1 mT. The output impedance varies from one element to another and is about 5–200 ohms, depending on the material and size of the element.

7.5   ELECTROMECHANICAL TRANSDUCERS

7.5.1   Tachometers

Measurement of angular speed of rotating shafts assumes great importance in instrumentation, particularly in the case of automatic speed control systems. For analog indication of speed, dc and ac tachometers are in use since long.

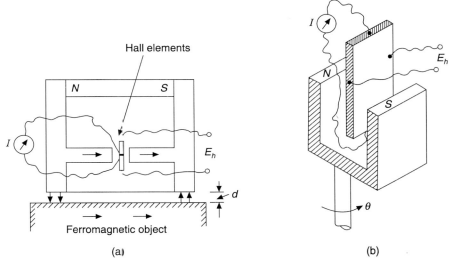

(a)   (b)

Figure 7.26   (a) Hall effect displacement transducer; (b) Hall effect proximity transducer.

A dc tachogenerator offers the simplest means of converting a shaft rotation into an electrical signal which is proportional to speed. It is basically a dc generator comprising a permanent magnet stator and a wound rotor. The emf generated, viz. E_g, is given by

$$E_g = \frac{NP\phi\Omega}{a}$$

where,

N = total number of conductors of the armature
P = number of field poles
ϕ = total magnetic flux per pole, Wb
Ω = speed of rotation, rad/s
a = No. of parallel paths in armature winding

The impedance of the windings is kept as low as possible, keeping in view its sensitivity. The inertia of the rotor is kept as small as possible. The commutator has a relatively large number of segments in order to provide as smooth an output voltage as possible. A light-weight wire brush assembly is used for good contact and minimum friction and noise. Even then a ripple content of 5% is noticeable. Sensitivity of the order of 5 V to 20 V per 1000 rev/min is obtained. Linearity as high as ± 0.01% is achievable. But the ripple content and the variation of ripple frequency with speed, render them unsuitable for situations involving feedback systems. For analog indication by dc or ac voltmeters, the dc tachometers offer the best solution. When high resolution of the order of $\pm 0.1\%$ to $\pm 0.01\%$ is desired, feedback measuring system using the dc tachogenerator or the ac pulse-generating tachometer is designed.

An ac two-phase induction motor having a squirrel-cage type rotor can be used as an ac tachogenerator, by exciting one phase with the reference ac supply and utilizing the second phase winding to provide output voltage related to the speed. With the rotor at rest, the output voltage is zero, as the coil axes are perpendicular to each other, as shown in Figure 7.27(a).

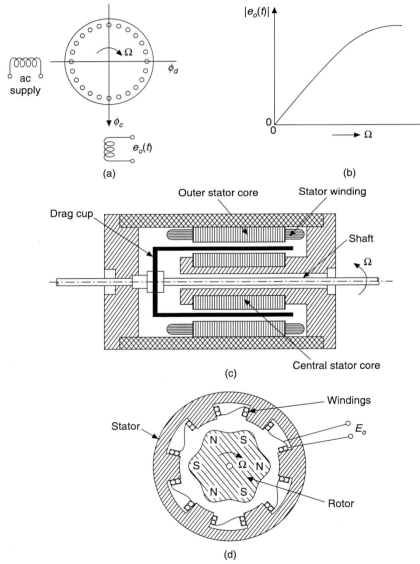

Figure 7.27 (a) AC tachogenerator; (b) its characteristic; (c) a drag cup type tachometer; (d) a multiple pole type ac tachogenerator.

Rotation causes dynamically induced emfs in the rotor, which are in phase with ϕ_d and are proportional to the speed of rotation. The resulting currents in the rotor conductors produce alternating magnetic flux ϕ_c at 90° to ϕ_d. It is this cross flux ϕ_c that induces voltage in the output winding by transformer action. This output voltage is of the same frequency as that of the supply and of an amplitude proportional to the instantaneous speed. It is in phase with the reference or excitation voltage, if the rotor behaves as a pure resistor. A phase reversal by 180° occurs, when the direction of rotation is reversed.

To reduce the mechanical loading of the rotor on the driving shaft, the rotor is made lighter by using copper or aluminium cup as the rotor. There will be no slot ripple in the output in the case of cup-rotor, as noticed in the case of squirrel-cage rotors. The air gap becomes necessarily large in the case of drag-cup type ac tachogenerators, and the exciting winding impedance becomes smaller. These factors ultimately influence the speed-voltage characteristic to the extent that linearity is lost at high speeds as shown in Figure 7.27(b). Figure 7.27(c) shows its constructional features.

The primary advantage of the ac tachometer is that it has no commutator ripple. But it should be noted that the accuracy of calibration is dependent on the excitation voltage as the output voltage is proportional to the excitation voltage. A polarity or phase discriminator is required to sense the direction of rotation. Normally ac tachometers are meant for excitation voltages of frequency of 50 Hz or 400 Hz. The sensitivity is less than that of dc tachogenerator and is generally about 2.5 V per 1000 rpm.

A slightly different version is shown in Figure 7.27(d) which does not need a commutator. The rotor is a multiple-pole permanent magnet structure. The stator coils are connected in series and the net induced emf is proportional to the speed.

Another type of tachometer that converts angular speed of rotation into angular displacement for purposes of analogue indication is the eddy current drag-cup type tachometer. The driving shaft rotates a permanent magnet inside a conducting cup made of nonmagnetic material. Eddy currents are set up in the cup and interact with the magnetic field in such a way as to follow the magnet. The torque developed is proportional to the relative velocity of the magnet and cup and when it is equal to the restraining torque, the cup comes to rest. Thus in steady state, the angular deflection is proportional to the angular speed of rotation.

The system behaves as a second-order indicating instrument. Resistivity variations of the cup-material with temperature limit the accuracy of indication. Accuracy of the order of 1% to 2% is attainable. With the provision of suitable gear-trains, the range is extended up to 50,000 rev/min. The spindle connecting the cup and control spring may be provided with a shaft-angle encoder for digital output apart from analog indication. For obtaining electrical output voltage, any low-torque angular displacement transducer may be used.

7.5.2 Variable-Reluctance Tachometers

In this type of tachometer, reluctance variation of a magnetic circuit is brought about by the rotating shaft and the consequent flux changes develop induced emf in the coils linking with the circuit. Figure 7.28 shows two such versions, one employing an electromagnet and the other a permanent magnet. The toothed rotor may as well be a gear wheel but should be made of a ferromagnetic material. The shape of tooth of the rotor and the shape of the pole piece of the

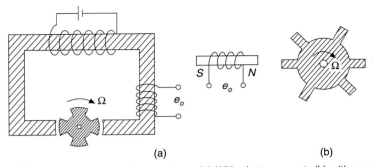

(a)   (b)

Figure 7.28   Variable reluctance type tachogenerators: (a) With electromagnet; (b) with permanent magnet.

magnet may be so designed, wherever necessary, as to develop a sinusoidal emf. The peak-to-peak value of the induced emf is proportional to the shaft speed. The number of pulses produced is proportional to speed. The rms value of the output voltage is dependent on the clearance between the rotor and the pole piece of the magnet. The pole piece of the magnet is known as perception head or proximity head. The rms voltage may be used for indication on a voltmeter and, in such cases, the tachometer is generally restricted to the range of speeds between 40 rpm to 6000 rpm. The frequency of the induced emf in the pick-up coil may also be processed so as to be indicated in analog fashion by a p.m. m.c. instrument. A circuit consisting of two Zener diodes and two diodes is shown in Figure 7.29, for such application and the average current $\overline{I}$ read by the meter is proportional to the frequency of the induced emf and is given by

$$\overline{I} = 2C(V_z + V_f - V)f \qquad (7.20)$$

where,

V = forward voltage of the D_1 and D_2 diodes
V_z = reverse voltage of Zener diodes
V_f = forward voltage drop of Zener diodes
f = frequency of the emf (speed in rev. per second $\times$ no. of teeth)

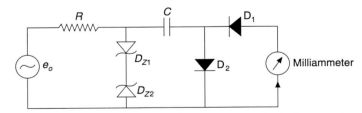

Figure 7.29   Circuit for use with frequency generating tachogenerator.

Such tachometers whose indication is proportional to frequency are useful for measuring speeds up to 100,000 rpm. The induced emfs of the coil are converted in their wave shape to look like pulses, by a pulse shaper before they are applied to the circuit of Figure 7.29. Variable reluctance tachometers with the permanent magnet perception heat have become very popular in the context of digital tachometry.

Two simpler versions of the variable reluctance transducer are shown in Figure 7.30. The armature in both cases is of a ferromagnetic material and its movement around its mean position induces emf in the pick-up coils. In each case the emf is proportional to the velocity of motion of the armature and hence the variable reluctance transducers are useful for the measurement of vibratory velocities, linear or angular. Integration of the output signals yields voltages proportional to the relative displacement of the armature with respect to the fixed magnet assembly. The reluctance transducers are suitable for measurement of dynamic displacements only and the range of frequencies for which each is suited, depends on the natural frequency of the system and the damping, fluid and electromagnetic. It is necessary to observe that no current is drawn from the coil due to the voltmeter or load connected to it.

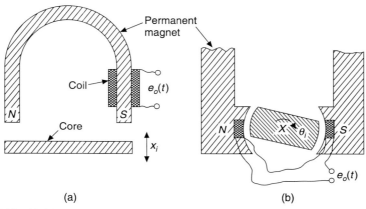

(a)   (b)

Figure 7.30   Variable reluctance transducers for small displacements: (a) Linear; (b) angular.

7.5.3   Electrodynamic Vibration Transducers

Electrodynamic vibration transducers are based on the principle of electromagnetic induction. The permanent magnet moving coil system is one such transducer in which a relative displacement of the moving coil with respect to the magnet induces emf in the coil and the emf is proportional to the angular velocity of its motion. The coil is lighter than the moving member of the variable reluctance type and the transducer is useful for the measurement of angular velocity and displacement around its mean position. The angular displacement is limited to $\pm 70°$ from its mean position, and the system is used for measurement of torsional vibrations and torsional vibratory velocities.

The corresponding version for the linear or transverse displacements and vibratory displacements is shown in Figure 7.31. The

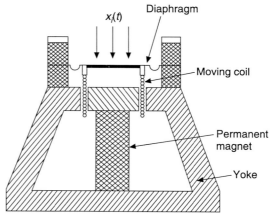

Figure 7.31   Electrodynamic vibration transducer.

system is equivalent to Hibbert's magnetic standard in which the light cylindrical hollow coil of *n* turns cuts the flux set-up in the annular gap of the magnet assembly. The soft iron yoke preserves the stability of the system and the transducer is used for the measurement of linear vibratory velocities from which the dynamic displacements of the coil are obtained around its mean position. Again, it should be noted that no current should be drawn from the coil while measuring the emf, since it entails the development of electromagnetic damping of the system. The natural frequency of the system depends on the mass of the moving system and the stiffness of the spring used to hold the coil in its mean position.

The system shown in Figure 7.31 cannot be used in moving vehicles as it requires a reference-fixed platform for the magnet assembly to be kept. In Section 5.10, a seismic system in conjunction with a relative velocity transducer is illustrated; it is intended for measurement of vibrational velocities encountered in moving vehicles and platforms. A system equivalent to the above is designed such that the mass of the seismic system carries the moving coil as shown in Figure 7.32(a). Other alternative arrangements make use of the magnet as the proof mass, with the coil remaining in fixed position. Figure 7.32(b) shows an arrangement in which the diaphragm, the linking rod, and the coil form the mass while the magnet is held rigidly in the housing. The desired lower cut-off frequency and the mass of the structure on which the transducer is to be planted decide the mass and the stiffness of the spring of the transducer.

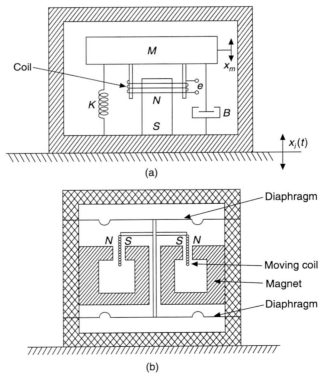

Figure 7.32 Seismic type vibration transducer: (a) With proof mass; (b) with coil and diaphragm as proof mass.

The transfer function relating the induced emf $E_o(s)$ and the input velocity $V_i(s)$ is given by

$$\frac{E_o(s)}{V_i(s)} = \frac{-BLs^2/\omega_n^2}{[s^2/\omega_n^2 + s(B_m + B^2L^2/R_t)/\omega_n^2 M + 1]}$$

and, when rewritten with B_m being negligible, this equation becomes

$$\frac{E_o(s)}{V_i(s)} = \frac{-BLs^2/\omega_n^2}{\left[\dfrac{s^2}{\omega_n^2} + s\left(\dfrac{B^2L^2}{R_t\omega_n^2 M}\right) + 1\right]} \tag{7.21}$$

where,

B = magnetic flux density in air gap, Wb/m^2
L = length of conductor (= πDn), m
D = diameter of coil, m
n = No. of turns of coil
R_t = total resistance of coil circuit, ohms
M = mass of the moving system, kg
K_s = stiffness of spring, N/m
ω_n = natural frequency, $\sqrt{(K_s/M)}$, rad/s
B_m = mechanical (viscous) damping coefficient, Nm s

A study of the transfer function reveals that the output voltage is zero when velocity is zero, and at frequencies well above its natural frequency, the transducer behaves as an ideal velocity transducer satisfying the fundamental relationship $E_o = BLV_i$. The damping present has considerable influence on the characteristic at frequencies below and around the natural frequency. In deriving the above transfer function, the self-inductance of the coil is ignored and its effect will be pronounced at very high frequencies at which the sensitivity will be lower than that indicated by the ideal relationship. The transducers are available with high sensitivity and for a lower cut-off frequency of about 3 Hz. Where necessary, oil damping is used; otherwise, electromagnetic damping is used by choosing the proper value for R_t.

7.5.4 Electrodynamic Pressure Transducer

The vibration transducer of Figure 7.31 can be modified to measure dynamic pressures and for converting sound pressure variations into voltage signals. Moving-coil microphones are not popular nowadays, but the performance of the system for dynamic pressure measurements is worth studying. The net force working on the diaphragm of area a is ΔPa, where ΔP is the pressure difference across the diaphragm. The performance of the system may be studied from the transfer function

$$\frac{E_o(s)}{\Delta P(s)} = \frac{sBLa/M}{s^2 + s(B_m + B^2L^2/R_t) + K_s/M} \tag{7.22}$$

At resonance, the response is governed by the damping term, and if damping coefficient is suitably designed, the transducer may have flat frequency response over a small range around its damped natural frequency. However, it will be seen that it does not offer much scope for wider application.

7.5.5 Electromagnetic Flowmeter

The electromagnetic flowmeter is based on Faraday's laws of electromagnetic induction. When a conducting fluid passes through a pipe of nonconducting and nonmagnetic material, it can be treated as equivalent to a set of moving parallel straight conductors lying in a plane perpendicular to direction of motion. The flow is allowed to take place in a steady magnetic field of density *B,* as shown in Figure 7.33. Thus there will be induced emf across the two electrodes as per Faraday's law of induction. The emf *E* is given by

$$E = BDV \text{ (volts)} \tag{7.23}$$

where,

B = magnetic flux density, Wb/m^2
D = diameter of the pipe, m
V = velocity of flow, m/s

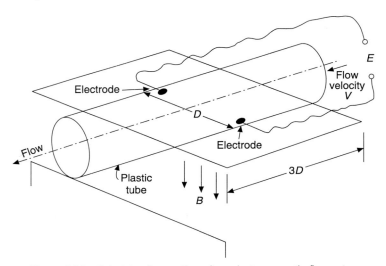

Figure 7.33 Principle of operation of an electromagnetic flowmeter.

Maximum emf is induced when the electrodes are placed across the diameter of the pipe and the direction of flow. The direction of magnetic field and the line joining the electrodes are all mutually perpendicular to each other. Since the liquid is in continuous contact with the element of conductor between the electrodes, there will be an equivalent of short-circuiting effect and thus the actual voltage across the electrodes will be less than *BDV*. If the field extends over a large area, this effect is less at the central portion of the field. The field is setup over a length of the pipe equal to thrice the diameter. It is better that a very high impedance type voltmeter is used for measuring the emf between the electrodes; otherwise, there will be a drop due to the internal resistance of the liquid. For a pipe of diameter of 0.65 cm, the tap water presents a resistance of about 7500 ohms.

Instead of a permanent magnet providing the magnetic field, ac electromagnets providing alternating magnetic field at 50 Hz are preferred since there will be no effect of polarization in the liquid. The small ac voltages can be amplified conveniently by ac amplifiers.

The electromagnetic flowmeter has several merits. It does not obstruct the flow and can therefore be used for measurement of velocity of flow of slurries and corrosive liquids. It is insensitive to viscosity, density and temperature of the fluid. It has no moving parts. Reversal of flow direction results in polarity reversal of the emf. It measures the average velocity only, and it is essential that the pipe should always be full. The fluid must have its conductivity greater than 20×10^{-6} mho/cm. It responds instantaneously to velocity changes, in the case of dc magnetic field system. The system is rugged and reliable. The sensitivity is low and the induced emf needs amplification. The system is also used to measure the velocity of liquid metals, using a stainless steel pipe.

7.6 PHOTOELECTRIC TRANSDUCERS

Transducers based on the effects of physical radiation on matter are extensively used in instrumentation. Of the entire gamut of electromagnetic radiation, optical spectrum is concerned with the range of wavelengths belonging to ultraviolet, visible and infrared radiation. The optical spectrum is mostly favoured because of the ease of emission and absorption of radiation and suitability for application in many situations. Opto-electronic devices are designed for the emission and absorption of this optical radiation. Photoelectric devices and transducers are meant for producing photoelectric effect, i.e. the effect of visible radiation. It is difficult to identify devices that are exclusively marked for application in the visible radiation with wavelengths ranging from 380 nm (violet) to 760 nm (red). Thus the choice of electro-optical and photoelectric sources and detectors is mostly influenced by the wavelength of radiation required. Some of the detectors may respond, though selectively, to the whole range of optical spectrum while some others may be useful only for visible radiation.

In this context, the definitions and terminology commonly used are recalled. The quantities and their units are different, depending on whether radiant energy or light energy is under consideration. The quantities are accordingly either radiometric or photometric. The radiant energy Q_e is expressed in ergs or Joules and the radiant flux $P(= dQ_e/dt)$ in Joules/s or watts. Other quantities are as follows:

$$\text{Radiant emittance } W = \frac{dP}{dA} \text{ W/m}^2$$

$$\text{Irradiance } H = \frac{dP}{dA} \text{ W/m}^2$$

Radiant emittance deals with sources of radiation while irradiance refers to incident radiation.

$$\text{Radiant intensity } J = \frac{dP}{d\omega} \text{ W/sr}$$

where ω is the solid angle through which flux from a point source is radiated.

$$\text{Radiance } N = \frac{dJ}{dA \cos \theta} \text{ W/sr-m}^2$$

where θ is the angle between line of slight and normal to the surface of area dA under consideration.

All the abovementioned units are known as radiometric units, while the following photometric units are used to convey the same but are limited to the energy contained within the visible region. Similarly, light intensity referring to the amount of power emitted from a source may be expressed in watts/sr, but is generally expressed in photometric units, as lumen/sr, lumen being the unit of luminous flux F.

The luminous efficacy K is used to refer to the amount of luminous flux available per unit of radiant flux and is given by

$$K = F/W \text{ lm/W}$$

$$\text{Luminous efficiency } V = \frac{K}{K_{max}}$$

$$\text{Luminous energy } Q_v = \int_{380 \text{ nm}}^{760 \text{ nm}} K(\lambda) Q_e \lambda \, d\lambda$$

where λ is the wavelength of the radiation and $K(\lambda)$ represents the luminous efficacy at the wavelength under consideration.

$$\text{Luminous emittance } L = \frac{dF}{dA} \text{ lm/m}^2$$

$$\text{Illuminance } E = \frac{dF}{dA}$$

$$\text{in lux (lm/m}^2\text{) or phot (lm/cm}^2\text{)}$$

L is used when referring to sources and E while referring to incident light.

$$\text{Luminous intensity } I = \frac{dF}{d\omega} \text{ lm/sr}$$

I is also known as candle power, expressed in candela (lm/sr = cd)

$$\text{Luminance } B = \frac{dI}{dA \cos \theta} \text{ cd/m}^2 \text{ (= nit) or cd/cm}^2 \text{ (= stilb)}$$

The radiant intensity of a source varies with wavelength, and hence for each source, spectral power density in watts per unit wavelength is plotted against wavelength. Similarly, all detectors are characterized by their selective spectral sensitivity. Sensitivity of the eye depends on the wavelength and is maximum at a wavelength of 550 nm (green).

Another way of measuring the illuminance is by specifying the number of photons emitted from incident power per unit time. The energy of the photon in Joules is given by

$$E_p = \frac{ch}{\lambda} \tag{7.24}$$

where,

h = Planck's constant (= 6.62×10^{-34} J s)
λ = wavelength of light, m
c = velocity of light (= 3×10^8 m/s)

At 555 nm, a light beam of 1 W corresponds to 2.8×10^{34} photons per second. All optical detectors are based on a quantum effect in which the photon energy is an important parameter. If the photon energy is expressed in electron volts, then the constant ch is equal to 1240 eV/nm. All electro-optical effects require that a single photon should have sufficient energy to free an electron by overcoming the bond energy or trap in a solid. Since most bond energies lie in the range of 0.5 eV to 5 eV, the minimum photon energy is within this range.

7.6.1 Photoelectric Phenomenon

If a light quantum of wavelength λ interacts with an electron bound in a metal surface, the entire quantum energy is converted into kinetic energy of the electron. The net energy with which the electron will be able to leave the metal surface is reduced because an amount of energy equal to $e\Phi$ is utilized to overcome the binding forces of the metal. Φ is the work function of the material and e is the charge of the electron. If $e\Phi$ is equal to hc/λ, then the net energy of the electron becomes zero. Hence for each material, there is a minimum wavelength λ_0, known as the threshold wavelength (= $1240/\Phi$), and for λ greater than this value, no electron will be released from this surface. The threshold wavelength of common photoelectric material surfaces is in the 600–1200 nm range. Also, the number of electrons emitted for a given number of absorbed light quanta, i.e., the quantum yield, increases with decreasing λ from zero at λ_0 in a monotonic fashion. It is this fundamental phenomenon that is known as *photoelectric effect*. The quantum yield on pure metal surfaces is of the order of 0.1% and is as high as 30% in some other materials. The photoelectric effect is quite pronounced in semiconductors. Depending on the way the photoelectric effect is recognized, the photoelectric devices are classified as (a) photoemissive, (b) photovoltaic, or (c) photoresistive (photoconductive).

The photoemissive type transducer consists of a metallic cathode and an anode in an evacuated tube. The emitted electrons are attracted towards the anode, thereby constituting an electric current proportional to the illuminance. The photocurrent depends on the wavelength of the radiation and the material of the surface. To increase the sensitivity, the tube is filled with an inert gas at a pressure of about 1 mm Hg. The vacuum and gas-filled type emissive cells are not much popular since the semiconductor type photocells have been found to possess several advantages. Only the photomultiplier type emissive photocell is still widely used for amplification of the small photocurrents. Commercial photocells of the emissive type are available with the cathodes coated with such compounds of alkali metals that respond to wavelengths of visible radiation only.

Solid-state semiconductor photoelectric devices are generally sensitive to not only optical radiation but also to X-rays, γ-rays, and other corpuscular radiation such as α- and β-rays. The process of energy absorption produces movable charges or a change of mobility of the charge carriers in the material. The effect of the optical radiation on the semiconductor may be observed as a change in either current, developed voltage or resistance. Photovoltaic cells are self-generating and are favoured for use in exposure meters. Photoresistive cells are passive and the changes in the resistance value should be measured by suitable circuitry. They are also known as light dependent resistors (LDR) and are widely used in instrumentation. Photojunction type cells are available as photodiodes and phototransistors. They are considered to function in both the photoemissive and photovoltaic modes. Their small size and fastness of response have contributed to the development of a variety of instrumentation schemes, apart from their application to detection and measurement of radiation in spectrophotometers.

The photoelectric phenomenon is not entirely a reversible one, but in certain cases, it is possible to observe the reverse effect. The light emitting diode (LED) is one in which energy is released in the form of photons, when a current flows through the junction.

In Sections 7.6.2–7.6.4, the more popular photoelectric transducers are described and their characteristics presented.

7.6.2   Photoconductive Transducers

For detection and measurement of optical radiation, photoconductive cells using cadmium sulphide as the material are described in Section 6.1.8 and the spectral response characteristics are presented in Figure 6.49. Cadmium sulphoselenide cell shows maximum response at a wavelength of 615 nm and can be used up to 1000 nm in the infrared region. Both of them show no response to wavelengths below 300 nm.

Lead telluride cells respond to the widest range of radiation, i.e. from the ultraviolet to about 5600 nm in the infrared region. They possess high sensitivity and shorter response times. Lead sulphide cells also respond to the infrared up to a wavelength of 3000 nm and are sensitive to weak sources of radiation such as from bodies at a temperature of 100°C. Their response time is about 0.1 ms.

Other commonly used materials are selenium, silicon and germanium with quantum yield of about unity. Response time of germanium is less than 10 µs and has high sensitivity.

Photoconductive materials respond to radiation over a spectral range extending from thermal radiation through infrared, visible, ultraviolet spectrum into the range of X-rays and γ-rays. Cadmium sulphide cell responds to bombardment with α-rays and β-rays. The spectral response normally falls sharply at longer wavelengths. Cooling of the cell generally improves its response to longer wavelengths.

The specific sensitivity S of the photoconductive cell is given by $S_{spec} = \Delta I/\Delta F$ when a potential difference of 1 volt is set up across the terminals of the cell. It is expressed in µA/lm. The specific sensitivity of the cells varies from 5×10^5 to 9×10^{16} A/lm-V.

The photoconductive cells are extensively used in pyrometry, and the detection of radiant heat and for quantitative spectroscopic measurements. With the help of semiconductor photoconductive cells, it has been possible to resolve the fine structure in the infrared absorption bands.

7.6.3 Photovoltaic Transducers

The photovoltaic or barrier-layer cell consists of three layers, as shown in Figure 7.34, a base plate B, a semiconductor layer S, and a thin transparent metal layer T. Proper processing produces a thin insulating barrier layer between the semiconductor and the metal layer T. If light is incident on the barrier layer, an emf is developed across the base plate and the top metal layer, with the base plate being the positive terminal. A thin film of gold or platinum is deposited on crystalline selenium with a base plate of iron. Presently, cells with silicon and germanium for semiconductor layers are available (they are also known as photojunction diodes).

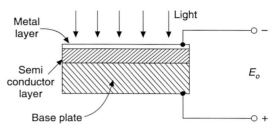

Figure 7.34 Barrier-layer photoelectric cell.

When the incident quantum of light is absorbed at the barrier layer, formation of an electron and hole-pair takes place. The electric field in the barrier layer causes separation of the carriers and hence a charge is generated between the semiconductor and the metal layer. The charge can either flow back through the barrier layer (setting up internal current) or can cause flow of current in the external circuit connected to the terminals.

Since all the energy of the current is derived from the light source and the cell acts as an energy converter, the current, though of measurable magnitude, represents very small power. The short circuit current is proportional to the area of the cell and increases linearly with illuminance as shown in Figure 7.35(a). The characteristic becomes nonlinear when (a) illuminance is high, (b) only a part of the surface is exposed to light, and (c) the load resistance is not small as compared to the internal resistance of the cell. The sensitivity is of the order of 1 mA/lm. The open circuit voltage increases in approximately logarithmic fashion, as shown in Figure 7.35(b), and is independent of the cell area. When the incident light flux is increased by 1000 times, the resistance of the cell falls by 100 times. The cell is considered as a source of current and used with external circuits of very low resistance. If amplification is desired, amplifiers with very low input impedance are to be used.

The spectral response of the selenium cell is almost similar to that of the human eye and extends from about 250 nm to 750 nm, with a maximum response at about 570 nm, as shown in Figure 7.35(c).

The response of the germanium cell is primarily in the infrared region responding to wavelengths from 200 nm (near ultraviolet) to 2000 nm, with a maximum sensitivity around 1500 nm. The silicon cell has its threshold wavelength at 1200 nm. All the above cells are also sensitive to X-rays, α- and β-rays and gamma radiation.

The generation of emf within the cell takes place within a microsecond after exposure to light, but the build-up of the output voltage is delayed by the large internal capacitance of the barrier layer system. Germanium cells are faster than selenium cells in their response.

All the above cells are affected by temperature. The voltage and the current decrease rapidly with increase in temperature.

The above cells are stable and are capable of retaining their calibration constant for reasonably long periods, if the temperature of the environment is held constant.

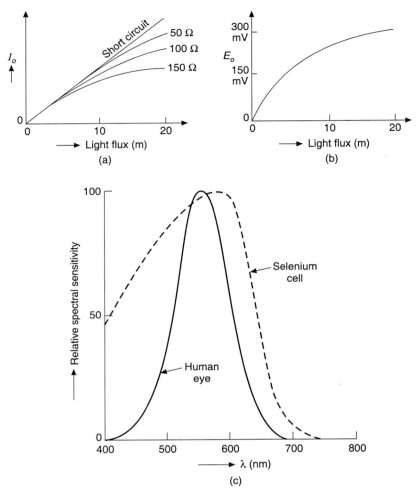

Figure 7.35 (a) Light-flux current characteristic; (b) Light-flux voltage characteristic; (c) spectral response of selenium cell.

7.6.4 Photoemissive Transducers

If a photojunction cell (described in the Section 7.6.3) is applied with a voltage E, and a load resistor is connected in series, then the photosensitive barrier layer changes the operation of the cell from that of a photovoltaic cell to that of a photodiode. It may be considered to behave as a photoemissive type transducer with the photocurrent being a function of the illuminance. As a photovoltaic cell, it has the internal current flowing from the metal layer to base plate. The cell has very low forward resistance in this direction, and hence an external source of supply E_a connected to it through a load resistance, as shown in Figure 7.36(a), results in a current flow through the load. The characteristic curves between the applied voltage and current are shown in the first quadrant of Figure 7.36(b). It is observed that the cell acts as a rectifier without

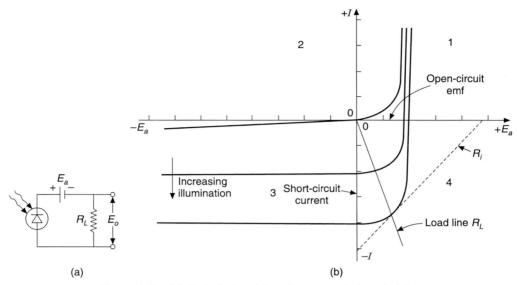

(a)   (b)

Figure 7.36   (a) Photodiode; (b) its voltage current characteristics.

illumination. The effect of light has negligible impact on the forward current. With no illumination, the current in the reverse direction is very small as in normal diodes, but the reverse current increases sharply with increase in illumination. The characteristics in the third quadrant indicate that the photodiode has its applicability for measurement of illuminance with the reverse current as a photocurrent. The short circuit current and the open circuit voltage which are obtained when the photodiode functions as a photovoltaic cell are shown in the Figure 7.36(b).

The sensitivity of the photodiodes varies from 10 to 50 mA/lm. The spectral response extends from the visible to the infrared region up to 2200 nm with the peak occurring at about 1500 nm. The rise in temperature causes increase in the dark current and photocurrent.

Photodiodes are constructed to have a much smaller area than the photojunction type. The capacitance of the cell becomes much smaller, thereby enabling higher frequency response. The construction of two types of germanium photodiodes is shown in Figure 7.37. The sensitive area is usually about 0.5 to 1.0 nm^2. A silicon solar cell consists of a thin slice of single crystal p-type silicon up to 2 cm square into which a very thin (0.5 μ) layer of n-type material is diffused.

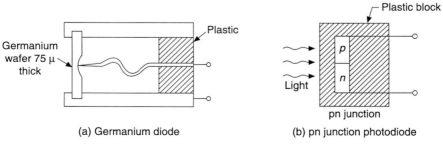

(a) Germanium diode   (b) pn junction photodiode

Figure 7.37   Germanium diodes: (a) Point contact type; (b) junction type.

Gold-doped germanium cells are used for high frequency applications possessing high spectral sensitivity in the infrared region. The build-up time and the decay time of the cells range from 10 µs to 25 µs. They are faster in response than the photoconductive cells.

The small size of the photodiodes has enabled the development of integrated circuits with the combination of photodiodes with circuits. Apart from application to sound pick-up from motion picture films, they are used to constitute a linear array of photodiodes on a single integrated circuit chip that can be used for edge and width scanning. The photodiode centres are spaced 25 µm apart, and 1024 diodes in a line give a chip of about 2.7 cm in length. The diodes in the array can be sequentially addressed at a rate up to 10 MHz so that a voltage level dependent on the light intensity falling on each diode is available in serial form. Specially controlled geometry of photodiodes allows measurement of position or null of position.

7.6.4.1 Phototransistor

Two photodiodes arranged back-to-back form a phototransistor which consists of three semiconducting layers in a p-n-p or n-p-n arrangement as shown in Figure 7.38. The exterior appearance of this transistor is the same as that of conventional transistors except for phototransistors without bases, in which only the emitter and the collector have leads. Normally the base is illuminated in phototransistors. The dark current flowing through the phototransistor not illuminated is governed by the charge carriers which reach the collector via the base. Although the base emitter is biased in the forward direction of the diode, the transistor has only little dark current. Light produces electrons and holes in pairs and the holes diffuse into the emitter and the collector. The holes entering the collector increase the collector current and the electrons induce noncompensated negative charge, thus reducing the threshold potential of the boundary between emitter and base. This allows more holes to flow from the emitter to the base. The holes passing through the base enter the collector and further increase the current.

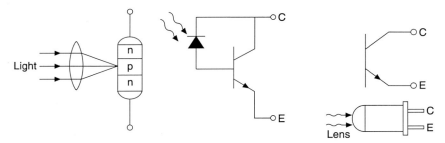

Figure 7.38   Phototransistor and its representation with a lens arrangement.

Phototransistors are mostly used in earthed emitter circuits, the role of the control base current being taken by external light signals sent to the base. The frequency response is poorer than the photodiodes. The upper frequency limit values are nearly the same as those for photoconductive cells. Their merit lies only in the amplification of the photocurrent.

7.6.4.2 Photomultiplier

The photomultiplier is, in fact, a vacuum photoemissive cell with a large number of anodes and one cathode. The merits of the cell with one anode and one cathode are (a) satisfactory stability and (b) very low threshold level for illuminance. The photocell cathode may be so chosen as to be sensitive to the visible range of frequencies. But the photocurrent for a certain illuminance is very small and it cannot be increased by raising the anode voltage beyond 20 V. Moreover, it cannot detect the lower energy photons of the red and infra-red end of the spectrum as the work function of the available materials is well above 1 eV. Hence the multiple anodes called by nodes arranged as shown in Figure 7.39 are used so that electrons released from one dynode are attracted to the neighbouring dynode because of its higher positive potential and, in that process, a larger number of secondary electrons are released. Two or three low-energy secondary electrons are emitted for each incident electron, thus producing an electron multiplying effect. The electrons from the ninth dynode are collected by the tenth and final anode and the final anode current I is a measure of the illuminance at the cathode. The ratio of the number of secondary electrons released at each dynode for each primary electron is represented by g. If n is the number of dynodes used, the net gain of the photomultiplier is given by g^n. For n value lying between 7 and 10, net gain can be obtained around 10^5 to 10^7.

Thus the sensitivity can be as high as 20 A/lm which, in other words, enables measurement of light flux less than 10^{-6} lm since the output current of photo-multiplier is limited to 10^{-5} A for continuous operation. Sensitivity can be increased by application of larger voltage; however, for constant sensitivity, highly stabilized supply voltage is essential.

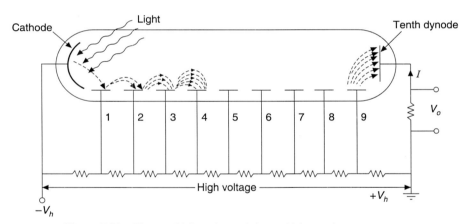

Figure 7.39 Photomultiplier tube and the multiple anode arrangement.

An end-on photomultiplier tube is provided with a semitransparent photocathode film on the end window. Light passing through the glass causes an electron to be released from the backside of the film. It is attracted by the focusing electrodes for subsequent multiplying effect.

Where necessary, further amplification may be attempted by using suitable amplifiers when measuring light flux ranging from 10^{-8} to 10^{-11} lm. At much lower levels of illuminance, the pliotomultiplier is cooled so as to minimize the dark current and noise components in the current. At light levels below 10^{-14} lm, only counting of the amplified pulses caused by the incidence of single quanta at the cathode enables the measurement. Operation of the photomultiplier at higher light levels in excess of 10^{-5} lm is not recommended as the tube may suffer fatigue effect or even permanent damage.

The response time of the photomultiplier is of the order of 10^{-8} s and it can be used to measure very short light pulses of duration less than 10^{-9} s though as a broadened output pulse. In spite of the size and fragility of the device, the photomultiplier happens to be the ultimate choice, when considering its speed of response and its sensitivity at very low levels of illuminance.

7.6.4.3 Photocounter

A schematic diagram of the photocounter is shown in Figure 7.40. The anode of the counter is a fine wire held in the centre of the cylindrical tube. A window is provided to admit ultraviolet and visible radiation into the tube. The counter is filled with a suitable gas plus a quenching agent at a low pressure of a few torr. Geiger-Muller counter for nuclear radiation measurements has similar construction, as can be seen from Section 7.7. A photoemissive layer is held opposite the window. The electrons emitted from the layer move towards the anode and cause ionization of the gas on their way to the anode. The ion current thus obtained is a measure of the illuminance. It is primarily used for the detection of ultraviolet radiation. The quantum yield is low, i.e. of the order of 10^{-3} to 10^{-4}. It is found to measure as little as 12 quanta/cm²-s, in an observation time of 10 to 12 min.

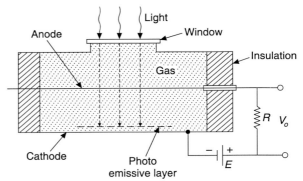

Figure 7.40 Photocounter.

7.7 IONIZATION TRANSDUCERS

Ionization transducers are based on the process of ionization of a gaseous medium contained in a chamber at low pressure. The atoms of the gas are split into ions and electrons which are drawn towards the respective electrodes to constitute current. The ionization of the gas may be effected either by fast moving electrons, exposure to high-energy radiation of X-rays and γ-rays or by bombardment of the gas by particulate matter such as α- and β-particles and neutrons. The ultraviolet region of the electromagnetic spectrum extends from 380 nm down to small wavelengths overlapping the soft X-ray region. Gamma rays have wavelengths from 0.1 nm and below, and hence they are considered to be high-energy electromagnetic waves. α-particles, β-particles, neutrons and γ-rays are emitted by radioactive materials (called isotopes), and they are known as nuclear radiation. An α-particle is a helium nucleus having a positive charge of 2 and a relative mass of 4. A β-particle is an electron having a relative mass of 0.000549. The neutron has zero charge and a relative mass of unity. Alpha particles are rather readily absorbed in many materials, while β-particles are usually more penetrating. Neutrons and gamma rays are the most penetrating types of radiation. The process of interaction of these nuclear radiations with materials is complex, but our interest in Sections 7.7.1–7.7.6 centres on the detection and measurement of the nuclear radiation and application of the ionization process for measurement of certain physical quantities such as thickness, displacement and pressure. Nuclear radiation detectors are normally known as ionization chambers, Geiger counters or scintilation counters and are based on the interaction of the radiation with the detecting system that produces an ionization process. A variety of nucleonic instruments are developed in spite of the hazards posed by the radioactive materials. In most of the instruments, the sources are shielded heavily, thereby making them very safe for the operating personnel. The nucleonic instruments are developed for application to situations where other methods fail as also for effecting noncontact and noninvasive measurements.

Though the utilization of the ionization effect does not yield a self-generating or an active transducer, the ionization transducers are described here to present a state of continuity for the detection of radiation.

7.7.1 Ionization Vacuum Gauges

The vacuum gauge with heated cathode resembles a thermionic valve as shown in Figure 7.41(a). The chamber is connected to the vacuum source, the pressure of which is below 1 μm when the mean free path of the electrons becomes fairly long. The anode is held at a negative potential of about 25 V, whereas the grid is at a positive potential of 100 V. Some of the electrons emitted by the cathode strike the gas molecules during their transit towards the grid. The secondary electrons thus released are, in fact, a small fraction of the electrons emitted from the cathode, as the pressure of the gas is very low. The number of positive ions formed is directly proportional to the gas pressure and also proportional to the number of primary electrons emitted. The positive ions are collected by the negatively charged anode and the electrons by the grid. The number of primary electrons and hence the grid current I_g is generally held constant and so the anode current I_a is a measure of the number of gas molecules per unit volume and hence of the pressure. The sensitivity S of the gauge is given by

$$S = \frac{I_a}{P I_g} \tag{7.25}$$

where, P is the absolute pressure of the gas.

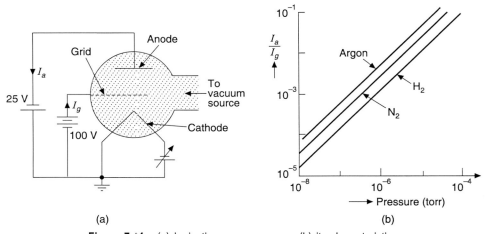

| (a) | (b) |

Figure 7.41   (a) Ionization vacuum gauge; (b) its characteristics.

The sensitivity varies with the construction of the gauge and is different for different gases as shown in Figure 7.41(b). The gauge is used for the measurement of pressure over a very wide range from 1 torr to 10^{-10} torr. The merits of the gauge are linearity in the range of pressure and high speed of response. It is the most accurate of all vacuum gauges and the calibration constant and sensitivity remain constant for some time. The drawback of the gauge is that the filament burns out if accidentally the pressure of gas becomes higher than 1 torr. The life of the cathode and gauge is affected by the temperature at which the cathode operates and hence the gauge needs calibration at regular intervals.

The cold cathode gauges overcome the problems associated with the high temperature filament. But the accuracy and linearity are poorer than the hot cathode gauge. The range of pressure over which cold cathode gauge is used, is only from 10^{-5} to 10^{-3} torr. The gauge consists of two electrodes only, a cold cathode and an anode held at a high potential of about 1–3 kV, as shown in Figure 7.42(a). Due to the high electric field between the electrodes ionization of gas molecules takes place and the ionization current, I_a is a measure of the low gas pressure. I_a varies approximately linearly with pressure only in the range of 10^{-5} to 10^{-3} torr as shown in Figure 7.42(b). Different electrode arrangements exist but with no improvement in linearity.

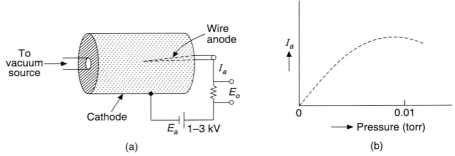

Figure 7.42 (a) Cold-cathode vacuum gauge; (b) its characteristics.

The Phillips–Penning ionization gauge is a modification of the cold cathode gauge yielding much larger anode current. The constructional details are shown in Figure 7.43. The anode A is held at a potential of about 2–4 kV and is kept between two outer cathodes. A magnetic field is applied in a plane perpendicular to the electrodes to maintain the discharge at lower pressures down to 10^{-5} torr. The electron traverses up and down inside the anode ring before striking it due to the presence of the magnetic field, and thus the pathlength of the electron becomes longer, enabling more collisions with gas molecules for maintaining the discharge. The discharge current is not linear with pressure for most gases and typical curves for common gases are shown in Figure 7.44. The gauges are applicable for pressures ranging from 10^{-2} to 10^{-5} torr. Various modifications of the gauge have resulted in extending the range up to 10^{-12} torr.

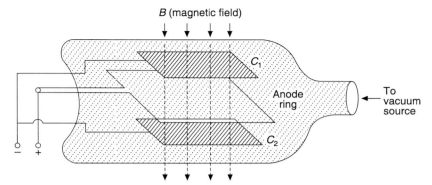

Figure 7.43   Phillips–Penning ionization vacuum gauge.

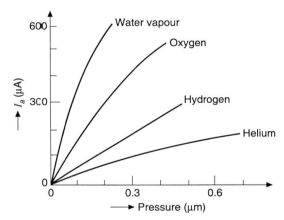

Figure 7.44   Characteristics of Phillips–Penning vacuum gauge.

7.7.2 Ionization Displacement Transducer

Ionization displacement transducers are based on the ionization of a gas when exposed to intense electric field of a high frequency of about 0.5–1 MHz. The gas is enclosed inside a small tube at a pressure of about 10 torr, and the tube is provided with two electrodes, as shown in Figure 7.45(a). Under the intense electric field, a glow discharge occurs in the space and the electrodes serve to indicate the potential difference due to the discharge. The tube is known as glow discharge tube. Their potential is determined by the space potential of the plasma surrounding each electrode and by the radio frequency potential induced due to their capacitative coupling with the external plates. The potential difference E_0 is zero when the tube is lying at the centre and symmetrically with respect to the plates, and is proportional to x when displaced from the central position. The dc voltage thus developed changes polarity when moving from one side to the other side of the central position. The characteristic between displacement and E is linear over a large part of the range about the centre as shown in Figure 7.45(b), and the sensitivity can be as high as 1 kV per mm of displacement, depending on the design of the system.

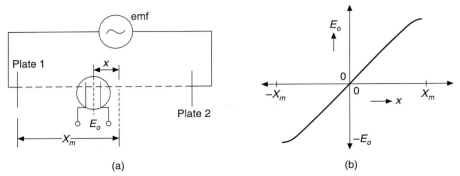

(a)   (b)

Figure 7.45   (a) Ionization displacement transducer; (b) its characteristics.

If the range of displacement is about 1 cm, the above system requires very large voltage across the plates. So a modified version, consisting of a long tube provided with a movable metal ring placed over it, is developed. The radio frequency generator is connected to the electrodes of the tube, as shown in Figure 7.46(a). A discharge takes place between the ring and each electrode. The direct voltage E_0 across the electrodes of the tube is a measure of the position of the ring, and is proportional to x over a major length of the tube (see Figure 7.46b). The sensitivity is of the order of 1 to 10 V per mm.

Both the above displacement transducers are suitable for measurements of steady displacements as well as dynamic displacements of frequencies up to 50 kHz, with an accuracy of ±0.1% and linearity of ±0.1% over a range up to 15 cm.

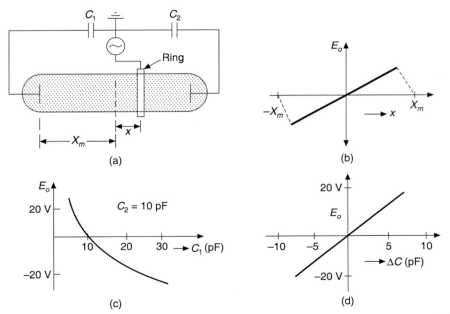

(a)   (b)

(c)   (d)

Figure 7.46   (a) Ionization displacement transducer; (b) characteristics of (a); (c) characteristics of (a) with C_1 on displacement transducer; (d) characteristic of (a) with C_1 and C_2 as push–pull displacement transducer.

The above arrangement can also be used for measurement of displacements, treating one of the two capacitors shown in Figure 7.46(a) as a capacitive displacement transducer. The ring is held centrally and the value of C_2 is fixed at, say, 10 pF. The variation of C_1 due to displacement gives rise to a variation of E_0 as illustrated in Figure 7.46(c).

If C_1 and C_2 are combined to constitute a differential capacitor with its central plate being subjected to displacements, then the E_0 variation due to capacitance variation of each half is shown in Figure 7.46(d). Unless displacement is very small, E_0 will not be proportional to displacement.

The output impedance of all the above types of ionization transducers is very high (of the order of 1 MΩ), and hence suitable signal processing circuits or voltmeters are to be used with the transducers, although the sensitivity is sufficiently high.

7.7.3 Nuclear Radiation Transducers

As already seen, ionization vacuum gauges and displacement transducers consist of an ionization chamber or a tube in which ionization of the enclosed gas is brought about by means other than a nuclear radiation. It can be seen that the output current is a function of the applied voltage in the case of vacuum gauges and in the displacement transducers, with the applied voltage adjusted to result in corona discharge.

Thus, when the effects of nuclear radiation on a gas enclosed in a chamber are under consideration, the influence of the applied voltage on the behaviour of the gas should be properly understood. The nuclear radiation may be detected and measured in two ways: by recognizing the total effect of radiation or by identifying the number of interactions of the radiation on the gas. The first type of measurement involves the mean energy level of the irradiation, whereas the second type is concerned with the counting process. Ionization chamber so far described is meant for energy-level measurement, and the applied voltage is held at far lower values, yielding an average anode current for measurement. As the voltage is increased, a stage is reached when counting of the interactions becomes possible. Geiger Müller counter is typically used for nuclear counting operations. For a clear understanding of the process, the operation of the chamber or tube with increasing voltages is divided into four regions, as described now.

7.7.3.1 Ionization Chamber

When ionization chamber is used for the detection and measurement of nuclear radiation, the chamber is normally filled with a suitable gas such as Argon at atmospheric pressure and sometimes at higher pressure to increase the absorption for incident high-energy particles. Air, or hydrogen or nitrogen is also used. The electrode system may consist of either two parallel plates, as shown in Figure 7.47(a), or a hollow cylinder with a tungsten wire at the centre serving as the anode. An electric field is set up between the electrodes by application of dc voltage. When radiation is incident on the chamber, the gas is ionized, and the electrons and ions thus formed are attracted towards the respective electrodes, thereby constituting the output current. For small applied voltage the current is very small because the positively charged ions move slowly and some of them will recombine with electrons. As the voltage is increased, the ions move faster and the current increases until a condition is reached when all the ions and

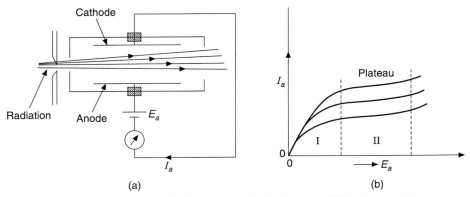

Figure 7.47 (a) Ionization chamber as radiation detector; (b) its characteristics.

electrons are collected at the electrodes before they have a chance to recombine. The two regions are shown in Figure 7.47(b) and it will be observed that the current is constant in region II and is independent of the applied voltage. The length of the plateau depends on the nature of gas inside the chamber and on the specific ionization S (defined as the number of ions formed per centimetre of length and per atmosphere of pressure), also known as ionization density. For a constant intensity radiation or for a steady flux N_0 of ionizing particles arriving each second into the space of the chamber, the output or the ionization current is given by

$$I_s = N_0 SLPe \tag{7.26}$$

where,

L = active length of the chamber, cm
P = pressure of the gas in the chamber, atmp
e = charge of an electron

For higher-intensity radiation, the value of I_s is larger. The ionization current is very small in magnitude and of the order of 10^{-10} A. It is read on a galvanometer, or the voltage drop due to it across a 1-MΩ load resistor is amplified by special amplifiers.

The shape and dimensions of the electrodes are of importance as a large volume is required for high sensitivity, and small electrode spacing develops intense electric field. The ionization chamber may also be used as a pulse-type transducer for the measurement of the number and energies of the particles.

7.7.3.2 *Proportional Counter*

If the voltage applied is increased further, the output pulse voltage rises as shown in region III of Figure 7.48. Due to the larger mobility and higher acceleration of the primary electrons, secondary ionization takes place. This phenomenon is known as gas multiplication or amplification. If the critical electric field for this effect occurs n mean free paths, away from the central wire anode, a gas amplification factor A of 2^n is obtained. As the applied voltage increases, A too increases in magnitude, and can reach values from unity at the end of region II to 10^4 at the end of region III. In this range, the output voltage pulse is proportional to the number of ion-pairs formed initially by the incident-ionizing particle. A high-intensity α-particle gives larger A than a β-particle and A depends, of course, on the nature of gas inside and the geometry of the chamber and its electrode system. All the positive ions excepting the few primary ones are produced close to the anode wire: the output pulse caused by the subsequent movement of the ions formed in the avalanche has an amplitude proportional to the primary ionization and independent of the particle track in the counter. The output pulse remains truly proportional to the number of ion-pairs initially formed over the first half of region III, and the system is used as a proportional counter. The system is thus useful for the accurate determination of the amount of energy absorbed from an ionizing particle by the gas of the chamber, so long as the total number of ion pairs in the avalanche is less than about 10^7. Above this value, space charge effects destroy the true proportionality. During the second-half of region III, the proportionality becomes limited and therefore, it is not of much consequence for measurement.

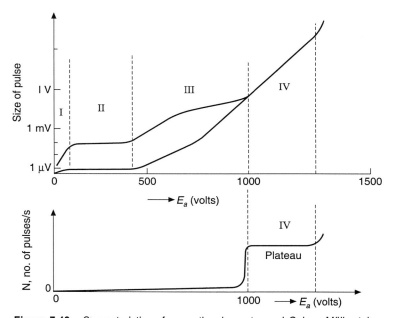

Figure 7.48 Characteristics of proportional counter and Geiger–Müller tube.

The gas chamber operating in region II (ionization chamber) region III (proportional counter), and region IV (Geiger–Müller counter) is considered as a counter-detector responding separately and producing an electrical discharge for each individual particle or quantum detected. In region II, ionization chamber is essentially used for the measurement of mean current which is proportional to the rate of energy absorption and, in general, the particle arrival rate. Pulse ionization chambers are useful when specific ionization is as large as in the case of α-particle incidence. In such cases, the source of radiation is located inside the chamber. Fast counting is not possible with ionization chambers.

7.7.3.3 Geiger–Müller Counter

A study of Figure 7.48 reveals the fact that the amplitude of the pulse increases with applied voltage except in region II. It is also observed that the pulse rate, i.e., the number of pulses per unit time for a given constant primary ionization, increases with applied voltage except in region IV which is known as the *Geiger region*. The characteristic property of a counter operated in this region is the appearance of output pulses which are all of the same size, independent of the number of ions formed initially by an incident-ionizing particle or radiation. In this region, a second phenomenon occurs before the electrons and ions formed in the avalanche are collected at the electrodes. Some of the electrons and ions recombine, resulting in the emission of light. The photons thus emitted have sufficiently high energy level, thereby releasing new electrons from the atoms of the gas due to photoionization. This is known as *Townsend discharge* which occurs in the vicinity of the central electrode. The discharge spreads along the wire until the entire anode is surrounded by a narrow cylinder of ions and the discharge becomes self-sustaining. It ceases when the positive ion sheath around the anode moves towards the cathode and virtually increases the effective radius of the anode. The consequent fall in the electric field intensity quenches the discharge.

When ions reach the cathode, they may liberate new electrons which may reignite new discharges. To prevent recurrence, a quenching mechanism is provided by means of a large resistor R in the circuit or other electronic quenching circuits. Self-quenching counters use an admixture of a poly-atomic gas (methane or alcohol vapour).

The counter operated in the Geiger region is known as *Geiger–Müller counter*, and the counting rate is independent of the applied voltage. Other merits are that it can be used to detect different types of radiation and that it is very sensitive. Output pulse amplitudes may be up to 100 V. The counter tube may be designed in various shapes and sizes as shown in Figure 7.49 and are provided with suitable windows. The counter shown in the figure is primarily used for γ-radiation, and if the walls are thin, for the more energetic β-radiation. The end-on type counters are used for the counting of α- and β-rays and soft X-rays, which can penetrate the thin windows made of mica or mylar.

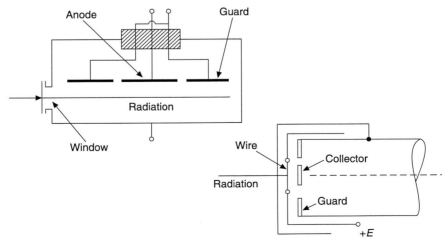

Figure 7.49 Two types of Geiger–Müller counter tubes.

When the incident particle causes a discharge or pulse, there is a time delay before the tube can detect another particle and register another pulse. This delay is approximately the time required to recharge the electrodes and establish a new space charge in the gas. Thus the counting rate is limited to 10^4 counts per second.

7.7.3.4 Scintillation Counters

The scintillation counter is the most versatile of the conventional devices explained so far for the measurement of nuclear radiation. In a scintillation counter, the particle energy is consumed in a phosphor to produce photons which are in turn detected by the photosensitive cathode of a photomultiplier tube, as shown in Figure 7.50. The phosphor material used depends on the application; however, it has to be matched with the spectral sensitivity characteristic of the photocathode material. For α-particles, a thin layer of phosphor is enough, but for γ-rays, thick blocks are necessary. Thallium-activated potassium iodide is used for α-, β- and γ-radiation, and nickel-killed zinc sulphide is used for α-rays only. Anthracene and naphthalene are useful for β- and γ-rays. Scintillation counters have the advantage of high counting efficiency of nearly 100%, whereas it is far less in the case of Geiger–Müller counters.

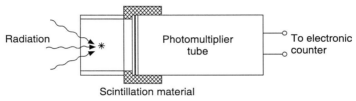

Figure 7.50   A scintillation counter.

7.7.3.5 Solid-state Transducers for Radiation

The solid-state transducer consists of a slice of a crystal such as silver chloride on which thin metal electrodes are deposited as shown in Figure 7.51. The applied voltage E_a causes an intense electric field of the order of several kV per cm in the crystal. Radiation incident on the crystal will release free electrons in the vicinity of the path in the same way as happens in the case of a gas. Accelerated electrons are collected by the anode, thus forming a current and a voltage pulse. If an ionizing particle releases n_0 electrons in the vicinity of the negative electrode, the migration of these electrons develops the voltage pulse with its magnitude proportional to n_0. If the crystal is thinner, the pulse

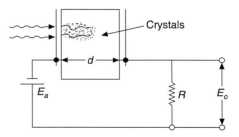

Figure 7.51   Solid state radiation detector.

height is more. The pulse rises first with applied voltage and then becomes nearly constant. The characteristic depends on the material. Other common materials are silver bromide and thallium bromide–thallium iodide at the temperature of liquid air and diamond and cadmium sulphide at room temperature.

The transducers can be used for measuring the energy of the incident particle and for counting the number of incident particles.

The resolving time of the solid state counter is considerably shorter than that of the ionization chamber or Geiger counter. The resolving time is about 0.2 μs for silver halides and is much less for diamond counter.

7.7.4 Radioactive Vacuum Gauge

Radioactive vacuum gauge is a cold cathode gauge, but it is based on the effect of ionization of the gas, whose pressure is under measurement, by α-rays emitted from a radioactive source kept inside the gauge. The arrangement is shown in Figure 7.52(a) and it is essentially the same as the ionization chamber except that the walls must be of sufficient thickness and of a material that no radiation leaks out. The number of ions formed is proportional to the gas pressure as long as the range of the α-rays exceeds the dimensions of the chamber. The ion current does not increase any further when the α-particles emitted are totally absorbed by the gas inside. The ion current is proportional to the pressure and the range of pressure is the largest with this gauge. It can be used from 10^{-4} torr to 10^{-3} torr. The input-output characteristic is different for different gauges and for some gases the approximate characteristics are shown in Figure 7.52(b). The radioactive decay of the radiation source limits the lifetime of the gauge and most of them have a lifetime of about five years. Hence the calibration need not be done as often as is the case with other gauges.

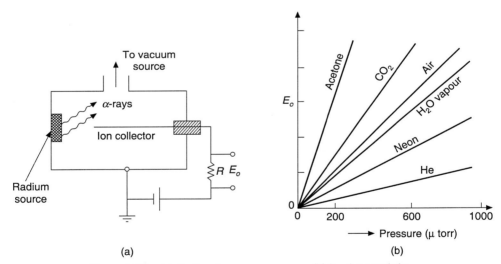

(a)   (b)

Figure 7.52   (a) Radioactive vacuum gauge; (b) its characteristics.

7.7.5 Radioactive Thickness Gauge

The thickness gauge consists of a shielded source of nuclear radiation and a radiation detector of either an ionization chamber, a Geiger counter or a scintillation counter. The source of radiation is shielded except in the direction required for absorption or penetration through the material whose thickness is under measurement.

If a β-, γ- or X-ray beam having an initial intensity I_0 penetrates an absorbing medium of thickness t, it will emerge with the intensity I given by

$$I = I_0 \exp (-\mu t) \tag{7.27}$$

where, μ is defined as the linear absorption or attenuation coefficient. The coefficient is proportional to the density ρ of the material and hence μt may be also written as

$$\mu t = k\rho t = k\frac{m}{a} \tag{7.28}$$

where,

m = mass of the material

a = area of surface exposed to radiation

Since μ changes with density, the measurement of I (with I_0 being constant) should be corrected for temperature (or pressure) variation of the material. The quantity μ/ρ is known as *mass absorption coefficient*.

Gamma and X-rays are highly penetrating and are used for heavy metals and thick specimens. Beta particles are much less penetrating and hence suitable for measurements of thickness of metallic foils and thin deposits of metals on paper, rubber or plastics. The α-rays are used only for very thin foils of a few microns thickness. Gamma and X-rays are normally composed of radiation of different wavelengths and hence thickness gauges have to be calibrated for each radiation and for each material. The primary advantage of such thickness gauges is that they are fast and need not be calibrated everyday. No mechanical contact between the gauge and test piece is necessary.

The arrangement for measurement of thickness by absorption method is shown in Figure 7.53(a). The attenuation characteristics of all radiations except those for α-rays follow the pattern of variation with thickness as shown in Figure 7.53(b) while for α-rays it is given in Figure 7.53(c). The intensity of the α-rays remains unchanged up to a minimum thickness of R for air or any other gas and then falls sharply. So, the introduction of a foil reduces the value of R by ΔR as shown, which, in other words, means that the film in its thickness is equivalent to that of the gas layer of thickness ΔR. Thicker foil of varying thickness results in a characteristic of a less rapid fall.

The value of R is about 5 cm for dry air at normal pressure and at 25°C, when the source of α-rays is polonium. On-line measurements of thickness of polystyrene and such others having 5–25 μm thickness can be measured with an accuracy of 2–4%. A higher accuracy in measurement is possible using the other radiations.

The sources of nuclear radiation are the radioactive isotopes which are the members of a family of nuclides. Isotopes of an element are characterized by different atomic weights due to the differences in the number of neutrons present. Some of the isotopes are unstable and so they undergo decay, during which process the nucleus changes to a different form by emitting particles or rays. The rate of decay is proportional to the number of nuclei, n, present at any instant and is given by

$$n = n_0 \exp (-\alpha t) \tag{7.29}$$

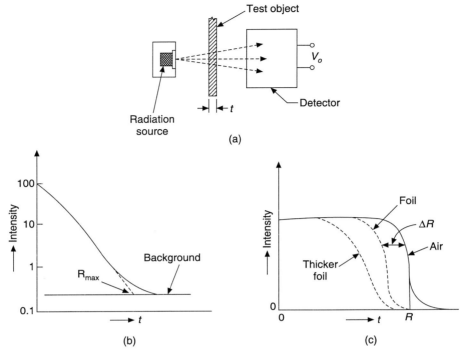

Figure 7.53   (a) Radioactive thickness gauge; (b) and (c) attenuation characteristics.

where n_0 = the number of nuclei at time $t = 0$ and α is a constant. Half-life of an isotope is the time by which n/n_0 becomes half. For industrial instrumentation, the commonly used sources are cesium-134 and cobalt-60; the former has a half-life of 2–3 years while the latter has 3–5 years. The heavy, naturally occurring radioactive elements such as thorium and uranium emit α-particles. They have very high energies of the order of 10 MeV and high ionizing power. But they have only slight penetrating power, being stopped by thin foils of solid materials, and they can penetrate only 5–7 cm of air. A 5-MeV α-particle which is stopped in 3.5 cm of air produces about 25,000 ion-pairs per cm of travel, whereas a 0.5 MeV β-particle produces about 60 ion-pairs per cm of its path. Gamma rays are high energy photons, having much greater penetrating power than that of either α- or β-particle, but the ionizing power is much less.

Each ion-pair formed during the passage of radiation through air represents an average loss of about 35 eV. The number of ion-pairs formed per cm of travel is known as specific ionization. One roentgen is the amount of radiation that will, on passing through 1 g of air under standard conditions, produce 1.6×10^{12} ion-pairs/g of air.

The strength of radioactive material is expressed in curie, its activity producing 3.70×10^{10} disintegrations per second. The unit of specific gamma emission is the roentgen per millicurie hour at one cm. The maximum thickness of a material in which the entire radiation is absorbed is known as the *range of the particle*, as shown in Figure 7.53(b).

For measurement of thickness by absorption method, a suitable source and a detector are connected as shown in Figure 7.53(a) with the test material held between them. Cobalt-60 and cesium-134 can be used as sources for γ- and X-rays, whereas strontium-90 and cesium-137 can be used for β-gauges, both having half-life of more than 20 years. Beta gauges can be used for aluminium sheets of 4–5 mm thickness.

Figure 7.54 shows the method of back-scatter applicable for cases where there is access on one side only. The amount of radiation incident on the detector varies with the attenuation in the sheet and hence with its thickness. The test sheet is kept in contact with a backing material of different atomic number. It should be noted that the amount of back-scattered radiation from the test sheet and the backing material depends on the amount of scattering and absorption of radiation in the test sheet and the backing material, and varies with the thickness of test sheet.

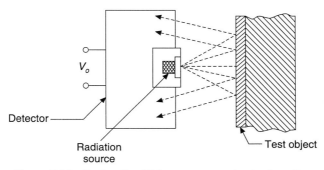

Figure 7.54   Radioactive thickness gauge using back-scatter.

All thickness gauges require calibration and the calibration characteristics are mostly nonlinear.

7.7.6   Radioactive Level Gauges

The level gauge has a column of liquid between the radioactive source and detector. The length of the path determines the intensity of radiation received by the detector. As is the case with sheet materials, the characteristic is nonlinear. Where linearity is required, a good number of strip sources and strip detectors may be located along the sides but on opposite faces so that the total output of the detectors when added makes up the output signal.

7.8   DIGITAL TRANSDUCERS

Transducers dealt with so far are analog transducers whose output signals are in analog form. The ease and versatility provided by digital signal processing circuits and digital computers necessitates the development of digital transducers providing digital output signals directly. As there are only a few such digital transducers, the analog outputs of analog transducers are converted into digital signals using analog-to-digital converters. With the increasing application of digital computers, digital transducers that are compatible with the digital nature of the computer are under development. Direct digital transducers provide output signals in the form of rectangular pulses of constant duration and amplitude, the presence or absence of which in its time slot is taken to stand for either 1's or 0's. However, transducers are treated as digital type, if they provide pulses whose pulse rate is counted. Similarly, transducers whose output signals are sinusoidal and the frequency of which is related to measurand are considered to be digital type when working in combination with digital frequency measuring system. Such transducer systems may be treated as indirect digital type. A few direct and indirect digital transducers are briefly described in the following sections.

7.8.1   Digital Displacement Transducers

One of the direct digital transducers is the digital encoder for linear and angular displacements. It is also known as linear or angular digital encoder (LDE or ADE). Such transducers are available in different sizes with differing resolution and accuracy. Basically, they are divided into two types: incremental and absolute encoders. Incremental encoders require a counting system which adds increments of pulses generated by an encoder, a sensing system, and some datum from which increments are added or subtracted. Absolute encoders present a digital readout for each angular position and do not require a datum.

All encoders require a sensing system of either the contacting type using brushes or the noncontacting optical technique.

The encoders shown in Figures 7.55 and 7.56 consist of two distinct regions signifying the two logic level signals, 0 and 1. The linear encoder of Figure 7.55 for the contacting type has a pattern of metallic areas on a matrix of nonconducting areas. All the metallic areas get connected together and energized through a fixed brush that rests on a continuous track and is in contact for all positions. The encoder shown has four tracks, resulting in digital output in four bits. The length of the encoder is 16 times the length of traverse required for a change of the digit in the least significant bit. If a resolution of 0.2 mm is desired in the measurement, a maximum traverse of the encoder in the direction shown is 3 mm. The brush makes contact over an area of each track much less than 0.2 × 0.2 mm. As the scale of encoder moves under the four tracks, the respective lamp circuits are connected or broken so that the position of scale is read out by noting the lamps under glow. Thus, 0000 stands for no flow of any lamp, and 1111 for glow of all lamps. If travel longer than 3 mm is desired, the number of tracks must be increased. If a higher resolution is needed, the number of tracks must be increased, in which case, much finer brushes are required. The limitation is generally due to the sensing or scanning system.

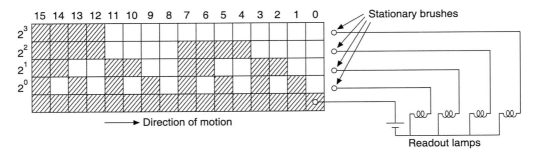

Figure 7.55   Linear digital encoder (LDE).

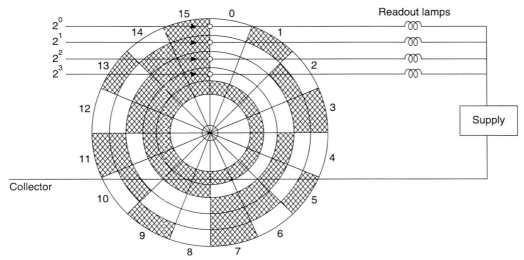

Figure 7.56 Angular digital encoder (ADE).

The disadvantages of brush pick-ups are normally associated with brush contact, brush bounce, and brush friction. Due to wear of brushes and contact surface of the encoder, life of the encoder is reduced. Also, arcing as brushes move across segments and mechanical vibrations are some of the problems in operation. The scales and discs shown in Figures 7.55 and 7.56 are encoders providing digital outputs in four bits. The angular digital encoder of Figure 7.56 is also known as shaft angle encoder and is normally meant for a total angular displacement of 360°. Both the encoders shown are absolute encoders. Discs with up to 10 tracks each, having a resolution of 1 in 1024 are available. However, by using more than one disc and having internal gearing, overall resolutions up to 1 in 10^5 can be obtained.

Linear digital encoders (LDE) may also be obtained by converting linear motion into rotary motion through a rack and pinion or some such arrangement and using the shaft-angle encoders. The accuracy of these systems depends on the linearity of the linear-to-rotary converter. Simple arrangements using a pulley or a cable are shown in Figure 7.57. With all such systems, the precision of the mechanical components sets the limit on the accuracy of measurement.

Incremental encoders are single track discs or scales provided with alternating conducting and nonconducting areas as shown in Figure 7.58. All incremental encoders are designed to generate a fixed number of pulses for each unit of angular or linear displacement of the encoder. The accuracy thus depends largely on the encoder disc or scale and the associated mechanical assembly used. The sensing systems for the incremental encoder may be the same as those adopted for the absolute encoders. Incremental encoders are used along with suitable signal processing circuits to count the pulses generated by the encoder and convert the same into a digital output in the required code. All incremental encoders have to indicate the datum point to the signal processing circuits. The direction of motion must be sensed and used to provide the corresponding output.

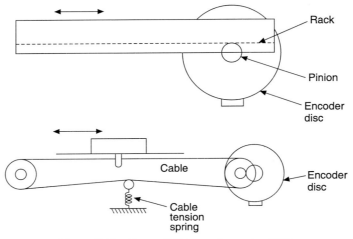

Figure 7.57   Linear digital encoders using ADE.

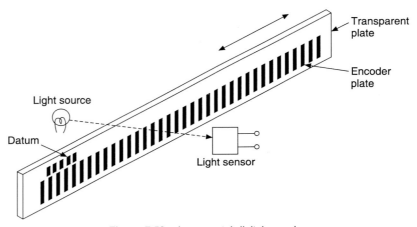

Figure 7.58   Incremental digital encoder.

7.8.1.1 Optical Encoders

The majority of shaft-angle encoders use noncontacting type sensing systems so as to make the measurements free from the problems of the brush contact, stated above. Optical encoders use optical and photoelectric sensing systems. The linear and angular encoders have a pattern of transparent and opaque areas corresponding to the conducting and nonconducting areas respectively of the contacting brush type. The sensing system consists of light sources, each provided with a focusing lens and an equal number of photoelectric devices, and receiving the light beam from its corresponding light source. The light sources are kept on one side and the photosensors on the other side of the encoder as shown in Figure 7.59(a). The sharpness of focusing and the special techniques adopted in the manufacture of the encoders have enabled achievement of an accuracy of better than 1 in 10^8. Most of the optical encoders are produced

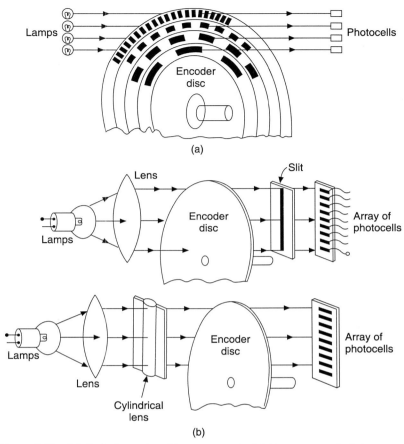

Figure 7.59 (a) Optical encoder; (b) arrangement of light sources and photosensors.

photographically, using a contact printing method from a master disc or scale, accurately machined to the requirements. The transition between a transparent and an opaque area must be sharp and well defined so that there is less noise in the sensors during the transition between the logic '0' and '1' and sure decision. In this respect, optical encoders are specially favoured for application in high-accuracy systems.

Instead of having a large number of light sources, a single lamp and a lens is used as shown in Figure 7.59(b) to flood the encoder on one side, while the sensors receive light through a narrow slit located accurately with respect to the reference line. Alternatively, a cylindrical lens produces a single line beam which is so projected on to the reference line of the disc as to be incident on the sensors, after passing through the disc. Using a fine filament lamp, up to 5000 bits can be detected on discs of about 15 cm diameter. The silicon type photovoltaic cells, developing an output voltage of 20–40 mV in a load resistor of 10 kΩ, are used as sensors. Amplifiers and other processing circuits develop the logic level signals in required code format. Light-emitting diodes (LED) may be used instead of filament lamps but their light output efficiency is far less than that of filament lamps. LEDs have longer life and greater immunity against mechanical disturbances.

7.8.1.2 *Magnetic Encoders*

The conducting areas of the contacting type encoders are replaced by a coating of magnetic material powder as is done in the case of magnetic tape, or it may be raised portions of suitably magnetized material while non-conducting portions are at a lower level with no magnetization. The sensing heads in each case consist of toroidal cores, each provided with two coils, and they are placed close to the pattern of the encoder, but not in contact with it. The proximity of the magnetized and raised portion saturates the core, and the coil system is utilized to enable detection by generating suitable output signals. One of the coils is known as *reading coil* and the other as *interrogate coil*. The reading coil develops output signals, only when the interrogate coil is energized with a constant voltage signal of 200 kHz, as shown in Figure 7.60. The output signal is low if the toroidal core is over the magnetized portion, as the core is saturated. The output voltage is high when the core is over the nonmagnetized portion as the toroidal core and coils operate as a transformer. Thus the output voltages of the reading coil are of 200 kHz but of amplitudes governed by the presence or absence of the magnetized portion of the encoder. Low output signal signifies logic level '0' and high output signal, '1'. The signals require demodulation and shaping so as to be distinct in their status as 0 or 1.

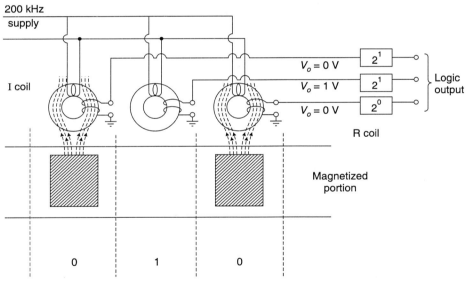

Figure 7.60 Magnetic encoder.

Magnetic encoders are seen to be more complex than the contacting type or the optical encoders. But they are more reliable in operation than the contacting-type encoders.

When used as an incremental encoder, a series of pulses are produced which are added and indicated by an external binary counter. The resolution of the encoder is limited by the gap length of the toroidal core head, which may typically be 0.01 mm. For sensing direction of motion, four heads are arranged to read consecutively phased signals that are decoded by external logic circuits to provide 'clockwise' pulses on one output line and 'anticlockwise' pulses on a second output line. A reversible counter integrates the total shaft rotation at any instant.

7.8.1.3 Scan Problems

All the encoders so far described have patterns related to binary output. The reliability of indication is naturally limited by the ambiguity of indication whenever brushes or sensors change from regions pertaining to '1' to regions pertaining to '0' or vice versa. For example, an encoder having a binary output 0111 has to indicate 1000 on advancing to the next increment. Such a change in reading necessitates the accurate and simultaneous change of all the four brushes. It may be difficult to achieve such a change in many encoders, and hence there is a chance for a large error in indication.

To minimize errors in this regard, additional brushes are located on the disc, one displaced ahead (leading) and another behind (lagging); the required but least significant bit is retained as a single brush and is located in the position to be measured. Each of the other tracks have a pair of brushes, one lagging and the other leading the position to be measured as shown in Figure 7.61. The gaps between the pairs of brushes are x, $2x$, $4x$, $8x$ ..., where x is the width of one LSB increment of a linear encoder. The arrangement goes with an associated logic and the logic circuitry has to identify that when the true output brush of a pair on a track reads '1'; then the lagging brush of the next higher-order track is to be read as the true output. If the true output from a given track is a '0', then it is the lead brush of the next higher-order track which should be read as the true output. Thus the changeover of the true output from all tracks occurs only when there is a change on the LSB track. The logic circuitry selects that brush on each track which has to be selected as representing the true output. This V-scan is adaptable for digital computer interrogation as the brush selection logic can be part of the computer program.

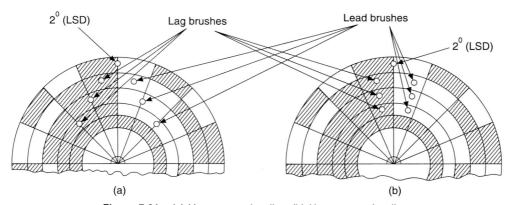

Figure 7.61   (a) V-scan encoder disc; (b) U-scan encoder disc.

For the position shown in Fig. 7.61(a), the LSB brush is reading '0' and the leading brush is read to give an '0', and so on for successive brushes. Thus although the LSB brush is on a point of transition, the total encoder output is stable on all 0's or 1's without ambiguity. Figure 7.62 shows the logic circuit for the V-scan encoder.

An alternative to V-scan is U-scan in which the brushes are spaced one least significant digit apart, as shown in Fig. 7.61(b). The tracks for each bit are electrically insulated from each other. A common collector track is used with each set of tracks. Figure 7.61(b) shows the two contact areas for the second LSB (2^1), one for the leading brush and the other for the lagging

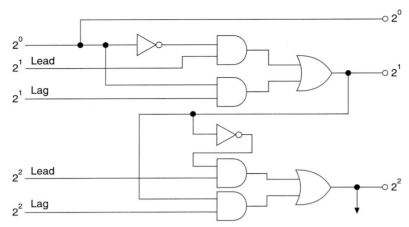

Figure 7.62   Logic circuit for V-scan encoder.

brush, located against two collector tracks. Similar arrangement exists for the contact areas of the next LSBs. Diodes between the two brushes are needed for each bit (lagging and leading brush) to prevent accidental energization of the collector tracks and hence the conducting areas of other bits.

There are other methods of scanning used with suitable logic circuitry.

7.8.1.4 Codes and Decoding

The other means of overcoming the risk of ambiguity is by using a 'reflected' or 'progressive' code in which only one digit is changed at a time. The Gray-code is one such in use. The encoder using Gray-code is shown in Figure 7.63(a); it can be seen from the figure that for any transition point between 0 to 15, the readout encounters a change in one brush position only. The code requires conversion to pure binary and involves the utilization of a Gray-to-binary decoder, based on exclusive OR gates as shown in Figure 7.63(b). It can be seen that the state of each brush is either confirmed or reversed, depending on whether the sum of the previous more significant '1' is even or odd respectively. Hence the output of each gate is '1' when only one input is present.

An effective technique for eliminating the ambiguity and minimization of noise in brush type encoders is by MR scan which employs two Gray-coded discs, each being a complement of the other and driven from a common shaft. A Schmidt trigger is used to deliver output.

7.8.2 Digital Tachometers

All the electromagnetic transducers described in Section 7.5.2 develop output signals whose frequency is proportional to the rotational speed. The transducers may be designed in such a way that the wave shape of the output signals is similar to pulses; this enables the pulses to be counted over a given time. The time between pulses may also be measured.

Wherever it is possible to convert the measurand into a rotational speed, the counting mechanism may be employed to deliver digital outputs. Turbine flow meters enable flow-velocity measurements in this fashion. One of the turbine blades carries or is embedded with a small permanent magnet or a small quantity of radioactive device. The sensing head consists

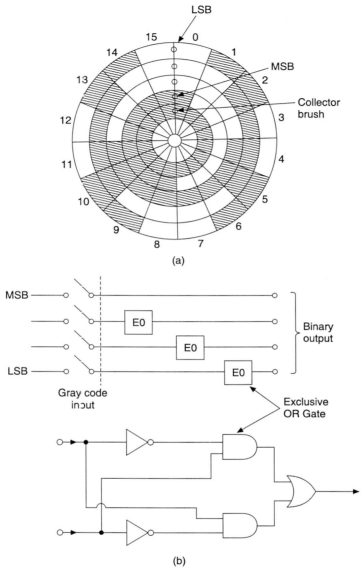

Figure 7.63 (a) Encoder using Gray code; (b) Gray-to-binary decoder.

of a coil and a core so as to develop the pulse whenever the magnet sweeps the region of the head. Or else, a Geiger–Müller counter counts the number of sweeps the blade makes near the detector.

The capacitance variations between a probe plate and a serrated (toothed) rotor may be used to generate pulses whose rate is proportional to rotor speed. The capacitor system is excited by a high frequency source so that the output is effectively a pulse modulated signal. The gap between the tooth and probe surface must be very small.

Incremental encoders described in Section 7.8.1 can be used for the measurement of either linear velocity over limited distances or for rotational speeds. The number of pulses in the output are counted over a period of time.

The pulse counting method provides measurement of the average speed only. The accuracy depends on the accuracy of the clock period, and the resolution varies with speed. Measurement of very low speeds is not possible. The basic stages for the measuring system are shown in Figure 7.64(a). The clock provides the pulses to open the gate for the prescribed period to reset the counter before each count and simultaneously update the digital output. Sensing the direction of motion and indication of the same is possible. The arrangement is known as *linear function* tachometer in contrast with the inverse function tachometer.

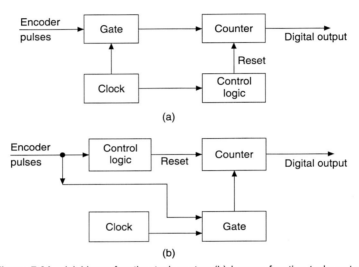

Figure 7.64   (a) Linear function tachometer; (b) inverse function tachometer.

The *inverse function* tachometer system is shown in Figure 7.64(b). It consists of a clock whose pulses are gated by each incremental pulse of the encoder. Control logic is used to reset the counter and update the output at each pulse from the incremental encoder disc. The technique offers high resolution at low speeds and low resolution at high speeds. An encoder disc having 100 segments and running at 12 rpm gives a resolution of 1 in 50,000 and at 12,000 rpm, a resolution of 1 in 50. The merit of this system is that the instantaneous speed of rotation at any instant during a revolution can be reckoned from the counter and that the transient changes in speed can be detected.

Absolute encoders may also be used for the measurement of rotational speeds but they require elaborate processing circuitry to develop digital readout for the speed.

Incremental encoder discs are light and hence the inertial load on the shaft of the machines and motors is practically negligible if the space permits the addition of encoder disc and its sensing system. Also, there is no magnetic drag or friction drag if the sensing system is optical.

7.8.3   Transducer Oscillators

Analog transducers whose outputs are changes in resistance, capacitance or inductance, have been dealt with in Chapter 6. These changes can be made use of to alter the frequency of oscillation of electronic oscillators, thereby converting signals from analog domain to frequency (or time) domain. The number of cycles per second or the time period can be measured and converted into a digital signal. It is an indirect conversion of analog signal into a digital signal but offers solution to situations where there are no means available for direct conversion.

Quartz crystals, stretched wires and, elastic members, such as cantilevers and diaphragms, may be made to remain in a state of oscillation whose frequency is dependent on a quantity pertaining to the oscillating system and has to be measured. Such systems are also considered to be classified as transducer oscillators and are used to develop output signals in frequency domain.

The frequency of oscillation of all such systems is also likely to be affected by several disturbing inputs, as is true of all analog transducers, and hence the basic accuracy is limited in this respect. The interest lies in developing digital output signals, while trying to identify oscillating systems that are sufficiently stable and reliable. Some systems of this kind, already developed and being used, are described briefly in the remaining part of this Section 7.8 while the electronic oscillators, using R, L and C elements are covered in Section 9.2.

7.8.3.1   Quartz Temperature Transducer

Quartz crystal oscillators are known for their stability in the frequency of oscillation. However, the frequency of oscillation f_T at any particular temperature T is governed by the relation

$$f_T = f_0(1 + aT + bT^2 + cT^3 + \ldots) \tag{7.30}$$

where, f_0 = fundamental frequency at 0°C.

When the quartz crystal is cut in a specific direction, the relationship becomes simpler and linear, with the coefficient a being 35.4×10^{-6} per °C. The fundamental frequency f_0 depends on the thickness of the crystal and can therefore be adjusted to a value that yields a change in frequency of 1000 Hz for a temperature change of 1°C. Using the coefficient of 35.4×10^{-6}, the value of f_0 is found to be about 28 MHz. The diameter of the crystals is about 6.5 mm and the crystal is kept sealed in an inert atmosphere of a small container. Detection of a change in frequency of oscillation by 1 Hz means a resolution of ±0.001°C. The difficulty in obtaining the resolution of this order lies in having another quartz oscillator operating at a fixed but known temperature and comparing the two frequencies by heterodyning. A block diagram of the scheme is shown in Figure 7.65, which is self-explanatory. Temperatures in the range of −40°C to +230°C can be measured precisely and accurately by this method.

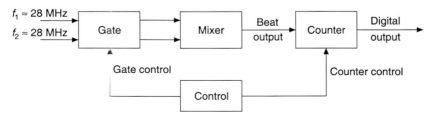

Figure 7.65   Principle of heterodyning two close frequency outputs.

7.8.3.2 Quartz Pressure Transducer

Quartz crystals are used for force measurements. When they are subjected to high pressure, the crystal gets strained and the thickness is slightly reduced. If the crystal is used as a resonator for the oscillator, a variation in the frequency of oscillation by 2 Hz is noticeable when the applied pressure changes by 1 atmosphere. The temperature of the crystal is maintained constant and high pressure is applied to the crystal.. The consequent change in the frequency is measured by methods similar to those employed for quartz temperature transducer. Pressures ranging from 0–700 atmospheres are measured by such transducers with a resolution of 1 in 10^6 for a one-second counting period.

7.8.3.3 Vibrating-diaphragm Pressure Transducer

Vibrating diaphragms are used to measure several quantities, particularly pressure in the range of 1–100 atmospheres. The diaphragm needs to be maintained in a state of vibration by means of positive feedback which necessitates again a pick-up and an amplifier. But in practice, the distortion or deformation suffered by the diaphragm due to application of pressure is utilized to change the tension and the angle of support of another thin diaphragm (membrane) as shown in Figure 7.66. This membrane is kept under constant vibration by means of a sensor and forcing coil energized by an amplifier. Whenever the diaphragm is subjected to a pressure change, the frequency of vibration of the membrane changes. The membrane is of a suitable material so as to reduce the effects of ambient temperature on the frequency of vibration.

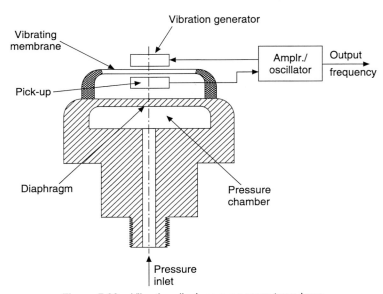

Figure 7.66 Vibrating diaphragm pressure transducer.

7.8.3.4 Vibrating-cylinder Pressure Transducer

Vibrating-cylinder transducer is based on the same principles as those of vibrating diaphragm type except that the frequency of the vibration of the cylinder itself is measured, at pressures ranging from 1–700 atmospheres. The cylinder is normally made of a special nickel–iron alloy so that changes in elasticity due to temperature variation have an effect which is equal and opposite to the changes that arise from dimensional variations. It is the satisfaction of this condition that allows the cylinder to be used for pressure measurements. The cylinder is designed to have stiff ends and thin flexible walls so that the modes of vibration are limited to the modes shown in Figure 7.67. The cylinder is vibrated in the required mode by suitably locating the pick-up coil and forcing coil, both inside the cylinder. The cylinder is about 5 cm long, 2 cm in diameter, and of wall thickness of 0.075 mm. The test pressure is admitted to the inside of the cylinder while its outside is held at a known reference pressure or vacuum. The nominal frequency of the cylinders varies from 5–15 kHz, depending on the range of pressure, and it is possible to have a 20% frequency change over the full range. The pressure-frequency characteristic is nonlinear, as shown in Figure 7.68.

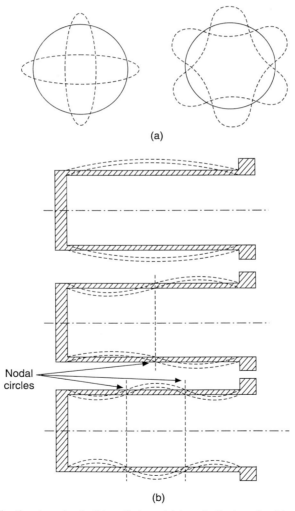

(a)

Nodal circles

(b)

Figure 7.67 Vibrational modes in thin cylinders: (a) Longitudinal mode; (b) meridional mode.

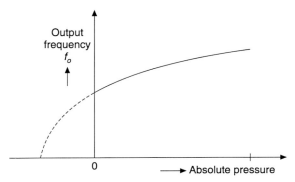

Figure 7.68   Pressure–frequency characteristic of a vibrating-cylinder pressure transducer.

However, a modification in the characteristic is essential due to the effects of variation in the density of gas. Temperature of the gas admitted also alters the density of the gas and hence the frequency of vibration. This variation is given by

$$\Delta f = \frac{k}{R} P \left(\frac{1}{T_1} - \frac{1}{T_2} \right) \tag{7.31}$$

where,

k, R = constants
P = pressure
T_1, T_2 = temperatures in degrees Kelvin
Δf = change in frequencies of vibration due to change in temperature from T_1 to T_2

For air, the total variation in frequency is less than 2% of the mean frequency over a temperature range of –55°C to +125°C. For accurate measurements, the correction is applied by measuring the temperature of the gas by means of a thermistor located at the base of the transducer through which gas is admitted. The overall accuracy of measurement is said to be within ± 0.1%.

7.8.3.5   *Vibrating-tube Gas-density Transducers*

The system described above is applicable for gas-density measurements, if the pressure difference across the walls of the tube is held constant. For effecting on-line measurement of the density, the gas is allowed to flow along the walls of the tube so that no pressure difference exists. The cylinders are normally smaller in size and their natural frequencies are lower than those of pressure transducers. A particular tube is seen to vary its frequency from 3.9 kHz to 4.9 kHz for a change of gas density from 0 to 60 kg/m³.

The gas density–frequency characteristic is nonlinear and is of the form

$$\rho = k_0 + \frac{k_1}{f} + \frac{k_2}{f^2} \tag{7.32}$$

where,

ρ = gas density
k_0, k_1, k_2 = constants of transducer
f = frequency of vibration of cylinder

7.8.3.6   Vibrating-string Force Transducer

The transducer is basically a metallic wire which is kept under tension by the addition of load as shown in Figure 7.69. Adjacent to the wire, a pick-up and an electromagnetic vibration generator are kept so as to keep the wire in continuous vibration. Once plucked, the wire will continue to vibrate at its natural frequency. The natural frequency depends on the length of the wire and the force or weight applied. The amplifier output frequency is measured, which indicates the force or weight applied. The frequency of vibration is given by

$$f_0 = \frac{1}{2L}\sqrt{\frac{T}{m}} \text{ Hz} \tag{7.33}$$

where,

L = length of wire
T = force or weight
m = mass per unit length of wire

A titanium wire of length 3.2 cm and of 0.063 mm diameter has a natural frequency of about 3700 Hz for a weight of 0.81 N. The frequency of vibration varies as the square root of the force and hence suitable electronic circuits are needed before a straightforward counting system can be used for a linear calibration of force.

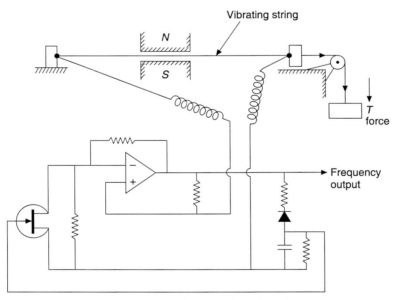

Figure 7.69   Vibrating string force transducer.

7.8.3.7   Vibrating-tube Density Transducer

The density of a flowing liquid can be measured by a vibrating-tube transducer, shown in Figure 7.70. The flowing liquid is divided into two paths and the two parallel tubes are held together securely at either end, leaving a free central span. It is the central span of both the tubes

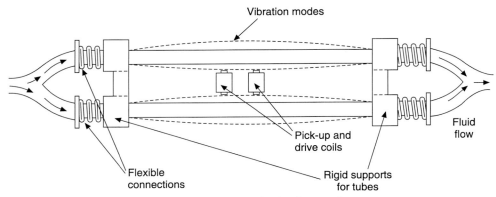

Figure 7.70 *Vibrating-tube density transducer.*

which is kept in vibration by using the pick-up and drive coils located in between the tubes. Flexible connections to the ends of the tubes prevent all disturbances from affecting the tube vibrations. The tubes vibrate in a lateral mode, and the natural frequency depends on the mechanical stiffness and on the mass per unit length of the tube along with its contents. The relationship between the liquid density ρ and the output frequency f is given by

$$\rho = k\left(\frac{f_0}{f}\right) - 1$$

where,

k = a constant

f_0 = frequency of vibration of the tube with no liquid flowing

The dimensions of the tubes are always chosen such that there is about 20% change in frequency for a density change from 0 to 1 kg/m^3 with a maximum frequency of about 1350 Hz with only air in the tubes.

The calibration is found to remain within 0.3×10^{-4} kg/m^3 over a range of flow rates from zero to 75 litres/min. The flow is in vertical direction from top to bottom so that no sediment accumulates in the tubes.

The frequency is affected by the pressure of the liquid and compensation is required to correct this error.

7.9 ELECTROCHEMICAL TRANSDUCERS

In the field of chemical analysis, electrochemical techniques based on various phenomena occurring within an electrochemical cell are very much favoured for application in laboratory and industry. The concentration of a component in solution is measured in terms of the effect it has on the electrical properties of the cell. The effect of the component on an electrode introduced into the solution forms the basis of the electrochemical transducer of the active type and the emf developed between the two electrodes of the cell constituted for this purpose is used to signify the concentration level.

There are two types of electrochemical cells: galvanic and electrolytic. A galvanic cell is used as a converter of chemical energy into electrical energy, whereas an electrolytic cell is supplied with electrical energy from an external source. At the exact point at which the galvanic emf is balanced by the applied emf, no current flows through the cell, and under these conditions, the potential of each electrode reflects the composition of the solution. The theory of electrodes and the principles that govern their design are a little involved and require an understanding of electrochemistry.

Electrochemical transducers are specially designed for use in biomedical instrumentations. In carrying out their various functions, certain systems of the body generate their own monitoring signals, conveying useful information about the functions they represent. Such signals are known as bioelectric potentials and are related to nerve conduction, brain activity, heart beat, etc. Bioelectric potentials are, in fact, ionic potentials developed as a result of electrochemical activity of certain types of cells formed in the tissues of the body. Electrochemical transducers enable presentation of these ionic potentials into a suitable form so that the output signals are properly related to the basic biochemical phenomena. In Sections 7.9.1–7.9.5, some electrode systems for the measurement of certain chemical properties of solutions with which the electrodes come in contact are described. These sections also show that care has to be taken in carrying out measurements. The basic mechanism and the relevant theory concerning the development of electrode potentials are reviewed briefly at the outset.

7.9.1 Basics of Electrode Potentials

When a metallic electrode (wire) is placed in a solution, a reversible chemical reaction takes place at the interface between the solution and the metal. Positive ions of the metal enter the solution, leaving the electrode negatively charged and the solution will acquire a positive charge. If the solution already contains ions of the metal, there is a tendency for ions to be deposited on the electrode, thereby making the electrode positively charged. The electrode potential is the result of the difference in diffusion rates of ions into and out of the metallic electrode. Equilibrium is finally established by the formation of a layer of charge at the interface. The interface may, in fact, be treated as consisting of two layers, the layer nearest the metal being of one polarity, while the layer next to the solution is of opposite polarity. The magnitude and sign of the potential depends on the concentration of metallic ions in the solution and the nature of the metal. The entire process of exchange requires an infinitesimal amount of ion transfer. The metallic electrode and the solution surrounding it constitute a 'half-cell' only as the electrode potential E cannot be measured unless a second electrode is used. The test electrode M is called *indicator electrode* and the second electrode R the *reference electrode,* while each constitutes a half-cell with the solution. The net emf E_0 across the two electrodes is the algebraic sum of the potentials of both half-cells, i.e.,

$$E_0 = E_{M/solution} + E_{solution/R} \qquad (7.34)$$

where,

$E_{M/solution}$ = indicator electrode potential

$E_{solution/R}$ = reference electrode potential

The indicator electrode potential can be obtained from the measurement of E_0, provided the reference electrode potential remains constant. All electrode potentials are given as relative values and are stated in terms of the potential of reference electrode. By international arrangement, the normal hydrogen electrode (NHE) is chosen as the reference standard and arbitrarily assigned an electrode potential of zero volts.

The half-cell potential E between an electrode and the surrounding solution is expressed by the Nernst equation and is given by

$$E = E^0 + \frac{RT}{nF} \ln (C_M^+ f) \text{ volts} \tag{7.35}$$

where,

E^0 = standard (reference electrode) potential

R = the gas constant (8.314 Joules per mole/degrees Kelvin)

T = the absolute temperature, K

F = the Faraday constant (96487 Coulomb per mole)

n = the valence of the ion (no. of electrons participating in the reaction to ionize the atom)

C_M^+ = concentration of the ions in solution

f = activity coefficient of the ions in solution

The activity coefficient f depends on such factors as the temperature, concentration, ionic strength, and the distance between ions. The product $C_M^+ f$ *is* called the *activity* of the ion responsible for the electrode potential. In a very dilute solution, the value of f is nearly taken as unity, and hence E is governed by concentration C_M^+.

For high concentrations, the measured half-cell potential differs from that computed from the Nernst equation because it depends on the activity rather than the molar concentration.

At 25°C, the half-cell potential E becomes

$$E = E^0 - \frac{0.05912}{n} \log C_M^+ \text{ volts} \tag{7.36}$$

For $C_M^+ = 1$, $E = E^0$.

By using NHE for reference electrode in a solution of unit activity ($C_M^+ f = 1$), E^0 is taken as zero at all temperatures, and all electrode potentials are measured relative to that of NHE.

The standard hydrogen electrode consists of a platinum electrode coated with platinum black, half-immersed in a solution of hydrogen ions at unity activity (1.228 M HCl at 20°C) and half in pure hydrogen gas at one atm pressure. Due to the inconvenience of setting up NHE for each instance of measurement, other reference electrodes are used, whose potentials relative to the NHE arc accurately determined. Examples of these are the silver/silver chloride electrode (Ag/AgCl), and the mercury/mercurous chloride (Hg/HgCl, calomel) electrode.

Only cells constituted by two electrochemically reversible half-cell reactions provide reproducible results. A criterion for reversibility is the reaction of a half-cell to a small current input from an external source in either direction. If the reaction proceeds with equal effects in both directions and if the current has negligible effect on the half-cell potential, the reaction is said to be *reversible* and the electrodes used are known as *reversible electrodes*. A reversible reaction does not exist if the solution reacts spontaneously with the electrodes or the presence of the externally imposed current results in formation of gas, oxidation or any other chemical effects.

The Nernst equation also suggests that if two identical electrodes are immersed in the solution, E^0 gets cancelled and the net potential difference (or emf) across them will also be zero unless the concentrations are not identical at the points where the electrodes are immersed in the solution. The concentrations may be made to differ by insertion of a porous barrier in the solution as shown in Figure 7.71(a).

The net potential difference between the electrodes can be sufficiently altered by the presence of a diffusion potential due to a membrane separating the solution as shown in Figure 7.71(b). If the concentrations of the solution in the half-cells are C_1 and C_2, then the net potential difference E_0 is given by

$$E_0 = \frac{0.05912}{n} \log \frac{C_1}{C_2} + E_m + E_1^0 - E_2^0 \tag{7.37}$$

where,

$$E_m = \text{membrane potential}$$

$$E_1^0 \text{ and } E_2^0 = \text{respective standard potentials of the half-cells}$$

Practically all biological potentials are associated with membranes, especially the membranes surrounding the cells. Certain types of cells within the body, such as nerve or muscle cells, are encased in a semipermeable membrane that permits some substances to pass through, while others are kept out. Surrounding the cells of the body are the body fluids which are conducting solutions containing mostly ions of sodium (Na^+), potassium (K^+), and chloride (Cl^-). Potentials develop across membranes when the concentrations of a permeable ion are unequal on either side of the membrane. A positive ion (cation) diffusing from the higher to lower concentration side leaves the anion behind since it is prevented by membrane, and hence the resulting charge separation develops the membrane potential. As the various ions seek a balance between the inside and outside of the cell (see Figure 7.71(c)), a net membrane potential E_m exists across the two sides of the membrane and is termed as the *resting potential* of the cell.

Since at equilibrium the chemical potentials of the two sides of the membrane are equal, the membrane potential E_m is given by

$$E_m = \frac{0.05912}{n} \log \frac{C_1}{C_2} \tag{7.38}$$

If the membrane is permeable to more than one ion, the above equation becomes invalid and E_m is governed by the relative ion permeabilities.

Biological reactions upset the equilibrium thus established and variations in E_m can be noticed. A wide variety of electrodes are used in biomedical instrumentation for the analysis of bioelectric events. A distinction should be made between the electrode systems and transducers developed for such studies. For electrodes used for the measurement of bioelectric potentials, the electrode potential occurs at the interface of a metal electrode and an electrolyte, whereas in biochemical transducers, both membrane barriers and metal-electrolyte interfaces are considered.

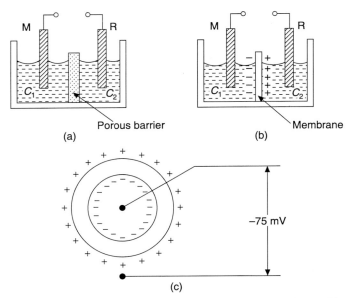

Figure 7.71 Electrochemical cells: (a) With porous barrier; (b) with membrane; (c) with bio-tissue.

The problem of measurement of ion concentration in electrolytes is to find an indicator electrode selective to the desired ion species and its concentration in the solution and a reference electrode whose potential remains unaffected by the variations of concentration of the same species in the electrolyte. No electrode by itself is an indicator electrode or a reference electrode. Many electrodes have been developed to respond to a wide range of selected ions.

7.9.2 Reference Electrodes

In practice, as it is inconvenient to set up the normal hydrogen electrode (NHE) other reference electrodes are developed and used.

The silver/silver chloride electrode shown in Figure 7.72(a) consists of a silver wire or plate coated with silver chloride and kept in contact with a saturated solution of potassium chloride and silver chloride. The half-cell potential of this reference electrode is rendered independent of the variation of the electrolyte surrounding the indicator electrode by separating the electrolytes as shown in Figures 7.72(b) and 7.72(c). A permeable membrane such as the fritted glass disc separates the electrolytes in Figure 7.72(b), whereas in Figure 7.72(c) a salt bridge consisting of saturated KCl solution with excess gelatin is used. In each case a potential difference exists at the boundary and the net emf E_0 between the indicator and reference electrodes becomes

$$E_0 = E_M + E_j + E_R$$

where,

E_j = the junction potential
E_M = indicator electrode potential
E_R = reference electrode potential

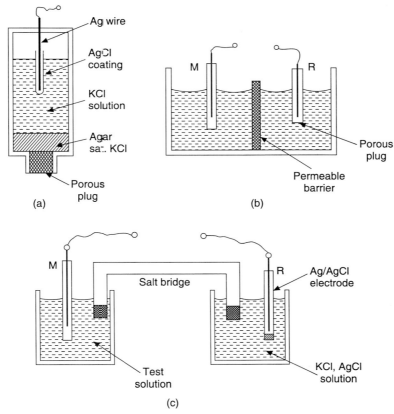

Figure 7.72   Reference electrodes: (a) Ag/AgCl chloride electrode; (b) with porous plug; (c) with salt bridge.

The junction potential E_j of the salt bridge is kept within ±2 mV by suitable choice of concentration of the salt solution. The Ag/AgCl electrode is used as a reference electrode for pH measurements. It can also be used as an indicator electrode for chloride ion concentration.

The other commonly used reference electrode is the calomel electrode shown in Figure 7.73. The metal used for electrode is mercury which is in contact with a paste containing equal weights of mercurous chloride (calomel) and potassium chloride. A column of saturated KCl solution over the paste enables contact with the test solution through a salt bridge of KCl whose concentration is kept about 3.8 mol per litre. Contact with mercury pool is by means of a platinum wire which may be amalgamated. The calomel electrode is used as a reference electrode for pH measurements.

The porous plug or the salt bridge serves the function of connection between the reference and indicator electrodes without allowing the solutions to mix with each other. However, the passage of ions takes place, though the unequal rates of diffusion develop some junction potential.

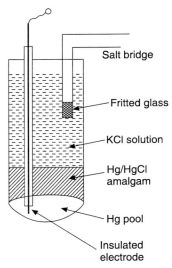

Figure 7.73 Calomel electrode.

7.9.3 Indicator Electrodes

The most commonly used indicator electrode is the glass electrode selective to hydrogen ions in a solution and used for the measurement of pH. The concentration of hydrogen ions is commonly expressed as

$$pH = \log \frac{1}{C_H} \tag{7.39}$$

where, C_H = concentration of H ions.

In pure water the concentration of H^+ ions and OH^- ions is alike and is given by

$$[C_{H^+}] = [C_{OH^-}] = 10^{-7} \text{ mol/litre at } 25°C$$

The product of both the concentrations remains constant. Thus pure water has C_{H^+} of 10^{-7}, and a pH of 7. If the pH is less than 7, the solution is acidic, if it is greater than 7, then it is alkaline.

The glass electrode is a membrane electrode and its constructional features shown in Figure 7.74 confirm that a thin ion-selective glass membrane is sealed onto the end of a glass tube that has no ion-selective properties. The bulb at the end of the tube contains an internal buffer solution of known hydrogen ion concentration of pH_i, (say, 0.1 M HCl solution). An internal reference electrode (either Ag/AgCl type or the calomel type) is immersed in the buffer solution and is connected to a screened lead for connection to external circuit.

Another reference electrode is used as an external reference electrode in the solution with unknown hydrogen-ion concentration (pH_x).

The entire electrode assembly can be represented as

external reference electrode	test solution of pH_x	glass membrane	buffer solution of pH_i	internal Ag/AgCl electrode

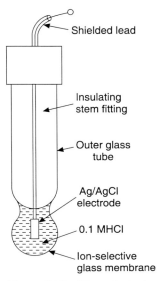

Figure 7.74   Glass electrode.

resulting in the net potential difference given by

$$E_0 = E_g^0 + 0.05912\,(\text{pH}_x - \text{pH}_i) \qquad (7.40)$$

where, E_g^0 is the equivalent of membrane potential, also known as asymmetry potential.

Due to the presence of E_g^0, empirical calibration of the glass electrode with a solution of known pH is always required.

Due to the presence of the glass membrane, the resistance of the glass electrode is always very high and is of the order of 10 to 10^3 MΩ. Hence, measurement of the emf of the pH electrode system is done by special measuring systems possessing much higher internal impedance (as explained in Section 3.5.2).

The composition of the glass used to serve as the membrane varies with the type of ion species under measurement. Glass electrodes selective to sodium, potassium, ammonium or silver are available. The glass membrane should be practically impermeable to all other species of ions, except the one for which it is meant and designed.

There are other types of ion-selective electrodes, being known as solid-state electrode, heterogeneous membrane electrodes etc., which are selectively permeable to a wide range of ion species.

Microelectrodes are specially designed glass reference electrodes for use in biological experiments. The microelectrode shown in Figure 7.75(a) consists of a hollow tip tapered to a fine point with the tip diameter of about 1 µm. The hollow of the electrode is filled with KCl solution into which an Ag/AgCl reference electrode is inserted. The fine tip is pushed into the nerve cell as shown in Figure 7.75(b), while a second electrode with a coarse tip is positioned outside the cell in the vicinity of the first electrode contact point. The potential difference between the two jumps from zero to the resting potential of about 70 mV as soon as the fine

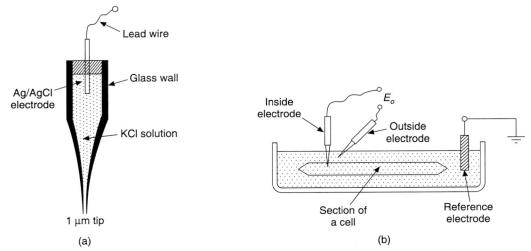

Figure 7.75   (a) Microelectrode; (b) insertion of microelectrode into nerve cell.

tip breaks through the cell wall. The whole process is undertaken while observing the same under a microscope.

Skin electrodes are placed on the skin to measure skin potentials originating from the electrical activity of the heart (electrocardiogram or ECG), brain (electroencephalogram or EEG) or muscle (electromyogram or EMG). The electrode must be securely fastened to the skin so as to make good contact with the skin to convey these small ac signals to the external measuring system.

A typical skin electrode is shown in Figure 7.76, which is held against the skin by a suction cup. An electrolytic conducting paste is smeared on the skin to provide better contact. The electrode is, in fact, an Ag/AgCl electrode, with the difference that the electrode is a silver disc coated with AgCl. The contact resistance between the electrode lead wire and the tissue underlying the surface of the skin is comparatively high and variable compared with the actual tissue resistance, which is of the order of 10–1000 Ω. Hence the measuring system should be of high input impedance.

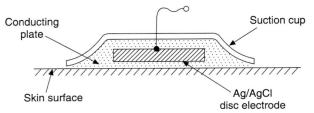

Figure 7.76   Skin electrode..

7.9.4   Measurement of pH

Measurement of pH is undertaken to determine whether a particular solution is acidic, neutral or alkaline and to find out the hydrogen ion concentration of the solution. The electrode system consisting of a glass electrode and a reference electrode (Ag/AgCl electrode or calomel electrode) provides the emf which is in turn amplified by means of high input impedance dc amplifiers. The measuring system consists of an amplifier and a microammeter, and using Eq. (7.40), the scale of the microammeter is calibrated in pH. To match with the high resistance of the electrode system, an FET-input op-amp is used to amplify the signal voltage. The gain of the amplifier is made adjustable over a narrow range to provide for temperature compensation.

7.9.5 Measurement of Bioelectric Signals

Skin electrodes are located on the body of a patient when ECG (or EEG/EMG) is taken either for monitoring the condition of the patient or for diagnostic purposes. In all cases, three electrodes are attached, one of them serving as ground.

An ECG signal is about 1 mV in its amplitude with its frequency content lying in the range of 0.2 to 100 Hz when used for diagnosis, and of 0.5 to 50 Hz when used for monitoring.

An EEG signal is in the range of 1–10 µV and of a frequency range 3–30 Hz.

EMG signals vary in amplitude from 0.1 to 10 mV, depending on electrode location and are of the frequency range from 5 to 500 Hz. Normally, the presence or count of impulses is only required and hence the gain of the amplifier used need not be known precisely.

However, for all the above measurements, it is important to see that dc potentials due to membranes are blocked and the ac signals of interest are amplified properly without any unduly large errors in the measured outputs due to hum or pick-up voltages of 50 Hz from the electric fields in the vicinity. 50 Hz components may be eliminated to a large extent by using 50 Hz notch filters in the case of EEG signals, but in the other two cases, these filters introduce distortion of the output as the desired signals have also a component of 50 Hz. Hence, enough attention should be paid while choosing the amplifiers for these purposes. Difference (balanced) or differential amplifiers having high common-mode rejection ratio are to be used. High input impedance for the amplifier is the main requirement in all cases. A slightly modified version of instrumentation amplifier is shown in Figure 7.77, with the provision for connection of the ground electrode to the input-side ground terminal. The circuit is suitable for dc signals also, but for the above purposes, it is necessary to use filters to block the dc.

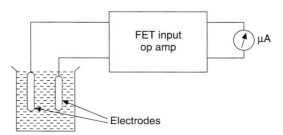

Figure 7.77 Measuring system of a pH meter.

A grounded electrode attached to the body of the patient always poses the problem of an electric shock, as it provides one of the two contacts to the body necessary for shock. The problem is created by the ground lead or any current path to ground connected to the subject through the amplifier input-leads under possible overload conditions. It is necessary to provide ground isolation to eliminate such shock hazards.

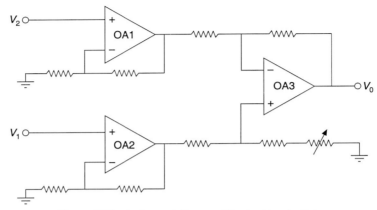

Figure 7.78 Instrumentation amplifier for biopotentials.

NUMERICAL EXAMPLES

7.1 The series connected thermopile is made up of copper–constantan thermocouples with T_1 at 150°C and its net output emf is 3.3 mV for the arrangement of three junction pairs, as shown in Figure 7.5(a). Calculate the value of temperature T_2, taking the sensitivity of each junction as 50 μV/°C.

Taking all the junctions as matched, the emf developed by each junction pair is given as 3.3/3 = 1.1 mV.

The sensitivity is 50 μV/°C. So the temperature difference is $\dfrac{1.1 \times 10^3}{50} = 22°C$. Hence T_2 is 150 ± 22 = 172°C or 128°C, depending on the polarity of emf.

7.2 A quartz crystal of thickness 2 mm is used for pressure measurement and is subjected to a static pressure of 1.5 kN/m². If the voltage sensitivity is 0.05 V-m/N, determine the output voltage of the crystal. If the absolute permittivity of quartz is 4.06 × 10⁻¹¹ F/m, determine the quantity of charge developed on the crystal.

Using Eq. (7.5),

$$V_x = 0.05 \times 2 \times 10^{-3} \times 1.5 \times 10^6 = 150 \text{ V}$$

Using Eq. (7.6),

$$d = \varepsilon_0 \varepsilon_r g = 4.06 \times 10^{-11} \times 0.05$$
$$= 2.03 \times 10^{-12} \text{ C/N}$$
$$= 2.03 \text{ pC/N}$$

7.3 A Hall-effect element is used for the measurement of magnetic field of 0.8 Wb/m^2. The thickness of the element is 2.5 mm and is of Bismuth material. If the current passed through the element is 4 A, calculate the Hall emf developed.

Using Eq. (7.18) and taking the Hall coefficient from Table 7.3 as 5×10^{-7}, the Hall emf is

$$E = \frac{K_H BI}{t} = \frac{(5 \times 10^{-7}) \times (0.8) \times 4}{2.5 \times 10^{-3}} = 6.4 \times 10^{-4} \text{ V}$$

EXERCISES

1. Explain the thermoelectric phenomena and show how a temperature measuring system can be built.

2. Making use of the thermoelectric lines of Figure 7.2 for the materials of copper and iron, obtain the temperature emf characteristic for the junctions, with the reference temperature maintained at 0°C. Determine the temperature at which inversion takes place and establish the approximate linear range.

3. Suggest the merits of a thermocouple system for the measurement of temperature, when compared with the wire resistance thermometer and the thermistor.

4. What is meant by cold-junction compensation and how is it effected in industrial environments?

5. What is a 'thermopile' and where is its application?

6. Explain how thermocouples may be used for measurement of surface temperature of bodies and indicate the characteristics of the thermocouple, which make them suitable for such measurements.

7. Why are thermocouples provided with an air-tight jacket and what is the effect of the jacket on its performance?

8. What is meant by 'dynamic compensation' and how and when is such compensation applied for a thermocouple system?

9. Explain the piezoelectric phenomenon and suggest the materials that exhibit this phenomenon.

10. Describe how a piezoelectric crystal is cut to serve as a transducer and show that the crystal is most suitable for force measurements.

11. Show the nature of output signals obtainable from a crystal, when subjected to mechanical force; show also how they can be processed.

12. Define the *d, g* and *h* coefficients by which the sensitivity of a crystal is identified and obtain the inter-relationship between them.

13. What are natural and synthetic piezoelectric materials? State their important and distinguishing properties by which their application is determined.

14. Distinguish between piezoelectric materials and ferroelectric materials. Mention some of the important ferroelectric materials and their distinguishing characteristics that render them widely applicable in instrumentation.

15. Explain the 'anisotropic stress effect' and show how force and pressure measurements are made possible by using this effect.

16. What are 'bimorphs' and 'multimorphs' and how do they enable application and measurement of force?

17. Show how piezoelectric transducer can be used to measure (i) torque, (ii) pressure, and (iii) acceleration. For each case identify the primary transducers required and describe the constructional features of such systems.

18. Explain the phenomenon of magnetostriction and how it lends itself for measurement of force. Is it suitable for measurement of static forces?

19. Describe the constructional features of a magnetostrictive transducer and obtain the input-output relationship.

20. Explain how acceleration and torsion can be measured by applying magnetostriction phenomenon.

21. Explain 'Hall Effect' and suggest its applicability for measurement of physical quantities.

22. Define 'Hall coefficient' and state the factors on which it depends.

23. Show as to how a proximity meter can be made by using a Hall element.

24. Describe the constructional features of a dc tachometer and an ac tachometer and distinguish their performance characteristics.

25. Show how variable reluctance transducers can be used to measure (i) linear and (ii) angular velocity over short ranges of motion.

26. Describe the construction of an electrodynamic vibration transducer and show how the range of vibration frequencies which it can measure accurately, depends on the design constants.

27. Describe the moving-coil microphone system and identify the primary and secondary transducers of the arrangement.

28. Explain the principle of operation of an electromagnetic flow meter and suggest the nature of fluids that can be used with the system.

29. Discuss the relative merits and demerits of an electromagnetic flow meter as compared to mechanical flowmeters.

30. Explain 'Photoelectric effect' and show how the phenomenon enables measurement of electromagnetic radiation.

31. Distinguish between 'photovoltaic', 'photoemissive', and 'photoconductive' cells and suggest the range of wavelengths of radiation for which they are suitable.

32. Explain the operation of a photojunction cell and show how it can be used as a photovoltaic and photoemissive cell.

33. Distinguish the construction and operational features of (i) a photo transistor and (ii) a photomultiplier, and indicate their application.

34. Explain the basic principles of operation of ionization transducers and show how the phenomenon of ionization enables measurement of certain physical quantities.
35. Describe an ionization vacuum gauge and explain its operation. Show how the system can be modified to yield higher sensitivity.
36. Explain the techniques by which ionization phenomenon may be used for measurement of mechanical displacements.
37. Explain what are nuclear radiation transducers and how they are used for measurement of vacuum and pressure.
38. Distinguish the operational features of an ionization chamber, proportional counter and a Geiger–Müller counter. Identify the nature of output signals obtainable from each.
39. Describe a scintillation counter and its importance in instrumentation.
40. Explain the reasons for the popularity of radioactive thickness gauges and show how they are usually constructed.
41. Indicate the commonly used radioactive sources and their half-life.
42. What are 'direct' digital transducers? Give an example and show how the output signals are obtained.
43. Describe the construction of shaft angle encoders and show how the angle can be measured with higher precision.
44. Explain how encoders are designed for measuring linear displacements.
45. Explain the problems faced in encoding the angular displacements and show the means adopted to avoid the same.
46. Explain the 'Gray' code adopted in shaft-angle encoders and indicate its merits over the direct simple binary code used.
47. Explain the techniques adopted for conversion of angular velocity into digital signals.
48. What are 'transducer oscillators' and how are they normally made up?
49. What are 'indirect' digital transducers? Give a few examples?
50. Show how quartz crystals may be used to make up indirect digital transducer systems for measurement of temperature and pressure.
51. Explain how vibrating tube and diaphragm systems are designed for measurement of pressure and force.
52. Explain how electrode systems are designed for measurement of hydrogen-ion concentration and indicate the problems associated with the measurement.
53. Explain why pH measurement is done with special amplifiers when using glass electrode.

SUGGESTED FURTHER READING

Doebelin, E.O., *Measurement Systems: Application and Design*, McGraw-Hill, New York, 1966.

Gibson, John E. and Tuteur, Franz B., *Control System Components,* McGraw-Hill, New York, 1958.

Holzbock, W.G., *Instruments for Measurement and Control*, Reinhold Publishing Corp., New York, 1955.

Jones, Barry E., *Instrumentation, Measurement and Feedback*, Tata McGraw-Hill, New Delhi, 1973.

Kuo, B.C., *Automatic Control Systems*, 3rd ed., Prentice Hall, Englewood Cliffs, New Jersey, 1975.

Nagrath, I.J. and Gopal, M., *Control System Engineering*, Wiley Eastern, New Delhi, 1982.

Neubert, H.K.P., *Instrument Transducers*, 2nd ed., Clarendon Press, Oxford, 1975.

Ryder, John D., *Electronic Fundamentals and Applications*, Prentice-Hall of India, New Delhi, 1974.

Wightman, E.J., *Instrumentation in Process Control*, CRC Press, International Scientific Series, Columbus, Ohio, 1978.

CHAPTER 8

Feedback Transducer Systems

8.0 INTRODUCTION

All the electrical measuring systems described in Chapter 3 and all the transducers dealt with in Chapters 5–7 are physical systems consisting of a combination of different elements and are characterized by an input-output relationship. Each element may be represented by means of a block with its transfer function relating its input and output variables. The blocks, when connected together, enable formulation of the functional relationship of the entire system. Block-diagram representation of physical systems is very helpful in analyzing the behaviour of the system and in effecting modifications in the design to conform to the desired specifications. All the systems described so far can be represented by a single block as shown in Figures 8.1(a) and 8.1(b) and are described as open-loop systems. Introduction of a link between the input and output quantities constitutes feedback and enables the system to assume considerable significance in its behaviour.

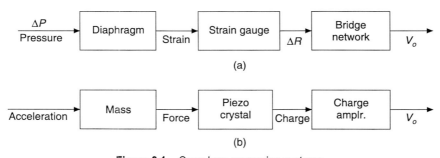

Figure 8.1   Open-loop measuring systems.

Feedback is extensively applied in the design of oscillators, amplifiers, voltage and current regulators, servomechanisms and automatic control systems. The block diagram of a feedback system is shown in Figure 8.2, and the magnitude and phase of the signal fed back to be added to or subtracted from the input signal, makes all the difference in the behaviour of the system. If the two are added, the system is described as positive feedback system and oscillators employ positive feedback. If the signal which is fed back is subtracted from the input variable, it is known as negative feedback system. Most of the amplifiers and automatic control systems employ negative feedback. When the input variable $x_i(t)$ and feedback signal $x_f(t)$ are of the same dimensions, and nearly equal in magnitude, as is the case with all automatic control systems, the difference is known as *error (deviation) $e(t)$* and is given by

$$e(t) = x_i(t) - x_f(t) \qquad (8.1)$$

If $x_f(t) = y_0(t)$, the system is described as a unity feedback control system.

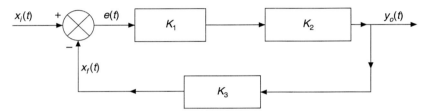

Figure 8.2   Block diagram of a closed-loop system.

In instrumentation, the measurement of the primary quantity is carried out by transducer systems and the output signals are processed suitably for purposes of indication, recording, telemetry and control function generation. A large number of electronic circuits and physical systems utilize the principles of negative feedback to satisfy the demands of the modern instrumentation systems. In Section 8.1, the merits of employing negative feedback to measuring systems are explained and later transducer systems using negative feedback are described. In Chapters 9 and 11 relevent data recording and telemetry systems, employing feedback are explained indicating the merits of its application.

8.1   FEEDBACK FUNDAMENTALS

A simple feedback system representing an amplifier of gain A in the forward path and a network in the feedback path of gain β is shown in Figure 8.3. Treating A and β as constants and the input

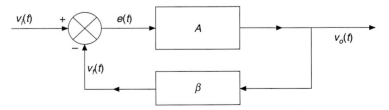

Figure 8.3   Block diagram of a feedback amplifier.

and output quantities as voltages, it is seen that

$$\text{error } e(t) = v_i(t) - v_f(t)$$
$$= v_i(t) - \beta v_o(t)$$
$$v_o(t) = A\, e(t)$$

The effective voltage gain of the system is

$$A_f = \frac{v_o(t)}{v_i(t)} = \frac{A}{1 + A\beta} \tag{8.2}$$

A_f is known as closed-loop gain and $A\beta$ as open-loop gain. If $A\beta \gg 1$, then $A_f \approx 1/\beta$ and becomes independent of A.

If the amplifier is an operational amplifier of high gain and the feedback network is a simple resistive potential divider, the voltage gain of the feedback amplifier is almost equal to $1/\beta$. A_f is as accurate as β, the ratio of the resistors making up the potential divider.

The amplifier gain A is likely to change due to aging effects, temperature, nonlinearity, or any other reason. If it increases by ΔA, it can be shown that the fractional change in output voltage is governed by

$$\frac{\Delta v_o(t)}{v_o(t)} \approx \frac{\Delta A/A}{1 + A\beta} \tag{8.3}$$

A change in amplifier gain has far less effect on the output voltage as the open-loop gain $A\beta$ is much greater than unity.

The feedback amplifier system is considered to be equivalent to a voltage source v_o with an output resistance R_o. In case the load current changes by ΔI, there will be a reduction in output voltage by Δv_o. Since A, v_i and β are constants and Av_i is given by

$$Av_i = v_o(1 + A\beta)$$

the feedback system provides a change in output voltage given by $\Delta v_o(1 + A\beta)$ to counter the influence of a change in the load current ΔI. Hence

$$R_o\, \Delta I = \Delta v_o(1 + A\beta)$$

$$\frac{\Delta v_o}{\Delta I} = \frac{R_o}{1 + A\beta} \tag{8.4}$$

Thus the effective output impedance is that of the amplifier divided by the open-loop gain. Application of negative feedback has considerably reduced the output impedance of the amplifier.

The voltage gain with feedback has become $1/(1 + A\beta)$ times the gain without feedback. The input impedance of the amplifier has increased by $(1 + A\beta)$ times with feedback.

If the amplifier is characterized by the transfer function given by

$$A(s) = \frac{A}{1 + sT}$$

where T is its time constant, the voltage gain of the feedback amplifier is represented by the transfer function

$$\frac{v_o(s)}{v_i(s)} = \frac{A}{(1 + A\beta)[1 + sT/(1 + A\beta)]} \tag{8.5}$$

The effective time constant is reduced to $T/(1 + A\beta)$, and hence the bandwidth of the amplifier becomes larger. The product of voltage gain and bandwidth remains the same.

It is interesting to study the effect of noise voltages N_1 and N_2 added at the input and output terminals respectively as shown in Figure 8.4. The equation for V_o is given by

$$V_o = \frac{V_i A}{1 + A\beta} \pm \frac{N_1 A}{1 + A\beta} + \frac{N_2}{1 + A\beta} \tag{8.6}$$

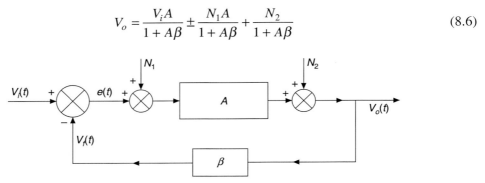

Figure 8.4   Feedback amplifier with noise inputs.

It is observed that the effects of interference due to noise and hum added at the output terminals of the amplifier are reduced by the open loop gain. But the noise added at its input terminals is as effective as the error voltage or the input voltage.

From the above analysis it is further clear that the application of negative feedback results in several advantages except that the gain becomes reduced. Application of negative feedback to measuring systems and transducer systems has been found to be equally important. The scheme applicable for such systems is shown in Figure 8.5. The primary transducer converts the measurand $r(t)$ into $x_i(t)$, and the secondary transducer converts the difference of $(x_i - x_f)$ into a voltage signal v_i. The output of the amplifier is a dc current i_o. The feedback path consists of a system that converts the current into a proportionate signal x_f. The comparator compares the two quantities and the difference $(x_i - x_f)$ is treated as the error signal.

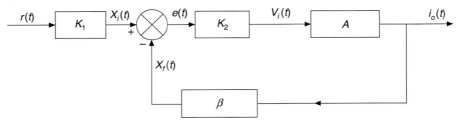

Figure 8.5   Block diagram of a feedback measuring system.

Following the same procedure of Eq. (8.2), it can be shown that

$$\frac{I_o(s)}{R(s)} = K_1 \left(\frac{K_2 A}{1 + K_2 A \beta} \right)$$

where K_1 and K_2 = gains of transducer 1 and 2 respectively.

Making the gain A of the amplifier very high such that $K_2 A \beta \gg 1$, the relation becomes

$$\frac{I_o(s)}{R(s)} \approx \frac{K_1}{\beta} \tag{8.7}$$

The output current has become independent of K_2 and A. The feedback system is now immune to variations in the values of K_2 and A due to their nonlinearity and other disturbing inputs such as temperature, and supply voltage fluctuation. By making $K_2 A \beta$ very high, I_o is made proportional to X_i. To make the relationship between X_i and I_o fixed and constant, it is necessary that β be constant. The physical system or the component in the feedback path should be stable, linear and belong to a zero-order system. Under these conditions, the feedback system has become a measuring system which has all the desirable characteristics such as high input impedance, low output impedance, linearity, stability, and immunity from interference which affect the parameters of the forward path components. The necessary output power is drawn from the source of supply of the amplifier. The feedback system does not load the primary transducer or the system under test. Accuracy is improved and is limited by the accuracy of the value β. Performance specifications pertaining to the second transducer and amplifier need not be rigid. It is only necessary that A be very high and the system in the feedback path be somewhat ideal. If the second transducer is a slow device, the speed of response of the entire system is improved because of feedback.

In certain cases of signal processing, it is desired that the output signal be related to the input signal in a desired manner, since both are functions of time. If the relationship is expressed as

$$\frac{I_o(s)}{X_i(s)} = H(s)$$

it is only necessary to identify a suitable physical system or a network in the feedback path such that

$$\frac{X_f(s)}{I_o(s)} = \frac{1}{H(s)} \tag{8.8}$$

Based on these principles, a wide variety of feedback measuring systems and function generators, both of electrical and mechanical type, are designed and made available commercially by manufacturers. The measuring systems employing negative feedback are known as *force balance* or *torque balance* systems, depending on the nature of the quantities X_i and X_f. The feedback system attempts always at equalizing the two of them and making the error negligible, but never zero. Since a zero error signal cannot provide an output signal, it is understood that a small error signal responsible for the output should be always present. As the gain is large, a small value is sufficient. The higher the value of K, the smaller the error E. The accuracy of the system expressed as a percentage of the full scale value, i.e., $100E/X_i$, where

X_i is the full scale value. Though there is additional complexity and increase in size due to the addition of components, the advantages gained indicate the popularity and increased attention paid to the development of such systems. Care must be taken to see that the increased gain of the amplifier does not lead to instability in the performance. If the amplifier and other components of the forward path act like a third-order system, then there is every possibility that the system will become unstable at some critical value for *A*. However, it may be noted that in most cases, the system is of either first or second order, as the following sections indicate.

8.2 INVERSE TRANSDUCERS

Transducers that function in reverse manner are known as *inverse transducers*. All those transducers used for measurement of nonelectrical quantities which exhibit excellent characteristics of operation when the input and output quantities are reversed, can be used as inverse transducers. They are also known as *output transducers*.

With the increasing demand on the number of variables of a process or plant to be measured and the necessity of converting them into electrical signals of suitable range for purposes of telemetry or for use with data processing computers, it has become necessary to standardize the range of electrical output signals as dc signals of range 4–20 mA or 0–1 V. Some other nonstandard feedback systems are used as recording instruments (instrument servomechanisms) having linear or angular displacement as output signal which, in fact, is easily converted into corresponding digital signals for display purposes. In the case of pneumatic instrumentation system, the output signal is standardized as a pressure of range equivalent to 3–15 psi for a long time.

We have already seen that the accuracy of feedback measuring systems depends on the characteristics of the physical system which is used to convert the output signal into a signal of desirable form and dimensions for comparison by the comparator. For each one of the output signals, a suitable system for use in the feedback path is to be chosen such that the quantity to be measured and the quantity fed back are of the same type. The quantities usually compared are torque, force, position or pressure and, accordingly, the feedback systems are designated as *torque-balance, force-balance* or *position-balance* systems. In electronic signal processing amplifiers and other circuits, they are either voltages or currents. It has been indicated in the last or the previous Section that the physical system constituting the feedback path should possess excellent stability, high degree of linearity, and belong to the class of zero-order systems. The choice of such systems thus becomes limited to a few inverse transducers. The permanent-magnet moving coil system, the parallel plate capacitor, and the piezoelectric crystal are exploited for their application in the feedback measuring systems with electrical output and the pneumatic bellows with mechanical output in pneumatic systems.

The permanent magnet moving-coil system is used to convert dc current into proportional torque in torque-balance systems. The torque T_f developed is given by

$$T_f = BANI_o = \beta I_o$$

where, B = flux density in the gap
A = area of moving coil
N = no. of turns of the moving coil

The effect of temperature on the permanent magnet is seen to change the flux density in the gap by -0.02% per $1°C$. For higher accuracy, temperature compensation is effected by providing a temperature-sensitive shunt across the gap or by employing swamping resistors along with the coil. For use in force-balance systems, an equivalent system, consisting of a cylindrical coil working in the radial magnetic field of a magnet assembly as shown in Figure 8.6 is used. The coil is wound on a cylindrical former and is free to move in the annular gap between the yoke and magnet. The force F_f produced by the coil is given by

$$F_f = \pi DNBI_o = \beta I_o \qquad (8.9)$$

where, B = flux density in the gap
$\quad\quad\quad D$ = mean diameter of coil
$\quad\quad\quad N$ = number of turns of the coil

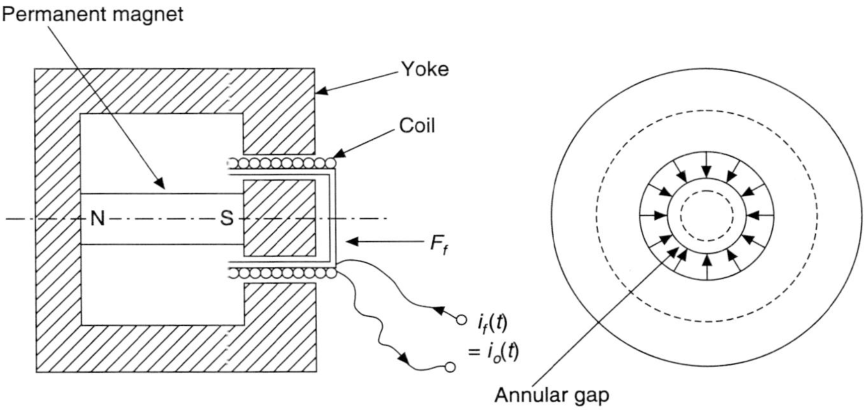

Figure 8.6 Pmmc type inverse transducer.

They may be used in position-balance systems, when the feedback system is designed to convert linear or angular displacements into proportional dc currents.

When a feedback system is designed to convert displacements into proportional voltages, the parallel plate capacitor or the piezoelectric crystal is used. As in the case of pmmc systems, they are, in principle, converters of voltage into force or torque. In the case of capacitor, the torque or force developed is not proportional to the voltage applied, as seen from the following equations:

$$F_f = \frac{V_0^2}{2}\frac{dC}{dx}, \qquad T_f = \frac{V_0^2}{2}\frac{dC}{d\theta} \qquad (8.10)$$

Hence, their application is limited to measurement and indication of quantities that yield very small force or torque for comparison with F_f or T_f. The basic nonlinearity prohibits their application to measurement of time varying quantities. The piezoelectric bender bimorph of the cantilever type shows a high degree of linearity between the voltage applied and displacement developed. Their size is small and they too possess high sensitivity in volts per displacement as in the case of electrostatic capacitor transducer. The piezoelectric bimorph can be used to have output voltage of one volt per μm of displacement under measurement. They are therefore used in displacement balance systems.

For pneumatic systems, the bellows element is considered to possess the desired characteristics of the feedback element. It is used as pressure-to-force converter in the feedback path. The relationship is governed only by the effective area of the bellows which is rendered sufficiently immune to temperature effects by a proper choice of the material for the bellows.

8.3 TEMPERATURE BALANCE SYSTEM

The internal (core) temperature of a hot solid body is difficult to measure and hence it is extrapolated from the knowledge of the surface temperature obtained by measurement and the estimated temperature gradient between the centre and surface of the body. The temperature gradient is effected by the conditions of the surrounding medium and hence the estimated internal temperature is likely to vary widely. To render the measurement more accurate, the feedback is employed as shown in Figure 8.7(a). A thermal insulation block of small size is placed on the body, and two thermocouples are located on the block such that one of them reads

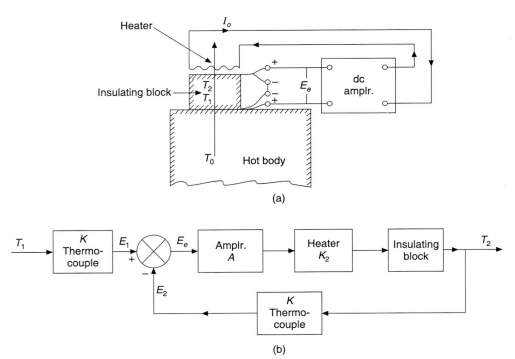

Figure 8.7 (a) Temperature balance system; (b) the block diagram.

the surface temperature T_1 of the body, and the other, surface temperature T_2 of the block. The surface of the block, exposed to the surrounding environment, is provided with a small heater of fine wire, so as to equalize T_2 and T_1. When $T_2 = T_1$, there is no heat flow across the insulation block and hence across the body from its centre to surface, along the direction shown.

Under these conditions, the internal temperature T_0 is equal to T_1. The feedback system utilizes the difference in the thermo emfs of the two thermocouples, and the high gain amplifier is used to drive the current through the heater for effecting the temperature balance.

From the block diagram of the system shown in Figure 8.7(b), it may be noted that the measurement of E_2 renders the measurement of T_0 possible and satisfactory provided the two thermocouples used are matched for their sensitivities. The system thus compensates for errors due to conduction and any loading due to the thermal capacity of the measuring system. This example is suggestive of the scope provided by feedback technique although not favoured for use in actual practice.

8.4 SELF-BALANCING POTENTIOMETERS

The merits of dc potentiometer for accurate measurement of dc voltages have prompted the development of self-balancing (automatic) potentiometers for the measurement of dc voltages. They are widely used for indication and record of voltages developed by various transducers. The automatic balancing is provided by an ac servomotor as shown in Figure 8.8(a). The servomotor drives the tapping point of the potentiometer, the writing mechanism, and the pointer for indication. The motor comes to rest when the unbalance voltage becomes zero. The block diagram is shown in Figure 8.8(b). The output displacement and voltage feedback are related by the calibration constant of the potentiometer wire. The amplifier is tuned to 50 Hz and has high gain. The unbalance dc voltages are converted into proportional ac voltage of 50 Hz by means of a synchronous chopper as shown in Figure 3.8. The forward path consists of the servomotor and the inertial load of the writing pen. By virtue of the negative feedback, the bandwidth of the system is increased, and the system is used to record faithfully voltages of

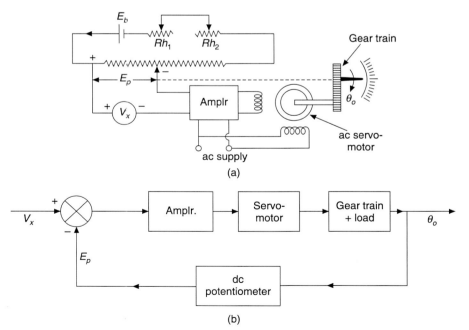

Figure 8.8   (a) Self-balancing servo-operated potentiometer; (b) the block diagram.

frequency range from dc to 5 Hz. The system is considered to function as voltage-to-displacement converter. The output shaft can be coupled to a shaft angle encoder if, in addition to analog indication, digital display is desired.

8.5   SELF-BALANCING BRIDGES

Based on the above principle, self-balancing bridges are developed. Most of the dc and ac bridges that can be balanced by adjusting one element utilize the servomotor to balance the bridge. The system thus serves as resistance-to-displacement converter. The scale is calibrated in terms of the quantities to which the resistance transducer of the bridge network responds. Where an output current related to the resistance of a transducer in the bridge is desired, electronic self-balancing is adopted. Two such systems are shown in Figure 8.9(a) and 8.9(b). In the circuit of Figure 8.9(a) the high gain amplifier drives the bridge with dc current and the unbalance voltage is applied to the amplifier. In fact, the amplifier needs a small input voltage to be able to deliver the output current I_o, and the bridge must be under slight unbalance to develop the required unbalance voltage. Any variation in resistance of the transducer affects the unbalance voltage in such a way that the change in the output current will bring back the bridge to balance conditions. The change in the output current is proportional to the change in resistance. If $R_3 = R_w$, the bridge is exactly balanced, but to drive the bridge with an excitation current of say, 12 mA dc, the value of R_3 or R_w must be slightly altered by ± 0.05% or so. After this initial adjustment, the bridge is let to self-balance. The design of the amplifier is done such that for the expected range of variation of R_w, the value of I_o goes through a change of ± 8 mA, so that a full-scale variation of I_o will be from 4 to 20 mA. I_o can be read or indicated at a distance from the bridge, as any small increase in line resistance does not affect the value of I_o required for balance. The circuit of Figure 8.9(b) uses a fixed compensating resistor R_c in series with R_w. The value of R_c is kept so small in comparison to R_w that the unbalance voltage developed will be enough to make the amplifier develop the desired output current I_o when $R_w = R_3$. For a change of R_w by ΔR_w, the change in I_o will be ΔI_o, and is given by

$$\Delta I_o = \frac{E_b}{2R_3^2} \tag{8.11}$$

These two circuits may be considered to serve as resistance-to-current converters and are useful for telemetering resistance variations to a distant location, and hence quantities that effect R_w.

8.6   HEAT-FLOW BALANCE SYSTEMS

Constant resistance type hot-wire anemometer can be constructed by using negative feedback for effecting the balance between the heat inputs to the hot wire due to changes in the velocity of flow and the current in the hot wire. The net change in the resistance of the hot wire is measured by means of the self-balancing bridge system shown in Figure 8.9(a).

The initial adjustment of the system is done when the fluid flow velocity is zero. With no current through the hot wire, R_w will be at the fluid temperature. The bridge is held under

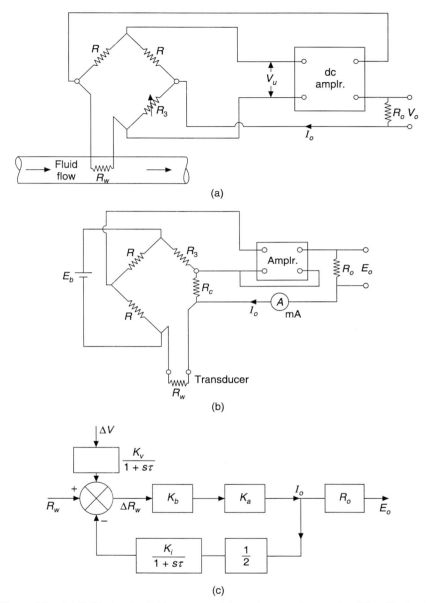

Figure 8.9 (a) Self-balancing bridge as a resistance-to-current converter; (b) self-balancing bridge for fluid flow rate measurement; (c) the block diagram for (b).

unbalance by keeping R_3 higher than R_w. With the application of excitation, I_o and the unbalance voltage V_u will cumulatively increase first and finally I_o will assume a steady value at which the resistance of hot-wire due to its being heated will be almost the same as R_3. A very small difference in their values exists because the finite-gain amplifier requires a small value for developing V_u.

When the fluid-flow velocity is brought to some steady value, the hot wire gets cooled and its temperature falls. The consequent rise in the unbalance voltage alters the value of I_o which compensates the fall in the value of R_w, and hence a state of equilibrium is established with the new value for I_o. The characteristic between the velocity of fluid flow and the value of I_o^2 will be as given in Eq. (6.19).

The system is also suitable for measurement and display of the fluctuating components of velocity about any steady flow velocity. If the equilibrium point for steady flow conditions is represented by steady flow velocity V_o, wire resistance R_{wo}, and current through hot wire I_{wo}, then the effect of small perturbations in the velocity and wire current can be obtained by linearized analysis as

$$\left.\frac{\Delta R_w(s)}{\Delta V(s)}\right|_{I_w = I_{wo}} = \frac{K_v}{1 + s\tau}$$

$$\left.\frac{\Delta R_w(s)}{\Delta I_w(s)}\right|_{V = V_o} = \frac{K_i}{1 + s\tau} \tag{8.12}$$

where K_v and K_i are the sensitivities of the hot wire for the changes in V_o and I_{wo} respectively and τ its time constant at the operating point, as given in Eqs. (6.22) and (6.23).

The gain of the dc amplifier is given by

$$I_o = K_a V_u$$

Assuming the Wheatstone bridge to be excited by a constant current source, the unbalance voltage V_u of the bridge can be obtained by using Eq. (3.15), as

$$V_u = I_o \left[\frac{m}{2(m+1)}\right] \Delta R_w = K_b \Delta R_w$$

where $m = R/R_w$.

The excitation current of the bridge divides equally into the paths and hence $\Delta I_w = \Delta I_o/2$ with $m = 1$.

Using the relationships, the overall transfer function relating the output voltag $\Delta E_o(s)$ (= $R_o \Delta I_o(s)$), and the change in the velocity $\Delta V(s)$ can be derived from the block diagram shown in Figure 8.9(c) and is given by

$$\frac{\Delta E_o(s)}{\Delta V(s)} = \frac{2K_v R_o}{K_i \left(1 + \dfrac{2s\tau}{K_a K_b K_i}\right)} \tag{8.13}$$

By making K_a very large, the dynamic performance of the system can be improved very much. The time constant of the hot wire may be reduced by 100 to 1000 times.

8.7 BEAM BALANCE SYSTEMS

Beam balance systems are a replica of the conventional mechanical type analytical balance used for comparison of forces in the context of weight determination. Beam balance system is a feedback measuring system used for comparison of the moments of two forces from which one physical quantity is converted into another physical quantity. The two physical quantities are required to set up forces proportional to them. Three typical examples are presented below.

Figure 8.10(a) shows the scheme of an electromagnetic balance in which the beam assumes horizontal position when the moments due to unknown and standard masses are equal. The horizontality of the beam is detected by the photoelectric cell and restored by the permanent-magnet moving coil system. The output of the photocell is processed suitably by the amplifier which in turn drives the current I_o through the moving coil. By suitably adjusting the initial coil position and gain of the amplifier, the system is made to assume null position when two masses are equal. If there is any difference in the two masses, the horizontality of the beam is disturbed and can be restored by the torque developed by the moving coil due to a change in I_o. This change in I_o is proportional to the difference in the two masses. Thus the system enables weighing very light objects apart from the ease of remote indication and applicability for continuous recording. Figure 8.10(b) shows the block diagram of a beam balance system when used for comparison of the moments due to two forces, one due to the error between the reference and measured pressure and the other due to the electromagnetic type force servo when carrying an output current I_o. As in all such feedback systems, the gain of the amplifier is kept high so that when the beam is horizontal, the moments are balanced. The block diagram is given in Figure 8.10(c) and the relationship in the null position can be shown to be

$$(P_r - P_m)A_b L_1 = K_f L_2 I_o \qquad (8.14)$$

The above relationship holds good under steady-state conditions.

Under dynamic conditions the time constants of the displacement transducer and the amplifier have negligible effect due to the negative feedback and high gain of the amplifier. The negative feedback has no effect on the time constant of the input bellows. The output current is generally of a range 4–20 mA and may be transmitted to a distance without having any effect of the line resistance and external electrical and magnetic fields on the indicated value. Where necessary, the drop across R may be converted into a digital voltage for digital read-out. The speed of response is sufficient for use in many process control schemes.

Figure 8.11(a), p. 473 shows a similar scheme but employing pneumatic null detector and developing output signals of air pressure variation equivalent to 3–15 psi for transmission over short distances of 50 metre. A flapper-nozzle valve detects the deviation of the beam from horizontality and the effect of the output pressure change on the feedback bellows is utilized to restore the beam to horizontality. The output pressure is proportional to $(P_r - P_m)$ when the beam is horizontal. The effect of the nonlinearity and time constant of the flapper-nozzle valve on the input-output characteristic is minimized to a large extent by increasing the stiffness K_b of the bellows-linkage combination and the sensitivity of the flapper-nozzle valve as an amplifier. To increase the gain of the forward path, a second stage of pneumatic amplification is used before the output pressure is fed back through the feedback bellows to restore the null conditions. The block diagram is shown in Figure 8.11(b), and it will be observed that the forward path belongs to a second-order system and hence the overall system is unlikely to become unstable for large gain values of the forward path.

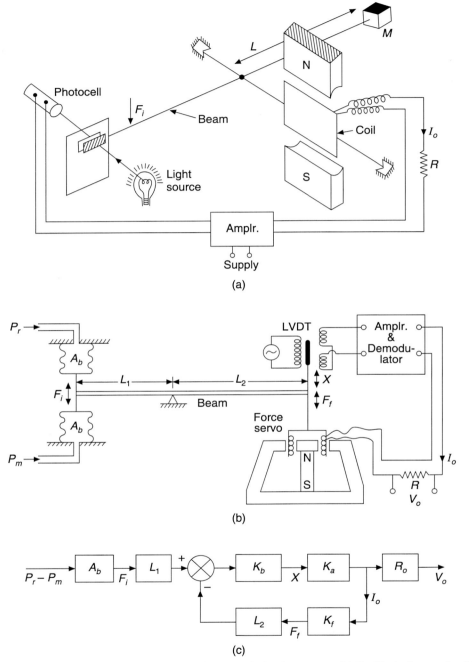

Figure 8.10 (a) Electromagnetic balance; (b) a beam-balancing system; (c) the block diagram for (b).

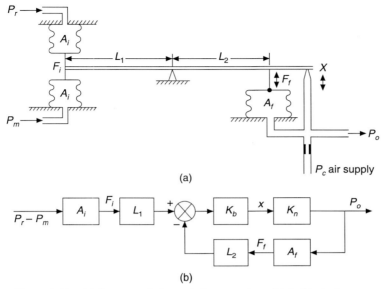

(a)

(b)

Figure 8.11 (a) A pneumatic beam-balance system; (b) its block diagram.

8.8 SERVO-OPERATED MANOMETER

The level of the liquid in manometers may be read automatically for indication and record purposes by making up a feedback measuring system. The system shown in Figure 8.12 is one of such servo-operated recorders, in which the position of a float resting on the surface of the manometric liquid is tracked by a linear variable differential transformer. The float has a small

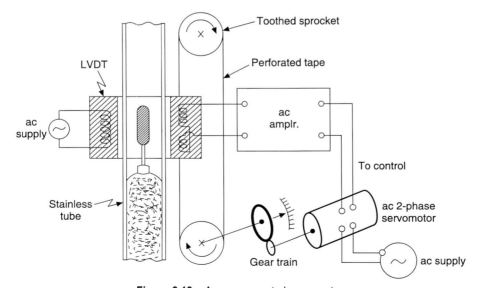

Figure 8.12 A servo-operated manometer.

iron core mounted on it which, when displaced from the null position of the LVDT, results in the development of output voltage signals. The signals are amplified by an amplifier and used to drive the two-phase ac servomotor in such a direction as to bring the coil system of LVDT back to the null position. In the null position, the LVDT output voltage is zero and the servomotor is at rest. The mechanical coupling of the servomotor with the toothed sprocket enables the perforated steel tape to move the LVDT into a new position along with a pointer and pen of an indicator/recording system.

For application to high pressure measurement, the manometer tube may be of stainless steel, which is nonmagnetic. But for low pressure measurements, a glass or transparent plastic tube may be used along with a photoelectric detector similar to the one employed in electromagnetic balance of Figure 8.10(a).

When the levels of liquid in the two legs of a manometer are to be read, two such systems are employed to follow the levels, and the difference in the levels is obtained by means of a gear differential coupled to the shafts of the two servomotors.

Such servo-operated manometers are used in process control systems, and they are found to possess high accuracy and resolution. The bandwidth of such recording systems is limited to 5 Hz and is satisfactory for pressure measurements arising in process control operations.

8.9 FEEDBACK PNEUMATIC LOAD CELL

Measurement of very large force and weight is normally undertaken by pneumatic and hydraulic load cells. These cells are small in size and are safe for use where fire hazards exist. The load cell is shown in Figure 8.13, and it simply consists of a suitably sized diaphragm and a flapper-nozzle valve. It is equivalent to a self-regulating system and does not possess a high gain amplifier in the forward path as can be observed from the figure. Applied force is balanced by the force due to output pressure on the underside of the diaphragm. The cell can be used either with air supply or hydraulic supply at a suitable high pressure of about 20 kg/cm^2. The gauge is used to read the output pressure. Hydraulic cells are faster in indication and can be used for measurements of weights up to 5×10^6 kg.

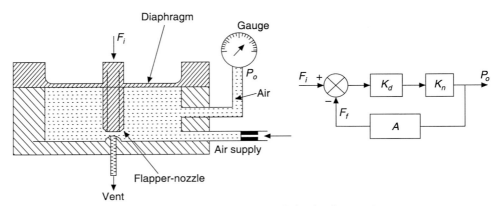

Figure 8.13 A pneumatic load cell.

The relationship between the input force F_i and output pressure P_o under steady state condition is given by

$$P_o = \frac{F_i}{A + (1/K_d K_n)} \qquad (8.15)$$

where,

K_d = the compliance (inverse of the spring rate) of the diaphragm

A = the area of the diaphragm

K_n = sensitivity of the flapper nozzle system

If $K_d K_n$ is sufficiently large, the output pressure P_o is linearly related to F_i. K_n is not really a constant and hence the linearity of the cell is affected by the nonlinearity of the flapper-nozzle system. It is seen that the input force F_i is limited to $P_o A$, and hence for measurement of large forces, A must be large, having chosen a convenient value for the supply pressure.

8.10 SERVO-OPERATED ELECTROMAGNETIC FLOWMETER

In Section 7.5.5, an electromagnetic flowmeter suitable for measurement of the flow rate, or the average velocity of slurries and conducting liquids with particulate matter, is described. The output signal E of Eq. (7.23) is seen to be proportional to the flux density B. If 50-Hz ac electromagnets are used, the B-value may fluctuate with supply voltage, resulting in error in measurement. It is possible to overcome this difficulty by application of feedback. A two-phase servomotor is used to complete the feedback system so that output angular displacements proportional to the average velocity may be obtained. The schematic diagram of the feedback system is shown in Figure 8.14. The field coils carry the exciting current of the electromagnet in whose field the pipe carrying the liquid is placed. The induced emf $e(t)$ between the two electrodes placed in the pipe is balanced by a feedback voltage $e_f(t)$, which should be proportional to the output angular displacement conveyed by the servomotor to the indicating pointer and the recording pen.

Any variation in the value of B affects $e(t)$, and hence it may be observed that $e_f(t)$ also has equal effect, thereby compensating the supply voltage fluctuation. For achieving this, the excitation current I_p of the electromagnet is passed through the primary winding of a current transformer, the secondary of which is kept almost short-circuited by a low-resistance potentiometer. The potential difference across this potentiometer is proportional to I_s and hence to I_p. Any swing in the value of I_p results in corresponding swing in the value of B and $e_f(t)$, assuming that B and I_p are linearly related in the operating region. A current transformer of suitable ratio is chosen so that $e(t)$ and $e_f(t)$ are equal. The servomotor drives the tapping point of the potentiometer until $e(t)$ is balanced by $e_f(t)$. Writing $B = I_p/K_i$ and $e_f(t) = I_s R_p \theta_o/\theta_F$, the system can be analyzed for its transfer function between output displacement θ_o and average flow velocity v_i. The forward path gain is made high by means of a high gain amplifier. Under steady state conditions, the error voltage is negligible and hence the relationship is given by

$$BDV_i = BK_i \frac{R_p}{\theta_F} \frac{N_p}{N_s} \theta_o$$

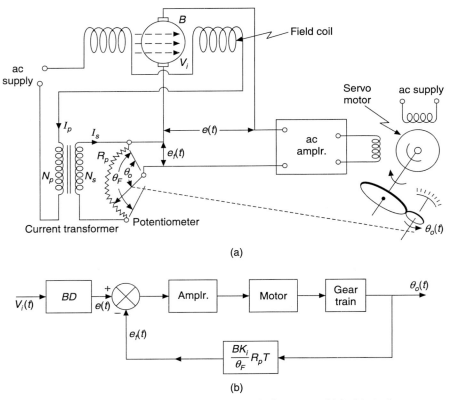

Figure 8.14 (a) A servo-operated electromagnetic flowmeter; (b) its block diagram.

$$\theta_o = \frac{\theta_F D V_i}{T R_p K_i} \tag{8.16}$$

where,

$T = N_p/N_s$, the ratio of primary to secondary turns of the C.T.

θ_F = full traverse of the potentiometer

R_p = resistance of the potentiometer

K_i = constant relating the primary current I_p with the flux density B

D = diameter of the pipe

The feedback system also has the merit of not drawing any current from the electrodes of the flowmeter. Due to the presence of the servomotor in the forward path, the measuring system behaves as a second order system for indication and recording of velocities within the bandwidth of 0–5 Hz.

8.11 FEEDBACK ACCELEROMETER SYSTEM

In all the open-loop type accelerometers, the relative displacement of the proof mass and the housing of the seismic system is measured by means of a displacement transducer. To improve the accuracy of measurement and the bandwidth of the system, feedback technique is applied as shown in Figure 8.15(a). The usefulness of the permanent magnet moving-coil system in the scheme can be seen as it provides both the feedback torque and the necessary damping to the system. The spindle of the coil carries an arm of length, r, at the end of which a proof mass M is connected. For acceleration inputs in the x-direction, the coil turns through an angle θ due to the inertial torque T_i given by

$$T_i = Mr\,\ddot{x}(t)$$

A feedback torque T_f equal to T_i should be developed by the coil under steady state conditions. To obtain the feedback torque, an inductive transducer or a linear variable differential transformer is used such that its output signals are amplified and processed to drive the d.c. current in the moving coil. Initially the system is zeroed such that a certain bias current of, say 12 mA, passes through the coil and the arm carrying M is held along a direction normal to the direction in which acceleration is desired to be measured. The core of the LVDT lies in the null position under these conditions. When the system is subjected to acceleration, the output signals of the LVDT are of such polarity as to develop feedback torque in opposition to the input torque T_i. If the output current of the amplifier is I_o, then under steady state conditions,

$$Mr\,\ddot{x}_i(t) = BANI_o = K_f I_o \tag{8.17}$$

where K_f is the torque constant of the moving coil system. An output signal of voltage may be obtained by using R, for purposes of analog indication and conversion into a digital signal. The transfer function of the pmmc system enters the forward path along with the transfer functions of the LVDT and amplifier. If the gains of the displacement transducer and amplifier are represented by K_p and K_a as shown in Figure 8.15(b), the system behaves as a second-order system of much larger bandwidth than the bandwidth of the moving coil system alone. If necessary, additional liquid damping is employed. The accuracy and stability of the system are very much improved. The gain of the amplifier cannot be increased at will, as the other stages of the forward path are likely to possess certain time constant. Very high value of gain renders the system unstable.

Such feedback or servo-operated accelerometers are widely used in inertial navigation and they are available in several ranges from ±10 g to ±100 g, possessing a natural frequency of 100 Hz to 250 Hz. Full-scale output current may be adjusted to 20 mA for use in industries, whereas for inertial navigation systems, it is held within ±1.2 or ±12 mA for acceleration inputs in both the directions.

In some cases, the knowledge of jerk, which is the rate at which acceleration is changing, is desired. A modification of the scheme of Figure 8.15(c) is possible so as to measure both the jerk and acceleration. The primary criterion in the design of the feedback measuring systems is to employ components in the feedback path, which constitute a transfer function that should be the inverse of the desired input–output transfer function of the measuring system. An electrical integrator is added to integrate the output voltage $v_o(t)$ and feed the current to the moving coil to develop the feedback torque (see Figure 8.15(c)). Under steady state conditions, the output voltage is proportional to jerk. But what is desired is the dynamic performance.

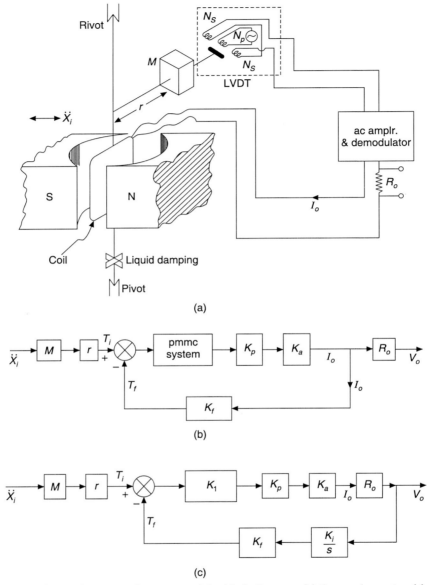

Figure 8.15   (a) A feedback type accelerometer and the block diagrams; (b) the accelerometer; (c) a jerkmeter.

Simplification of the figure reveals that the transfer function of the system belongs to a third order system and the gain of the amplifier cannot be increased beyond a critical value. Thus the bandwidth of the system is limited to very low frequencies only. Jerkmeters of this type are available for full scale jerk values ranging from ±0.5 g to 20 g/s and for acceleration range of ±1 g to ±30 g.

8.12 INTEGRATING SERVO

A motor which has its speed proportional to its applied voltage may be considered to serve the function of an integrator of the input voltage. The total angular displacement may be measured by a counter and the total number of revolutions will be proportional to the integral of the input voltage with time. A permanent magnet dc motor whose armature is excited by $e_i(t)$ develops a torque T_d given by

$$T_d(s) = \frac{[E_i(s) - K_m s\theta_o(s)]}{R_a} K_m \tag{8.18}$$

where,

$K_m s\theta_o(s)$ = back emf of motor
K_m = motor torque constant
R_a = armature circuit resistance
θ_o = output displacement

Considering the load on the shaft of the motor to be purely inertial, the load torque T_o becomes $Js^2\,\theta_o(s)$.

Equating the two torques and rearranging, the transfer function of the motor becomes

$$\frac{s\theta_o(s)}{E_i(s)} = \frac{1/K_m}{1 + s\tau_m} \tag{8.19}$$

where,

$$\tau_m = \text{motor time constant} \left(= \frac{R_a J}{K_m^2}\right)$$

J = inertia of the motor and load

The dc motors meant for such purposes are generally of small size having a τ_m of 0.01 s designed to operate on input voltages ranging from 1.5 V to 24 V. But at low speeds the performance is poor because of friction, iron losses and brush-contact voltage drop.

For greater accuracy and ability to deal with larger inertial loads, the feedback system shown in Figure 8.16 is used. A high gain amplifier drives the dc servomotor, and a dc tachogenerator is connected in the feedback path. If the amplifier gain K_a is made very large, it will be seen that

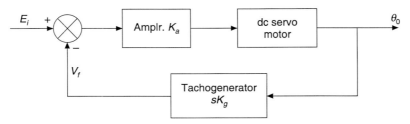

Figure 8.16 Integrating servosystem.

$$\frac{\theta_o(s)}{E_i(s)} \approx \frac{1}{K_g s} \qquad (8.20)$$

where, K_g = constant of tachogenerator.

If the time constant of the motor is taken into consideration, the transfer function is given by

$$\frac{\theta_o(s)}{E_i(s)} = \frac{K_a / K_m}{s[(1 + K_a K_g / K_m) + s\tau_m]} \qquad (8.21)$$

It is observed that the effective time constant is reduced sufficiently due to the large value of K_a. The accuracy of integration is improved to a large extent by employing feedback.

In aircraft instrumentation, speed measurement is required to high accuracy and resolution. An ac tachometer of the type shown in Figure 7.28 is used along with the smoothing circuit of Figure 7.29, and the smoothened dc is measured by a self-balancing servo-operated potentiometer of Figure 8.17. The output shaft of the servomotor drives the pen and the pointer in most cases. For obtaining high resolution in the speed measurement, an auxiliary multiturn helical potentiometer is used with its contactor coupled to the shaft of servomotor.

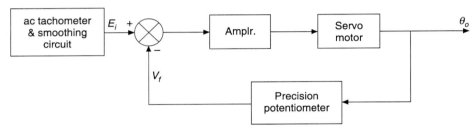

Figure 8.17 An accurate speed measuring system.

8.13 AUTOMATIC MEASUREMENT OF DEW POINT

Application of negative feedback may be used to eliminate the manual effort in several measurements, especially in the field of electrochemistry and analysis instrumentation. To determine the relative humidity, measurement of dew point is carried out by using feedback as shown in Figure 8.18. Dew point is the temperature at which the moisture in the atmosphere can saturate the air and is estimated by noting the temperature of a smooth polished surface of a mirror or a silver thimble when the moisture starts settling on it. Thus the temperature of the mirror has to be continuously and automatically adjusted such that the surface of the mirror starts becoming misty due to moisture settling on it. A small mirror is held at one end of a hollow metallic tube, the other end and part of it is in a cold bath of acetone and carbon dioxide. The tube is heated by an induction coil energized by a radio frequency oscillator. Light from a source is made to fall on two similar photo cells, one cell receiving the light directly and the other receiving it after reflection from a mirror. Whenever the mirror surface becomes misty, the unbalance in the illumination of the photo-cells is utilized to increase the heating of the induction coil. When the surface becomes bright, the heating level is reduced so that the cold bath starts cooling the mirror. Thus the system adjusts the temperature of the mirror such that

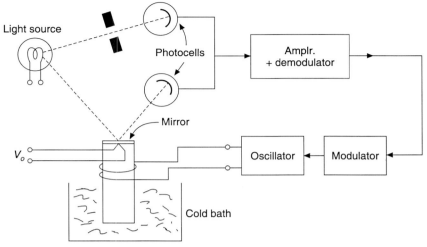

Figure 8.18 Automatic measurement of dew point.

its surface is just on the verge of being misty. Initial adjustment of the heating and illumination of the photocells is adjusted by means of the diaphragm. The temperature of the surface is measured by a thermistor or a thermocouple and the scale is calibrated in terms of the relative humidity of air. Thus the dew point is automatically determined with an accuracy of about ±1°C over the range of environment temperature from –25°C to +25°C.

8.14 NON-CONTACT POSITION MEASUREMENT

The cathode-ray oscillograph, with its electron beam controlled by the application of voltages to the deflecting plates, can be considered to function as an inverse transducer, in a feedback system for the measurement of position. A schematic diagram of such a system is shown in Figure 8.19, in which the amplitude of vibrations of an object in motion in a particular direction is measured without establishing physical contact with the object. The object is provided with a reference line above which the surface is silvered so as to function as a mirror, while the surface below is totally blackened. Light beam from a cathode-ray oscillograph is passed through a beam splitter and focussed onto the object by a lens. Reflected light is deflected by the beam splitter towards the photodetector. In case the spot is focussed on the blackened surface, no light is received by the photodetector. The output voltage of the photodetector is amplified and fed to the deflector plates in such a way as to shift the spot until it is focussed on the reference line serving as the contrast line. The system is so arranged as to see that the spot is locked on/to the reference line which, in other words, means that the spot is in correspondence with the motion of the object. Variations of the deflecting voltage with time are a measure of the amplitude of vibrations of the object from its mean level. Vibration studies of very light members, vibrating at reasonably high frequency, can be measured by the system.

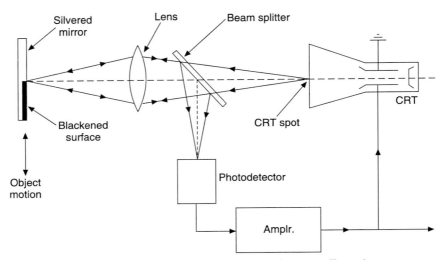

Figure 8.19 A position-follower-system using an oscillograph.

8.15 BIMORPH POSITION-CONTROL SYSTEM

As already explained, the piezoelectric bimorphs can be used as inverse transducers. A scheme showing the application of the bimorph for development of displacements proportional to command signal voltage is shown in Figure 8.20(a). When a voltage is applied to the bimorph, the upper element expands and the lower element contracts with the result that the cantilever bimorph bends and its tip is displaced from its neutral position: strain gauges mounted on either side of the cantilever bimorph sense the strain of each element, and the corresponding unbalance

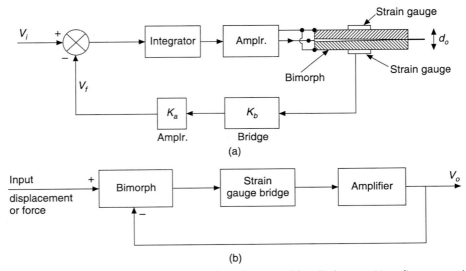

Figure 8.20 (a) A position control system using a bimorph; (b) a displacement-to-voltage converter.

voltage is amplified and fed back for comparison with the command signal. Any error signal present is integrated and amplified to drive the cantilever into such a position as to reduce the error. The linearity between d_c and v_i is mostly governed by the linearity of the strain gauges and the bridge circuit. The displacement d_o can be generated, with some amount of force, and the necessary power is drawn from the drive-amplifier of the forward path.

The entire scheme is reversible in operation so that input displacements and feedback displacements from the bimorph can be compared. The scheme is shown in Figure 8.20(b). When the input displacement is zero, the bimorph cantilever is flat and the strain gauge output voltage is zero. The output voltage is used to drive the bimorph in such a way as to annul the input displacement when the input displacement strains the bimorph. The output voltage necessary to hold the bimorph in horizontal position without any strain is proportional to the input displacement. The system may as well be used for the measurement of forces both in static and dynamic conditions.

8.16 OTHER APPLICATIONS OF FEEDBACK

In all the above examples, negative feedback is utilized to result in measurement systems with considerable improvements in linearity, speed of response, stability, accuracy and resolution, apart from making the system automatic. It has been shown that the provision of a high gain amplifier in the forward path can be utilized to develop an input–output relationship which is the inverse of the transfer function of the system connected in the feedback path. The technique is utilized widely in many signal processing circuits with the operational amplifier in the forward path. In Chapter 9, several such circuits are briefly described, indicating the extent to which the processing of the electrical signals of the transducers by means of electronic circuits render the instrumentation systems effective and versatile.

Automatic test systems and telemetry systems are based on the application of negative feedback. In the field of automatic control of industrial processes, standard units of control function generators developing output signals related to error signals are essential. Several such controllers of electrical, electronic, hydraulic and pneumatic are designed on the principles of feedback and are commercially made available.

EXERCISES

1. Distinguish the effects of positive and negative feedback in the performance of a closed-loop system.

2. Explain how negative feedback is applied to a measuring system and discuss the merits of applying the same.

3. Show that the product of overall gain and bandwidth of a closed-loop system having a single time constant for its open-loop gain is constant.

4. Discuss the effects of noise added (i) at the output terminals of comparator, and (ii) at the output terminals of the closed-loop system, on the overall performance of the negative feedback system.

5. Describe how a feedback transducer system is made up and explain how its performance is improved by using negative feedback.

6. What are force-balance and torque-balance systems and how do they enable accurate measurements?

7. Define *inverse transducers* and explain their role in feedback measuring systems.

8. What are the desired characteristics of inverse transducers? Suggest a few inverse transducers popularly used.

9. Show how a self-balancing dc potentiometer can be constructed by using negative feedback. How is its accuracy estimated?

10. Show how a self-balancing dc potentiometer may be made to indicate the unknown voltage, both in analog and digital fashion. Show how it will serve as analog indicator as well as recorder.

11. Explain how automatic balancing of dc Wheatstone bridges can be achieved by using negative feedback. Show how self-balancing is done (i) mechanically, and (ii) electronically.

12. Explain the principles of operation of a constant resistance type hot-wire anemometer and show how the speed of response of the system is increased by using negative feedback. Derive the necessary relationship and method of indication and recording of the flow rate of fluid.

13. What are beam-balance systems and what physical quantities can be measured by them? Describe the system for each.

14. Describe the constructional features of a servo-operated manometer. Is it suitable for use with any liquid in manometer?

15. What is a pneumatic load-cell and what is it used for? Can it be used with liquids in place of air?

16. Explain the basic principles of operation of an electromagnetic flow meter and show that the measurement may be made automatic.

17. Explain how a basic seismic system may be used for linear acceleration measurements. Show how a pmmc system enables automatic measurement of the same. What are the advantages of the system when it is used as a recorder?

18. What is an integrating servo and how is the system built?

19. Define *dew point*; what does it signify? Show the technique of measuring the same automatically.

20. What is piezoelectric bimorph and what is it used for? Does it serve as an inverse transducer? If so, illustrate its use in a feedback measuring system.

21. Discuss the advantages of making an automatic measuring system with the output quantity being a small dc current.

22. Indicate the standard output signal ranges employed for the pneumatic and electrical feedback measuring systems.

23. Show how current-to-voltage and voltage-to-current conversion is made possible by using a high gain amplifier and suitable feedback.

24. Discuss the nature of problems faced while designing a feedback measuring system. Is it worth increasing the complexity of the measuring system by use of feedback? If so, justify the same.

SUGGESTED FURTHER READING

Cromwell, L., *Biomedical Instrumentation and Measurements*, Prentice Hall, Englewood Cliffs (N.J.), 1973.

Doebelin, E.O., *Measurement Systems: Application and Design*, McGraw-Hill, New York, 1966.

Kurt, S. Lion, *Instrumentation in Scientific Research*, McGraw-Hill, New York, 1959.

Mansfield, P.H., *Electrical Transducers for Industrial Measurement*, Butterworth Scientific Publications, London, 1973.

Neubert, H.K.P., *Instrument Transducers*, 2nd ed., Clarendon Press, Oxford, 1975.

Noltingk (Ed.), Jones' *Instrument Technology*, Vol. 2, *Measurement of Temperature and Chemical Composition*, Butterworth Scientific Publications, London, 1985.

Norton, H.N., *Handbook of Transducers for Electronic Measuring Systems*, Prentice Hall, Englewood Cliffs (N.J.), 1969.

Oliver, F.J., *Practical Instrumentation Transducers*, Pitman, London, 1972.

Wightman, E.J., *Instrumentation in Process Control*, CRC Press, Columbus, Ohio, 1972.

Willard, H.H., Merrit, L.L. and Dean, J.A., *Instrumental Methods and Analysis*, 4th ed., Affiliated East-West Press, New Delhi, 1965.

Woolvet, G.A., *Transducers in Digital Systems*, Peter Peregrinus, London, 1977.

CHAPTER 9

Signal Processing Circuits

9.0 INTRODUCTION

Sophisticated instrumentation systems rely heavily on the application of a wide range of electronic circuits. Signal processing circuits constitute the essential link between the transducers and the final output equipment such as readout systems, computers and other devices. The signals obtained from transducers may be in analog or digital domain; very few instrumentation systems are fully digital by nature. Signal processing and conditioning is carried out in order to bring up the output signal to the desired description and standardized levels. In most cases, it is essential to see that the signal processing circuitry chosen preserves the desired functional relationship between the input and output signals and does not in any way impair the basic accuracy with which the measurement is carried out. The functions of the circuitry may be conversion of analog signals into digital signals and vice versa, conversion of voltages into currents and vice versa or conversion of voltages into frequency and vice versa. The power levels at the input and output ends are sufficiently low throughout except at the final stage where it may be necessary to drive the end device with a little power. Control windings of servo-motors, solenoids and relays may usually be the end devices.

Data acquisition systems and other processing circuitry may consist of several small plug-in modules, each effecting the desired function on its input signal. The availability of solid state circuits in the form of IC chips makes the job of system designer simpler.

In the following sections, some of the electronic circuits are briefly presented with a view to revealing the scope of their application in instrumentation. Excellent books are available on Analog and Digital Electronics and the reader is advised to go through the Suggested Reading for additional information.

9.1 DC POWER SUPPLIES

A large number of transducer systems and electronic circuits designed for discrete functions are energized by dc power supplies, usually within the voltage range 6 V–30 V. The sources of supply may be a bank of batteries (or dry cells) or full-wave rectifier circuits operated from ac supply through transformers. The transformer provides the proper voltage and, in addition, the more important isolation between the instrument ground and the power line.

Ripple reduction, to a large extent, and good voltage regulation are highly desired in almost all cases. Integrated circuit type voltage regulators of differing designs and specifications are available for selection and application. The more important specification is related to regulation against variations in load and line voltage.

Regulation against load refers to the change in output voltage since load current is varied. It is usually specified as the fractional change in output voltage when the load current is increased from zero to the maximum value of its range. Most of the regulators are designed to be within ±0.2% for regulation against load. Regulation against line voltage fluctuations is the percentage change in output voltage for a given percentage change in line voltage. It is about ±0.2% for a 10% change in line voltage.

Precision voltage regulators are those that develop output voltage much closer to true voltage and are required where a reference voltage is assumed as in some digital voltmeters and AD converters.

Zener diodes constitute the heart of all the precision voltage regulators, and a variety of IC regulators are available with differing specifications. The output voltage is compared with the reference voltage obtained across the Zener diode and the difference is amplified and used to correct the output voltage. The two simple circuits shown in Figure 9.1 illustrate the use of Zener diode. Voltage regulators incorporating op-amps perform much better.

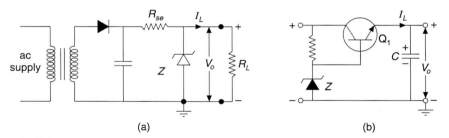

(a)	(b)

Figure 9.1 Voltage regulators using: (a) A simple Zener diode, and (b) an emitter-follower Zener diode.

The stability of the output voltage is entirely dependent on the stability of the Zener diode. The temperature coefficient of the Zener diode is small and is ignored when it is used at voltages between 6 V and 9 V. When excellent stability over a long time of usage is desired, it is kept in a temperature-controlled casket.

The circuit of a typical series-type voltage regulator using an op-amp for error voltage amplification is shown in Figure 9.2. The op-amp drives the base of the power transistor Q_1. The drop across R_{se} is small with Q_1 effectively working as an emitter follower. Hence the output voltage is nearly equal to the base voltage less the usual base-emitter diode drop. The function of θ_2 is to provide short-circuit protection. Normally, Q_2 is off. Only when a high load current flows, the voltage drop across R_{se} exceeds the emitter-base diode voltage. The collector current flows through Q_2 and Q_1, causing reduction in output voltage.

IC voltage regulators are sometimes provided with an overvoltage protection, by means of a silicon-controlled rectifier circuit as shown in Figure 9.2.

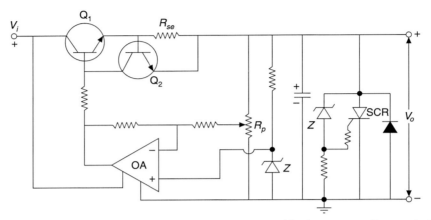

Figure 9.2 Series type voltage regulator using an op-amp and having an overvoltage protection.

In addition to the above regulated power supplies, a precision variable voltage-reference source is required for use in instrumentation systems. The circuit shown in Figure 9.3 provides variable output voltage that can be adjusted to give the desired value by comparison with a known standard voltage. The Zener diode is usually operated around 6 volts and is temperature controlled. Fine adjustment is possible by means of a 10-turn potentiometer for R_p. Long-term stability of about 10 ppm is achievable.

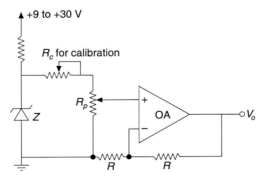

Figure 9.3 A precision variable voltage-reference source.

Smaller reference voltages less than 2.5 V can be obtained by using a series combination of several forward-biased silicon diodes or LEDs.

9.2 OSCILLATORS AND SIGNAL GENERATORS

Certain transducers and Wheatstone bridge networks with one or more arms consisting of transducers are excited by ac supply voltages, mostly sinusoidal in waveform. Low distortion is one of the essential requirements for these applications. The amplitude and frequency of the voltage, in all such cases, should be stabilized to a high degree of accuracy. The frequency of the voltage may be chosen at any value within the range of 0.001 Hz–1 MHz, depending on the actual application. The sources are invariably treated as fixed-frequency oscillators, with no provision for tuning, as is normally found in variable frequency oscillators.

When the dynamic behaviour of physical systems has to be experimentally determined, signals generators whose frequency can be varied over the desired range are used. The waveform may be sinusoidal, square wave or saw tooth, but the amplitude of the output voltage should be held constant when the frequency is varied.

Equally important are the signal generators whose output signals are either a series of unipolar pulses or bipolar rectangular waves. They are very much used for pulse modulation and digital applications. It is possible to convert waveforms of one shape to another once the frequency of the generated signals is accurately stabilized and known.

Conversion of voltage to frequency has a particularly important role in A-D converters and other data acquisition systems. The circuits performing this conversion are known as voltage-controlled oscillators (or V–F converters) and are used in conjunction with a transducer, usually of resistive type, to generate output voltages of a frequency proportional to the amplitude of the input voltage. The output signals thus brought out into frequency (or time) domain enable counting operations for digital readout as also transmission of the signal to a remote location, without entailing any error due to the transmitting link.

Similarly, tuned *LC* oscillators enable the conversion of a measurand into frequency, when an inductive or capacitive transducer forms one of the elements of the *LC* circuit. Such tuned-frequency oscillators are known as subcarrier oscillators when their output signals are used for modulation of a carrier signal for telemetry and transmission purposes. The resonant frequency of the *LC* shunt circuit, when the amplitude of the measurand is zero, is given by $f_0 = \dfrac{1}{2\pi\sqrt{LC}}$.

Variation in the values of *L* or *C* occurs in such a way that their fractional changes ($\Delta L/L$ or $\Delta C/C$) are proportional to the instantaneous amplitude of the measurand, and hence the instantaneous frequency deviation Δf *is* directly proportional to the fractional change, provided $\Delta C/C$ is less than 0.05. The frequency deviation Δf is so approximated that

$$\Delta f = f_0\left(1 \pm \frac{\Delta C}{C}\right)$$

or

$$\Delta f = f_0\left(1 \pm \frac{\Delta L}{L}\right) \tag{9.1}$$

Some of the popular oscillator circuits are briefly described in the remaining part of this Section 9.2.

9.2.1 Wien-Bridge Oscillators

One of the most commonly used sine-wave generators is the Wien-bridge oscillator for generation of signals in the low frequency range and up to audio frequency range. The circuit has the merits of simplicity, amplitude stability, and ease of frequency variation.

Two simple circuits are shown in Figure 9.4, showing the technique of obtaining positive feedback for building up oscillations, by means of an op-amp and a feedback network consisting of resistors and capacitors. The frequency of oscillation is precisely governed by the values of R and C and is given by

$$f = \frac{1}{2\pi RC} \tag{9.2}$$

The amplifier gain is adjusted so that the closed-loop gain is slightly greater than unity for small signals. The nonlinear element (the diode) provides the means for reducing the gain at high signal amplitude. The amplitude of output signals increases until it self-limits. The limiter of circuit of Figure 9.4(a) enables the waveform to become reasonably sinusoidal in waveshape and, without it, the waveform approaches a square wave. Figure 9.4(b) has the provision for automatic gain control by means of an FET. It is possible to achieve generation of sinusoidal signals of negligible distortion and precise amplitude control by means of additional circuits.

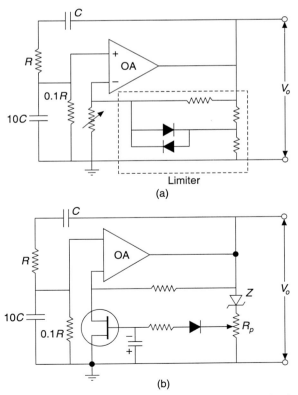

Figure 9.4 Wien-bridge oscillators using an op-amp with positive feedback.

RC phase-shift oscillators constitute alternative sources of sinusoidal signals at low frequencies. The frequency of oscillation is precisely governed by the values of *R* and *C* used. Its principle of operation is similar to that of the Wien-bridge oscillator.

9.2.2 Crystal Oscillators

Crystal oscillators have an important role in instrumentation systems as they are often employed as accurate time references. Most stable oscillators have the crystal contained in a temperature-controlled oven. However, for common applications, the stability in frequency is sufficiently high as the temperature coefficient of the quartz crystal is very small. Crystal oscillators are normally for high frequency within the range 100 kHz–10 MHz.

The simple circuit of Figure 9.5 is used when the output signals are desired to be nearly sinusoidal and is known as Pierce oscillator. The constants of the crystal and the value of *C* determine the frequency of oscillation. The frequency stability is of the order of 10^{-6} to 10^{-10}.

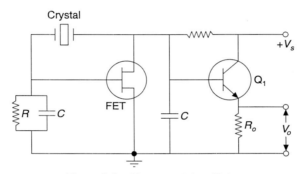

Figure 9.5 Pierce crystal oscillator.

The circuit shown in Figure 9.6 is particularly convenient for digital applications, where the signals are required to be square wave type. These circuits are usually built from integrated circuits of transistor-transistor logic (TTL) or complementary metal-oxide-semiconductor (CMOS). The principle of operation is the same as in all oscillators, i.e. building up of oscillations using positive feedback. The circuit shown has two inverters in cascade to provide the positive feedback. Such circuits are very much needed for driving counters and frequency dividers.

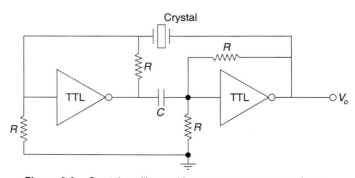

Figure 9.6 Crystal oscillator with square-wave output voltages.

9.2.3   LC Tuned-circuit Oscillators

Tuned-circuit Oscillators consist of resonant circuits of a self-inductance L in parallel with a capacitor C and amplifier whose output is fed back to provide for continuous oscillations. The commonly known versions are Hartley and Colpitts oscillator circuits.

The simple circuit of a Hartley oscillator using a transistor is shown in Figure 9.7(a). The tuning capacitor (or the capacitive transducer) is shunted across the coil with a tap. The oscillator can be operated at high frequencies up to a few MHz.

The Colpitts oscillator provides tuning over a wider range at frequencies above 30 kHz. The circuit shown in Figure 9.7(b) consists of a FET and an emitter follower to serve as a buffer between the oscillator and the load. The frequency of oscillation is given by $1/(2\pi\sqrt{LC})$, and tuning is usually done by variation of L. The circuit is useful as a subcarrier oscillator with the self-inductance type transducer in place of L.

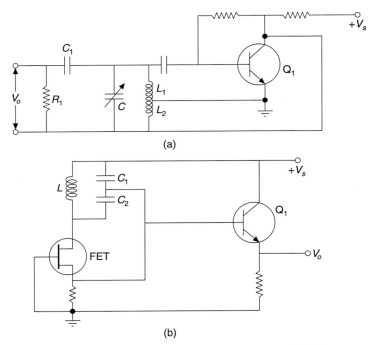

(a)

(b)

Figure 9.7   LC tuned-circuit oscillators: (a) Hartley type; (b) Colpitts type.

9.2.4   Square-wave Generators

An astable (free-running) multivibrator using an RC circuit and an op-amp is the most commonly used form of square-wave generator. The simple circuit is shown in Figure 9.8 with the frequency variation brought about by variation of R. The frequency of oscillation is approximately given by $1/RC$. These circuits are useful for frequencies up to 30 kHz.

There are several methods of obtaining square-wave generation. Pulse generators, triangular-wave generators and sine wave oscillators may also be used to provide square-wave output.

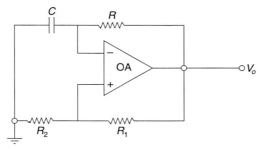

Figure 9 8 An astable multivibrator for square-wave output.

9.2.5 Pulse Generators

Pulse generators and square-wave generators differ in respect of the mean level of the pulse train. The square-wave generator has the mean level of zero and the positive and negative pulses are each of duration equal to $T/2$, where T is the period of the wave. In a pulse generator, the pulse is on for a time T_1 and off for a time T_2, and the pulse rate is given by $1/(T_1 + T_2)$, as shown in Figure 9.9(a).

A simple pulse generator using a unijunction transistor is shown in Figure 9.9(b). The capacitor C is charged from V_s through R and the charging time can be varied by R. When the voltage across the capacitor is equal to the threshold voltage of the unijunction transistor (UJT), the transistor conducts and partially discharges the capacitor. The output voltage across R_L is processed suitably to constitute the pulse train of desired on-time and pulse rate. The pulse rate is proportional to $1/RC$.

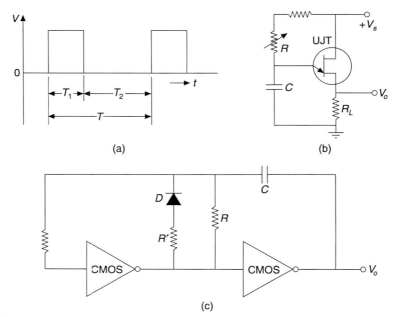

Figure 9.9 (a) Pulse generator waveform; (b) simple pulse generator using a UJT; (c) pulse generator using two inverters.

Another simple version consists of two CMOS used as two inverters as shown in Figure 9.9(c). They alternately charge the capacitor C and force the output of the following inverter to reach high state. The diode D and resistor R are used to provide the mean level of pulses at zero and vary the off-time and pulse rate. The on-time is approximately equal to RC and off-time, $R'C$. The circuit becomes a square-wave generator if the diode and R' are switched off from the circuit.

9.2.6 Function Generators

Low-frequency oscillators are designed by using op-amps based on the method of building-up oscillations satisfying the necessary conditions of a second-order differential equation. Simulating the equation by means of two integrators and an inverting amplifier is possible; a simple circuit is shown in Figure 9.10. The frequency of oscillation is given by $f_o = 1/(2\pi RC)$ and sinusoidal waveshapes can be generated at frequencies above 10 Hz. The potentiometer R_p serves to vary the damping of the circuit from positive values to negative. Such sine-wave generators enable the variation of the frequency by variation of V_0 and V_1 in a suitable way from an external voltage, and constitute frequency modulation.

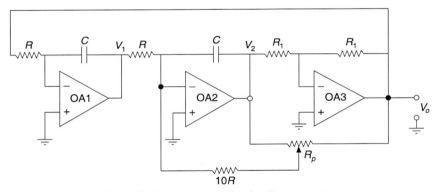

Figure 9.10 A second-order function generator.

Apart from the above method, function generators can be designed by using a Schmitt trigger and an integrator to generate a triangular output voltage. The simple scheme shown in Figure 9.11 enables generation of not only rectangular and triangular waveshapes but also sinusoidal outputs by making use of sine-function networks. The frequency of the output voltage is given by

$$f = \frac{1}{4RC}\frac{R_2}{R_1} \tag{9.3}$$

The scheme is useful as a function generator with controllable frequency or a voltage-to-frequency converter. It is possible to vary the input voltage of the integrator and thereby modulate the frequency of oscillation. The circuit shown in Figure 9.12 may be used as a voltage-to-frequency converter by varying the amplitude of input voltage, and without going into the details of operation, it can be shown that the frequency of oscillation is given by

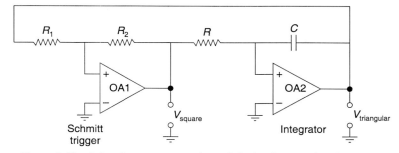

Figure 9.11 A function generator using a Schmitt trigger and an integrator.

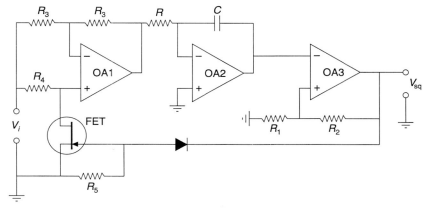

Figure 9.12 A voltage-to-frequency converter.

$$f_o = \frac{R_1 + R_2}{4R_1 RC} \frac{V_C}{V_{i\,max}} \tag{9.4}$$

The analog switch using the FET is satisfactory for generating voltages of a frequency up to 10 kHz. For generating higher frequency outputs, the circuitry may be modified by using a transistor switch in place of the FET. For high stability of amplitude and frequency, *OA* 3 may be replaced by a precision type Schmitt trigger.

It is convenient to generate square wave voltages of variable duty cycle by effecting simple modifications in the circuit shown and addition of a circuit for splitting the input control voltage for altering the duty cycle.

9.3 HIGH FREQUENCY AMPLIFIERS

For industrial instrumentation systems, dc and low-frequency amplifiers are commonly required (Some of the amplifier configurations have been described in Section 3.5). They also form the heart of many signal conditioning circuitry. They are all suited for handling and amplifying signals of frequencies up to 100 kHz. When amplification of signals of a higher frequency is desired, extreme care is needed in component choice, component placement, wiring, and minimization of unwanted couplings. Although there may not be many situations in instrumentation warranting the use of high frequency amplifiers, a brief introduction to some of them is considered desirable.

9.3.1   Tuned RF Amplifiers

In FM telemetry, high frequency amplifiers are required to amplify radio-frequency signals of a frequency range 100 MHz–2000 MHz. They are designed to amplify signals of a narrow bandwidth, and are therefore tuned to a fixed frequency. The amplifying devices may be electron tubes with *LC* tuned circuits or, if the frequency is lower, special IC networks are used as tuned circuits.

9.3.2   Wideband Amplifiers

For use in electronic instruments, such as cathode ray oscillographs, a fixed-gain wideband amplifier is required. Integrated circuits are available for use at frequencies up to 100 MHz. The gain is not very high and is usually limited to 100–300. Wideband amplifiers are often single-ended.

9.3.3   High Frequency Op-amps

By using internal or external compensation circuits, it is possible to have the frequency range of a conventional op-amp extended to a higher limit. Some of the op-amps have a bandwidth of 15 MHz and a slew rate of 50 V/μs. Amplifiers with gain-bandwidth products in the gigahertz region are also available. In all such cases, circuit layout is important.

9.4   COMPARATORS

An analog comparator is equivalent to an analog switch that initiates the switching process when a continuous input voltage exceeds or falls below a certain predetermined level. An op-amp without feedback can perform this function satisfactorily as its output voltage attains the positive or negative saturation value, depending on whether the input voltage difference, $(v_+ - v_-)$, is positive or negative. Owing to its high gain, the op-amp responds to a very small voltage difference and hence it is suitable for comparison of two voltages.

However, at zero crossing of the input voltage v_i, the output voltage does not immediately reach the saturation value, because the transition is limited by the slew rate which is ≈ 1 V/μs. An additional delay is incurred due to the recovery time needed for the amplifier, after it has been saturated. These difficulties do not pose any problem when dealing with voltage signals obtained from transducers, as their frequency is not very high.

In certain situations, it is necessary that the output voltage of the comparator accurately assume two well-defined levels. To achieve this, the output voltage of the comparator is fed back to operate the switch. In the circuit shown in Figure 9.13, two Zener diodes are connected in the feedback path so that the output voltage levels are governed by the Zener diode characteristics.

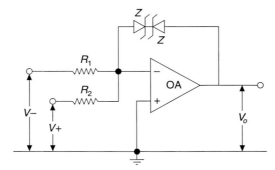

Figure 9.13 A Zener diode type analog comparator.

9.4.1 Window Comparator

A window comparator is the circuit that is used to determine whether or not the value of an input voltage lies within two reference voltage levels. The circuit thus requires two comparators connected to a logic circuit as shown in Figure 9.14. The output of the logic circuit is indicative of 'yes' or 'no'. The combination is available as an IC chip.

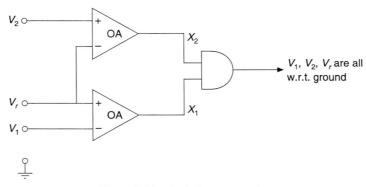

Figure 9.14 A window comparator.

9.4.2 Schmitt Trigger

A Schmitt trigger is a comparator for which the positive and negative transitions of the output occur at different levels of the input voltage. It is similar to a comparator with hysteresis. Schmitt trigger circuits can be realized in several ways, but their complexity is due to the function for which they are designed.

The difference between the levels of the input voltage that causes the transitions is characterized as *hysteresis*. In the circuit shown in Figure 9.15(a), the hysteresis is accomplished by means of positive feedback through the potential divider arrangement. The input–output characteristic is given in Figure 9.15(b).

The threshold voltage for switching on the circuit is determined by R_1 and R_2 and is given by $V_2 R_1/(R_1 + R_2)$, and similarly, the voltage for switching off is $V_1 R_1/(R_1 + R_2)$. Thus the width of the hysteresis is governed by R_1 and R_2 for a particular op-amp.

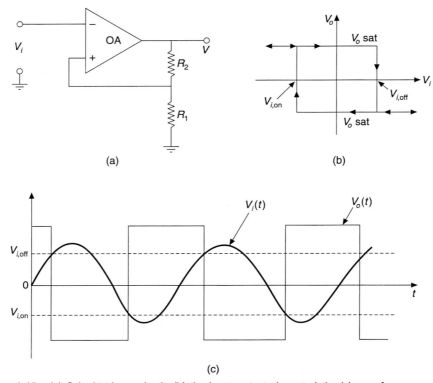

Figure 9.15 (a) Schmitt trigger circuit; (b) the input–output characteristic; (c) waveform conversion.

The circuit has an important application in instrumentation for conversion of voltage signals of any waveform into square-wave signals. From Figure 9.15(c), the operation of the waveform converter can be understood. The zero crossings of the input voltage signal and the rectangular wave generated do not occur at the same instant. The square-wave output voltage has the same frequency as the fundamental of the input voltage, and the amplitude of the square wave is rendered fixed, in spite of fluctuations in the input voltage due to noise.

Precision Schmitt triggers are those in which either the trigger levels or the output voltage levels are known accurately, instead of being determined by the voltage levels at which the op-amp saturates.

Schmitt triggers are available in IC chip form with different specifications and for different functions.

9.5 ACTIVE ELECTRICAL FILTERS

In Section 4.7, frequency-selective systems are introduced, indicating their role as filters to segregate signals of certain frequencies from the rest. Electrical filters play an important role in all the modern data transmission and signal processing fields, and it is difficult to conceive design of any electronic measuring system without the presence of an electrical filter. Analog filters are used for processing of analog signals, whereas digital filters process digitized continuous signals.

Simple first- and second-order filters employing resistors, capacitors and inductors are introduced in Section 4.7.1, and they are known as *passive filters*. The major limitations in their performance are:

1. Signal source should be of constant-voltage type.
2. Load impedance should be nearly infinite.
3. Voltage gain of the filter cannot be higher than unity unless it has an inductor in the circuit, which resonates with a capacitor.
4. Amplitude-frequency response characteristic is not truly flat in the pass-band region.
5. Phase-frequency characteristics are not linear with frequency in the passband.
6. In the stop (reject)-band region, the attenuation in not infinite.
7. The rate of attenuation of frequencies beyond cut-off frequency is usually unsatisfactory.
8. The presence of an inductor in the filter network is a source of noise and interference.

Depending on the requirements, it is possible to design passive filters using modern theories of approximation and realization having the desired amplitude-frequency characteristic or the phase-frequency characteristic. But in such a case, the filter becomes a higher-order filter, requiring a large number of components. Such filters are available as Butterworth, Chebyshev or Bessel type and involve inductors in the circuit. At times, it is also desired to have an 'all-pass' filter that passes all frequency components equally well but with the phase being a function of frequency.

In the design of filters for use at low frequencies, say from 0.1 Hz onwards up to 0.5 MHz, inductors become impractical for incorporation in the filter circuit because of their size and large deviation from ideal behaviour. They are not adaptable to integrated circuit techniques; which have now become very common due to their miniature size, low weight and reliable performance. Therefore, filter designs are modified so as to incorporate active elements such as transistors in the integrated circuit. Thus modern filters are seen as *active filters* because other circuits functioning as operational amplifiers, gyrators or negative impedance converters can be integrated with resistors and capacitors. Such designs have yielded considerable versatility in the performance characteristics of the filter. The commonly used active filters are of the *RCA* type, where *A* refers to the high-gain amplifier, and they have considerably enlarged the scope of application in the field of instrumentation.

Some of the more common configurations of the active filters of the first-, second-, and higher-order type are briefly described in the remaining part of this Section 9.5. Relevant theoretical background for the design of these filters is available from literature on Passive and Active Network Design.

9.5.1 Simple Active Filters

Simple versions of active filters may be designed by the addition of a dc high-gain, single-ended amplifier or a differential operational amplifier to an *RC* circuit.

An active low-pass filter of first order is shown in Figure 9.16(a) using an operational amplifier. The transfer function relating the output and input voltages may be obtained as

$$\frac{V_o(s)}{V_i(s)} = \frac{-R_b}{R_a(1 + sCR_b)} \tag{9.5}$$

The filter network is acting as an inverting amplifier with a dc gain of R_b/R_a.

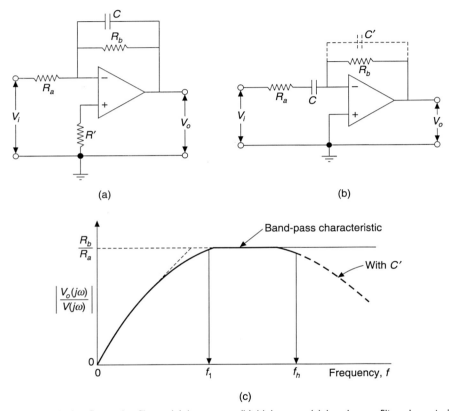

(a)   (b)

(c)

Figure 9.16   Active first-order filters: (a) Low pass; (b) high pass; (c) band pass filter characteristic.

Although the frequency response characteristics are not different from those of the equivalent passive network of Figure 4.10, the active filter enables one to make the gain variable and greater than unity by adjustment of R_b. The filter can be used with finite impedances on both the source and load-side, without much effect on the filter characteristic.

The inverse equivalent of the low-pass filter is the high-pass filter shown in Figure 9.16(b), with its TF being given by

$$\frac{V_o(s)}{V_i(s)} = \frac{-sCR_b}{1 + sCR_a} \tag{9.6}$$

The voltage gain at very high frequencies is $-R_b/R_a$.

In the event of a capacitor C' shunting R_b, the TF gets modified as

$$\frac{V_o(s)}{V_i(s)} = \frac{-R_b sC}{(1 + sCR_a)(1 + sC'R_b)} \tag{9.7}$$

The gain at zero frequency and at very high frequencies becoming zero, it can be said that the network behaves like a band-pass filter for some intermediate frequencies between the lower and upper cut-off frequencies, as shown in Figure 9.16(c). The value of C' can be so chosen as to obtain a gain of R_b/R_a for some frequencies within the pass band.

As indicated in Section 4.7 (Figure 4.29), it is possible to obtain band-pass and band-reject filters by cascading the active low-pass and high-pass filters. In such a case, a buffer amplifier between the two filters is not needed, as the first filter network is not loaded by the next stage.

A first-order *all-pass type filter* having its TF given by the equation

$$\frac{V_o(s)}{V_i(s)} = \frac{1 - sT}{1 + sT} \tag{9.8}$$

can be obtained by using the simple RC low-pass and high-pass filters along with a difference amplifier as shown in Figure 9.17. Considering the gain of the amplifier to be K, $V_o(t)$ can be

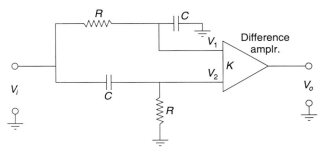

Figure 9.17　An all-pass filter using simple RC circuits and an amplifier.

obtained in terms of $V_1(t)$ and $V_2(t)$. Using the transfer functions of $V_1(s)/V_i(s)$ and $V_2(s)/V_i(s)$, $V_o(s)$ is obtained as

$$\frac{V_o(s)}{V_i(s)} = K\frac{1 - sT}{1 + sT} \tag{9.9}$$

The all-pass filter has a constant magnitude of K for all frequencies but the phase shift between output and input signals changes from $0°$ to $180°$.

The twin-T network of Figure 4.31(a) can be used with a high-gain amplifier, resulting in certain specific advantages. A rejection amplifier having steeper slopes for the characteristic of the twin-T notch filter can be obtained by using the twin-T network in the forward path of the unity feedback system, as shown in Figure 9.18(a). An analysis of the feedback system indicates that the higher the gain of the amplifier, the sharper is the null.

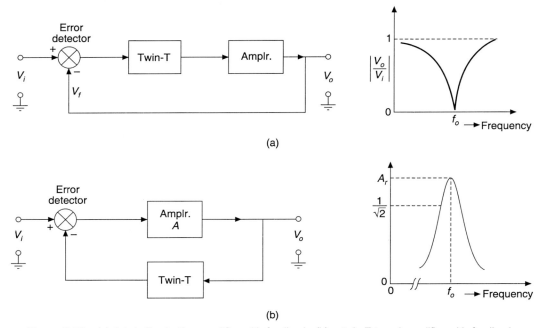

Figure 9.18   (a) A twin-T rejection amplifier with feedback; (b) a twin-T tuned amplifier with feedback.

If the position of the twin-T network is changed to the feedback path, as shown in Figure 9.18(b), while retaining the high gain amplifier in the forward path, it can be shown that the overall characteristic will have a gain of unity at very low frequency and very high frequency, but at the null frequency of the twin-T, the gain is the same as the gain of the amplifier in the forward path. This is because at null frequency, the feedback path is equivalent to an open circuit, thereby making the overall system behave like an open-loop-system. The entire system, thus, behaves like a narrow band-pass filter, having very high gain for notch frequency, while not rejecting totally any other frequency component.

9.5.2   Second-order Filters

Ideal filter characteristics as indicated in Figure 4.26 are physically unrealizable. So a response, which is practical and realizable and which approximates to the ideal response within some set of tolerances specified for the pass-band, reject-band and transition region of the filter, is required. Approximating the gain-frequency characteristic of these three distinct regions and realizing the corresponding filters makes filter design varied and extensive. Butterworth filters are the outcome of a realizable approximation of the gain-frequency characteristic of an ideal low-pass filter. The magnitude of the voltage gain of the filter is expressed by

$$\left| \frac{V_o(j\omega)}{V_i(j\omega)} \right| = |H(j\omega)| = \frac{A_{\text{dc}}}{(1+\omega^{2n})^{1/2}} \quad \text{for } n = 1, 2, 3, \ldots \tag{9.10}$$

The characteristic resulting from the above approximation is known to be of the nth-order Butterworth low-pass filter.

A_{dc} may be equal to unity if it is realized as a passive filter: otherwise it may be made higher than unity, if synthesized as an active filter. At $\omega = \omega_c = 1$ rad/s, the value of $|H(j\omega)|$ $= A_{dc}/\sqrt{2}$, and hence the filter realized is said to be frequency-normalized. Frequency scaling may be utilized to denormalize the filter with $\omega_c = 1$ to obtain any desired cut-off frequency. Also, transformation techniques enable conversion of the normalized low-pass filters into high-pass, band-pass and band-reject filters.

Although it is possible to synthesize low-pass filters by using resistors, capacitors and inductors, active filters using resistors, capacitors and op-amps are synthesized and are commercially made available as integrated circuit devices. Active filter networks offer several other advantages, the more important of them being adjustable gain and low output impedance.

Another important approximation due to Chebyshev resulted in the formation of Chebyshev filters with the magnitude of the gain given by

$$\left| \frac{V_o(j\omega)}{V_i(j\omega)} \right| = |H(j\omega)| = \frac{A_{dc}}{[1 + \varepsilon^2 C_n^2(\omega)]^{1/2}} \tag{9.11}$$

where ε is a design constant and $C_n(\omega) = \cos(n \cos^{-1} \omega)$, which is known as Chebyshev polynomial of the first kind of degree n.

The Chebyshev filter is also known as *equiripple filter* because, in the pass band, the value of $|H(j\omega)|$ exhibits ripples of equal magnitude, with the ripple width RW given by

$$RW = 1 - \frac{1}{\sqrt{1 + \varepsilon^2}} \tag{9.12}$$

The ripple width is often expressed in dB and is equal to 0.5 dB for $\varepsilon = 0.3493$.

Figure 9.19(a) shows the response characteristics of the Butterworth low-pass filter, for n values of 1 to 5 (i.e. second order to 10th order). It is seen that as n increases, the sharpness of the cut-off and the approach to the ideal response increase.

Figure 9.19(b) shows the response characteristics of the Chebyshev low-pass filter. It is seen that for fixed values of ε, there are more ripples in the passband with the increase in n and any increase in n results in sharper cut-off. Also, it may be seen from Eq. (9.11) that increasing ε for a fixed n improves the characteristic in stop-band region.

From the above response curves, it is seen that a Butterworth filter, by contrast, has a monotonic decrease in gain with frequency and a maximally flat response up to the cut-off frequency.

For applications where phase is important, a minimal phase-shift filter, known as Bessel filter, is similarly realized.

In all the above cases, the transfer function relating the input and output voltage is obtained from the approximation adopted and then the filter network synthesized. Second-order filters are realized first as active versions and are used in cascade to obtain higher-order filters. Given the transfer function of a normalized second-order filter, it is possible to realize the filter in various configurations, depending on the criteria chosen.

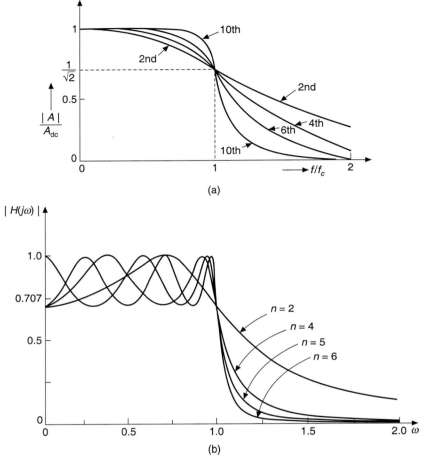

Figure 9.19 Gain frequency characteristics of higher order filters of (a) Butterworth type, and (b) Chebyshev type.

The second-order Butterworth low-pass filter is described by the transfer function

$$\frac{V_o(s)}{V_i(s)} = \frac{K}{s^2 + \sqrt{2}s + 1} \tag{9.13}$$

The TF is realized by using one op-amp with feedback and a few resistors and capacitors as shown in Figure 9.20(a).

The op-amp with its feedback resistors provides a finite gain of 1.586 (= K) while presenting infinite impedance at its input terminals. Hence the filter network is known as VCVS (voltage-controlled voltage source) low-pass filter. The cut-off frequency is equal to $1/RC$.

The same TF may be obtained by using an infinite-gain op-amp using multiple feedback as shown in Figure 9.20(b). It is an inverting unity gain filter (gain = R_2/R_1) frequency normalized filter with C_1 and C_2 chosen such that $C_1C_2 = 1$. The values of resistors are so chosen as to make it a Butterworth filter.

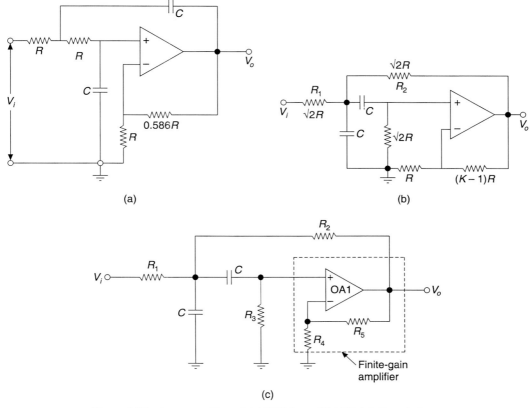

(a)

(b)

(c)

Figure 9.20 Low-pass filters: (a) VCVS type; (b) MFB infinite gain type; (c) band-pass filter using a finite gain amplifier.

The above two configurations of the low-pass filter are the result of matching the coefficients of the given function with the coefficients of the TF of the filter. Applying a similar procedure, second-order high-pass, band-pass and band-reject filters may be synthesized so as to possess the required voltage gain. If it is an inverting-gain configuration, as is the case with Figure 9.20(b), another op-amp is used to invert the non-inverting gain.

A second-order VCVS band-pass filter, obtained by matching the coefficients, is shown in Figure 9.20(c). The TF of the filter is represented as

$$\frac{V_o(s)}{V_i(s)} = \frac{Ks/Q}{s^2 + s/Q + 1} \tag{9.14}$$

where,

Q = the quality factor of the filter, $1/b$

(= reciprocal of the bandwidth)

The voltage gain K of the filter and the op-amp gain μ may be so chosen as to obtain the desired value of $Q(= KR_1/\mu)$. Q values up to 25 may be easily obtained by the above configuration.

Also, by utilizing $RC:CR$ transformation, a high-pass filter may be obtained from the low-pass filter by interchanging resistors with capacitors and vice versa and leaving the amplifier part unchanged.

9.5.2.1 Biquad Filters

A biquad network is a four-terminal network designed to realize the general biquadratic transfer function of the form

$$H(s) = \frac{V_o(s)}{V_i(s)} = \pm \frac{a_2 s^2 + a_1 s + a_0}{s^2 + b_1 s + b_0} \qquad (9.15)$$

where a_0, a_1, a_2, b_0 and b_1 are all real constants and b_1 and b_0 are each greater than zero. The product $a_0 a_2$ should also be less than zero so that the entire circuit can be made up of ideal (infinite gain and infinite input impedance) amplifiers, resistors and capacitors. Using the state-variable technique for synthesizing of networks, the biquad filter can be realized by using two integrating amplifiers and the required number of inverting amplifiers. As it is a second-order system, the network is stable, provided the conditions related to coefficients are satisfied. A circuit that realizes the TF of Eq. (9.15) is shown in Figure 9.21 with inverted gain. The choice of values for all the resistors and capacitors is done on the basis of values specified for the coefficients. Such a circuit is known as all-purpose biquad. By making suitable coefficients zero, thereby rendering the values of some resistors and capacitors zero or infinite impedance, low-pass and other second-order filters can be obtained.

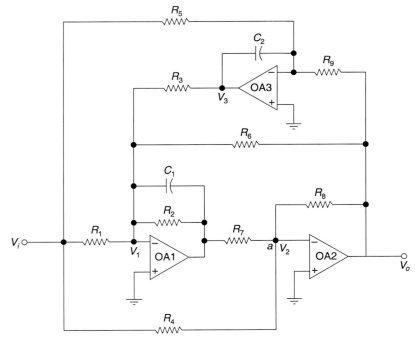

Figure 9.21 General purpose biquad circuit.

The circuit shown in Figure 9.22 is a low-pass filter, synthesized on the lines suggested for the TF given by

$$\frac{V_o(s)}{V_i(s)} = \frac{K}{s^2 + b_1 s + b_0} \qquad (9.16)$$

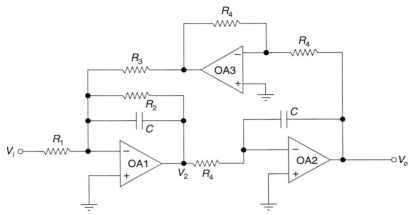

Figure 9.22   Low-pass biquad filter.

Apart from its excellent stability, the merits of the circuit lie in its tunability for satisfying the characteristics of the filter. In the given circuit, R_1 can be adjusted to result in the required gain of the filter, R_2 for the coefficient b_1, and R_3 for b_0.

Analyzing the circuit, it can be shown that the TF between $V_2(s)$ and $V_i(s)$ is given by

$$\frac{V_2(s)}{V_i(s)} = -\frac{Ks}{s^2 + b_1 s + b_0} \qquad (9.17)$$

showing that the same filter circuit can be used as band pass filter, by considering $V_2(s)$ as the output voltage. Adjustment of b_1 by R_2 provides the necessary bandwidth and Q.

Although the above biquad filters require more resistors and op-amps, the operational features render the circuit more popular. Biquad band-pass and band-reject filters may be designed for Q values up to 100.

A biquad all-pass filter having its TF given by

$$\frac{V_o(s)}{V_i(s)} = K \frac{s^2 - b_1 s + b_0}{s^2 + b_1 s + b_0} \qquad (9.18)$$

is realized from the general purpose biquad. The circuit shown in Figure 9.23 provides for the adjustment of b_0, b_1 and K. It is necessary to make $R_4 = R_5 = 1/K$, and in such a case, $R_1 = R_2/2K$, $R_2 = 1/b_1$ and $R_3 = 1/b_0$.

All the second-order filters can be obtained in *IC* form and provision is made for obtaining the necessary tuning of the filter. The drift in the values of the components may make the filter characteristics differ from the desired characteristics, but can be corrected to some extent by tuning it.

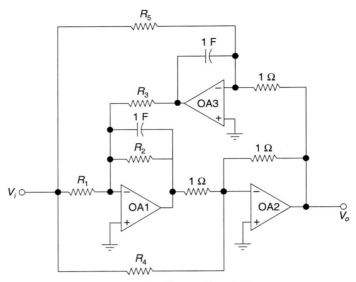

Figure 9.23   All-pass biquad filter.

9.5.3   Higher Order Filters

There are two general methods of synthesizing higher-order filters. The given transfer function may be synthesized by using the state-variable technique and the corresponding network realized. The second and more common method is to split the given TF into second-order functions, realize each function as a subnetwork, and then cascade the subnetworks to obtain the overall network.

The given $H(s)$ may be represented as

$$H(s) = \Pi H_i(s) \tag{9.19}$$

where $H_i(s)$ is realized by a quadratic filter in case the order of the overall filter is even. Where the order of the filter is odd, a first-order subnetwork is necessary. By realizing all subnetworks as active networks, it is possible to cascade them without the need for buffer amplifiers. After the given $H(s)$ is split into its factors, it is enough to identify suitable quadratic filters with tunability for matching the corresponding coefficients of each subnetwork. For example, a third-order low-pass Butterworth filter, with its TF split into the factors is given as

$$\frac{V_o(s)}{V_i(s)} = \frac{4}{(s^3 + 2s^2 + 2s + 1)} = \left(\frac{2}{s+1}\right)\left(\frac{2}{s^2 + s + 1}\right) \tag{9.20}$$

The corresponding filter is shown in Figure 9.24 with two VCVS filters connected in cascade.

Similarly, a fourth-order low-pass Butterworth filter has its transfer function given by

$$\frac{K}{(s^4 + 2.61313s^3 + 3.41421s^2 + 2.61313s + 1)} = \frac{K_1}{s^2 + 0.76537s + 1} \frac{K_2}{s^2 + 1.84776s + 1} \tag{9.21}$$

Two biquad filters may be synthesized and connected in cascade to obtain the fourth-order low-pass filter.

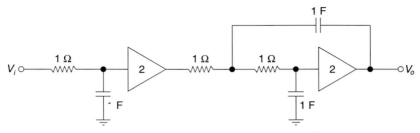

Figure 9.24 Butterworth third-order low-pass filter.

Other higher-order filters are obtained in a similar fashion using the basic transfer function and its factored form.

Even though the cascaded configuration of the higher-order filter possesses the advantages of tunability and stability, it suffers from the drawback of drift of the overall gain due to the drift of each one of the subnetworks. The overall characteristic of the higher-order filter of the cascaded type cannot be accurately specified for these reasons.

The higher-order filter of the multiple-feedback type as obtained by the first method possesses the merit of compensating the effect of drift due to the feedback from the output to the input terminals. A third-order low-pass filter, with its TF given by

$$\frac{V_o(s)}{V_i(s)} = \frac{-K}{s^3 + a_2 s^2 + a_1 s + a_0}$$

can be synthesized by using the state variable technique, and the circuit synthesized is shown in Figure 9.25. It should be observed that the output voltage is fed back to the input terminals, apart from the feedback from the output terminals of the other integrating amplifiers. A comparison of the third-order filters shown in Figures 9.24 and 9.25 indicates the number of components and op-amps required to realize the TF in each case.

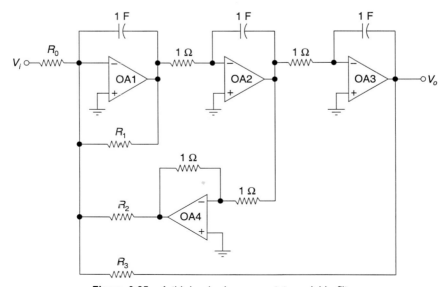

Figure 9.25 A third-order low-pass state variable filter.

9.6 ANALOG MODULATORS

As explained in Section 2.6, the process of modulation of analog signals is adopted in signal processing and signal transmission systems. In each case, the interest lies in preserving the desired relationship between the analog signals at the input and output end. When measured data is transmitted from the transducer end to the receiving end, there is every likelihood of the data being adulterated by noise all along the path. It is necessary to identify the proper technique of modulation and suitable equipment so that the effect of noise is minimized. In this respect, frequency modulation technique is seen to be far superior for analog transmission of measured data. Amplitude and frequency modulators are briefly described in the remaining part of this Section.

9.6.1 Amplitude Modulators

The essential operation for obtaining an amplitude modulated (AM) signal is the multiplication of the data signal $m(t)$ with a higher frequency sinusoidal carrier signal, $E_c \cos \omega_c t$ and development of a signal proportional to their product as per Eqs. (2.33) and (2.35). There are not many devices that can carry out the operation of multiplication of two electrical signals as observed in the case of Hall-effect transducer (Section 7.4). However, all electronic devices that possess a nonlinear characteristic between the input voltage and output current can be utilized to produce the desired output. If a diode is applied with a small voltage $v_i(t)$, the output current $i_o(t)$ is nonlinear and the voltage $v_o(t)$ across a resistor is given by

$$v_o(t) = R i_o(t) = a v_i(t) + b v_i^2(t) + c v_i^3(t) \tag{9.22}$$

where a, b and c are constants of sufficient value (other higher-order terms being neglected). Treating $v_i(t)$ as a sum of $m(t)$ and $E_c \cos \omega_c t$, the output voltage $v_o(t)$ can be written as

$$\begin{aligned} v_o(t) = {} & a E_c \cos \omega_c t + a m(t) + b m^2(t) + 2b m(t) E_c \cos \omega_c t \\ & + b E_c^2 \cos^2 \omega_c t + c m^3(t) + 3c m^2(t) E_c \cos \omega_c t \\ & + 3c m(t) E_c^2 \cos^2 \omega_c t + c E_c^3 \cos^3 \omega_c t \end{aligned} \tag{9.23}$$

Of all the terms, the $2b m(t) E_c \cos \omega_c t$ is of interest and hence some means should be adopted to eliminate all other components by proper filtering. The problem becomes acute when a modulating signal consists of components of two frequencies. So by using two matched diodes, it is possible to cancel some of the terms, as can be seen from Figure 9.26(a). The output voltage $v_o(t)$ is proportional to the difference of the two diode currents. Using Eq. (9.22), it can be shown that

$$v_o(t) = K[2a m(t) + 4b m(t) E_c \cos \omega_c t + 6c m(t) E_c \cos^2 \omega_c t + 2c m^3(t)] \tag{9.24}$$

where K is a constant.

The number of unwanted terms is reduced and the term $4Kb m(t) E_c \cos \omega_c t$ is the DSBSC output, which is extracted from $v_o(t)$ by filtering the rest.

An alternative and a better method is by operating the diodes as switching devices which are driven into the forward and reverse regions of the characteristic with the bias voltages removed. The output voltage then consists of the product of $m(t)$ and the Fourier components of square wave representing the switching action of the diodes. A balanced modulator of the type shown in Figure 9.26(a) may be used along with filters.

A modified version of the modulator known as double-balanced modulator is shown in Figure 9.26(b) whose output is of the form

$$v_o(t) = m(t)\left[\frac{4}{\pi}\left(\cos\omega_c t - \frac{1}{3}\cos 3\omega_c t + \frac{1}{5}\cos 5\omega_c t \dots\right)\right] \qquad (9.25)$$

The merit of double-balanced modulator is that it does not have the modulating signal in the output as observed in the case of balanced modulator.

Balanced modulators in the form of ICs using bipolar transistors, based on the above principles of operation, are available for use with large input signals. For small signal operation, ICs using matched FETs are chosen. These devices display characteristics that are nearly of the square law type, and hence the components of unwanted frequencies have much less amplitudes and become insignificant at higher frequencies.

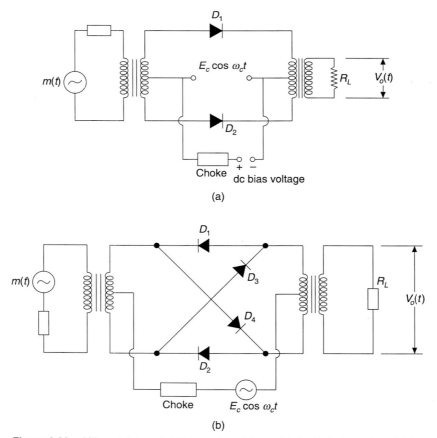

Figure 9.26　AM modulators: (a) Balanced modulator; (b) double-balanced modulator.

9.6.2 Frequency Modulators

A variety of frequency modulators exist, depending on the unmodulated (centre) frequency f_c and the bandwidth of the modulated signal, Δf. FM generators for radio communication and broadcasting usually have the centre frequency contained within 10–100 MHz allotted for such work, whereas for space communication systems, the centre frequency is very high. Space telemetry systems meant for transmission of measured data demand stability of the centre frequency to a very high degree. Also, the generation of the final FM signal is brought about in two stages. Thus the centre frequencies of FM generators of the first stage (sub-carrier oscillators) may be within the range of 400 Hz to 70 kHz and of the second stage far above the frequencies allotted for message communication, broadcasting of music or television. Also, the power levels of FM signals of space telemetry having centre frequencies above 200 MHz is far lower than those meant for entertainment broadcasting systems. Hence a proper circuit for frequency modulation may have to be chosen, depending on whether the data transmitted is obtained from transducers and measuring systems or from microphones, TV cameras and other such equipment.

The frequency modulation techniques may be broadly classified as (i) direct and (ii) indirect.

9.6.2.1 Direct FM Generators

The generator that produces a carrier of a frequency up to 500 MHz is, in many cases, an LC tuned circuit oscillator. It provides a sinusoidal waveform for the FM signal, over the range of variation of the centre frequency. The modulating signal $m(t)$ is usually applied in such a way as to vary the capacitance or the self-inductance of the LC tank circuit, but proportional to the instantaneous amplitude of $m(t)$. The variation of L or C may also be brought about electrically or electronically, apart from L or C becoming part of a transducer system. The electrical or electronic technique of effecting a variation of L or C is used when the modulating signal is a voltage or current. The self-inductance of a saturable reactor can be varied by changing the dc biasing current through the bias winding, and thereby changing the core permeability. Similarly, the dc biasing voltage of a voltage variable capacitor can be varied, bringing about a variation in its capacitance. Semiconductor diodes, when operated with a reverse bias, have characteristics suitable for use as voltage variable capacitors.

A voltage variable capacitor, commonly known as varicap or varactor diode, is kept in parallel with a fixed capacitor C of the tank circuit as shown in Figure 9.7(a). The capacitance C_v of the varicap changes in proportion to the amplitude of the modulating voltage $m(t)$, and hence causes a corresponding variation in the oscillator frequency.

At high frequencies, diodes behave as capacitors and the variation of the depletion region of the diode with voltage results in the variation of its capacitance. The capacitance change per unit voltage is highest at the lower values of dc bias voltages and the relationship is nonlinear. Usually, amplitude of the modulating voltage is so chosen that the diode operation is always within the nonconducting region. The capacitance of the diodes is in the range of 10–100 pF at −4 V bias, and the modulating voltage is coupled to the diode as shown in Figure 9.27. The coupling of the diode with the tank circuit should, in no case, result in the loading of the tank circuit by the source of $m(t)$ or vice versa. For centre frequencies about 100 MHz, the circuit forms part of a Colpitt's oscillator.

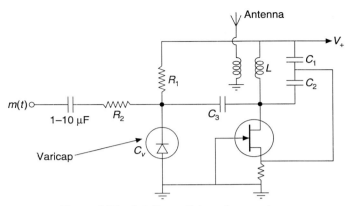

Figure 9.27   Colpitts oscillator using a varicap.

Frequency modulation may also be achieved by the variation of the reactance of a reactance valve or the resistance of a field-effect transistor or a PIN diode.

For situations in which the FM signal is desired to be rectangular for its waveform, the multivibrator circuits can be used to serve as carrier signal generators. The frequency of the carrier depends on the supply voltages used for biasing as well as on the values of the passive components of the circuit. Hence frequency modulation is effected by using the modulating signal to control these biasing voltages or by using resistive transducers in place of passive transistors.

FM signals at frequencies higher than 100 MHz can be generated directly at the carrier frequency by using thermionic devices such as klystron.

All the above techniques have one common drawback, i.e. the carrier signal frequency drifts when $m(t) = 0$. The FM generator has to possess long-term stability in its frequency of generation, especially when used for transmission of measured data. When the frequency modulators are used for the first stage of modulation, for telemetry of measured data, they are commonly known as *sub-carrier oscillators* (SCOs). The required degree of constancy of frequency should also be ensured; otherwise, a continuous check and adjustment of the same has to be undertaken.

The indirect method of FM signal generation, as described in Section 9.6.2.2, provides frequency constancy of the unmodulated carrier, by using a crystal oscillator as the basic carrier signal generator.

9.6.2.2   *Indirect FM Generators*

FM signals can be generated by making use of crystal oscillators which provide high stability for the frequency of oscillation. As it is not possible to control the frequency of the resonant circuit by direct application of the modulating signal, an indirect means is used. The most popular and commonly used technique is due to Armstrong and is known as *Armstrong modulator* which makes use of two crystal oscillators, a narrow-band FM generator, frequency multipliers and a frequency changer (mixer). Both the broadcasting and data-telemetry systems employ this technique. Although the technique seems to be more complex, it has the additional merit of providing FM signals of desired frequency deviation (and hence wide-band FM signals), at the desired value of unmodulated frequency.

The block diagram of the Armstrong modulator is shown in Figure 9.28. The first stage of the modulator consists of a narrow-band FM signal generator shown in Figure 9.29 and is based on Eq. (2.42). With m_f of about 0.2, Eq. (2.42) simplifies to

$$e_m(t)\big|_{\text{FM}} = E_c\left[\cos \omega_c t - \sin \omega_c t\,(m_f \sin \omega_m t)\right] \tag{9.26}$$

as $\cos \theta \approx 1$ and $\sin \theta \approx \theta$ for small values of θ.

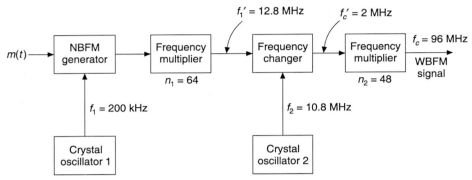

Figure 9.28 Armstrong wideband frequency modulator.

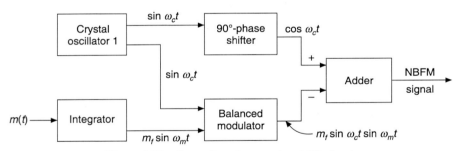

Figure 9.29 Generation of narrow-band FM signal.

The two components of the signal represented by Eq. (9.26) are produced separately and added together as shown in Figure 9.29, to form the narrow-band FM signal. The DSBSC output of a balanced modulator is added to the quadrature component of the output of the crystal oscillator. As the signal applied to the balanced modulator is obtained after integrating the modulating signal, the circuit is equivalent to a narrow-band FM generator with the modulation index $m_f (= \Delta f_c/f_m)$ limited to a value of 0.2 rad. For music broadcasting, modulating signal frequency components lie within the range of 50 Hz to 15 kHz, and hence a choice of Δf of 25 Hz makes m_f of 0.5 at 50 Hz and of much smaller value at the highest frequency. For such situations, the crystal oscillator 1 is chosen with its frequency of oscillation at 200 kHz. At this frequency, crystal oscillators are highly stable, and balanced modulators are conveniently designed. However, for FM/FM telemetry, as can be seen in Chapter 11, the modulating signal for the second stage of frequency modulation consists of components of frequency ranging from 370 Hz to 80.5 kHz. In such a case, a crystal oscillator of 1 MHz may be chosen and a Δf of 75 Hz.

For both the music broadcasting and radio telemetry systems, it is desired that the unmodulated centre frequency for transmission through space be as high as possible while conforming to the bands of frequencies provided for such purposes.

For commercial broadcasting systems, the unmodulated carrier may be about 96 MHz with Δf of about 75 kHz. For radio telemetry systems, the centre frequency for transmission through space is usually very high (216–235 MHz) with Δf of 125 kHz. Thus, it is necessary to bring up the value of the unmodulated frequency as well as the frequency deviation, by a process of frequency multiplication.

When a frequency multiplier circuit is used, the instantaneous frequency of the input signal is increased by, say, n times which, in other words, implies that both the unmodulated carrier frequency and the frequency deviation are increased in their values by n times. However, it may be noticed from the requirements of either the broadcast systems or the telemetry systems that the same factor of multiplication is not suitable.

Δf of 25 Hz can be increased to 75 kHz by having n equal to 3000, but the centre frequency of 200 kHz becomes 600 MHz, which is not what is desired. Likewise, Δf of 75 Hz requires a value of 1667 to be made up to 125 kHz while its centre frequency goes up to 1667 MHz. Hence, a frequency changer is used to obtain the desired centre frequency having the desired bandwidth of $\pm \Delta f$. The multiplication factor of n is split up such that $n = n_1 n_2$ so that the above requirements are satisfied while at the same time, n_1 and n_2 are brought down to reasonable values, as seen from Figure 9.28.

A frequency multiplier is a combination of a nonlinear element and a band-pass filter as shown in Figure 9.30. The transistor is operated in the class C mode so that it operates in the highly nonlinear region of its characteristic. The input signal is a sinusoidal signal of frequency, say f, in which case the collector current waveform is highly distorted. The LC parallel resonant circuit acts as a narrow band pass filter, with its resonant frequency equal to the nth harmonic of the frequency f of the input signal. The output signal of LC filter is nearly sinusoidal in waveform at the frequency of nf. The circuit is suitable for values of n-ranging from 2–5 only. If higher multiplication factors are required, multiplying circuits are cascaded.

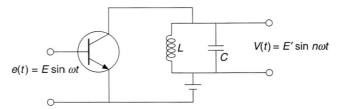

Figure 9.30   A frequency multiplier circuit.

Considering the narrow band FM signal as a sinusoidal signal, whose frequency changes from instant to instant, it can be said that the instantaneous frequency of the output signal of the multiplier is n times the instantaneous frequency of the input signal.

If the multiplication factor of multiplier 1 is chosen as 64, the centre frequency becomes 12.8 MHz (= 64 × 200 kHz) and Δf = 1.6 kHz (25 × 64 Hz).

For the FM broadcasting system, the final value of the unmodulated frequency is required to be 96 MHz and the value of Δf, about 75 kHz. This is achieved by a process of frequency changing, followed by multiplier 2. If the multiplication factor of multiplier 2 is chosen as 48, to make Δf equal to 76.8 kHz, the centre frequency becomes 614.4 MHz (12.8 × 48 MHz). Therefore, a second crystal oscillator is used to bring down the centre frequency of 12.8 MHz to a value of 96/48 MHz (= 2 MHz) by heterodyning the signal of 12.8 MHz with a signal of 10.8 MHz from the crystal oscillator. By such a process, the final FM signal is made to have an unmodulated frequency of 96 MHz with a Δf of 76.8 kHz. The modulation index m_f for a 50 Hz modulating signal becomes 1536, and for the highest frequency component of 15 kHz, m_f becomes 5.12.

For heterodyning, the signal of frequency of 10.8 MHz can as well be obtained from the crystal oscillator 1, by adopting frequency multiplication. In the above case, the multiplication factor needed is 54. The elimination of the second crystal oscillator possesses the merit of adding additional frequency stability for the Armstrong modulator.

9.7 ANALOG DEMODULATORS

A modulated high frequency signal is used to convey analog data over a distance either by wires or by radio waves by means of a transmitting antenna. The analog data is contained either in the amplitude or the frequency of the carrier signal, and is required to be reconstructed at the receiving end, as faithfully as possible. The process of extraction of the analog signals from the modulated carrier signal and their reconstruction is known as *demodulation*. Amplitude demodulators recover the analog signal from the AM signals and frequency demodulators from the FM signals.

9.7.1 Amplitude Demodulators

Amplitude demodulation is effected by simple circuits consisting of a diode and a low-pass filter as shown in Figure 9.31(a). The rectified voltage across the RC parallel circuit contains the envelope of AM signal along with some dc voltage. A capacitor blocks off the dc and makes the analog signal available at the output. The values of R and C are so chosen that the carrier frequency signal is filtered out, but the modulating signal frequency is passed on.

The circuit shown in Figure 9.31(b) uses a transistor that functions as a rectifier in place of the diode. The tank circuit is tuned to the carrier frequency and the received AM signals are amplified and demodulated.

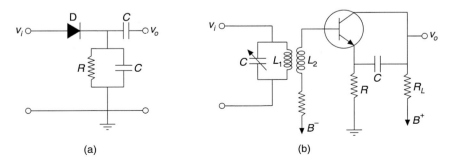

Figure 9.31 Amplitude demodulators using (a) simple RC circuit; (b) a transistor.

9.7.2 Frequency Demodulators

Frequency-modulated signals may be either sinusoidal in waveform or a series of rectangular pulses. Frequency demodulators or discriminators are designed to convert the FM signals into analog signals and the systems are commonly known as FM detectors. The carrier frequency f_c may be low when the FM signal is derived from a frequency-generating transducer and is very high when it is used to convey measured data and other information from a remote space vehicle or a satellite. A variety of circuits are developed for frequency demodulation, and circuits such as those possessing high accuracy and linearity are preferred for use in instrumentation.

Demodulation of FM signals consists of two operations: conversion of frequency changes into proportional amplitude variations, while its frequency remains at the centre frequency (i.e., FM signals into AM signals) and demodulation of AM signals. In simple circuits, the two processes are effected in a single circuit.

9.7.2.1 Tuned Circuit FM Detector

A simple FM detector using an LC tuned circuit and a diode amplitude demodulator is shown in Figure 9.32(a). It is usually known as slope detector since it is based on the slope of the characteristic when it is detuned (see Figure 9.32(b)). The resonance frequency is so chosen that it is slightly higher or lower than the unmodulated carrier frequency f_c of the FM signal. The linear portion of the resonance curve is selected for application and it is essential that the changes in frequency of the carrier signal be limited to this linear portion. The impedance of the tuned circuit changes by $\pm\,|\,\Delta Z\,|$ and in proportion to Δf, and hence the amplitude variations of the voltage across the tuned circuit are proportional to Δf. However, the detector is affected by fluctuations in the amplitude of the FM signal itself, and to reduce this effect, the detector is applied with FM signals whose amplitude is held constant by limiter circuits.

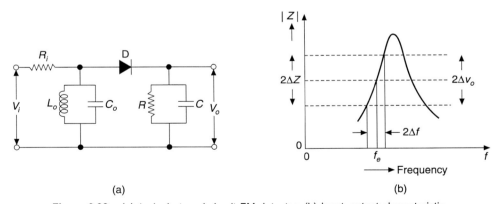

(a)   (b)

Figure 9.32   (a) A single tuned-circuit FM detector; (b) input–output characteristic.

The resonance characteristic of a single-tuned circuit is linear over a small range of frequency and hence is limited for application to narrow band FM signals only. However, by connecting two tuned circuits, in a push–pull fashion, a balanced demodulator with linearity over a much wider range of frequency can be constructed. Figure 9.33(a) shows the circuit employing the two tuned circuits: one tuned to a frequency f_1 which is slightly higher than the unmodulated carrier frequency f_c and the other to f_2 which is lower than f_c by the same amount. The outputs of the tuned circuits are amplitude-demodulated and connected in phase opposition. The resultant output v_o is zero as can be seen from Figure 9.33(b) when the input voltage is of frequency f_c. The range over which the demodulator is linear is thus made larger, and the circuit can be used to deal with wide band FM signals having a frequency deviation of $\pm 15\%$.

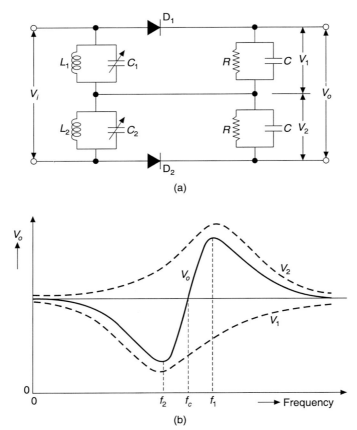

Figure 9.33   (a) FM detector using two tuned circuits; (b) input–output characteristic.

9.7.2.2 Foster–Seely Detector

Another similar FM demodulator, known as Foster–Seely circuit, is shown in Figure 9.34(a). The circuit is more complex but is widely used at carrier frequencies above 1 MHz. The FM input signal v_i is passed through a limiter and is then impressed on the two halves of the diode amplitude demodulator in two ways. Both the tank circuits consisting of L_1C_1 and L_2C_2 are tuned to the centre frequency f_c of the FM signal. The magnetic coupling between L_1 and L_2 enables the two halves of L_2 to be impressed with induced emfs e_1 and e_2. Also, the centre-tap of L_2 is brought to a potential e since the output voltage of the limiter is applied across the series combination of L_0 and C_0. The reactance of the *RF* choke L_0 is far higher than that of C_0, and hence e may be considered to remain fixed in phase and amplitude.

As C_1 and C_2 present negligible reactance to carrier frequency voltages, the voltages across the diodes will be governed by $(e + e_1)$ and $(e + e_2)$. At centre frequency, the tank circuit L_1C_1 is at resonance, and e_1 and e_2 are both in quadrature to e. The net voltages across diodes are equal in amplitude as seen from Figure 9.34(b), and hence v_o is zero. But when FM signal changes its frequency from f_c, the phase angles of e_1 and e_2 with reference to e change with the result that the net voltages across the diodes differ in amplitude and the difference sets up the output voltage v_o. Thus, $v_o(t)$ is proportional to the instantaneous frequency deviation.

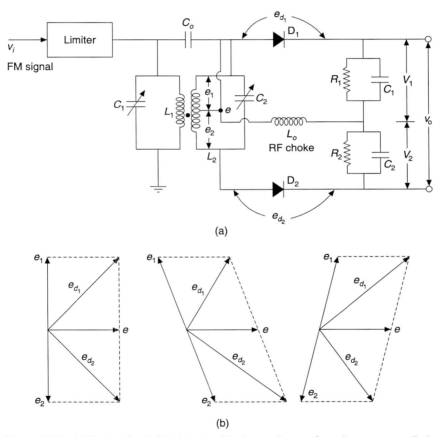

Figure 9.34 (a) Foster–Seely FM detector; (b) phasor diagram for voltages across diodes.

9.7.2.3 PLL Detector

The basic scheme of a phase-locked loop (PLL) is shown in Figure 9.35; it consists of three important stages. It may be noted that it is basically a negative feedback system, with the voltage-controlled oscillator in the feedback path. The error detector is the phase sensitive detector (PSD), whereas the low-pass filter constitutes the forward path and provides the

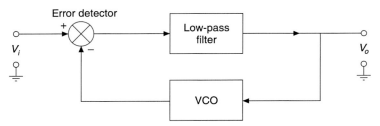

Figure 9.35   Basic scheme of a phase-locked loop.

necessary gain. The phase lock is said to have been established when the phase difference between the input voltage and the output voltage of the VCO is zero. It can only happen when the frequency of the input signal is equal to the VCO frequency. Under these conditions the VCO frequency is at its centre frequency f_c, which is just the same as the frequency of the unmodulated FM signal acting as the input signal.

Whenever the frequency of the input signal changes, it amounts to a phase difference between the input signal and the output voltage of the VCO, and hence the output voltage of the PSD changes. This change results in a new value for the output voltage v_o of the PLL system and acts on the VCO to make its frequency of oscillation correspond with the frequency of the input signal.

The instantaneous frequency of the feedback signal, $f_f(t)$, is given by

$$f_f(t) = f_c + Kv_o(t) \tag{9.27}$$

where K is the sensitivity of the VCO and $v_o(t)$ is the output voltage of the PLL.

If $\theta_i(t)$ is the instantaneous phase of the input signal and $\theta_f(t)$ is that of the VCO output voltage, then the difference $\theta_e(t) = \theta_i(t) - \theta_f(t)$ constitutes the error in phase. $\theta_e(t)$ results whenever a deviation occurs in phase lock which, in other words, means that the input signal has changed its frequency from that of unmodulated conditions to that of modulation. Thus, $\theta_e(t)$ is proportional to the modulating signal $m(t)$ as per Eq. (2.38). The PSD output $v_1(t)$ is proportional to $\theta_e(t)$ and is given by

$$v_1(t) = K_1\theta_e(t) \tag{9.28}$$

It is desired that the low-pass filter has its frequency response characteristic and low frequency gain so adjusted that the feedback system is stable. The new value of the output voltage of the PLL, when fed back to VCO, modifies its frequency of oscillation until the phase lock occurs once again. In this way, the VCO is made to track the variation in the frequency of input (FM) signal by effecting continuous equality of phase of the input and the feedback signals. The output voltage $v_o(t)$ is thus a replica of the instantaneous frequency deviation of the FM signal from its unmodulated value.

The entire PLL system is available as an integrated circuit version and is highly suitable for use as FM demodulator. Of course, it has its many applications in other electronic instruments.

9.7.2.4 *Pulse Counting Demodulator*

Sometimes a frequency demodulator that can convert FM signals whose frequency varies over far wider limits is required, and the methods so far described may not be suitable. This demodulator is based on the principle of converting the FM signals into a train of constant amplitude and constant duration pulses, whose pulse rate depends on the instantaneous frequency of the FM signal. The block diagram of the demodulator is shown in Figure 9.36. A Schmitt trigger is used to generate a square wave from the FM signal and by differentiating the square wave, two sharp pulses of short duration are obtained. One of them is used to trigger a monostabie multivibrator. Since its output is a continuous train of pulses of constant height and duration, the time average of the pulses is proportional to the instantaneous frequency. Averaging is accomplished by an *RC* network and amplifier connected as a low-pass filter. A zero control is used to provide zero output when the input signal is at the centre frequency.

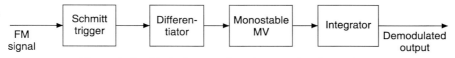

Figure 9.36 Block diagram of a pulse counting demodulator.

The demodulator is used to process FM signals having the centre frequency lying within the range of 200 kHz–10 MHz. The merit of the demodulator lies in its linearity for wide-band FM signals.

9.8 SAMPLING CIRCUITS

In signal processing circuits, at times it is necessay to pass a signal as it is, only when it is commanded. The command is in the form of an electrical voltage signal applied to a 'GATE' terminal, and it may, in most cases, be a short-duration pulse. An input signal appears at the output terminal when the gate pulse appears and the output voltage is zero in the absence of the gate pulse. Thus a *gating circuit,* also known as *linear gate,* is equivalent to a switch operated by application of an electrical voltage. Circuits performing this function may be realized in many ways and they become part of sampling circuits.

A diode bridge linear gate is shown in Figure 9.37(a) which makes the input signal available at the output terminal, when a gate voltage appears at the terminals A and B. If V_{AB} is zero, diodes D_1 and D_4 block the input signal from appearing at the output terminal. The

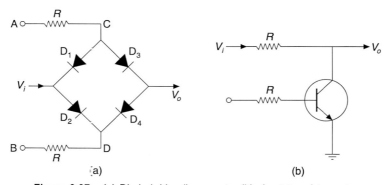

Figure 9.37 (a) Diode-bridge linear gate; (b) shunt transistor gate.

circuit operates as a switch for gate voltages of both polarities. It is enough to ensure that the voltage drop across CD terminals is negligible. In this circuit, the V_{CD} is due to the forward resistance of the two diodes forming the path of output current.

Another gate circuit for input signals of one polarity is the shunt transistor gate shown in Figure 9.37(b). When the gate signal is zero or negative, the transistor does not conduct and so the input signal appears at the output; the output signal is less than the input signals the difference being voltage drop across R. The transistor has to be brought to saturation by means of a large positive gate voltage so that the output voltage is nearly zero.

Linear gates are also fabricated in IC form with different specifications for gate voltage, switch resistance and output current.

The output voltage thus obtained from a linear gate may be used to charge a capacitor so that the voltage across the capacitor may be treated as representing the input voltage. It is desired that the gate signal be present and the gate be on for the time duration required for the charging of the capacitor to the value of the input signal. The charging time is thus determined by the value of the capacitor and the source resistance of the circuit that charges the capacitor. The capacitor is known as *HOLD capacitor* and the circuits as *sample-and-hold* or *track-and-hold circuits*.

Figure 9.38(a) shows the distinction between the two circuits. The input signal is commanded or gated to appear at the output terminal, for such a small interval of time, so that the capacitor is fully charged to the input voltage. Even after the gate voltage is withdrawn, the voltage across the capacitor is treated as equal to the value of the input voltage and is used for different purposes such as analog-to-digital conversion. In such a case, it is a sample-and-hold circuit. If the gate voltage is on for a longer interval, the output voltage follows and 'tracks' the input voltage faithfully as long as is desired. If at any instant of time during the tracking operation the magnitude of the input is desired, the command signal is switched off so that the voltage across the capacitor represents the value of the input voltage which it has at the instant of withdrawal of the command signal. Then the circuit is known to serve as track-and-hold circuit. The circuit is said to operate in 'sample' mode or 'track' mode, if the capacitor is connected to the input signal and 'hold' mode when the capacitor holds and retains the value of the voltage to which it is charged.

For the sample-and-hold circuit, the *acquisition time* T_{ac} is the time required for the capacitor to charge up to the value of the input voltage after the switch is first started. The *aperture time* T_{ap} is the time required for the switch to change from 'on' state to 'off' state. The *holding time* T_h is the duration of time for which the circuit can hold the charge on the capacitor and hence its voltage without dropping more than a specified percentage of its initial value. A sample-and-hold circuit is designed after careful consideration of the maximum rate of variation of the input signal (or frequency), the time required for processing of the voltage across the capacitor (say, for analog-to-digital conversion), and the impedance levels of source and load lying on either side of S-and-H circuit. The circuits are usually built with fast acting FET switches and op-amps so as to meet the requirements of charging current for the capacitor, and isolating it from the effects of its load impedance.

The circuit shown in Figure 9.38(b) uses two op-amps and an FET switch. The hold capacitor is isolated from both the source and loadside alternately during sampling and holding times.

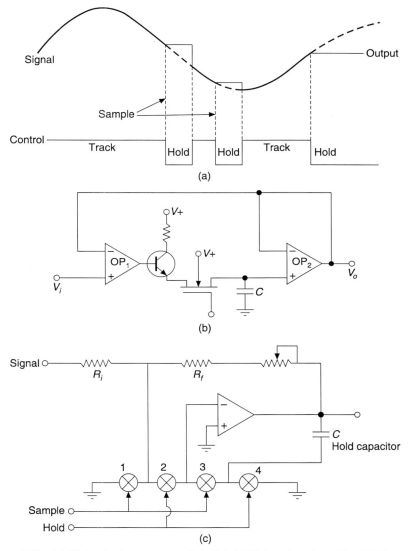

Figure 9.38 (a) Distinction between sample and hold (SH) and track and hold (TH) circuits; (b) SH circuit using two op-amps; (c) SH circuit using one op-amp.

The circuit shown in Figure 9.38(c) uses only one op-amp while the gate pulses operate the switches in a sequential manner. The sample and hold gate pulses are complementary. While the sample gate pulse puts switches 1 and 3 in 'off' state, switches 2 and 4 are 'on', and hence the capacitor gets charged through the op-amp. When the hold pulse occurs, switches interchange their states, resulting in the capacitor going into the feedback loop of the op-amp while R_i and R_f get connected to ground. R_i and R_f are made equal by means of the trimming resistor so as to make V_o equal to V_i in sample mode. In the hold mode, V_o equals the voltage across the capacitor, with the op-amp acting as a voltage follower.

9.9 BASIC DIGITAL BUILDING BLOCKS

A variety of digital IC devices are available for application in instrumentation systems and microprocessor-based instrumentation. They are mostly either the transistor-transistor-logic (TTL) circuits or the complementary metal-oxide-semiconductor (CMOS) circuits in a chip form. They are mostly based on the common and conventional binary positive logic and their input and output voltage levels are referred to as logic levels. Digital devices have two stable states or logic levels: a higher positive voltage state defined as *high* or *one* and a lower voltage state as *low* or *zero*. It is essential that all the digital devices of either TTL family or CMOS family have the same voltages for each of the two states so that they can be interconnected in any possible manner.

The digital IC devices may be broadly classified according to the number of input and output signals they handle. A brief introduction to such devices is considered relevant for a better understanding of digital instrumentation systems. Specification of the logic outputs for various combinations of the inputs is done by means of a truth table. Figure 9.39 presents the devices with a single output along with the corresponding truth tables and symbols.

Inverter is the simplest device, it being the digital equivalent of a unity-gain inverter.

Driver is the digital equivalent of a unity-gain amplifier, which simply presents the logic state of the input at the output terminal. Two inverters may constitute a driver, resulting in larger output current and power.

An OR gate has its output HIGH if any one of its inputs is HIGH.

A NOR gate is the equivalent of an OR gate followed by an inverter.

A XOR (exclusive OR-gate) is similar to OR gate except that its output is LOW if all its inputs are HIGH.

An AND gate has its output HIGH only if all inputs are HIGH.

A NAND gate is the equivalent of an AND gate followed by an inverter.

An INHIBIT gate is an AND gate with one of its input terminals being preceded by an inverter. The output assumes HIGH state only when all its inputs except the negated input assume the HIGH state.

Although digital IC circuits of the TTL and CMOS families are superior in performance, simpler logic circuits consisting of diodes and op-amps are used when only a few simple digital operations are required. With op-amps, a LOW state is defined as negative saturation and HIGH state as positive saturation. An OR gate is shown in Figure 9.40(a), using diodes only and in Figure 9.40(b) using an op-amp along with diodes. Similarly, an AND gate consisting of diodes only is shown in Figure 9.41(a), and in Figure 9.41(b) with an op-amp added to the simple diode gate.

A transmission gate is equivalent to a voltage-controlled switch, which allows bidirectional flow of pulses when it is on and no flow at all when off. These gates are symmetric, and either terminal may be connected to the input. Analog as well as digital signals can be switched, but the input and output voltages must be within a certain range. They are widely used in multiplexing circuits.

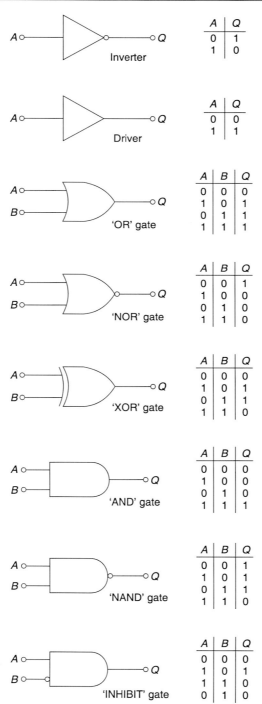

Figure 9.39   Basic logic devices and their truth tables.

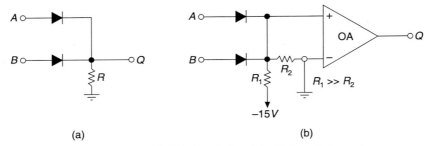

(a)   (b)

Figure 9.40   An 'OR' gate: (a) With two diodes; (b) with two diodes and an op-amp.

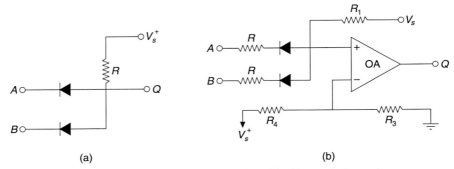

(a)   (b)

Figure 9.41   An 'AND' gate: (a) With two diodes; (b) with two diodes and an op-amp.

9.10   FLIP-FLOP CIRCUITS

Storage is one of the most important functions in data processing. Digital devices with two stable states that can be made to undergo a change from one state to another by application of a trigger pulse to one of its inputs, constitute the heart of the binary counting systems.

A flip-flop is a circuit with two stable states, which is capable of remaining in either state indefinitely until triggered by an external signal. It can be said to have a 'memory' as it will remember the last signal it received until a new signal is applied to the circuit.

A variety of flip-flops designed to carry out different functions are available in IC form.

The basic flip-flop is the Set-Reset (SR) flip-flop with two inputs and two outputs as indicated in Figure 9.42. An emitter-coupled bistable multivibrator can be used as an SR flip-flop. In most cases a pair of two-input NOR or NAND gates are connected to act as a flip-flop.

For the SR flip-flop of Figure 9.42(a), the initial states of S and R are made LOW (0– no pulse), with Q and $\bar{Q}$ remaining in NO CHANGE condition. Application of a pulse to input S brings S momentarily to HIGH (1) state which in turn renders the output Q to go to HIGH state while ensuring $\bar{Q}$ to be in the inverse state of LOW (0). The reset action of bringing R to HIGH state by application of a trigger pulse renders Q to go to LOW (0) state and $\bar{Q}$ to HIGH state (1), as seen from the corresponding truth table. If both inputs are taken to the HIGH state, the outputs may go to either state and hence become indeterminate (?). In a properly designed circuit, both inputs will not be brought to HIGH simultaneously.

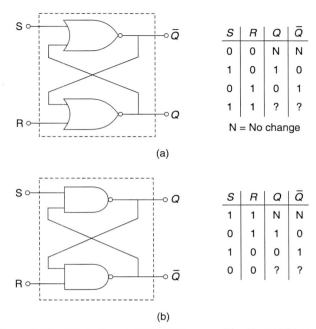

N = No change

(a)

(b)

Figure 9.42   SR flip-flop: (a) With 'OR' gates; (b) with 'NAND' gates.

The SR flip-flop consisting of two NAND gates is shown in Figure 9.42(b) with its corresponding truth table. The input logic sense is inverted.

If HIGH-LOW transitions are applied alternately to the S and R inputs, then the states of Q and $\overline{Q}$ reflect the input which last had LOW (0) applied to it. Thus the circuit acts as a memory and is usually referred to as *latching circuit*.

In order to provide a common triggering point for counting applications, circuits are often based on the "master-slave" principle, with two ranks of SR flip-flops and auxiliary gates for the routing functions. A JK or master-slave flip-flop has two logic inputs J and K, two outputs Q and $\overline{Q}$, and additional control inputs such as a clock (CK), a clear (CLR), and a preset (PR), as indicated in Figure 9.43(a). The outputs of JK flip-flop do not change until a pulse is applied to the clock input. In microprocessor-based systems, the clock synchronizes the various logic operations.

The operation of the JK flip-flop may be understood by referring to Figure 9.43(b) which, for convenience, is shown to consist of two SR flip-flops along with other gate circuits. When the clock is in LOW state, the input AND gates allow the JK input signals to the master flip-flop and so follow the inputs. But the slave flip-flop does not follow the master because the transfer AND gates do not pass the signal until the clock goes HIGH. The JK inputs are first disconnected to prevent the master from changing during the clock pulse. Next, the transfer AND gates connect the slave to the master and remain connected with the clock in HIGH state. If the clock returns to LOW state, the slave does not change its state even though the master will again follow the input. At any instant, the slave can be reset (Q to LOW state) by bringing the CLEAR input to HIGH state, by means of a pulse. Usually, the JK flip-flop is set to LOW (0), of cleared before starting an operation.

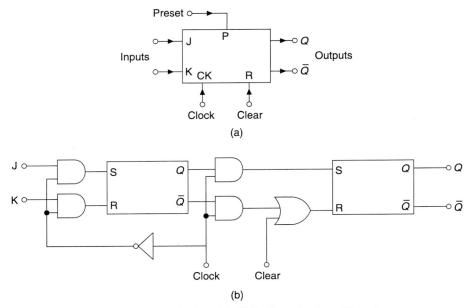

Figure 9.43 (a) JK flip-flop; (b) JK flip-flop using two SR flip-flops.

The JK flip-flop can be operated as a binary counter because it reverses states on application of a clock pulse when both JK inputs are at HIGH (1) state. JK flip-flops are available with several modifications of the basic scheme shown and are used in complex counters.

A more popular building block for use in more complex sequential logic circuits is the D flip-flop which is similar to the JK flip-flop except that it has only one data input D, as shown in Figure 9.44(a). The D input is the same as the J input and K input connected to clock (CK). In a synchronous sequential logic system, the states of latches change only when the particular logic signal 'CLOCK' makes a transition, and thus all the memory elements change together. In the D flip-flop, the state of the latch can change only at the clock transitions; then the latch is set to HIGH (1) if $D = 1$ and to LOW (0) if $D = 0$.

The operation of the D flip-flop may be understood from the scheme shown in Figure 9.44(b) when CLOCK = 0, $R = S = 1$, and the latch remains in whatever state it was in. If $D = 1$, $A = 0$, $B = 1$, whereas if $D = 0$, then $A = 1$, $B = 0$. This means that with CLOCK at 0, A and B are free to respond to changes in D.

Now when the clock makes a 0-to-1 transition, S assumes the state of D at the time of transition and R the opposite state. Thereafter, R and S do not respond to changes in D. Thus the D flip-flop remembers the state of the D input at the time of a 0-to-1 clock transition.

In certain D flip-flops, an additional unlocked PRECLEAR input is added to set $Q = 0$ and $Q = 1$, regardless of the states of the DATA and CLOCK inputs.

The D flip-flop may be used in shift registers and other sequential logic circuits.

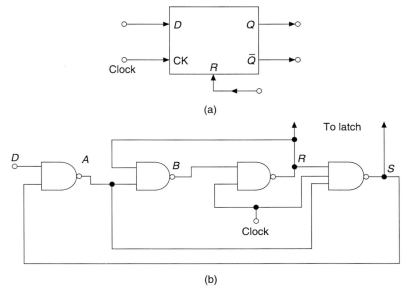

(a)

(b)

Figure 9.44 *D* flip-flop.

9.11 SHIFT REGISTERS AND COUNTERS

Binary data can be stored in flip-flops. In the case of JK flip-flops, the input data is transferred to an internal store at one edge of the clock pulse and transferred later to the output. New data is taken in only at the next edge of the same type.

9.11.1 Shift Registers

Four flip-flops are cascaded as shown in Figure 9.45 to form a shift register in which the data is shifted with each clock pulse from one flip-flop to the next. It is referred to as clocked-shift register. With four flip-flops forming the circuit, it can store only four bits. The read-out of data may be either serial or parallel, whereas the read-in of data can only be serial.

There is often parallel data which must be read into the shift register, and in such cases, loading of parallel data is best done by using *D* flip-flops and a suitable switching system.

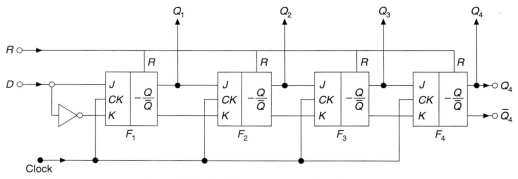

Figure 9.45 A shift register using four flip-flops.

9.11.2   Binary Counter

Another important application of logic circuits is the counting of pulses. A counter is the simplest sequential circuit which has a defined relationship between the number of input pulses and the states of outputs.

The JK flip-flops can be used to constitute a binary counter because it reverses states on application of a clock pulse when both JK inputs are at 'ONE'. The scheme shown in Figure 9.46 consists of three JK flip-flops, in which the clock becomes the input and the JK terminals are permanently wired to a source of supply, making them stay at ONE. Before counting, a pulse is applied to all the clear inputs, so that all the outputs are brought to ZERO. If a series of pulses is applied to the clock (CK) input, its output reverses each time the clock pulse goes through one complete cycle.

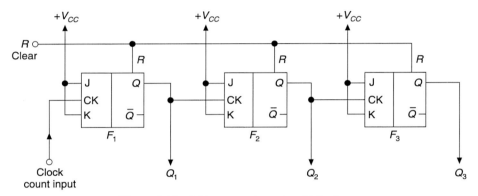

Figure 9.46   A binary counter using JK flip-flops.

A set of four flip-flops with the JK inputs internally connected to ONE is referred to as a 4-bit binary or *ripple* counter with the clock input designated as the *toggle* or count input T. If indicator lamps are connected to indicate the states of the outputs, the system can be treated as a binary counter. They are provided with CLEAR input and ENABLE input which can inhibit the count. In most of the counters the output changes on falling edge of the clock pulse. The counter may be extended to any size and can count up to 1023 with 10 flip-flops.

Every such counter may be treated as a frequency divider. The frequency at the output of flip-flop A is half the counter frequency. A quarter of the input frequency is the frequency of the output of B and so on.

9.11.3   Decimal Counters

A binary coded decimal (BCD) counter is identical to the 4-bit binary counter except for an internal arrangement by which all the flip-flops are set to "ZERO" on the tenth count which is 1010 in binary system. A 'carry' bit is produced at the end of the tenth count and this carry bit controls the decade counter for the next higher decimal digit. Each 4-bit binary counter is equivalent to one decimal digit and is separately decoded from the binary as a digit from 0 to 9.

The decimal digit can be read on a seven-segment display device (described in Section 10.3) after proper decoding.

The ripple counter of Section 9.11.2 is known as *asynchronous* counter, implying that the clock pulse is applied only to the input of the first flip-flop while the remaining are indirectly controlled. This means that the input signal of the last flip-flop arrives only after all previous stages have changed state. Thus the counter exhibits delay and for long chains and high counter frequencies, this is objectionable. Additional provision must be made to eliminate the propagation delay.

'Synchronous' counters do not have this drawback as the clock pulses are applied simultaneously to all clock inputs. For example, the three digits of a decimal counter (i.e. all the 12 bits) while going from 099 to 100 will change at the same time.

Complex counters are designed with additional provisions for (i) presetting the counter to a desired initial value, (ii) counting down as well as up, and (iii) dividing by a factor.

9.11.4 Frequency Counters

A frequency counter is designed to indicate the number of cycles during a fixed time interval, usually 1 s. The input signal is a train of rectangular pulses having a pulse height treated as HIGH or ONE. The block diagram of the frequency counter is given in Figure 9.47, where the input pulses are allowed to pass to the counter during the desired time interval T. A flip-flop controls the NAND gate and is turned on by a start-pulse generator. Simultaneously, the time-base counters connected to a crystal clock of 1 MHz are started. When the time-base counters reach 10^6 pulses, the carry output turns off the flip-flop. During this interval the input pulses are counted so that the frequency is equal to the number of pulses registered. A brief clear pulse is generated as the gates are switched on.

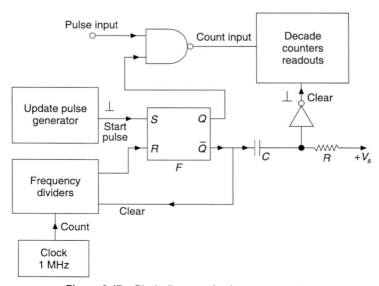

Figure 9.47 Block diagram of a frequency counter.

9.12 ENCODERS AND DECODERS

Encoding is the operation of converting an input data into a code consisting of a series of logic level signals of ZERO and ONE, thereby forming a binary number or a BCD number. Encoders are devices or circuits that enable the conversion.

In Section 7.8.1 shaft-angle encoders are described, showing how the angular position of the shaft is represented in the binary number system. Another simple common example may be found in the thumbwheel switch encoder which enables conversion of switch positions from 0 to 9 into corresponding numbers in the BCD code, and entry into the circuitry of the instrument. Figure 9.48 shows the arrangement in which the four output lines corresponding to the BCD code are brought into the states of ONE or ZERO, when the four contactors ganged together are moved from one position to the other. Logic level ONE is obtained when the supply line is contacted and ZERO when the switches are open by providing resistors connected to ground.

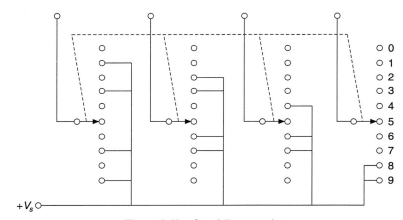

Figure 9.48 2-to-4 line encoder.

Decoders may be treated to enable the reverse operation. However, they are normally used to convert information obtained over a certain number of lines into a corresponding information over a larger number of lines. They may be known as 2-to-4-line, 4-to-16-line, etc. depending on the number of input and output lines.

In Figure 9.49, a 2-to-4 line decoder is shown, having only one of the output lines at ONE at any instant. The decoder is constructed from inverters and AND gates, and the decoders are available in integrated units. From the truth table it is seen that one of the output lines goes to ONE if the corresponding binary number appears at the input.

Similarly, a 4-to-7-line decoder is used to light up the light-emitting diodes of the seven-segment display system. Decimal numbers from 0 to 9 are brought out into a BCD code constituting four input lines of the decoder, whereas its seven output lines are used to drive the display system in such a manner as to display the decimal number.

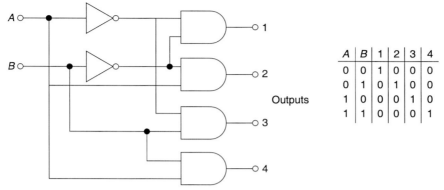

A	B	1	2	3	4
0	0	1	0	0	0
0	1	0	1	0	0
1	0	0	0	1	0
1	1	0	0	0	1

Figure 9.49 2-to-4 line decoder.

9.13 DIGITAL-TO-ANALOG AND ANALOG-TO-DIGITAL CONVERTERS

Conversion of an analog signal into a digital signal appropriate to a digital device assumes great importance and application in instrumentation systems. In Section 3.2.2.2, simple techniques of analog-to-digital (A-D) conversion of an analog voltage are described briefly, for application in the design of digital voltmeters.

Equally important is the reverse process of converting the digital signal into its equivalent analog since the fastest method of AD conversion utilizes a digital-to-analog converter (DAC). Also, most of the final control elements are analog devices, requiring analog excitation from DAC.

DACs and other powerful ADCs are described briefly in Sections 9.13.1 and 9.13.2.

9.13.1 Digital-to-Analog Converters

The digital signal is generally available as a number in straight binary form or binary-coded-decimal form. If the relative weight of each binary digit is independent of the other digits, the digital to analog conversion is simple. Thus, in the 8421 BCD representation, a 1 in the 8-bit must represent eight times the analog signal which is represented in the 1-bit and so on.

An op-amp can satisfy the requirements of the above process, if connected as an adder as shown in Figure 9.50. The resistors on the input side are in the ratio of 1, 2, 4, and the output voltage, V_o is, therefore, proportional to the BCD input number since it is given by

$$V_o = 8B_8 + 4B_4 + 2B_2 + B_1 \tag{9.29}$$

where B_1, B_2, B_4 and B_8 may be at 'ONE' or 'ZERO', depending on the logic state of each bit.

The constant of proportionality may be set by R_f/R.

BCD signals with several digits can be converted to analog form by using a DAC of the type shown in Figure 9.50 for each BCD digit and then adding the outputs in another op-amp.

Accuracy depends not only on the number of bits but on the precision of the reference voltages that stand for 'ONE' and 'ZERO' states and the values of resistors used in the circuit.

So for higher accuracy, it is necessary to use a transistor switch or an FET switch which connects the resistors (Rs) to a reference voltage when the input logic is at "ONE" and to ground if at "ZERO"

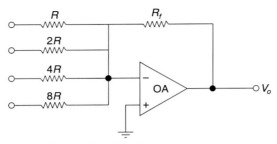

Figure 9.50   Digital-to-analog converter.

To avoid the use of a large number of precision resistors of differing values in the op-amp adder circuit, a precision ladder network of resistors as shown in Figure 9.51 is used. It can be shown that the output voltage V_o is given by

$$V_o = \frac{V}{2}(8B_8 + 4B_4 + 2B_2 + B_1)$$ (9.30)

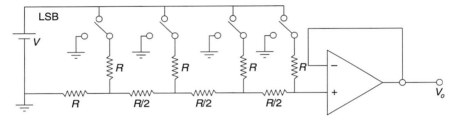

Figure 9.51   A precision ladder network for DAC.

Extension to 8, 10 or 12 bits can be done in a similar fashion and IC chips are available for direct connection to binary inputs and reference voltage. The input is desired in parallel binary form.

Most DACs are 8-, 12- or 16-bit type with correspondingly higher resolution.

The speed of a DAC is generally expressed in terms of settling time required for a large step in voltage to settle within half the value of the least significant bit of the final value. Higher resolution thus means larger settling time, ranging from 0.1 μs to 50 μs.

9.13.2 Analog-to-Digital Converters

Analog-to-digital conversion of an analog signal can be effected in a variety of ways. Some of the techniques are presented in Section 3.2.2.2 while dealing with digital voltmeters. The cost and complexity of the ADC increases with the speed and accuracy required. The speed of an ADC is expressed as *conversion time* which is the interval of time between the 'start' command and the instant at which the conversion is completed. Typical speeds are 1 to 100 μs. The ADCs are usually designed to provide both serial and parallel output.

Conversion speed for computer applications is necessarily higher than that for digital voltmeter applications. The indirect conversion methods are meant for relatively slow changing signals and are particularly suited for digital panel meters. The direct method of conversion is the successive approximation method, the block diagram of which is given in Figure 3.16. The direct method of *A*-to-*D* conversion is based on utilizing the feedback technique and the DAC forms the feedback path. The details of functioning are as follows: The basic scheme of a DAC is redrawn in Figure 9.52, and the technique consists of comparing the unknown analog voltage against a precisely generated internal voltage at the output of the DAC, the input number of which is the output of the ADC. When the conversion starts, the most significant bit of a digital register is turned on and the analog equivalent of this trial number v_f is compared with the input signal v_i. If $v_i > v_f$, the trial "1" in the MSB is retained; otherwise it is erased. The same procedure is adopted on the next most significant bit and so on until a "1" has been tried in the least significant bit and either retained or erased. The erasure of a bit always coincides with the setting of the next less significant bit. Conversion time is thus fixed for a given number of bits as all must be tried and is independent of the magnitude of the input voltage. The converted digital result is ready for use as soon as the last comparison has been made, being already present in the registers. Each conversion is unique and independent of the result of the previous conversion since all the logic is cleared at the start of the process.

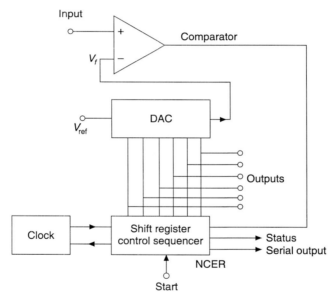

Figure 9.52 Basic scheme of a DAC using successive approximation technique.

Clearly, the above process is much faster than counting pulses, especially if the number of bits is large. The method affords both accuracy and fastness, the conversion time for 10-bits being 5 μs.

The successive-approximation type ADCs are normally available as complete modules. They are also implemented by microprocessor software. They may have an external clock, which may be synchronized with other digital circuits.

During the time of conversion process, a 'STATUS' or 'BUSY' output goes to '1' and the digital output is not read until the 'STATUS' output returns to '0'. Serial output is also available for interfacing with other equipment.

For non-critical low-cost and low speed applications, a slightly different version of the successive approximation converter known as counter-type converter is used. It also has a DAC, but is driven by an increasing count generated in a counter, as shown in Figure 9.53. The result is that the output of DAC is a staircase output which forms v_f for comparison with v_i. The count is stopped when v_f crosses the v_i value. The count is then stored and the corresponding digital output is displayed.

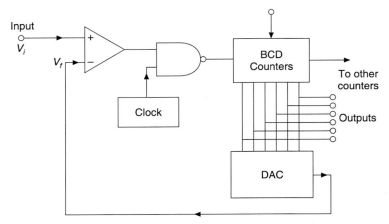

Figure 9.53 Basic scheme of a counter-type ADC.

9.13.3 Voltage-to-Frequency Converters

A voltage-to-frequency converter (VFC) develops an output frequency which is proportional to the analog input voltage. It can be used as an important stage of the ADC as stated in Section 3.2.2.2 and, in such a case, the output waveform is a pulse train.

In Section 9.2.6, a function generator is shown to work as a VFC. There are integrated circuit versions of the VFC using a comparator-triggered one-shot for frequency outputs ranging from 1 Hz to 100 kHz.

One of the common circuits of VFC uses an integrator that automatically resets itself when the output reaches a certain threshold. The scheme is shown in Figure 9.54. The input voltage v_i is integrated, resulting in a negative-going ramp at the output of OA1, with a slope proportional to v_i. A comparator connected to the integrator has an output v_o, the polarity of which is dependent on the relative levels of the threshold voltage v_T and the output voltage v' of the integrator. When v_o is positive, the FET is switched on, causing the capacitor C to discharge. Capacitor C' controls the output pulse length, t_p, which is made long enough to allow C to fully discharge.

A narrow output pulse is required for a reasonably high conversion linearity. If the operation at frequencies higher than 1 kHz is desired, the design of the circuit and the time constants become critical in order to preserve the linearity between v_i and the period of the output pulse T. The conversion factor of the VFC is defined by the ratio of v_i to f_o, where f_o is the frequency of the output pulse train.

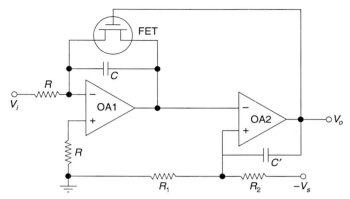

Figure 9.54   Circuit of a voltage-to-frequency converter.

When VFC is used to function as an ADC, the output pulses of the VFC are counted over a particular time interval. Thus the addition of a time-interval generator and a pulse counter makes the scheme complete. The important stages of this ADC can be IC versions, and hence the assembly becomes easier.

9.14   MULTIPLEXERS AND DEMULTIPLEXERS

Multiplexing is the process of transferring several independent signals through a common line or circuit. A multiplexer is thus a device or system that has several inputs and one output. Both analog and digital signals can be multiplexed. In telemetry and data transmission, a large number of signals have to be transmitted over a single wire or radio link to a remote station. Multiplexing of digital data is also done in order to reduce the number of decoders and ADCs and such other complicated circuits, or to reduce the number of lines or pins that must be passed in or out of an integrated or printed circuit.

Multiplexing operation can be carried out either in time domain or frequency domain. In time-division multiplexing (TDM), the analog input signals are sampled and transmitted sequentially through the common circuit. In frequency-domain multiplexing, analog input signals of widely differing centre-frequencies, whose bandwidths to not overlap each other, are mixed up together to form a composite signal and transmitted over the wire or radio link. Analog multiplexers are circuits used for multiplexing operation, and demultiplexers are used for reverse operation.

Digital data is normally presented in parallel form and the number of lines over which the data is available is equal to the number of bits of data. The states of all the bits are available simultaneously. In serial form the data is transmitted sequentially, one bit at a time over the single line or link. Usually one data bit is transmitted per clock pulse. Conversion to and transfer of serial data is equivalent to multiplexing operation and is implemented by means of *D* flip-flops. Similarly, demultiplexing or conversion of serial data back into parallel form is done by a set of *D* flip-flops.

A multiplexer may also be provided with additional inputs known as *ADDRESS,* whose state decides which one of the inputs can be connected to the output. Similarly, in a demultiplexer the separate ADDRESS inputs determine which of the outputs is connected to the input. All of the nonselected outputs may remain, say at logic "ONE", and the selected one at "ZERO".

Some multiplexers may have an additional 'ENABLE' input which, in one state, connects the input(s) to the output(s) and in the other state disconnects them from the outputs.

Latches built from *D* flip-flops transfer data from the input to the output, usually on the rising edge of the clock. For multiplexing, a tristate latch consisting of a standard latch with a transmission gate connected to the output is usually preferred where the outputs of two or more latches are to be connected in parallel.

9.14.1   Analog MUXRs and DEMUXRs

An analog multiplexer (MUXR) consists of a set of analog switches which can be switched 'ON' in sequence by a counter and decoder. They are used at the inputs to ADCs so that one converter is only needed for conversion of several inputs.

A demultiplexer (DEMUXR) at the receiving end is identical to MUXR.

When the rate of change of the analog signals is low, mechanical switching is used with proper synchronization of the switching system at both the transmitting and receiving ends.

The analog switch is an electronic switch of the CMOS transmission gate and is used when the analog signal frequency is high. When transmission gate is 'ON' the equivalent resistance of the switch may be within 100–400 Ω, and hence a MUXR using transmission gates requires signal amplifiers and line drivers.

A four-channel MUXR and DEMUXR is shown in Figure 9.55. At the receiving end, capacitors are used to hold the voltage for the time between samples. 2-to-4 line decoders are

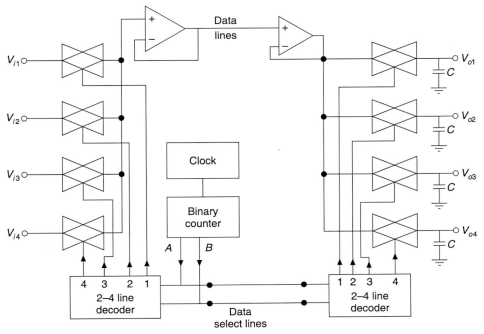

Figure 9.55   A four-channel multiplexer.

used to select the data lines (as data-selector) and they are driven by the same counter to obtain synchronization of the input and output lines at the transmitting and receiving ends. In the scheme shown, the total number of lines running from transmitting end to the receiver end is three, but the saving in the number of lines is considerable in the case of a larger number of input channels. The two data-select lines may be reduced to one, if one clock is used at one of the ends and a single line known as *synchronization line,* to furnish CLOCK inputs to two counters located one at each end.

9.14.2 Digital MUXRs and DEMUXRs

Digital data may be switched by transmission gates or by a combination of gates. The output lines may have to be brought to a third state of 'OFF' apart from the HIGH (1) and LOW (0) states. In such situations, a tristate latch, which is a combination of a standard latch and a transmission gate connected to the output as shown in Figure 9.56(a) is used. The DISABLE input provides the facility bringing the output lines into the 'OFF' state.

In Figure 9.56(b), an arrangement is shown in which the appropriate data line is switched on by an AND gate controlled by a decoder. The OR gate brings the output lines to 'OFF' state when all AND gate outputs are at "ZERO"

Seven-segment LED display units require a different approach for multiplexing the data obtained from a BCD counter. Multiplexing is done in order to obtain the display by using as many drivers as the number of display digits and one BCD-to-7 segment decoder. The four BCD data lines representing each decimal digit are switched on simultaneously by the 2-to-N-line decoder, and N decimal digits are scanned in sequence at a fast rate of about 100 Hz so that no flicker can be noticed on the display digit. Only one BCD-to-seven-segment decoder is used as can be seen from the scheme. All the seven-segment LED common anode displays are connected in parallel, but only the anodes of a particular display unit selected by the 2-to-N-line decoder get energized with the positive supply. The segment that goes to HIGH state from the BCD-to-7 segment decoder goes to lighted condition. Thus the switching of the anode driver and the latch output is synchronized by the data-selector decoder which is controlled by the scan clock and counter. The scan frequency may be chosen at will and it may range from 0.1 kHz to 10 kHz. Other details about LED display units are presented in Section 10.3.

9.15 MICROPROCESSORS

Digital devices described so far in the preceding sections have proved their versatility to meet the requirements of instrumentation systems. Moreover, they can all be built in the form of integrated circuits which are reliable in respect of their small size. The latest and most powerful device added to them is the microprocessor chip.

The *D* flip-flop is shown to have one-bit data input, whereas the JK flip-flop has a 2-bit data input. Digital devices having inputs and outputs of 4-bits, 8-bits or 16-bits along with the capacity of remembering previous inputs and outputs (being known as registers) are in greater demand. They are provided with external or built-in clocks so that at CLOCK transitions the outputs would assume states, depending on the inputs and the contents of the internal registers. The input data, otherwise known as *word,* consists of a set of bits and the word lengths are

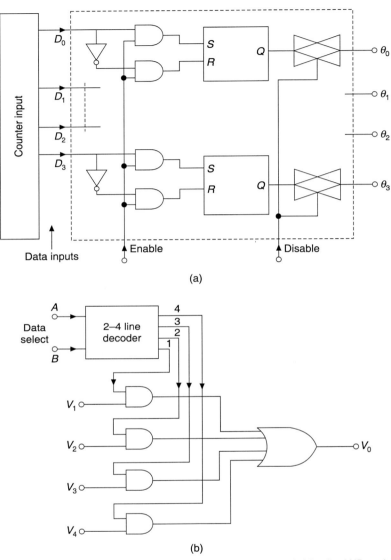

Figure 9.56 (a) A four-bit latch with tristate output; (b) digital data switching by AND and OR gates.

usually of 4-bits, 8-bits or 16-bits. The microprocessor is a large scale integrated-circuit chip and operates as a generalized sequential device usually on word lengths of 8-bits. An 8-bit word is called a *byte*. Large digital computers are based usually on 32-bit words. The microprocessor is capable of performing arithmetic and logic functions as defined by a given program and can function as a microcomputer, provided it is supported by the addition of memory circuits, input/ output devices and other components for interfacing it with the system. In other words, the microprocessor is the central processing unit (CPU) of the microcomputer.

The block diagram of a microcomputer is shown in Figure 9.57, and the entire unit is built as a large scale integrated circuit. The memory stores in coded form the operations to be performed and the information or measured data to be processed. The information is transferred from the system to the computer and vice versa by the input/output devices. The microprocessor (CPU) controls the input/output devices and transfers the data to and from the data memory. The program memory stores the 'program' which is a sequence of instructions, each instruction being a suitably coded word. Each instruction consists of two parts, an operator and an operand. The operator specifies the operation to be performed by the processor and the operand specifies the 'address' of memory location to be operated on.

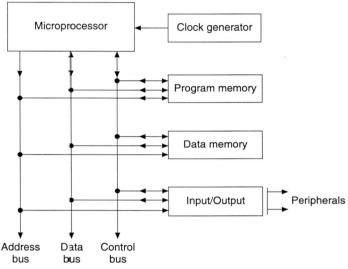

Figure 9.57   Block diagram of a microcomputer.

Each byte in memory is labelled by an address, which itself is a set of binary digits. If the memory contains 65,536 (2^{16} or 64 K) words, then a word length of 16 binary digits (2 bytes) is required to address any word in memory. The word length of data is usually 8-bits and can be 4 or 16 bits. The 'bus' is the bunch of conductors carrying signals from one or more sources to several other units. The address and data 'buffers' are storage devices used to compensate for short-term differences in information transfer rate between a source and receiver of information. Figure 9.58 presents the block diagram of the important stages incorporated in a microprocessor.

The program counter stores the 'location' in the memory of the next instruction to be executed by the computer. The instruction register stores the next instruction to be executed by the computer. The accumulator is a register that receives and stores the data used in or obtained from the arithmetic or logic operations. The arithmetic logic unit (ALU) performs the specified arithmetic and logic operations conveyed in the instruction. The instruction decoder is typically an internal read-only memory (ROM) which translates the machine instruction code into instructions that are executed by the processor. Transfer of data, addresses and control signals between the system elements are carried by the time-shared buses.

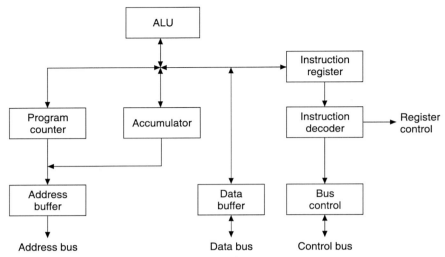

Figure 9.58   Block diagram of a microprocessor system.

Other important auxiliary hardware that go with the microprocessor are the memory systems. They are commonly known as RAM, ROM and PROM. RAM refers to a memory described as random access memory, in which arbitrary binary data may be written into or read from it. RAM is volatile as the memory loses the stored data when the power supply goes off. Whatever data is desired to be stored should be written into again after power supply is resumed. In a read-only-memory (ROM), information is initially stored at the time of manufacture and it is for purposes of reading only. It retains the stored information, independent of the power supply. ROMs are used for supplying dot-matrix patterns for use in character generation on a CRT alphanumeric display. EPROM is an erasable programmable ROM whose stored data can be erased and a new bit pattern written into. A transparent window is provided in the sealed package over the chip and exposure of the surface through the window to ultra-violet light results in erasure.

Microprocessors have invaded the field of instrumentation in a big way, where measurement and control operations involving reasonable amount of data and microcomputers using microprocessors have become the choice.

EXERCISES

1. Stress the role of electronics when dealing with measured data in the field of instrumentation.

2. Indicate the basic characteristics of an operational amplifier and show how it has enabled the processing of measured data obtained from transducers.

3. Briefly describe the following types of oscillators and explain their importance in instrumentation systems:

(a) Wien-bridge oscillator

(b) Crystal oscillator

(c) LC tuned-circuit oscillators.

4. Explain why pulse generators find increasing application in instrumentation.

5. Draw a simple circuit of a Schmitt trigger and suggest its applications.

6. Distinguish between passive and active filters and discuss the relative merits of each.

7. Show how simple low-pass and all-pass filters of the active type are built and obtain their transfer functions.

8. Distinguish between Butterworth and Chebyshev filters and show their characteristics.

9. Using operational amplifiers, show how the second-order filters are made?

10. What are 'biquad' filters and how is the choice of the number of op-amps made?

11. Describe the techniques of obtaining third and other higher-order filters.

12. Indicate the merits of frequency modulation when compared to analog modulation.

13. Draw the circuits of balanced and double balanced modulators and show how amplitude modulation is effected by them.

14. Distinguish between direct and indirect frequency modulation.

15. Explain the techniques of obtaining direct and indirect frequency modulation.

16. Draw the scheme of Armstrong frequency modulator and explain its functioning. What are its merits?

17. Distinguish between narrow-band and wide-band frequency modulation.

18. Draw the circuit of a simple amplitude demodulator.

19. Explain the operation of the (i) tuned-circuit FM detector, and (ii) Foster-Seely detector, and indicate the merits of each.

20. Draw the basic scheme of a phased-locked loop and show how it enables frequency demodulation.

21. Draw the scheme of a sample-and-hold circuit and explain its operation and application.

22. Describe a flip-flop circuit and its application. Show how it is built, by using logic gates?

23. Show how the following systems are constructed by using flip-flop circuits:

   (i) Shift register

   (ii) Binary counter

   (iii) Frequency counter.

24. Draw the circuit of a simple digital-to-analog counter and explain its operation.

25. Discuss the types of analog-to-digital converters available and the criteria adopted for the design of each.

26. What is the difference between a voltage-controlled oscillator and a voltage-to-frequency counter?

27. Explain the operations of 'multiplexing' and 'demultiplexing' and indicate their application in instrumentation.

28. Draw the basic schemes of the analog and digital type multiplexers and explain their operation.
29. Draw the block diagram of a microprocessor system and explain the function of each stage.
30. Briefly discuss the ways in which a microprocessor can be used to aid modern instrumentation activity.

SUGGESTED FURTHER READING

Betts, J.A., *Signal Processing, Modulation and Noise*, ELBS, London, 1975.

Carrick, Alan, *Computers and Instrumentation*, Heyden, London, 1979.

Cooper, William, D. and Helfrick, Albert, D., *Electronic Instrumentation and Measurement Techniques*, Prentice-Hall of India, New Delhi, 1970.

Graeme, J.G., *Applications of Operational Amplifiers—Third Generation Techniques*, McGraw-Hill, New York, 1973.

Graeme, J.G, Tobey, G.E. and Heulsman, L.P. (Eds.), *Operational Amplifiers: Design and Applications*, McGraw-Hill, New York, 1971.

Johnson, David, E., *Introduction to Filter Theory*, Prentice Hall, Englewood Cliffs (N.J.), 1976.

Kurt, S. Lion, *Elements of Electrical and Electronic Instrumentation*, McGraw-Hill Kogakusha, Tokyo, 1975.

Oliver, B.M. and Cage, J.M. (Eds.), *Electronic Measurements and Instrumentation*, McGraw-Hill, New York, 1971.

Rangan, C.S., Sarma G.R. and Mani, V.S.V., *Instrumentation Devices and Systems*, Tata McGraw-Hill, New Delhi, 1983.

Richard, Henry, W., *Electronic Systems and Instrumentation*, John Wiley, New York, 1978.

Taub, H. and Schilling, D.L., *Principles of Communication Systems*, McGraw-Hill, New York, 1971.

U. Tietze Ch., Schenk, *Advanced Electronic Circuits*, Springer-Verlag, Berlin, 1978.

Wobschall, Darold, *Circuit Design for Electronic Instrumentation*, McGraw-Hill, New York, 1979.

10

Data Display and Recording Systems

10.0 INTRODUCTION

A large number of sensors and transducers continuously furnish information about the condition of a plant, process, vehicle or patient, to a central point to provide real-time display of the quantities measured. These quantities may be recorded, plotted or printed for further scrutiny at a later date or stored in memory in such a way as to reconstruct, if necessary, the signal with time for visual display and analysis. The raw data thus obtained from transducers has to be processed into a suitable form so that necessary mathematical operations are carried out, if necessary, by digital computers. All the data is thus pooled at the centralized control room where the digital computer is located. The digital computer may as well be used for on-line control of the plant or process; otherwise it may serve as a data logger. It has become possible to effectively monitor or control a complex system only after the full capabilities of the digital computer are exploited. Along with the measured data, a large amount of computed data and associated decisions are communicated by the digital computer for the benefit of the operator, through a larger number of indicators, display systems, printers, recorders and plotters. Data acquisition systems (DAS) acquire the data from transducers and convert the same into digital form so as to be processed and presented suitably by the data logger. Simple data acquisition systems may at times be totally analog in nature, but in such a case, high accuracy is not demanded. When the number of data channels is low and their bandwidth high, analog data acquisition systems offer sufficient versatility. But when the data from transducers is slowly varying and the number of data channels is very high, digital DAS are invariably preferred due to their higher accuracy and low cost per channel. Once the data is in digital form, a large number of computations can be carried out on the input data, rendering on-line instrumentation and control possible. A large amount of data can be stored if it is in digital form. Magnetic tape recorders and floppy disks are used to store the data. In the following sections, some of the common devices used for data display record and printing are briefly described. For details, literature on electronic display systems may be consulted.

10.1 DATA LOGGERS

For an effective condition-monitoring of a plant or vehicle, a large number of transducers are used for obtaining the necessary data. Some of the variables may be measured at several points of the system simultaneously for checking its performance for uniformity. All the data is mostly obtained in real time and processed for further understanding of the system performance. Where necessary, the data is utilized for control of the process or the system. In all the cases, the large amount of raw data obtained from the transducers have to be processed into a form suitable for analog or digital display and recording. The electrical signals obtained from transducers and associated circuits may be dc or ac, varying widely in their bandwidth and levels. While some of them have to be displayed and recorded continuously, others may be sampled at discrete intervals of time and presented. Data acquisition systems are intended to collect the output signals from the transducers, process them suitably, and provide output signals of a suitable range linearly related to the measurand. The DAS has to provide the necessary power supplies for the transducers, amplifiers and other stages. The several stages normally used for conditioning the input signal to the desired dc level may be understood from a scrutiny of two examples shown in Figure 10.1. For dealing with the large number of slowly varying input signals, they are sampled at regular intervals of time and the subsequent stages are thus time-shared for further processing of the signal. An analog multiplexer is used to sample the analog signals, and an analog-to-digital converter is used to convert the signals into digital form.

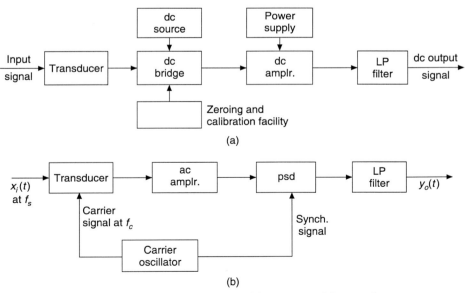

Figure 10.1 Signal conditioning systems: (a) dc system; (b) ac carrier system.

The timing of all the necessary operations is usually accomplished by a control unit. The control unit may be controlled by signals released and programmed from a clock incorporated in the data logger. The block diagram of a typical data logger employing a microcomputer is shown in Figure 10.2. The clocking system provides the necessary control signals for selection of the input data channel and getting it logged along with the time at which the logging is done. The channel selected is also logged for identification. The output data may be displayed, printed on a print-out system, and recorded on a tape recorder. The data loggers may also be provided with memory-devices for storage of data. A data logger essentially deals with digital signals and hence all the input data is brought out into digital form by ADC. The control signals (or the ADDRESS) of the microcomputer are synchronized in such a way that the input data channels are scanned one after another and fed to the ADC. The microcomputer may be provided with an operating console and keyboards for the entry of necessary command instructions.

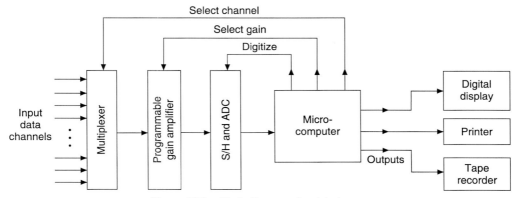

Figure 10.2   Block diagram of a data logger.

An analog DAS handling 6–12 analog signals may be provided with the facility of analog recording on a magnetic tape recorder. The complexity of DAS increases with the number of input signals to be processed, and in such cases, a microprocessor-based data logger similar to that of Figure 10.2 becomes essential. It is necessary to preserve the accuracy of the input data, while getting processed and conditioned through all the stages of the data logger.

10.2   ANALOG INDICATORS

Analog indicators are mostly meters that indicate the value of the measurand on a calibrated scale by means of a pointer. The permanent magnet moving coil indicators described in Section 3.1.1 are used as end devices in instrumentation systems for the indication of the amplitude of the primary measurand. The scale calibration may not always be linear and hence accuracy in reading becomes poorer in portions of the scale where it is cramped. Indicating instruments are helpful when static measurements are carried out. A variety of electrical indicating instruments are available for measurement of electrical quantities, and most of them are current-operated systems and meant for use on dc and ac at frequencies below 250 Hz.

Several electronic instruments are developed, using the basic pmmc system of range 0–50 µA as the end device and they are useful for measurement of dc and ac quantities of a wider range and frequency. The basic distinction between the electrical and electronic instruments lies in the fact that the electronic instrument constitutes negligible loading of the system under test. Nevertheless, the user should exercise considerable caution while selecting an instrument for measurement of a particular quantity.

In the field of instrumentation, it is possible to use electrical and electronic instruments as end devices. However, certain mechanical indicating systems reign supreme in areas where fire-hazards exist. They are mostly mechanical systems that develop linear or angular displacements as output signals which, after mechanical amplification by means of levers or gears, actuate the pointer for indication. In pneumatic instrumentation systems, air pressure variations constitute the output signals and hence Bourdon tube type pressure gauges are the end devices for indication of the value of the primary measurand.

All the indicating systems are designed in such a way as to respond as fast as possible and settle at the final value without any undue number of oscillations. When they are meant for use in industries, additional care is taken to provide them with vibration-free mountings. However, it must be said that the analog indicators are mostly used for laboratory work and their place is usurped by digital indicators, and that too, in much smaller numbers since data loggers enable the display of the value of any desired primary variable on command by the operators.

10.3 DIGITAL READOUT SYSTEMS

The outputs of the digital systems are usually presented in visual form by means of various types of display systems. The outputs of measured and computed quantities are displayed in numerals and associated instructions by means of letters from A to Z and by punctuation marks. Alpha numeric devices enable generation of all the required numerals, letters and other symbols. For a long time, the cold cathode neon lamp was used in digital voltmeters for display, but today, the solid-state devices using light-emitting diodes (LEDs), liquid-crystal displays (LCDs) and other hot filament devices constitute the common display systems. The development of the visual display unit (VDU), a modified version of the cathode ray tube, has provided great scope for both readout and graphic display.

10.3.1 Alphanumeric Devices

The neon lamp simply consists of a metallic anode and a cathode in a glass envelope which is filled with neon gas at a suitable pressure. When the voltage applied between the electrodes exceeds the breakdown voltage, the gas becomes ionized, resulting in a red glow around the cathode. The colour of the glow depends on the type of gas. The numerical indicator tube (NIT) has stacked cathodes formed in the shape of numerals giving good character shape. Due to their placement at differing depths, the numbers displayed are not on the same plane. Seven-segment displays of the neon lamp are also available with cathodes formed as seven independent segments, as shown in Figure 10.3(a), presenting the numerals in one plane. The requirement of a high anode potential prohibits them from being applied in instruments.

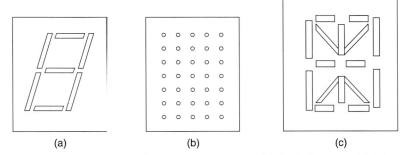

Figure 10.3   Digital display devices: (a) Seven-segment array; (b) 5 x 7 dot array; (c) 16-segment array.

LEDs are semiconductor diodes cut from gallium arsenide phosphide or gallium phosphide crystals containing certain impurities. They emit electromagnetic radiation in visible region (red) under forward bias conditions. Variation in the materials used for impurities provides emission in other colours such as green, yellow and amber. Each diode requires only about 1.5 V for forward biasing, drawing a current of about 5 mA only. However, it should be protected from the effects of voltage fluctuations, by means of a current-limiting resistor, since a small change in voltage can result in a large change in brightness. The merits of LEDs are their small size (almost a point source of light) and the small 'turn-on' and 'turn-off' times (about 50 ns). Hence they are preferred for forming an array of diodes to make up the seven-segment display for numerals from 0 to 9 and a 5 × 7 dot array for alphanumeric readout, as shown in Figure 10.3(b). The 16-segment alphanumeric display system is shown in Figure 10.3(c), which is used to generate additional characters such as those required in a typewriter-style keyboard. Because it is impractical to bring out one lead per dot, the dot matrix units have a built-in decoder or, more commonly, have one lead for each row and column. A dot at the intersection of a specific row and column is illuminated by applying a voltage between the specific row and column leads. To produce a complete character, the row or column is scanned rapidly. In Figure 10.4 is shown an LED display system with an external decoder. All anodes are connected together and the decoder output is seven lines, one for each segment. Decoder input is supplied in 4-bit BCD form. A transistor driver within the decoder/driver is switched on to turn on a segment. A suitable decoder is used when all cathodes are connected together. Neon lamp displays are used when brighter display is required, but they require a high-voltage drive per segment as shown in Figure 10.5(a), LED characters are only about 7 mm high, and when characters of 10–30 mm height are desired, neon lamp, hot-filament (incadescent) lamps or LCDs are used. They can be read from a distance.

Incandescent lamps are operated on 5 V supply and each segment draws a current of 10 mA. They are normally used in the seven-segment array. They are used alongwith the transistor segment drivers as shown in Figure 10.5(b). They are slow to respond and hence the resistor R_w is kept for holding the filament warm.

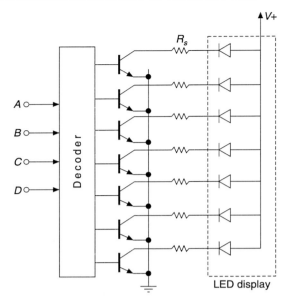

Figure 10.4   External decoder for LED display.

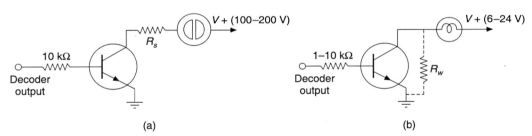

Figure 10.5   Segment drive circuits: (a) For neon lamp; (b) for incandescent lamp.

The LCDs are preferred when power requirements have to be kept lower. They are passive devices and consist of a film of liquid crystal about 15 μm thick sandwiched between two transparent electrodes. When exposed to electric field, the liquid crystal changes from being transparent to becoming milky opaque. Light from a separate light source passes through the crystal, and the brightness of the display depends on the intensity of the light passing through it. The operation of the LCD is based on the way the light is allowed to enter. To avoid electrolytic action, ac excitation of about 10 V at 50 Hz is used. The liquid crystal has high resistance and hence the current drawn is only about 250 μA. The LCDs are slow to respond with response time in the range of 20–100 ms. Their short operating life and low reliability render them unsuitable for purposes of instrumentation.

An array of LEDs or LCDs may be arranged in an $n \times m$ matrix form so as to have the amplitude of a signal indicated in a linear or angular format. Such a display is known as bar-graph display and the LED bar graph usually consists of a number of LED arrays (row-wise) arranged in a single column or multiple columns.

10.3.2 CRT Readout Systems

The cathode-ray tube (CRT) provides the scope for presentation of signals in digital fashion in addition to its conventional display. The digital readout cathode-ray oscillograph basically consists of a high-speed oscilloscope along with an electronic counter. The circuitry of both is connected by means of a display-logic control, allowing measurements to be made with great speed and sufficient accuracy. The digital readout oscilloscope presents not only the amplitude of the signal but also its rise-time and other information about the difference in time of two signals under comparison. Counter-circuits driving nixie tubes display the measured quantity.

Figure 10.6 shows the block diagram of the scheme used for measuring the amplitude of a voltage. The gated clock pulses are counted by the digital counter and the number of pulses is directly proportional to the voltage between the selected measurement points of the waveform displayed on the screen and it is readout in millivolts or volts by the nixie tube display.

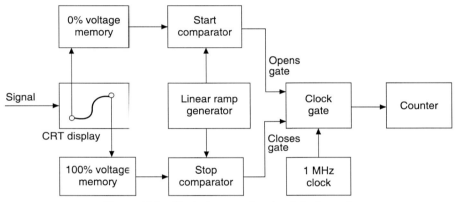

Figure 10.6   Block diagram of a visual display unit.

Viewing distance for display with cathode-ray tubes can be increased to 9 metres if the screen size is 75 cm. Black and white screens are visible at any light level if reflected light is eliminated by hood. Colour CRTs require larger lighting level.

A visual display unit (VDU) is a device displaying alpha-numeric, graphic and pictorial data, generated electronically as shown in Figure 10.6. For computers they have become essential for use as input/output devices for display of data and instructions flowing in and out of the computer. The cathode-ray tube is designed to suit the display of characters and is made flat with its size matching that of the TV screen. The CRT thus provides the alternative for the display of a large number of characters. The systems built up with the alpha-numeric devices are slow and cannot be used for a large number of characters as the drive logic circuitry becomes very complex. A variety of VDUs are available as terminal devices for use with computers. Characters are generated by a line raster technique similar to the technique used in TV tubes. The tube employs a pattern of 20 horizontal lines, and characters are made from a 5×7 dot matrix caused by unblanking the electron beam sweep at the proper instants of time. The character generator determines the display capabilities of the tube because it contains the information or memory required to produce the characters. It is either self-contained or provided by the computer memory. The decoder is connected to the character generator and accepts the

B-C-D inputs which are converted to unblanking the control grid signals that generate the desired characters.

Versatile storage type VDUs have a basic format comprising 39 rows of 85 characters, each of them generated in the form of 9 × 7 dot matrix. Graphic modes allow 1024 (*x*) by 780 (*y*) points on the screen to be individually energized straight lines of chosen length to be drawn between any two points, curves and complex graphics to be drawn from point by point.

With VDUs connected to digital computer, the need for many analog or digital indicators is eliminated. The operator selects the information. The alpha-numeric and graphic capabilities of the CRT are fully exploited and used, implying that they are found to be more efficient and economical than the other usual modes of display when large data is involved.

10.4 ANALOG RECORDERS

Analog recorders are used to provide a trace of the amplitude variations of a signal as a function of time. The trace is obtained in most cases either in ink on the recording paper or by means of a light beam on a light sensitive paper. Recording mechanisms differ widely in their bandwidth and the recorders are selected for use, depending on the bandwidth of signals to be recorded. The signal can be recorded and stored in a magnetic tape recorder so that it can be played back any number of times at a later time and at slow speed, if necessary. A brief description of some of the more popular recorders and their important features are now presented.

Direct pen recorders are the conventional indicating instruments, the moving system of which is provided with the ink-writing system. The power for driving the writing system is invariably to be provided from the source of signal. All such recorders record very slowly varying signals whose frequency in no case exceeds 0.1 Hz. Circular chart recorders record the signal on circular charts, each of which shows the record of variation of the measurand over a period of 24 hours. They are normally found in many process industries. Bourdon-tube type pressure gauges are used to record pressure variations which are proportional to the measurands.

Some.of the permanent magnet moving coil type galvanometers are used as direct writing recorders for signals lying within a frequency of 100 Hz. The pointer of the recorder is provided with a heated stylus that marks on a heat sensitive chart paper. To minimize the loading of the source of signal, an optical or an ultra-violet beam is reflected from the mirror attached to the moving coil on to the light sensitive paper. The photographic paper is developed for obtaining the record and, in certain cases, the trace is visible as soon as it is exposed to day-light. By increasing the spring constant and decreasing the moment of inertia of the moving coil, the undamped natural frequency of the galvanometer can be brought up to 10 kHz so that signals having bandwidth from dc to 4 kHz are recorded faithfully under a damping factor of 0.707.

Servo-operated self-balancing potentiometer recorders described in Section 8.4 are used to record low-level dc voltages. Their main advantage lies in the fact that all the power necessary to drive the pen is obtained from the power supply. But the inertia of the motor and the mechanical system is so much that its application is limited to recording of signals of a frequency below 5 Hz. The accuracy of recording renders them more suitable for use in instrumentation systems.

The *X-Y* plotter is a variant of the servo-recorder and is used to record one variable on one coordinate and another variable on the other coordinate. The plotter consists of two independent self-balancing potentiometers and the resulting motion of the recording pen provides a plot of the variation between the variables on a cartesian coordinate graph paper. By using a linear time base for the *X*-input of the recorder, the *X-Y* plotter can be converted to function as the *Y-T* plotter. The normal size of the chart paper is either 22 cm × 28 cm or 28 cm × 44 cm.

10.4.1 Cathode-ray Oscilloscope

The cathode-ray oscilloscope is one of the most versatile instruments used for visual indication and record of electrical voltages varying with time. Waveforms of complex signals and voltage transients can be studied for their variation with time. Repetitive waveforms can be brought to steady display and can be photographed for record and analysis after magnification. Similarly, non-repetitive transients can be photographed. Response of the cathode-ray oscilloscope for the fast transient studies is limited by the persistence of the phosphor coating provided on the screen.

The cathode-ray tube can be considered to function as a voltage-to-displacement converter as the position of the bright spot on the screen reflects in real-time the variations of voltage signal with time. The accuracy of the display is limited by the size of the screen and constancy of the gain of the amplifiers used. The relationship between the linear deflection *d*, and the voltage applied to the deflecting plates, *v*, as shown in Figure 10.7, is given by

$$d = \left(\frac{L_p L}{2 D_p V_a} \right) v \tag{10.1}$$

where,

d = deflection, m
L_p = length of the deflection plates, m
L = distance between the deflection plates and the screen, m
D_p = separation between the deflecting plates, m
V_a = accelerating potential, V

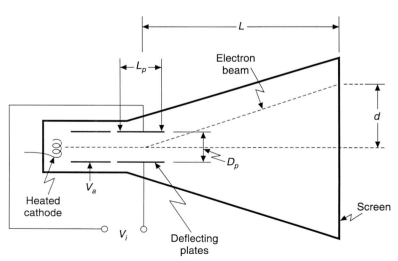

Figure 10.7 Deflection of a cathode-ray beam.

The test signals are usually amplified by means of amplifiers and v is the amplified voltage in volts. The other details about oscilloscope may be obtained by the reader from standard textbooks. They may be designed to provide bandwidth from dc to very high frequencies of the order of 10 MHz. The input impedance of most of the oscilloscopes is about 1 MΩ in parallel with an input capacitance of 30 pF. Some of the oscilloscopes are provided with two vertical beams and two amplifiers so as to enable comparison of their waveforms and measurement of phase difference between them.

The storage oscilloscope enables the capture and storage of a transient signal that occurs only once. The trace of the signal can be reproduced later for sufficiently long time so as to be visually analyzed or for photographic recording. A special cathode-ray tube is used for such oscilloscopes with provision for display of the signal over a period of time that can be adjusted as desired. When the signal is no more desired, it can be erased.

A different form of the storage oscilloscope is the transient recorder that captures the transient signal or a desired portion of the same for display at a later time on the screen of the cathode-ray tube. The storage facility is provided by means of a high speed analog-to-digital converter and a digital memory which can retain the signal truthfully for a very long time without any loss. The signal can be retrieved from the memory either for visual display in analog form or for further processing and printout by a digital computer or data logger. It can also be connected to a conventional pen recorder for getting the trace of the signal on chart paper run at the desired speed. The bandwidth of the transient recorder extends up to a frequency of 1 MHz, as is the case with the storage oscilloscope, but the accuracy and resolution of the reproduced record of the signal is higher because of the digital memory.

The fibre-optic oscilloscope is basically a single channel cathode-ray oscilloscope with fibre optics to enable the recording of the signal on direct-print paper. The photographic recording is thus eliminated. The light output from the cathode-ray tube is focussed on a continuous strip of direct-print paper by the fibre-optics system. Photosensitized paper is held in close contact with the fibre strip and an optical transfer of the CRT image occurs directly on the paper. The paper is run continuously and a skew correction is provided to correct for the motion of the paper. All other advantages and provisions of the cathode-ray oscilloscope can be utilized with the speed of the paper adjustable from 0.1 cm/s to 250 cm/s.

10.5   MAGNETIC TAPE RECORDERS

So far, recording systems providing visual record of the variations in the measured variable with time have been described. Magnetic tape recorders are unique in their application in the sense that the data is recorded for storage. The stored data can be reproduced at any subsequent time for visual display by the conventional analog recorder or on the screen of a cathode-ray oscillograph. The stored data may be played back or reproduced any number of times and that too at slow speeds to match with the speed of reading or recording system. The magnetic tape that used to store the data may be reused after erasing the existing data. The tape may have as many tracks as 16 so that 16 variables can be simultaneously recorded on the tape. The reproduce signals are electrical voltages which can be analyzed conveniently by spectrum analyzers.

The versatility of the tape recorder lies in its suitability for recording of analog signals of frequency ranging from dc to 1 MHz and digital signals of bit rates up to 10^5 per second. Although the home entertainment type tape recorder is based on the same principles of operation as the instrumentation tape recorder, the latter possesses high accuracy, higher bandwidth, high signal-to-noise ratio, and high dynamic range due to the techniques of recording the signals and higher tape speeds.

The tape recorder is widely used for recording of data in data acquisition and telemetry systems. The digital tape recorder assumes great importance due to its direct compatibility with computer systems, allowing direct transfer of data to and from the computer.

The magnetic tape recorder consists essentially of three different systems: (a) electromagnetic systems consisting of the record/reproduce/erase heads and the magnetic tape, (b) electronic circuitry to process the input and output signals, and (c) the mechanical system to run the magnetic tape at the desired speeds.

The record head is similar to the reproduce or erase head and it consists of a laminated toroidal core of a high permeability material, with a coil wound on it. The toroidal core is provided with a narrow gap filled with a nonmagnetic material as shown in Figure 10.8 and the magnetic tape is run close to and across the gap at the desired speed. The magnetic tape consists of a ribbon of tough plastic material, on which small magnetic particles of iron oxide (Fe_2O_3) are securely embedded. The particles bridge the gap and complete the magnetic circuit of the

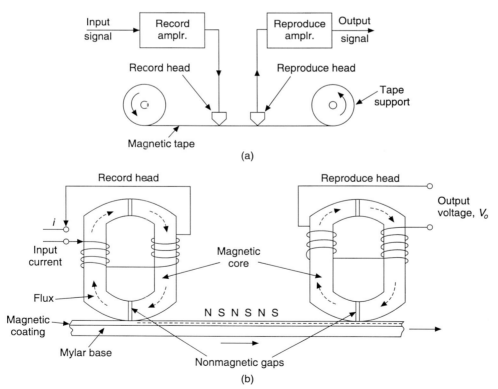

Figure 10.8 (a) Basic systems of a magnetic tape recorder; (b) record and reproduce heads and tape.

record or reproduce head. A current passing through the coil of the record head sets up magnetic field in the core, and the particles located in front of the gap are magnetized by the concentrated magnetic field set up across the gap. All the particles lying within the gap length at a given instant have nearly the same degree of magnetization.

If the magnetizing current through the coil is sinusoidal with time and the tape is moving across the gap at a speed v, then the length of the tape that would have passed the gap in a time interval equal to the period T of the current is given by

$$\lambda = vT = v/f \qquad (10.2)$$

where f = the frequency of the sinusoidal current signal and λ is known as the recorded wavelength. As the tape leaves the gap, each magnetic particle retains the state of magnetization that was last impressed on it by the magnetic flux bridging the gap, and hence the actual recording occurs at the trailing edge of the record head gap.

When the recorded tape passes across the gap of the reproduce or playback head, the magnetized particles bridge the gap of the head and magnetic flux set up through the core. The variation in flux linkages with time, due to the variation in the magnetization of the particles, causes induced emfs in the coil of the reproduce head.

If the magnetization curve of the magnetic particles is linear, magnetization of the particles is proportional to the current in the coil of the record head, but the induced emfs of the reproduce head coil are proportional to the time-derivative of magnetization. The induced emf which is the output signal of the reproduce head is thus seen to increase linearly with frequency of the record signal current. To compensate for this effect, an equalization amplifier is used to provide output voltages independent of frequency of the record signal over the range of frequencies of interest. Figure 10.9 shows the effect of equalization amplifier. It should be observed that the induced emf of the reproduce head becomes zero, if the record signal current is dc. Also, there is an inherent noise level of the recorder system, which sets the limit for the lowest frequency at which the reproduce signal bears correspondence with the record signal. Response below 50 Hz is generally unsatisfactory.

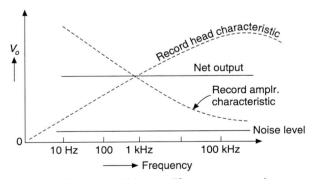

Figure 10.9   Effect of equalizing amplifier on tape recorder response.

The high frequency response is limited by the gap length of the reproduce head. As stated earlier, the induced emf of the coil is proportional to the rate of change of flux linkages with time. Hence a change in the magnetization of the particles (as seen from the trailing edge from instant to instant, while the tape moves across the gap) is essential for developing the induced emfs. From Figure 10.10(a) it can be seen that the net magnetization of the particles is zero, if the length of gap L is equal to λ. If $L = \lambda/2$, the induced emf is maximum and the variation of the induced emf with λ in terms of L is shown in Figure 10.10(b). Thus the choice of tape speed for a particular gap length determines the high frequency limit.

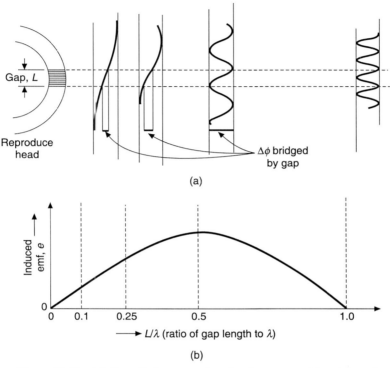

(a)

(b)

Figure 10.10   (a) Gap effect on reproduce head response; (b) variation of induced emf with λ in terms of gap length.

The entertainment type tape recorder is usually run at tape speeds $1\frac{7}{8}$, $3\frac{3}{4}$ and $7\frac{1}{2}$ inch per second, whereas the instrumentation tape recorder is run at higher speeds up to a maximum of 304.8 cm per second (120 ips). All these speeds are standardized for use in instrumentation by Inter Range Instrumentation Group (IRIG). If the tape is run at 120 ips, a gap length of 1.27 μm enables recording and reproduction of signal currents of 1.2 MHz. In such a case, the tape recorder is said to hold 10^4 cycles per inch and is considered to serve as index of its resolution. With the same gap length, if it is run at lower speeds, the high frequency limit is reduced correspondingly.

10.5.1   Direct Recording

The direct recording technique simply consists of converting the electrical voltage signals into proportional current signals which are, in turn, passed through the record head coil. But due to the nonlinearity in the B-H characteristic of the magnetic particle material, the magnetization of the particles will not be proportional to the record signal currents. Hence the reproduced signals will be distorted in waveform while record signal currents are sinusoidal. To overcome the problem, an ac bias current is added to the record signal current so that only the linear portions of the magnetization curve are utilized as shown in Figure 10.11. The bias signal is of such a high frequency as cannot be reproduced in the reproduce signal.

The bias frequency is usually 100 kHz for the entertainment tape recorder, and for others it is set at least five times the highest signal frequency. The amplitude of the bias current is also higher than the amplitude of the signal current and is adjusted in such a way as to reach from the origin to the centre of the linear portions as shown in Figure 10.11. Too high a value of bias current results in loss of high frequency response as it saturates the magnetic particles.

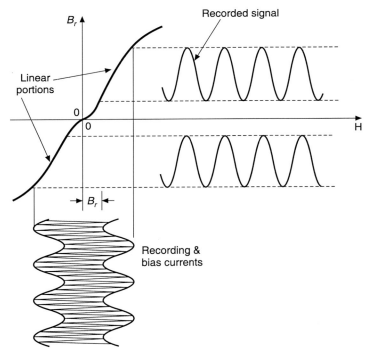

Figure 10.11   Effect of bias current on the output signal.

Although direct recording technique provides maximum band width, it is still unsatisfactory when high accuracy in recording is desired. The amplitude of reproduce signals is affected by presence of dust particles across the gap and on the tape, and also due to the uneven distribution of the magnetic particles on the tape. Fluctuations in tape speed is a common problem, and while recording and reproducing, tape speeds may not remain exactly the same. 'Flutter' is due to sudden changes in tape speed while 'wow' is due to speed changes over a long term. Accumulation of dust particles results in 'tape dropout' and also in loss of sensitivity of reproduction in course of time. Direct recording technique is thus restricted to recording and reproduction of music. Accurate recording of dc and low frequency signals obtained from transducers cannot be done by direct recording technique. To overcome this drawback, and to improve the signal-to-noise ratio of the recording system, frequency modulation recording technique is employed.

10.5.2 Frequency Modulation Recording

The data signal is used to frequency-modulate a high frequency carrier signal and the modulated carrier signal is directly recorded on tape. There is no need of a biasing signal. The reproduced signal is frequency-demodulated and the data signal is reconstructed. Thus the effect of amplitude variations in the reproduce signal is eliminated by the frequency modulation recording. A low-pass filter is used finally to remove noise and traces of carrier frequency signals.

To obtain high accuracy by the FM recording technique, the tape speed is controlled by means of high accuracy speed control systems. Any variation in tape speed directly effects the amplitude of the data signal when reproduced. In situations where the tape speed is not rigidly under control, it is better to make the percentage deviation of the carrier at 40% instead of the normal 7½% found in telemetry systems. Modulation index of 5 is used in most cases. When the tape recorder is used for recording of data obtained from telemetry, the tape speeds, centre frequencies of the carrier, and modulation index are set at values specified in IRIG standards.

It is possible to use one carrier signal and record more than one data signal on a single track. The record and reproduce circuits make the tape recorder more complex and costly in comparison to the direct recording system. FM recording technique enables dc signals to be recorded, but the highest frequency that can be recorded is limited to 40 kHz.

10.5.3 Digital Recording Technique

Direct and FM recording techniques are suitable for recording and reproduction of analog signals. Digital data are recorded directly on tape by means of the same record heads but the coils are driven by means of suitable electronic circuits to deliver current pulses of such amplitude as to saturate the magnetic particles with one polarity for one logic level and reverse polarity for the other level. This is known as industry-accepted IBM format of NRZ (non-return-to-zero) recording. In this way, the tape is magnetically saturated at all times and the change in flux direction on the tape is used to indicate a 1-bit and no change as a 0-bit (see Figure 10.12). Errors are likely to result because of tape drop-out and hence most tape systems employ parity checks.

The digital tape recording systems are operated in two ways. The incremental system advances its tape by one increment for each digital character to be recorded. Hence all the characters are uniformly and precisely spaced from each other. In the synchronous system, recording is done after the tape is brought to the correct high speed. After the recording is over, the tape is quickly brought to stop. All the characters are uniformly spaced in each bunch and there is an erased portion of tape between one bunch of characters to another.

Erasing the recorded data on the tape is done by the same record head or an equivalent head additionally used. The erase head is operated at a frequency higher than the upper limit of its bandwidth, and an intense magnetic field is set up across the gap so as to magnetize the particles to the maximum extent. Thus the magnetization due to previous record is destroyed, leaving the particles demagnetized.

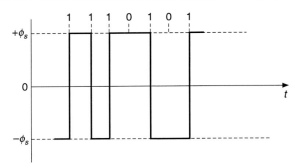

Figure 10.12   Digital tape recording by NRZ format.

10.5.4 Floppy Discs

A floppy disc is a magnetic recorder consisting of a disc coated with a magnetic material instead of the tape of the tape recorder. The disc forms the base and support for the coating while providing mechanical strength. It is about 25 to 50 cm in diameter and is coated on both sides in such a way as to form about 100 to 1000 tracks on each side. The read (reproduce) and write (record) heads are positioned on movable arms that are controlled to select the track required. The disc is run at constant speed, and data can be recorded at the rate of 10,000 to 40,000 bits/s. The storage capacity of the disc is much higher than the tape and can hold as many as 4×10^8 characters.

The diskette is similar to the floppy disc except that it is smaller in size. It is a disk of thickness of 0.125 mm and is made of a tough plastic material called mylar.

Both the disc and diskette store data and instructions in a way compatible with the digital computers.

10.6 DIGITAL INPUT–OUTPUT DEVICES

Where the digital computer is used for data processing, it has to be coupled with a few devices that accept the data from the computers and its operators and transmit the data for read-out, print-out, plot and record. Both the input and output devices handle data that represent numbers, symbols and letters. Between the input and output devices, there is a storage unit that stores the data before utilization by the output device.

The on-line operation of the input and output devices is directly commanded by the computer and is considered to be integral with the computer. There are various other peripheral equipment that go with the computer, but the most common of them are the punched-card or paper-tape equipment, magnetic tape recorder or floppy discs, and the visual display unit. The VDU serves the purpose of plotting the data for visual display and analysis. The production of graphic images on VDU is known as *Computer graphics.*

For low speed computers, punched card and paper tape equipment are used, such that information is conveyed by the presence or absence of punched holes in the card or paper. The paper tape punches are slower and have a speed of 100 to 500 characters per second.

Print-out systems constitute one of the important peripheral equipment of a computer. Printers are synchronous equipment used for the registration of graphic characters, mostly numerals and alphabetic letters on paper. A teleprinter is a serial printer which prints out one character at a time at a rate of 5 to 40 characters per second. Parallel printers print all the characters on a given line simultaneously and hence are faster in operation.

Line printers are parallel printers having a speed range of 600 to 1200 lines per minute. Most of the line printers are impact printers that can print all the numerals, letters and some symbols. The print-out has normally 120, 132 or 144 character positions in a line and the printer prints about 5 to 6 lines in a vertical span of 2.5 cm with 0.5 cm line spacing. There are a variety of line printers based on different mechanisms for production of the characters and they are known as bar-, wheel-, chain-, cylinder-, and matrix-printers.

EXERCISES

1. What are data acquisition systems and what is their role in the field of instrumentation?

2. Describe briefly, by means of a block diagram, the various stages of a data logger and indicate the function of each stage.

3. Distinguish between 'analog' and 'digital' indicators and discuss the means adopted for the indication of the measured value.

4. Briefly describe the types of alpha-numeric devices available and indicate the merits and demerits of each.

5. Describe a 'VDU' and its basic mechanism. Explain the reasons for its popularity and use in modern data acquisition systems.

6. Classify the various types of analog recorders and distinguish the limitations of each when recording analog data.

7. Describe the various stages of a cathode-ray oscilloscope and derive from fundamentals the relationship between the deflection of the spot and the voltage applied to the deflecting plates.

8. Explain the basic mechanism of recording and reproduction of an analog voltage signal by means of a magnetic tape. Describe the important stages that make up a magnetic tape recorder.

9. What are the merits of a magnetic tape recorder when used for recording measured data?

10. Distinguish the requirements of an 'entertainment' tape recorder and an 'instrumentation' tape recorder and explain why the latter is operated at very high tape speeds.

11. Explain how the gap of the recording head limits the range of frequencies of signals to be recorded.

12. Briefly explain the analog and digital recording techniques employed and show that higher percentage deviation of the carrier frequency is essential in FM recording.

13. What is a 'floppy disc' and what is its application?

14. Briefly describe the types of printers used for the print-out of data in digital form.

SUGGESTED FURTHER READING

Carrick, Alan, *Computers and Instrumentation*, Heyden, London, 1979

Cooper, William, D. and Helfrick, Albert, D., *Electronic Instrumentation and Measurement Techniques*, Prentice-Hall of India, New Delhi, 1978.

Kurt, S. Lion, *Elements of Electrical and Electronic Instrumentation*, McGraw-Hill Kogakusha, Tokyo, 1975.

Magrab, Edward, B. and Blomquist, Donald, S., *The Measurement of Time-varying Phenomena: Fundamentals and Applications*, Wiley-Interscience, New York, 1971.

Richard, Henry W., *Electronic Systems and Instrumentation*, John Wiley, New York, 1978.

Sheingold, Daniel H., *Transducer Interfacing Handbook*, Analog Devices Inc., Massachusetts, 1980.

U. Tietze, Ch. Schenk, *Advanced Electronic Circuits*, Springer-Verlag, Berlin, 1978.

Wightman, E.J., *Instrumentation in Process Control,* International Scientific Series, CRC Press, Columbus, Ohio, 1972.

Wobschall, Darold, *Circuit Design for Electronic Instrumentation*, McGraw-Hill, New York, 1979.

CHAPTER **11**

Data Transmission and Telemetry

11.0 INTRODUCTION

Most of the measurement and instrumentation systems discussed so far are shown to consist of all its stages interconnected with each other by means of short connecting wires. Wherever the connecting wires posed problems in processing the signals, care has been taken to overcome the same, as in the case of signals obtained from high-impedance sources such as piezoelectric transducers. A large variety of situations, however, exist as detailed in Chapter 1, wherein measurement is carried out at one location and its value is required at a remote location for indication, record or control-action initiation. The distances involved may be small in the case of instrumentation inside a laboratory, a few hundred metres in case of industries and a few thousand kilometres in case of aircraft and spacecraft instrumentation. In all these cases, the primary interest lies in reproduction of the measured data at the remote location without entailing any more loss of accuracy in the course of transmission over the distance. Special attention needs to be paid towards the choice and design of a suitable interconnecting channel or link along with other associated equipment at either end. For the data available in electrical form, the link is invariably the electrical wires or lines for reasonable distances, and electromagnetic waves for long distances. In areas where fire hazards exist and electrical systems pose problems, signals are obtained from pneumatic sources and transmitted over pneumatic pipelines for distances up to 250 m. Once in the safe region, the pneumatic signals are converted into electrical signals for further transmission. These signals may be transmitted directly as they are, over electrical lines, for short distances up to a few kilometres but are subjected to the process of modulation, for purpose of transmission over long distances by means of wires and electromagnetic waves. The modulated signals are demodulated at the receiving end and such a process offers several advantages besides preservation of accuracy.

Telemetry (telemetering) is thus taken to mean presentation of measured values at a location remote from the site of measurement. It is quite different from remote measurement which implies that the measuring equipment is located remotely from the object or system under investigation. Measurement of the position of an aircraft by means of radar located at ground station constitutes a case of remote measurement. Telemetry systems may be treated as a class of instrumentation systems since the difference lies only in respect of the distance between the sites of the measured quantity and displayed quantity.

The advent of real-time computation and control by means of digital computers is primarily responsible for the development of telemetry systems. The amount of measured data that can be handled by the computers and data loggers has become exceedingly large, with the result that it has become imperative to identify some means of conveying the large number of measured variables effectively, accurately and in real-time for condition monitoring and automatic control of systems, whether it be a process, plant or a moving vehicle. Whatever may be the distance involved, the computer needs to be fed with the measured data in the desired form along with a set of instructions to the computer. In course of developments in this regard, it has come to be understood that the data transmission systems deal with the signals to and from the computer, and data includes all information constituting the input and output signals of the digital computer. Thus data transmission systems deal mainly with digital data, while telemetry systems are concerned with the transmission of both the analog and digital signals. Both the telemetry and data transmission systems may be considered to fall into the same category in respect of preservation of accuracy of data in course of transmission, regardless of the distance involved.

Information is a general term and is used to convey an idea of all the relevant details pertaining to a system or source of information. Voice message, music, pictures—static or moving—may be treated as information, but qualitative in nature as the exacting demands on accuracy of reproduction are reasonably relaxed for reasons of inability of the human sensory organs, ears and eyes, to detect the deviations in the reproduced information. But for purpose of communication over distance, the above information needs to be converted into electrical signals by means of transducers and such data when reproduced may be allowed deviation, to some extent, from the original information. Communication system design differs mainly in this respect from that of data transmission and telemetry systems. Although communication theory lends itself all the principles for the development and design of data transmission and telemetry systems, communication system design permits the incorporation of the facility for the flow of information from any one point to another (or in both the directions for the communication link). The data transmission and telemetry system allows the flow of data in one direction only and it is always from the source of data to the receiving system. The measured data is processed for transmission at the measuring end, and the entire system is known as transmitter, whereas receivers such as read-out devices, recorders, data loggers and computers may constitute the receiving systems. Both the wire and radio links allow the transmission of analog and digital data, but the distinct feature of digital telemetry and data transmission methods is that they are inherently capable of not only reproducing the data with the desired accuracy but are also capable of detection and recognition of an error along with attendant correction.

Simple telemetry systems of electrical and pneumatic type exist in certain industrial processes and power generating stations and they usually consist of a transmitter, a receiver and associated cables or pneumatic piping. Each system is designed for a particular variable of a certain range and some of them need recalibration if the length of the lines is altered. The basic scheme of a telemetry system is shown in Figure 11.1.

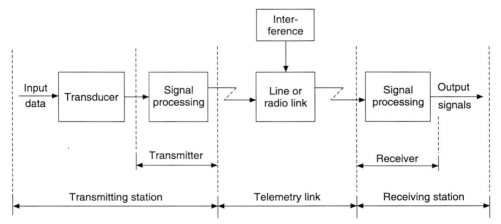

Figure 11.1   Block diagram of a telemetry system.

Both the wire and radio links allow transmission of a group of measured values provided the basic data consisting of these values is used to modulate a high frequency carrier. Telemetry of a small group of variables is possible through the existing telephone and power lines. The process of utilizing one data link for transmission of a group of variables is known as *multiplexing*. Demodulation and demultiplexing of data are the reverse processes of modulation and multiplexing respectively.

Once data acquisition through sensors has gained significant status and sophistication, efforts to utilize the data in a big way to transfer it over distances not just confined to the area of the industry, but over long distances, encountered in guidance of an unmanned aircraft or remote control of a space craft. For the computer controlled automated production systems, it is desired that a standard field instrumentation bus be designed, so as to effect a two-way communication between the digital sensors and actuators. Factors leading to such designs are indicated in Section 14.7.

Consequent to the growth of the signal transmission system, the need for remote control has arisen. Based on the signals received at the ground station, a control action should be initiated at the ground station for the automatic control of the variables pertaining to a vehicle in space. In such situations, a continuous dialogue exists between the spacecraft and the ground station. The subject of "Telemechanics" deals with such operations as desired for remote control (control from a distance) and telemetry.

11.1   CHARACTERISTICS OF A TELEMETRY SYSTEMS

The performance of a telemetry system is mostly dictated by the characteristics of the transmission medium. The transmission media may be wire or radio link as stated already, but it may be noted that optical, ultrasonic and laser beam links are also used for transmission of data. For the present study, wire and radio links are only considered and electrical signals invariably constitute the input and output signals of the link. The signals transmitted over the link usually carry data obtained from more than one data source.

Direct telemetry systems are those in which the data from one transducer alone is transmitted over the link. Data from a number of transducers is grouped together for transmission purposes, and is termed as 'raw data that consists of a number of base band signals'. If all the base band signals do not overlap on each other in their frequency spectrum, they are mixed together and transmitted directly over the data link. The range of frequencies contained in this composite signal is referred to as the bandwidth of the baseband-signals. It is essential that the data link be capable of transmitting all these frequency components faithfully for reproduction at the receiving end, and in such a case, it is clear that the bandwidth of the transmitted data falls within the bandwidth of the data link. If the bandwidth of data is small, the baseband signals are utilized for amplitude or frequency modulation of a high frequency carrier signal so as to transpose (shift) the spectrum of the baseband signals to a region around the frequency of the carrier signal. The bandwidth of the data link, if large, is usually divided into a number of channels so that each channel is used for carrying either one baseband signal or a mixture of them. The link is thus put to optimum usage and its data-handling capacity is increased. Electrical wires, power lines, telephone lines and coaxial cables are studied for their information handling capacity and each one of them is considered for the merits of its bandwidth.

Similarly, radio telemetry provides facility for transmission of a large amount of data by the choice of a high frequency carrier in the spectrum of electromagnetic waves. The channels for radio (space) telemetry are allotted by the International Frequency Registration Board (IFRB) of the International Telecommunication Union (ITU) so that there is no interference from one channel to another.

Based on the above considerations, the important characteristics of a telemetry system may be written as:

1. As the telemetry system is primarily equivalent to an instrumentation system, the primary criterion for choice and design is accuracy.

2. The system is decided, depending on whether the transmitted variable is in analog, time or digital domain.

3. If digital data is transmitted, error detection and recognition capability as well as correction capability, make the system highly accurate.

4. Each data link should be studied for its information carrying capacity, and optimum use should be made of its bandwidth.

5. As telemetry systems deal with signals, the power levels of data links and associated equipment should be as low as possible to reduce the noise generated.

6. Bandwidth of the data channel and the data link should be so chosen as to avoid crosstalk from one to another.

7. Signal-to-noise power level should be as high as possible; choice of the transmission method is dependent on the specifications of telemetry error.

8. The data link should be reliable for operation.

9. The size and weight of the transmitting equipment should be low when it is meant for use on air-borne vehicles.

10. Wherever possible, existing telephone lines, power lines and microwave links may he utilized for economical reasons.

11.2 LAND-LINE TELEMETRY

Direct transmission of data from the signal sources through the physical medium of electrical and pneumatic lines constitutes land-line telemetry. In electrical wire telemetry systems, transmitted variables are electrical voltages or currents. When the data received at the receiving terminal is in analog form, it constitutes analog data transmission system. In such a case, the voltages or currents transmitted and received are in analog domain and the data is contained either in their amplitude or frequency levels. Pulse telemetry systems utilize electrical voltages in pulse form to convey the data and it is possible to convert the data accordingly. Any one of the pulse characteristics represents the amplitude of the data. For digital telemetry, the pulses are encoded in the desired format after obtaining the samples of data at regular intervals of time. While all these methods are suitable for conveying the data when it is no more than a single variable, some of them can be utilized for transmission of data consisting of more than one variable by resorting to processes of modulation and multiplexing.

Analog telemetry systems for a single variable are commonly found in laboratories, hospitals and industries, where the distances and conditions permit the use of electrical lines and pneumatic tubing. In most cases, the data is received in analog form. The entire system consisting of the transmitter, receiver and the transmission medium or link may be of either open loop or closed loop type. The closed loop type is equivalent to the automatic measuring system employing negative feedback. Feedback type telemetry systems are preferred for reasons of accuracy, linearity and certain other advantages.

11.2.1 Open Loop Type Telemetry Systems

The simplest method of conveying an electrical signal from the transmitter to the receiver is by means of a pair of wires, as in the case of power transmission and telephone lines. Cables may be used for short distances up to 2 km to transmit even low level signals of the order of 10 mV. But the received signals are attenuated and shifted in phase due to the impedance of the lines. Moreover, the received signals are corrupted very much due to the electromagnetic interference, noise voltages and cross-talk from the adjacent lines. Hence cable telemetry systems,where the data resides in the amplitude levels of the transmitted variable, are not favoured.

However, a pair of lines or a cable may be used to convey ac voltage signals whose frequency is a function of the measured variable. It presents no difficulty to the receiving system in identifying the frequency of the signal, as long as the received signal is of satisfactory amplitude. The voltage signals transmitted are periodic functions of time, (may be sinusoidal in waveform) or a train of unipolar rectangular pulses, the pulse rate of which is a function of the measured variable. In such cases, signal sources such as the frequency generating transducers and other oscillator type transducers are favoured for use as transmitters. Otherwise, the analog electrical signals obtained from transducers have to be converted into signals in time domain for purposes of transmission by using voltage controlled oscillators and multivibrators. At the receiving end, the receiving system may consist of equipment that converts the time-domain signals into either analog domain or digital domain.

Encoded data obtained from a single digital transducer such as the angular displacement encoder may also be directly transmitted over a pair of wires. When the data is to be transmitted for connection to data loggers and computers, special cables are used. Usually the distances involved may not be more than one km. In such cases, the systems are simply referred to as data transmission systems and constitute special equipment at the transmitting and receiving end so as to ensure the desired accuracy, integrity and, in some cases, electrical isolation. The cables are designed for use at data rates as high as 10 MHz. Optical links by use of optocouplers enable transmission of data at higher data rates. Additional details related to transmission of encoded digital data may be obtained from treatises on the subject.

To serve as a guideline, the attenuation and phase characteristics of a pair of 20 SWG polyethylene insulated wires and a coaxial cable are presented in Figures 11.2(a) and 11.2(b), respectively. It may be observed that both the attenuation α and phase-shift ϕ are functions of the frequency of the voltages and are expressed to represent the same for a length of one kilometre. Where the phase shift and time delay are not prohibitively large, the wire telemetry systems may be used for transmission of the signals containing the data in the time domain. Signals of a frequency far below 50 Hz, and a frequency far higher than 50 Hz may be transmitted over the existing power lines. Similarly, telephone lines may be used for transmission of data by coupling the same to the line once the contact is established by dialing. In all these cases, special equipment is required at either end of the line for proper channelling and reception of data.

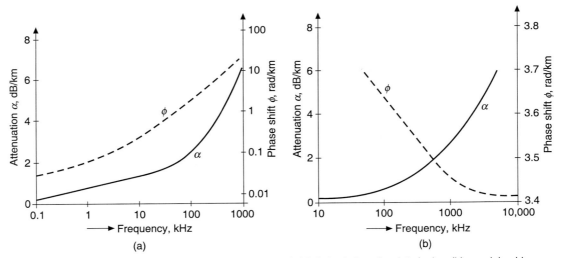

Figure 11.2   Attenuation and phase characteristics of: (a) Polyethylene insulated wire; (b) coaxial cable.

Simple analog telemetry systems for transmitting liquid level over a distance of a few hundred metres are shown in Figures 11.3 and 11.4. In each case, the receiver is a crossed-coil pmmc system and the transmitted variable, current. The scheme of Figure 11.3 employs a dc

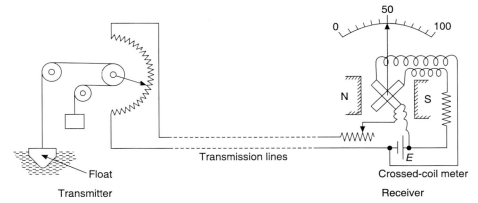

Figure 11.3   Liquid-level telemetry system using two wires.

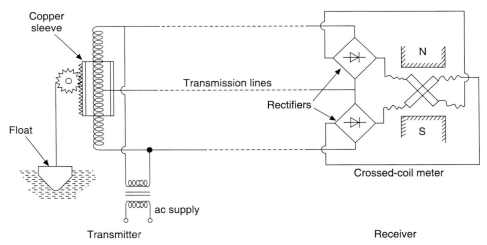

Figure 11.4   Liquid-level telemetry system using three wires.

source and the resistance variation due to level change controls the amplitude of the current. The scheme of Figure 11.4 uses ac supply at 50 Hz and the induced emfs, and hence the currents in the secondary coils are controlled by the position of the copper sleeve enclosing the coil system. Both the systems are designed for indication only at the receiving end.

Similarly, for transmission of angular displacements over a distance, a pair of synchros may be used. They are commonly used in many applications where the distance is less than a kilometre. The synchro-generator is used at the transmitting end, and the synchro-repeater, at the receiving end as shown in Figure 6.61. It is necessary to use a 3-wire cable and inequalities in the impedances of the three wires cause error in the indicated value. It has been stated in Section 6.2.2.2 that the load on the shaft of the synchro-receiver is also a cause of error in indication. The error in indication may not be more than ±10 min., if the synchro-receiver has only a light pointer attached to it. Synchro-pairs are available for use with ac power supply of 50 Hz and 400 Hz only. An additional pair of lines is necessary for exciting the synchro-pair, and a separate power supply system is usually employed for such situations.

Figure 11.5 shows another simple telemetry system for transmission of linear displacements obtained from primary mechanical transducers. The induced emfs of the two secondary windings are transmitted over the wires to the receiving end. The receiver consists of a servo-operated self-balancing system providing indication of input displacements. The arrangement is more or less free from the effect of loading the line due to the receiving system. It is equivalent to an ac bridge network ivith one-half located at the transmitting end and the other half at the receiving end. It may be noted that the accuracy of the system is limited due to the fact that the servo-amplifier cannot discriminate the 50 Hz components of the unbalance voltage and 50 Hz noise signals induced in the wires. A pair of screened wires may be used with their metallic screens connected together and grounded so that the effect of noise signals is minimized.

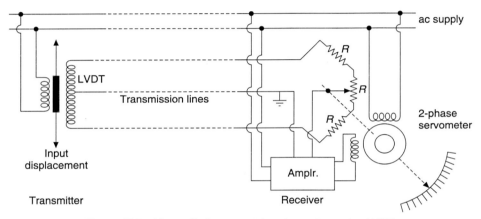

Figure 11.5 Linear displacement telemetry system using LVDT.

11.2.2 Feedback Type Telemetry Systems

Feedback type telemetry systems for a single variable are designed primarily to overcome the problems of the open loop type telemetry systems and to ensure the equality in the values of the transmitted and received signals. They are similar in principles of operation to the feedback type measuring systems dealt with in Chapter 8. Pneumatic telemetry systems develop output signals in the range of 3–15 psi in FPS system and electrical systems, in the range of 4–20 mA dc. Manufacturers of telemetry equipment have been designing them since long and the output-signal range is treated as standardized. The pneumatic systems are exclusively for use in areas where fire hazards exist and for distances up to 250 metres only, whereas the electrical type are suitable for distances up to 30 km. Process variables such as fluid flow-rate, liquid level and temperature are first measured by primary transducers, and their corresponding output signals of force are, in turn, converted into proportional pneumatic or electrical signals by means of the force-balance systems described in Section 8.6.

Figure 11.6 shows the scheme of a pneumatic telemetry system in which the beam balance system serves as the transmitter. The output pressure signals are transmitted over the lines of copper tubing to pressure gauges for indication. Crossed-leaf springs are used as pivots for the beam balance system. Input and feedback forces are developed by diaphragms. The beam is

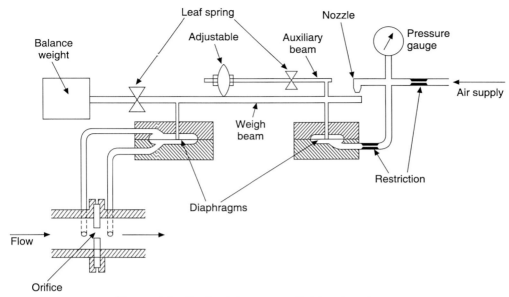

Figure 11.6   Feedback type pneumatic telemetry schemes

initially balanced in horizontal position with the output pressure adjusted at 3 psi for zero flow rate. The feedback provides linearity between the output pressure and input differential pressure, but the receiver pressure gauge is calibrated in terms of the flow rate of the fluid. In case the differential pressure developed is lower, bellows are used in place of diaphragms.

The main drawback of the pneumatic telemetry systems is that they are slow to respond. It may take as much as 4 seconds to receive a signal at the end of a pipeline of length of 250 m and of 1 cm outer diameter. The application of negative feedback improves the speed of response of the system but the time taken for the signal to travel the length of the line cannot be overcome.

The scheme of an electrical feedback type telemetry system with output signals in the range of 4–20 mA dc is shown in Figure 11.7. It is seen that it is akin to the system shown in

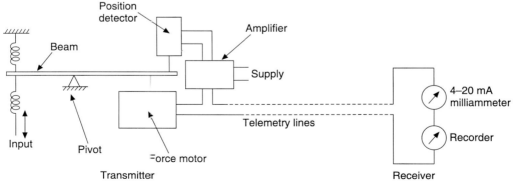

Figure 11.7   Feedback type electrical telemetry system.

Figure 8.10 except for the distance of separation between the indicating meter and the transmitter. The current may be transmitted over a two- or three-wire system over distances, depending on the line resistance. Telemetry systems are designed to operate at such power levels as to meet the line loss up to certain limits. Line resistance values up to 600 ohms do not affect the value of indicated current in the case of two-wire systems. In a three-wire system, two wires are used for power lines connecting the transmitter and receiver, and the output signal current is transmitted over a third wire with the return path provided by one of the power supply lines. The zero value of the measured variable and the break in the transmission lines are distinguished by the reading of the mA meter. A break in the lines is signified by the reading falling from 4 mA to zero value. These systems are faster than the pneumatic type and are very much used in industries.

11.3 RADIO TELEMETRY

The use of a radio link is not always meant for telemetry of data over a considerable distance. There are many situations where the data sources are located in inaccessible regions and the measuring and other recording equipment are located only a few metres away. In radioactive areas, the transmitter and receiver may be separated only be the thickness of a screening wall. During the check-out and countdown operations, the guided missile or the satellite and their launchers are not very far away from the receiving equipment. But after the satellite is launched, the distance between the transmitter of the satellite and ground-station receiving equipment becomes considerably large. Similarly, small transmitter is planted in a heart patient, which transmits signals related to the patient's heart condition, and the monitoring equipment may be located in the doctor's room only a few metres sway. While testing land-borne or air-borne vehicles, with or without pilot, the transducers used for measuring the operating conditions of the vehicle and the transmitters used are within a distance of 1–100 km from the receiving equipment. In all these and other similar situations, the transmitter and receiver are kept tuned to the operating frequency allotted for each set. Two-stage modulation may be employed if the number of measured variables is large and the frequency of the radio-carrier signal is so chosen as not to interfere with the frequencies of the radio-broadcast bands. As the carrier frequency is always sufficiently high, and the measured data from transducers is usually of low frequency, multiplexing of data becomes the rule, so that data from a good number of transducers may be obtained either on a continuous or time-shared basis.

In certain industries, one may come across a radio telemetry system in which a single measurand is conveyed over a radio link, by means of a radio frequency carrier even though the distance is not considerable.

The present-day user of a telemetering system has the choice between using a radio or a wire link. Radio links may prove very advantageous for quick erection and usage in situations where the installation of land lines would prove very expensive and time-consuming. But, due to the ruggedness and reliability of the transmission towers used for power lines, the conductors of the power lines are used for telemetry of data between generating stations and load despatch centres. Efficient operation of the electrical power systems is rendered possible by use of the power-line carrier current (PLCC) facilities. Multiplexing is possible, but the carrier frequency is limited to frequencies appropriate to the conductor size. Data signals of a frequency bordering on the frequency of the power supply and its harmonics should be avoided.

For space telemetry and satellite communication systems, it is essential to see that the carrier frequency is very high so that the length and weight of the antenna of the transmitter is considerably reduced. The frequency lies in the range of 216–235 MHz. Two-stage modulation of either FM/FM or PDM/FM is usually recommended. The rf carrier must remain stable to within ±0.01 per cent. The transmitters are usually allowed half a megahertz bandwidth for FM/FM and 0.2 MHz for PDM/FM system. A reasonable spacing between adjacent channels is maintained in order to prevent interference. The rf carrier deviation in the case of FM/FM system ranges between ±75 kHz and ±125 kHz and for PDM/FM system, it lies in the range of ±25 kHz to ±45 kHz. The output power of the transmitter is usually kept at low values consistent with the requirements and distances involved and it may be in the range of 2–100 watts. Other rf bands are also employed for nonstandard systems from 70–85 MHz to 2200–2300 MHz. The transmitting antennas are usually either projecting vehicle surfaces such as wings or tail fins or spike antennas that serve the double function of a pitot tube. For receiving the rf carrier, a wide variety of antennas are used, if the receiving station is land based. The receiving antenna may be designed to track the transmitter automatically.

A microwave telemetering system, employing transmitting and receiving parabolic reflecting antennas, is used to telemeter data from one station to another when both stations are land based. As the frequency is very high, the distance between the transmitting and receiving stations cannot, in any case, be more than 75 km, and such systems may be used for telemetry as well as digital data transmission.

With the increased awareness of the powerfulness of the pulse-code-modulation (PCM) technique, digital telemetry and digital-data transmission are receiving greater attention. Only when a few base-band signals of widely differing frequencies ranging from dc to 1 kHz are desired to be telemetered simultaneously and continuously in time domain without loss of information at any instant, analog telemetry systems are used. For such requirements, FM/FM systems using frequency-division multiplexing (FDM) are recommended with appropriate choice of the rf carrier frequency. In other cases, where the measured data consists of low frequency signals, each one of a frequency not widely different from that of the other, sampling process is used to obtain PAM signals first and then generate coded formats for transmission by wire or radio link. The process of obtaining signals for pulse and pulse-code modulation is referred to as time-division multiplexing (TDM). Both the FDM and TDM techniques are explained in the following sections.

11.4 FREQUENCY-DIVISION MULTIPLEXING

Frequency-division multiplexing (FDM) is always associated with FM/FM telemetry systems, in which a number of analog signals obtained from transducers are transmitted to a far-off station by using one single high-frequency carrier signal. Depending on the frequency of the high frequency carrier, the telemetry link may be wire or radio waves. In power line carrier current (PLCC) system, the hf carrier is of a frequency within the range of 20–300 kHz, depending on the characteristics of the wire-link, and in the case of radio telemetry a radio frequency carrier of a frequency lying in the range of 100–1000 MHz is chosen. The FDM technique is invariably the choice when primary (raw) data consists of 12–18 analog signals of differing bandwidths and when they have to be reproduced at the receiving station, simultaneously and continuously with time. The FM/FM telemetry system is an analog telemetry system utilizing the FDM technique; the various stages of the system are shown in Figure 11.8. The base-band signals chosen for the radio telemetry system using the FDM technique may have a frequency content from dc to a frequency up to 70 kHz. But the upper frequency limit of each signal varies

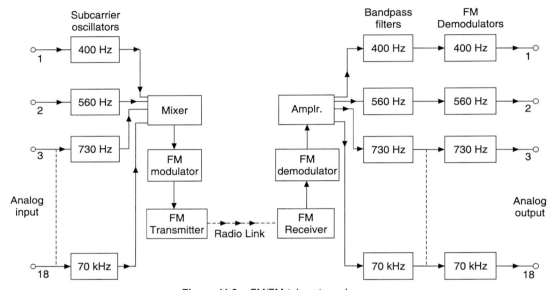

Figure 11.8   FM/FM telemetry scheme.

widely. Thus the bandwidth of each primary signal is specified and standardized by the Inter-Range Telemetering Working Group as presented in Table 11.1. Each one of 18 base-band signals is frequency-modulated by means of a voltage-controlled oscillator (VCO), otherwise known as subcarrier oscillator (SCO). The centre frequencies of the 18 SCOs are specified and standardized and are spread over the frequency range of 400 Hz to 70 kHz. Each subcarrier signal is modulated such that its centre frequency f_c deviates by ± 7.5 per cent. The frequency modulation index m_f for each SCO is also standardized at 5, taking into consideration the upper frequency limit of each base-band signal. Taking the case of the lowermost and first base-band signal, it is seen that its frequency content extends from dc to 6 Hz, and with an m_f of 5, the frequency deviation (Δf) becomes 30 Hz, and hence the first subcarrier signal of centre frequency of 400 Hz swings by ± 30 Hz. It may also be recalled that the modulated signal contains 8 sidebands for an m_f of 5 (see Section 2.6.3), and hence the bandwidth of the modulated subcarrier extends by ± 48 Hz on either side of 400 Hz. The subcarrier frequency of the next SCO is so chosen that their sidebands do not interfere with each other. With the second SCO frequency at 560 Hz, it may be seen that its 8th sideband becomes 50.4 Hz (= 6 × 8.4), and hence there is a clear gap between the outputs of the two SCOs. Each subcarrier frequency is about 1.3 to 1.4 times the subcarrier frequency of the next lower SCO. The 18th data channel is seen to have its bandwidth from dc to 1050 Hz, and the 18th SCO has a centre frequency of 70 kHz. The modulated outputs of all the 18 SCOs are combined to constitute a composite signal which, in turn, is used to modulate a radio frequency carrier of a frequency not less than 100 MHz, for transmission over long distances.

Table 11.1   Standard telemetry channel frequencies for FDM

Band	Centre frequency f_o, Hz	$\pm$ Full-scale frequency deviation, %	Overall frequency response, dc (0) $-f$ (f-values, Hz)
1	400	7.5	6.0
2	560	7.5	8.4
3	730	7.5	11.0
4	960	7.5	14
5	1,300	7.5	20
6	1,700	7.5	25
7	2,300	7.5	35
8	3,000	7.5	45
9	3,900	7.5	59
10	5,400	7.5	81
11	7,350	7.5	110
12	10,500	7.5	160
13	14,500	7.5	220
14	22,000	7.5	330
15	30,000	7.5	450
16	40,000	7.5	600
17	52,500	7.5	790
18	70,000	7.5	1,050

Optional bands

	Band	Centre frequency f_o, Hz	$\pm$ Full-scale frequency deviation, %	Overall frequency response, dc (0) $-f$ (f-values, Hz)
Omit 15 and B	A	3,300	15.0	660
Omit 14, 16, A, C	B	4,500	15.0	900
Omit 15, 17, B, D	C	6,000	15.0	1,200
Omit 16, 18, C, E	D	7,880	15.0	1,600
Omit 17 and D	E	10,500	15.0	2,100

Sometimes, the last five bands may be used with a frequency deviation of ±15.0 per cent (m_f of 10). In such a case, certain other adjacent bands are prohibited from the rf channel in order to allow the spread.

When FM/FM systems are designed for use on a wire link, as in PLCC, the centre frequency of the hf carrier and the centre frequencies of the SCOs may be chosen at will, but conforming to the choice of m_f of 5 and the frequency deviation of $\pm7.5\%$.

At the receiving end of the FM/FM system, an FM receiver receives the rf signal. The demodulated output is equivalent to the composite signal and hence it is passed through a set of band-pass filters, each having its centre frequency equal to that of SCO. The output of each band-pass filter is frequency demodulated and then passed through the low-pass filter. The bandwidth of each low-pass filter is the same as the bandwidth of the base-band signal.

The output of any of the band pass filters may also be used for recording on a magnetic tape, after choosing the correct speed. Otherwise, the composite signal, which is the output of the FM receiver, may directly be used for tape recording, with the speed set at the maximum value of 300 cm/s.

The problem with FM/FM telemetry system is mostly due to the drift of the centre frequency of the SCOs and the rf oscillator. The telemetry error is very much dependent on the frequency stability of the oscillators.

11.5 TIME-DIVISION MULTIPLEXING

Time-division multiplexing (TDM) is concerned with the usage of a single wire or a radio link for transmision of analog data obtained from a large number of data channels. The data is obtained from a set of transducers which may develop analog voltages. These voltages are periodically and sequentially sampled by means of a switching device and the sampled data thus obtained constitutes the pulse-amplitude modulated (PAM) signal. When the signals to be multiplexed vary slowly with time, mechanical switches known as commutators may be used for sampling the input data. At the receiving end, a similar device known as decommutator is used to deliver the respective samples to the corresponding output terminals. Both the commutator and decommutator should rotate in synchronism as shown in Figure 11.9(a) so that they make contact simultaneously at terminals bearing the same channel index. If the multiplexed signals are transmitted directly over a pair of wires, no additional processing is done. At the receiving end, the output pulse of each channel is passed through a low-pass filter, so that its output is treated as the reconstructed signal resembling the original signal.

When the input signals to be multiplexed are of a higher frequency, higher switching speeds are required, which may be beyond the range of mechanical commutators, and in such a case, electronic switching systems are employed. The train of PAM pulses becomes the base band signal, when the data is to be transmitted over a radio link, thereby constituting either a PAM/FM system or PAM/FM/FM system. The carrier or the sub-carrier signal is frequency modulated, in most cases, by the PAM pulse train.

The PAM pulse train thus consists of a series of rectangular pulses, regularly spaced at intervals of T_s as shown in Figure 11.9(b). Each pulse is flat-topped, signifying the fact that the analog signal sampled has not changed in its amplitude during the period of sampling and that it has not changed much during the period between any two of its successive samples. The amplitude of the pulse is a representation of the amplitude of the signal at the instant at which it is sampled and the duration of the pulse is so adjusted that it is between 40–65 per cent of the time interval T_s between any two pulses of the pulse train. A synchronizing pulse is added at the beginning of each series of the pulses so as to signify the starting of the sampling process from input channel with index number one. The synchronization (synch) pulse is of full-scale amplitude and its duration is equal to $1.5T_s$ with 'off' time of $0.5T_s$, as shown in Figure 11.9(b).

A frame is said to consist of the pulses obtained from all the input channels once, along with the synch pulse. The frame rate is the number of times the entire group of input channels get sampled per second. Thus the number of pulses (samples) obtained per second is the product of the frame rate and the number of samples per frame. In each frame, two more pulses are usually added, one representing the zero level and the other the maximum value which a pulse is likely to assume and which may be used to calibrate the system at the receiving end.

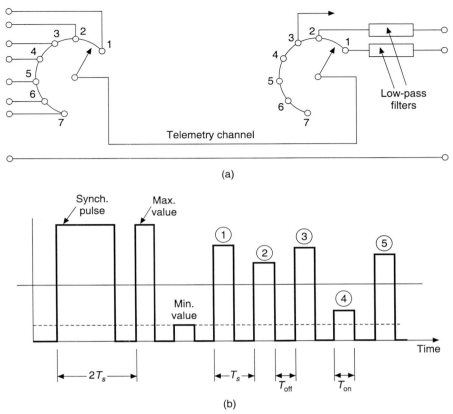

Figure 11.9 (a) Simple TDM scheme; (b) PAM pulse train.

The number of samples per frame is normally chosen as 18 or 30 and the frame rate is standardized such that the commutation rate for mechanical sampling varies from 75 to a maximum of 900 samples per second. For PAM/FM/FM telemetry system, the lowest subcarrier frequency recommended for a train of 75 samples per second is 14.5 kHz, whereas for the case of 900 samples per second a subcarrier of 70 kHz is used. But for electronic switching, the number of channels sampled per second may go as high as 200 and the number of frames per second may be increased to the desired level, depending on the limitations of the link and the technique of telemetry chosen.

Thus it is seen that in time-division multiplexing, a regularly occurring time slot is allocated to each data input channel and the PAM pulse train is the consequence of amplitude modulation of a pulse which is located within the specified time slot. Other possibilities of modulation such as PWM, PPM or PCM are possible, once the PAM train is obtained, as can be understood from the following sections.

11.5.1 Pulse-Time Modulation

The amplitude of each pulse of the PAM train conveys the amplitude of the particular channel sampled. Instead of the amplitude of the analog signal at any instant modulating the amplitude (or height) of a pulse, it can be used to modulate the duration of the rectangular pulse, thereby constituting a train of PDM (or PWM/PLM) pulses. In other words, the amplitude of an analog signal at the sampled instant is converted from analog domain to time domain. In such a case, it is broadly referred to as pulse-time modulation (PTM). Other alternatives are pulse-position modulation (PPM) or pulse-frequency modulation (PFM).

In PDM systems, it is necessary to design the modulation in such a way as to have the maximum duration of the pulse generated smaller than the duration of the time slot T_s allocated for each sample so that a 'guard time' is provided between the termination of one pulse and the starting of the next pulse. It is equally important to see that the minimum duration of the pulse is finite when the amplitude is zero. The PDM train is shown in Figure 11.10(a). The commutation rate is standardized at 900 samples per second while the number of samples per frame being set at either 30, 45, 60 or 90. Within the T_s of 1110 µs the minimum pulse duration is specified as 90 ± 30 µs to represent zero amplitude and a maximum pulse duration of 700 ± 50 µs to represent the maximum amplitude. The pulse rise and decay times should not be greater than 20 µs. When the PDM train is used to frequency-modulate, the frequency deviation of the subcarrier frequency is chosen at ±15%.

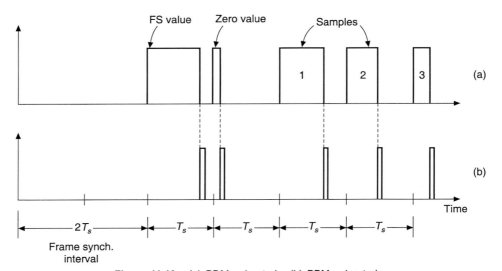

Figure 11.10 (a) PDM pulse train; (b) PPM pulse train.

Once the PDM signal is obtained, it can be used to generate the PPM signals. The PPM train consists of very narrow pulses of constant height, the pulse rate being the same as the pulse rate of PDM signals. The occurrence of the narrow pulse in its time slot signifies the amplitude of the sample. It is easier to understand the generation of the PPM signals from Figure 11.10(b), in which the trailing edge of the PDM pulse is used to trigger the generation of a narrow pulse. If PDM pulse is of maximum duration, the narrow PPM pulse occurs at a later instant in the time slot of T_s. Similarly, the narrow pulse signifying the zero amplitude of the sample occurs at the later instant than the instant of sampling.

One method of generating the PTM signals from PAM signals is shown in Figure 11.11. In synchronism with the PAM samples, a linear ramp-type (or reversed saw-tooth) pulses are generated and added to the PAM signals. The sum of these two signals is applied to a comparator circuit, the output of which assumes only two levels, depending on whether the sum is larger or smaller than its reference level. The first crossing of the reference level by the input signal generates the leading edge of the PDM pulse, while the second crossing generates the trailing edge. Thus the duration T_d of the pulse is proportional to the amplitude of the PAM pulse. The trailing edge of the PDM pulse is used to trigger a pulse generator (a monostable multivibrator) which generates pulses of fixed amplitude and duration. Thus PPM signals are generated.

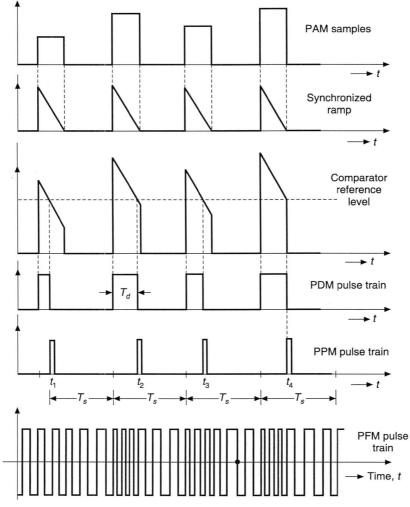

Figure 11.11 Generation of PTM signals from PAM samples.

It is also possible to generate PTM signals directly from the analog signal, without generating the PAM train at the start.

A pulse frequency modulated (PFM) signal consists of a series of rectangular pulses, the frequency of the pulse train at any instant being related to the modulating voltage. A PFM signal may be obtained by generation of a rectangular pulse for every complete oscillation of an FM wave. The sinusoidal unmodulated carrier of a frequency f_c yields a pulse train of the same frequency when the positive halves are clipped after rectification, or the process of detecting the FM wave by a cycle counter is equivalent to conversion of FM wave to PFM train and selecting the low-frequency components of the PFM signal by a low-pass filter. Such a process lends itself for application to a single modulating signal. A similar process may be considered applicable for a PAM train, but the technique does not have any particular merit.

However, pulse frequency modulation technique is applied for a PCM signal with considerable complexity in circuitry. PCM/PFM system is identified to be a high efficiency telemetry system.

11.5.2   Pulse-Code Modulation

Each sample of the PAM train may be quantized into a certain discrete number of levels and each quantized level may be represented by a code number. The code number is expressed in the straight binary system and each digit of the binary number (bit) is identified by the presence or absence of a pulse in the allotted time slot. Such a system constitutes (binary) pulse-code modulation (PCM). The modulating signal $e_m(t)$ in a binary PCM system is thus a sequence of 1's and 0's, which are grouped together in binary words describing the value of the particular data channel sampled at the instant of sampling.

When angular displacements are encoded by shaft angle encoders, the Gray code shown in Section 7.8.1.4, is used.

Two-level analog signals corresponding to 1's and 0's may be used to represent the on- and off-conditions of switches, valves and so on.

The pulse train thus obtained from ADCs, shaft-angle encoders etc. may be treated as a digital signal, as the train is composed of pulses in their time slots. It is sufficient to distinguish the presence or absence of a pulse in the time slots, to recover the original signal. The exact height or amplitude of the pulse is not important. There is an advantage in raising the pulse width as wide as possible within the limits prescribed by the telemetry channel. Higher width for the pulse means higher pulse energy, thereby facilitating easier recognition of the pulse against the background noise.

Figure 11.12 shows a PAM signal and the pulse representation of the binary numbers used to code the samples of the PAM train. The pulse pattern shown is the binary PCM waveform that is transmitted over the telemetry channel either directly or by subsequent modulation. For short distances and low bit rate, it is transmitted by wires. In other cases, the PCM signal is subjected to further modulation such as FM or PFM. To eliminate the dc component present in the PCM signal, the excursions of the pulse are brought into two levels ranging between $-V$ and $+V$ volts.

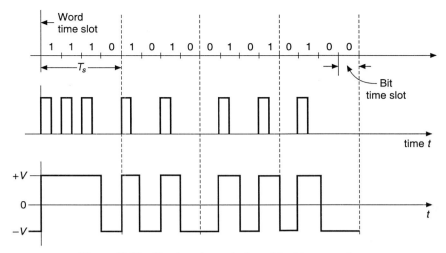

Figure 11.12   Simple pulse-code format for binary numbers.

The binary number can be encoded in several formats, each having a particular merit. Figure 11.13 shows two such pulse-encoding formats, known as 'return-to-zero' (RZ) and 'nonreturn-to-zero' (NRZ) formats. The return-to-zero method produces a pulse for each 'ONE' data pulse. It is satisfactory for remote aperiodic pulse-counting applications. It can be unipolar or bipolar. The method is very popular for slow speed transmission and is used for speeds up to 600 bits per second.

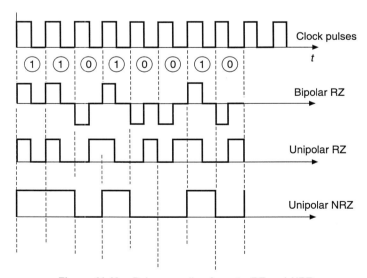

Figure 11.13   Pulse encoding formats: RZ and NRZ.

Unipolar NRZ format is most popular for computer applications and for data transmission speeds up to 2,400 bits per second. In this method, pulse level transitions occur only when there is a change from '1' to '0' or '0' to '1'. There is no change between successive 0's and 1's. The bandwidth of the channel can be efficiently used by using this format.

A block diagram of the PCM signal generator is shown in Figure 11.14. The multiplexer (MUXR) is controlled by the programmer so that the analog data channels are sequentially sampled. The signal is then fed to an ADC, the output of which is fed in turn, to a set of digital gating circuits. Additional digital signals obtained from a synch generator and other data sources are also fed to the digital gating circuits. The PCM system programmer programs all timing and control functions of the system. The final output signal is a PCM signal containing synchronization words and digital data. Final transmission by wire or radio link may be invariably accomplished after a second modulation process, frequency modulation being the choice.

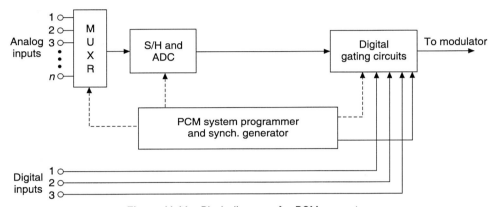

Figure 11.14   Block diagram of a PCM generator.

At the receiving station, the function of data recovery equipment is to detect the incoming data bits, determine the location of the first bit in the frame, then by means of a counter with programmed control, to decommulate and demultiplex each of the data words to its appropriate output system, which may be a display recorder or a printer system.

PCM telemetry systems are widely used for missile and space missions. The primary merits of the PCM telemetry system are its capability of transmitting data with higher accuracy and its flexibility as regards the choice of the number of channels and the channel sampling rates. The system possesses very high noise immunity.

11.5.3   PCM/FM Systems

The PCM signal is now seen to be in any one of the two levels for its amplitude, and hence PCM train may be considered as the basic modulating signal for an rf carrier wave. In such a case, FM signal may be characterized as at any one of two frequencies, one representing the '1' level and the other the '0' level. Such a process is known as *frequency-shift keying*. The transmitted FM signal is of a fixed amplitude and of a frequency $f_c + \Delta f$ or $f_c - \Delta f$ representing the two levels. Demodulation of such a signal is fairly simple. PCM/FM telemetry is used in many applications.

A high efficiency telemetry system is one in which the binary number representing the amplitude of a sample is used to directly generate a pulse train of a certain pulse rate. If a three-bit word represents the amplitude of a signal, then there are eight possible levels into which the signal amplitude is discretized, and hence for each level a certain frequency is allotted. For each binary number of the three bits, there is a frequency allotted and the pulses are generated during the sampling period T_s at that frequency. A typical case is shown in Figure 11.15, and it should be observed that there is a guard time of 10 ms between each series of pulses. Such a process of modulation is known as pulse-frequency modulation (PFM) and is considered applicable for both wire and radio telemetry over short distances.

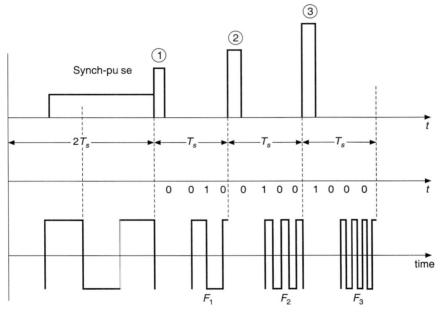

Figure 11.15   Formation of PFM signals from PCM signals.

EXERCISES

1. Define the term 'telemetry' and explain in what way it differs from 'remote measurement'. Illustrate both by means of examples.

2. What are the important characteristics of a telemetry system and distinguish the criteria on which the telemetry and communication systems are based.

3. Explain land-line and radio telemetry systems and the means adopted for each.

4. Briefly explain a closed-loop telemetry system and explain the merits in comparison to an open-loop type. Suggest an example for each.

5. Explain why digital telemetry and data transmission systems are considered superior to analog systems.

6. Indicate the situations in which analog and digital telemetry systems are preferred.

7. Explain the frequency-division and time-division multiplexing and the situations in which each is recommended.

8. Show how a frequency-division multiplexing system is constituted and state the criteria adopted for its design.

9. Describe the various stages of a FDM system and a TDM system, by means of block diagrams and explain the functions of each stage.

10. Explain two-stage modulation and show how FM/FM systems are superior to others when designing an analog telemetry system.

11. Explain pulse-time modulation and show how pulses can be utilized for transmission of measured data.

12. Distinguish the PAM, PWM, PPM, PFM and PCM systems and show that PAM is the basic requirement for the PCM system.

13. Briefly explain the PCM telemetry system and show that it can be used for both land-line and radio telemetry.

SUGGESTED FURTHER READING

Foster, Leroy, E., *Telemetry Systems*, John Wiley, New York, 1965.

Gruenberg, E.L. (Ed.), *Handbook of Telemetry and Remote Control*, McGraw-Hill, New York, 1967.

Herbert, Taub and Schilling, D.L., *Principles of Communication Systems*, McGraw-Hill, New York, 1971.

Ilyin, V., *Remote Control Systems*, Mir Publishers, Moscow, 1973.

McKay, F.S., *Biomedical Telemetry*, John Wiley, New York, 1968.

Nicholas, M.H. and Rauch, L.L., *Radio Telemetry*, 2nd ed., John Wiley, New York, 1956.

Perry, A.B. and Wilfrid, J. Mayowells, *Telemetering Systems*, Reinhold Publishing Corp., New York, 1959.

Podszeck, H.K., *Carrier Communication over Power Lines*, Springer-Verlag, Berlin, 1972.

Mischa Schwartz, *Information Transmission, Modulation and Noise*, McGraw-Hill, New York, 1970.

Mischa Schwartz, Bennett, W.R., and Seymour Stein, *Communication Systems and Techniques*, McGraw-Hill, New York, 1966.

Swoboda, G., *Telecontrol Methods and Applications and Telemetering and Remote Control*, Van Nostrand Reinhold, New York, 1971.

Young, F.E., *Supervisory Remote Control Systems*, Peter Peregrinus, London, 1977.

Young, R.E., *Telemetry Engineering*, Iliffe, New York, 1968.

CHAPTER **12**

Developments in Sensor Technology

12.0 INTRODUCTION

All living species are endowed with sensory organs and actuating devices for carrying out their routine activities. Sensing precedes actuation and the signals obtained after sensing, may undergo some processes. The first one after sensing is the measurement process, when a man in interested in making use of the sensed signal for all activities conceived by him. To gain knowledge and organize several activities, a large number of sensing systems are needed and they are known as 'transducers' and lately as 'sensors'. There are subtle distinctions in the way the sensing systems are developed, but the extraordinary capabilities of the present-day sensors in sensing systems without any interference, have made them unique in many respects. A sensing system of the present day is termed and identified as 'SENSOR' no matter how many components make up the system. In the twenty years past, extensive research has yielded a galaxy of sensors based on several phenomena not only from Physics and Chemistry but also from Biology. Chapters 5 to 8 deal with those discrete physical devices, but the new breed of sensors, described in this chapter may be classed as 'hybrid' in the sense that they employ a variety of components to deliver sensed information, in a way convenient and comfortable for the man to organize his technological effort. Present day achievements in all walks of life starting from home to transportation systems by land, air, water and space, could not have been what they are without the service of sensors. The productivity and product quality of the automated industry is entirely due to the presence of 'SMART SENSORS' in the production processes. Today, the healthcare activities in hospitals; and safety and security systems for protection of personnel and properties from fire, pollution and poisoning in industrial areas in big cities, make use of a large variety of sensors and sensing or monitoring devices. Such development in sensor technology has been possible because of the simultaneous research effort in production of a large number of alloys and compounds and identification of appropriate manufacturing processes for production of integrated-circuit (IC) chips and micro-components.

The microprocessor, the microcomputer and the communication technology have sufficiently contributed to the development of sophisticated instrumentation. The understanding that 'the small the beautiful' or 'the small the powerful', is fine but when dealing with the micro-size signals from the microsensors, one needs to tbe cautious, especially, if they are electrical in nature, because of the effects of noise in electrical components and the electromagnetic interference. Under these circumstances, interest shifted to utilization of sound waves (ultrasonic), heat waves (Infra-red) and light waves to constitute sensing systems, as they are considered to be safe and versatile in many ways. Attention is paid to the utilization of energy present in mechanical and electromagnetic waves, the nuclear radiation, and electric and magnetic fields, resulting in development of a variety of sensors for application in several situations, hitherto considered difficult.

Solid state devices and opto electronic components embedded on chips have extended their application to constitute several complex fibre-optic systems for developing imaging systems and sonography. Vocabulary like thermal imaging (thermography) and tomography have come to the knowledge of common people. Biomedical sensors have been rendering services in the medical diagnostics while the biosensors based on the knowledge of biological phenomena and properties of biological species are found to serve the needs of foods and drugs industries, medical diagnostics and pollution studies.

Fabrication and commercial production of sensors is still a guarded process. A sensor, marketed hints probably the basics of operation while claiming the merits and application potential. The users and operators do invariably look to the manufacturers for their services in regard to assurance of performance and maintenance. With the advent of new generation of sensors, the solution seems to be discarding the faulty and replacing it with a new one.

12.1 SEMICONDUCTOR SENSORS

12.1.1 Basics

Semiconductor materials, such as silicon and germanium and other semiconductor compounds have been extensively used, since long, for the production of physical transducers as discrete elements for measurement of temperature, strain, etc. Their properties, configurations and applications are presented in Chapters 6 and 7. The favourable status of electrical and electronic systems for measurement of several nonelectrical quantities, reinforced the quest for sensing systems with improved performance and more importantly applicability in situations considered inaccessible all along. Scientists and metallurgists have identified certain materials with certain unique properties while the manufacturing processes simultaneously enabled fabrication of sensors in smaller sizes as thin films on substrate materials, thereby enabling them get integrated with electronic processing circuits on a single chip. The discrete transducers using silicon and germanium in pure or doped form detailed in Sections 6.1.1.2 and 6.1.4.7 and the junction diodes and transistors in Section 7.2.2.2 have provided the base for developments in Sensor Technology.

A semiconductor material is known to behave as an insulator at certain temperatures and as a conductor at other temperatures. Other substances and compounds that display similar characteristics are also classified as 'semiconductors'. The ceramic materials and likewise the ferroelectric materials are some of them. Likewise, sevaral new materials have been identified, thereby enlarging the scope of application to detection and measurement of several non-electrical quantities. The following sections present some of the developments in this direction, paving the way for a new generation of micro-sensors and enabling the sensor technology as a field of considerable importance. It is possible to see extremely small size cells through powerful microscopes but now the sensor technology helps in detection and measurement of minute quantities and all the more in situations never conceived as practicable earlier.

12.1.2 Materials and Techniques

Silicon is the most favoured and largely utilized element with all its physical properties extremely suited for development of smart sensors.

Silicon possesses certain unique properties both mechanical and electrical, resulting in its popularity for embedding IC circuits on the transducer chip. It has high tensile strength, Young's modulus hardness and strength-to-weight ratio. It has high fatigue strength, negligible hysteresis, consistent quality and high corrosive resistance. It has the exceptional property of accepting high surface finish.

Planar silicon integrated technology is rendered possible by applying photo lithographic method and an oxide window masking. On one side of a silicon wafer, it is possible to incorporate passive and active components and their interconnections. It is very easy to deposit a metal layer on to a silicon wafer with suitable etching techniques. Laser techniques and advanced manufacturing technology enable the production of accurate resistors on a silicon chip. The resistors are of permalloy (nickel–iron alloy) with high temperature coefficient of resistance and hence suitable for temperature measurement. The chip can be 1 mm^2 and of thickness only of 10 μm. These chips are available for measurement of temperature and are nowadays found to be stable over long-term and at the same time cheap as no precious metals are used. Temperatures below 150°C can be accurately measured.

For pressure measurements, the mechanical properties of silicon have been proved to be extremely suitable (see Section 5.2.4) and its combination with strain gauges for pressure measurements is described in Section 6.1.4.19. Modern pressure sensors are now available in two forms: one with piezo-resistive elements integrated with the silicon diaphragm and the other silicon diaphragm itself forming as one plate of a capacitor system. Both types are available as smart sensors for a wide range of pressures. Likewise strain bars of silicon for pressure and cantilever beams of silicon for acceleration measurements are available as integrated resistive pressure sensors using the piezoresistive elements. A novel device is the tiny electret microphone in which the silicon diaphragm in combination with a polarized electret, thereby makes up a self-generating sensor

The piezo-resistive effect noticed in silicon and germanium, when doped, has been successfully applied for measurement of mechanical strain. The description of the system is given in Section 6.1.4.6 to Section 6.1.4.9. Presently such systems are available as smart sensors.

In Section 7.4, Hall-effect transducers are described and it is shown that semi-conductors yield large values for Hall coefficients. A large number of Hall-effect sensors are brought out as microsensors for measurement of a large number of variables including electrical quantities.

Magneto-resistive materials such as Bismuth and Indium antimonide are shown to be sensitive to magnetic fields, resulting in changes of electrical resistivity. The effect is very small in silicon. Details of the traditional transducers are presented in Section 6.1.7.

Micromachining techniques enable production of microsystems with silicon being configured as 1. cantilevers, 2. diaphragms and 3. microtuning forks. All these and other configurations have their own natural frequencies of oscillation. A resonant micro sensor is one of such members along with its associated sensors, operating at its resonant frequency. Any change in the resonant frequency changes the output of the sensor and such changes may be brought to work on the resonator either by physical or chemical input variables. The input variables may be addition of mass, stress or deformation of the resonator. The systems are kept in a state of oscillation by using positive feedback. Resistors can be integrated by diffusion or ion implantation for processing the signals.

In a microsensor, a cantilever is like a small tongue of fine silicon strip with its resonant frequency of oscillation given by

$$f = 0.16 \frac{t}{l^2} \left(\frac{E}{\rho} \right)^{1/2} \tag{12.1}$$

where, ρ is specific mass, kg m^{-3}
  l, length of the tongue, m
  t, the thickness of the tongue, m
  E, modulus of elasticity, Pa.

Piezoelectric semiconductors are dealt with in Section 7.2.2.2 describing the anisotropic stress effect in *p-n* junctions, wherein the mobility of charge carriers is detected and measured when a small force is applied by means of a diamond stylus (with sharp tip). Based on this phenomenon a large number of electronic devices such as diodes, transistors, FETs and MOSFETs are effectively used for force and pressure measurements.

They have been encased and provided with outfits relevant to the particular quantity under detection and measurement as identified from the Table 12.1. A typical *pn* diode is also used as a position sensor by making use of lateral photoeffect. When a light spot illuminates a silicon solar cell, an emf is developed across the ohmic contacts as a function of the position of the light spot. Using this effect, two-dimensional light spot sensitive photodiodes are fabricated.

Making use of two matched transistors and effecting a difference between their emitter voltages due to the quantity under measurement has been a popular and common technique. The possibility of integration of desired signal processing circuits on board the chip has considerably enhanced the versatility of the technique. For example, a sensor system of this kind for temperature measurements over a range of −25°C to +85°C, with a Zener diode voltage reference and built-in operational amplifiers, is commercially available with a sensitivity of 10 mV/°C, being linear over its range. An integrated silicon flowmeter is designed on the samelines using matched pairs of piezotransistors for flow measurements at temperatures between −40°C to +85°C.

Table 12.1   Sensors using semiconductor materials and silicon

Energy domain	Self generating	Passive (R, L.C type)	Diode type	Transistor type	Smart sensors
1. Radiation	Solar cell	Photo conductor	Photo-diode	Photo transistor	Photo IC–CCD
2. Mechanical	—	Piezo resistors	Piezo junction diode	Piezo-transistor	Accelerometers piezo ICs
3. Thermal (for temp)	Thermocouple	Thermistor semiconductor resistor	Reverse biased diode	Forward biased transistor	Temperature IC
4. Electrical	—	Electric field MOSFET	Electric field MOSFET	Dual gate-MOSFET	IC chips for all
5. Magnetic	Induction coils	Magneto resistor	Magneto diode	Hall effect	Hall effect ICS
6. Chemical	Galvanic	Ion-concentrator	—	Ion selective FET	Electronic nose

Developments in material science research to meet the demands of instrumentation industry in bringing out materials that can be brought out as thin films, sensitive to some measurement and admission for integration on a silicon chip have been successful enough.

Aluminium nitride thin films have been grown by reactive magnetron sputter technique using pulsed power supply. The highly (002)-textured columnar films deposited on platinized silicon substrates exhibited quasi-single-crystal piezoelectric properties. The effective d_{33} is 3.4 m/V and effective e_{31} as 1.0 C/m^2. The pyroelectric coefficient is seen to be positive at 4.8 $\mu Cm^{-2}K^{-1}$, due to the dominating piezoelectric contribution. Thin film bulk acoustic resonators (TFBAR) with fundamental resonant frequency of 3.6 GHz have been fabricated for adoption in some integrated-sensor systems.

Likewise polymer thin films are made from polyvinylidene fluoride (PVDF) which exhibit strong piezoelectric effect and strong pyroelectric effect with piezoelectric coefficient of 15 pCN^{-1} (or 14.4 VN^{-1}) and pyroelectric coefficient of 8 VK^{-1}.

Photo-sensitive detectors or photoreceptors are used in xerography and they are needed to be sensitive to the entire visible spectrum. A new class of amorphous semiconductors are available and these materials are found to be mechanically and chemically stable, when used as sensors for the visible radiation.

Amorphous materials can be insulators, semi-conductors or superconductors at low temperatures. a–Se is amorphous selenium is a chalcogenide glass, used in xerographic industries. The addition of tellurium (Te) to the a–Se film improves sensitivity in the red region while addition of Arsenic (As) improves both sensitivity and stability. It can also retard the crystallization rate in the presence of sufficient As, allowing the crystallization to take place above 100°C. Then the a–Se film becomes conductive and allows built-up charge dissipation faster. It has been observed that a–Se with 40% of As, makes it a–As_2–Se_3 (arsenic triselenide). Addition of chlorine (30 ppm) improves its speed of response.

Also hydrogenated amorphous silicon films (a–Si : H) with resistivity of 10^{13} Ω/cm and high photo-sensitivity are also found to be suitable for use in optical tomography and other scanning systems for use on transparent foods and vegetables.

A new family of electrically-active ceramic materials are brought out by Elmwood Sensors Ltd., replacing the conventional metallic compounds in a wide range of applications. They are used as intrinsically-safe heaters and temperature sensors. They combine the electrical conductivity of metals with the durability and stability of ceramics. They possess high electrical conductivity.

Physical and electrical properties of the ceramic enables it ito be used as a filler material for the manufacture of electrical contacts for medium and low voltage equipment. Oxides can not form on the surface of material.

These materials can be sputtered on to a substrate in the same way as the existing compounds but provide a more stable product. They can replace the existing thin film technology in light-sensitive electro-chromic electrodes, thus affording significant advantages in electrical and optical performance.

They can also be used for producing a new solid-state rechargeable battery replacing the conventional one, there by eliminating the use of hazardous cadmium metal in the cell.

12.2 SMART SENSORS

12.2.1 Definition of a Smart Sensor

In traditional systems, each transducer is specified by its sensitivity, input and output signal ranges, offsets, linearity hysteresis, etc., and interest has been always to compensate for some or all deficiencies, somewhere in the system or technique. Linearity between the input and output signals is always coveted. Provision of electronic signal processing by means of an integrated circuit (IC) chip, with the output signals processed so as to bring out them in a standard range, linearly related to the measurand is considered a welcome development. When the traditional transducer, preferably miniaturized so as to be considered as a single device along with the IC chip, it has been treated as a 'smart' sensor. The micro miniaturization of the entire system, if implemented on a single chip it is considered 'smartness' of the higher order. In course of time, additional provisions have been incorporated to the extent that they are produced and commercially made available with standard output signals and be replaceable without the need of calibration. It is for these provisions and many more, that they are classed as 'SMART SENSORS' of the present day. Smart sensors may in short be defined as those sensors in which much of the signal conditioning is carried within the transducer housing and which will provide standardized output signals in digital form, and suited for transmission via a communication bus to the central control room. They will linearize their own output, compensate for environmental changes, and include self-calibration and diagnostic functions, both for themselves and for the systems to which they are applied.

A 'soft' sensor is a sensor that handles mathematical operations necessary to deliver the output signals in the desired dimensions, after effecting computation on the measured data obtained from other sensors. So soft sensor may also be treated as a sensor. For example, when a tank is getting filled with a liquid, liquid level is measured and from the level values, flow-rate of the liquid can be calculated. The calculator is the soft sensor.

In this respect, it is nice to define a 'smart' field device also. It is defined as a microprocessor based process transmitter or actuator that supports two-way communications with a host, digitizes the transducer signals and digitally corrects its process variable values to improve system performance. The value of a 'smart' field device lies in the quality of the data it provides.

12.2.2   Configuration of a Smart Sensor

A typical smart sensor is configured with the three following elements:
1. a physical transducer
2. a network interface, and
3. a processor and memory core.

The signal from the transducer is fed to an A/D converter so as to be acceptable to the microprocessor as shown in Figure 12.1. The processor will perform some processing on the digital data depending on how it is programmed. The network receives the data from the processor and the network interface block handles the transactions between the processor and the network field bus.

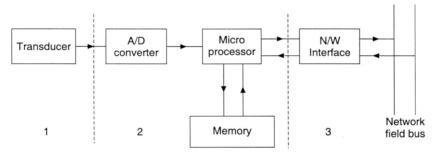

Figure 12.1   Configuration of a smart sensor.

It is possible that all the above are implemented on a single chip, thereby attributing the qualification of 'smartness' to the sensing system. All such sensing systems, in common terminology are taken as sensors and the modern materials and manufacturing techniques enabled production of such sensors commercially to meet the demands of the industries (see Fig. 12.2). The network bus, known as field instrumentation bus or in short field bus offers immense advantages, the foremost being, easy and quick replaceability of a sensor, if found

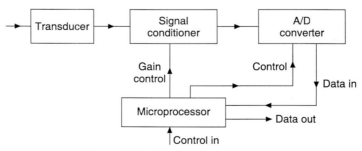

Figure 12.2   IC chip of a smart sensor.

defective. There may be some variations in the above configuration with some additional sensors on board, permitting compensation for ambient temperature changes and such others.

12.3 MICROSENSORS

12.3.1 Introduction

A microsystem is an assembly of many small elements coupled together in a way to effect a desired function and the microsensor is one of them, revolutionizing the development and design of the present day instrumentation systems. Mechatronics is known to constitute successful integration of mechanical and electronic devices and the microsystems belong to the same family with the difference that the microsystems are the smallest and lightest, ever known to be capable of serving the interests of man in his technological achievements. The microsystems, otherwise known as Micro Electro Mechanical systems (MEMs), are successful outcomes of the advances made in manufacturing processes, such as planar treatment, micromachining, deposition of extremely thin layers of substances on to substrates, etc. such systems are found to perform effectively and efficiently not only in air transportation and space navigation systems but also in medical diagnostics and robotics.

The development of microsensors going along with the processing power of the Intel P_6 and P_7 chips with about 3×10^6 components on board of a single chip is an important milestone in the field of instrumentation and thereby automatic control. The decrease in line width in integrated-circuit technology is made possible by application of deep ultraviolet, electron or X-ray photolithographic techniques. The single crystal silicon has immensely contributed to the development of microsensors, not only by its mechanical properties but also by its excellent micromachining admissibility. It is known to serve as a sensing element or as its complement, in most of the microsensors. MEMs naturally require actuators, in some cases, extremely small in size, and successes in this front are proved by an exhibit, working like an artificial butterfly.

The microsensors are in most cases, a combination of sensing systems and electronic circuits, integrated to silicon substrates. The successful production of microsensors is possible because of the production processes in nanotechnology which is the art, science and technology dealing with micron-size objects. In the following sections, a few typical microsensors are briefly introduced.

12.3.2 A Microsize Microphone

A novel but popular device, used as the tiniest microphone, consists of an equivalent bimorph structure as described in Section 7.2.3 and shown in Figure 7.13(a). It is useful as an acoustic pressure sensor or alternatively as a microphone. As an inverse transducer system, it can be used as a microspeaker. Usually, a single piezoelectric wafer is used on a silicon nitride diaphragm. For use with the bimorph, parylene-D is preferred for reasons of high thermal stability and flexibility. A couple of transverse expander mode piezoelectric films are used to serve as a bimorph and they are connected so as to develop output signals of voltage between the two outer electrodes, when the diaphragm is strained due to applied acoustic pressure. ZnO serves as substrate material, sputter-deposited at 250°C on the diaphragm. The entire micromachined system is shown to be better in sensitivity as well as the signal-to-noise ratio, when compared to the conventional system using a single film (or uni-morph system). It is extensively used in hearing aids and other diagnostic systems in medical profession.

12.3.3 Inertial Sensors

Micromachined inertial sensors, such as accelerometers and gyroscopes are much coveted for use in all navigation systems and missile guidance systems. Information about the angular and linear motion of a body in its six degrees of freedom is required. The technology of surface-micromachining enables the integration of mechanical structures and the integrated circuit chips. Thin films of poly-silicon and SiO_2 are grown on a wafer and the oxide layer surface is removed by a release step by a wet-etchant of hydrogen fluoride. Several sensors based on this technique are brought out as monolithically integrated devices.

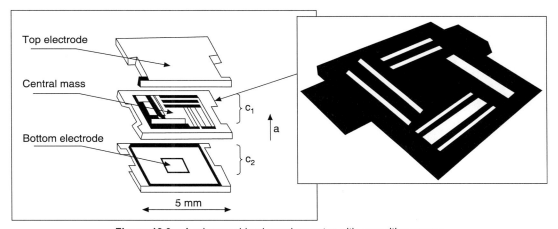

Figure 12.3   A micromachined accelerometer with capacitive sensor.

The piezoelectric accelerometers described in Section 7.2.7 are popular for use while the arrangement shown in Figure 7.19 is brought out as a microsensor by diffusing the piezo-resistive material into the elastic member that carries the proof mass. The diffused films constitute the four arms of a wheatstone bridge network, developing output voltage of 1–3 mV/g with about 10 V supply voltage.

Another popular version of the accelerometer makes use of the proof mass serving as a middle plate of a differential capacitor, with the metallic body on either side being the fixed plates as shown in Figure 12 3. The differential capacitor akin to the one described in Section 6.3.2.5 and shown in Figure 6.77, is used as two arms of a wheatstone bridge network for sensing the deflection of the diaphragm, when subjected to acceleration inputs. Converting the system into a feedback type measuring system results in higher resolution in the 1 µg range along with improvement in bandwidth.

Multi-axis accelerometers based on this principle with a proof mass of about 0.4 µg are available from Analog Devices of USA, with a bandwidth of 100 Hz and a range of ± 11 g.

Likewise, micromachined gyroscopes based on Coriolis acceleration, requiring no rotational parts with bearings can be miniaturized and get integrated with the accelerometers.

12.3.4 Hall Effect Sensors

While introducing the Hall effect transducer in Section 7.4, it has been suggested that the semiconductors are extremely fast and require very little space (see Figure 12.4(a)). Intensive search is on for identifying sensors that can be produced on a silicon wafer using the new technologies while at the same time lending scope for incorporation of electronic circuits on board. Hall effect transducers are well suited for adoption and mass production at very low cost.

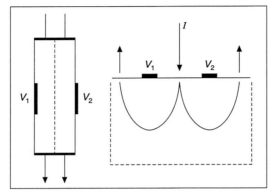

Figure 12.4(a)   Versions of Hall effect configuration.

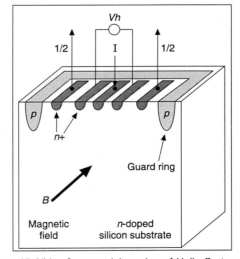

Figure 12.4(b)   Commercial version of Hall effect sensor.

The structure of the modified Hall effect transducer is shown in Figure 12.4(b) in which the contacts for admission of current and pick-up of output voltage are all on the silicon wafers or films deposited on the *n*-doped silicon substrate. The sensor is sensitive to magnetic fields oriented horizontally in the plane of the wafer. To enhance the sensitivity to magnetic fields, concentrators of magnetic field are located directly on the chip. This is known as vertical Hall sensor, useful for measurement of variables that affect the intensity of magnetic field.

The unidirectional vertical Hall sensor is shown in Figure 12.5(a) and the two-dimensional one in Figure 12.5(b). Three legs of the unidirectional one can be used when connected in a star-fashion and can serve measurements of angular position of shaft under rotation. It is only necessary to have a small magnet attached to the rotating axis.

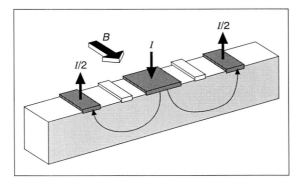

Figure 12.5(a)   Unidirectional Hall effect sensor.

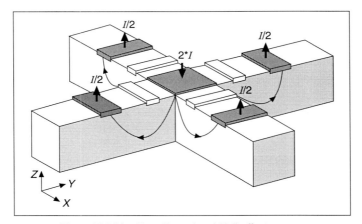

Figure 12.5(b)   Two-dimensional Hall effect sensor.

Several possibilities exist for developing microsensors, based on the above techniques of fabrication. Inductive proximity sensors are integrated with the electronic circuits on a silicon chip using CMOS-compatible technologies. World's smallest passive infrared (PIR) motion sensor is produced for use in security systems, lighting control amusement and vending machines. It has an onboard amplifier, comparator and output circuits which enables the device to be driven directly by microprocessor circuitry. Standard range is 5 metres with 64 detector zones and the slight motion unit has 104 zones over a range of 3 km. Output is 300 mA at 3 V dc.

12.3.5   Polymer Sensor

A large number of polymer sensors are being developed using a polymer thick film paste on a wide variety of substrates. The polymer thick film acts as a sensing element, when fabricated and printed on silicon. The paste of polymer film consists of a polymer matrix such as epoxy, silicon or phenolic resin that binds together filler particles of differing physical properties. Copper or silver particles are for conductive pastes, carbon particles for resistive pastes, with mineral addition to improve the electrical and mechanical properties. Finally solvents are added for use as ink for screen printing. The pastes are cured after the solvents are dried off at the required temperatures by convection or infrared heating. The carbon paste cured at 220°C behaves like a thermistor film while copper paste behaves like a linear resistor.

Polymers absorb water, swell the polymer binder and cause increase in resistance and hence finds application as humidity sensor.

Adding lead zirconate titanate (PZT) grains in the polymer matrix, to make a polymer PZT, behaves like dielectric and serves in its application as a piezoelectric sensor with a relative permittivity of 240. The other piezoelectric coefficients are d_{33} of 10 pC/N and g_{33} of 25×10^{-3} Vm/N. With the film provided with electrodes, serves as a force sensor, developing electric field on open circuit and a charge on short circuit.

The piezo resistive properties of carbon filled polymer is one leading to the development of strain gauges with gauge factors up to 60.

12.4   IR RADIATION SENSORS

12.4.1   Basics

Optical instruments, utilizing the effects of electromagnetic radiation, containing the ultraviolet, visible and infra-red spectra, on matter are in use since long, normally in scientific laboratories for research purposes. Recognising their potential for application and extension to situations arising in industry, simpler and modified versions are developed for on-line and real-time instrumentation of several quantities. Present day concern about healthcare of patients as well as the condition-monitoring of machinery and systems in industry, has resulted in development of several infra-red instrumentation systems such as thermometers, thermographs and other pollution monitoring equipment.

Considerations such as dangers from exposure to ultra-violet and effects of interference from sources of visible radiation, prompted the choice of near infra-red radiation with wavelengths ranging from 700 to 2,500 nm. It is well known that the techniques of using this radiation are based on emission and absorption with likewise classification of instruments also. A single wavelength of the spectrum or a bunch of them may be used depending on the application and instruments such as dispersive and non-dispersive type of analysers are known since long enabling detection of impurities and trace elements and analysis for composition of substances. Traditional physical transducers, used in such instruments have been replaced wherever possible, by the present-day miniaturized versions of solid-state sensors, to enable the presentation of data compatible with the display devices and communication systems. The following sections reveal some of the unique systems using the infra-red radiation.

The infra-red (IR) radiation sources are classified as thermal and non-thermal (quantum) types. As IR radiation is identified as heat waves, every hot body emits radiation, characteristic of the source, differing in intensity at each wavelength of the emitted spectrum. Every hot body at temperatures above 0 K, emits IR radiation purely by virtue of its temperature and in such a case it is known as 'Thermal Source'. It is this phenomenon, that has enabled development of IR-thermometry. The characteristics of thermal sources, briefly dealt with in Section 6.1.1.5, indicate the distinction between IR-thermometry and pyrometry. The non-thermal (quantum) sources operate by some quantum mechanical activity and they may be simple tungsten filament lamps, light emitting diodes (LEDs) or lasers. They need external sources for excitation and thereby the emission and it is usually electrical excitation. They may be emitting over a narrow range of wavelengths (λs) (lasers and LEDs) and possess high intensity at each λ, and in some cases, they may be modified to emit coherent beams at one or two λs. The filament lamps provide radiation of λs over a wide range and are used in IR analysers. IR lasers and LEDs with their unique characteristic of being turned on or off at high speeds, have been popular in the development of sophisticated instrumentation, high speed communication and condition-monitoring of processes.

The distinction is equally to be noticed in the principles of operation of the detectors for IR radiation. IR thermal detectors, sense temperature changes brought about in them by absorption of heat energy by some sensing elements, known already as thermocouples, thermopiles, bolometers, etc. The thermocouples and thermopiles are introduced in Section 7.1.2 and the bolometer in Section 6.1.1.5. The sensitivity of a detector is the ratio of the signal obtained to the incident power, usually expressed in volts/watt or amps/watt. The characteristics of the radiation source should be specified and the detector area is specified for each so that radiation flux can be calculated. For certain applications, the sensitivity at each wavelength is specified as monochromatic sensitivity. For quantum or photoelectric sensors, the sensitivity is necessarily specified as luminous sensitivity in amps/lumen. With no radiation falling on, the detector still generates a small random voltage due to various electrical noise within the detector iteself. The amount of incoming radiation required to produce a signal just equal in strength to the inherent electrical noise level is called noise equivalent power, NEP. NEP indicates the smallest power detectable. Spectral response curves, for all the photon detectors are generally available. Other data regarding, response time, time constant, impedance etc. are also specified for each.

12.4.2   Thermal Detectors

In Section 6.1.1.5, thermal radiation detectors are shown for use in radiation pyrometry enabling the measurement of the temperature of the outer surface of hot bodies from a distance. Study of the energy distribution pattern with wavelength shown in Figure 6.8 indicates that for temperatures below 650°C, all the energy is contained in the infrared region only. As temperature increases from 650°C, it is seen that the energy content creeps into the visible region and that more energy is emitted at progressively shorter wavelengths as the temperature is increased. Such an observation led to the development of pyrometry wherein total radiation pyrometers enable measurement of very high temperatures, responding to the total energy in the visible and infrared regions. Infrared thermometers are specifically meant for measurement of temperatures below 650°C right down to −100°C making use of only thermal detectors for detection of emitted IR energy.

A slightly different thermal detector, preferred for use in IR gas analyzers, is the Golay thermal detector, based on pneumatics. Incident radiation falls on a metallised collodial film having a resistivity matched to the impedance of free space, so that much of the radiant energy is absorbed by the film. The cell consists of a cylinder filled with Xenon and sealed with one side being the blackened metal plate and film, and the other side by a flexible metal diaphragm. When the radiation falls on the film, heat is conducted to the gas inside the cylinder causing it to expand and deform the diaphragm which separates the two chambers, like in the case of the differential capacitor transducer shown in Figure 6.77. Light from a lamp inside the instrument can be focussed on the diaphragm which deflects the light onto a photocell. Thus motion of the diaphragm moves the light beam across the photocell surface and changes the photocell output. This system is as good as thermocouples with a time constant of 4 ms.

Some of the characteristics of thermal detectors are given in Table 12.2.

Table 12.2 Thermal detectors

	Sensitivity (VW^{-1})	Response time (ms)	Area (mm^2)
1. Thermistor Bolometer	1.3×10^3	16	0.5×0.5
2. Pt-metal Bolometer	4	16	7.0×0.30
3. Thermopile (Schwartz vacuum)	30	100	0.5×9.0
4. Golay detector (including amplifier)	25×10^4	15	3 mm dia
5. Pyroelectric detector	0.25×10^3	20	2.5×6

The pyroelectric detector is essentially a capacitor that can be charged by an influx of heat. It consists of a crystalline substance capable of generating electric charge in response to heat flow (i.e. in response to change in its temperature). The substance is kept between a pair of electrodes and a heat absorbing layer is provided on one electrode and the IR radiation is incident on this layer. Since the pyroelectric detector generates charge related to the change in temperature only, it requires a reference ambient temperature sensor and a chopper mechanism kept between the pyroelectric detector and radiation source.

The net thermal power P received by the detector is given by

$$P = \sigma A \varepsilon_a \varepsilon (T^4 - T_a^4) \qquad (12.2)$$

where σ, the Stefan-Boltzmann constant
   A, the area of the target
   ε, the emissivity of the target
   ε_a, the emissivity of the surroundings
   T, the temperature of the target
   T_a, the temperature of surroundings

The net radiated flux, F (= P/A) is sensed by the detector. Using the above eqn., T can be given by

$$T = \sqrt[4]{kF + T_a^4}$$

where $k = \dfrac{1}{\sigma\varepsilon\varepsilon_a}$, a constant

The charge developed at the electrodes is suitably processed by electronic circuits. The detector response is highly linear and is very sensitive. Typical application is seen in measurement of temperature of the human body, by using it to measure the radiation from the inner surface of the auditory canal and from the tympanic membrane. The so-called tympanic temperature is considered the most accurate and reliable indicator of the body temperature. It takes only 3 seconds to measure while conventional ones take more than 1 minute.

12.4.3 Quantum Detectors

The quantum detectors (also known as photon detectors), on the other hand, are wavelength-dependent and are not limited by the thermal effects. They operate in a manner that is characterized as photoelectric effect described in Section 7.6.1 where one quantum of radiation releases a quantity of electric charge. They are normally non-uniform in their spectral response, which increases with λ and then falls to negligible levels at a λ known as cut-off wavelength λ_o (or threshold wavelength). Each detector has its own peak wavelength at which it exhibits maximum response and its own λ_o. The photoelectric effect is mainly characterized by the energy of the incident quantum, which shows up as thermal energy, only when it is greater than the energy needed for excitation i.e., the energy needed to release an electron or enable it to jump from one discrete energy level to a higher one. That is why the efficiency of detection is less for shorter wavelengths than the cut-off. There are a large number of photon detectors, smaller in size, faster in response but are normally known as photoelectric transducers as given in Table 12.3, (as detailed in Sections 7.6.2, 7.6.3 and 7.6.4) or simply optical sensors.

Table 12.3 Photon Detectors

	Wavelength range µm		Sensitivity	Response time (ms)	Area mm^2
	λ_{min}	λ_{max}			
1. Silicon photovoltaic	0.35	1.13	70×10^3 V/W	5	10 mm dia
2. Selenium photovoltaic	0.25	0.80	0.015 A/W	10	40 mm dia
3. Silicon Avalanche diode	0.35	1.13	0.5 A/W	0.4×10^{-6}	0.10 mm dia
4. Silicon photodiode	0.35	1.13	0.45 A/W	2.0×10^{-6}	7.5 mm dia
5. Photoemissive diode (AgOCs cathode)	0.3	1.25	0.045 A/W	0.5×10^{-6}	25×12 mm^2
6. Photoemissive diode (SI cathode)	0.3	1.25	0.001 A/W	0.1×10^{-6}	20 mm dia
7. Photomultiplier (SI cathode)	0.3	1.25	0.5×10^3 A/W	1.0×10^{-6}	36 mm dia
8. Cadmium sulphide	0.3	0.9	0.3×10^6 A/W	100	15×10 mm^2
9. Lead sulphide	0.4	3.5	0.1×10^6 A/W	0.15	6×6 mm^2
10. Indium antimonide	1.0	7.2	1 V/W	0.1×10^{-3}	6.0×0.5 mm^2
11. Cu doped Ger (4.2 K)	2.3	25	0.5×10^3 V/W	0.5×10^{-3}	4×4 mm^2

12.4.4 IR Thermometry

Infrared thermometers consist of thermal detectors sensitive to the wavelengths of radiation from 0.72 to 20 μm. The radiation is brought to focus by means of lenses, the lens materials being transparent to the spectrum of frequencies of the above region, as shown in Table 12.4.

Table 12.4 Lens materials

Material	Lower limit (μm)	Upper limit (μm)
Lithium fluoride	0.11	7.0
Calcium fluoride	0.125	10.0
Fluorite, CaF_2	0.13	9.5
Fused quartz, SiO_2	0.17	4.0
Saphire, Al_2O_3	0.18	6.1
Air	0.19	—
Barium fluoride	0.20	13.5
Potassium Bromide	0.20	30
Sodium chloride	0.20	17
Pyrex glass	0.31	4.0
Arsenic trisulphide	0.72	10.0
Irtran-2, ZnS	1.0	15
Germanium	1.7	20
Cesium iodide	1.0	40
Diamond	5.8	7200
Silver chloride	10	25
Thallium bromide	0.7	60

No single material is available for the entire region and so choice is limited to those that cover the maximum range of λs in each application. The detectors need to be small in size and of low thermal capacity. The thermal detectors develop output signals that are solely functions of the absorbed thermal energy received by them and not on the energy spectrum. There are a large number of radiation detectors for application in the visible and ultra-violet regions apart from infrared with sensitivities, not totally restricted to any one but to wavelengths overlapping in most cases.

For detection of infrared radiation and for use in the IR thermometers, thermal detectors are best suited as they can be obtained with essentially uniform response for all wavelengths and in the smallest sizes desired. Among all the thermal detectors, the thermocouple and the thermopile are widely used. A blackened gold foil of size 2 mm × 0.2 mm and of thickness of 1 μm is supported by quartz fibres. The 'hot' junction made by welding together wires of two different semiconducting materials at their ends, is fastened to the gold foil. The corresponding cold junction is kept darkened and at a constant temperature in the receiver cell, an evacuated steel casing with a Potassium bromide window. The time constant may be large but it is most sensitive. Thermopiles with sensitivity of 200 V/W with sizes of 0.4 mm^2 and of NEP of about 10^{-11} W are available. Bismuth-alloy and antimony-alloy combinations are also preferred. The main advantage of the thermopile over the bolometers is its good zero-point stability.

A bolometer can be made up of either a thin film of a noble metal or a semiconductor flake known as thermistor. Traditional types are used in vacuum cells to reduce noise. The time constant may be brought down to 4 ms with sizes of 0.5 mm^2. They are invariably chosen to make up IR thermometers. Nanotechnology application yielded microbolometers of extremely small sizes permitting a large number of them to constitute an array for scanning purposes in thermography. Some of the thermal sensors given in Table 12.2 are brought out in a similar manner in reduced sizes as microsensors yielding faster response.

Photoconductive cells can be used for radiation in the near IR region as well as in the visible region. Lead sulphide cell is the most popular one for use in the visible region. This material can be coated with a thickness of 0.1 μm on glass backing and can be used up to a wavelength of 3.5 μm while lead telluride has a range up to 6 μm. These materials can be developed as microsensors for use in scanning systems and enable high enough scanning rates.

A novel sensor known as photoelectromagnetic detector utilizes the Hall-effect. A semiconductor crystal is subjected to a strong magnetic field and radiation is applied to one side. A potential difference is developed across the ends of the crystal.

12.5 ULTRASONIC SENSORS

12.5.1 Introduction

The increasing interest in having sensor systems that are immune to electro-magnetic interference, noninvasive in nature and fast in response has led to the adoption of sound waves as means for the development of sensing systems. Sound waves are pressure waves that require a medium for propagation but do not alter its characteristics in any way. Other attractive features of sensing systems using sound waves are that they pose no health hazards to the operating personnel while at the same time, they are safe for use in harsh and hot environments, prisonous, explosive, inflammable or corrosive in nature. With no moving parts the sensor systems are non-contacting type affording ease of installation and maintenance. Instrumentation systems using ultrasonic (US) waves, with frequencies far above the audible range, have been finding extensive application in traditional industries as well as the present-day food and pharmaceutical industries. Their effectiveness is very much realized in medical diagnostics and condition monitoring of engineering systems.

12.5.2 Basics of Ultrasonics

Sound waves are known normally in three categories; subsonic, sonic and ultra-(or super)-sonic waves, the latter usually of the frequency range of 0.5 MHz to 10 MHz. Ultrasonic waves are the invariable choice for use in instrumentation so as to avoid the interference from the sonic sound waves, always present in the surroundings.

The important property of sound waves is that they behave like light-waves as far as reflection and refraction are concerned. They are also both longitudinal and transverse in the media. The velocity of sound waves is least in air, being 350 m/s and going up to 6,000 m/s in metallic substances. Sound energy attentuates highly in gases and to some extent in viscous matter. The volume density of the material is responsible for the attenuation, the higher the volume density, the larger the impedance offered for the transmission of the waves. The choice of the frequency of the ultrasonic (US) waves (usually interpreted as ultrasound) is dependent on the material through which the waves are transmitted and the path length of the beam in the material.

12.5.3   Sonar

In naval warfare, the role of *sonar* is well known since long, in detecting and tracking submarines in the sea bed. The *Sonar* (SOund NAvigation and Ranging) system is unique in its application for activities related to under-water acoustics and exploration of seabed contents. The piezoelectric and magnetostrictive type generators of sound waves are used, the latter normally chosen for larger power signals at audible sound frequencies. The generating systems are so designed to transmit a parallel beam of waves into the sea, similar to that of a searchlight. The system is made up of a large number of vibrating elements in a pattern known as 'mosaic' shown in Figure 12.6 resulting in certain advantages in the formation of the wavefront.

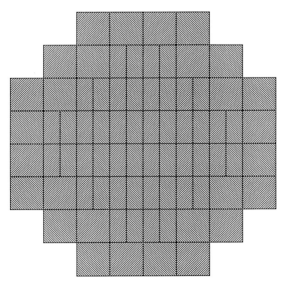

Figure 12.6   Mosaic array of crystal elements.

While a single piezoelectric crystal stuck to the outside surface of a tank is sufficient to measure the level of a liquid in the tank, the sonar transmitter consists of an array of such crystal elements as shown in Figure 12.7. The elements of the mosaic are cemented to a metal (or rubber) backing support to hold them firmly and conform to the desired alignment. The frequency of the waves generated by the transmitter and other characteristics are determined by the material used for support and the way the elements are attached. All the elements are driven electrically, in parallel, by an electrical source of supply. The vibrator dimensions may vary with their thickness being either quarter or half of the wavelength, λ of the waves. The crystals of the piezoelectric materials used are so cut as to have maximum coupling coefficient between mechanical strain and electrical polarization directions. For use in sonar, the ADP or Rochelle salt crystals are used with the elements sealed in oil as the crystals are water-soluble.

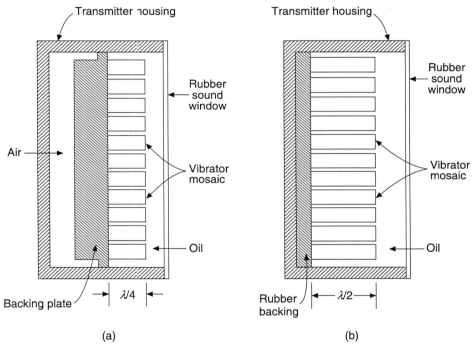

(a)   (b)

Figure 12.7   Typical mosaic transmitters. (a) Quarter-wave vibrators with metal backing plate.
(b) Half-wave vibrators with pressure release rubber backing.

The magnetostrictive type transmitter, based on the phenomenon of magnetostriction explained in Section 7.3 employs longitudinal vibrations of a number of nickel tubes, each of length equal to $\lambda/4$, attached to a driven plate as shown in Figure 12.8. Each tube is excited by an exciting coil, carrying the same alternating current. The permanent magnets provide the bias-field. The entire system is sealed and made watertight. In place of tubes, laminations can be used.

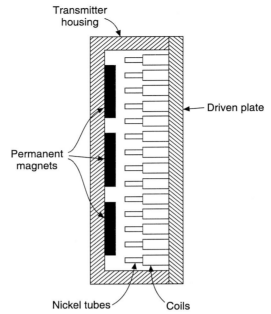

Figure 12.8   Cross-section of a typical magnetostrictive transmitter.

12.5.4   Ultrasonic Sensing System

The development of the ultrasonic sensors is based on utilization of the effect of the measurand on the properties of the wave. Reduction in the intensity of the wave leads to the attenuation of the mechanical energy of the waves in the medium. The reduction is due to the impedance of the path of the US beam, usually of a liquid under test for its flow rate and the walls of the pipe through which the liquid is flowing. The impedance (otherwise known as acoustic impedance) of a path in a medium is proportional to the product of the three quantities; (1) path length (2) velocity of the wave in the medium and (3) volume density of the medium.

Each US sensing system comprises of a source and a detector with the frequency of the waves chosen to suit the application. They are usually referred to as transmitter and receiver and they are in most situations piezoelectric transducers as explained in Section 7.2. The transmitting transducer is energized by a tuned oscillator for generation of continuous waves and for pulse generation a capacitor discharging circuit is used.

The size of the quartz crystal used in the transmitter being small, sufficient care is taken to establish good coupling of the crystal with the walls of the tank or pipe through which the US beam requires to be propagated. The crystal is attached to the walls simply by sticking or gluing with a coupling compound of known acoustic properties. At low temperatures, water or grease can be used. It is essential to see that the path of the beam comprises of solids and liquids but not air as the impedance offered by it will be large, variable and unpredictable. It is for this reason that the ultrasonic sensors are limited to measuring levels of liquids in containers and liquid flow rates. They are also finding extensive application in patient monitoring and medical diagnostics.

Ultrasonic techniques use continuous waves or a single pulse of short duration for measurement of distances of targets from a spot or levels of liquids in containers. A matched pair of piezoelectric transducers is used to serve as a transmitter–receiver combination when the ultrasonic beam is transmitted through a liquid column. When a single pulse or a train of pulses are used, reflected beam from the surface of the liquid is received by the same transducer. It is usually known as pulse-echo technique being found effective in most of the sonars. The time delay in the detection of the reflected pulse determines the distance of the target in sonar-detection, as the velocity of sound in sea-water is known. For other cases, such as the finding the depth of water surface in a well, velocity of sound in air enables the estimation of depth with the use of a single pulse. The accuracy of estimation depends on the value V, of the velocity used. If the delay or the transit time of the pulse is T s, the depth or the distance, D is given by half of VT.

For situations encountered in industry, it is either the measurement of the liquid level or its control. It is essential to establish perfect coupling between the transducers and the tank surfaces by means of couplant, as shown in Figure 12.9. The continuous waves from the transmitter are received by the receiver transducer, and the significant change in the detector output signals due to the change in the path of the beam affords scope for using the system as part of an automatic measuring system in the way described in Section 8.8. The system can also be treated as an on-off level-detector for level-control schemes.

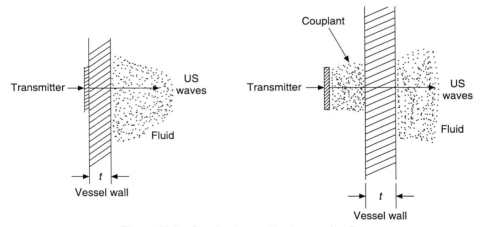

Figure 12.9 Coupling transmitter to vessel walls.

If the path of the beam is transmitted all the time through the liquid medium, then the arrangement is used for qualitative studies only such as concentration of solutions, identification of different liquids and presence of bubbles and dust. Measurements cannot be very precise due to limitations of media making up the entire path of the beam. However, due to the unique advantages of these sensors, claimed earlier, the sensing systems enable some other measurements. Velocity of sound is highest in heavy metals and lowest in gases. The attenuation of the beam is highest in gases and lowest in metals. When a plane wave is incident normally to the surface and passes into water, the intensity of the wave is reduced by a factor of 0.00112. The velocity of sound in water is about 1480 m/s and those in steel and quartz are 5,050 m/s and 5,750 m/s respectively. The continuous wave and pulse-echo techniques can as well be utilized to determine the density of the medium and concentration of the solutions. In such studies, the waves pass through the medium over a fixed length, while the medium is at rest. Measurement of the velocity of the wave or its transit time enables identification of the medium, its density and concentration and permits monitoring of deposits formed due to precipitates, dust or rust. Likewise, the position of the interface between two immiscible liquids can also be determined by any of the two techniques as shown in Figure 12.10.

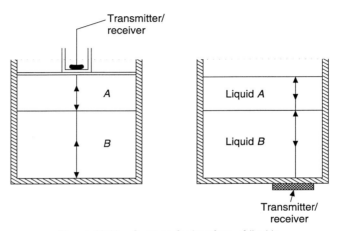

Figure 12.10   Systems for interface of liquids.

12.5.5   Ultrasonic Flow Meters

Measurement systems for the velocity of fluids flowing through pipes, by using the ultrasonic waves, constitute a large variety and are mostly preferred in industries due to the many advantages claimed earlier. Ultrasonic flowmeters are successfully applied for situations where other systems cannot work, such as those where the medium is highly corrosive, nonconducting or slurry type and where pulsating flow is to be monitored. Simple to complex flow meters are available for selection by the user, providing accurate measurements for flow-rates over a wide range. The principles on which the flowmeter is made to work, may be measurement of transit time of pulsed beams, or use of Doppler effect, where clean fluids are the flowing media. Cross-correlation techniques are utilized for situations where the flowing medium presents problems.

Transit-time type flowmeters using one pair of transmitters and one pair of receivers are common and constitute simpler versions of flowmeters. The transmitters and receivers may as well be placed inside the flowing medium whether it be in an open channel or pipe and where flow is laminar and the medium clean and free from particulate matter. The velocity of the US waves, c in the medium should be known as the transit time t_o of the pulses from the transmitter at A to the receiver at B is given by L/c where L is the length of the path from A to B, as shown in Figure 12.11, when the fluid is at rest. If the flow-rate of the fluid is V, then the transit time, t is $L/(c + V)$, for flow in the same direction as the pulses, i.e. from A to B. As the fluid flow-rates are very small compared to c, the difference between t and t_o, Δt is given by

$$\Delta t = \frac{LV}{c^2} \tag{12.3}$$

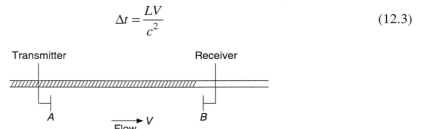

Figure 12.11   Ultrasonic flowmeters

If the temperature of the fluid is constant, and the value of c, if known accurately, the measurement of V from the above Eqn. (12.3) is satisfactory. Otherwise, an additional error creeps in if V is as much as 1% of c. In such a case, the use of another transmitter at C and receiver at D, results in reduced error as well as enabling the measurement, as difference in the transit times, given by twice that of the above Eqn. (12.3).

When dealing with clean and passive liquids, flowing in open channels or through large diameter pipes, the above arrangement can be used, though invasively, by suitably selecting two lines of flow, spaced apart from each other and ensuring that the paths of the beam are parallel to the direction of flow.

To eliminate the effect of temperature on c and thereby on V, a similar arrangement but with the ultrasonic beams transmitted into the stream at an angle to the direction of flow as shown in Figure 12.12. It is used for the flow through pipes with the transmitters and receivers, stuck to the outside surface of the pipe walls. The transit times of the beam for upstream and

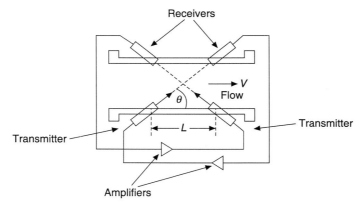

Figure 12.12   'SING AROUND' flowmeter.

downstream directions are measured over the same path. If the θ be the angle between the path of the beam and the axis of flow, the velocity of flow V is given by

$$V = \frac{L(T_u - T_d)}{2T_u T_d \cos \theta} \qquad (12.4)$$

where T_u and T_d are the transit times of the beam for upstream and downstream respectively. It may be noted that the relationship has become independent of the velocity of waves in the fluid.

If the transmitter–receiver of each path is converted into a self-exciting oscillator system as shown in Figure 12.12 by using the received pulses to trigger the transmitter pulses as a positive feedback, the frequency of oscillation of one path differs from that of the other by $2V \cos \theta/L$. The system is known as 'sing around' flowmeter. This frequency difference is measured by multiplying the two signals to produce the beat frequency. Even though the flow is streamlined, the velocity calculated by using the value of beat frequency, will be the average velocity of flow only. As the measurement is effected non-invasively, the system can be used for flow rate of liquids containing particulate matter and those of variable density.

The 'sing around' type of flow meter also yields the value of the sound in the fluid. If the variation of the velocity of sound in fluids with temperature is known and available prior to measurement, then it is possible to estimate the temperature of fluid also, provided flow rate is constant.

12.5.6   Doppler Flowmeter

Doppler effect is utilized as the basis for measuring flow-rate of fluids containing particulate matter. The particles in the medium are necessitated as they are to reflect or scatter the ultrasoninc beam directed into the fluid as shown in Figure 12.13(a). The velocity of the particle and that of the fluid should be the same, which in other words dictates the size and character of the particulate matter. Bubbles or droplets will also help the measurement. The receiver picks up the reflected or scattered waves and compares its frequency with that of the transmitter, as shown in Figure 12.13(b). The velocity of flow is proportional to the 'Doppler Shift' in frequency given by Δf. Using Snell's law of refraction, applicable to sound waves, the velocity of flow, V is given by

$$V = \frac{c \Delta f}{2f \sin \theta} \qquad (12.5)$$

where c, the velocity of soundwaves in the fluid

f, the frequency of the transmitted waves

and θ, incident angle at which the waves from the transmitter enter the fluid.

The frequencies of the transmitted and received beams are then 'beat' together to generate Δf and thereafter processed to obtain output signals in the desired form. Δf may be around 100 Hz for f value of 5 MHz. Accuracy is poor ($\pm 10\%$) because of variation in velocity of fluid from centre to the walls of the pipe.

Pulse echo technique also can be used by using the Doppler Principle, for determining the velocity distribution across the pipe. When combined with computer modelling of the flow, the temperature distribution can also be determined.

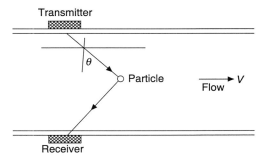

Figure 12.13(a)   Flowmeter for liquids with particles.

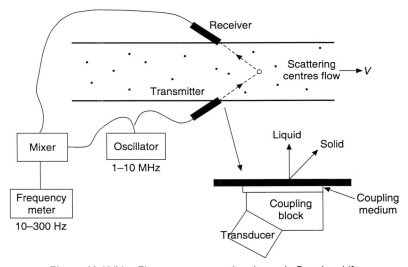

Figure 12.13(b)   Flow measurement by ultrasonic Doppler shift.

In nuclear industry, temperature distribution of liquid sodium, flowing with constant velocity, is studied non-invasively.

In process industries temperature does not change appreciably and so sonic velocity change becomes a measure of the solute concentration.

Concentration of particles can also be determined by making attenuation measurements across the pipe, by directing the beam in a direction perpendicular to the flow towards the receiving transducer. As particles much smaller than the wave length of sound scatter sound disproportionately less. So by performing the measurements at different frequencies, it is possible to measure size distribution. It is to be noted that gas bubbles behave differently.

Immersion type transducers are specially made by encapsulating them entirely out of nickel, stainless steel or special alloys for use in corrosive liquids and for use at temperatures around and up to 500°C and for liquid nitrogen temperatures right down to –195.8°C, at pressures up to 20 bars.

12.5.7 Cross Correlation Flowmeter

The metering of multiphase flow, usually problematic, is nowadays effected by application of a mathematical tool known as 'correlation technique' and utilizing the electronic support for the signal processing and data display. The popularity and powerfulness of the technique is seen in such other applications where measured signals are weak and infested with random noise and the ultrasonic cross-correlation flowmeter is recognized as the most suitable one for flow metering using simple transducers enabling online, real-time non-invasive measurement.

The principle of the cross-correlation technique rests on processing of signals derived from two sensors, located at a distance of L from each other as shown in Figure 12.14. A pair of ultrasonic transmitter–receiver combination provides two signals, one from the upstream pair and the other from the downstream pair, differing in pattern not only because of the time delay T s between them due to the flow-rate but also from any randomness due to the type of flow.

If $x(t)$ and $y(t)$ are the two signals with $y(t)$ being the delayed one by τ s, the cross-correlation function $R_{xy}(\tau)$ relating these signals in terms of τ is given by the expression

$$R_{xy}(\tau) = \lim_{T \to \infty} \frac{1}{T} \int_0^T x(t - \tau)\, y(t)\, dt \qquad (12.6)$$

where T is the integration time.

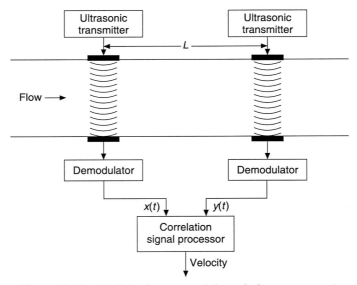

Figure 12.14   Principle of cross-correlation velocity measurement.

It is to be noted that the integration process is the statistical technique used to find the best fit by adjusting τ until a maximum correlation is found as shown in Figure 12.15. Measurement of τ_m by electronic signal processing enables calculation of the flow velocity, V from L/τ_m.

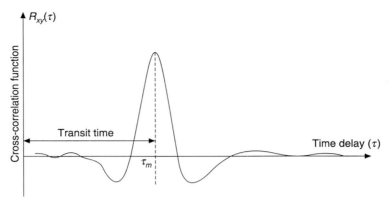

Figure 12.15   Cross-correlation function.

The randomness in the flow pattern due to turbulence of flow or due to presence of another phase fluid, does not pose problems and flow rates obtained in this way are reliable and more accurate when compared to other methods. The measurement technique is applicable to different sizes of pipes, as the velocity is dictated only by L and τ_m. It goes without saying that the accuracy is dependent on the value of L and so the attachment of the transmitters and receivers is given sufficient attention. Clamp-on arrangement requires to be avoided, as it cannot ensure exact geometry of spacing. The ultrasonic waves get affected in intensity and phase in gases and multiphased liquids but in low velocity single-phase fluids, intensity modulation is so small as to be unusable. The technique is favourable for use over a velocity range of 0.1 to 80 m/s for a sensor spacing of 40 mm.

Applications of this technique range from the velocity measurements of solid/liquid mixtures, water/gas mixtures and oil/gas mixtures. The technique when dealing with oil/gas mixtures, requires the mixture to be well dispersed. Such dispersion is found in oil wells, where oil from the depths is obtained and hence the technique is preferred in oil industries. With slight modifications and in association with measurements of density or water content, additional data can be obtained.

Solid metal waveguides known as 'stand-off' waveguides are developed for use on tanks and applied to the tank wall in a kind of welding operation. Also for use in areas of difficult access, strip waveguides are fabricated. They are applied to the outside of a vessel just like normal transducers.

12.5.8 Surface Acoustic Wave (SAW) Sensors

A novel type sensor known as surface acoustic wave sensor (SAW) consists of inter-digitated electrodes deposited on a piezoelectric plate. The plate when excited, develops surface waves that can be propagated in the surface plane normal to the overlap of the electrodes at a frequency dependent on the propagation velocity and spacing of the electrodes shown in Figure 12.16. Its centre frequency can be altered by stress when used as a pressure sensor. These sensors can be applied to produce acoustic beams which can diffract light in Brigg mode.

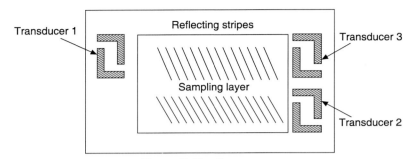

Figure 12.16   Saw sensor.

One of the applications is for monitoring of nitrogen dioxide.

Such sensors constitute the new breed of micro-sensors where combination of optics, acoustics and electronics plays an important role in the hybrid sensor technology.

12.5.9   Coriolis Type Flowmeter

Of all the meters for fluid-mass flow measurement, the Coriolis type stands unique and has gained prominence by virtue of its rangeability, repeatability, accuracy and stability. Apart from its noninvasive nature, other advantages claimed are insensitivity to a wide range of effects such as pulsations, viscosity, flow profile, two-phase flow and fluid density. Considering all the above factors, the pressure drop due to its introduction is tolerated in most of the applications. As can be seen, the meter should be protected from any external mechanical vibrations.

The principle of operation is based on the generation of a force known as Coriolis force on a particle whenever it is present in a rotating body and moves relative to the body in a direction either away from or towards the centre of rotation. In Figure 12.17(a) the particle of mass M slides with constant mean velocity v in the tube which is rotating with an angular velocity ω about a fixed point P. The body experiences two components of acceleration and force:

1. a radial acceleration, $A_r = \omega^2 r$, the centripetal acceleration, towards the axis.
2. an angular acceleration, A_t perpendicular to the above 1 equal to $2\omega v$ where v is the radial velocity.

To impart the Coriolis acceleration, a_t to the particle, a force of magnitude $2\omega vM$ is required in the direction of a_t. This comes from the tube. The reaction of this force back on the tube is the Coriolis force $F_c = 2\omega vM$.

Applying the above understanding to the case of a fluid of density ρ flowing at a constant velocity v along the tube which is kept in rotation as shown before. The elemental length of Δx of the tube experiences a transverse Coriolis force of magnitude $\Delta F_c = 2\omega v\rho A\Delta x$ where A is the area of cross-section of the tube. Since the mass flow rate is given by $m = \rho vA$, the Coriolis force is given by

$$\Delta F_c = 2\omega\Delta x \cdot m$$

If this Coriolis force exerted by the flowing fluid on the rotating tube can be measured then the force becomes a measure of the mass flow rate m.

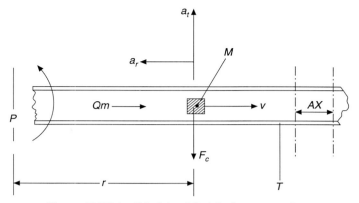

Figure 12.17(a) Principle of Coriolis force generation.

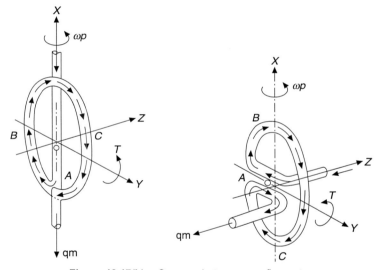

Figure 12.17(b) Gyroscopic type mass flowmeter.

As the tube cannot be rotated in actual case of measurement, it is subjected to vibration and the driving force required to keep the tube in constant vibration is to be directly or indirectly measured. This force is smallest when it is vibrated at its natural frequency. In all meters, the flow tube is anchored at two points and vibrated at a position midway between the anchors, thus resulting in opposite oscillatory motions of the two halves of the tube. The tube looped as shown in Figure 12.17(b) functions like a cantilever, is vibrated at its free end. Fluid entering at A, experiences an increasing forced amplitude of vibration (A to B) and the amplitude then decreases as the fluid along the outlet limb (from B to C). These forces are at right angles to the axial motion of the particles which is constant relative to the tube. The fluid exerts an opposing force to resist the imposed perpendicular motion. As these forces are in opposite directions, in the inlet and outlet limbs, a mechanical couple is set up that causes the tube to twist through an angle of θ that is proportional to the mass flow rate. Position sensors measure the amount of tube twist by means of timing the passage of the two limbs of the tube past a reference line. The time interval is a direct function of mass flow as given by

$$Q_m = \frac{K_s \Delta t}{8r^2} \tag{12.7}$$

where, K_s is a constant for the tube material r, the radius of the tube and Δt, the measured time interval.

The tube movements are usually of the order of 25–50 μm full scale and can be measured.

The commercial versions of the Coriolis meter are of the three different designs shown in Figure 12.18 where one is a U-tube, the second a loop and the third a straight pipe.

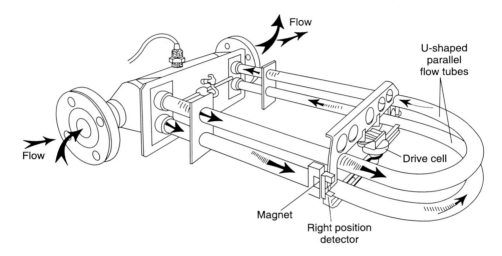

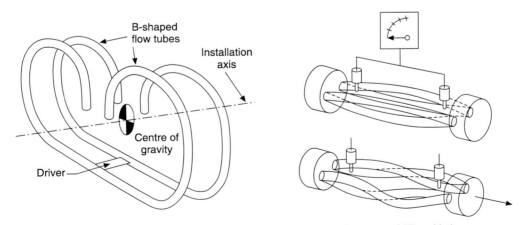

Figure 12.18 Three designs of Coriolis mass flowmeters (courtesy of Micro Motion, K-flow and Endress and Hauser respectively).

12.6 FIBRE OPTIC SENSORS

12.6.1 Introduction

During the last two decades, fibre optic technology has contributed immensely for modernizing the instrumentation systems in almost all industries, more so in those where safety and security of personnel and property are of utmost priority. It goes without saying that the fibre optic sensors are favoured for use in medical equipment for diagnostics as well as biological investigations and research. The fibre optic technology has emerged as a contender to the solid-state electronics, in several ways, notably to the development of fibre optic sensors, that are immune to electromagnetic interference and radiation hazards and suitable for use in harsh and inaccessible environments. Fibre-optic sensors and systems have been recognized to have the application potential for a large number of variables both in the conventional localized and distributed fashion for their measurement. Their versatility is proved by their application to measurement of not only physical and chemical quantities but also those arising in biomedical and biological research. They have also satisfied most of the rigorous demands of the industry, in regard to accuracy, repeatability, long-term stability and long life, apart from their access for on-line control and real-time monitoring, telemetry and compatability for integration with electronic circuits. The principles of operation on which the fibre optic sensors are based, render them near ideal in their sensing capability, small in size and weight and fast in response. No wonder, they have been found to be exceptionally suited for use in aircrafts, defense systems and space navigation systems. Some of the simple sensors and sensing systems are described in the following sections revealing the extensive scope of application and lending credibility to the above observations.

12.6.2 Basics

Since long, a large number of simple to sophisticated instruments are developed and used, making use of rays and beams of *lightwaves*. They have all been designed with the medium of air for the path of the rays or vacuum where necessitated. The possibility of providing guided path through glass fibres, has extended the modern technology to exceptional heights, not only in the field of communication but also in the present day instrumentation. Development of instrumentation systems, with the degree of sophistication, as seen today, could not have been possible without the corresponding sources of light rays, the semi conductor lasers and the light-emitting diodes (LEDs), essential for use in the instrument. The exceptional nature of immunity provided by the materials used for the fibres, essential for the guidance of rays of light, has widened their suitability for application in several situations. The optical beam for use in fibre optic sensors is usually chosen from the visible region, though the ultraviolet and the infra-red permit their application is certain situations, through the use of relevant sources of radiation, detectors and other auxiliary devices. The materials used for fibres, should be low-loss and low dispersion type for the wavelengths of the spectrum chosen. The absorption characteristics of silica glass limit its use for a spectral range of about 200 nm to 2,200 nm for wavelengths. Hence optical quartz fibres are preferred. Two types of silica glass are also available for use, one with high hydroxyl (OH) concentration useful for 250–800 nm and that with low OH for 400–2100 nm wavelengths. Flouride glass is chosen for infra-red beam for wavelengths up to

3,200 nm. The optical fibre cable consists of several thousand filaments inside a cladding sheath while a stainless-sheath may be provided as outermost sheath for obvious reaons. The filaments or the fine fibres can be bunched in random fashion or grouped to form different arrangements. Most of the glass fibre-optic cables will need a polished epoxy seal at the ends of the bundle to provide a liquid and air-tight-seal.

Plastic optical-fibre cables having acrylic monofilaments of different diameters can be enclosed in polyvinyl chloride (PVC) or stainless steel sheaths and used. The dielectric properties of the optical fibres, offers a number of advantages over the electrical sensing systems thereby being most suitable for use in aerospace transportation systems. Plastic optical cables are usually meant for short lengths only. For long-distance use, plastic cladding over the core is used to afford flexibility and an additional armour for mechanical protection. Glass fibres have losses of 3 dB/km while plastic fibres have much higher losses of 100–1250 dB/km. Plastic fibres enable short distance telemetry with information capability of about 6 Mbits/s. They are better suited for measurements at low temperatures over −30°C to +70°C and are also easy to establish coupling with end devices.

The transmission of the rays through the glass fibres along the axis is made possible by use of the cladding around the core, where the cladding has lower refractive index than that of the core. In step-index fibres the light is guided by total reflection at the core-cladding interface. Core diameters range between 50 and 400 μm. It is necessary to ensure that the light power that can be coupled into a fibre is maximum. This is dependent on the refractive index of core, n_1 and that of the cladding, n_2. If θ_m is the maximum acceptance angle, then the parameter known as numerical aperture (NA) is given by

$$NA = \sqrt{(n_1^2 - n_2^2)} = \sin \theta_m \qquad (12.8)$$

If the light beam incident on the endface of the fibre is to enter the fibre, its incidence angle must be smaller than θ_m. Most common value for θ_m is between 8° and 13°, and for NA it is 0.14 to 0.22.

The sources of radiation are laser diodes and light emitting diodes (LED). The laser diode emits coherent light at a wavelength of about 0.82 μm with a bandwidth of 2 nm. Due to the small emitting region extending from 1 μm to 20 μm, light signals get coupled into the fibre more efficiently, without much loss. The LEDs emit incoherent light and can be coupled into the fibre, but with greater losses than the earlier case. The emission bandwidth is also larger i.e. at about 30 nm. To obtain a parallel beam lenses are necessary. The power delievered by LED is smaller than that of laser diode and is about 1 mW.

In a simple fibre-optic sensor, the intensity of the light waves, modified or modulated due to the influence of measurand is detected and converted into an electrical signal. Silicon PIN photodiodes and the avalanche photodiodes serve as detectors at the receiving end. Pin diodes, described in Section 7.6.4, possess higher bandwidth while photo transistors of Section 7.6.4.1 have higher sensitivity.

12.6.3 Typical Fibre-optic Sensors

Fibre-optic sensors are treated as belonging to two distinct principles of operation:
  1. intrinsic and   2. extrinsic

Intrinsic sensors are those in which the quantity under measurement causes a change in the optical properties of the fibre itself. The fibre itself is the sensing device.

In extrinsic type sensing system, the fibre serves merely as a carrier or guide for the light from the source to a sensing device and back. It is the light waves from the sensing device that has undergone a change in its parameters due to the interaction at the sensing device.

The sensing system shown in Figure 12.19(a) utilises a fibre cable incorporating fibres at the centre for the transmission of light on to the target, along with fibres around the central core fibres, for the reception of the reflected light from the target. The displacement, t, between the

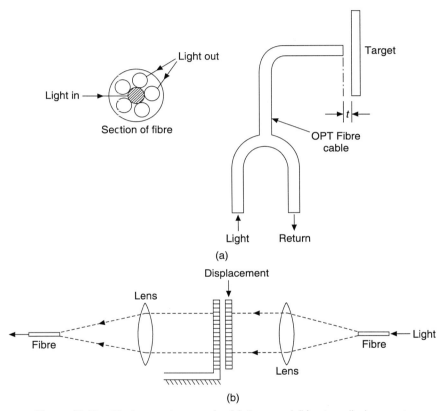

Figure 12.19 Displacement sensor for (a) linear and (b) rotary displacement.

target and the tip of the fibre optic cable determines the intensity of the reflected light. Angular displacements can be measured by using Moire fringe modulator shown in Figure 12.19(b) wherein the amount of light received by the receiver fibre is a measure of the angular displacement of the movable plate such as the rotor movement in a turbine flow meter.

For level measurements, an optical dipstick of the shape shown in Figure 12.20 with a conical tip is placed near the surface of the liquid whose level is required to be monitored for actuating a switch as shown in figure. The intensity of the beam, undergoing total internal

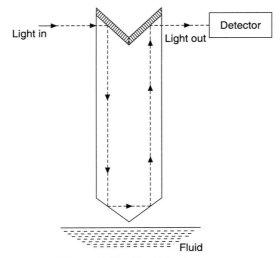

Figure 12.20   Liquid level sensor.

reflection at the tip of the dipstick gets altered as soon as the tip is in contact with the liquid. A feedback measuring system of the type described in Section 8.8 can be employed for continuous measurement of level.

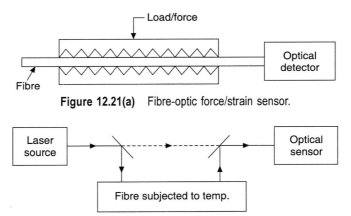

Figure 12.21(a)   Fibre-optic force/strain sensor.

Figure 12.21(b)   Fibre-optic temperature sensor.

A microbend sensor is constituted to communicate strain to the fibre by application of mechanical force as shown in Figure 12.21(a). The system serves as a point sensor, sensing force or strain of a member at a point. The system may be used along with an interferometer for measuring the phase variations of the waves passing through the fibre, when subjected to external forces or even temperature as shown in Figure 12.21(b).

While point sensor enable measurements of quantities at certain points only, it is possible to effect distributed sensing where the variable under investigation can interact at any position along the length of the fibre. In such a case it is known as a distributed sensor.

It is like-wise possible to have a quasi-distributed sensor, where certain points of fibre are sensitive to respond to the measurand and then yield a value of the measurand at these known and predetermined points.

12.6.4 Fibres Optic pH Sensor

It is surprising to realize that pH can be measured by optical fibre sensor. In the simple case, the emission of a fluorescent dye coating at the end of the fibre when dipped into the solution under test is made use of. The emission characteristics change with pH and the change has to be detected.

pH is also measured spectro-photometrically by using fibre-optic sensor. A pH sensitive indicator dye such as phenol red is immobilized inside a cuprophan membrane at the end of a pair of fibre optic cables. Light at two wavelengths is transmitted to the dye. Reflectance at one wavelength is pH dependent while the same at the other wavelength is independent of pH. Reflected light at both wavelengths is collected by the output optical fibre cable and processed for indication. But the system output is proportional to pH over a limited range only.

12.6.5 Fibre-optic Humidity Sensor

An intrinsic optical fibre sensor for humidity measurement is shown in Figure 12.22 wherein a single optical fibre of hard clad silica (HCS) with 600 μm core is used. At the sensing point along the fibre, the cladding is removed over 5 cm length and replaced with cobalt chloride

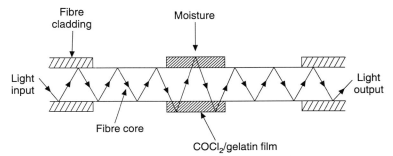

Figure 12.22   Fibre-optic humidity sensor.

gelatine thin film. It is immobilised on the surface of the core in a film such that light in the optical fibre travels through the film. Since the refractive index of the film is greater than that of the fibre core, practically all the light-launched through the fibre will pass through the film resulting in interaction with the cobalt chloride molecules. This anhydrous salt is blue in colour and turns pink as the water of hydration converts the salt to hexahydrate. As the external humidity level changes, the moisture content of the film changes and so the light passing through the film will be absorbed in the sensitive wavelength λ region. Thus the light output from the fibre will be found to have been attenuated at that λ. Whitelight is launched into the fibre and the intensities of spectra of the collected light are measured, using a spectrum analyser. Light over a wavelength region of 600–740 nm is found to be absorbed. A second wavelength outside this region at 850 nm is used to provide an intensity reference to enable rejection of changes in common mode light intensity variations (or temperature effects). The light attenuation of the sensor at 850 nm is seen to be 0.2 dB. The relative intensity plotted against relative humidity does not yield a linear relationship and hence the graph is split to consist of three distinct ranges; 0–40, 40–70 and 70–100% relative humidity.

It may be noted that more than 10 such sensors may be sited over a single multimode optical fibre of about 1 km length, so that humidities at those 10 points may be measured. An optical time domain reflectometer (OTDR) technique is thus used as a distributed sensing network. A pulsed laser diode is coupled to the fibre via a directional coupler which allows the return backscattered light to be collected by a photo-diode detector.

Fully distributed sensor systems are developed for monitoring stresses in structures such as bridges, buildings, etc. They are very much preferred especially where security checking for safety of personnel and property are considered. Likewise, monitoring of hot spots in a system, such as a transformer (with the fibre wrapped around it) is done both in terms of temperature and its spatial location.

Considerable research is undertaken to exploit as much as possible the potential of optical fibre-sensors for application to situations where even the microsensor is kept off for fear of having the EMI affecting its performance. Their long term stability and repeatability is very much desired by industries and as it is, their usage is high in scientific research, pollution monitoring, condition monitoring and medical diagnostic efforts.

12.7 CHEMICAL SENSORS

12.7.1 Introduction

Chemical sensors are normally meant for measurement of chemical properties of substances and compounds in solid, liquid and gaseous states. Taken together with the measurement of some physical properties, the data obtained may be utilized for analysis of some that are simple or binary mixtures. Industries use a great variety of raw materials to produce their end-products and are interested in on-line analysis of the substances at both ends. The products may be organic or inorganic in nature and may be fertilisers, pesticides, herbicides, food products, wines, beverages, scents, cosmetics or petrochemicals. An optimistic estimate of the number of chemical substances making up such products stands at 10,000. It has become a more challenging activity to identify and develop proper sensing techniques for their analysis, more so with the intent of assuring the quality of the products. Intense research is undertaken in the past two decades, to develop sensing systems for composition studies, arising out of the demands in the areas of medical diagnostics and pollution monitoring. The range of modern materials under the flag of ceramics, semiconductor materials and polymers, has contributed largely to the development of sensors much smaller in size and much faster in response. The developmental effort is considerable and rather involved due to the following reasons:

1. interaction between sensor and sensed media,
2. sensors are usually non-specific,
3. physical state of the sensed substance,

4. requirement of detecting and measuring the degree of concentration of the components (sometimes ppb in pollution studies), and

5. requirement of the sensor to be noninvasive, non-consuming or non-interfering type.

Of all the chemical sensors, those meant for gas analysis and detection of a particular gas in a mixture, vary widely, not only due to the principles of their operation, but also due to the need of isolating the gas when it is present as a very small quantity in a mixture and that too when it is known to be poisonous or toxic. The necessity is well realized as the sources of some dangerous gases are kitchens, land and air transportation vehicles and all industries including power plants. In most cases, real-time and on-line detection and analysis is desired, while gas analysis by procuring samples is done in laboratories using sophisticated techniques such as gas chromatography, spectrophotometry or mass spectrometry.

A different class of sensors, known as 'Bio-sensors', have been developed lately and are meeting the special demands of gas detection in chemical industries, pollution monitoring and medical diagnostics. These sensors use enzymes, organisms or biological tissues for their operation.

The necessity of guarding the health and safety of people both inside and outside of the working environment, and incorporating measures to prevent fire accidents in the industries has evoked great interest in developing gas sensors for all toxic gases and vapours and more importantly the flammable gases. Increased reliability and computer-compatibility are given prime attention in the designs of all sensors. Flammable gas sensors have to be provided with a rugged mechanical enclosure as their principles of operation are based on a heater inside to ignite the gas. The heater and the temperature of the gases should not be a cause for explosion of flammable gases in outside environment. The mechanical outfit of the instrument should be provided with shields inside so as to contain the hot gases quite well contained inside only. The entire outfits should conform to internationally approved safety standards.

As in all sensing systems, the electromagnetic compatibility has become a major issue in the gas sensor designs. The sensors should also be immune to the effects of external electric and magnetic fields. In the following sections the techniques used for gas-sensing are presented.

12.7.2   Semiconductor Gas Detectors

The behaviour of semiconductor materials when subjected to physical and chemical stimulii has been opening up opportunities for development of sensors and sensing systems for detection and measurement of flammable gases such as hydrocarbons.

Metal oxide semiconductor materials, particularly, those of transition and heavy metals, such as tin, zinc and nickel have been observed to change in their electrical conductivity when a gas molecule is adsorbed on the semiconductor surface. The process of adsorption involves in the formation of bonds between the gas molecule and the semiconductor by transfer of electrical charge. The charge transfer affects the electronic structure of the semiconductor thereby changing its conductivity. The change in conductivity is related to the number of gas molecules adsorbed on the surface and hence to the concentration of the adsorbed species in the surrounding atmosphere.

Organic semiconductors, e.g. polypyrrole can also be used instead of metal oxides. A typical semiconductor sensor consists of a bead (2 mm diameter) of the material between two small coils of platinum wire; one for raising the temperature of the bead and the other facilitating the measurement of the resistance of the bead. With the rise in temperature of the bead, adsorption of the molecules of the component gas in the gas mixture takes place resulting in a change in the resistance of the bead. The bead is mounted in a stainless steel gauge enclosure so as to ensure that molecules diffuse to the semiconductor surface. The gauge-enclosure brings the molecules to a stand-still state.

Better versions make use of alumina substrate, containing the heater together with the electrodes onto which the metal oxide is deposited.

Thick film processes are also commonly used like those in thick film hybrid technology. The heater may be of 500 mW power.

Micro-machined semiconductor sensor shown in Figure 12.23 requires only 100 mW. These are better suited for detection of flammable gases, such as the LPG domestic gas. Their susceptibility to humidity changes and the sensitivity drift with time renders them unsuitable for measurement.

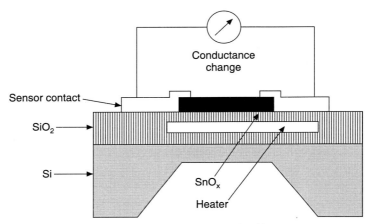

Figure 12.23 Micro-machined metal oxide sensor.

12.7.3 Ion Selective Electrodes

Ion selective electrodes (ISE) are electrodes meant for responding to a wide range of selected ions. The one meant for measuring hydrogen ion activity (pH) is the glass electrode described in Section 7.9.3 and is provided with glass membrane sensitive to hydrogen ions. It has been shown that the composition of glass used for membrane varies with the type of ion species under measurement and glass electrodes selective to sodium, potassium, silver, ammonia and other univalent cations are made. Water purity is tested by pH meters.

In solid state electrodes, the membrane consists of a single crystal or a compacted disc of the active material. The membrane isolates the reference solution from the solution under measurement.

Figure 12.24 shows the version of one type in which the membrane is sealed with a metal backing by means of a solid metal connection. A solid state electrode selective to fluoride ions employs a membrane of lanthanum fluoride (LaF). One which is selective to sulphide ions has

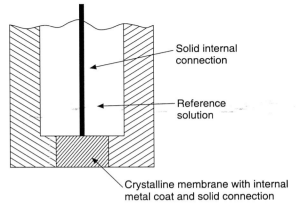

Figure 12.24 Solid state electrode.

a membrane of silver sulphide. There are other electrodes available for measurement of Cl^-, Br^-, I^-, Ag^+, Cu^{2+}, Pb^{2+} and Cd^{2+} ions.

Heterogeneous membrane electrodes are similar to the solid state electrodes but differ in having the active material dispersed in an inert matrix. Electrodes of this type are available for Cl^-, Br^-, I^-, S^{2-} and Ag^+ ions.

Another version, known as liquid-ion exchange electrode shown in Fig. 12.25 has the internal reference solution and the measured solution separated by a porous layer containing an organic liquid of low water solubility. Dissolved in the organic phase are large molecules in which the ions of interest are incorporated. The most important electrode is calcium electrode but others are also available for Cl^-, NO_3^-, Cu^{2+}, Pb^{2+} and BF_4^- ions. These have certain limitations in their application but may be suited to measure ions that cannot yet be measured by means of solid state electrodes.

A different version of gas sensing membrane electrode is shown in Figure 12.26. It makes use of the sensing surface of a flat-ended glass pH electrode which is pressed tightly against

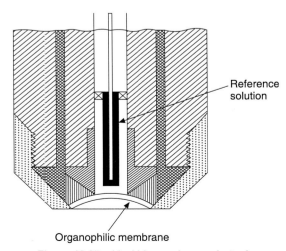

Figure 12.25 Liquid ion exchange electrode.

a hydrophobic polymer membrane. The membrane serves as a seal for the end of a tube containing NH_4 solution. A silver/AgCl electrode is immersed in the bulk solution. The membrane permits the diffusion of free ammonia NH_3 but not ions between the sample solution and the film of ammonium chloride solution (see Figure 12.26). The introduction of free ammonia changes the pH of the internal ammonium chloride solution which is sensed by the internal glass pH electrode.

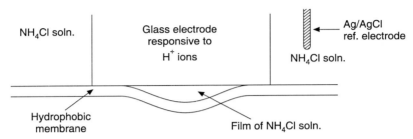

Figure 12.26   Gas sensing membrane electrode.

12.7.4   Conductometric Sensors

Electrical conductivity transducers or cells for mesurement of water purity and concentration of electrolytic solutions are treated in Section 6.1.6.2. When air-pollutants in gaseous form, such as CO_2, SO_2, ammonia and hydrogen sulphide are to be detected and estimated, they are absorbed in suitable electrolytic solutions and the variation in electrical conductance of the solution is taken to represent the amount of pollution. Simple portable instruments are available for such air pollution studies.

If the test gases are in vapour phase, polymer absorption sensors, known as chemiresistors are used for detection and estimation. A chemically sensitive absorbent is chosen and deposited onto a solid phase that acts as an electrode. When vapours are admitted absorption occurs resulting in swelling of the absorbent. The consequent change in the resistance of the absorbent is a function of the concentration of the vapour. A hand-held artificial nose uses an array of such 32 chemiresistors.

A slightly different system, patterned on the basis of a hot-wire pressure transducer described in Section 6.1.2.2, with the difference that one of the filaments is coated with a catalyst and is allowed to come in contact with the combustible gas under test. The other filament is enclosed in a chamber and is maintained at the same temperature (anywhere between 300°C to 800°C) as the other before the test gas is admitted. The test gas after admission into the chamber, goes into combustion, and affects a rise in the filament temperature. The rise in temperature results in the unbalance of the Wheatstone Bridge network. The filaments can be replaced by thermistor beads as well, for efficient and sensitive detection. If the concentration of the test gas in a mixture needs to be estimated, the temperature, pressure and rate of flow into the test chamber require to be controlled at set values.

Photo-ionization and flame-ionization techniques are favoured for detection of many organic gases. They are time-tested techniques and are fast in response. They are very sensitive and can detect 0.1 to 1 ppm of the test gas in a mixture. The dynamic range is also high. They are not sensitive to CO, CO_2 and highly halogenated compounds.

Photo-ionization system uses a high energy discharge lamp (11.7 eV) emitting ultraviolet radiation. The radiation from the lamp is allowed to fall on the sample gas, thereby ionizing the molecules with ionization potentials less than the lamp energy. Electrons are collected on the electrode and an electrometer is used to measure electronic current. The system is contained in a compact outfit along with the necessary signal processing circuitry.

Flame ionization is similar to the above except the source of ionization. A flame produced from a hydrogen–air mixture, ionizes the test gas and the ions generated are detected and measured as current at the collector-electrode. Many organic and inorganic gases and vapours are detected by making use of infrared (IR) radiation in the range of wavelengths from 3–14 μm. When the radiation passes through the sample cell it is absorbed by the test gas. Measurement of the intensity of radiation after absorption is a means of detecting and measuring the concentration of the test gas in the mixture. The system has its own IR source inside the instrument known as IR nondispersive analyser (described in next chapter) useful for detection as well as measurement of the test for its concentration. They are well suited for monitoring hydrocarbon gases and vapours. The system adopts itself for the open-path gas monitoring technique, in which the path up to 150 metres can be monitored in an open area where the toxic gas or vapour is found to exist. For pin-pointing the spot where the gas is located, other techniques making use of optical fibres is employed.

For detection of organic compounds and vapours generated in work places, such as aromatic amines, ethers and ketones, the instrument, ion mobility spectrometer as shown in Figure 12.27 presents very low detection limit of even 0.01 ppm and possesses very fast response. It is selective to certain components in the gas mixture. A radioactive source such as Ni^{63} beta particle emitter is used to ionize the test gas. Ions generated pass into a drift tube where they migrate in air under an electrostatic potential towards the collector. The current is related to the concentration. The instrument can be used as a handheld toxic gas monitor in work places. The ion species can be separated according to their mobility which is a function of their mass charge and ionic state. It can also be used for detection of certain inorganic compounds and some environmental contaminants.

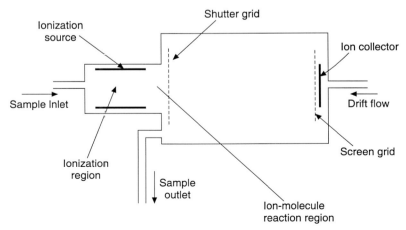

Figure 12.27   Ion-mobility spectrometer sensor.

12.7.5 Mass Sensor

A novel sensor known as mass sensor is developed for real-time detection and measurement of many inorganic and organic compounds. It is very small in size and is available in a chip form with on board processing circuitry and/or along with a microprocessor to constitute the modern smart sensor. It is, in fact, a combination of sensors under the name of surface acoustic wave (SAW) sensor (described in Section 12.5.8). It consists of a piezo electric substrate that serves as a transmitter and receiver of a surface acoustic wave at 100 MHz. The waves pass on the chemically adsorbed film resulting in mass change at the surface of the crystal. The attenuation of the wave is a measure of the gas mass adsorbed by the film. The schematic diagram shown in Figure 12.28 uses a piezopile. The sensor is useful for detection and measurement of organo phosphates chlorinated hydrocarbons, ketones, alcohols and aromatic hydrocarbons. The generated waves can be propagated through the bulk also in which case they are known as BAW devices.

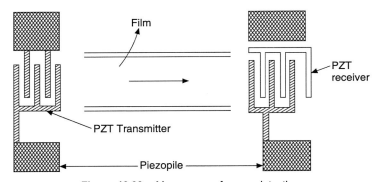

Figure 12.28 Mass-sensor for gas detection.

12.8 BIO SENSORS

12.8.1 Introduction

The latest member of the multi-disciplinary engineering programmes is 'BIOTECHNOLOGY'. This new discipline is defined and accepted as the one dealing with the application of biological systems and organisms to manufacturing processes. Bioprocesses and associated operations are based mostly on biological sciences apart from physical sciences. Investigations utilizing the discipline are directed to three areas:

1. preservation of biological components in food processing
2. biosynthesis of useful products for the pharmaceutical industries, and
3. biodegradation of unwanted materials and waste such as those faced in sewage treatment.

A distinctly different approach is necessitated in the support technology to deal with the above operations.

The role of microorganisms in the production of bread, cheese, beer and wine is known and the experiences have to be translated so as to be applicable for fermentation process and such others when producing antibiotics, amino acids, enzymes and other food products.

Pharmaceutical science utilizes the techniques for detection, isolation, and characterization of the different proteins and other biological products that regulate the biological functions vital to health and human life. Intense research in this direction has been enabling splicing of DNA molecules derived from different sources. Gene manipulation and selection can be used to increase the synthetic capabilities of microorganism to produce large amounts of harmones and proteins. This recombinant DNA technology extends itself a proven path for successes on various fronts right from health of human beings to that of production of vaccines by the drug industries. The sensor technology has assumed a new dimension to enable the above activities, and such sensors are now known as 'biosensors'. Advances in biosensor development, alongwith those in areas of microelectronics, computer technology, fibre optics and nanotechnology have been of great help to the medical sciences. Sensors meant for medical profession are designated as 'biomedical sensors'.

A major activity of modern bio-analytical chemistry is conversion of sample into invasive contacting sensors. All chemical sensors have limited lifetimes due to oxidation. It is difficult to prevent biofilm formation due to bacteria on surfaces exposed to active media such as bacterial or eukaryotic cell culture. Growth of biofilm degrades the performance of sensor.

The sensor performance is also degraded by continuous contact with biological tissues. The life time of a sensor may become shorter, if the risk to the health of the patient increases with time for which the sensor is in place. It is essential to ensure bio-compatibility of the chemical sensor before it is used.

Bio-sensors are becoming increasingly important bioanalytical tools in the pharmaceutical, biotech, food and other consumer-oriented industries. An array of biosensors are available for use making use of the modern technologies referred above. Choice is becoming enlarged with time.

12.8.2　Biosensor Structure

The biosensor is thus seen to be distinct in its design features, in the sense that it uses a combination of a recognition step and a transducer step. The first step involves a biological sensing element called biocatalyst, on the surface so as to recognize the biological or chemical analytes in solution or in its environment. The sensing element may be antibody, enzyme, tissue slice, organelle or a whole cell. This element is immobilised on the surface of a transducer and hence is in close contact with it. The analyte-sensing element reaction is converted by the transducer into a quantitative electrical or optical signal which may be further processed suitably. The transducer needs to be highly specific for the analyte of interest. A membrane is used to protect the transducer and provide the biocompatibility. It is also helpful in the immobilization of the biocatalyst. The analyte-sensing element reaction is evinced either by production or consumption of species (electrons, or specific gases, e.g. O_2, H_2 or NH_3) that can be easily detected by transducer.

12.8.3　Biomedical Sensors

Biomedical sensors are intended to measure: 1. physical quantities (such as pressure, temperature, blood flow, etc.), 2. chemical quantities (such as concentration of chemicals in a volume of gas, tissue, etc.), and 3. biological quantities (such as biomolecules, H^+ ions, bacteria, etc.). The first type is common and relatively simple. The second type is not much necessitated. But it is the third type that demands care and attention in the design and fabrication. To detect and measure concentration of the chemical or bio-species, some other biological molecules, tissues or organisms are borrowed from nature so that this borrowed biological entity interacts with the sample species and enables the measurement. Such a step is essential if the sample species is complex and sensors based on this principle are 'biosensors'. The most important advantage of using biochemicals for sensing is that despite the disadvantage of using chemically unstable (labile) components in a sensor they allow measurement of chemical species that cannot otherwise be sensed. Sensors have been fabricated that incorporate small biochemicals such as antibodies, enzymes and other proteins ion channels, whole bacteria and eukaryotic cells (both dead and alive) and even plant and animal tissues. The difficulty arises in immobilization of the enzyme or other materials. The enzyme solution is trapped between semipermeable membrane and a metal electrode.

The biomedical sensors can be contacting/noncontacting type, or invasive/noninvasive type. Non-contacting type may be electromagnetic or radiation type. All invasive sensors do damage the biological system as known from the use of 1 mm dia pH electrode. The sensors may be consuming/nonconsuming type. In the nonconsuming type, there is no net transfer of matter or energy between sensor and its environment as observed in the use of temperature sensors, ion-selective electrodes or antibody sensors. But in the use of glucose sensor for monitoring blood glucose of diabetics, glucose is destroyed to some extent and measured value may become slightly less. Such effects are not favoured in biometry.

Essentially the biosensor consists of a physical transducer, a membrane and a biocatalyst. The combination of biocatalyst and membrane serves the purpose of receptor of the sample. The biosensor needs to be highly specific to the sample species so as to enable its use in complex media such as blood, serum, urine, food, etc., with minimum sample treatment. It has been possible to estimate a desired protein among a host of them with a discrimination ratio of 10^7 to 10^8.

Another important aspect governing the biosensors is the use of a proper material for the protection of the sensor from the biological environment and also protection of the recipient. Silicon elastomers, polyurethranes, plasma deposited polymeric films, polyimides and a variety of metals such as stainless steel and titanium are in use.

Likewise a large number of transducers developing output signals, electrical or optical in nature, have been applied to constitute the bio-sensors and optical sensors based on surface plasma resonance offer high sensitivity. The modalities of developing the output signals in biosensors, are based on either physical or chemical phenomena. Microminiaturization techniques presented a chemical field-effect transistor. While some chemical sensors using pH and ion selective electrodes are used, mechanical and acoustic sensors, like the surface acoustic wave (SAW) or bulk acoustic wave (BAW) detectors are offering several advantages. Many combinations are possible from among the range of modern physical and chemical sensors.

Where fast determination of the metabolic status of a patient is desired, the coupling of biological assays with sensitive electro-chemical detectors, has provided the effective means for application in biomedical field. Glucose biosensors are already popular for use on the diabetic patients. While a range of other sensors are aiding the industry apart from environmental pollution studies.

A simple arrangement of a biosensor is shown in Figure 12.29 with the location of the transducer.

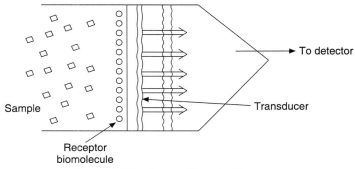

Figure 12.29   Composition of a bio-sensor.

12.8.4   Quartz Crystal Microbalance

Piezoelectric immuno-sensors are the developed biosensors using the quartz crystal microbalance (QCM) as shown in Figure 12.30. It consists of a thin quartz disc with electrode plates on it. As the QCM is piezoelectric, an oscillating electric field applied across the device induces an acoustic wave that propagates through the crystal and meets minimum impedance when the thickness of the device is a multiple of a half wavelength of the acoustic wave. QCM is a shear mode device in which the acoustic wave propagates in a direction perpendicular to crystal surface. Deposition of a thin film on the crystal surface decreases the frequency in proportion to the mass of the film. Changes in the resonant frequency are related to the mass accumulated on the crystal due to the medium adjacent to it.

QCM and surface acoustic wave (SAW) sensors are much preferred for use as they are offering impressive sensitivities. (Ex.: 4×10^{-2} g/cm^2). A SAW sensor with LiNBO$_3$ resulted in a sensitivity of 0.3 ppb when NO$_2$ is monitored.

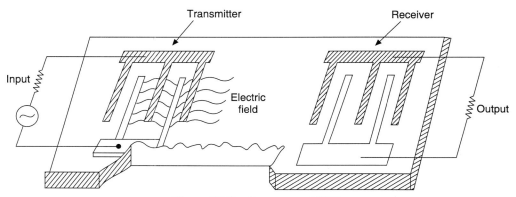

Figure 12.30   Saw sensor (QCM).

Another novel sensor is a magnetic biosensor for DNA detection. The scheme shown in Figure 12.31 comprises of an array of wire like magnetic field microsensors. These sensors are coated with simple stranded DNA probes specific for a gene from a bioagent. Once a strand of bioagent DNA in a sample binds with a probe, the resulting double strand binds a single microbead. When a magnetic bead is present above a sensor, the resistance of the sensor decreases in proportion to the number of microbeads.

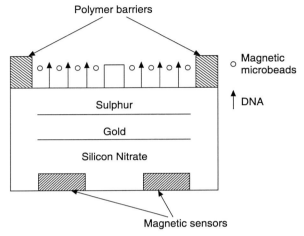

Figure 12.31   A magnetic biosensor.

Some sensors use fluorescent antibodies to bind the bacterial cells. The area of biosensor technology is embracing all aspects of the earlier sensor techniques and at the same time presenting access for data modification, processing and presentation by using computers.

EXERCISES

1. How can a sensor be distinguished from a transducer?

2. Distinguish the processes of sensing and measuring, illustrating the same with relevant example for each.

3. Classify the range of semiconductor materials and indicate the typical characteristics of each or of each group.

4. Discuss the important properties of silicon that makes it a favoured material for sensor development.

5. Explain the phenomenon of piezoelectric effect and indicate the groups of materials exhibiting this effect.

6. Critically examine the versatility and utility of the piezoelectric sensors.

7. What are the additional attributes of a 'smart sensor' and how they are usually configured?

8. What is a 'microsensor' and what are the means employed for production of the same?

9. What is a 'button' microphone and what are the principles of operation? Comment on their sensitivity and fidelity.

10. Describe the technique that enables production of a micromachined accelerometer.

11. Show how the basic 'Hall' element is modified to make up a range of microsensors for sensing different variables.

12. Suggest a few polymer materials and show how they constitute sensor systems for a few variables.

13. Explain how radiation from hot bodies is composed of and how their temperatures can be sensed.

14. Explain how IR radiation is considered most helpful for thermometry while presenting the types of detectors available.

15. Distinguish 'thermal' detectors and 'quantum' detectors and suggest their applications range.

16. Explain in what ways radiation thermometers differ from the commonly used ones and present the merits of the former type.

17. How are mechanical waves produced in air and how are they classified for use in instrumentation?

18. Explain why ultrasonic sensors and sensing systems are gaining popularity for application in measurement.

19. Describe a 'Sonar' and explain its functioning.

20. Indicate the laws governing reflection and transmission of ultrasonic waves when going through material media.

21. Show how liquid flow rate is measured by means of ultrasonic waves when liquid is flowing in (1) open channels and (2) long pipes.

22. What is a 'sing around' flowmeter and explain its operation?

23. What are the problems faced in measurement of fluid flow rates and explain why volume flow rate of fluids is important in many cases.

24. Explain the 'Doppler' flowmeter and indicate the reasons for its popularity.

25. Define 'Cariolis' force and show how it enables measurement of fluid flow rate.

26. What are 'optical fibres' and how they are found useful in instrumentation?

27. Describe the sources and detectors available for use with fibre optic sensors.

28. Show how the fibre optic sensors are constituted for measuring the following variables: (a) level, (b) temperature, (c) strain, (d) humidity, and (e) pH.

29. Discuss the nature of problems encountered while making measurement of physical and chemical quantities.

30. Define 'ion selective electrodes' and show a typical one for use in (1) solutions, and (2) gases.

31. Describe the ion-mobility spectrometer sensor and show how it enables gas detection successfully.

32. Describe the basics of a mass sensor and explain its usefulness in chemical analysis.

33. What is a 'biosensor' and in what respects it is different from the industrial versions?

34. Suggest an example of a biosensor and show how it is constituted for applications.

SUGGESTED FURTHER READING

Armitage, A.F., Radiation Thermometers—An Introduction Theory, Technology and Techniques, *Measures & Controls*, Vol. 28, No. 8, Oct. 1995, p. 238.

Armitage, A.F., The Sensors of the Future, *Measures & Controls*, Vol. 29, No. 2, March 1996, p. 46.

Armitage, A.F., Neural Networks in Measurement and Control, *Measures & Controls*, Vol. 28, No. 7, Sep. 1995.

Asher, R.C., Ultrasonic Sensors for the Process Variables, *Measures & Controls*, Vol. 30, No. 5, June 1997, p. 138.

Atherton Paul, D., Nanometric Precision Mechanisms, *Measures & Controls*, Vol. 31, No. 2, March 1998, p. 35.

Baltes, H., *Enabling Technology for MEMS and Nanodevices,* Vol. 1, Wiley, 2004.

Barry, C., and Brackbenbury, A., Online Gas Analysis, *Measures & Controls*, Vol. 24, No. 8, Oct. 91, p. 231.

Bearman, K.R. and Carter, T.J.N., A New Environmental Monitor for NO_2, Ozone and Aldehyde, *Measures & Controls*, Vol. 34, No. 2, March 2001, p. 40.

Besse, P.A., and Schott, C. and Others, Realized Examples of Microsystems and their Application, *Measures & Controls*, Vol. 33, No. 9, Nov. 2000, p. 261.

Boby, J., and Terralt, S. et. al., Vehicle Weighing in Motion with Fibre-Optic Sensors, *Measures & Controls*, Vol. 26, No. 2, March 1998, p. 35.

Bradshaw, A., Sensors for Mobile Robots, *Measures & Controls*, Vol. 23, No. 2, March 1990.

Cahn, T.M., *Biosensors*, Chapman and Hall, France.

Chetwynd, D.G., *Nanometrology by X-ray Interferometry*, *Measure & Controls*, Vol. 31, No. 2, March 1998, p. 42.

Cobb, J., What is in it For Me—Field Bus, *Measures & Controls*, Vol. 23, No. 1, Feb. 1990, p. 7.

Coulthard, J., and Yan, Y., Ultrasonic Cross-Correlation Flowmeters, *Measures & Controls*, Vol. 26, No. 6, Aug. 1993, p. 164.

Ewyk. R. Van, Flammable Gas Sensors and Sensing, *Measures & Controls*, Vol. 29, No. 1, Feb. 96, p. 13.

Ferbes, A.M., POIC Kersen, A Miniature Silicon Photoacoustic Detector for Gas Monitoring Application, *Measures & Controls*, Vol. 34, No. 2, March 2001, p. 40.

Furio, Cascetta, Experimental Inter Comparison of Medical Thermometers, *Measures & Controls*, Vol. 26, No. 9, Nov. 1995, p. 267.

Gaussorogues, G., *IR Thermography*, Chapman and Hall, France.

Grattan, K.T.V., '*The Season of Light*': Optics in Measurement and Control, *Measures & Controls*, Vol. 34, No. 1, Feb. 01, p. 3.

Grattan, K.T.V., New Developments in Sensors Technology Fibre and Electrooptics, *Measures & Controls*, Vol. 22, No. 6, July/Aug 1989.

Greengrass, S.M., and Cunningham, Endoscopy, *Measures & Controls*, Vol. 26, No. 4, June 1993, p. 109.

Halit, Eren, *Electronic Portable Instruments Design and Application*, CRC Press, 2003.

Hencke, H., The Design and Application of Honey Well's Laser-Trimmed Temperature Sensors, *Measures & Controls*, Vol. 22, No. 8, Oct. 1989, p. 37.

Hon-Lin, Zhu and ZouMin, A New Simple Noninvasive Method for Flow Measurement, *Measures & Controls*, Vol. 32, No. 6, July 1999, p. 178.

Johnston, Jim, Sensors and Instrumentation for Energy Efficiency, *Measures & Controls*, Vol. 29, No. 5, June 1996.

Jones, T.A., Trends in the Development of Gas Sensors, *Measures & Controls*, Vol. 22, No. 6, July/Aug 1989, p. 176.

Kakolaki, J.B., and Henry, R.M., Ultrasonic Flow Measurement Using Two Asymmetric Sing Around Paths, *Measures & Controls*, Vol. 34, No. 10, p. 306.

Kharaz, A. and Jones, B., A Distributed Fibre-Optic System for Humidity Measurement, *Measures & Controls*, Vol. 28, No. 4, May 1995, p. 99.

Krohn, D.A., *Fibre Optic Sensors—Fundamentals and Applications*, Inst. Soc. of America, 1988.

Lawrence, E., Kinsler and Austin R. Frey, *Ultrasonic Flow Meters*, 2nd ed., Wiley Eastern Ltd., New Delhi, 1978.

Lindner, K.P., A Milestone in Field Instrumentation Technology, *Measures & Controls*, Vol. 23, No. 9, Nov. 1990, p. 272.

McKnight, J.A. and Clare, A., Using Ultrasonics to Measure Process Variables, *Measures & Controls*, Vol. 23, No. 7, Sep. 1990, p. 208.

Med Lock, R.S., Recent Advances in Industrial Measurement Techniques, *Measures & Controls*, Vol. 20, No. 4, April 1987, p. 8.

Medlock, R.S., and Furness, R.A., Massflow Measurement—State of the Art Review, *Measures & Controls*, Vol. 23, No. 4, May 1990, p. 100.

Michael Kraft, Micromachined Inertial Sensors—State of the Art and a Look into the Future, *Measures & Controls*, Vol. 33, No. 6, July 2000, p. 164.

Papakostas, T.P., and White N.M., Thick Film Polymer Sensors for Physical Variables, *Measures & Controls*, Vol. 33, No. 4, May 2000, p. 99.

Richard, S., and Cobbold, C., *Transducers for Biomedical Measurements—Principles and Applications*, John Wiley & Sons, 1974.

Togawa, Tat svo, and Tumura, Toshiyo, *Biomedical Transducers and Instruments*, CRC Press, New York (1997).

Trent Wood, An Introduction to the Use of Hall Effect Devices, *Measures & Controls*, Vol. 21, No. 4, May 1988, p. 109.

Twalsh, Peter, Toxic Gas Sensing for the Workshop, *Measures & Controls*, Vol. 29, No. 1, Feb. 96, p. 5.

Van Cam, P.D., Neurons for Computers, *Scientific American*, Sept. 1992.

Whitehouse, D.J., Nanotech Instrumentation, *Measures & Controls*, Vol. 24, No. 2, March 1991, p. 37.

13

Sophistication in Instrumentation

13.0 INTRODUCTION

It is well known that all living species are endowed with the necessary instincts for their survival and their sensory perceptions aid them sufficiently in matters related to the choice of food and water as well as safeguards from the effects of natural disasters such as fires, earthquakes, volcanic eruptions and tsunamis. Present day civilized world has been contributing immensely to the comfortable life of man (probably at the cost of havoc to other species) in many ways such as speed of transportation, electric power and communication systems and of course never-visualized amusement through gadgets such as T.V., cellphone with camera and personal computer. The motto of the twenty first century, being 'speed, safety and security' which are highly contributory to the achievements of man as well as his welfare, has been exciting to the democratic governments and of course, the galaxy of intellectuals and professionals in the fields of basic sciences, engineering and technology. Man's instincts, backed by his intelligence and ingenuity are driving him with inspiration to search for information devices that act as extensions to his sensing capabilities, for fulfillment of the motto. It has become all the more necessary due to the impact of population explosion, heavy industrialization, the atmospheric pollution and probably inadequacies in man's approach to live in balance with nature. The personal computer and the high-speed communication enable him to converse with the world, educate himself, entertain himself while at the same time provide and procure many day-to-day services. But today's needs point out the necessity of providing safety and security from health risks by providing him with accompaniments such as blood-pressure monitor, temperature monitor and such others encased or embedded in his dress itself apart from pocket-size glucose monitor and gas masks in his head-gear to warn him of toxic gases around him. It is to be understood that sensor technology that is at the root of the present day civilization and the sophistication in instrumentation consisting of the composite of smart sensors and microcomputers is becoming the saviour of life, apart from all other services provided to the man from day-to-day work schedules to that of intensive research in scientific laboratories.

Miniaturization and microminiaturization of sensors through the use of modern materials and the modern manufacturing processes, has extended considereable opportunity for the design of sophisticated instrumentation. In the following sections, a few typical situations have been briefly presented to enable the reader understand the techniques that have been incorporated in the instrumentation systems, to process the signals obtained from an array of sensors to the extent that the data displayed is in the suitable form for the operator. Sophisticated instrumentation provides the scientist, engineer or operator, with data obtained after a test, analysis of a product, scanning of a patient, or condition monitoring of machinery, road and rail bridges and over huge mechanical structures. The data obtained may be helpful for identification of a fault, prevention of an accident or adoption of a treatment procedure or corrective action. In fact, there is a danger of one kind or other, at every nook and corner, right from the residential house to the large industry. Safety and security checks are desired in a variety of ways and sensors have to be incorporated at crucial positions. Checking the high speed of motor cars on highways is as much desired as the checking of an intruding thief into the house. Infra red sensors using unseen beams of infra-red radiation help the detecting agencies. Pollution monitoring of cities due to exhaust fumes of motor cars is becoming essential as much as the environment around the chemical factories and nuclear power plants. Quality of life comes to the forefront in future and assurance of quality of essential foods and water is an obsession of everyone.

Control instrumentation as desired by industries is different in certain respects. Control instrumentation system is part of the entire control scheme, set up with the objective of production at the desired rate while conforming to the specifications framed for the end-product.

13.1 THERMOMETRY AND THERMOGRAPHY

13.1.1 Basics

Radiation thermometers are designed to measure the temperature of an object without any physical contact. It has been explained in Section 12.4.1 that all objects with a temperature lying anywhere between 0 K to 650°C emit electromagnetic energy in the infrared region at wavelengths ranging from 0.72 μm to about 1000 μm. The total energy emitted within this spectrum of wavelength increases with temperature but with the striking difference that the peak of the emitted energy distribution curve shifts towards shorter wavelengths as the temperature rises, as shown in Figure 6.8. This pattern of variation necessitates selection of proper detectors having higher sensitivity at longer wavelengths for measurement of low temperatures. The wavelength at which the peak is observed depends on the temperature and is given by $2891/T$ where T is the temperature expressed in absolute scale. At a temperature of 1000°C, 90% of the radiant energy is contained in the wavelength region from 1.5 μm to 10 μm; while at 100°C, from 5 μm to 25 μm; at 0°C from 7 μm to 45 μm and at –100°C from 11 μm to 70 μm. Thermal detectors are invariably chosen to enable measurement of temperature of any of these bands of wavelength as they are unique by their property of uniform response for all the wavelengths of

IR region. Yet, all radiation thermometers are designed suitably for each of the above ranges, so as to obtain optimum response from each band of wavelengths. For measurement of high temperatures, a photo conductive cell is used in place of a thermal type, as this type of photon detector is highly sensitive in the region of near IR where the peak of energy occurs. To capture maximum energy, all sensors are blackened with sizes reduced to minimum for fast response.

Even though the technique seems to be straightforward for measurement of temperature of remotely located objects, the following difficulties need to be given attention in the design of the radiation thermometers.

1. It is essential to effect calibration against a black-body reference, for each application situation.

2. The need to focus the radiant energy from the body under measurement, to a point, through the application of filters, lenses and mirrors, which may not be working without any loss in the energy.

3. Objects under consideration, vary in their size, shape, smoothness of surface and the material of the object, thereby having differences in their sensitivities for reception.

4. The path of the radiant beam cannot always be air. Water vapour and other gases such as CO_2 in the path may result in loss of energy, while dust, smoke or other vapours related to the object or its (environment) may also be present in the path.

5. The proximity of other objects at varying temperatures, to the object under observation causes error as their radiation interfares with the measurement.

6. The object may not always be a stationary one and if it is moving, the time duration of exposure needs to be considered.

7. Attainment of same accuracy in measurement is difficult even if the temperature is same.

8. The same object does not have the same emissivity as it varies with its temperature and wavelength of the radiation.

13.1.2 Radiation Thermometers

In spite of all the above difficulties, radiation thermometry is unique in its capability and flexibility extended to remote measurement of temperature and also because there is no other equally effective technique around the corner. Corrective measures to surmount some of the above problems are incorporated in the instruments. To ensure better accuracy in measurement, frequent calibration against a black-body is considered essential. A convenient black-body reference for on-the-spot calibration, is a simple blackened cavity of about 15° cone angle. It is usually kept in a temperature controlled oven with a window for access and maintained at temperatures required for calibration. Wherever possible, it is kept in the same location as that of the test body and calibration of the thermometer is effected by comparison with a platinum resistance thermometer.

The lenses and mirrors used in the instrument for focusing purposes entail in loss of some energy on the way while mirrors used do not pose a major problem, lenses do because of the materials they are made of. Infrared rays are governed by the same laws of reflection and refraction as those of visible radiation. The lens materials do have different transmission characteristics for different wavelengths. Glass materials used for visible region are opaque to infrared wavelengths while some like arsenic trisulphide is opaque to visible region and transparent to infra-red wavelengths up to 10 μm. Thallium bromide is transparent up to a maximum λ of 30 μm and like zinc sulphide up to 15 μm, both being favoured for low-temperature thermometry. Fused silica and sapphire are transparent to infrared λs up to 6 μm but crown glass is useful for a narrow band of frequencies around 0.75 μm (see Table 12.4).

The difficulty posed by the presence of CO_2, O_2 and water vapour in the path of the IR beam from the test object, is overcome in some radiation thermometers, by sensing the temperature of the object, at wavelengths or wavelength bands in which the absorption of radiation by those interfering gases in the path are minimum. There remains only one major problem and that is due to the dependence of emissivity of an object on λ and its temperature. It is normally higher at shorter λs than at longer λs. Using the basic relations governing the dependence of measured temperatures on emissivity changes, computations quantifying the error in °C for a 1% change in emissivity at different values of λ and temperature were carried out during the development period of radiation thermometry and they are reproduced in Table 13.1 for a better understanding. The error in measurement of temperature is lower at lower values of λ and at lower temperatures. A study of the error pattern reveals the necessity of minimizing the error and bringing it into acceptable limits whatever be the temperature of the object under measurement. The standard technique of holding all variables constant and studying the effect on the output due to the one variable of interest is done at two values close to each other. In this case, there is variability in the influence of the intervening medium between object and detector and also the one due to nature of the object and its proximity. The hot body under test and its temperature being kept the same, and the path of the beam made clean and constant, the output signals for radiant energy at two wavelengths λ_1 and λ_2 are measured. The closeness is so chosen as to permit the responsivity of the detector to be the same at λ_1 and λ_2. Then the ratio of the energies at λ_1 and λ_2 and thus the ratio of output signals becomes a function of the temperature of the object only. The ratio is calibrated against temperature and such thermometers are known as ratio-thermometers or 'two-colour' thermometers. Thermal detectors are used in infra-red thermometers where λ_1 and λ_2 are obtained by use of filters in the path of the beam. Two colour thermometers use photon detectors having selective response to λ_1 and λ_2. The independence of the ratio from the effects of emissivity of the test body, shows up as an advantage due to the cancellation of the effect of environment on the ratio of the output signals.

Table 13.1 Temperature effects due to 1% change in emissivity

Target Temp (°C)	*Effective wavelength in* μm						
	0.65	0.9	1.6	2.3	3.4	5.0	10.6
100	0.06	0.08	0.15	0.22	0.33	0.49	1.00
500	0.27	0.37	0.68	0.96	1.40	2.10	3.60
1000	0.74	1.00	1.80	2.60	3.70	5.10	7.80
1600	1.60	2.20	4.00	5.50	7.50	9.60	13.00

13.1.3　Radiation Pyrometers

Realizing the importance of remote measurement of temperature of hot bodies, and with no other effective means available, radiation thermometers (also known as pyrometers) of various descriptions, are designed and made available in the market. The nomenclature signified, at times, the region of radiation used, with total radiation pyrometers, implying the application of both the visible and infra-red regions. With the use of thermal detectors, the instrument is known as thermal or broadband radiation thermometer, whereas the use of photon detectors make them known as restricted (or selective) band radiation thermometers.

Infrared thermometry has become a powerful and popular technique due to the developments in sensor technology. Nanotechnology has presented microsensors for use as thermal detectors, having the desired sensitivity and speed of response. Infrared thermography, an extension of infrared thermometry, has become a powerful condition-monitoring technique in industry and medical diagnostics.

Infrared (IR) thermometers vary in their design depending on the application-needs. Dealing with stationary objects is a relatively simpler affair as the choice of the detector rests on its suitability for the temperature of the object. It is a common practice to provide, internal or external to the instrument, a black body reference source for frequent calibration. The output signals, are processed by electronic circuitry for display after affecting compensation inside the instrument for interfering inputs such as, ambient temperature.

It is the needs of the industry that have to be taken care of, whether it involves condition monitoring or process control. It is invariably, a hot body in motion, such as the ingot of steel or a glass sheet under production. The nature of gases, vapours emanating from the hot object as well as the smoke, dust and flake formation on the object, have to be considered, so as to ensure reasonable accuracy apart from the speed desired in the measurement.

The most accurate one is the disappearing filament type optical pyrometer for measurement of temperature (above 700°C) of stationary objects. The human eye judges the equality of brightness of a hot filament and the image of the hot object. The current used to heat the filament is indicative of the temperature.

All IR instruments necessitate the use of lenses, mirrors and filters for proper measurement. The techniques of dealing with the IR beam, like filtering and chopping the beam vary from one instrument to another and descriptions of such instruments may be found in standard literature. Aperture adjustments may be provided in certain instruments to deal with large-sized objects at a distance. In the context of measuring temperature in the range of 600°C to 3000°C, by means of a silicon sensor sensitive to λ of 0.7 to 1 μm, the diameter of the test object can be up to 1% of its distance from the sensor. Likewise, when a thermopile is used for temperatures low enough ranging from −50°C to +100°C, it needs to have objects with diameters of 2.5% of the distance. Specifications concerning the range of the instrument and other provisions vary from one to another. The detector and associated optics are usually enclosed in a proper outfit, sometimes temperature controlled.

At times, difficulty is posed because of the object under study cannot come in the line of sight of the instrument. Hence the thermometer is provided with a fibre optic light-guide instead of a lens. The inaccessibility of the hot body has never been a problem with infra-red thermometry. Electronic signals processing, being known for what it can do, in the present day, it is the array of micro sensors, the variety of lens materials in association with computational techniques by means of computers, that has rendered scanning applicable in industry and commercial production of infra-red cameras. In manufacturing industries, a variety of situations demand the remote measurement of temperature of bodies, stationary or moving, of different sizes and properties, having temperatures in the range of –100°C to 3500°C. Where materials like plastics, polymers, polyesters, ceramics, paints and waxes are involved, the temperatures encountered are low and object sizes are small and of low thermal capacity. They may be thin strips or films. While on the other hand, glass and steel industry requires scanning the surface of the materials in sheet or slab form, for ensuring the uniformity of temperature at several points along a line or over the entire surface. Table 13.2 summarizes some of the detectors and their temperature ranges along with the lens materials used for the wavelengths, suited to the application.

Table 13.2 Temperature sensors—Characteristics

Sensor	Wavelength (in μm)	Temperature range (°C)	Response time (ms)
Silicon	0.7 to 1.1	400–4000	10
Lead sulphide	2.0 to 2.6	100–1400	10
Pyroelectric	3.4 to 14	0–1500	100
Thermopile	1 to 14	–150–500	100

13.1.4 Infrared Thermography

Infrared thermography, sometimes known as thermal imaging is based on the techniques used in radiation thermometry. Infrared thermograph also known as thermogram is a photograph showing the surface temperature of a scene at several points and it is facilitated by the equipment or instrument called Infrared (IR) camera. The thermograph enables the identification of points or regions of the body or even a machine, with increased or reduced heat emission due to a fault in the body or malfunction of a part of the body. The camera provides the thermograph constructed from the output signals from the array of detectors used for scanning the temperatures of a large number of points along a line of the surface of the body. It is akin to the operation of a closed circuit television system with the difference that the operation is not continuous. Figure 13.1(a) shows the outfit of an IR camera with an array of detectors in a square matrix and provision for cooling the detectors and Figure 13.1(b), a thermogram taken by the camera. IR cameras have normally fewer pixels than a typical video camera as the resolution requirements are less. The microbolometer, of the array is the thermal detector which is very fast in response because of its extremely small size (made possible by nanotechnology micromachining technique). Figure 13.2 shows the layout of a microbolometer array. Vanadium oxide pixels are connected to the processing circuitry by means of metal tracks, all on a silicon substrate with integrated circuit as well for processing the output signals of the array. The pixels are thermally insulated from the substrate by thin support legs keeping the bolometer slightly above the substrate.

Photon detectors may also be used provided their sizes are small enough to constitute the array. Well defined images are obtained when a large no. of detectors are used in the array.

Figure 13.1(a)   An infrared camera sealed to IP65 for use in hostile environments
(photograph courtesy of Land Infrared)

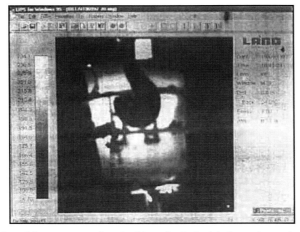

Figure 13.1(b)   Infrared thermogram of a steel transfer ladle (photograph courtesy of Land Infrared).

Unusual and unwanted temperatures of the bodies, if detected in time, helps prevention of failures, fires and disasters. With the top priority accorded to energy conservation measures everywhere, IR thermography presents the most powerful technique for inspection of systems whether they be buildings, machinery, reactors or vehicles in motion. An important condition monitoring tool of the present day is the IR camera.

Likewise, IR thermography is useful in medical diagnostics. A number of circulatory disorders and inflammations are diagnosed by inspection of a thermogram.

Other important applications include missile guidance in which the hot jet exhaust of an aircraft engine serves as the hot body for tracking and guiding; also the satellite attitude sensing system making use of earth or another planet as the hot body for reference.

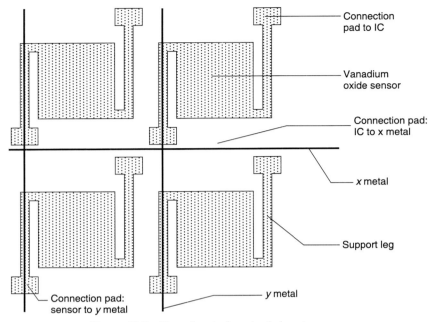

Figure 13.2   Plan of part of a microbolometer array.

13.1.5   Thermal Flowmeter

An interesting application of heat transfer principles is found in a simple noninvasive thermal flowmeter where temperature measurements enable flow-rate determination. The metallic pipe through which the liquid is flowing is heated by means of a clamp-on heater located on the outer surface of the pipe. Two matched thermal detectors of the platinum-film thermistors (2 mm wide, 0.3 mm thick and 1 cm length) are glued on the outer surface of the pipe, symmetrically placed on either side of the heater as shown in Figure 13.3. With no flow, their resistance values

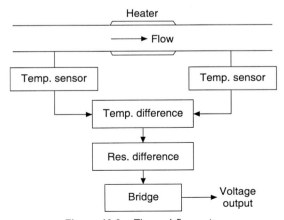

Figure 13.3   Thermal flowmeter.

are equal. With flow, the temperature difference of the pipe wall brings in resistance changes that are measured by means of a simple bridge network. For each case of the pipe and the liquid-nature, calibration is essential. Even though the detecting system is stable and accurate in measuring temperature, the small variations, if any, in the density, viscosity, specific heat and thermal conductivity may cause errors in the final measurement. Two-phase (liquid/solid) fluids can as well as measured, if of constant composition, and with no deposits being formed on the inside surface of pipe. The system can be used for gas-flow rate measurements also.

13.1.6   Thermal Mass-flow Meters

Mass-flow meters are important when mass and energy balances are required in industrial processes. Fluids are generally sold industrially on a weight rather than a volume basis and energy conversions are best measured in mass units because of the constancy mass has in spite of variations in composition, temperature and density.

The thermal mass-flow meters of the following types are mostly meant for gas flows, some of them can be extended to liquid flows also. But the types are all invasive type and it is necessary to ensure that the effects of heating directly or indirectly on the flowing medium are trivial. Although they do respond to the mass-flow rate, most are sensitive to the variations in thermal conductivity, specific heat and density of the flowing fluid. They are usually meant for gases flowing at low rates of mass flow.

The principles of operation are the same as those presented in Section 6.1.4 where hot-wire and hot-film anemometers of the constant-current and constant-temperature types are discussed. The sensor in these types can be a heated wire, a thermistor, a thin film, a resistance thermometer or a thermocouple. The response of the heated wire which is inserted into the flowing medium is nonlinear being typically of the form:

$$Q_m = K \left[\frac{H}{\Delta T} \right]^{1.67} \tag{13.1}$$

where Q_m is the mass flow rate

   H, the rate of heat loss

   ΔT, the temperature difference between the fluid and sensor and K, a constant.

A second unheated sensor is used to measure the fluid stream temperature, thus determining ΔT. Figure 13.4 shows the systems usually employed and they can be based on the constant power (current) type or the constant temperature type (ΔT maintained constant). The hot-film resistor if used, is a deposit on a ceramic base and is provided on a stainless steel body for protection purposes.

A slightly different technique shown in Figure 13.5 uses a constant power source to heat the entire stream. The temperature difference ΔT, between the upstream and downstream sensors is measured and Q_m is calculated from $H/C_p \Delta T$ where C_p is the specific heat at constant pressure.

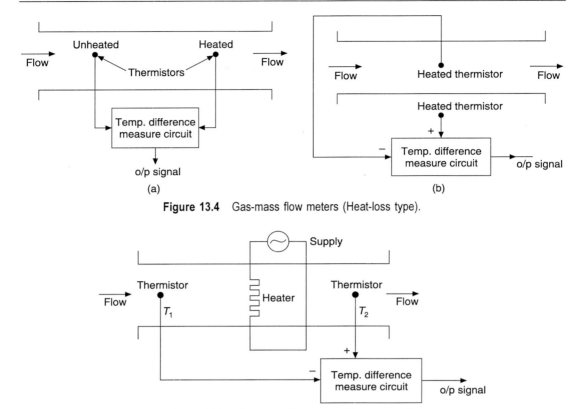

Figure 13.4 Gas-mass flow meters (Heat-loss type).

Figure 13.5 Gas-mass flow meter (Thomas type).

13.2 NANO INSTRUMENTATION

13.2.1 Introduction

It is the realization of man that 'Nature is his best teacher' and that 'the small is not only beautiful but powerful', that has enthused him to focus his attention for the study and exploitation of the power encased in the living microorganisms and tiniest particles of matter. One of the successful accomplishments is the development of computers with higher and higher speeds of operation, by putting together a large number of micron-size components on a single chip. Biological science has enabled him understand natures' biological processes, better than ever, and other secrets too governing the biomaterials. Inspired by the fact that the smaller the system, the faster its performance, considerable research is undertaken to utilize suitable transduction techniques and structure them as new sensing systems, much smaller in size and being called 'microsensors'. Traditional precision engineeering techniques could not serve the objective of production of these micro-sensors. Successful integration of mechanical components and electronic devices is covered in the subject of 'Mechatronics'. But due to the problems faced in the integration of a large number of elements of extremely small size, some mechanical and some electronic in nature, new methods of fabrication and manufacture are necessitated. The activities pertaining to the production of these and other microsystems constitute the subject of 'Nanotechnology'. The term nano is known to mean 10^{-9} and all those systems consisting of components of sizes from 1 μm to 1 nm, and fabricated by the new technology are termed as microelectromechanical systems (MEMs). Several sophisticated instrumentation systems are necessary for checking dimensions of components during manufacture and later calibration and product quality checking. Nanotechnology relies heavily on the effective measurements of the surface properties of components, during and after production, by means of highly sophisticated instruments. The fabrication and manufacturing techniques of the microcomponents and MEMs are different and equally sophisticated. The range of measurements is around 1 μm and the resolution desired is around 0.1 nm or in other words at atomic scale. Investigation of the surface properties demand different techniques as they depend on the material, its state, the on-going reactions if any on the surface and the presence of ions and molecules chemical or biological in character. The measurements are intended to establish the topographic features, shapes, edge sharpness, detection of defects, movement of atoms and structural characterization. Selection and application of instruments require thorough understanding of the processes. Nanotechnology is treated as the 'ultimate', and is termed as 'enabling technology' implying that all the facets of science and technology lend to their application in an interdependent fashion. A brief introduction of the types of instrumentation systems, commonly known as 'Nano Instruments', is justified to draw the attention of the reader to the elaborate assembly in the instrument while appraising him of the inter-disciplinary character of the same. It is one technology in which the exceptional role of instrumentation systems is unique, leading to the successful realization of man's dreams. The instruments may be grouped as 1, stylus type 2, optical type and 3 scanning electron-beam type.

The traditional precision engineering techniques and the instruments aiding the operations are not effective in nanotechnology and hence require suitable modification to meet the exacting demands. Precision positioning and scanning systems with higher resolution and repeatability are essential. It is the materials such as silicon and procedures and processes to make silicon chips with line widths below 200 nm, that demand new approach in the design of systems for use. The ultra-precise positioning systems make use of the piezoelectric transducers and capacitive sensors that are known to possess near-ideal or infinite resolution apart from other worthy characteristics. The positioning mechanism consists of several stages, known as nano-mechanisms using the above devices along with flexures for guidance and positioning with subnanometric accuracy. They are designed to work either as single axis or *x-y* stages and also to provide even six degrees of freedom, in certain cases. The positioning system is a closed-loop control scheme, wherein the precision attainable is totally decided by the capabilities of the sensor which is invariably the capacitive type. Hence exceptional care is taken in the choice of the materials used for the plates. Thermal gradients can cause structural distortions which are minimized by the use of Super Invar and Zero-dur, known to be of very low thermal expansion coefficient, for the plates of the sensor and other components of the system. The interference effects of temperature and some parasytic motions need to be suppressed as far as possible, in design of such nanometre positioning mechanisms.

13.2.2 Nano Stylus Instruments

Conventional stylus methods involving contact with the surface under investigation using a sharp diamond stylus are seen to result in slight surface damage. But it has become possible to produce stylus with tips of size as small as that of an atom, for use of nanostylus instruments and the techniques such as ion beam milling has enabled such tip production. The instrument primarily consists of a sensor and a scanning system, aided by a miniature interferometer for attaining the desired high resolution (see Figure 13.6). However, newer designs, employing components of zero thermal expansion, have come up for application, with better performance in all respects.

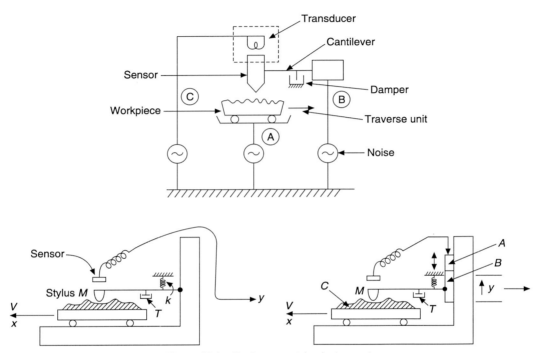

Figure 13.6 Basic nano stylus instrument.

13.2.3 Optical Instruments

Optical instruments using light or laser beams, are known to counter the effects of mechanical type instruments and at the same time offer higher resolutions. They offer the considerable advantage of being non-contacting type and hence superior to stylus-type. The optical types have some differences in their basic methods of beam generation and application. The methods are based on either focused systems or diffraction, interference and gloss techniques. Detailed descriptions are not attempted here. One parameter usually quoted in these types, is the numerical aperture (N.A.). Better profiling of surfaces is possible with highly focused systems yielding large N.A. But lower N.A. systems, are faster and less accurate. Another important factor in nano-instrumentation, is the ratio of range to resolution and this factor is limited in the case of optical instruments. A laser scanning system is shown in Figure 13.7 providing constant numerical aperture. An optical instrument based on interference method is shown in Figure 13.8 to indicate the complexity of the system and versatility offered for use along with computers as part of a sophisticated position control scheme.

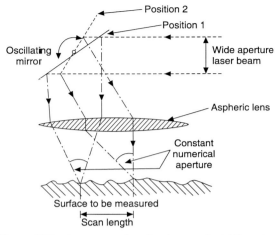

Figure 13.7 Laser scanning showing constant NA system.

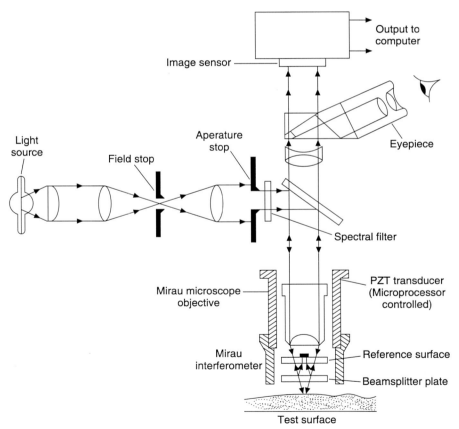

Figure 13.8 Interference methods.

13.2.4 Scanning Electron Beam Type Instrument

The scanning electron microscope is totally an electronic instrument in which an electron beam is focused onto the surface of the specimen, instead of an optical beam (see Figure 13.9). The sharp beam is incident on the test surface at a point wherefrom the secondary emission takes

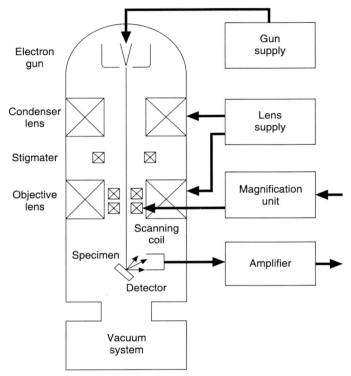

Figure 13.9 M block diagram of a scanning electron beam microscope.

place as shown in figure. The emitted electrons strike the collector and the resulting current is amplified and used to modulate the brightness of the cathode ray tube (CRT) screen spot. The brightness corresponds to the number of secondary electrons collected from a point on the surface. Scanning coils placed before the final lens cause the electron spot to be scanned across the specimen surface in the form of a square raster, similar to that on a television screen. The current passing through the scanning coils is made to pass through the deflecting coils of the CRT so as to produce a larger raster on the viewing screen in a synchronous fashion. Thus an image of the surface is progressively built up on the screen. The image magnification is solely dependent on the ratio of the sizes of the rasters on the CRT screen and on the specimen surface. Reduction of the currents in the scanning coils increases the magnification. The instrument yields high resolution in the lateral direction but not sufficiently in the vertical direction. This drawback is overcome in the subsequent versions of micrscope. However, the instrument is favoured for visualization of topographic features of surfaces, edges and slopes look brighter than reality.

The newer designs, being known as scanning tunnelling microscope (STM) and the atomic force microscope (AFM) enable scanning to the desired atomic scale, while having compatible resolutions in the vertical and lateral directions. For details of these instruments, the reader is referred to consult references provided.

13.3 CONDITION MONITORING

13.3.1 Introduction

An industrial establishment is mainly characterized as an investment on machinery and materials, housed in suitable layouts. Machinery may be for generating electric power or production of goods using electrical energy. Materials may be stored in open or closed tanks or stocked in store houses. All rotating machines develop heat while in operation and radiate heat to their surroundings and incidentally emit a lot of noise due to friction, wearout of contacts or minor faults. Undue and unexpected rise in temperatures of machinery or, in fact, the store houses of stock, leads to fire accidents and loss of property. Diagnosis of sources of such dangers is essential and diagnostic techniques to detect and raise alarm, whenever a deviation is noticed in the normal conditions of their existence or operation, are highly desirable. The source of danger may be a small device, a spot or an area of a stock-yard. Any device used to warn the onset of fire-breakout, is required to monitor the rise of temperature continuously with time (real-time) of all the machinery of the industry and, in fact, all building complexes, residential or shopping. Condition monitoring of any system, in fact, is necessitated from considerations of safety, security of personnel and property apart from the desirability of health-care measures in medical hospitals. Because of the vast differences in the systems under diagnosis, condition monitoring of industrial systems is known as industrial diagnostics and that of patients as Medical diagnostics.

Industrial diagnostics is an area of science and engineering, the wider area being known as 'Nondestructive Testing' (NDT). The philosophy of approach to this branch of testing or monitoring is based on the measurement of the condition without affecting or altering the properties, conditions or functioning of the system under study, i.e., noninvasively without any interaction, interference, destruction, or consumption of material pertaining to the system.

13.3.2 Vibration Monitoring

All structures and fixtures vibrate during motion and rotation and the intensity of the vibrations is a cause of failure in most cases. So for a long time, monitoring the amplitude and frequency of the vibrations is considered relevant and satisfactory. But mechanical vibrations of bodies and structures, can create mechanical waves i.e., pressure waves in the medium of air surrounding them. The pressure waves set in are sound waves, with frequency ranging from very low to very high frequencies, i.e. subsonic, sonic (audible), ultrasonic or supersonic when classified. Sensing sound waves, may be treated as sensing vibrations without establishing a physical contact with the body under vibration. Monitoring the vibrations of the body of an aircraft or spacecraft or its parts, require the use of accelerometers necessitating the physical contact between them. Seismic instruments assess the vibrations of the regions of earth during an earthquake, but it is observed that birds and animals sense the onset of earthquake, much better due to their sensitivity to frequencies of sound waves beyond the audible range. Bats sense the ultrasound in air while ants and snakes sense the vibrations through their contact with the ground. Accelerometers are used to monitor the vibrations of the aircraft floor or some parts, to enable measurement of the absolute amplitudes of vibrations while remaining in contact with the floor. But contacting sensors for the condition monitoring of machinery, are not popular, in the context of other non-contacting and noninvasive techniques, having been proved versatile and powerful in many ways. Presently, the sound waves emitted by the machine (known as acoustic emission) and the radient energy of the bodies with rise in temperature from the normal one, due to some fault or malfunction, provide the means of monitoring the machinery. The technique that provides the image of the body, with its temperature distribution profile of its surface, is known as 'thermography' as explained in Section 13.1.4 and the process is akin to that of a video-camera.

13.3.3   IR Camera

There are several situations in industries where infra-red monitoring offers excellent service in preventing failure of operations. Rotating electrical machines are known to fail due to rotor body defects, rotor winding faults, water coolant faults, winding faults and winding insulation defects. Defective bearings, rotor unbalance and misalignments of components cause failures, and in most cases result in developing heat at some spot or in some area of the machines. Spot radiometers are excellent instruments for remote measurement of possible hot spots only. With infra-red camera a large area can be covered with the additional advantage of recording. Infra-red cameras have fewer pixels than typical video camera and hence there is a likelihood of missing a hot spot in examination. If inspection is carried out by going round installations, then the camera operator needs to understand the problem sites and adjust the camera settings accordingly as not to miss hot spot. Foggy, misty or smoky atmosphere may effect quality of detection.

Fixed location of cameras help continuous monitoring with time over 24 hours and the scanned images are subjected to identification of bright spots over a background. Presence of a bright spot leads to triggering alarms when the signals are processed and fed to programmable logic controllers. For example, the infra-red imagers can be mounted on pan and tilt heads at the top of pylons, to survey large stock yards.

The use of infra-red thermography to monitor the quality of building insulation is widespread. Building Management Systems take care of buildings through the use of imagers for locating areas with the rise of temperature due to failure of utilities of the building.

13.3.4   Acoustic Emission Technique

The discovery regarding the behaviour of a wide range of materials when stressed provided the ground for an effective condition-monitoring tool for detection of defects in mechanical structures. The phenomenon of producing sounds over a wide range of frequencies, by materials of structures, when they are mechanically stressed, is known as Acoustic Emission (AE). Today, the AE-sensors along with associated electronic data processing resulted in development of ready-to-use instruments for structural integrity assessment. The AE emission from machinery is due to processes such as friction, vibration, impacts, turbulence and cavitation. The signals at the high detection frequencies possess much higher energy than the signals due to any defect growth in the system (categorized as background noise).

The application of AE techniques has been firmly established as an important condition monitoring technique, for industrial machinery. The AE technique is treated as 'Passive' as it deals with listening to and recording the sounds generated by a process or a machine and then using frequency analysis as the tool for identification of the abnormal conditions or an event. 'Active' technique for condition monitoring necessitates the application of sound waves, but of a range of frequencies well above the audible range (ultrasonic frequencies), obtained from ultrasound generators (piezoelectric type transmitters). Active techniques are equally preferred for use in certain situations, such as detection and location of defects in structures and assessment of corrosion effects in pipelines and containers. Active systems are commonly known as 'ultrasonic condition-monitoring' technique.

The technology of AE has become an established NDT technique and it is extremely suited for condition monitoring of all sizes of machinery: gear boxes of different configurations, and bearings heavily/lightly loaded, all being important in the continuous operation of an industry.

All AE activity originates from one or more localized source positions as a momentary transient. The nature of the source processes which are present and their relative participation decides the nature of the AE signal. As a result, the AE signals can take the form of discrete transients, quasi-continuous or a mixture of the two. The amplitudes span over a large dynamic range, generally high at lower frequencies. Utilization of data at lower frequencies is made difficult due to some proximate sources, by their contribution and corruption of the desired signals. This difficulty is overcome by detection at high frequency by using highly sensitive sensors for sensing the low-amplitudes. Narrow-band detection is preferred for use along with the choice of piezoelectric sensor of the ceramic type that is resonant at a frequency in the range of 90–100 kHz. The signal-to-noise ratio (SNR) is improved by positioning the sensor so that it is physically close to AE source and far away from any background noise source. The merit of AE technique lies in its inherent high SNR compared to the simple vibration-based monitoring technique. It is the frequency based approach in the vibration monitoring system that leads to identification of defect whereas it is the SNR with AE technique that enables easy detection of the abnormal conditions in the machinery.

The extension of this technique for process monitoring is riddled with a large number of problems such as the specific nature of the information that each application needs. The simplicity of the technique as well as the data processing method, makes the instruments get located permanently for continuous monitoring of essential machinery.

13.3.5 Ultrasonic Scanning Technique

For monitoring the effect of corrosion on the pipelines, a single transducer coupled to the outside surface by means of a couplant will not be sufficient. A flexible transducer array is developed so as to be strapped around the pipe, like a belt, consisting of thin strips of a film made from either Poly Vinylidine Fluoride (PVDF) or Copolymer, the film thickness being only between 50–100 μm. Copolymer is applicable for temperatures up to 145°C whereas PVDF is limited for use below 70°C. Both the materials are piezoelectric in nature and the thin film is manufactured with the provision of metallic conducting electrode layers on either side, the electrode material being either silver ink or nickel-copper alloy. A suitable size of the strip needs to be divided into small transducer elements of 12–15 mm square with each one of them provided with a conducting link-track to a distribution box at the end of the strip. The number of elements and the length of the strip are determined by the size of the pipe. At the end of the strip, a temperature sensor may be added to enable correcting the data due to temperature variations. The surface of the pipe is cleaned and polished to accept a thin coating of the couplant and thereafter the transducer elements. The strip is provided with a double-sided adhesive tape by use of which the transducer strip is firmly pressed on to the prepared surface from one end to the other. The elements are used like a single transducer, serving both as a transmitter and receiver. They are applied with ultrasound pulses of a few µs duration and the functioning is as per the common Pulse Echo technique. The released pulse travels through the pipe wall and is reflected at the back surface or earlier if it encounters an object in its path. The trace that shows the rectified envelope of the signal against time base is known as 'A scan'. It shows the recovery from the excitation pulse followed by multiple echoes from the back of the pipe. A single-chip microprocessor is programmed to enable each transducer element to be automatically connected in turn, via the coaxial connecting cable to a computer that can compute wall-thicknesses automatically without an operator. The set-up may be extensive enough to permit or deal with similar data from other transducer, located suitably along the pipe line, and provision to download wall thickness and transducer data via a field bus interface to a remote PC. Such flexible array transducers are being used on oil/gas extraction industries.

13.4 BIOMEDICAL INSTRUMENTATION

13.4.1 Introduction

Lately, one has been witnessing the escalating efforts of all agencies towards healthcare activities through an array of sensing and scanning systems. The gains in the application potential of sensor technology have targeted the human body for monitoring and controlling of its functions, after reaping sufficient successes in the improvement of human habitation, environment and the industrial scene. The complexity of the functioning of the human body, though small in structure, with no major differences from one to another, has been baffling to the medical scientists and associated professionals, probably due to the fact that it is not a creation of his skills. The body is seen to possess a good number of self-regulating mechanisms between the so-called energy sources, chemical factories, communication and computing techniques control function strategies and most internal diagnostic and heating mechanisms. In spite of such safeguards provided inside the body, human beings suffer from various kinds of diseases partly due to their style of living and reasonably due to invasive interference due to external agencies such as bacteria, bacilli and viruses and pollutants in air, water and food on the other side. Accidents due to fast and adventurous life and mechanised transportation are increasing day by day. To meet all such eventualities and render assistance through medication and hospitalization, a variety of healthcare measures are undertaken by private and Government agencies. Timely and effective treatment necessitated application of a large number of sensors and instruments to aid the decision of medical professionals. Biomedical instrumentation is the field of activity playing a very critical role in augmenting the efforts of doctors in diagnosis and extension of treatment to patients.

Sensors and sensing systems meant for the same measurand and of the same range, differ considerably from each other, due to the nature of the system under investigation and its location too. Biomedical instrumentation deals with measurements related to the human body, primarily, and so instrumentation systems designed for use in industry, need to be judged for their compatibility for use on the living species. Exceptional attention and care is exercised to ensure that there is no interference to the on-going bio-processes of the body and that there is no reaction and interaction between the materials of the sensor and chemicals of the tissues of the body, when in contact with each other. For a better understanding of the extent to which the biomedical instrumentation differs from industrial instrumentation, it is considered proper to picturize the human body by a short description.

The human body, though conceptualized as a machine, is sufficiently complex as it is a hybrid system in a single outfit, comprising of electrical, mechanical and chemical systems all working together in an interdependent fashion. It is akin to a chemical factory with its inputs of raw materials of air, water and food and outputs of not only organic and inorganic compounds but also tissues, bone, skin and blood through the bioprocesses, contributory to the growth and health of the body. The cardiovascular system is a closed hydraulic system with its pump (heart) and elastic tubing (arteries, veins) along with the automatically operated valves, located suitably along the forward and return paths of blood. The four-chamber pump is systematically excited by electrical signals for its reception and delivery of blood to the body parts. The respiratory system (lungs) serves the functions of providing oxygen to blood from air-intake and exiting CO_2 to atmosphere in a controlled fashion depending on the needs of the body. It is a pneumatic system with an automatic control function. The nervous system is the communication network extending from brain down to all corners of the body. Central to the functioning of the entire system is the tiny cell that responds to any stimulus or disturbance and develops electrical signals for communication and command operations. The pacemaker, known as a biological clock, releases the necessary synchronizing signals for operation of the heart. The brain functions like the central processing unit of a computer, carrying out computations on data obtained from all the five sensors of the body, storing relevant data in memory and releasing command signals for actuation of the body parts, not only eyelids but all others including hands and legs. The much-talked about sixth sense is the extraneous power of the brain, acquired in the course of life by mind through the exercise of thinking right from childhood. It is thus clear that the physiological system of the body is extremely complicated linking all the basic disciplines of science and engineering (evolving into a new discipline of Bio-medical Engineering). Man can only attempt to understand it to extend attention and care for its proper functioning. Sensing systems of various types are required and they may be used in contacting, non-contacting, invasive or noninvasive manner. Analysis instrumentation is very much needed alongwith techniques for imaging the internal parts of the body. The needs of biomedical instrumentation pose a challenging task for instrumentation personnel dealing with design, operation and maintenance while medical personnel and specialists apply their expertise in interpreting the data and images successfully leading to correct diagnosis and effective treatment in time.

13.4.2 Sensing Physical Variables

Most people suffer from lung and heart problems as they are mostly linked with blood flow patterns in the body. Preliminary studies are conducted for obtaining data of flow rates of air in and out from lungs and the analysis of inhaled and exhaled air. The electrical activity of the heart is studied from the electrocardiogram (ECG) obtained through the application of electrodes attached to the chest area. The bioelectric potentials between electrodes are measured by special amplifiers shown in Section 7.9.5 and the record of these potentials with time reveal the nature of the problem of the heart. Likewise EEG and EMG obtained through electrodes located in other regions help preliminary diagnosis. Blood pressure measurement is undertaken to study the functioning of the cardio-vascular system. pH of blood is analysed in certain situations, by obtaining samples from the patient and testing them in laboratory. Some simple sensors and instruments commonly used in biomedical instrumentation are shown in Table 13.3 along with the ranges of the physiological variables and their bandwidth.

Table 13.3 Physiological variables and sensors

Variable	*Range*	*Sensor*	*Instrument*
1. Temperature	35°C–38°C	Thermocouple Thermistor IR Techniques	Thermometers Thermograph
2. Heart Sounds puse rate	30–150 Hz 0.1–40 Hz	Capacitive	Microphone
3. (a) Blood Pressure arterial	30–400 mm Hg 0 to 30 Hz	Diaphragm (Capacitive, strain gauge)	Sphygmomanometer
(b) Veinous	0–15 mm Hg 0–30 Hz	Capacitive	
4. (a) Blood flow rate	0.5 to 1650 ml/s 0–50 Hz	Electromagnetic Ultrasonic (SAW) Doppler sensor	Flow-meters
(b) Flow volume	0–40 Hz	Use of dyes	
5. (a) Biosignal (Heart)	0.1 to 4 mV p-p 0–100 Hz	Micro Electrodes	ECG
(b) –do– (Brain)	10–100 µV 0–100 Hz	Surface Electrodes	EEG
(c) –do– (Muscle)	50–1000 µV 10–3000 Hz	Surface Electrodes	EMG
(d) –do– (Nerves)	1000Ω–50 KΩ	Surface Electrodes	Skin Resistance Unit
6. Mechanical Movements	0.1 to 5 µm 0–40 Hz	LVDT	Ballisto Cardio Graph
7. Breathing rate	250–300 ml/s 0–20 Hz	Tachometer, hotwire	Breathalyzer
8. Composition of Breath		Hotwire for CO_2 gas	Gas chromatograph
9. Ear	10 Hz to 20 kHz	Microphone	Audiograph

Ultrasonic (ultrasound) are finding extensive application due to their inert nature (when passing through the muscle tissues and blood vessels) and capacity to enable measurement non-invasively. Structural and blood flow information regarding the performance of the heart and peripheral vessels supplying blood to the head and legs can be obtained by using ultrasonic waves of frequencies ranging from 5 to 10 MHz. Systems described in Section 12.5.3 using waves or pulses are common for studying interfaces between blood and tissue and are visualized enabling location of vessels and heart chambers. Blood flow patterns are studied by using Doppler Effect technique of Section 12.5.6. Ultrasonic scanning and image presentation is common in many laboratories. Doppler technique using low frequency waves is used to examine blood-flow patterns inside the bony skull. A modified version known as Duplex scanning includes screening for angiography and pre- and post operative assessment.

Audiograms are graphical representations showing the frequency response of ear to judge its sensitivity for audio frequency sound waves. Audiology tests enable fitting of an appropriate hearing aid.

Analysers based on absorption of optical radiation at certain wavelengths are developed for measuring the oxygen concentration of the blood, known as Oxymetry. The visible red and the near-infrared part of the electromagnetic spectrum is chosen and the differences of absorption spectra of oxyhaemoglobin and haemoglobin are measured leading to the desired estimation.

13.4.3 Radioactive Techniques

Radioactive materials are used in hospitals for purposes of diagnosis of some diseases and treatment of patients. The radioactive material is usually Technetium-99 m (Tc 99 m) and is a source of gamma rays. It has a half-life of only 6 hours. For diagnostic imaging work, a chosen chemical is tagged to the radioactive isotope and then administered to the body, usually in the form of an intravenous injection. The chosen chemical is different for different diseases and is treated with Tc 99 m in a laboratory just before the administration through injection. Such activated chemicals are known as 'radio-pharmaceuticals' in hospitals, Tc 99 m is a weak radioactive isotope and the level of radiation from the injected radiopharmaecutical is kept within safe limits to ensure safety and avoid untoward effects on health of patient and others around. The distribution of the chemical within the patient's body is imaged by a 'Gamma camera'. In this equipment an array of radiation detectors are located close to the patient emitting gamma rays from the injected radiopharmaceutical. The spacial distribution of the detectors produces a display as an image of the uptake. From the image it is possible to observe asymmetry in organ uptake as well as 'hot spots' with additional information available on the function of the organ by examination of the 'washout' curves of the radio-active material.

Utilization of X-rays for getting images of bones is known to yield information regarding bone density, and its structure and fractures. The instrument is known since long as Dual Energy X-ray Absorption meter.

13.4.4 X-ray and Radiotherapy

The other important and common application of X-rays is in radiotherapy, essential for the treatment of deep-seated tumours causing cancer, by irradiating it with X-ray beams. Elaborate attention and care is exercised while deciding the required dose of radiation to work on the tumour and the mode of administration without causing any damage to the surrounding healthy tissues. The planning and execution of such treatment is carried out with the help of computers. The computerized planning system is enabled with prior 'education' and 'training' regarding the relevant procedures to be adopted first on a 'phantom' or fictitious tumour before exposing the patient. The CT scan (computerized tomography scan) slices of the patient are fed to the computer which with its stored-planning system computes the required dose of radiant energy, and the associated distribution through a combination of X-ray beams proper to the size and location of the tumour.

Linear accelerators are routinely used in clinical practice to provide beams of X-rays and electrons. The equipment is not only sufficiently complex but also requires to conform to regulations for quality assurance (QA) to ensure that the treatment routines are accurate leading to neither under—nor overexposure of patient to radiation. The energies of X-rays are usually in the range of 4–20 MeV. The QA of clinical radiation is concerned with the quality and quantity. Beam quality is decided by the size, shape and direction of the beam as well as the beam energy. Alignment of the movable collimators inside the instrument is microprocessor controlled and is checked and cross-checked for ensuring the distribution of radiation on the tumour. Measuring the quantity of the beam is a check of absolute calibration by dose measurement, using an ionization chamber.

13.4.5 Chemical Analysis

Optical methods offer an alternative to the commonly used electrochemical sensing, for analysis of samples consisting of serum electrolytes, metabolites enzymes, coenzymes, immunoproteins and inhibitors. The use of fluorescent techniques (in a fibre-optic configuration) is recognised as offering a very sensitive method for measuring the low concentrations of hormones, steroids and thyroid function constituents (see Figure 12.6.4). A micro-sensor (device of size in mm only) has been developed for monitoring pH, CO_2 gas concentration and oxygenation, all of them using the fluorescent technique. The sensor, suitable for *in vitro* application is a powerful device as it can measure temperature also along with all the three. Oxygen is sensed by its quenching effect on an aromatic hydrocarbon, pH, by the change in fluorescence of an immobilized indicator and CO_2 via optically detected changes in the pH of a buffer solution entrapped in a polymer.

The sensing of glucose in blood is an essential requirement in the present days and a sensor based on the fluorescence quenching, is a novel development. A lot of effort is on for bringing out suitable sensors for use *in vitro* and *in vivo*, utilizing a variety of chemical substances, biomaterials and membranes, for analysis of bio-samples.

13.4.6 Endoscopy

Endoscopy is an exercise undertaken by doctors to gain an insight into the gastrointestinal tract, the delicate subdivisions of the lung and other internal human organs and cavities, normally inaccessible to direct examination. Likewise, endoscopy in industry enables examination of structures and assemblies without expensive and destructive dismantling. It is a powerful tool in the hands of medical professionals considering the nature of the human body, and the essentiality of inspection of the inside before deciding on the surgical operation. The instrument, as a whole, is known as endoscope, permitting visual inspection of the internal parts, through a 'keyhole' for introduction of the optical-fibre cable, being flexible for use where needed. The popularly known 'laproscopic surgery', is the precision keyhole surgery using the endoscope and other associated devices. Laproscopes used with tiny CCD cameras allow several surgeons to view and control the procedure from TV monitors and perform operations with precision using only 'keyhole' incisions. A complete examination of the lung and airways can be accomplished within a short time say 15–30 mins, enabling the patient as well to participate in viewing the parts on the colour monitor of a video endoscope. Surgeries, supposed to be time-consuming and involved, are carried out comfortably in a short time, sometimes without the use of anaesthetics, resulting in considerable saving of time, energy and cost to the patient. The optical components of an endoscope are split into two categories, the light transmission system and the image transmission system. A flexible fibrescope contains both an incoherent bundle for the transmission of light and a coherent (image guide) bundle to relay an image from the distal tip (tip near the point of study) to proximal eye piece. Ultra compact CCDs (3–4 mm square) are incorporated into the distal tip of the flexible endoscope.

13.5   TOMOGRAPHY

13.5.1   Introduction

Simple formulations of measuring systems of oldertimes, necessitating physical contact, interaction or invasion with the system under observation have, in the course of the recent 20 years, evolved into a hierarchy of highly sophisticated instrumentation systems. Electronics and computer technology have their exceptional role and application in this evolution. The enthusiasm to sense what is inside an enclosure and to know what is happening inside, and that too non-invasively without disturbing its normal function, has always been high and it is this that made Radon in 1917 conceive a technique known as 'tomography'. Tomos, in Greek means section or slice and graphy, to picture or image. It was in 1971, that Hounsfield announced the development of the tomographic body scanner and was awarded Noble Prize. Medical tomography, also known as computerized tomography (CT) scanning utilizes X-rays for acquiring cross-sectional images of the human head and body for clinical purposes. Sensor system is rotated relative to the human body to obtain 'tomograms' with high resolution.

During the past decade the interest centred around the application of tomography technique for nonintrusive visualization of multiphase flows through pipelines and multiphase processes within reactors and other vessels. X-ray scanning cannot serve the objectives and so a different sensing system with high speed of response has become necessary. Process tomography, designed to deal with situations in industry, is identified by the type of sensing system. It can be electrical, ultrasonic, Gamma ray or optical type and the choice depends on the application situation.

13.5.2   Electrical Capacitance Tomography

Electrical techniques for process tomography do not present any of the type of hazards noticed in X-ray and gamma-ray tomographs. They are straightforward to implement and fast with potential for high resolution in time domain. Electrical capacitance tomography (ECT) system is suitable for systems with variable permittivities, either as mixtures of phases showing permittivity contrast or spatial variation due to composition or temperature.

The basis of process tomography is to put a number of sensors around an object, collect measured data from the sensors and reconstruct cross-sectional images from the data. The ECT system consists of a set of capacitance electrodes mounted equally around the cross section of an object to be imaged and the capacitances of all possible combinations are measured. This data is then processed to reconstruct an image of the contents of the pipe. These electrodes can be mounted either inside or outside the pipe or vessel depending on the wall material. Figure 13.10 shows the block diagram of the ECT system, wherein charge/discharge capacitance measuring circuits provide data, a PC, a multi-sync monitor and an array of transputers. The software is based on the popular linear back-projection (LBP) algorithm. Typical capacitance values are in the range of 0.1 to 500 femto-Farads (a femto-Farad is 10^{-15} Farads). The number and size of the electrodes used depend on the specific application, the choice being larger number for obtaining higher resolution images. The DC signals proportional to the measurand capacitances with a resolution of 0.1 fF are fed in turn to a common conditioning circuit via a multiplexer. The processed signals, as shown, in digital domain are fed to the transputer array via a serial data link. The simplicity of the LBP algorithm is

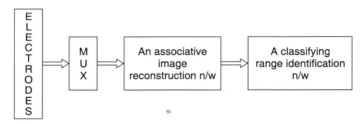

Figure 13.10   Use of neural networks for ECT.

explained by considering the simple 9-element square grids shown in Figure 13.11, that results when three pairs of electrodes in the X and Y directions are used. The shaded areas represent a fluid or solid in the vessel. In case 1, row sensors X_2 and column sensors Y_1 detect an object and so an unambiguous image for this situation can be deduced from the capacitance measurement. Likewise in case 2, but in case 3, there is no unique solution offered from the simple measurement of the capacitances between pairs of opposing electrodes. Hence, it is necessary to look at an object in more than two directions to pinpoint the location. In the case of a circular pipe, having twelve electrodes, distributed uniformly around its diameter, it is possible to measure 66 unique inter-electrode capacitance values (see Figure 13.12). To obtain images on a 32×32 pixel grid, it is necessary to augment the measured data with a second set of data known as the sensitivity map that is to be made available before the start of any measurement. The sensitivity map is a set of coefficients which define how the interelectrode capacitance values are modified, when the value of permittivity of each pixel within the circular cross-section is changed from the minimum to the maximum value in turn with another pixels set to the minimum permittivity value. The image is then calculated by superimposing the interelectrode capacitance measurements, using the sensitivity map coefficients as weighting factors for each pixel. The hardware necessary for the entire processing can be seen in Figure 13.13. The LBP algorithm, possessing the knowledge of the capacitance sensitivity

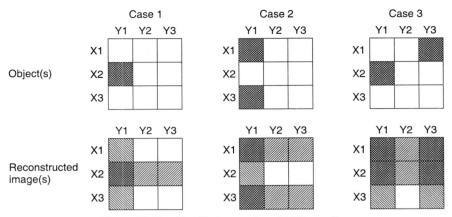

Figure 13.11 Simplified examples of LBP algorithm.

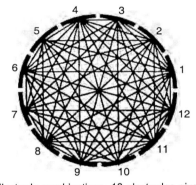

Figure 13.12 Electrode combinations, 12 electrodes give 66 combinations.

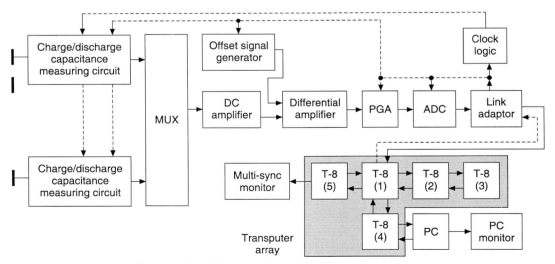

Figure 13.13 Schematic diagram of ECT system hardware.

distribution (i.e. sensitivity map) for all combinations of electrodes enables the image reconstruction. For the 12-electrode sensor there are 66 inter-electrode combinations and these 66 interelectrode capacitances when viewed against a square grid of 32 × 32 pixels, will be inside 812 pixels. It is required to calculate the value of permittivity of each of the 812 pixels from only 66 capacitance measurements. That is where the map comes useful. Online imaging rate of 25 frames per sec is used for 32 × 32 definition or 40 frames per sec if 20 × 20 definition. The data collection rate without reconstruction and display is usually 350 frames per sec for 6-electrode system, 240 frames per sec for 8-electrode system and 100 frames per sec for 12-electrode system. In view of the small values of capacitance to be measured, the sensor electrodes must be enclosed inside an earthed screen to minimize noise and the influence of external noise and objects.

The medium of flow through the pipe needs to be a dielectric essentially, like grain, powder (under pneumatic conveyor pipe), or two-phase flow of gas and oil from an oil well. But if it is a mixture of oil, gas and water, it is difficult to discriminate between three phases. For determination of the velocity of flow in the pipe, two or more sets of sensor electrodes are fitted to the pipe and spaced a short distance apart. By using correlation technique on the two sets of images obtained as detailed in Section 12.5.7, axial velocity-of-flow profile of the flowing medium can be obtained.

13.5.3 Electrical Resistance Tomography

Electrical resistance tomography (ERT) is used for the measurement on the aqueous-based slurries consisting of a conducting medium and a nonconductive solid phase, if they are of dielectric type (such as gas bubbles in an aqueous solution). The arrangement requires a short section of non-conducting pipework with equally spaced metallic electrodes radially positioned around a section of interest. The electrodes, normally 16 in number are invasive in many cases, but non-intrusive. The way the data is collected is slightly different from the ECT. The system normally injects a small constant alternating current to flow between two adjacent electrodes, while monitoring the potential difference between all adjacent pairs of the remaining electrodes. This process is repeated with all possible electrode pairs being used as excitation electrodes in turn. The reconstruction process is similar to earlier ECT system. Acquisition rates of between 40 and 100 frames per sec have been used with parallel processing techniques.

Processes that have mixtures of ferromagnetic and/or conductive materials can be applied with the recent electromagnetic tomography (EMT). It is based on the measurement of complex mutual inductance by means of sets of coils of a 16 + 16 system.

For a thorough and complete understanding of the tomography systems, references may be studied.

12.5.4 Fibre Optic Tomography

An interesting tomography system, though complex-looking at first sight, is conceived by making use of optical-fibres, external to the pipe, with the flowing medium inside, for on-line monitoring of particles and droplets having low concentration in the flowing medium. The first job is concerned with the provision of proper terminations at both ends of the optical fibre by means of lenses with suitable radii as shown in Figure 13.14. All the fibres placed around the

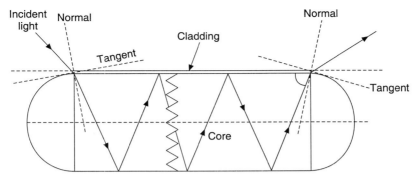

Figure 13.14 A side view of the fibre-optic model.

pipe, to constitute an array, are required to have smooth surfaces and same radii on each side, though different from one side to the other, so that all fibres used for admission and reception of light energy are of identical characteristics.

Light waves get modulated in intensity while passing through the pipe (transparent) and enable the recognition of the constituents of the medium. Each sensor (fibre in this case) provides information relating to particle concentrations within its sensing zone. Particles or droplets of different sizes and shapes flowing inside cause varying levels of attennuation and scattering and are detected by the optical-fibre sensors. Attenuation of light received by the photodiode is a measure of the particle size. It is observed that particles of size down to 100 μm could be detected.

The light source consists of four halogen bulbs providing collimated beams. The light receiver consists of an array of optical-fibre sensors arranged in a way to constitute two orthogonal and two rectilinear projections consisting of an 8 × 8 array of optical fibre sensors whereas the rectilinear projections consist of an array of 11 × 11 sensors as shown in Figure 13.15(a) and (b). As such there are 38 sensors in each plane.

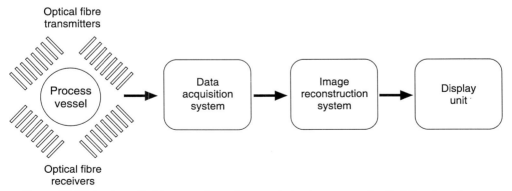

Figure 13.15(a) An optical tomography system employing two arrays of optical-fibre transducers.

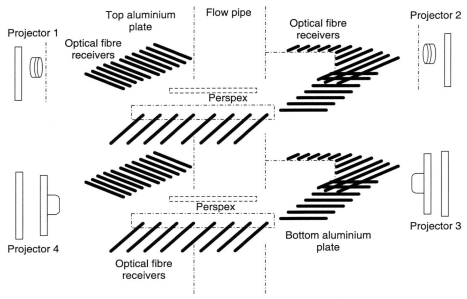

Figure 13.15(b)   An isometric view of the arrangement of optical fibres around flow pipe.

13.6   ANALYSIS INSTRUMENTATION

13.6.1   Basics of Analysers

Quality assurance measures in industry to assess the closeness of the end product to the desired specifications, include determination of the chemical composition as well as the physical properties. On-line instrumentation systems are invariably preferred with provision for display of data in real-time. Analysis instrumentation systems for ascertaining the chemical composition of the end product are highly complicated and sophisticated, and rely heavily on chemical phenomena and reactions. What used to be simple techniques for laboratory instrumentation, are studied and given effect to development and production of sensors suitable for each application. It is considered beneficial if a single system, though complex and sophisticated, is designed to effect the chemical analysis, thereby dispensing the use of multiple sensors. Traditional spectro chemical analysis techniques have assumed prominence in this context and are made available as elaborately designed commercial instruments known as spectrographs.

It is desirable to recall briefly the principles governing spectroscopy and spectrometry to recognise the potential of these techniques for use in industry as well as in research. Electromagnetic radiation extending from microwaves to Gamma rays at the other end of the electromagnetic spectrum, has been fully explored for use in instrumentation for chemical composition data apart from physical-variable measurements through interferometry and chromatography. All these traditional instruments have been considerably modified to be in tune with the times, in regard to the compatibility to go with computers, computational requirements and communication capability. Spectroscopy mainly utilizes the optical region of the electromagnetic radiation, consisting of infrared, visible and ultraviolet regions. Of these three, it is infrared region that is widely applied for analysis of gas mixtures. It has the advantage of being least influenced by light in the visible region. It has been known that assessment of the product quality cannot be complete in most cases unless the chemical composition is ascertained. Moreover, the range that are considered to be end-products of chemical, food, and drugs and pharmaceutical industries are increasing day by day and the look-out of industries for such on-stream analysis instrumentation systems resulted in adaptation of only certain powerful techniques that are accurate, reliable and safe for their environment. Analysis may also be desired in some industries, at all the various stages of their production processes. Analysers suitable for the test materials, which may be solids or fluids, usually of known composition but of varying concentrations, are desired. Situations abound in reality, such as those in pollution monitoring, where the composition is not known and with component concentrations going down to such low levels around 1 ppm. It is this extent of variability in the demands that is to be met with by properly identifying the technique for each situation. Adulteration and contamination of all the essential requirements of man, i.e., food, water and air, is the root cause of slow poisoning and deterioration in health. Combating the menace by suitable measures is being accorded highest priority by Governments in all countries. A large number of portable instruments for pollution monitoring have been developed. Obtaining samples of the test materials or gases compounds is also a tricky affair and the attention needed to effect the analysis in a non-destructive, non-invasive and non-interfering manner depends on the particular situation, even though the spectro-chemical analysis techniques conform nearly to the above requirements in most cases.

Simple chemical effects have been utilized in certain sensors employing certain compounds and polymers for measurement of the contribution of a single component in a gaseous mixture. Chemical science is increasingly being explored with the intent of developing fast-acting sensitive and reliable sensors to meet the demand for analysis of simple gaseous mixtures.

Research activity related to the advancement of chemical science depends a lot on the insight provided by the large number of instruments designed for spectro-chemical analysis. Extensive literature is available about these instruments, yet a brief introduction of those popular versions, useful to process industries is given in the following sections.

13.6.2 Emission Spectrography

All material bodies behave as sources of electromagnetic radiation when excited by external means and also as absorbants of some or all of the incident radiation. The spectrum of frequencies of the above phenomena are designated as 'Emission Spectra' of the emitted radiation and 'Absorption Spectra' of the absorbed radiation. Spectrography reveals the discrete frequencies of these spectra and enables computation of relative energies of all the frequency components. Spectrograph is the instrument that disperses the radiation into its frequency components and displays them as sharp lines on the screen with their brightnesses as indicative of their intensities. It is these lines that are designated as 'Spectra', and they are none but the images of a fine slit, kept in the path of the radiation. The dispersion is effected by means of prisms, gratings or filters and the line spectra can be photographed and/or displayed on computer screen. Analysis by emission spectroscopy requires a few milligrams of the sample.

Qualitative as well as quantitative analysis of solids and solutions is possible. The technique is capable of detecting 1–10 ppm of the metallic ions and of some non-metals like phosphorus and silicon in a small sample. Electrical methods of excitation are by dc arc or ac arc or spark by means of electrodes of the material of the test sample if it is electrically conducting type. Otherwise the sample is placed in a small core in a carbon or graphite electrode, with the upper electrode of carbon or graphite ground to a blunt point. Graphite is preferred as it is a better conductor than carbon. If the test sample is a solution, it is placed in a similar manner, but in a small hole in the centre of the lower electrode. The lower electrode is usually made the positive electrode. An optical ruby laser can also be used to excite samples of even nonconducting materials. The concentrated laser beam vaporizes the sample by absorption of the energy of the laser beam and intense heat generation. Samples of solutions can also be studied by spraying the solution through an atomizer on to a flame of a burning gaseous fuel. But the flame serves only as a low energy source and excites only a few lines of the spectrum.

The emission spectra thus obtained are of three kinds: continuous spectra, band spectra and line spectra. The continuous spectra are emitted by incandescent solids and are characterized by the absence of any sharply defined lines. Band spectra consists of groups of lines that come closer and closer together as they approach a limit, the head of the band. Band spectra are caused by excited molecules. Line spectra consist of definite, usually widely and somewhat irregularly spaced lines. Line spectra is the characteristic of atoms or atomic ions that are excited and are emitting their extra energy in the form of light of definite and discrete wavelengths.

Every spectrograph requires dispersion of the radiation either by prism or a grating. Lenses of proper materials are essential for collimation and focusing and be transparent for the wavelengths in the spectrum. The emission spectra are studied to identify all the elements of the sample qualitatively, however small they may be in the sample and quantitatively as well by studying the intensity of the line at the chosen wavelength. To enable such studies, additional systems for direct reading and/or obtaining a record of the spectra on a photographic film and photometers to read the intensity of the lines, are used in the instrument. As the quantitative analysis requires the measurement of the intensity of the lines of the unknown element, attention is paid to the control of the excitation conditions, nature of the exposure and the type of photographic film. To eliminate the effects of such variations, comparison methods are adopted, as in other situations, by using the spectrum of a reference standard with known concentration, in justaposition with that of the unknown sample. Once the sample is inserted for excitation, it takes only 2–3 minutes for the results to be available.

13.6.3 Absorption Spectrography

When a beam of radiant energy is incident on a substance, it may be totally transmitted with no absorption at all, with the direction of propagation getting altered, while in some cases part or total of the incident energy gets absorbed. Absorption of energy means transfer of energy to the medium and this absorption process is a specific phenomenon related to the characteristic molecular structures of the medium or substance. Molecules possess three types of internal energy, electronic, vibrational and rotational. Absorption of radiation occurs when a quantum of radiant energy coincides with an allowed transition to a higher energy state of the atom or molecule under study. If this amount of energy is supplied from an external source of radiation, the external energy required to bring about this change of state will be provided by the photons of one particular wavelength which may then be selectively absorbed. Study of the frequencies of photons which are absorbed would thus indicate a lot about the nature of material. Also the number of photons absorbed provides information about the number of atoms/molecules of the material present in a particular state. It thus provides a method for qualitative and quantitative analysis of substances. Electronic transitions correspond to ultraviolet (UV) and visible regions of radiation, vibrational transitions to near infrared and mid infrared (IR) and rotational to mid IR and far IR. IR radiation interacts with all molecules (except the homonuclear diatoms O_2, N_2, H_2, Cl_2, Argon and Helium) by exciting molecular vibrations and rotations. IR absorption spectroscopy assumed a special status as the IR absorption pattern is unique and distinct for each compound and is treated as its IR 'FINGERPRINT'. Very few compounds absorb UV radiation.

The absorption of radiation may be continuous meaning that the whole range of frequencies are reduced in intensity. Radiation absorption may be seen over a band of λs, which is common in many cases.

It will be wise to realize that there are particular groups of atoms absorbing radiation at the same λ with very little influence from the rest of the molecule. Also similar molecules have similar IR absorption spectra. So IR absorption spectroscopy is most straightforward when the component molecules of the sample have significantly different groupings. IR spectrum ranging between 7 μm to 15 μm wavelengths offers the best discrimination between molecules and hence identified as the 'FINGERPRINT' region. Absorption curve showing the reduction in intensity at each λ for a substance is typical for each substance. Beer's law enables determination of the concentration, in most cases and is given by

$$I_t = I_i 10^{-abc} \tag{13.2}$$

where I_i is the energy reaching the detector with no sample in the beam path,
I_t is the energy reaching the detector with sample in the path,
b, the path length of sample, cm
c, the concentration of sample component, and
a, the absorption coefficient of pure component of interest.

If c is expressed in g/litre, then a is called specific absorptivity and with c in moles per litre, then a is known as molar absorptivity. The product abc is the absorbance of the component and its variation with c is drawn for known samples so as to analyze the unknown ones.

In place of IR energy, UV energy can be used to obtain absorption spectra of certain substances available as aqueous samples.

UV analyzers are more sensitive but not as specific as IR analyzers. Both types are used choice depending on the type of sample. The spectrophotometers are all similar in operation except that the sources of radiation, lenses and detectors are different. IR analyzers ignore the presence of gases, such as O_2, H_2, N_2, Cl_2, Argon and Helium as they do not absorb IR energy. The analyzers are known as 'dispersive' type, as absorption is studied at discrete wavelengths of the incident radiation. It is essential to control the humidity and temperature of the instrument. The presence of CO_2 and water vapour in the optical path is detrimental as they are strong absorbers, and hence double beam instruments are preferred in practice to effect compensation.

13.6.4 Nondispersive IR Analyzer

The above dispersive type of IR spectrophotometer has been simplified to suit the conditions of process industries. The optical system requiring lenses or gratings is dispensed with and thus the version known as 'nondispersive' type of analyzer is designed wherein the total radiant energy from the IR source is transmitted through the sample. Two simple schemes as shown in Figure 13.16 are in use, the difference between them being the way in which detection takes place. The incident beam is split into two; with each half being received by the detector cells 1 and 2. In the case of (a) cell 1 is filled with the pure form of the gas, A being determined while cell 2 is filled with a nonabsorbing gas. If gas A concentration in sample is 100%, then there is no left-over energy reaching the cells 1 and 2 and so the difference in their output signals becomes zero. But if gas A is absent, all the energy received by cell 2 results in the maximum signal while pure gas A in cell 1 absorb all of energy resulting in zero signal. Thus the difference in output signals is indicative of gas A concentration in the test sample. The arrangement in Figure 13.16(b) has both detector cells containing gas A in pure form, while one half of the beam passes through the test sample and the other gas through a reference cell. The difference in output signals is indicative of the concentration of gas A.

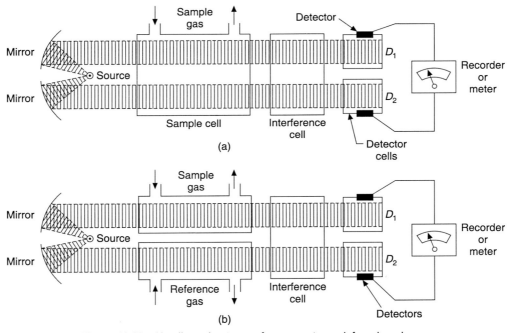

Figure 13.16   Nondispersive types of process-stream infrared analyzers.

These analyzers have become popular as they possess high selectivity and high resolving power. The test gas through the sample cell may be obtained from the main stream through a by-pass arrangement. Materials for the sample cell body pose a great problem as no material is transparent for the entire IR region. The alkali halides are widely used; sodium chloride is transparent up to a λ of 16 μm, Potassium bromide up to 26 μm and cesium iodide up to 40 μm.

Gaseous samples are measured in a cell about 10 cm long. For samples in liquid form, thin layers of the liquid with thickness between 0.01–0.05 mm are kept sealed with the provision of windows on either side and enclosed securely by means of gaskets. For solids, it is necessary to identify proper solvent which is nonabsorbing in regions of interest and dissolve the solid to make a dilute solution. Carbon tetrachloride and carbon disulphide are two such solvents commonly used. Powders or solids reduced to particles can be examined as a thick slurry.

13.6.5 Mass Spectrometry

In the emission spectrograph, radiation produced by a spark between electrodes of the sample is used to analyze the sample substance, while in mass spectrometer, it is the ions that are produced that render the analysis of test substances. Even though the first mass spectrometer devised by F.W. Aston in 1919, enabled the analysis in laboratory, it is the recent modified and sophisticated versions that enable quick and accurate analysis of a wide variety of gas and liquid mixtures, in a way suitable for use in process industries. Techniques of dealing with solid substances have been identified and used. Ions are formed when an electron is removed from a neutral atom or molecule and this ion has its own mass/charge ratio. Thus for each susbtance, a spectrum known as 'Mass Spectrum' is obtained showing a record of the numbers of different kinds of ions. The component is identified by the relative numbers of each ion and therefrom the quantitative and qualitative analysis of the substance made possible. The mass of the sample required for analysis can be as low as 100 micrograms. Mass spectra obtained under constant conditions of ionization depend on the structure of the molecules from which the ions originate. Each substance has its own characteristic mass spectrum, just like the finger print, and the mass spectrum of a mixture may thus be analyzed in terms of the spectra of pure components, and the percentage of the different substances in the mixture calculated.

Analysis of the mass spectrum of a mixture may involve the solution of a large number of simultaneous equations by using a computer. The principal difference between the various types of spectrometers lies in the method of separating the ions according to their mass/charge (m/e) ratio. Considering the nature and essentiality of information provided by such instruments, the complexity of the associated parapharnalia apart from the spectrometer is understandable and tolerable.

To handle all types of substances, different inlet systems are designed; (1) to ensure admission of the same quantity of sample each time, to the reservoir, (2) to expand the gas from reservoir into its expansion vessel and (3) to pump out the remnant sample at the end of each analysis. After expansion, the pressure of the gas is about 100 μm of Hg. In continuous monitoring the inlet systems the gas sample must be admitted at or near atmospheric pressure. Solid samples may be painted on a thin filament and heated.

The neutral molecules enter the ionization chamber of which pressure and temperature are controlled at 5 μm of Hg and 200°C respectively. Electrons produced from a hot tungsten filament are accelerated towards an anode kept at a fixed potential. These fast-moving electrons flying past in a direction perpendicular to that of the neutral molecules of the sample, bombard them thereby ionizing them to release positive ions, characteristic of the molecular structure of the sample. The rate of formation of these ions needs to be maintained at a constant value. The ions get accelerated in an electric field through a system of electrodes and are made to enter the analyzer tube at a very high velocity through a narrow slit of width of 0.075 mm. The analyzer tube is the heart of the mass spectrometer, through which the ions are directed to travel and emerge through an exit slit at the other end of the tube. In the magnetic-deflection type mass spectrometer of Figure 13.17 the accelerated ions are deflected by application of a magnetic field through an angle of 60° (even higher at 90°/120° or 180°). If V be the accelerating voltage of the ionization chamber, the velocity with which the positively-charged ions enter the analyzer tube is given by $v = \sqrt{(2eV/m)}$, where e and m are the charge and mass respectively of the ion. The ion beam is deflected from its straight path due to the presence of the strong magnetic field provided by the permanent magnet and finally get collected at the collector plate at the other end of the analyzer tube through an exit slit. The collected ions constitute a current of the order 10^{-10} amps and is processed electronically. The analyzer tube is controlled at near vacuum condition. The radius r of the path of the ions is governed by the relationship given by

$$\frac{mv^2}{r} = Bev$$

Using the value of $v (= \sqrt{(2eV/m)})$, r is given by

$$r = \left(\frac{2V}{B^2} \cdot \frac{m}{e}\right)^{1/2}$$

where B is the flux density of the magnetic field.

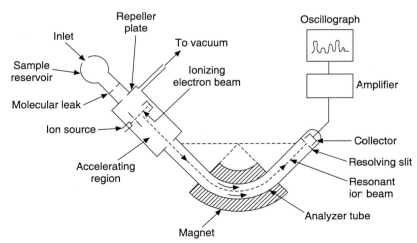

Figure 13.17   Schematic diagram of a Nier 60° sector mass spectrometer.

So it can be seen that only those ions that follow the path which coincides with the arc of the analyzer tube will be brought to focus on the collector slit and fall on the detector kept behind the slit. Other ions will strike the analyzer tube at some point, be neutralized and get pumped out of the system. The magnetic field thus classifies and segregates the ions into beams each of a different m/e given by $B^2r^2/2V$.

To obtain a mass spectrum, either the voltage or the strength of the magnetic field can be varied continuously at a constant rate. The potential attained by the collection plate is a measure of the number of ions striking it, provided they each carry the same charge. The output signals properly processed, are plotted against values of m/e ranging from 10 to 60 normally showing sharp peaks and making up the mass spectrum equivalent to the frequency spectrum. The mass spectrum becomes a sort of 'fingerprint' for each compound as no two molecules will be fragmented and ionized in exactly the same manner on electron bombardment. For example, from the simplest hydrocarbon, methane, CH_4, positive ions are produced corresponding to structures CH_4^+, CH_3^+, CH_2^+, CH^+, C^+, H_2^+ and H^+.

In identification of a compound, the most important single quantity of interest is the molecular weight. The mass spectrometer is unique among the analytical methods in providing this information very accurately. The mass spectrometer has been aiding the scientific research in a spectacular fashion, and its potential is utilized in process industries, nowadays, more so in dealing with a large number of organic compounds.

13.7   ELECTRONIC NOSE

13.7.1   Introduction

The physical transducers, described in earlier chapters, sensitive to visible and infrared radiation and pressure waves are, in a way, aids to human sensory systems of eyes, skin and ears respectively. They are useful in responding to light rays, temperature and sound waves. But when it is to aid the other two human sensory systems, i.e., nose and tongue, to sense odours and tastes, no simple or complex sensor system with compatible performance, could be developed. Mimicing the human olfactory system is tough as it involves the interaction of the complex chemical mixtures with the biological sensors of the nose and tongue. A food product is enjoyed not by taste alone but also by smell. The human sense of smell is still the primary means used to qualify and classify the raw materials or finished products. Present day industrial scene is bearing heavily on processing and production of goods related to human needs, such as milk, cheese, packed foods, wine, beverages, cosmetics, medicines and drugs. Other utilities include a large number of chemicals manufactured by chemical industries such as fertilizers, pesticides and other petro-chemicals. A great demand has arisen for the detection and estimation of many gases and vapours for their proportions in their mixtures. While the industries are interested in the principal component analysis of the gaseous and vapour mixtures to organize their production processes, the environmentalists and medical personnel also desire the same in the context of their safety and healthcare activities. The components of the mixtures may be organic or inorganic in nature and some of them are identifiable by their characteristic smell. All the organic materials, fruits, vegetables and even the processed foods release vapours that characterize their odours. As they are all normally meant for human consumption and other needs, extensive effort has gone in to develop a sensing system that mimics and serves as an alternative to the human nose. Such a system is expected to enable the on-line monitoring of the end-products and assessment of their quality. Human expert panels, such as those engaged in quality-checking of coffee, tea or wine, may be dispensed with, as they cannot serve continuously at the desired speed.

All raw materials such as fruits, vegetables and others and also the end-products of the food industry possess unique odours and tastes recognisable by human sensory system. The odour released by each substance contains vapours of many chemical components and is known to consist of a set of molecular components of varying concentrations. It is not the simple chemical reactions that decide the odour, but it is also the consequence of the biological activity of the microorganisms.

The fermentation process is known to be due to a kind of biological activity and all food and drug industries have their specific fermentation processes. Bacteria and fungi can contribute to the flavour of a food-product and at times they improve the taste. The same microorganisms also cause the decay of the food-stuffs and spoil their taste. Fruits and particularly, oranges are susceptible to fungal attack producing esters and alcohols. The vapours from bacteria generally come from the breakdown of protein as well as carbohydrates and consist of the products of decarboxylation and deamination of aminoacids that have been formed by the proteolytic degradation of proteins. These vapours are generally methylamine, dimethylamine, trimethylamine and ammonia. They can also be hydrogen sulphide, dimethyldisulphide and such others derived from sulphur-containing amino acids. Amines such as histomine are released from bacterial metabolism of the aminoacid histidine. Bacteria tend to produce vapours that are abnoxious to humans and they are usually sulphides and amines. The bacterial breakdown of proteins does not necessarily produce obnoxious smelling compounds. Some sulphur compounds, although obnoxious, at very low concentrations can results in a desirable flavour to foods. The fungi tend to attack food stuffs with a high carbohydrate content producing alcohols and esters giving pleasant smell. Components such as ethane or methane have no odour but contribute to the odour of the substance. The discriminatory capability of the human nose enables the grading of the substances as good, bad, fruity, floral, musky or earthy, and to be a substitute for the nose, one has to bank on a large number of sensors. It is not necessary to know the volatiles that are being emitted for assessing the quality of the foodstuff as long as one has standards for reference. It is exceedingly difficult to know all the volatile vapours and other molecular components that constitute the unique odour of a food-stuff, and its quality. Some of the natural products, such as vegetables and fruits have the concentration of each component varying with time during their last stages of growth and it is required that a follow-up of the varying concentrations are sensed by the sensors for the final assessment of the product. An array of sensors, some being specific in nature, is chosen so that their combined or processed output is classified against the reference provided by the expert-panel for that product. Thus, a pattern recognition system is provided for each product. Electronic noses based on this principle, can save lot of time and speed up analyses of the products considerably.

Exclusive application of gas chromatographs and mass spectrometers may possibly help the situation but the analysis takes sufficiently long time. They may be used to identify the major components of the vapours from a product thereby knowing the most significant volatiles from among them.

13.7.2 Structuring of Electronic Nose

A sensor array is necessary to respond to the various molecular components. Certain active materials and sensor types are available for making up the array. They may be metal-oxide thin film semiconductor (MOS)/conducting polymer (CP)/chemical resistors/coated quartz crystal microbalance (QCMs) and MOS field effect transistors (MOSFETs) with catalytically active gates. The tin-oxide sensor is sensitive to limonene, ethanol, ethyl acetate, acetaldehyde and methanol. Zinc-doped tin-oxide sensor has higher sensitivity to limonene while palladium doped one has much larger response to ethanol. Organic conducting polymers exhibit the desired rapidity in adsorption and desorption and their merit lies in their inertness to reactions with sulphur-containing compounds.

The coated quartz crystal balance of Section 12.8.4 is preferred for use in the array. Usually most of the sensors are nonspecific and their use in the array results in certain advantages. The schematic diagram of Figure 13.18 indicates the manner in which the outputs of the sensor array are processed using the pattern recognition system such as an artificial network. Artificial nose, commercially available from Arom a Scan uses polymer sensors and that of Alphamos uses ceramic heated sensors. They are used extensively in industries for quality assurance of the products, with the number of sensors coming to 32 in the array.

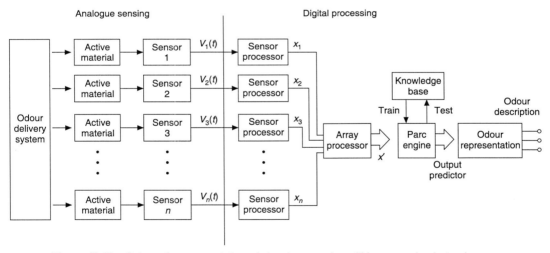

Figure 13.18 Schematic representation of signal processing within a generic electronic nose.

Increase in the number of sensors of an array burdens the processing circuitry and hence an alternative approach is through an optimization technique by using a much smaller number of sensors. For a certain application domain, the sensors are chosen to respond to the most significant volatiles and by optimization of their number, a certain confidence level is attained in the performance for ensuring the quality of the particular application. It is possible to reduce the cost and enable a hand-held instrument to be marketed.

A miniaturized conducting polymer odour sensor for monitoring *in vivo* food flavour release is developed exhibiting adequate sensitivity towards food flavour compounds and fast in response for the purpose of breath-by-breath analysis. Of all the materials available, conducting polymers are considered most suitable for use as odour sensors, with the sensor array made up of different sensing materials responding to a particular odour differently. The system thus mimics the human nose and responds to the air from the nasal tract during eating.

The sensor substrates are integrated structures fabricated on silicon wafers using standard photolithographic and micromachining techniques. A schematic cross-section through the sensor substrate is shown in Figure 13.19 along with one electrode pair, which when coated with conducting polymer behaves as a single sensor element in the array. The working electrode area of each sensor is 0.0625 mm^2 and six of them are fabricated on a single piece of silicon of 15 mm × 4 mm size. After proper cleaning and deposition of electrodes, the polymers are deposited. The polymers deposited are a series of poly (pyrrole) alkyl sulphonate (*n*-alkyl chain length 4 to 9) films. Each sensor is deposited with a different polymer substance. The sensor array is inserted into a printed circuitboard suitable for interfacing the polymers to a computer through relevant signal processing circuitry. The resistance variations of each sensor are monitored continuously and finally the custom-designed software enables the data acquisition, processing and display. The array is incorporated into a system capable of sampling air from the nasal cavity. It is fitted with a stainless steel lid with inlet and outlet ports of 3 mm diameter giving a volume of the inside around 500 ml. Once the response is tested for individual's breath, further analysis is undertaken to show the effects of eating on the array. All foods consumed, possess their typical tastes and odours, when consumed. The array system is designed to identify food through the odour sensed while breathing with food being chewed.

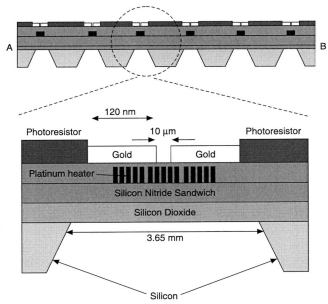

Figure 13.19 Schematic cross section through the 6 electrode pairs on a single sensor substrate (not to scale) for odour.

13.8   ENVIRONMENTAL POLLUTION MONITORING

13.8.1   Introduction

Nature's endowments for the environment on earth has a sort of balance considered congenial for sustenance of life on earth. The living species on earth apart from the human beings constitute animals on the ground, birds in the air, aquatic animals under the water as well as all kinds of microorganisms. Man has realized long ago, the interdependence of each other, for continuity of existence. In the interest of providing better standards of life through a lot of amenities such as comfortable housing, speedy transportation and fast communication, he has inadvertantly landed in a serious problem known as 'POLLUTION'. Extensive industrialization is noticed to meet the demands of goods and services, as seen presently, while elaborate efforts are on to unearth and consume nature's resources from underground, such as coal and crude oil and undersea, aquatic species such as fish. In this process, being very optimistic about the kind of life in future and with his convenient philosophy, he has been dumping indiscriminately enormous quantities of unwanted matter into the earth's environment, the philosophy being, 'out of sight and out of mind'. In course of time, the disturbance and degradation of the essential requirements of man, i.e., air, water and food products, are noticed as 'POLLUTION'. The state of affairs has provoked the environmentalists and the Governments of several countries to institute mechanisms for identification of the degree of pollution, its sources of origin, its effects on living matter and conditions and of course for corrective measures.

Continuous monitoring of air and water resources is considered a necessity while all industries are directed to look back and avoid releasing of materials to outside, considered objectionable. At the same time, all such dangerous wastes are required to be recycled or to enable production of useful goods, so as to minimize pollution. Development of relevant sensing systems with the desired characteristics of high sensitivity and high speed of response, has been on a war-footing and the designers and manufacturers, apart from the instrumentation specialists, are all sufficiently stressed.

13.8.2   Air Pollution Monitoring

The so-called 'Green House Effect' is increasingly being talked about, and the 'Green House' gases, e.g., methane, carbon dioxide and such others trap heat and increase the temperature of the earth. Chloro-Fluro Carbons (CFCs) destroy the layer of ozone in the stratosphere, at high altitudes and expose the humans to the cancerous levels of ultraviolet radiation. The atmospheric boundary layer, near to earth, descends down to earth during nights. Pollutant gases and vapours get mixed up in the air below this boundary layer and adversely affect the life on ground. Several other gases such as oxides of sulphur, carbon and nitrogen, HCl, NH_3, CFCs and hydrocarbons, released from industries and various other vehicles of transport, get mixed up and find their way into the atmosphere. Flue gases, stack gases and some particulates may be poisonous and allergy-causing. Toxic gases such as CO, H_2S and nitrogen oxides cause immediate harm to the health of people, while some others are known to poison the body slowly. Asthma is a common health disorder in big cities. High density of population and the high-rise buildings affect the health of the residents considerably. City environment is prone to pollution, because of cooking-gas usage, high vehicular traffic and on-and-often occurring accidental fires and explosions. Each one of the constituent components of the gaseous mixture not only needs to be detected, but also be known for its concentration level in the mixture. For some of the components, medical establishments have to declare minimum levels for each gas. Gas masks, worn by workers in industries are designed to warn them, once the minimum level is crossed. Workers in mines are known to protect themselves against the dangers posed by the inflammable gas, methane. Traditional ultraviolet or infrared analyzers for analyzing gas mixtures of unknown composition from samples obtained from test environment, do not serve the requirements. Search for sensors and sensing systems specific to a particular component and sensitive to that component alone without being affected by the presence of other components, goes on with funding by Governments. They are in most cases, required to be small in size, highly sensitive (for dangerous gases) and fast in response. To some extent, semiconductor sensors are found to meet these demands. Chemical science, mathematical tools and the development of polymers have provided the base for the necessary sophistication for the analysis of gas mixtures. An array of solid-state sensors, as many as ten, along with a Fourier Transform Spectrum Analyzer is used to analyze the poorly mixed gas mixtures and present the concentration of each component in real-time. Portable instruments for on-spot measurement of oxygen in exhaust gases of automobiles and measurement of alcohol content in the exhaled air of a driver are commercially produced.

It has become clear that pollution poses a great problem and that pollution is severe in cities and regions around the industrial belts. Sensing systems located at one spot, do not yield truthfully the extent of pollution occurring in the area around the spot. The closely-knit poorly ventilated high-rise buildings of cities are potential candidates for extension of pollution monitoring systems in order to ensure safety and security of residents and their properties. Likewise, areas surrounding heavy industries and power plants, need to be monitored for pollution. All sensors and instruments are designed for point monitoring and are suited for use in residential and office rooms. It will be proper to recall that office equipment, furniture, paints used and decorative material emit volatile organic compounds (VOC) and particulates which are equally harmful to health. Particulates transfer viruses, bacteria and other bio-effluents. These are all difficult to monitor. It is to be noted that the CO_2 level in excess of 300–350 ppm of air is not conducive to good health and that it reaches levels of 1000 ppm of air in kitchens of dense localities.

A new generation of Building Management Services (BMS) is concerned with pollution problems arising in each big building and is incorporating systems for monitoring and control of all pollutants from smoking lounges, office equipment, laboratories and/or kitchens along with other systems for lighting, heating and airconditioning. BMSs are established to provide healthy and comfortable environment in the building and effect optimization of energy consumption in the building. BMS systems are akin to the industrial instrumentation and control systems and are increasingly favoured for use in large office buildings of industries and business establishments and shopping malls. The fifth generation of BMS systems is a centralized system consisting of computers for communication and control, backed by relevant sensors for all conditions of interest.

When gaseous pollution is noticed in a large area single-point monitoring is not a solution. The area to be covered may be in and around an industry or localities around multistorey buildings. A technique known as 'long-path optical measurement' using a beam of infra-red or ultraviolet radiation over a path length from a few metres to a few kilometres as shown in Figure 13.20, is identified and found appropriate for application. The wavelength of the radiation chosen is such that the beam is absorbed by the gas component under investigation and not by any other in the path of the beam. It may be construed that is a limitation on the application of this technique, but it is justified when a culprit gas is noticed after leakage or explosion in an industry. Different gas components may be sensed, by changing the wavelength of the radiation in stepwise or continuous mode.

The long-path monitoring technique is based on the spectroscopic principles in which the optical radiation interacts with gases via Raman scattering, fluorescence or absorption processes. Absorption process is a natural choice as it is simple to setup and much less costly. Ultraviolet radiation is absorbed by oxygen while water vapour and carbon dioxide absorb infrared radiation. The amount of attenuation in the intensity of the beam is a measure of the average concentration of the gas along the full path of the beam. The usual configurations adopted are shown in Figure 13.20(a) and (b) with each one suited to a particular situation. The systems are developed by the National Physical Laboratory (NPL) of UK, for use in developed countries. The systems shown are for straightline paths. It is possible to deal with curved paths by using optical fibre cables, as sensors, with its cable cladding exposed to atmospheric gases to be investigated as explained in Section 12.6.5. The technique is seen to possess the required rapidity of detection where continuous vigilance is a must.

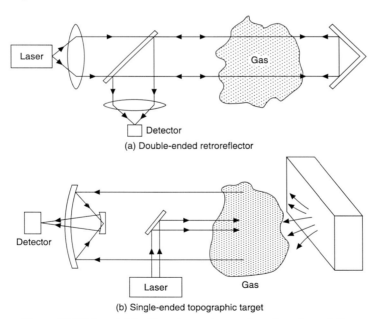

(a) Double-ended retroreflector

(b) Single-ended topographic target

Figure 13.20(a) and (b)   Remote sensing configurations for pollution.

All pollutants shot into the atmosphere find their place ultimately on ground, when carried down along with raindrops. Heavy rain lasting for less than 30 min is supposed to carry down half of the mass of air borne gases and particulates. Acid rain is the result of sulphur dioxide in the air.

13.8.3 Water Pollution Monitoring

Likewise industrial effluents, organic and inorganic, pollute all water resources and the soil coming in contact with the polluted water. The major effluent of food and brewery industries is yeast waste while paper and fabric industries discharge a large volume of dyes into streams of fresh water. In fact, every industry releases waste water containing a large variety of chemicals thereby contaminating fresh water. It is no wonder that a majority of Indians really get affected by using polluted waters and suffer from allergy and other diseases. Much worse damage to the health of people, cattle and poultry, is caused by the extensive application of chemical fertilizers, pesticides and herbicides with the motive of benefiting from higher agricultural production. All such organic and inorganic disease-causing and poisonous substances sink sufficiently into the soil strata and pollute the underground water flowing in the aquafers. Investigations reveal the extent of damage but control measures are totally inadequate, more so in countries like India.

Spilled oils from oil tankers and petroleum industries constitute the major oil pollution of seawater. Dissolved organic compounds such as proteins, carbohydrates, lipids, etc., found in freshwater streams due to waste discharge in towns and cities pose a lot of problems causing sufficient damage to life, sometimes causing death of fish and other aquatic species.

Analysis of water for determining the type of pollutant and its concentration in water, is quite cumbersome and involved. In many cases, samples drawn are analyzed by using the ultraviolet, visible and infra-red spectrophotometers. There are some electrochemical sensors useful for online testing of polluted water, but they are all invasive type. Some sensors using electrolytes confined to membranes and some biosensors do find application in testing of polluted water.

Oil pollution manifests as very thin film on water surface. These films are opaque to ultraviolet. Surveilance of surface water pollution is carried out by laser remote sensing systems. Other sophisticated systems make use of laser induced fluorescence and Raman scattering techniques in laboratories, by obtaining samples. That there is urgency in containing the manmade pollution of water, is realized but scientific investigations and associated effort in developing suitable sensing systems for monitoring demand conscious commitment and financial funding in a big way for limiting the hazards due to water-pollution.

13.9 ROBOTIC INSTRUMENTATION

13.9.1 Introduction

A robot is an essentially mechanized system, static or mobile, so as to carry out automatically certain jobs and behave like a human operator. In industry, they may be located in inaccessible or potentially harmful environments and are designed to carry out one or other of the jobs like welding and cutting, paint spraying and moving objects around the factory floor. Present day applications range from garment cutting to fire fighting outside and patient handling inside hospitals. They are provided with arms-equivalent to do some jobs while the mobile robots necessitate use of ranging systems, platforms and power supply for mobility. They are fully automated systems, whose control functions are computer-controlled aided by a variety of sensors for its operation. The microprocessors in association with newly developed microsensors and microactuators enable several operations on board the robot. The robot is an example of an intelligent system wherein the decision making capability of its computer, located inside the robot, rests squarely on the accuracy of information provided by various sensors. At times, externally located sensors feed the information, similar to the remote control signals of children's toys. The internal sensors are intended to sense the operations inside the robot, like positioning of its arms while the external sensors are meant for the interaction with its environment and monitor the position, location, colour, etc. of the objects proximate to the robot and located in its field of view. For stationary robots with an assigned job, the designing of the relevant feedback system is relatively simple using the available sensors and actuators with its on-board power sources of batteries.

For designing mobile robots, a deeper understanding of the job assigned to it and the kind of environment and obstacles faced by it during movements from a known position to a new location is essential. It is not enough to detect the presence of an obstacle; the data obtained from the scanning of the environment is fed to the computer for steering the robot in the proper direction. At the same time, provision is necessary to see that the robot does not collide with any obstacle. Multiplicity of sensors is needed to ensure such motion of the robot in its environment, where there may be more than one obstacle. Non-contacting type sensors are a natural choice but those meant for steering the robot towards its target, require high angular and range resolution than those meant for avoidance of collision. Likewise, the field of view will be appropriate for each case.

The choice of proper non-contacting type sensor systems and their application technique require sufficient attention, as can be seen from the following attempts.

13.9.2 Ranging Techniques

Proximity is usually taken as nearness, but in the present situation, non-contacting proximity sensors of the inductive or capacitive type are unsuitable. As the distances involved are not so small, other alternative methods using ultrasonic waves or electromagnetic waves, are considered. Transmitter–receiver combination can be used for detecting obstacles and measuring the distance of a target from the robot. Continuous waves or pulses are used as explained in Section 12.5.4. Prior knowledge of the properties, size, colour and the environment are necessary before choosing the energy waves. Sufficient signal strength is essential at the receiver end if the object serves as reflector. Interference from ambient light, temperature and humidity as well as the noise level has to be taken into account, before choosing the source of waves. For successful performance of a mobile robot, the principal requirement is its ability to detect all significant objects in its vicinity.

A proximity switch is energized by the proximity-sensing system as soon as an obstacle is detected, so as to avoid a collision. For an effective detection arrangement of emitter and detector shown in Figure 13.21 is used, so that their beams intersect over a defined range. When the obstacle is in this area only, the proximity switch is energized.

Sensor systems for ranging of the target can be simpler versions based on the 'time of flight' technique, as used in sonars and radars. A little involved system makes use of the triangulation method in which distances a and b are measured from the point of location of the obstacle, shown in Figure 13.22. It requires use of a transmitter–receiver pair at each point on

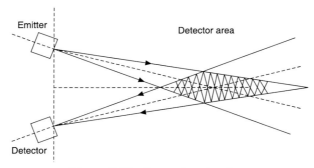

Figure 13.21 Beam aligned system.

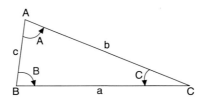

Figure 13.22 Triangulation method.

the baseline AB of length c. Distance of the robot from target is therefrom calculated. The system can be modified to obtain a three-dimensional image of the target, by adjusting the incidence angles of the detectors and/or transmitters in a raster sequence, so that several points related to the target are known for their location with respect to the robot.

For use in mobile robots, radar technique comes handy provided the antenna size and weight are brought within the permitted limits. If the wavelength is kept within 1–10 mm, the antenna becomes small enough for use in robots. The choice of high frequency for the radar, results in greater sensitivity and resolution and improved security, suiting the demands of military services.

If ultrasonic or optical techniques are employed for locating or tracking the targets, the cost of the robot becomes less. Optical techniques overcome the specular reflection problems encountered by acoustic ones except in the case of polished surfaces and transparent materials. Both light and infrared lasers are used in a wide variety of ranging systems.

13.10 NEURAL NETWORKS

It is the revolution in the information technology that has been altering the modalities of life. The successful integration of microprocessors, microcomputers and the so-called PCs in all systems of utilities apart from those related to engineering has been felt at all levels. Their strategic application to provide solutions for intricate situations in the areas of instrumentation and control is a great and welcome development. The artifacts devised in this context are known as 'Artificial Neural Networks', in short, neural networks, architectured relevantly to effect the desired function. While dealing with a large number of signals provided by the smart sensors, the data-processing operations by the neural networks are conceived and conceptualized as that of 'brain-like'. What were seen as unsurmountable and arduous tasks, have been solved and rendered possible by the application of neural networks. Extraction of desired signal from a highly corrupted and noise-infested signal and automatic control of industrial processes, characterized as complex, nonlinear and/or difficult to model, are examples of their application. Better performance is realized by replacing the traditional pid-control in electrical systems, with the neural networks. Likewise, in the image-processing area, neural networks in place of the standard image-processing algorithms, yielded superior results. It goes without saying that the main requirement for their application is the presence of a large number of signals at the input port, called layer in the present terminology. The neural networks are usually software simulations of networks running on a single processor or PC. Hardware implementations of neural networks are available but difficult to apply.

The key concept in neural networks is to have a large number of simple processing elements (PEs) called neurons, with each PE gathering signals again from a large number of other PEs. These signals are weighted, added and then put through a nonlinear threshold before being output as shown in Figure 13.23. A neural network is composed of a large number of PEs connected in different ways with or without feedback between them. The architecture imposes certain properties on the network. Then the network undergoes through a learning stage during which the connection weights are adjusted. The network can then be used in its final form, usually by being fed with input signals from sensors and producing outputs that indicate certain conditions. For introduction purposes, the network is visualized as a parallel network. One of the architectures, that is simple to understand, is the feed-forward network otherwise, known as Multi-Layer Preceptron (MLP).

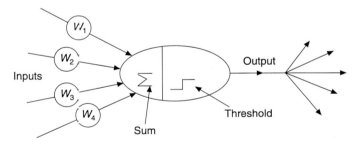

Figure 13.23 A typical processing element.

The MLP shown in Figure 13.24 consists of three layers: 1. input layer, 2. hidden layer and 3. output layer. The central layer, the hidden one, has no direct connection to any other PEs except to those PEs in layers adjacent to it. There can be more than one hidden layers. Data from external devices is received at the PEs of the input layer. The input layer PEs serve as buffers that distribute values to the next layer. PEs of hidden layer perform weighted sums and thresholdings. There will be no interconnection between PEs in any one layer. The PEs have multiple inputs a bias (acting as a threshold) and an output that can be either 1 or 0. The PEs of these layers constitute the MLP which deals with weighted sum of the inputs and presents outputs either as 1 or 0 depending on whether the sum is above the threshold or below. The simple perceptron is trained by supervised learning. The popularity of MLP has led to a lot of work on improvements.

Calculated outputs

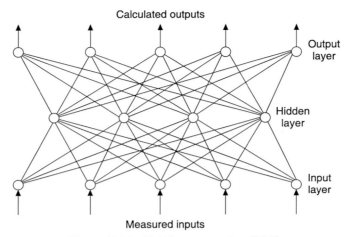

Output layer

Hidden layer

Input layer

Measured inputs

Figure 13.24   Multi layer perceptron (MLP).

There are other alternatives to MLP with certain modifications, obviously resulting in certain desired objectives. They are known as Radial Basis Function (RBF) and the Hopfield Network (HN) and in each there are different architectures. Deciding on the actual architecture to be used is largely a matter of experience.

Software packages are available making the process of developing networks a lot easier. A typical high-end package is NWorks (distributed by Scientific Computers). The majority of users who persevere with neural networks for real applications end up with an implementation running on a PC.

The methodology to be adopted for designing architecture of a neural network depends on the model in hand of the system or other data pertaining to it. A simple but interesting exercise showing the use of neural networks is given (Ref.: Patra, A.) wherein a sensor validation scheme and its application to a realistic model of a variable air volume air-conditioning system are described in detail.

Neural networks are applied to form models of some industrially important bioprocesses for which a complete mechanistic model is not available. A process for recombinant protein production uses fermentation processes that are highly nonlinear. Because of this, it is difficult to use traditional statistical methods to form a model that can be used for design decisions during the process of expanding from the initial small scale experimental stage up to full-scale production facilities. Having developed the neural network models with difficulty, however, they can then be used for on-line estimators of biomass and recombinant protein levels.

EXERCISES

1. Discuss the factors that interfere with the correct measurement of the temperature of hot and inaccessible objects from a distance.

2. Present the extra provisions incorporated in the radiation thermometers to ensure correct estimation of the measured temperature of an object at a distance.

3. What are 'two colour' or 'ratio' thermometers and where are they preferred for use?

4. Describe the basics of operation of a disappearing filament type pyrometer and indicate the range of temperatures for which it is designed.

5. List what sensors are incorporated in the pyrometers and the basis on which they are selected.

6. Explain what is 'thermography' and describe the instrument designed for such activity.

7. Suggest the areas of use where thermography is found relevant.

8. Describe how thermal flowmeters are designed for measurement of flow rate of fluids.

9. Explain what does 'nano-instrumentation' imply and explain the extent of sophistication in the design of instruments for such an activity.

10. Classify the distinctly different principles of operation of the instruments useful for nano-instrumentation.

11. Distinguish between 'nondestructive testing' and 'condition monitoring' techniques and explain the essentiality of the latter for services in industrial processes.

12. Describe how an IR camera works and show how such instruments enable prevention of disasters in industry.

13. Explain why vibration monitoring of mechanical structures is essential and show the means available for the same.

14. Indicate the role of ultrasonic waves in nondestructive testing and condition monitoring in industrial diagnostics.

15. Explain the 'acoustic emission' technique and show the situations where it is found appropriate for applications.

16. Indicate the additional considerations and provisions necessary for medical diagnostics in comparison to industrial diagnostic techniques.

17. Explain what is meant by 'ultrasonic' or 'ultrasound' scanning and how it is effectively carried out by instrumentation systems.

18. What is 'endoscopy' and how it is effectively implemented in medical diagnostics?

19. What is meant by 'CT Scanning' and what techniques are available for the same?

20. Describe the system used in electrical capacitance tomography technique and explain how the the entire system is configured.

21. Explain why fibre optic tomography is preferable for medical diagnostics and show the basics of operation of the same.

22. List the techniques available for tomography and discuss the merits of each.

23. Consider the role of 'analysis instrumentation' in chemical industries and discuss the considerations behind the choice of a technique.

24. Distinguish between the spectrography techniques of 'emission' and 'absorption' and indicate the areas of application in analysis.

25. Describe the mass spectrometer and show how the technique enables analysis of compounds otherwise considered insufficient.

26. Explain what is meant by 'finger print' of chemical substances and suggest the powerful means available for the same.

27. Describe an 'odour' sensor and explain the difficulties in the choice of techniques and implementation of the same for each odour.

28. What is an 'electronic' nose and what is its role in industry?

29. Distinguish the techniques of air pollution monitoring to be extended to building management systems and areas around industries considered to be releasing poisonous gases.

30. Discuss the range of sensors necessary for use in a mobile robot.

31. What are 'neural' networks and how they enable the design of electronic nose.

32. Indicate some powerful proximity sensors explaining the means adopted for their power.

SUGGESTED FURTHER READING

Amavasai, A. Bala, and Others, Vision control for a MEMS robotic system, *Measures & Controls*, Vol. 33, No. 9, Nov. 2000, p. 269.

Asher, R.C., Ultrasonic sensors for the process industry, *Measures & Controls*, Vol. 30, No. 5, June 1997.

Barry, C. and Bruckbenbury, A., Online gas analysis, *Measures & Controls,* Vol. 24, Oct. 1991.

Bartlett, P.N. Elliott, J.M., and Gardener, J.W., Integrated sensor arrays for the dynamic measurement of food flavour release, *Measures & Controls*, Vol. 30, No. 9, Nov. 1997, p. 273.

Corcoran, P., *Optimal Configuration of Sensor Arrays for Odour Classification.*

Coulthard, J. and Yan, Y., Ultrasonic cross correlation flowmeter, *Measures & Controls*, Vol. 26, No. 6, Aug. 1993.

Cowell, David, and Radcliffe, Norman, Detection of microorganisms in foodstuffs by the use of vapour sensors, *Measures & Controls*, Vol. 30, No. 2, March 1997, p. 35.

Cunningham, A.J., *Introduction to Bioanalytical Sensors,* John Wiley & Sons, New York, 1998.

Dyne, S.J.C., Satellite vibration monitoring, *Measures & Controls*, Vol. 26, No. 9, Nov. 1993.

Ewyk, R. van, Flammable gas sensors and sensing, *Measures & Controls*, Vol. 29, No. 1, Feb. 1996.

Flynn, A.M., Combining sonar and IR sensors for mobile robot navigation, *Int. Jou. of Robotics Res.*

Gaisford, W.C. and Rawson, D.M., Biosensors for environmental monitoring, *Measures & Controls*, Vol. 22, No. 6, July/Aug. 1989, p. 183.

Green, Steve, Radiation thermometry, *Measures & Controls*, Vol. 28, No. 8, Oct. 1995.

Henoke, H., The design and application of Honeywell's laser trimmed temperature sensors, *Measures & Controls*, Vol. 22, No. 8, Oct. 1989.

Holroyd, Dr. Trevor, Acoustic emission–The condition monitoring tool, *Measures & Controls*, Vol. 34, No. 4, May 2001.

Holroyd, T.J., *The Acoustic Emission and Ultrasonic Monitoring Handbook*, Coxmor Publishing Co., Oxford, 2000.

Ibrahim, S., Yunos, Y.M. and Green, R.G., A tomography system using optical fibre sensors for measuring concentration and velocity of bubbles, *Measures & Controls*, Vol. 34, No. 2, March 2001.

Jenner, R.P., and Vaezi-Nejad, S.M., Application of amorphous semiconductors for optical tomography, *measures & Controls*, Vol. 33, No. 6, July 2000, p. 175.

John Saffell, Environmental measurement and control in buildings, *Measures & Controls*, Vol. 29, No. 10, Dec./Jan. 1996/1997, p. 298.

Jones' *Instrument Technology,* Volumes 1–5, English Language Book Society, 1986, Butterworths Pubs.

Kakolaki, J.B., and Henry, R.M., Ultrasonic flow measurement using pure asymmetric sing around paths, *Measures & Controls*, Vol. 34, No. 10, Dec. 2001.

Kraft, Michael, Micromachined inertial sensors the state-of-the-art and a look into the future, *Measures & Controls*, Vol. 33, No. 6, July 2000.

Lawrance, E., Kinsler and Austin, R. Frey, *Ultrasonic Sensors*, 2nd ed. (1978), Wiley Eastern Ltd., Delhi.

McKnight, J.A., and Clare, A., Using ultrasonics to measure process variables, *Measures & Controls*, Vol. 23, No. 7, Sep. 1990.

Medlock, R.S., and Furness, R.A., Mass flow measurement a state-of-the-art review, *Measures & Controls*, Vol. 23, No. 4, May 1990.

Partridge, R.H., Longpath monitoring of atmospheric pollution, *Measures & Controls*, Vol. 23, No. 10, Dec./Jan. 1990/91, p. 293.

Patra, A. and Mukhopadhyay, S., Sensor Validation via Artificial Neural Network, Monograph on 'Frontiers of Measurement and Instrumentation, (FOMAI) IIT Kharagpur, May 2002, p. 414.

Persaud, K.C., Kaffel, S.M., and Pisanelli, A.M., Measurement of sensory quality using electronic sensing system, *Measures & Controls*, Vol. 29, No. 1, Feb. 1996, p. 17.

Ratner, B.D., *An Introduction to Materials in Medicine,* 2nd ed., Elsevier Academic Press, New York.

Robert, B. Northrop, *Instruments and Measurements*, CRC Press, Philadelphia, 1997.

Sadana, Ajit, *Engineering Biosensors: Kinematics and Design Applications,* Academic Press, New York, 2002.

Smith, F.B., Air pollution and the weather, *Measures & Controls*, Vol. 22, No. 9, Nov. 1989, p. 272.

Thomas, R.A., Jones, N., and Donne, K.E., Infrared thermography in industrial diagnostics, *Measures & Controls*, Vol. 34, No. 4, May 2001.

Thomas, R.A., *Thermography*, Coxmoor Publishing Co., Oxford, 1999.

Walsh, Peter T., Toxic gas sensing for the workplace, *Measures & Controls*, Vol. 29, No. 1, Feb. 1996.

Wood, T., An introduction to the use of Hall effect devices, *Measures & Controls*, Vol. 21, No. 4, May 1988.

Yang W.G., Beck M.S., and Byars, M., Electrical capacitance tomography from design to applications, *Measures & Controls*, Vol. 28, No. 9, Nov. 1995.

CHAPTER 14

Process Control Instrumentation

14.0 INTRODUCTION

Sensor development has been drawing a much greater attention than ever before and more so to comply to the demands of the industry. Sensing is an independent process like thinking, while control function is a follow-up activity to sensing. Initially automatic control systems rendered higher productivity possible, but in course of time, the urge and the necessity of producing a large variety of goods and products of enhanced quality, has been responsible for developments in sensor technology along with automated instrumentation, communication and condition monitoring systems. Every cause and effect suggested in the scientific world, is exploited to constitute a sensing system, to assist the industry in various situations. Identification and production of a variety of engineering materials and chemicals goes unabated while the nanotechnology and modern manufacturing processes have accelerated and intensified the research for developing more and more powerful sensors, yet extremely tiny in size.

The automated industry has its telescope focussed on production of food products and pharmaceutical drugs, apart from all other utilitarian gadgets for the man. The nuclear industry for production of the essential requirement, electricity and the oil industry to harness the nature's potential have been given greater attention. Each industry may bank on the established range of sensors for the measurement of process variables, but the qualifications, configurations and mechanisms of the sensors have also been going through a revolution and continuous evolution. It is this peculiar situation that draws closer all scientists, engineers, technologists as well as bioengineers, to pool their knowledge and sustain the development of sensors for the present day needs. In this context, it is considered desirable to acquaint the reader with some typical situation arising in industries and how they necessitated development of suitable and customer-specific sensors and sensing systems. The following sections are only intended be reveal the concern and attention extended to sensor development.

14.1   PHARMACEUTICAL   INDUSTRIES

14.1.1   Introduction

In many industrial processes, the water content of solvents is an important factor in determining the quality of the final product. The need for accurate on-line measurement and control of residual water in industrial solvents is felt very high, especially in the cosmetic and pharmaceutical industries, where methyl ethyl ketone is continuously monitored for water content. In textile industries, cloth finishing is very much dependent on the moisture content, as well as the water content in perchloroethylene when polyester cloth is produced. Likewise, in distilling industry, continuous monitoring of ethanol and methanol in their mixtures with water is very essential. Accurate measurement of moisture in industrial solvents, not only improves the quality of the end-products, but also reduces production costs due to the efficient and economic operation of continuous distillation columns.

14.1.2   Measurement of Water Content

Unlike the conventional potentiometry, a modified potentiometric method based on the ion iso-concentration technique is used for the determination of water in organic solvents. In this technique, potentials of cells with ion-selective electrodes vary systematically and reproducibly with variation in water content and solvent concentration respectively.

The development of the above technique is to a large extent, found effective considering the nature of moisture problems when dealing with industrial solvents. The solvent has to be known for its (1) water miscibility, (2) hygroscopicity, (3) polarity, (4) density and (5) viscosity.

Perchloroethylene is practically immiscible in water while ethanol and acetone are miscible in all proportions. Some others are partially miscible.

The ionizing power of solvents is zero for tetra methylsilane and unity for water. Density variation between the solvents is not much but viscosity variations are very high. Monoethylene glycol is highly viscous.

After extensive search, two types of ion-selective electrodes are found useful for application in the ion iso-concentration technique and they are:

(a) the fluoride electrode for the determination of solvent in largely aqueous solvent/water mixtures using the fluoride iso-concentration technique (FICT) and (b) the pH glass electrode for the determination of residual water in organic solvents using the proton iso-concentration technique (PICT).

Of the two, PICT provides high sensitivity to small changes in water concentration at high solvent content. It is useful for a 0–10% water concentration range in simple and complex solvent mixtures and at the same time be applicable to industrial online moisture monitoring and/or effective automatic control of the process. The change of cell potential under ion iso-concentration technique is a function of the property of the indicator ion in solution.

A typical system constituting the above measuring system along with other electronic circuits is shown in Figure 14.1 which makes use of the IEEE–488 bus for making the instrumentation get connected to the computer. The inputs and outputs from the computer system to the measuring equipment are under the direct control of a slave microprocessor system based on an 8-bit Intel 8085. This has 56 K–bytes of ROM/RAM and 8K-bytes of

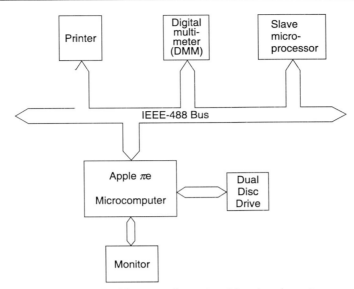

Figure 14.1 Microcontroller and peripheral equipment.

memory-mapped I/O including the IEEE–488 interface. The sequence of commands to control the equipment is downloaded to the slave from the Apple Ile. Programs appropriate to the particular liquid to be measured are loaded into the system via the $5\frac{1}{4}$ in floppy disk-drive. Under the computer control, a valve selects the input either from one of the calibration solutions or the sample stream under monitoring. Large number of readings from the electrodes are processed to present the average value after every 60 s.

14.2 PAPER AND TEXTILE INDUSTRIES

14.2.1 Introduction

Paper and Paper board manufacturers usually claim the quality of their products by printing on the outer wrapper labels that the paper is made to be used in conditions of environment with a relative humidity (RH) of 55% at 20°C. The paper is expected to be stored, unpacked and used under these conditions for better conformity with the printing and other applications. There are, likewise, food, pharmaceutical and textile industries which value moisture measurement in their production schedule for many reasons. The measurement and control of water activity (WA) and equilibrium relative humidity (ERH) also concern safety of their products in use and presentation of a unified value to the marketeers and users.

Removal of moisture from a solid material or a product online, can be effected to a certain extent only, as it depends on the RH of the air or gas in the vicinity of the product. The minimum of moisture in the product that cannot be drawn off is known as equilibrium moisture content which is a function of the humidity, temperature and nature of the solid or liquid. This data is important for products that get stored. Declaration of moisture content of the end-product does not mean much when assessing the performance for its application or usage.

14.2.2   Measurement of Relative Humidity

Water activity (AW) can be defined as the RH value achieved from measuring a moisture containing sample kept in a sealed container along with a sensor that can accurately measure the RH. The sample, container and the sensor, all are kept under the same temperature. The tests are done for products ranging from paper, food or pharmaceutical products. Water activity (AW) is expressed as

$$AW = \frac{\text{Water vapour pressure over sample at C}}{\text{Water vapour pressure over water at C}}$$

$$= \frac{\text{relative equilibrium moisture at C}}{100}$$

It is realized that water contained within a food product or paper exists in two states: bound or unbound. Whatever be the causes, the condition of water in the sample will be either fixed or environmentally influenced. The fixed or bound moisture can be driven off by heating with most of it becoming vapour initially and finally to escape into air after sometime.

While AW is seen from the above that it is expressed as a fraction, say 0.85 if the RH is 85%. But it is to be declared, when testing is done as specified already, under strictly controlled temperatures.

Where standards exist, it is essential that RH instruments are calibrated, and the standards came into being in 1985. The provision of convenient standards providing calibrations traceable to National Standards has greatly contributed to the confidence and certainty in the measurement of both RH and ERH. Calibration points for RH are provided by NPL through a set of saturated salts such as Lithium chloride (for 11.3%); Magnesium chloride (for 33.1%) and Potassium chromate (98%), all in pure form. Test rigs for ERH, the national standards for RH, and the establishment of Calibration Laboratories have enabled the Industries to qualify their production control systems with higher accuracy. Moisture content in paper is conditioned by the composition of the paper, the finish, amount of loading of coating and hysteresis. ERH is unaffected by these factors and therefore, provides a more reliable link between the manufacturer and consumer bringing them closer in their understanding of the conditions under which the paper can be stored, unpacked and used.

Measurement of moisture in paper is a bit tricky in the sense that paper is highly hygroscopic and hence can increase weight by 1% in 10 s when exposed. However, the gravimetric method is the oldest laboratory technique for most materials. For online measurement, two systems are found convenient: (1) Infrared absorption method, and (2) Microwave loss method.

Infrared radiation between 1–3 μm wavelengths is considered suitable for moisture measurements, not only in paper but also in food industries dealing with fat- and protein-rich foods, or in tobacco leaves, cigarette-packets, etc. IR radiation from the source is directed to be incident on the paper through a set of filters, brought into the path by means of a rotating shutter. The absorption is measured by a detector, after receiving the back scatter radiation as shown in Figure 14.2. A second detector also receives the radiation, and serves to correct errors automatically coming up in the measurement. The output signals are processed electronically

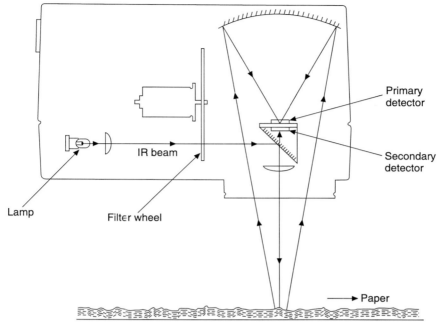

Figure 14.2 IR sensing head for paper.

and the moisture content is continuously recorded, as the paper passes through. Alternative arrangements use two sensing heads, one on each side of the dryer or one on either side of the paper. The only difficulty with the IR technique is that the scatter-radiation differs with the composition of paper material.

The above difficulty is overcome by using microwaves of much longer wavelengths around 3 cm. A rectangular waveguide, as shown in Figure 14.3 is used with the paper sheet, running between the two split halves of the waveguide. The planes of the paper sheet and the voltage vector coincide. The reduction of the wave power at the output end of the guide relates to the moisture content. The method is fundamentally measuring the power-loss in the paper material and it generally relates to paper condition right through the material without any hysteresis effect.

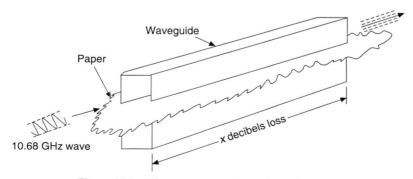

Figure 14.3 Microwave sensor for moisture in paper.

14.3 FOOD-PROCESSING INDUSTRY

14.3.1 Introduction

The food-processing industries are smaller in size when compared to some chemical industries, but are similar in the requirements of instrumentation and control systems for their production range. Yet they demand exceptional attention because of the nature of their raw materials and their end products. Ensuring the freshness and quality of the natural products of agriculture at the input end and bringing out hygienically processed and packed food products for human consumption. Labelling the product conveying data regarding its composition, calorific value, date of expiry etc., is considered essential.

All materials making up the storage tanks, processing vessels, pipelines, that come in contact with the processed material, should be carefully chosen so that nowhere there is interaction or reaction between them. Nowhere and at no time, the material should get trapped and the entire system should permit proper cleaning at regular intervals. 316 grade stainless steel and teflon coatings are generally used. Some food products using acetic acids, such as sauces and pickles, are surprisingly corrosive and in such applications, titanium is preferred. Glass is banned for use as it is brittle.

The moisture content of food products is very important as it affects storage life, the physical properties and taste. Water activity (AW) defined in Section 14.2.2 is measured in the food industry from the following considerations:

1. stability of composition
2. behaviour of product under environmental changes
3. shelf-life
4. sensory characteristics, like taste, odour and colour
5. biological activity of microorganisms
6. coagulation or physical separation of the constituent components of the product.

Additives to the food stuff may solve some of the above problems but not always found advisable. At the same time, it is noticed that at higher temperatures, the AW values are higher and at AW values ranging from 0.91 to 0.95, most bacteria are active for attack. The lower the temperature, the lower the AW value at which the biological activity is far less subdued and hence one finds the instruction on food packets: store the contents in refrigerator. As almost all food processing involves heating or cooling accurate measurement of temperatures ranging between $-50°C$ to $+150°C$ is very essential. An accuracy of $\pm 0.5°C$ is essential during sterilization of canned foods whereas in the dairy industry, maintenance of temperature to very close limits of $\pm 0.003°C$ over a very limited range of temperatures is very essential. Measurement of temperature is relatively easy to accomplish in comparison to moisture content and ERH of foodstuffs. ERH of a food stuff is measured as specified in Section 14.2 and the ERH value will indicate the storage life of the food under specified conditions. RH values can be measured by using the polymer thin film sensors (Section 12.3.5) which are regularly calibrated.

14.3.2 Measurement of Flow

Flow measurement in food industry is riddled with problems as most food liquids are viscous from about 100 to 10,000 centipoise and also non-Newtonian. The viscosity of chocolate is dependent upon many controllable factors like temperature, moisture content, and degree of stirring. Chocolate-coated biscuits and such others depend on viscosity and proper consistency. Another difficulty is due to the presence of particulate matter such as the sugar syrup used for confectionary. Volume as well as mass flow rates are of interest in the food industry whether the flowing medium be viscous like the above or particulate like rice curry and paper pulp or granular like coffee powder. At the final stage, the flowing media go directly for packaging and they have to be regulated at times for the flow rate to enable release of the substance in small quanta after correct measurement of weight/volume. The exacting demands of not only the quality of the end product but also the weight of the packed food, force the industry to look for accurate measurements through suitable instrumentation systems. Weighing by electronic means is invariably preferred but volume and mass flow rates pose considerable difficulty. Non invasive flowmeters of the electromagnetic or ultrasonic type are satisfactory, while the Coriolis mass flowmeter is the preferred choice.

The measurement of viscosity needs to be made over a wide range of shear rates. In most cases, it is enough if the technique is sensitive to the small changes in viscosity around an operating value and be reproducible over time. A viscometer, similar to the type explained in Section 5.7.2 is designed to suit the dynamic conditions of the industry rendering its application for continuous-flow medium. It consists of an inverted conical rotating bob that moves in an off-axis circular motion, inside an outer cylinder. The motion is such that the inverted cone precesses around the inside of the outer cylinder, drawing in and exiting the fluid medium as it does so. The drive is applied via a differential that acts as a torque transducer. The planetary gear cage of the differential is restrained by a spring, the extension of which is converted into a reading.

Any foreign body in the flowing medium is dangerous and should be detected before it enters into the package. Checking the medium at several stages is essential, apart from the end-stage. They may be metallic or any others like plastic or glass pieces or cotton threads. Detection of metallic bodies, while in stream is rendered easier by having a system of three coils surrounding the aperture through which the product passes. The three coils are patterned to function like those of a linear variable differential transformer (Section 6.2.3.1) when the central coil is excited from a radio-frequency source of voltage. The outer coils, present zero signal in normal conditions but, when a metallic body enters one coil there is an alarm signal working on the unbalance voltage of the two outer coils.

14.4 AEROSPACE INDUSTRY

14.4.1 Introduction

It is the air-transportation and space-navigation systems that have been continuously demanding better performance through the instrumentation and control strategies and techniques. The main areas of concern are: (1) flight control systems, (2) engine control systems and (3) structural monitoring systems. The vehicles in the air may be either for transportation or for warfare, with the former ensuring safety and comfort of passengers while the latter necessitating high manoeuvrability for combat operations. The present-day passenger aircraft is characterized by higher (supersonic) speeds, higher passenger capacity (going up to 800) and the long distance non-stop flights (not just Trans Atlantic). It is the combat aircraft that has been performing with its unique rapid manoeuvre potential offered by an unstable airframe. As ever, aircraft is becoming more complicated and complex while the pilot and the machine are getting stressed to do more through their vigilance and performance. Consider the objective of the designers to get the new novel (VSTOL) aircraft that is marked to fly at supersonic speed and its ability to assume a hovering condition. Whatever be the aircraft, its performance is based on a certain number of subsystems, each of which furnishes data from sensors to the pilot and communicates commands to the relevant systems from the pilot. A study of Figure 14.4 implicates that the pilot is the central processing and management system. The more the aircraft sophistication, the more the complicated and complex system it has in all its engines, airframe and avionics and weapon systems, thereby increasing the workload of the pilot. The proper decision by the pilot is based on the data provided to him by all the sensors, not only those enabling him with the control strategy but also all others that monitor the performance of the engines, airframe and the environment.

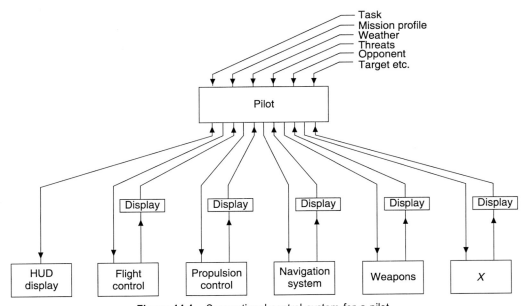

Figure 14.4 Conventional control system for a pilot.

It is easy to judge the thoughtfulness exercised in the choice of sensors, as they should be accurate, reliable, fast in response and rugged enough to withstand the vibrations, mechanical overlads, and high temperatures. Microsensors, dealt in Section 12.3.3 are favoured because of their size and weight but immunity to electromagnetic interference is of high priority. These demands of aircraft industry provoked professionals of all specializations, with the result that there are now the largest number of sensors based on different principles of operation. However, only those sensors, that have gone through on ground applications with satisfactory performance for sometime and found to be credible and compatible for use on board the aircraft do finally find place in combat aircraft. It is this high-tech industry that has enabled an all-round escalation in scientific research, production of new materials as well as revolution in manufacturing processes for production of MEMs.

During the early seventies, the smart and intelligent sensors, have become available for use and the avionics changed its structure from the earlier one to the integrated micromachined silicon sensors with the associated signal processing circuitry on board the sensor chip. Aircraft industry found them credible enough for measurement of pressure in all the areas of aircraft, i.e., engine performance and monitoring air frame and other systems both on ground and in-flight conditions. In particular, the introduction of fly-by-wire and closed loop control of hydraulic systems has also demands for pressure sensors of different ranges. Considerable improvements in the performance of aircraft were possible with the use of such silicon sensors. Good signal-to-noise ratio and high EMC/EMI immunity could be obtained. Size and weight reduction is a definite change for good.

14.4.2 Selection of Sensors

In spite of the perceived improvements, the quest for sensors that are non-electrical in nature and at the same time totally immune to electromagnetic interference, went on simultaneously. During that period, optical-fibre-based sensing techniques for measurement of various measurands were on development. Once they have proved their versatility and flexibility for applications in situations never considered possible, aerospace and defence establishments have targeted the optical fibres and optical fibre sensors to satisfy the stringent and new requirements of the advanced aircraft systems.

The major measurement requirement in flight control systems is for rotary and linear displacements. A fixed wing aircraft requires about 100 such sensors. Likewise, for engine control, optical fibre sensors, have proved effective with their accuracy and speed of response and more importantly suitable for the harsh environment of the engine. For the military and aerospace systems, accelerometers and gyroscopes are essential sensors and these requirements are met by the micromachined inertial sensors. Accelerometers of various sizes and specifications based on either capacitive sensing or piezo-resistive sensing are available for use in aircraft, but micromachined gyroscopes, based on different principles of operations (such as the one shown in Figure 14.5) are yet to establish themselves matching the stringent demands of the aircraft industry. Likewise, optical gyroscopes, such as Ring Laser Gyro and the Fibre Optic Gyro (FOG) are equally established systems with the required demands totally taken care of. They are: (1) no moving parts, (2) no start-up delay, (3) no 'G' sensitivity, (4) increased reliability and (5) reduced weight. The FOG consists of a coiled optical fibre, sufficiently long, with its length under continuous measurement by an optical interferometer. As the coil rotates the effective coil length, as seen by the interferometer, changes due to relativistic effects, enabling the rotation rate to be measured. A typical FOG has its size of 75 mm × 75 mm × 55 mm, and is linear in its response over a wide dynamic range. As said earlier such devices will be time tested on ground and also in unmanned vehicles or missiles before they become part of the present day aircraft.

Optical fibre sensors for monitoring the performance of the engines in aircraft are used in the context of measuring the temperature of the engine and specially the clearance between the rotating turbine blades and the shroud surrounding them.

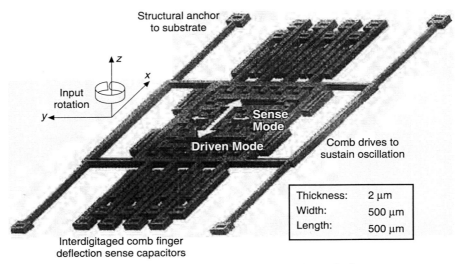

Figure 14.5 Micromachined gyroscope (magnified).

Likewise, optical fibre sensors for displacement and pressure have been found suitable in all respects, more importantly with immunity from EMI.

The in-service monitoring of airframe components offers the scope to fly an aircraft closer to its structural limits without causing damage and also facilitating better performance. Several other advantages can accrue by building optical fibres into the fabric of a composite/fibre-reinforced component during manufacture. The types of sensors incorporated into composite structures, fall into the four following categories:

1. strain measurement
2. temperature sensing
3. crack detection
4. vibration measurement.

Unless highly integrated aircraft flight control systems are realized through the reliable and accurate instrumentation, the workload of pilot cannot be reduced. Identification of suitable strategies for control functions and integration of accurate and fast-acting sensors with the control schemes is enabling the present day achievements in the aircraft operational capability alongwith its speed, size and safety.

14.5 NUCLEAR POWER INDUSTRY

14.5.1 Introduction

Nuclear Power Generation is treated as the only way to meet the demands of the standards of the present-day civilization, no matter whatever be the levels of risk which the society shudders. The safety of the nuclear power stations is accorded top priority by its nuclear safety department as well as the usual Government's Inspectorate branch of Health and Safety Executive. Acceptable low levels of risk is ensured and acceptable risk is expressed as a maximum frequency of accident for different levels of radiation release. Provision of instrumentation and control systems (I & C) of varying integrities are provided to limit the individual fault causation frequencies and to mitigate their effects. The safety of the plant and its working personnel as well as the people of the region in which the plant is located, is given topmost consideration. The following short account is intended to draw the attention of the reader towards the measures taken in Western World, of late, in effecting a modified approach in the I & C systems.

The totality of I & C systems comprises numerous systems which are engineered to satisfy integrity targets expressed as probability of failure on demand (pfd) ranging from 10^{-7} pfd for primary safety systems to 10^{-1} for safety related systems.

Normally I & C systems discourage use of software for safety applications because of the difficulties and high cost of qualification which is necessary to demonstrate a satisfactory safety case. However, for safety related systems, software is used. The totality of I & C systems comprises of certain categories arranged in descending order of integrity, starting with essential systems to emergency announcers.

14.5.2   Safety and Security Instrumentation

Essential systems either maintain or restore plant operation or initiate plant into an active state in response to an emergency situation. Highest integrity is demanded in this case. The last category is associated with the least integrity and is looked after by an independent team of operators and the system is made to announce alarms in the event of hazards or faults affecting the central control area or its operation (e.g. hot gas or steam release on cabling, earthquakes, fires, etc.).

The main control room is the central location for the operation of the power plant during all planned modes of operation. In the unlikely event of its becoming uninhabitable (say sudden smoke hazard) the reactor would be tripped manually and the operators would shift to the auxiliary shutdown room from where they would monitor the shutdown reactor and direct recovery operations. In the designs of these rooms, highest levels of integrated human-machine reliability and performance will be achieved under all conditions. Thorough and expert attention is paid to both the human and technological characteristics of the man machine interface (MMI). Computer-generated displays provide a compact and flexible way of viewing information. However, when things are happening rapidly, it can be difficult for an operator to build up quickly a mental picture of changing conditions. So a mix of computer-based displays and conventional I & C systems arranged on mimic control panels provides the most effective implementation of MMI.

In order to meet all the requirements specified for the I & C systems, robust system design principles are used to achieve high reliability and protection against the propagation of failures. These include segregation of major I & C systems functioning into separate electrical groups and the use of fault-tolerant systems. One change from the earlier times is the use of fibre-optic cables to provide the high data transmission rates required with the benefit of their inherent electrical isolation and freedom from the effect of EMI. A distributed computer architecture is used to provide a flexible and extensible system without overloading the data communication highways. Communication between remote processors takes place overlocal area networks (LANs) and data processing takes place as close as possible to the point of acquisition, with data only transmitted over the network if required by a remote processor.

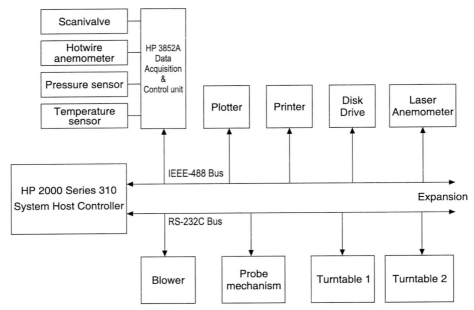

Figure 14.6   Wind tunnel measurement and control system using fieldbus.

14.5.3   Waste Disposal System

As times go by, instrumentation systems are seen as saviours of humanity from a variety of dangers posed by certain technological achievements. Nuclear industry is one of them and it presents special challenge to the instrument design engineers. The industry uses all the traditional systems used for measurement of process variables. It is the waste materials of the industry that have to be taken care of in all respects right from the release of the same from the plant to that of encapsulation and transfer to the store. Most industries have to comply to the rigid regulations framed for their waste products, gaseous or otherwise and mechanisms adopted are relatively simpler. But in nuclear industry, the waste disposal is a tough and complex process. The waste needs to be encapsulated and immobilized. The so-called intermediate level nuclear waste requires to be stored in retrievable form for a period of 50 years before it is transferred to a permanent underground repository. The waste is fed into stainless steel drums and filled with some filler material along with cement. The drums are then transferred to a suitably located store in which they may be stored for up to 50 years prior to removal to a permanent underground repository. A lot of ingenuity and knowledge of the various sensing systems resulted in the development and design of the detection and monitoring mechanisms in all the three important situations of the waste encapsulation plant.

First the monitoring and cleaning of feed lines needs a detection system for the pipelines through which grout flows from the plant down to a point where drum gets filled. Once the grout from the plant is completed for discharge, the feed pipes need thorough cleaning. Water is not permitted as it may be contaminated when let out. So a special pipeline inspection guide is designed and developed so that it can be pneumatically propelled forward for cleaning. A special magnetic assembly is designed and used.

The watered swarf (waste) from the plant when encapsulated in a monolithic form within a stainless steel drum is ready for transfer to the encapsulation plant store the swarf drum, when getting filled with swarf is required to be monitored for any fire accident. An instrument is developed for this aerial discharge monitor, by using a plastic scintillator and a photomultiplier system.

The drums need to be labelled and provided with date etc. and a means of recognition of the identification labels is also required. Image processing and recognition technique using radiation tolerant camera is applied to read drum labels in order to track the movement of material through the plant.

The provisions for remote detection, the degree of care exercised in monitoring the materials, for the safety of personnel, are indicative of the potential means presented by the variety of sensing systems.

14.6 BIOPROCESS INDUSTRY

14.6.1 Introduction

All the technological effort put in for industries towards enhancing their productivity and product quality of the variety of goods and products from micro to macro size, is being tested now for combating the complexities arising while processing of materials by biological agents to provide a new variety of goods and services. Bio-chemical engineering or in other words bioprocess engineering as indicated in Section 12.8.1 deals with the design, development and operation of processes involving the biological materials used in: (1) food processing, (2) manufacture of pharmaceutical drugs and (3) biodegradation of unwanted materials (sewage treatment). It may be reckoned as chemical engineering with all its techniques suitably modified or adapted to deal with the conditions imposed by the living cells i.e., microorganisms. These microorganisms are made to work in a manmade environment since long, in the making of bread, cheese, wine and beer. But now, they have been used to produce a wider range of products, necessitating large-scale fermentation processes for the production of antibiotics, aminoacids and enzymes. The youngest discipline of biotechnology has assumed great importance due to the development of an exciting range of techniques such as recombinant DNA manipulation, monoclonal antibody preparation, animal and plant tissue culture. The scientific studies pertaining to the above may be carried out in laboratories but considerable industrial activity is noticed in all western countries and India is no exception. The industries are concerned with food products from raw materials from nature, antibiotics and bio-medicines for healthcare, biofertilizers for agriculture and of much urgency for attention to waste treatment and utilization (ex. biogas from animal excreta). Day by day capital outlays for industries producing readymade food packs, drinking water, pasteurized milk, soft drinks and beer, is rising at phenomenal rate. Sugar syrups obtained by hydrolysis of cereals or potato starch with enzymes isolated from microorganisms are incorporated into many food products such as jam marmalade and concentrated fruit drinks. Healthcare products, such as vaccines and antibiotics are in great demand. The popular penicillin, known as Amoxil, is produced through three sequential processes, the first, a fungal fermentation, the second an enzyme conversion and the last a chemical step. Degradation by bacteria is prevented by including with the penicillin an inhibitor made by a separate fermentation process. For sure success in production of the above in industry, fermentation process is primary and the elaborate steps taken in this regard can be seen from the following short exposition. Fermentation provides not only antibiotics but other complex compounds as well useful for healthcare. Genetically manipulated vegetable products and such others are known and it is not intended to go into details of the same.

14.6.2 Fermentation Process

The design and construction of food processing plants are difficult as it is not possible to specify exactly the desired product and the subjective characteristics such as flavour and texture which are crucial for product acceptability. Introduction of a new process in the biological industry, especially if fermentation-based is not as easy as in conventional industries because the difficulty of striking at the right operating conditions really takes time after the construction is completed. The culture of microorganism for biomass or for the products that they make is indeed a tough and complex process that may take a long time in certain cases. The microbial cell is a complex system of many thousands of interacting biochemical reactions. To enable the reactions to proceed in the correct sequences and in the required amounts, the cell has evolved a number of internal control procedures. These procedures involve the production of enzymes that catalyze the biochemical reactions. The biosynthesis of an enzyme is in turn controlled at the genetic level. In addition, the cell requires a supply of energy obtained from the oxidation of carbon containing compounds and distributed around the cell. Thus the cell's activities are regulated by a delicate balance of nutrient components and energy availability. Most industrial fermentations are batch operations in which all the equipment and culture nutrients have to be sterilized before use. In addition, three requirements dominate fermenter design: good mixing of the fermenter contents, adequate oxygen supply to the growing microbial culture and removal of the metabolic heat produced. Most fermentations resulting in useful products are done in aerated stirred tanks of 100–250 m^3 in volume.

The temperature of the culture medium is the most essential state variable to be measured and controlled. Other variables to be measured and controlled are over pressure the air flow-rate and the stirrer speed for mixing. The formation of foam caused by the aeration has necessarily to be controlled for the process to operate satisfactorily. The rate of metabolism is controlled by the pH value of the culture medium and so pH measurement and control is another essential requirement. But the control of pH value does not in all cases assist the process. In complex batch culture processes, the progress of the process is indicated by the changing value of pH. Thus pH control is employed after proper understanding of its effects.

During the process, supply of nutrient components such as vitamins and trace elements as well as oxygen should be assessed properly and used so that growth of organisms continues to take place. Likewise carbon in the form of glucose, should be supplied at such slow rates so as to produce a number of enzymes. Measurements of outlet gas CO_2 concentration by infra-red analyzers, and estimation of the rate of growth, oxygen supply limitation and changes in metabolism by using robust dissolved oxygen electrodes are essential to regulate aeration rates and nutrient supply rates.

Thus the bioprocess control is really a tricky and complex affair. A number of process variables apart from those mentioned are to be identified and fed to the bioprocess operator on an online basis. Even then an effective control strategy may not be possible for all cases. Online measurement of glucose is rendered possible by means of biosensors, but for many products, complex off-line analyzers based on biochemical reactions and proper sensors are required. It is the physiology and amazing versatility of microorganisms, particularly in the process environment that makes matters complicated. A direct measurement of biomass, if enabled, is of great help to the bioprocess industries. The challenge lies in developing bioprocess sensing devices. Extensive research effort has yielded such devices and the subject of biotechnology has become a discipline.

Apart from the fermentation process, there are other equally tough processes, such as bio-catalysis for manufacture of sugar syrups and other products. The above introduction is intended only to reveal the nature and extent of care to be exercised in the control of bio-processes and bioreactors. It has become possible to achieve sufficient success in these operations by making use of the several sensors along with the biosensors and adoption of computer control. The essentiality of specialists in the disciplines of biology, chemistry, physics, technology and instrumentation, coming together is well recognised through the discipline of biotechnology and the new breed of bioprocess engineers are going to change the scene considerably in future.

14.7   FIELD INSTRUMENTATION—FIELDBUS

14.7.1   Introduction

In the past twenty years, digital techniques have been recognized to offer several advantages and their application range is getting extended. In the field of instrumentation, it is no exception as the all-pervading digital communication techniques, seem to present a successful scope for adoption in the instrumentation and control systems of the process industries. The 'smart' sensors have already replaced the traditional sensors with analog output signals. The standard output signals of 3–15 psi, found acceptable and convenient by many industries up to 1950, have become 4–20 mA with the successful performance of electrical and electronic techniques. The smart sensors with digital output signals, with the microprocessor chip on board, are treated as information sources, in the environment of the industry. The entire outfit, being miniaturized in their size, and located in or near the spots of origin of the measurands, has appealed and excited considerably, the minds of the operating personnel. If there are a good number of such information sources, confined to the area of the industry, it is possible to enable them 'converse' with each other, by means of a local area network, all digital in character. In the process control industry, there are other devices such as actuators and anunciators, which can participate in conversation with sensors, if all of them are digital. The central control room enabled the centralized control of all operations so far. It is considered to offer several advantages, if an all-digital communication network is designed, linking up all devices and the central control room and all essential requirements, already existing are also incorporated. Distributed digital control making use of the local area networks, has thrown open certain merits and inter-communication system, known as field-bus, has emerged as a possible solution for adoption of distributed digital control of industrial processes. All the devices that participate in the fieldbus, necessitated conformity to certain standards regarding the signals they have to deal with and their interoperability. The designs of the devices, also need to conform to certain standards regarding their specifications and other factors, so as to enable them get replaced, when necessity arises, no matter the country wherefrom they are bought. It means that the fieldbus design, while being standardized, for use anywhere in the world, dictate standards for the design of the devices constituting the fieldbus equipment. Initial enthusiasm has been responsible for the development of fieldbus by countries like France, Germany and Japan along with that of Instrument Society of America (ISA). Each one differed from the other in the philosophy of approach and an all-out effort to effect a world-standard has been attempted. The different versions are seen functioning in the countries, wherefrom they originated. All those interested groups developing fieldbus have joined together to form a working group in 1990, with the ultimate objective of gaining agreement for a single international fieldbus standard. The ISA SP50 Committee comprised of three major working subcommittees known as Physical Layer Subcommittee, Data Link layer subcommittee and the Application layer subcommittee.

The Physical Layer subcommittee defines the physical medium to be used and the encoding of the datastream on the medium. The second group deals with the issues of how devices gain access to the network, error checking, recovery, arbitration of conflicts, and message framing. The third committee deals with the interpretations of the fieldbus messages. In order to obtain interchangeable fieldbus devices the subcommittee has to determine what message must be supported by various classes of fieldbus devices.

14.7.2 Fieldbus Requirements

The fieldbus thus conceived makes use of the full potential of microprocessor based field instruments. Standardization of the fieldbus enables this technology to be applied in a broad line of products providing interchangeability of devices offered by different vendors. A fieldbus standard thus opens the way to a new generation of field instrumentation that will provide the means for new productivity enhancements in the process and manufacturing industries. It is not intended to present in detail all the versions proposed and used in those countries. To gain a reasonable understanding, a brief introduction to some approaches in the make up of the field bus are presented. The Fieldbus Foundation (FF) formed in 1994, has worked to develop a worldwide, unified Standard Fieldbus for process control and manufacturing automation. The definition of a single, IEC fieldbus (International Electrotechnical Commission) standard, accepted throughout the world but has not seen much progress. One has known the extent of delay when the MKS system of Units has to be applied throughout the world. Luckily, it was possible to effect the traditional 4–20 mA telemetry and communication system in industry, for all the obvious advantages over the earlier 3–15 psi standard. The changeover to the new fieldbus standard is expected to yield certain benefits such as:

1. Substantial saving on field wiring over the conventional 24 V, 4–20 mA system

2. Interoperability between devices from multiple vendors

3. Potential for many previously centralized functions to migrate back into the field, and

4. The ability to realize significant savings in commissioning, maintenance and trouble shooting of field mounted equipment.

The fieldbus may be simple in cases of its use as a multiplexing system or it can be larger in its applicability for process control. The number of devices, the power requirement and data transmission rates may make the differences.

On the other hand, testing the complete system of fieldbus will also be more difficult, as with many systems, the first test of multiple vendors can only be carried out when all the instrumentation is installed on site. The number of terminations on a single bus is also limited particularly when power is carried on the bus in hazardous areas.

The similarity of fieldbus and the large-scale digital communication systems should not be extended too much into the fieldbus instrumentation systems.

At higher levels of functionality are fieldbuses for process plant applications using digital control systems (DCS), that are formulated by Foundation FieldBus (Ref: Klaus-Peter Linder), WorldFIP (Ref: John Beeston and Masc Desjardins), ProfibusPA (Ref: Bob Squirreb) and LON. With experience in development of digital radio, a radio version of the fieldbus (low-level communication standard) particularly for use of the process industries and utility sectors is conceived with the radio serving as the physical layer.

The technical features and other aspects concerning each one of the versions, so far developed and on offer are to be consulted from papers under Ref.

14.7.3 IEEE-488 Fieldbus

While all the attention is paid to standardize the fieldbus design, situations abound in practice where the first generation buses which are IEEE-488 and the RS232C are intelligently applied for specific tasks in testing laboratories and industry. Usually, a centralized minicomputer is used to perform all the desired functions of data acquisition, control, data reduction and analysis. Of late, due to the decrease in cost of microprocessor-based systems coupled with significant improvements in performance has made popular the application of multiprocessor measurement systems. It is not unusual to find a multiprocessor system with at least one each for data acquisition and data reduction.

The IEEE-488 bus is one of the most popular interfaces supported by microprocessor based instrument manufacturers. The interface utilizes a party-line bus structure permitting byte serial bit parallel communication. The standard allows for 15 devices to be connected to one contiguous bus in either a star, linear or combinational arrangement.

A standard IEEE-488 is mostly supported by instrument manufacturers for use with intelligent controllers of the system. It comprises of 16 single lines and 8 ground lines allowing a maximum transmission rate of 1 megabyte per sec over limited distances. The typical transmission rate is between 250 and 500 kilobytes per sec, operating within a total network transmission path up to 20 metres. Where high speed communication is desired between intelligent instruments in close physical proximity this interface will be an ideal choice. But the cabling cost may be on the higher side.

The EIA RS-232C standard has a transmission rate lower than the above IEEE-488 bus but has no distance limit. It has not as much advanced capabilities as that of IEEE-488. It is primarily for use as an interface between a computer and a modem, but it has been widely adopted for the serial input/output port interface in a nonstandard way. It may be costing less than the IEEE-488. A judicious combination of the above is shown in the following application:

A real-time measurement and control system is set up for a low speed recirculating wind tunnel scheme as shown in Figure 14.6. It employs a dual bus distributed control architecture utilizing a number of low-performance microprocessors working in parallel to produce a combination the computing capability of which far exceeds that of a conventional centralized architecture. Incidentally, it is shown that it is cheaper. In this scheme, the data acquisition unit (DAU) communicates with the host on a relatively slower RS-232 network, only to provide in this case, a low-cost solution by matching bus specifications to data transfer requirements. The scheme shown in Figure 14.6 comprising the IEEE-488 and RS-232C interface, also supports the parallel centromic interface which is only used for a single point-to-point communication. The IEEE-488 bus caters to the needs of intelligent standalone dedicated controllers of the DAU and the laser Doppler anemometer. It satisfies the sophisticated communication protocols and high speed transmission requirements of these controllers. This bus also serves the mass storage device, digital plotter and printer. The parallel execution of data acquisition and control functions of the DAU frees the system host for real-time data reduction, analysis and display.

The slower RS-232C interfaces serve the less intense data traffic systems comprising the controllers, the blowers, turntables and probe positioning mechanisms that are remotely located.

It can be seen that the requirements of the application may necessitate the choice of the bus as it stands while the future architecture of the standard buses is directed for use in big industries.

EXERCISES

1. Explain the importance of
   (a) on-line instrumentation, and
   (b) real-time instrumentation in the present day automated industries.

2. Discuss the role played by sensors in achieving
   (a) higher productivity
   (b) superior quality of products
   (c) monitoring the environment of the industry, and
   (d) security of personnel and property.

3. In what ways control instrumentation differs from scientific instrumentation?

4. Explain why pharmaceutical and textile industries are more concerned with moisture and humidity measuring techniques.

5. What measuring techniques are found suitable for measuring the flow-rate of paper-pulp and food-slurry in paper and food industries.

6. For air transportation and space navigation vehicles, why the commonly used sensors are not recommended? What features of sensors are found most appropriate?

7. Explain why special attention is paid to safety and security aspects in the nuclear power plants.

8. What is understood by man-machine interface (MMI) and suggest the means adopted for effective implementation.

9. Discuss the nature of problems encountered when dealing with biological materials in bio-product industry.

10. Explain the fermentation process and show how the control steps are designed for dealing with the problems of the process.

11. What is understood by 'field bus' and discuss the scope of its applications.

12. What are the essential requirements of a 'field bus' system and suggest the kind of industries where is found most relevant.

13. What are the features of IEEE-488 field bus? Explain, by means of a typical situation where it is used.

SUGGESTED FURTHER READING

Bob Squirreb, ProfiBus Developments, *Measures & Controls*, Vol. 30, No. 3, April 1997.

Boettcher, D.B., and Hickling, E.M., Instrumentation and control systems for sizewell, B., *Measures & Controls*, Vol. 26, No. 5, July 1993.

Cobb, J., Field Bus: What is in it for me? *Measures & Controls*, Vol. 23, No. 1, Feb. 1990.

Crisp, R.J.T., Process instrumentation in the confectionary industry, *Measures & Controls*, Vol. 24, No. 7, Sept. 1991.

Desjardins, M., World FIP, *Measures & Controls*, Vol. 27, No. 2, March 1994.

Donald E., Brown, Bioprocess measurement and control, *Measures & Controls*, Vol. 25, No. 6, July/Aug. 1992.

Frank, R. Ried ijk and John., H. Huijsing, Sensor interface environment based on serial bus interface, *Measures & Controls*, Vol. 30, No. 10, Dec. 1997.

Greaves, Peter, Measurement of moisture in food pharmaceuticals and other products, *Measures & Controls*, Vol. 25, No. 4, May 1992.

Herschdoerfer, *Quality Control in the Food Industry*, Vol. 4, (1987), Academic Press.

Integrated Instrumentations and Control Upgrade Plan, Electric power research institute, EPRINP-7343, 1992.

John W. Beeston, World FIP advances in process control, *Measures & Controls*, Vol. 30, No. 3, April 1997.

Klaus-Peter Lindner, Fieldbus—a milestone in field instrumentation technology, *Measures & Controls*, Vol. 23, No. 9, Nov. 1990.

Lilly, M.D., Biochemical engineering—Its contribution to society, *Measures & Controls*, Vol. 21, No. 4, May 1988.

Neve, B.D., Progress on radio bus, *Measures & Controls*, Vol. 23, No. 1, Feb. 1990.

Norman, R.C., and Glease, I., Novel instrumentation in the nuclear industry, *Measures & Controls*, Vol. 24, No. 6, July/Aug. 91.

Richards, J.I., Principles and architecture of total control and instrumentation systems for Scottish nuclear power stations, *Measures & Controls*, Vol. 26, No. 3, May 1993.

Twork, J.V., and Yacynych, A.M. (Eds.), *Sensors in Bioprocess Control*, (1990), Marcel Dekker, New York.

Wood, T.J., Optical fibre based sensors for aircraft applications, *Measures & Controls*, Vol. 22, No. 7, Sept. 1989.

1

Some Physical Constants

Boltzmann constant	(k)	1.3806×10^{-23} J K^{-1}
Electric space constant (permittivity of space) $[(1/(\mu_0 c^2)]$	(ε_0)	8.854188×10^{-12} F m^{-1}
Magnetic space constant (permeability of space)	(μ_0)	$4\pi \times 10^{-7}$ H m^{-1}
Velocity of light in vacuum	(c)	2.99792×10^{8} m s^{-1}
Stefan–Boltzmann constant	(σ)	5.67032×10^{-8} W m^{-2} k^{-4}
Planck's constant	(h)	6.62618×10^{-34} Js
Molar gas constant	(R_0)	8.31441 J mol^{-1} K^{-1}
Gravitational constant	(G)	6.6720×10^{-11} N m^2 kg^{-2}
Faraday constant	(F)	9.6485×10^{4} C mol^{-1}
Rest mass of electron	(m_e)	9.10953×10^{-31} kg
Charge on electron	(e)	1.60219×10^{-19} C
Electron, specific charge on	$\left(\dfrac{e}{m_e}\right)$	1.7588×10^{11} C kg^{-1}
Acceleration of gravity	(g)	9.80665 m s^{-2}

2

Some Commonly Used Physical Quantities in SI and FPS Units

Velocity

Unit	metres per sec	miles per hour	Remarks
1 m s^{-1}	1	2.2369	1 knot
1 km h^{-1}	0.2777	0.6214	= 1 int nautical mile/h
1 mile h^{-1}	0.4470	1	= 1.852 kmh^{-1} = 0.5144 ms^{-1}
			= 1.1508 mh^{-1}

Mass

Unit	kilogram	pound (av)
1 kg	1	2.2046
1 lb (av)	0.4536	1
1 slug	14.5939	32.1740

Pressure

Unit	newton per sq. mm	pound force per sq. inch (psi)	
1 N mm^{-2}	1	145.038	1 Pascal = 1 Nm^{-2}
1 psi	6.8948 × 10^{-3}	1	= 10^{-5} bar

Unit	millibar	1 mm water gauge
1 m bar	1	10.1972
1 at (atmosphere)	980.665	10^4
1 torr (1 mm Hg)	1.3332	13.5951
1 mm w.g.	99.0665×10^{-3}	1

Force

Unit	newton	lbf
1 N	1	0.2248
1 lbf	4.4482	1

Power

Unit	watt	ft-lbfs^{-1}	
1 W	1	0.7376	1 kW = 1.341 hp
1 ft-lbfs^{-1}	1.3558	1	1 hp = 0.7457 kW

1 kjoule = 0.2388 kcal

1 kcal = 4.1868 kJ

Numerical Problems

1. A unity-ratio dc Wheatstone bridge network has two matched resistive transducers working in push–pull fashion, and making up the remaining two arms. Show that the bridge unbalance voltage is twice that of the bridge using only one transducer. Show also that the unbalance voltage is linearly related to the fractional change in the resistance of the transducer.

2. A unity-ratio Wheatstone bridge network consists of three 120-ohm resistors and a resistive strain gauge of 120 ohms under no-strain conditions. The strain gauge is limited for its current to 25 mA. Determine the maximum value that can be applied to the bridge for excitation. If the strain gauge is bonded to a steel structure, find out the unbalance voltage of the bridge, when the steel structure is subjected to a stress of 100 kg/cm². The gauge factor is 2 and the Young's modulus of steel is 2×10^6 kg/cm².

3. What is the effect of temperature rise on the bridge output voltage in the above problem? Determine the additional output voltage due to a rise of temperature of 30°C from the initial temperature. Thermal expansion coefficient of steel is $41.7 \times 10^{-6}/1°C$ and that of the material of the gauge, $27 \times 10^{-6}/1°C$. The temperature coefficient of resistance of the gauge material is $10.8 \times 10^{-6}/1°C$.

4. A load cell consists of a solid cylinder of steel 30 cm diameter with four strain gauges bonded to it and connected in a Wheatstone bridge arrangement. If gauges are 120 ohms each with a gauge factor of 2.0, what is the sensitivity of the load cell expressed in mV/N with the bridge excited by 1 V dc. The modulus of elasticity of steel is 200×10^9 N/m² and Poisson's ratio = 0.3.

5. A rectangular strain gauge rosette is mounted on a steel plate (Figure 6.32) and the three strains measured are as follows:

$$\varepsilon_1 = +500 \ \mu m/m$$
$$\varepsilon_2 = +400 \ \mu m/m$$
$$\varepsilon_3 = -100 \ \mu m/m$$

Calculate the principal strains and stresses and the maximum shear stress. Identify the axis of principal stress.

6. A diaphragm pressure gauge has a diaphragm of 5-cm diameter and is required to measure a pressure difference of 15 kg/cm^2. Calculate the thickness of diaphragm required so that it has a linear characteristic between pressure difference and deflection. What is the natural frequency of the diaphragm?

7. A pressure transducer employs a thin plate diaphragm on which four matched strain gauges are mounted so as to be connected suitably as a Wheatstone bridge network. The diaphragm is of steel and of radius of 7.5 cm. One pair of strain gauges located at 6.25 cm from the centre of diaphragm measures the radial strain. The second pair measures tangential strain and is located at 1.5 cm away from the centre. The thickness of the diaphragm is 1.25 mm and each gauge is of 120 ohms. If the bridge is excited from a battery of 5 V, determine the sensitivity of the arrangement in mV per N/m^2.

8. A cylindrical strain tube of steel is used for pressure measurement. The inside diameter of the tube is 5 cm and outside diameter is 5.25 cm. Calculate the maximum pressure that can be measured if the hoop strain is to be limited to 2×10^{-3} µm/m.

Also calculate the sensitivity in ohms per N/m^2, if the gauge resistance used is 360 ohm and the gauge factor is 2.

9. The output signals of a linear variable differential transformer are recorded by a self-balancing potentiometer recorder having its undamped natural frequency of 10 Hz and working at a damping factor of 0.707. The LVDT is excited by a supply voltage of 50 Hz. Calculate the maximum frequency of the displacement signal that can be recorded with an error of ±2%. How does the recorder respond to the 50 Hz carrier frequency voltages?

10. A galvanometer oscillograph of natural frequency of 1000 kHz working at a damping factor of 0.707 is used to record signals from an LVDT. The LVDT is excited from a source of 10 kHz and is used to measure displacements of a frequency range from zero to 500 Hz. From the frequency response calculations of the oscillograph, judge whether the entire system possesses the desired accuracy of measurement.

11. For the above LVDT excited from a source of 10 kHz, with displacements having a frequency content up to 500 Hz, a cathode ray oscillograph is used to record the output signals. If a diode-rectifier demodulator along with RC filters is used, suggest values for the resistors and capacitors to be used for a single RC filter, so that the response to carrier frequency voltages is contained within ±5%. If not found possible, use another properly designed RC filter in cascade to the first for satisfactory response. The input impedance of oscillograph may be taken as 10^6 ohm.

12. Five plates each of area 3 cm × 3 cm overlapping each other are arranged so as to serve as a displacement transducer. The plate spacings are 0.25 mm. The movable set is subjected to linear displacement along the plane of the plates. Calculate the sensitivity of the transducer in pF/cm.

Suggest suitable measuring network for delivering electrical voltages as output signals.

13. The thickness of paint coating on a metallic surface is measured by means of an electrode of area of 6.25 sq cm placed over the paint surface. The dielectric constant of the paint is 5. The capacitor thus formed is connected in one arm of an ac excited Wheatstone bridge having unity ratio made up of two resistors each of 1000 ohms. Determine the capacitance needed in the fourth arm such that the bridge becomes balanced for a nominal paint thickness of 0.05 mm.

    If the bridge supply is 1 V at 1 kHz, determine the smallest detectable variation in thickness from its nominal value, for an unbalance voltage of 0.1 mV.

    Suggest any alternative arrangement for the connection of the bridge network and comment on the arrangement in comparison to the earlier connection.

14. A quartz crystal transducer having a capacitance of 1000 pF and charge sensitivity of 4 μC/cm is connected to a charge amplifier. A fixed capacitor of 0.01 μF in parallel with a resistor of 10^6 MΩ is connected in the feedback path. Obtain the ultimate sensitivity of the combination, in volts/cm.

    What is the lowest frequency of the input displacement signal that can be measured satisfactorily? Is the system considered suitable for static measurements?

15. A piezoelectric transducer of sensitivity of 2 μC/N has a capacitance of 1000 pF. The connecting cable has a capacitance of 300 pF, and the input impedance of the oscilloscope connected is 1 MΩ in parallel with 50 pF. What is the sensitivity of the transducer alone in V/N, and what is it like with the connection to the oscilloscope?

    Determine the suitability of the scheme for the display of force signals of a frequency of 10 Hz. If not suitable, how do you modify the scheme for making it suitable?

16. The above transducer and the cable are used along with a charge amplifier having a capacitor of 0.01 μF in parallel with a resistor of 10^6 MΩ in its feedback path. Calculate the ultimate sensitivity of the entire system in V/N.

    Is the measuring sybtem suitable for static measurements? If not, calculate the lowest frequency at which the error is within ± 2%.

17. The feedback system for the measurement of flow rate of fluids through pipes is shown in Figure 8.14. Assume that the flux density produced is proportional to the primary current and the current transformer is ideal. Develop the overall transfer function of the system relating the flow rate and the indicated angular deflection. Comment on the effect of the supply voltage fluctuations on the overall sensitivity of the system.

    In a particular case the gain of the feedback path stage is 0.1 mV/l° and the motor constant, K_m, is 0.1°/mV. Suggest a suitable value for the voltage gain of the amplifier K_a to make it a satisfactory measuring system. If the motor time constant is 2 seconds, determine the natural frequency of the system and its damping conditions.

18. Signals having a bandwidth of 0–2 kHz have to be faithfully recorded on a tape recorder having a gap length of 6.4 μm for the recording and playback heads. Determine suitable values for frequency deviation and modulation index so that the signals can be frequency-modulation recorded. Suggest the centre frequency and the tape speed desired.

19. The fluctuations in the tape speed of a tape recorder are about ± 1%. What will be the error in the recording of signals if FM recording technique is used with a frequency deviation of (i) 7.5%, and (ii) 40%?

20. A feedback type accelerometer (Figure 8.15) with output signals of dc current is used with a proof mass of 1 g placed at the tip of the pointer at a distance of 3 cm from the pivot. The pointer is designed to deflect in vertical plane when subjected to vertical accelerations. The deviation from horizontality is detected by a sensor of sensitivity of 1 V rad, and the deflection sensitivity of the moving coil system is 10^5 rad/N-m. The amplifier gain is 100 mAV, and the torque constant of the pmmc system is 2×10^{-5} Nm/mA of coil current. The system is working under stable conditions. Calculate the coil output current when the accelerometer is subjected to vertical acceleration equal to g. (g = 9.81 m/s^2).

21. A servomotor–tachogenerator combination is used to serve as an integrating system for developing output displacements proportional to the integral of a dc voltage (see Figure 8.11). An amplifier of gain 1000 is used in the forward path. The moment of inertia of the servomotor is 0.2×10^{-6} kg-m^2. The armature circuit resistance is 500 ohm and the motor develops a torque of 0.1 N-m per amp. What should be the sensitivity of the tachogenerator to make the system an integrating system having a steady state accuracy of ± 0.2%? Calculate the time constant of the overall system.

22. A capacitive transducer is made up of a parallel plate capacitor, with a plate area of A and the separation between plates a. A dielectric material of nominal thickness d is introduced and is to be continuously monitored for its thickness variations. Taking the relative permittivity of the dielectric as ε_r, obtain the relationship between percentage variations in capacitance in terms of $\Delta d/d$ and other constants.

Comment on the relationship between the two for the sensitivity and linearity of the system. Plot curves relating the sensitivity factor with $d/(a - d)$ for values of ε ranging from 2 to 10.

23. A thin plate diaphragm type capacitive transducer is to be designed and used as a single-ended transducer for pressure measurements of the range of 1000 N/cm^2 at frequencies ranging from 0 to 5 kHz. The material of the diaphragm has Young's modulus of 2×10^6 kg/cm^2 and Poisson's ratio of 0.3. The diaphragm of the transducer should not exceed 1.25 cm. Calculate the thickness and initial separation desired for the system with air as the dielectric.

Estimate the sensitivity of the systems in terms of percentage changes in capacitance to pressure differences.

24. A capacitive transducer is formed of a thin plate diaphragm and a fixed plate for measurement of pressure. The diaphragm has a radius of 7.5 cm, and a thickness of 0.125 cm, and the separation between the plates is 0.2 cm. If the material used has its Young's modulus of 2×10^6 kg/cm^2 and Poisson's ratio of 0.3, calculate the percentage change in the capacitance when a pressure of 0.5 kg/cm^2 is applied to the diaphragm.

25. A membrane type capacitive transducer is used for measurement of differential pressure of the order of 10^{-4} kg/cm^2. The radial tension on the membrane is 25×10^{-3} kg/cm. The

radius of the membrane is 5 cm and the initial gap is 0.2 cm. Calculate the initial capacitance and the percentage change in capacitance when the differential pressure is applied.

26. A push–pull type resistance transducer having a resistance of 120 Ω for each half is connected in a transformer-coupled unity-ratio arm bridge network. The self-inductance of each half of the inductive arms is $0.1/\pi$ H, and the coefficient of coupling between the windings of the transformer is assumed to be unity. Work out the ratio of the sensitivities of the transformer-coupled bridge and the conventional Wheatstone bridge (with coefficient of coupling being zero) when the bridge is excited at supply frequencies of 50, 500, 5,000 and 50,000 Hz.

27. The resistive transducer of the above problem is replaced by the push-pull capacitive transducer of Problem 25. Obtain the unbalance voltage when the Tr-coupled bridge is excited with 10 V at 150 Hz for a differential pressure of 10^{-4} kg/cm^2.

    Will the bridge become more sensitive if the frequency of excitation of the bridge is increased to 500 Hz? If so, calculate the sensitivity.

28. A push–pull type self-inductance transducer is designed by using U and I stampings (as shown in Figure 6.50(a)) of a material having an initial (relative) permeability of 600. The length of the magnetic path of each half is 10 cm, while the area of cross-section of the U-core is 1 cm^2. Each of the two U-cores is provided with windings of 500 turns each. The initial gap of air of each half of the assembly is 0.5 mm.

    Obtain the sensitivity factor and the self-inductance value of each half for initial air gap values of 0.1, 0.2, 0.3, 0.4 and 0.5 cm when used as a single-ended transducer.

    Repeat the calculations when the system is used as a push–pull transducer.

    Estimate the range of input displacements that can be measured with a linearity of ± 2%, if the initial gap of each half is chosen as 0.25 mm.

29. If the rigid structure cn which the seismic system is fixed is subjected to vibration, determine the range of frequencies of vibration for which the flatness of response is within ± 5%. The given system has a stiffness of 100 N/m, a mass of 1 kg, and a damping coefficient of 12 N s/m.

30. If the above seismic system is used for measurement of translational accelerations, what is its static sensitivity? If an LVDT of 25 mV/mm is used to sense the displacements, what will be the output voltage of the LVDT, when the structure has an absolute acceleration of 10 m/s^2 against gravity. Is it a suitable system for acceleration measurements? If not, suggest the modifications necessary.

Index